EARTH SCIENCE

EARTH SCIENCE

Eleventh Edition

Edward J. Tarbuck
Frederick K. Lutgens

Illustrated by

Dennis Tasa

Upper Saddle River, New Jersey 07458

Library of Congress Cataloging-in-Publication Data

Tarbuck, Edward J.
   Earth Science/Edward J. Tarbuck, Frederick K. Lutgens ; illustrated by Dennis Tasa.— 11th ed.
     p. cm.
   Includes index
   ISBN 0-13-149751-0
   1. Earth sciences.   I. Lutgens, Frederick K.   II. Title.
QE26.2.T38
550—dc22   2006 2004060080

Executive Editor: *Patrick Lynch*
Editor-in-Chief, Science: *John Challice*
Executive Managing Editor: *Kathleen Schiaparelli*
Production Editor: *Edward Thomas*
Managing Editor, Science Media and Supplements:
  Nicole M. Jackson
Media Editor: *Chris Rapp*
Director of Marketing, Science: *Linda Taft-Mackinnon*
Manufacturing Buyer: *Alan Fischer*
Director of Creative Services: *Paul Belfanti*
Art Director: *Kenny Beck*
Interior Designer: *Dina Curro*
Cover Designer: *JMG Graphics*
Copy Editor: *Barbara Booth*
Editorial Assistant: *Sean Hale*
Marketing Assistant: *Larry Grodsky*
Photo Research Administrator: *Melinda Reo*
Photo Researcher: *David A. Tietz*
Proofreader: *Alison Lorber*

Image Permission Coordinator: *Debbie Hewitson*
Color Scanning Supervisor: *Joseph Conti*
Production Assistant: *Nancy Bauer*
Manager, Formatters: *Allyson Graesser*
Composition: *Jacqueline Ambrosius, ArtWorks*
Assistant Formatter: *Karen Stephens*
Senior Manager, Artworks: *Patty Burns*
Production Manager, Artworks: *Ronda Whitson*
Manager, Production Technologies, Artworks: *Matt Haas*
Project Coordinator, Artworks: *Jessica Einsig*
Illustrator, Artworks: *Mark Landis*
Illustrations by: *Dennis Tasa, Tasa Graphic Arts*
Cover Photo: *Clouds Over the Mitten and Merrick Buttes,*
  Monument Valley Tribal Park, Arizona/Utah
  (Photo by Tom & Susan Bean, Inc.)
Title Page Photo: *Delicate Arch in Arches*
  National Park, Utah.
  (Photo by David Muench Photography, Inc.)

10 9 8 7 6 5 4 3 2 1

ISBN 0-13-149751-0

Pearson Education Ltd., *London*
Pearson Education Australia Pty., Limited, *Sydney*
Pearson Education *Singapore*, Pte. Ltd
Pearson Education North Asia Ltd., *Hong Kong*
Pearson Education Canada, Ltd., *Toronto*
Pearson Educación de *Mexico*, S.A. de C.V.
Pearson Education—Japan, *Tokyo*
Pearson Education Malaysia, Pte. Ltd

To Joanne and Nancy

BRIEF CONTENTS

GEODe: Earth Science v.2

A copy of *GEODe: Earth Science*, v.2 is packaged with each copy of *Earth Science*, Eleventh Edition. This dynamic learning aid reinforces key concepts by using tutorials, animations, and interactive exercises.

Unit 1: Earth Materials

A. Minerals
  1. Introduction
  2. Major Mineral Groups
  3. Properties Used to Identify Minerals
  4. Mineral Identification
  5. Minerals Quiz
B. Rock Cycle
C. Igneous Rocks
  1. Introduction
  2. Igneous Textures
  3. Naming Igneous Rocks
  4. Igneous Rocks Quiz
D. Sedimentary Rocks
  1. Introduction
  2. Types of Sedimentary Rocks
  3. Sedimentary Rocks Quiz
E. Metamorphic Rocks
  1. Introduction
  2. Agents of Metamorphism
  3. Textural and Mineralogical Changes
  4. Common Metamorphic Rocks
  5. Metamorphic Rocks Quiz

Unit 2: Sculpturing Earth's Surface

A. Hydrologic Cycle
B. Running Water
  1. Stream Characteristics
  2. Hydrological Cycle and Running Water Quiz
C. Groundwater
  1. Groundwater and its Importance
  2. Springs and Wells
  3. Groundwater Quiz
D. Glaciers
  1. Introduction
  2. Budget of a Glacier
  3. Glaciers Quiz
D. Deserts
  1. Distribution and Causes of Dry Lands
  2. Common Misconceptions About Deserts
  3. Deserts Quiz

Unit 3: Forces Within

A. Earthquakes
  1. What is an Earthquake?
  2. Seismology: Earthquake Waves
  3. Locating an Earthquake
  4. Earth's Layered Structure
  5. Earthquakes and Earth's Interior Quiz
B. Plate Tectonics
  1. Introduction
  2. Plate Boundaries
  3. Plate Tectonics Quiz

C. Igneous Activity
  1. The Nature of Volcanic Eruptions
  2. Materials Extruded During an Eruption
  3. Volcanoes
  4. Intrusive Igneous Activity
  5. Igneous Activity Quiz

Unit 4: Deciphering Earth's History

A. Geologic Time Scale
B. Relative Dating
C. Radiometric Dating
D. Geologic Time Quiz

Unit 5: The Global Ocean

A. Floor of the Ocean
  1. Mapping the Ocean Floor
  2. Features of the Ocean Floor
  3. Floor of the Ocean Quiz
B. Coastal Processes
  1. Waves and Beaches
  2. Wave Erosion
  3. Coastal Processes Quiz

Unit 6: Earth's Dynamic Atmosphere

A. Heating the Atmosphere
  1. Solar Radiation
  2. What Happens to Incoming Solar Radiation?
  3. The Greenhouse Effect
  4. Temperature Structure of the Atmosphere
  5. Heating the Atmosphere Quiz
B. Moisture and Cloud Formation
  1. Water's Changes of State
  2. Humidity: Water Vapor in the Air
  3. Cloud Formation
  4. Moisture and Cloud Formation Quiz
C. Air Pressure and Wind
  1. Measuring Air Pressure
  2. Factors Affecting Wind
  3. Highs and Lows
  4. Air Pressure and Wind Quiz
D. Basic Weather Patterns
  1. Air Masses
  2. Fronts
  3. Introducing Middle-Latitude Cyclones
  4. Examining a Mature Middle-Latitude Cyclone
  5. Basic Weather Patterns Quiz

Unit 7: Earth's Place in the Universe

A. The Planets: An Overview
B. Calculating Your Age and Weight on Other Planets
C. Earth's Moon
D. A Brief Tour of the Planets
E. Solar System Quiz

 This *GEODe: Earth Science* icon appears throughout the book wherever a text discussion has a corresponding module on the CD-ROM.

Contents

5   Running Water and Groundwater   114

6   Glaciers, Deserts, and Wind   153

UNIT THREE

Forces Within   187

7   Earthquakes and Earth's Interior   187

8   Plate Tectonics: A Scientific Theory Unfolds   215

9   Volcanoes and Other Igneous Activity 249

10   Mountain Building 283

UNIT FOUR
Deciphering Earth's History 309

11   Geologic Time 309

12 Earth's History: A Brief Summary   335

UNIT FIVE
The Global Ocean   359

13 The Ocean Floor   359

14 Ocean Water and Ocean Life   383

UNIT SIX

Earth's Dynamic Atmosphere 435

16 The Atmosphere: Composition, Structure, and Temperature 435

15 The Dynamic Ocean 402

17 Moisture, Clouds, and Precipitation   465

18 Air Pressure and Wind 501

19 Weather Patterns and Severe Storms   527

20 Climate   559

UNIT SEVEN
Earth's Place in the Universe   585

21 Origin of Modern Astronomy   585

22 Touring Our Solar System   611

Preface

Earth Science Eleventh Edition, like its predecessors, is a college-level text designed for an introductory course that often has the same name as this text. It consists of seven units that emphasize broad and up-to-date coverage of basic topics and principles in geology, oceanography, meteorology, and astronomy. The book is intended to be a meaningful, nontechnical survey for undergraduate students with little background in science. Usually these students are taking an Earth science class to meet a portion of their college's or university's general requirements. In addition to being informative and up-to-date, a major goal of Earth Science is to meet the need of beginning students for a readable and user-friendly text, a book that is a highly usable "tool" for learning basic Earth science principles and concepts.

Distinguishing Features

Readability

The language of this book is straightforward and *written to be understood*. Clear, readable discussions with a minimum of technical language are the rule. The frequent headings and subheadings help students follow discussions and identify the important ideas presented in each chapter. In the Eleventh Edition, improved readability was achieved by examining chapter organization and flow, and writing in a more personal style. Large portions of the text were substantially rewritten in an effort to make the material more understandable.

Focus on Learning

When a chapter has been completed, three useful devices help students review. First, the **Chapter Summary** recaps all of the major points. Next is a checklist of **Key Terms** with page references. Learning the language of Earth science helps students learn the material. This is followed by **Review Questions** that help students examine their knowledge of significant facts and ideas. Each chapter closes with a reminder to visit the **Website** for Earth Science, Eleventh Edition (http://www.prenhall.com/tarbuck). It contains many excellent opportunities for review and exploration.

Illustrations and Photographs

The Earth sciences are highly visual. Therefore, photographs and artwork are a very important part of an introductory book. Earth Science, Eleventh Edition, contains dozens of new high-quality photographs that were carefully selected to aid understanding, add realism, and heighten the interest of the reader.

The illustrations in each new edition of Earth Science keep getting better and better. In the Eleventh Edi-tion more than 100 pieces of line art are new or redesigned. The new art illustrates ideas and concepts more clearly and realistically than ever before. The art program was carried out by Dennis Tasa, a gifted artist and respected Earth science illustrator.

Focus on Basic Principles and Instructor Flexibility

Although many topical issues are treated in Earth Science, Eleventh Edition, it should be emphasized that the main focus of this new edition remains the same as its predecessors—to foster student understanding of basic Earth science principles. Whereas student use of the text is a primary concern, the book's adaptability to the needs and desires of the instructor is equally important. Realizing the broad diversity of Earth science courses in both content and approach, we have continued to use a relatively nonintegrated format to allow maximum flexibility for the instructor. Each of the major units stands alone; hence, they can be taught in any order. A unit can be omitted entirely without appreciable loss of continuity, and portions of some chapters may be interchanged or excluded at the instructor's discretion.

Three Important Themes

The newly revised and expanded Chapter 1 "Introduction to Earth Science" presents students with three important themes that recur throughout the book—*Earth as a System, People and the Environment*, and *Understanding Earth*.

Earth as a System

 An important occurrence in modern science has been the realization that Earth is a giant multi-dimensional system. Our planet consists of many separate but interacting parts. A change in any one part can produce changes in any or all of the other parts—often in ways that are neither obvious nor immediately apparent. Although it is not possible to study the entire system at once, it is possible to develop an awareness and appreciation for the concept and for many of the system's important interrelationships. Therefore, starting with revised discussion of "Earth System Science" in Chapter 1, the theme of "Earth as a System" keeps recurring through all major units of the book. It is a thread that "weaves" through the chapters and helps tie them together.

Several new and revised special interest boxes relate to *Earth as a system*. To remind the reader of this important theme, the small icon you see at the beginning of this section is used to mark these boxes.

Finally, each chapter concludes with a section on *Examining the Earth System.* The questions and problems found here are intended to develop an awareness and appreciation for some of the Earth system's many interrelationships.

People and the Environment

Because knowledge about our planet and how it works is necessary to our survival and well being, the treatment of environmental issues has always been an important part of *Earth Science.* Such discussions serve to illustrate the relevance and application of Earth science knowledge. With each new edition this focus has been given greater emphasis. This is certainly the case with the Eleventh Edition. The text integrates a great deal of information about the relationship between people and the natural environment and explores the application of the Earth sciences to understanding and solving problems that arise from these interactions.

In addition to many basic text discussions, many of the text's special interest boxes involve the "People and the Environment" theme and are quickly recognized by the distinctive icon you see at the beginning of this section.

Understanding Earth

As members of a modern society, we are constantly reminded of the benefits derived from science. But what exactly is the nature of scientific inquiry? Developing an understanding of how science is done and how scientists work is a third important theme that appears throughout this book beginning with the section on "The Nature of Scientific Inquiry" in Chapter 1. Students will examine some of the difficulties encountered by scientists as they attempt to acquire reliable data about our planet and some of the ingenious methods that have been developed to overcome these difficulties. Students will also explore many examples of how hypotheses are formulated and tested as well as learn about the evolution and development of some major scientific theories. Many basic text discussions as well as a number of the special interest boxes on "Understanding Earth" provide the reader with a sense of the observational techniques and reasoning processes involved in developing scientific knowledge. The emphasis is not just on what scientists know, but how they figured it out.

Highlights of the Eleventh Edition

The Eleventh Edition of *Earth Science* represents a thorough revision. *Every* part of the book was examined carefully with the dual goals of keeping topics current and improving the clarity of text discussions. People

familiar with preceding editions will see much that is new in the Eleventh Edition. The list of specifics is long. Examples include the following:

- Chapter 1 "Introduction to Earth Science" is *new.* In previous editions the introduction did not have chapter status. This reorganized and expanded series of discussions provides a more complete introduction and overview of Earth science.

- Much is new in the chapters on Earth materials. Chapter 2 has a completely rewritten section on the geologic definition of minerals and an all new discussion of elements that more clearly explains atomic structure and bonding. Chapter 3 includes an expanded treatment of the rock cycle including new line art and also a significant revision of the section on igneous rocks.

- Among the changes in Chapter 4 is an all new section on "Classifying Soils" that includes a new table on world soil orders and a large new world map showing global soil regions.

- The portion of Chapter 5 that focuses on running water has been reorganized and almost entirely rewritten so that the discussion of streams progresses in a manner that is clearer and more logical for the beginning student. The portion of Chapter 5 on groundwater includes new sections on the storage and movement of groundwater.

- Chapter 8 on plate tectonics has been made even better! This chapter has been *extensively* reorganized, revised, and rewritten. Now more than ever, the chapter clearly summarizes and explains the most important unifying theory in the Earth sciences. The new title "A Scientific Theory Unfolds" serves to highlight the fact that a significant emphasis involves tracing the historical development of the theory of plate tectonics as a way of providing students with insight into how science and scientists work.

- Chapter 10, "Mountain Building," more effectively explains basic processes in the context of plate tectonics. Organizationally, the chapter has a more logical progression of topics and includes new discussions of the Himalayas and Appalachians.

- Unit 5, "The Global Ocean," which was substantially revised and expanded in the previous edition, has undergone additional improvements. Highlights include new information on mapping the ocean floor and the oceanic ridge system in Chapter 13. Chapter 15 contains an all new section on "The Coastal Zone."

- Unit 6, "Earth's Dynamic Atmosphere," has experienced many changes that have strengthened and improved discussions and updated coverage. Chapter 16 has an expanded section contrasting "Weather and Climate," and Chapter 17 has a revised discussion of "Water's Changes of State." Chapter 18 has easier to understand sections on "Understanding Air Pressure" and "Measuring Air Pressure," as well as two new special interest boxes. An expanded look at hurricanes and their formation can be found in Chapter 19. Chapter 20 concludes the unit with a clearer introduction to "The Climate System" and updated coverage of global warming.

- Chapter 22 "Touring Our Solar System," a part of the concluding unit on "Earth's Place in the Universe," includes a revised discussion of lunar history and clearer explanations of lunar features. Updated coverage of Mars includes results from recent Mars landers.

Additional Highlights

- "*Students Sometimes Ask . . .*," a feature that was new in the Tenth Edition, has been retained and improved in the Eleventh Edition. Instructors and students reacted favorably and indicated that the questions and answers that are sprinkled through each chapter add interest and relevance to text discussions.

- Although the total number of special interest boxes is slightly reduced, thirteen are totally new or substantially revised. As in the previous edition, most are intended to illustrate and reinforce the three themes of "Earth as a System," "People and the Environment," and "Understanding Earth."

- The text's interactive CD-ROM *GEODe: Earth Science*, v.2 is an even better learning tool than before. Each section now concludes with a review quiz consisting of randomly generated questions that help students test their understanding of basic facts and concepts. We have maintained the use of a special icon that appears throughout the book wherever a text discussion has a corresponding *GEODe: Earth Science* activity.

The Teaching and Learning Package

The challenge is fundamental and too often overlooked in what seems to have become a weapons race of resources supplemental to the text: *instructors need more time, students need more preparation*. With this as a credo, Prentice Hall has produced for this edition perhaps the best set of instructor and student resources ever assembled to support an introductory Earth science textbook. Not only are they of the highest quality, they are the most *useful*. Please see pages xx–xxii of this Preface for detailed descriptions.

Acknowledgments

Writing a college textbook requires the talents and cooperation of many individuals. We appreciate the aid of Professor Alan Trujillo at Palomar College. His contributions to the oceanography unit and to the "Students Sometimes Ask . . ." feature remain an important part of *Earth Science*. Working with Dennis Tasa, who is responsible for all of the outstanding illustrations, is always special for us. We not only value his outstanding artistic talents and imagination but his friendship. We are grateful to Professor Ken Pinzke at Southwestern Illinois College. In addition to his many helpful suggestions regarding the manuscript, Ken is responsible for the text's student study guide, and laboratory manual. Ken is an important part of our team and a valued friend as well.

Special thanks go to those colleagues who prepared in-depth reviews. Their critical comments and thoughtful input helped guide our work and clearly strengthened the text. We wish to thank:

David Brackney, *Rochester College*
Patricia Crews, *Florida Community College, Jacksonville–Kent Campus*
Susanne Holmes-Koetter, *Faulkner State Community College*
Christopher Hooker, *Waubonsee Community College*
Leslie Kanat, *Johnson State College*
Mark Micozzi, *East Central University*
Tina-Gayle Osborn, *Palm Beach Community College*
David Pitts, *University of Houston–Clear Lake*
Bethan Salle, *Collin County Community College*
W. Patrick Seward, *Rogers State University*
Lisa Vanderbloemen, *St. Petersburg College*
David Voorhees, *Waubonsee Community College*

We also want to acknowledge the team of professionals at Prentice Hall. We sincerely appreciate the company's continuing strong support for excellence and innovation. Special thanks to our executive editor Patrick Lynch. We value his leadership and appreciate his attention to detail, excellent communication skills, and easy-going style. The production team, led by Ed Thomas, has once again done an outstanding job. Thanks also to Barbara Booth for her excellent copyediting skills. The entire production team consists of true professionals with whom we are very fortunate to be associated.

Edward J. Tarbuck
Frederick K. Lutgens

Instructor Resources

Prentice Hall continues to improve the instructor resources in this edition with the goal of saving you time in preparing for your classes.

Instructor's Resource Center (IRC) on DVD:

The IRC puts all your lecture resources in one easy-to-reach place:

- Three **PowerPoint**® **presentations** for each chapter
- **43 animations** of Earth processes
- **All of the line art, tables, and photos** from the text in .jpg files (Are illustrations central to your lecture? Check out the *Student Lecture Notebook*.)
- *Images of Earth* photo gallery
- *Instructor's Manual* in Microsoft Word
- *Test Item File* in Microsoft Word
- **TestGenEQ** test generation and management software

PowerPoints®

Found on the IRC are three PowerPoint® files for each chapter. Cut down on your preparation time, no matter what your lecture needs:

1. **Art and Animations**—All of the line art, tables, and photos from the text, along with the animation library, pre-loaded into Power-Point® slides for easy integration into your presentation.
2. **Lecture Outline**—Authored by Stanley Hatfield of Southwestern Illinois College, this set averages 35 slides per chapter and includes customizable lecture outlines with supporting art.
3. **Classroom Response System (CRS) Questions**—Authored for use in conjunction with any of the new classroom response systems. These systems allow you to electronically poll your class for responses to questions, pop quizzes, attendance, and more.

Animations

Found on the IRC on DVD, the *Prentice Hall Geoscience Animation Library* includes 43 animations illuminating many difficult-to-visualize Earth science topics. Created through a unique collaboration among five of Prentice Hall's leading geoscience authors, these animations represent a truly significant leap forward in lecture presentation aids. Each animation is mapped to its corresponding chapter. They are provided as both Flash files and, for your convenience, pre-loaded into PowerPoint® slides.

- Earth-Sun Relations
- Convergent Margins
- Foliation
- P & S Waves
- Stream Processes
- Faults
- Transform Faults
- Angular Unconformity and Nonconformity
- Global Warming
- Beach Drift
- Folding
- Seismograph Operations
- Breakup of Pangaea
- Nebular Hypothesis
- Hurricanes
- Oxbow Lake Formation
- Crater Lake
- Igneous Features
- El Niño/La Niña
- Hydrologic Cycle
- Tidal Cycle
- Seafloor Spreading
- Glacial Processes—Ice Budget
- Relative Dating
- Coriolis Effect
- Tectonic Settings and Volcanic Activity
- Glacial Processes—Plucking and Moraines
- Water Phases
- Wave Motion
- Coastal Stabilization—Jetties, Groins, Breakwaters
- Ocean Circulation
- Accretion of Terranes
- Formation of Tornadoes
- Global Atmospheric Circulation Model
- Fronts
- Ozone Depletion
- Atmosphere Energy Balance
- Cyclones & Anticyclones
- Seasonal Pressure and Precipitation Patterns
- Jetstream and Rossby Waves
- Wind Pattern Development
- Mid-Latitude Cyclones
- Atmospheric Stability

"Images of Earth" Photo Gallery

Supplement your personal and text-specific slides with this amazing collection of over 300 geologic photos contributed by Marli Miller (University of Oregon) and other professionals in the field. The photos are available on the IRC on DVD and searchable by keyword.

Transparencies

Simply put: every Dennis Tasa illustration in *Earth Science*, Eleventh Edition is available on full-color, projection enhanced transparency—over 350 in all. (Are illustrations central to your lecture? Check out the *Student Lecture Notebook*.)

Instructor's Manual with Tests

Authored by Stanley Hatfield (Southwestern Illinois College), the *Instructor's Manual* contains: learning objectives, chapter outlines, answers to end-of-chapter questions and suggested, short demonstrations to spice up your lecture. The *Test Item File* incorporates art and averages 75 multiple-choice, true/false, short answer and critical thinking questions per chapter.

TestGenEQ

Use this electronic version of the *Test Item File* to customize and manage your tests. Create multiple versions, add or edit questions, add illustrations—your customization needs are easily addressed by this powerful software.

Test Item File in Blackboard and WebCT formats

Already have your own web site set up? Prentice Hall can provide you with the *Earth Science Test Item File* in formats suitable for importation into your Blackboard or WebCT course. Additional course resources are available on the IRC and are available for your use, with permission (see the ReadMe file on the IRC).

For the Laboratory

Applications and Investigations in Earth Science, Fifth Edition. Written by Ed Tarbuck, Fred Lutgens, and Ken Pinzke, this full-color laboratory manual contains 22 exercises that provide students with hands-on experiences in geology, oceanography, meteorology, astronomy, and Earth science skills. The lab manual is available at a discount when purchased with the text; please contact your local Prentice Hall representative for more details.

Student Resources

The student resources to accompany *Earth Science*, Eleventh Edition have been further refined with the goal of focusing the students' efforts and improving their understanding of the concepts of Earth science.

GEODe: Earth Science

Somewhere between a text and a tutor, *GEODe: Earth Science* reinforces key concepts using animations, video, narration, interactive exercises, and practice quizzes. A copy of *GEODe: Earth Science* is automatically included in every copy of the text purchased from Prentice Hall.

Study Guide

Written by experienced educator Ken Pinzke (Southwestern Illinois College), the *Study Guide* helps students identify the important points from the text, and then provides them with review exercises, study questions, self-check exercises, and vocabulary review.

Online Study Guide

www.prenhall.com/tarbuck Authored by Tina-Gayle Osborn (Palm Beach Community College), the *Online Study Guide* contains numerous chapter review exercises (from which students get immediate feedback). Links to other resources are also included for further study. Professors can utilize the quizzing modules in conjunction with a course management system to assess student progress.

Student Lecture Notebook

All of the line art from the text and transparency set are reproduced in this full color notebook, with space for notes. Students can now fully focus on the lecture and not be distracted by attempting to replicate figures. Each page is three-hole punched for easy integration with other course materials.

OneKey

OneKey is all you need.

OneKey offers the best teaching and learning online resources all in one place. Conveniently organized by textbook chapter, these compiled resources help instructors save time and help students reinforce and apply what they have learned in class. OneKey for convenience, simplicity and success. For more information contact your local Prentice Hall representative or visit **www.prenhall.com/onekey**.

For instructors:
- Electronic versions of the art
- *PH Geoscience Animation Library*
- PowerPoint™ presentations
- *On-Line Study Guide* self-tests
- The *Test Item File*
- Access to all OneKey student resources

For students:
- *Research Navigator*
- *On-Line Study Guide*
- *New York Times: eThemes of the Times for Earth Science*

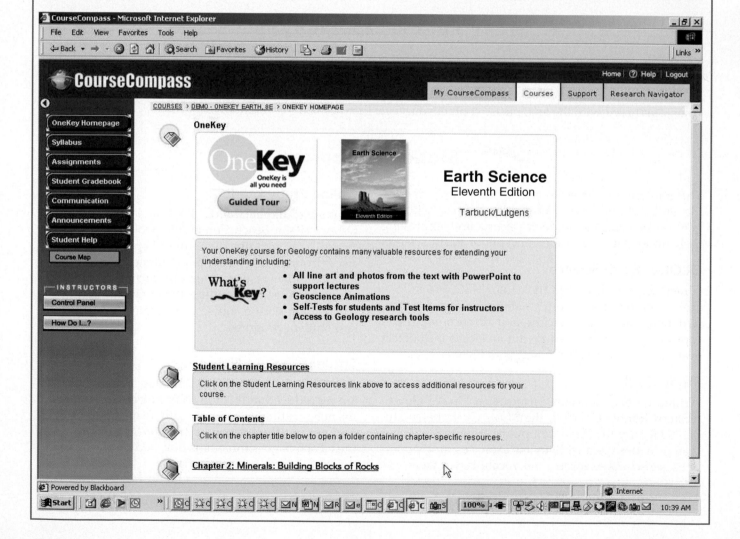

EARTH SCIENCE

Reflection Lake in Washington's Mount Rainier National Park. Glaciers are still sculpting this large volcanic mountain. (Photo by Art Wolfe)

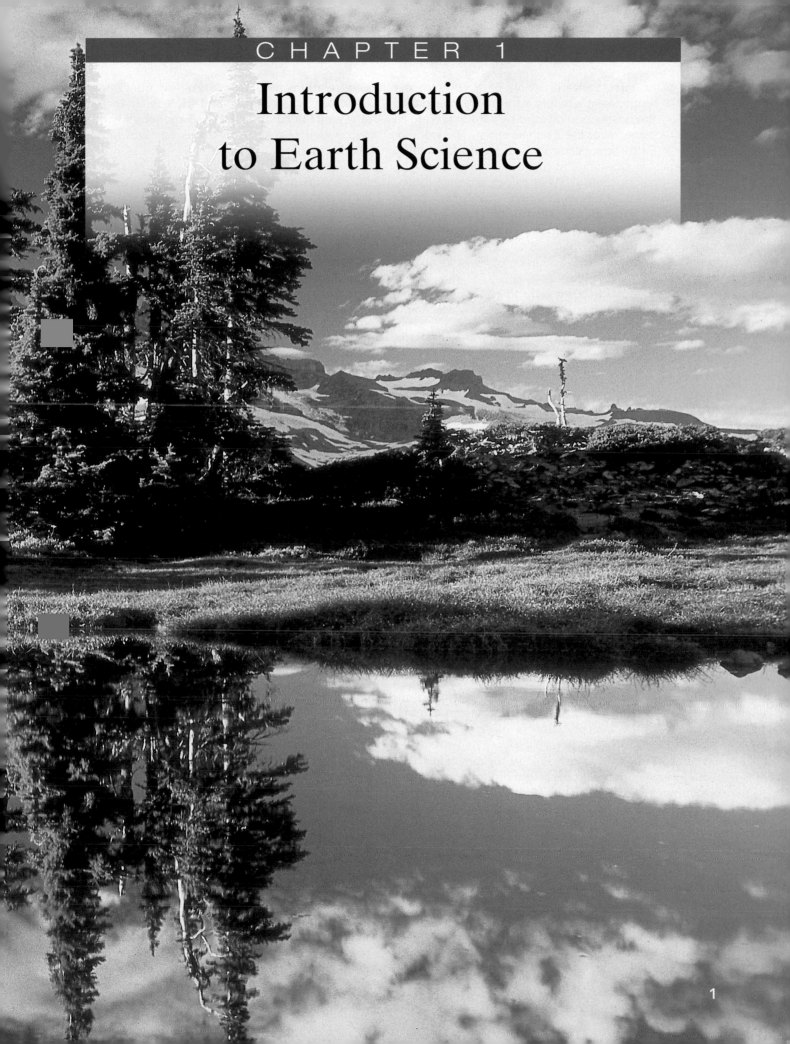

CHAPTER 1

Introduction to Earth Science

The spectacular eruption of a volcano, the magnificent scenery of a rocky coast, and the destruction created by a hurricane are all subjects for the Earth scientist. The study of Earth science deals with many fascinating and practical questions about our environment (see chapter-opening photo). What forces produce mountains? Why is our daily weather so variable? Is climate really changing? How old is Earth, and how is it related to the other planets in the solar system? What causes ocean tides? What was the Ice Age like? Will there be another? Can a successful well be located at this site?

The subject of this text is *Earth science*. To understand Earth is not an easy task, because our planet is not a static and unchanging mass. Rather, it is a dynamic body with many interacting parts and a long and complex history.

What Is Earth Science?

Earth science is the name for all the sciences that collectively seek to understand Earth and its neighbors in space. It includes geology, oceanography, meteorology, and astronomy.

In this book, Units One through Four focus on the science of **geology**, a word that literally means "study of Earth." Geology is traditionally divided into two broad areas—physical and historical.

Physical geology examines the materials composing Earth and seeks to understand the many processes that operate beneath and upon its surface. Earth is a dynamic, ever-changing planet. Internal forces create earthquakes, build mountains, and produce volcanic structures. At the surface, external processes break rock apart and sculpture a broad array of landforms. The erosional effects of water, wind, and ice result in a great diversity of landscapes. Because rocks and minerals form in response to Earth's internal and external processes, their interpretation is basic to an understanding of our planet.

In contrast to physical geology, the aim of *historical geology* is to understand the origin of Earth and the development of the planet through its 4.5-billion-year history. It strives to establish an orderly chronological arrangement of the multitude of physical and biological changes that have occurred in the geologic past. The study of physical geology logically precedes the study of Earth history because we must first understand how Earth works before we attempt to unravel its past.

Unit Five, *The Global Ocean*, is devoted to **oceanography**. Oceanography is actually not a separate and distinct science. Rather, it involves the application of all sciences in a comprehensive and interrelated study of the oceans in all their aspects and relationships. Oceanography integrates chemistry, physics, geology, and biology. It includes the study of the composition and movements of seawater, as well as coastal processes, seafloor topography, and marine life.

Unit Six, *Earth's Dynamic Atmosphere,* examines the mixture of gases that is held to the planet by gravity and thins rapidly with altitude. Acted on by the combined effects of Earth's motions and energy from the Sun, the formless and invisible atmosphere reacts by producing an infinite variety of weather, which in turn creates the basic pattern of global climates. **Meteorology** is the study of the atmosphere and the processes that produce weather and climate. Like oceanography, meteorology involves the application of other sciences in an integrated study of the thin layer of air that surrounds Earth.

Unit Seven, *Earth's Place in the Universe,* demonstrates that an understanding of Earth requires that we relate our planet to the larger universe. Because Earth is related to all of the other objects in space, the science of **astronomy**—the study of the universe—is very useful in probing the origins of our own environment. Because we are so closely acquainted with the planet on which we live, it is easy to forget that Earth is just a tiny object in a vast universe. Indeed, Earth is subject to the same physical laws that govern the many other objects populating the great expanses of space. Thus, to understand explanations of our planet's origin, it is useful to learn something about the other members of our solar system. Moreover, it is helpful to view the solar system as a part of the great assemblage of stars that comprise our galaxy, which in turn is but one of many galaxies (see Box 1.1)

To understand Earth science is challenging because our planet is a dynamic body with many interacting parts and a complex history. Throughout its long existence, Earth has been changing. In fact, it is changing as you read this page and will continue to do so into the foreseeable future. Sometimes the changes are rapid and violent, as when severe storms, landslides, or volcanic eruptions occur. Just as often, change takes place so gradually that it goes unnoticed during a lifetime. Scales of size and space also vary greatly among the phenomena studied in Earth science.

Earth science is often perceived as science that is performed in the out of doors, and rightly so. A great deal of what Earth scientists study is based on observations and experiments conducted in the field. But Earth science is also conducted in the laboratory, where, for example, the study of various Earth materials provides insights into many basic processes, and the creation of complex computer models allows for the simulation of our planet's complicated climate system. Frequently, Earth scientists require an understanding and application of knowledge and principles from physics, chemistry, and biology. Geology, oceanography, meteorology, and astronomy are sciences that seek to expand our knowledge of the natural world and our place in it.

Earth Science, People, and the Environment

Environment refers to everything that surrounds and influences an organism. Some of these things are biological and social, but others are nonliving. The factors in this latter category are collectively called our **physical environment**. The physical environment encompasses water, air, soil, and rock, as well as conditions such as temperature, humidity, and sunlight. The phenomena and processes studied by the Earth sciences are basic to an understanding of the physical environment. In this sense, most of Earth science can be characterized as environmental science.

However, when the term *environmental* is applied to Earth science today, it is usually reserved for those aspects that focus on the relationships between people and the natural environment (Figure 1.1).

We can dramatically influence natural processes. For example, river flooding is natural, but the magnitude and frequency of flooding can be changed significantly by human activities such as clearing forests, building cities, and constructing dams. Unfortunately, natural systems do not always adjust to artificial changes in ways we can anticipate. Thus, an alteration to the environment that was intended to benefit society may have the opposite effect.

Interactions between people and the natural environment is an important theme that recurs throughout this text and helps to tie its units together. The primary focus of this book is to develop an understanding of basic Earth science principles, but along the way, we will explore numerous important relationships between people and the natural environment.

Resources

Resources are an important focus of the Earth sciences that is of great value to people. They include water and soil, a great variety of metallic and nonmetallic minerals, and energy. Together they form the very foundation of modern civilization. The Earth sciences deal not only with the formation and occurrence of these vital resources but also with maintaining supplies and the environmental impact of their extraction and use.

Few people who live in highly industrialized nations realize the quantity of resources needed to maintain their present standard of living. For example, the annual per capita consumption of metallic and nonmetallic mineral resources for the United States is nearly 10,000 kilograms (11 tons). This is each person's prorated share of the materials required by industry to provide the vast array of products modern society demands. Figures for other highly industrialized countries are comparable.

Resources are commonly divided into two broad categories. Some are classified as **renewable**, which means that they can be replenished over relatively short time spans. Common examples are plants and animals for food, natural fibers for clothing, and forest products for lumber and paper. Energy from flowing water, wind, and the Sun are also considered renewable.

By contrast, many other basic resources are classified as **nonrenewable**. Important metals such as iron,

◄ **FIGURE 1.1** Air pollution episode in Santiago, Chile Air-quality problems affect many cities. Fuel combustion by motor vehicles and power plants provides a high proportion of the pollutants. Meteorological factors determine whether pollutants remain "trapped" in the city or are dispersed. (Photo by P. Baeza/Publiphoto/Photo Researchers, Inc.)

BOX 1.1 EARTH AS A SYSTEM:
Earth's Place in the Cosmos*

For centuries, those who have gazed at the night sky have wondered about the nature of the universe, Earth's place within it, and whether or not we are "alone." Today many exciting discoveries in astronomy are beginning to provide some answers concerning the origin of the universe, the formation and evolution of stars, and how Earth and its materials came into existence.

The realization that the universe is immense and orderly began in the early 1900s when Edwin Hubble and other dedicated scientists demonstrated that the Milky Way Galaxy is one of hundreds of billions of galaxies, each of which contains billions of stars. Today we understand that Earth, its materials, and indeed all living things are the consequence of a sequence of events, governed by natural laws, that occurred within the universe during the past 10–15 billion years.

Astronomical evidence supports the theory, called the Big Bang, that the universe began about 14 billion years ago when a dense, hot, supermassive concentration of material exploded with cataclysmic force (Figure 1.A). Within about one second, the temperature of the expanding universe cooled to ap-

proximately 10 billion degrees, and fundamental atomic particles called protons and neutrons began to appear. After a few minutes, atoms of the least complex elements—hydrogen and helium—had formed, and the initial conversion of energy to matter in the young universe was completed.

During the first billion years or so, matter (essentially hydrogen and to a lesser extent helium) in the expanding universe became clumpy and fragmented into enormous clouds and groups of clouds that eventually collapsed to become galaxies and clusters of galaxies. Inside these collapsing clouds, smaller concentrations of matter formed into stars. One of the billions of galaxies to form was the one we call the Milky Way.

The Milky Way is a collection of several hundred billion stars, the oldest of which is about 10 billion years. It is one of a cluster of approximately 28 galaxies, called the Local Group, that exist in our region of the universe. Initially, the oldest stars in the Milky Way formed from nearly pure hydrogen. Later, succeeding generations of younger stars, including our Sun, would have heavier, more complex atoms available for their formation.

After forming, all stars go through a series of changes before finally dying out. During the life of most stars, energy is produced as hydrogen nuclei (protons) fuse with other hydrogen nuclei to form helium. During this process, called *nuclear fusion*, matter is converted to energy. Stars begin to die when their nuclear fuel becomes exhausted. Massive stars often have indescribably explosive deaths. During these violent events, called supernovas, nuclear fusion advances beyond helium and produces heavier, more complex atoms such as oxygen, carbon, and iron. In turn, these products of stellar death may become the materials that make up future generations of stars. It was from the debris scattered during the death of a preexisting star, or stars, that our Sun, as well as the rest of the solar system, formed.

Our star, the Sun, is at the very least a second-generation star. It, along with the planets and other members of the solar system, began forming nearly 5 billion years ago, some 9 billion years after the Big Bang, from a large, interstellar cloud, called a *nebula*, which consisted of dust particles and gases enriched in heavy elements from supernova explosions. Gravitational energy

---

aluminum, and copper fall into this category, as do our most important fuels: oil, natural gas, and coal. Although these and other resources continue to form, the processes that create them are so slow that significant deposits take millions of years to accumulate. In essence, Earth contains fixed quantities of these substances. When the present supplies are mined or pumped from the ground, there will be no more. Although some nonrenewable resources, such as aluminum, can be used over and over again, others, such as oil, cannot be recycled.

Population Growth

The population of our planet is growing rapidly. Although it took until the beginning of the nineteenth century for the number to reach 1 billion, just 130 years were needed for the population to double to 2 billion. Between 1930 and 1975 the figure doubled again to 4 billion, and by about 2010 nearly 7 billion

people may inhabit the planet. Clearly, as population grows, the demand for resources expands as well. However, the rate of mineral- and energy-resource usage has climbed more rapidly than the overall growth of population. This fact results from an increasing standard of living. This is certainly true in the United States, where about 6 percent of the world's population uses approximately 30 percent of the annual production of mineral and energy resources.

How long will the remaining supplies of basic resources last? How long can we sustain the rising standard of living in today's industrialized countries and still provide for the growing needs of developing regions? How much environmental deterioration are we willing to accept in pursuit of basic resources? Can alternatives be found? If we are to cope with an increasing per-capita demand and a growing world population, it is important that we have some understanding of our present and potential resources.

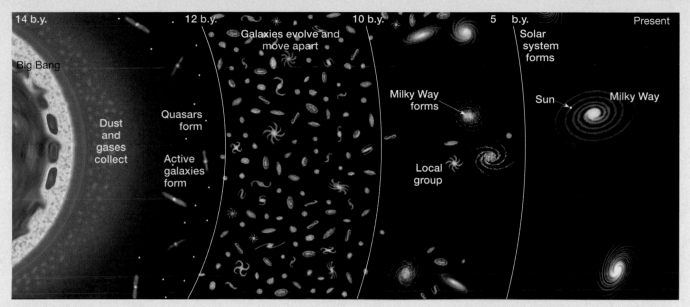

14 b.y.

12 b.y.

Galaxies evolve and move apart

10 b.y.

5 b.y.

Present

Big Bang

Dust and gases collect

Quasars form

Active galaxies form

Milky Way forms

Local group

Solar system forms

Sun

Milky Way

Figure 1.A About 14 billion years ago, an incomprehensibly large explosion sent all the matter of the universe flying outward at incredible speeds. After a few billion years, the material cooled and condensed into the first stars and galaxies. Because the universe is expanding, the evolving galaxies continue to move away from one another. About 5 billion years ago, our solar system began forming in one of these galaxies, the Milky Way.

caused the nebula to contract and, in so doing, it began to rotate and flatten. Inside, smaller concentrations of matter began condensing to form the planets. At the center of the nebula there was sufficient pressure and heat to initiate hydrogen nuclear fusion, and our Sun was born.

It has been said that all life on Earth is related to the stars. This is true because the very atoms in our bodies, as well as the atoms that make up every other thing on Earth, owe their origin to a supernova event that occurred billions of years ago, trillions of kilometers away.

*This box was prepared by Kenneth Pinzke, Southwestern Illinois College.

Environmental Problems

In addition to the quest for adequate mineral and energy resources, the Earth sciences must also deal with a broad array of other environmental problems. Some are local, some are regional, and still others are global in extent. Developed and developing nations alike face serious difficulties. Urban air pollution, acid rain, ozone depletion, and global warming are just a few concerns that pose significant threats. Other problems involve the loss of fertile soils to erosion, the disposal of toxic wastes, and the contamination and depletion of water resources. The list continues to grow.

In addition to human-induced and human-accentuated problems, people must also cope with the many natural hazards posed by the physical environment (Figure 1.2). Earthquakes, landslides, floods, and hurricanes are just four of the many risks. Others, such as drought, although not as spectacular, are nevertheless equally important environmental concerns. Of course, environmental hazards are simply *natural* processes. They become hazards only when people try to live where these processes occur. Figure 1.2 and 1.3 clearly illustrate this idea. In many cases, the threat of natural hazards is aggravated by increases in population as more people crowd into places where an impending danger exists.

It is clear that as world population continues its rapid growth, pressures on the environment will increase as well. Therefore, an understanding of Earth is not only essential for the location and recovery of basic resources but also for dealing with the human impact on the environment and minimizing the effects of natural hazards. Knowledge about our planet and how it works is necessary to our survival and well-being. Earth is the only suitable habitat we have, and its resources are limited.

▲ **FIGURE 1.2** Two geologic hazards are represented in this image. On January 13, 2001, a magnitude 7.6 earthquake caused considerable damage in El Salvador. The damage pictured here was caused by a landslide that was triggered by the earthquake. As many as 1000 people were buried under 8 meters (26 feet) of landslide debris. (Photo by Reuters/STR/Archive Photos)

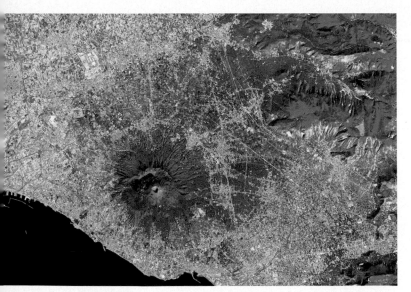

▲ **FIGURE 1.3** This is an image of Italy's Mount Vesuvius in September 2000. This major volcano is surrounded by the city of Naples and the Bay of Naples. In 79 A.D. Vesuvius explosively erupted, burying the towns of Pompeii and Herculaneaum in volcanic ash. Will it happen again? Such hazards are *natural* processes. They only become hazards when people try to live where these processes occur. (Image courtesy of NASA)

The Nature of Scientific Inquiry

As members of a modern society, we are constantly reminded of the benefits derived from science. But what exactly is the nature of scientific inquiry? Developing an understanding of how science is done and how scientists work is another important theme that appears throughout this book. You will explore the difficulties in gathering data and some of the ingenious methods that have been developed to overcome these difficulties. You will also see many examples of how hypotheses are formulated and tested, as well as learn about the evolution and development of some major scientific theories.

All science is based on the assumption that the natural world behaves in a consistent and predictable manner that is comprehensible through careful, systematic study. The overall goal of science is to discover the underlying patterns in nature and then to use this knowledge to make predictions about what should or should not be expected, given certain facts or circumstances. For example, by knowing how oil deposits form, geologists are able to predict the most favorable sites for exploration and, perhaps as important, how to avoid regions having little or no potential.

The development of new scientific knowledge involves some basic logical processes that are universally accepted. To determine what is occurring in the natural world, scientists collect scientific *"facts"* through observation and measurement (Figure 1.4). Because some error is inevitable, the accuracy of a particular measurement or observation is always open to question. Nevertheless, these data are essential to science and serve as the springboard for the development of scientific theories.

Hypothesis

Once facts have been gathered and principles have been formulated to describe a natural phenomenon, investigators try to explain how or why things happen in the manner observed. They often do this by constructing a tentative (or untested) explanation, which is called a scientific **hypothesis** or **model**. (The term

▼ **FIGURE 1.4** This geologist is taking a lava sample from a "skylight" in a lava tube. Near Kilauea Volcano, Hawaii. (Photo by G. Brad Lewis/Tony Stone Images)

model, although often used synonymously with hypothesis, is a less precise term because it is sometimes used to describe a scientific theory as well.) It is best if an investigator can formulate more than one hypothesis to explain a given set of observations. If an individual scientist is unable to devise multiple models, others in the scientific community will almost always develop alternative explanations. A spirited debate frequently ensues. As a result, extensive research is conducted by proponents of opposing models, and the results are made available to the wider scientific community in scientific journals.

Before a hypothesis can become an accepted part of scientific knowledge, it must pass objective testing and analysis. (If a hypothesis cannot be tested, it is not scientifically useful, no matter how interesting it might seem.) The verification process requires that *predictions* be made based on the model being considered and that the predictions be tested by comparing them against objective observations of nature. Put another way, hypotheses must fit observations other than those used to formulate them in the first place. Those hypotheses that fail rigorous testing are ultimately discarded. The history of science is littered with discarded hypotheses. One of the best known is the Earth-centered model of the universe—a proposal that was supported by the apparent daily motion of the Sun, Moon, and stars around Earth. As the mathematician Jacob Bronowski so ably stated, "Science is a great many things, but in the end they all return to this: Science is the acceptance of what works and the rejection of what does not."

Theory

When a hypothesis has survived extensive scrutiny and when competing models have been eliminated, a hypothesis may be elevated to the status of a scientific **theory.** In everyday language we may say, "That's only a theory." But a scientific theory is a well-tested and widely accepted view that the scientific community agrees best explains certain observable facts.

Theories that are extensively documented are held with a very high degree of confidence. Theories of this stature that are comprehensive in scope have a special status. They are called **paradigms** because they explain a large number of interrelated aspects of the natural world. For example, the theory of plate tectonics is a paradigm of the geological sciences that provides the framework for understanding the origin of mountains, earthquakes, and volcanic activity. In addition, plate tectonics explains the evolution of the continents and the ocean basins through time—a topic we will consider later in the book.

Box 1.2 Understanding Earth
Studying Earth from Space

Scientific facts are gathered in many ways, including laboratory studies and field observations and measurements. Satellite images like those in Figure 1.B and Figure 1.C are another useful source of data. Such images provide perspectives that are difficult to gain from more traditional sources. Moreover, the high-tech instruments aboard many satellites enable scientists to gather information from remote regions where data are otherwise scarce.

The image in Figure 1.B makes use of the Advanced Spaceborne Thermal Emission and Reflection Radiometer (ASTER). Because different materials reflect and emit energy in different ways, ASTER can provide detailed information about the composition of Earth's surface. Figure 1.B is a three-dimensional view looking north over Death Valley, California. The data have been computer enhanced to exaggerate the color variations that highlight differences in types of surface materials.

Salt deposits on the floor of Death Valley appear in shades of yellow, green, purple, and pink, indicating the presence of carbonate, sulfate, and chloride minerals. The Panamint Mountains to the west (left) and the Black Mountains to the east are made up of sedimentary limestones, sandstones, shales, and metamorphic rocks. The bright red areas are dominated by the mineral quartz, found in sandstone; green areas are limestone. In the lower center of the image is Badwater, the lowest point in North America.

The image in Figure 1.C is from NASA's *Tropical Rainfall Measuring Mission (TRMM)*. Rainfall patterns over land have been studied for many years using ground-based radar and other instruments. Now the instruments aboard the *TRMM* satellite have greatly expanded our ability to collect precipitation data. In addition to data for land areas, this satellite provides extremely precise measurements of rainfall over the oceans where conventional land-based instruments cannot see. This is especially important because much of Earth's rain falls in ocean-covered tropical areas, and a great deal of the globe's weather-producing energy comes from heat exchanges involved in the rainfall process. Until the *TRMM*, information on the intensity and amount of rainfall over the tropics was scanty. Such data are crucial to understanding and predicting global climate change.

Figure 1.C *TRMM* is a research satellite designed to expand our understanding of the water cycle and its role in our climate system. By covering the region between the latitudes 35°N and 35°S, it provides much needed data on rainfall and the heat release associated with rainfall. Many types of measurements and images are possible. Here the satellite captures rainfall rates in Cyclone Gafilo on the morning of March 6, 2004. Dark red indicates rainfall rates of up to 50 millimeters (2 inches) per hour. (NASA/*TRMM* image)

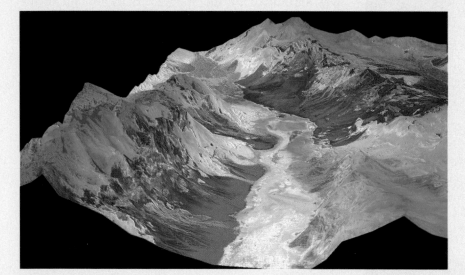

Figure 1.B This satellite image shows detailed information about the composition of surface materials in Death Valley, California. It was produced by superimposing nighttime thermal infrared data, acquired on April 7, 2000, over topographic data from the U.S. Geological Survey. (Image courtesy of NASA)

Scientific Methods

The process just described, in which researchers gather facts through observations and formulate scientific hypotheses and theories, is called the *scientific method*. Contrary to popular belief, the scientific method is not a standard recipe that scientists apply in a routine manner to unravel the secrets of our natural world. Rather, it is an endeavor that involves creativity and insight. Rutherford and Ahlgren put it this way: "Inventing hypotheses or theories to imagine how the world works and then figuring out how they can be put to the test of reality is as creative as writing poetry, composing music, or designing skyscrapers."*

---

*F. James Rutherford and Andrew Ahlgren, *Science for All Americans* (New York: Oxford University Press, 1990), p. 7.

There is not a fixed path that scientists always follow that leads unerringly to scientific knowledge. Nevertheless, many scientific investigations involve the following steps: (1) the collection of scientific facts through observation and measurement; (2) the development of one or more working hypotheses or models to explain these facts; (3) development of observations and experiments to test the hypotheses; and (4) the acceptance, modification, or rejection of the model based on extensive testing.

Other scientific discoveries may result from purely theoretical ideas, which stand up to extensive examination. Some researchers use high-speed computers to simulate what is happening in the "real" world. These models are useful when dealing with natural processes that occur on very long time scales or take place in extreme or inaccessible locations. Still other scientific advancements are made when a totally unexpected happening occurs during an experiment. These serendipitous discoveries are more than pure luck, for as Louis Pasteur said, "In the field of observation, chance favors only the prepared mind."

Scientific knowledge is acquired through several avenues, so it might be best to describe the nature of scientific inquiry as the methods of science rather than the scientific method. In addition, it should always be remembered that even the most compelling scientific theories are still simplified explanations of the natural world.

In this book, you will discover the results of centuries of scientific work. You will see the end product of millions of observations, thousands of hypotheses, and hundreds of theories. We have distilled all of this to give you a "briefing" on Earth science.

But realize that our knowledge of Earth is changing daily, as thousands of scientists worldwide make satellite observations, analyze drill cores from the seafloor, measure earthquakes, develop computer models to predict climate, examine the genetic codes of organisms, and discover new facts about our planet's long history. This new knowledge often updates hypotheses and theories. Expect to see many new discoveries and changes in scientific thinking in your lifetime (see Box 1.2).

? STUDENTS SOMETIMES ASK...

In class you compared a hypothesis to a theory. How is each one different from a scientific law?

A *scientific law* is a basic principle that describes a particular behavior of nature that is generally narrow in scope and can be stated briefly—often as a simple mathematical equation. Because scientific laws have been shown time and time again to be consistent with observations and measurements, they are rarely discarded. Laws may, however, require modifications to fit new findings. For example, Newton's laws of motion are still useful for everyday applications (NASA uses them to calculate satellite trajectories), but they do not work at velocities approaching the speed of light. For these circumstances, they have been supplanted by Einstein's theory of relativity.

Scales of Space and Time in Earth Science

When we study Earth, we must contend with a broad array of space and time scales (Figure 1.5). Some phenomena are relatively easy for us to imagine, such as the size and duration of an afternoon thunderstorm or the dimensions of a sand dune. Other phenomena are so vast or so small that they are difficult to imagine. The number of stars and distances in our galaxy (and beyond!) or the internal arrangement of atoms in a mineral crystal are examples of such phenomena.

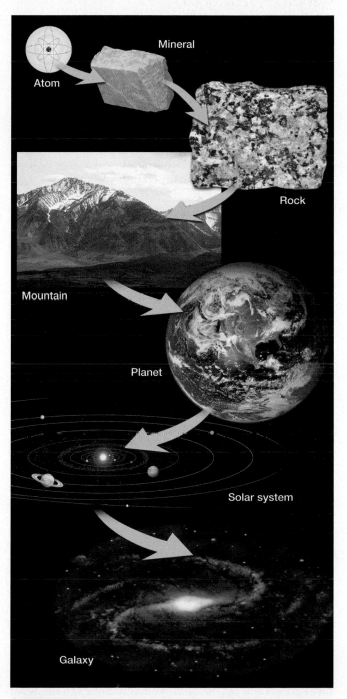

▲ **FIGURE 1.5** Earth science involves investigations of phenomena that range in size from atoms to galaxies and beyond.

Some of the events we study occur in fractions of a second. Lightning is an example. Other processes extend over spans of tens or hundreds of millions of years. The lofty Himalaya Mountains began forming about 45 million years ago, and they continue to develop today.

The concept of geologic time is new to many nonscientists. People are accustomed to dealing with increments of time that are measured in hours, days, weeks, and years. Our history books often examine events over spans of centuries, but even a century is difficult to appreciate fully. For most of us, someone or something that is 90 years old is *very old*, and a 1000-year-old artifact is *ancient*.

By contrast, those who study Earth science must routinely deal with vast time periods—millions or billions (thousands of millions) of years. When viewed in the context of Earth's 4.5-billion-year history, an event that occurred 100 million years ago may be characterized as "recent" by a geologist, and a rock sample that has been dated at 10 million years may be called "young."

An appreciation for the magnitude of geologic time is important in the study of our planet because many processes are so gradual that vast spans of time are needed before significant changes occur.

How long is 4.5 billion years? If you were to begin counting at the rate of one number per second and continued 24 hours a day, seven days a week and never stopped, it would take about two lifetimes (150 years) to reach 4.5 billion!

Over the past 200 years or so, Earth scientists have developed the **geologic time scale** of Earth history. It subdivides the 4.5-billion-year history of Earth into many different units and provides a meaningful time frame within which the events of the geologic past are arranged (Figure 1.6). The principles used to develop the geologic time scale are examined at some length in Chapter 11.

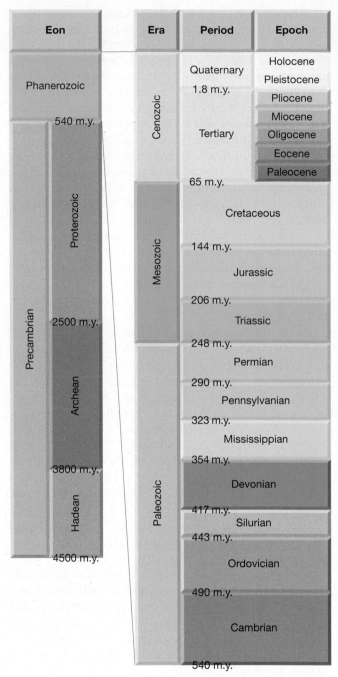

▲ **FIGURE 1.6** The geologic time scale divides the vast 4.5-billion-year history of Earth into eons, eras, periods, and epochs. We presently live in the Holocene epoch of the Quaternary period. This period is part of the Cenozoic era, which is the latest era of the Phanerozoic eon.

STUDENTS SOMETIMES ASK...

I've heard scientists use the term "light-year" when discussing astronomy. What is a "light-year?"

At first you might think that a light-year is some sort of time measurement. But, actually, the light-year is a unit for measuring distances to the stars. Such distances are so large that familiar units such as kilometers or miles are too cumbersome to use. A light-year is the distance light travels in one Earth year—about 9.5 trillion kilometers (5.8 trillion miles).

Early Evolution of Earth

This section describes the most widely accepted views on the origin of our solar system. Remember that it is a hypothesis, and like all hypotheses, it is subject to revision and even outright rejection. Nevertheless, it remains the most consistent set of ideas we have that explains what we know about the solar system today.

The orderly nature of our solar system leads most astronomers to conclude that the planets formed at essentially the same time and from the same primordial material as the Sun. This material formed a vast cloud of dust and gases called a *nebula*. The **nebular hypothesis** suggests that all bodies of the solar system formed from an enormous nebular cloud consisting mostly of hydrogen and helium, and a few percent of all the other heavier elements known to exist. The heavier substances in this frigid cloud of dust and gases consisted mostly of elements such as silicon, aluminum, iron, and calcium—the substances of today's common rocky materials. Also prevalent were other familiar elements, including oxygen, carbon, and nitrogen.

Nearly 5 billion years ago this huge cloud of gases and minute grains of heavier elements began to slowly contract due to the gravitational interactions among its particles (Figure 1.7). Some external influence, such as a shock wave traveling from a catastrophic explosion

▲ **FIGURE 1.7** Formation of the solar system according to the nebular hypothesis. **A.** The birth of our solar system began as dust and gases (nebula) started to gravitationally collapse. **B.** The nebula contracted into a rotating disk that was heated by the conversion of gravitational energy into thermal energy. **C.** Cooling of the nebular cloud caused rocky and metallic material to condense into tiny solid particles. **D.** Repeated collisions caused the dust-size particles to gradually coalesce into asteroid-size bodies. **E.** Within a few million years these bodies accreted into the planets.

(*supernova*), may have triggered the collapse. As this slowly spiraling nebula contracted, it rotated faster and faster for the same reason ice skaters do when they draw their arms toward their bodies. Eventually the inward pull of gravity came into balance with the outward force caused by the rotational motion of the nebula (Figure 1.7). By this time the once vast cloud had assumed a flat disk shape with a large concentration of material at its center called the *protosun* (pre-Sun). (Astronomers are fairly confident that the nebular cloud formed a disk because similar structures have been detected around other stars.)

During the collapse, gravitational energy was converted to thermal energy (heat), causing the temperature of the inner portion of the nebula to dramatically rise. At these high temperatures, the dust grains broke up into molecules and excited atomic particles. However, at distances beyond the orbit of Mars, the temperatures probably remained quite low. At −200°C, the tiny particles in the outer portion of the nebula were likely covered with a thick layer of ices made of frozen water, carbon dioxide, ammonia, and methane. (Some of this material still resides in the outermost reaches of the solar system in a region called the *Oort cloud*.)

The formation of the Sun marked the end of the period of contraction and thus the end of gravitational heating. Temperatures in the region where the inner planets now reside began to decline. The decrease in temperature caused those substances with high melting points to condense into tiny particles that began to coalesce (join together). Materials such as iron and nickel and the elements of which the rock-forming minerals are composed—silicon, calcium, sodium, and so forth—formed metallic and rocky clumps that orbited the Sun (Figure 1.7). Repeated collisions caused these masses to coalesce into larger asteroid-size bodies, called *protoplanets*, which in a few tens of millions of years accreted into the four inner planets we call Mercury, Venus, Earth, and Mars. Not all of these clumps of matter were incorporated into the protoplanets. Those rocky and metallic pieces that remained in orbit are called *meteorites* when they survive an impact with Earth.

As more and more material was swept up by the inner planets, the high-velocity impact of nebular debris caused the temperature of these bodies to rise. Because of their relatively high temperatures and weak gravitational fields, the inner planets were unable to accumulate much of the lighter components of the nebular cloud. The lightest of these, hydrogen and helium, were eventually whisked from the inner solar system by the solar winds.

At the same time that the inner planets were forming, the larger, outer planets (Jupiter, Saturn, Uranus, and Neptune), along with their extensive satellite systems, were also developing. Because of low temperatures far from the Sun, the material from which these planets formed contained a high percentage of ices—water, carbon dioxide, ammonia, and methane—as well as rocky and metallic debris. The accumulation of ices accounts in part for the large size and low density of the outer planets. The two most massive planets, Jupiter and Saturn, had a surface gravity sufficient to attract and hold large quantities of even the lightest elements—hydrogen and helium.

Earth's Spheres

A view such as the one in Figure 1.8A provided the *Apollo 8* astronauts as well as the rest of humanity with a unique perspective of our home. Seen from space, Earth is breathtaking in its beauty and startling in its solitude. Such an image reminds us that our home is, after all, a planet—small, self-contained, and in some ways even fragile.

As we look closely at our planet from space, it becomes apparent that Earth is much more than rock and soil. In fact, the most conspicuous features in Figure 1.8A are not continents but swirling clouds suspended above the surface of the vast global ocean. These features emphasize the importance of water on our planet.

The closer view of Earth from space shown in Figure 1.8B helps us appreciate why the physical environment is traditionally divided into three major spheres: the water portion of our planet, the hydrosphere; Earth's gaseous envelope, the atmosphere; and, of course, the solid Earth, or geosphere.

It should be emphasized that our environment is highly integrated and is not dominated by rock, water, or air alone. It is instead characterized by continuous interactions as air comes in contact with rock, rock with water, and water with air. Moreover, the biosphere, the totality of life forms on our planet, extends into each of the three physical realms and is an equally integral part of the planet. Thus, Earth can be thought of as consisting of four major spheres: the hydrosphere, atmosphere, geosphere, and biosphere.

The interactions among the spheres of Earth's environment are incalculable. Figure 1.9 provides us with one easy-to-visualize example. The shoreline is an obvious meeting place for rock, water, and air. In this scene, ocean waves that were created by the drag of air moving across the water are breaking against the rocky shore. The force of the water can be powerful, and the erosional work that is accomplished can be great.

Hydrosphere

Earth is sometimes called the *blue* planet. Water more than anything else makes Earth unique. The **hydrosphere** is a dynamic mass of liquid that is continually on the move, evaporating from the oceans to the atmosphere, precipitating to the land, and running back to the ocean again. The global ocean is certainly the most prominent feature of the hydrosphere, blanketing nearly 71 percent of Earth's surface to an aver-

A.

B.

▲ **FIGURE 1.8 A.** View that greeted the *Apollo 8* astronauts as their spacecraft emerged from behind the Moon. (NASA Headquarters)
B. Africa and Arabia are prominent in this image of Earth taken from *Apollo 17*. The tan cloud-free zones over the land coincide with major desert regions. The band of clouds across central Africa is associated with a much wetter climate that in places sustains tropical rain forests. The dark blue of the oceans and the swirling cloud patterns remind us of the importance of the oceans and the atmosphere. Antarctica, a continent covered by glacial ice, is visible at the South Pole. (NASA/Science Source/Photo Researchers, Inc.)

age depth of about 3800 meters (12,500 feet). It accounts for about 97 percent of Earth's water. However, the hydrosphere also includes the fresh water found in streams, lakes, and glaciers, as well as that found underground.

Although these latter sources constitute just a tiny fraction of the total, they are much more important than their meager percentage indicates. In addition to providing the fresh water that is so vital to life on the land, streams, glaciers, and groundwater are responsible for sculpturing and creating many of our planet's varied landforms.

Atmosphere

Earth is surrounded by a life-giving gaseous envelope called the **atmosphere**. When we view the atmosphere from the ground, it seems to be very deep (Figure 1.10). However, when compared to the thickness (radius) of the solid Earth (about 6400 kilometers, or 4000 miles), the atmosphere is a very shallow layer. One half lies below an altitude of 5.6 kilometers (3.5 miles), and 90 percent occurs within just 16 kilometers (10 miles) of Earth's surface. This thin blanket of air is nevertheless an integral part of the planet. It not only provides the air that we breathe but also acts to protect us from the dangerous radiation emitted by the Sun. The energy exchanges that continually occur between the atmosphere and Earth's surface and between the atmosphere and space produce the effects we call *weather* and *climate*.

If, like the Moon, Earth had no atmosphere, our planet would not only be lifeless but also many of the

processes and interactions that make the surface such a dynamic place could not operate. Without weathering and erosion, the face of our planet might more closely resemble the lunar surface, which has not changed appreciably in nearly 3 billion years.

Biosphere

The **biosphere** includes all life on Earth. Ocean life is concentrated in the sunlit surface waters of the sea. Most life on land is also concentrated near the surface, with tree roots and burrowing animals reaching a few meters underground and flying insects and birds reaching a kilometer or so above Earth. A surprising variety of life forms are also adapted to extreme environments. For example, on the ocean floor, where pressures are extreme and no light penetrates, there are places where vents spew hot, mineral-rich fluids that support communities of exotic life forms. On land, some bacteria thrive in rocks as deep as 4 kilometers (2.5 miles) and in boiling hot springs. Moreover, air currents can carry microorganisms many kilometers into the atmosphere. But even when we consider these extremes, life still must be thought of as being confined to a narrow band very near Earth's surface.

Plants and animals depend on the physical environment for the basics of life. However, organisms do more than just respond to their physical environment. Through countless interactions, life forms help maintain and alter their physical environment. Without

life, the makeup and nature of the geosphere, hydrosphere, and atmosphere would be very different.

Geosphere

Lying beneath the atmosphere and the ocean is the solid Earth or **geosphere**. The geosphere extends from the surface to the center of the planet, a depth of 6400 kilometers, making it by far the largest of Earth's four spheres. Much of our study of the solid Earth focuses on the more accessible surface features. Fortunately, many of these features represent the outward expressions of the dynamic behavior of Earth's interior. By examining the most prominent surface features and their global extent, we can obtain clues to the dynamic processes that have shaped our planet. A first look at the structure of Earth's interior and at the major surface features of the geosphere will come in the next section of this chapter.

Soil, the thin veneer of material at Earth's surface that supports the growth of plants, may be thought of as part of all four spheres. The solid portion is a mixture of weathered rock debris (geosphere) and organic matter from decayed plant and animal life (biosphere). The decomposed and disintegrated rock debris is the product of weathering processes that require air (atmosphere) and water (hydrosphere). Air and water also occupy the open spaces between the solid particles.

A Closer Look at the Geosphere

In this section we will make a preliminary examination of the solid Earth. You will become more familiar with the internal and external "anatomy" of our planet and begin to understand that the geosphere is truly dynamic. The diagrams should help a great deal as you begin to develop a mental image of the geosphere's internal structure and major surface features, so study the figures carefully. We begin with a

▲ **FIGURE 1.9** The shoreline is one obvious meeting place for rock, water, and air. In this scene along the Oregon coast, ocean waves that were created by the force of moving air break against the rocky shore. The force of the water can be powerful, and the erosional work that is accomplished can be great. (Photo by Galen Rowell)

▶ **FIGURE 1.10** This jet is flying high in the atmosphere at an altitude of more than 9000 meters (30,000 feet). To someone on the ground, the atmosphere seems very deep. However, when compared to the thickness (radius) of the solid Earth, the atmosphere is a very shallow layer. (Photo by Warren Faidley/Weatherstock)

look at Earth's interior—its structure and mobility. Then we will conduct a brief survey of the surface of the solid Earth. Although portions of the surface, such as mountains and river valleys, are familiar to most of us, those areas that are out of sight on the floor of the ocean are not so familiar.

Earth's Internal Structure

Early in Earth's history the sorting of material by compositional (and density) differences resulted in the formation of three layers—the crust, mantle, and core (Figure 1.11). In addition to these compositionally distinct layers, Earth is also divided into layers based on physical properties. The physical properties that define these zones include whether the layer is solid or liquid and how weak or strong it is. Knowledge of both types of layers is essential to an understanding of our planet.

Layers Defined by Composition.   The **crust**, the thin, rocky outer layer of Earth, is divided into oceanic and continental crust. The oceanic crust is only about 7 kilometers (4.3 miles) thick and is composed of the ig-

neous rocks *basalt* and *gabbro*. The continental crust averages about 40 kilometers (25 miles) thick but may exceed 70 kilometers in some mountainous regions. It consists of many rock types. The average composition of the upper continental crust is similar to a *granitic rock* called *granodiorite*. Continental rocks have an average density of 2.7 g/cm^3, and some are over 4 billion years old. The rocks of the oceanic crust are younger (180 million years or less) and have a density of about 3.0 g/cm^3. As you will see later, the greater density of oceanic crust as compared to continental crust has significant implications for the workings of our planet.

Over 82 percent of Earth's volume is contained in the **mantle**—a solid rocky shell that extends to a depth of about 2900 kilometers (1800 miles). The boundary between the crust and mantle represents a change in chemical composition. The dominant rock type in the uppermost mantle is *peridotite*, which has a density of 3.4 g/cm^3.

The **core** is a sphere composed mainly of an iron-nickel alloy. At the extreme pressure found in the center of the core, the iron-rich material has an average density of almost 13 g/cm^3 (13 times heavier than water).

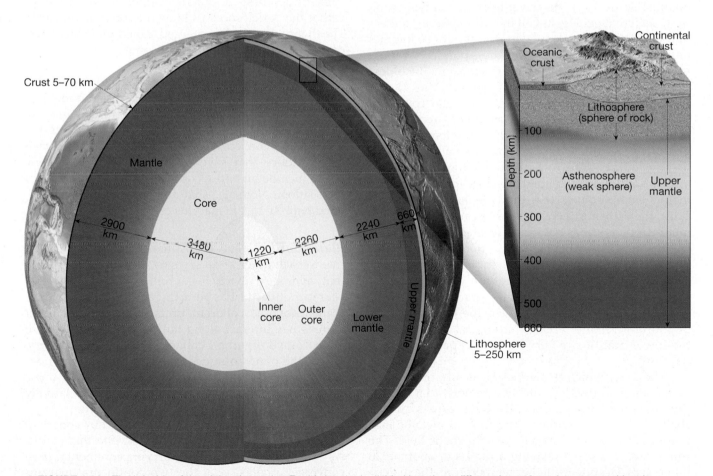

▲ FIGURE 1.11   The left side of the globe shows that Earth's interior is divided into three different layers based on compositional differences—the crust, mantle, and core. The right side of the globe shows the five main layers of Earth's interior based on physical properties and mechanical strength—the lithosphere, asthenosphere, lower mantle, outer core, and inner core. The block diagram shows an enlarged view of the upper portion of Earth's interior.

Layers Defined by Physical Properties. Earth's interior is characterized by a gradual increase in temperature, pressure, and density with depth. These changes in turn affect the physical properties and hence the behavior of Earth materials. In particular, as temperature increases, the atoms in a substance vibrate faster and the chemical bonds weaken, resulting in a loss of strength. Furthermore, if the temperature exceeds the melting point, the material's chemical bonds break and melting begins.

If temperature were the only factor that determined whether a substance melted, our planet would be a molten ball covered with a thin solid outer shell. However, pressure also increases with depth and counteracts the effects of temperature by increasing rock strength. Thus, depending on the physical environment (temperature and pressure), a substance may behave like a brittle solid, deform in a puttylike manner, or melt and become liquid.

Earth can be divided into five layers based on physical properties—lithosphere, asthenosphere, lower mantle, outer core, and inner core.

Earth's outermost layer, as defined by physical properties, consists of the crust and the uppermost mantle (Figure 1.11). This layer forms a relatively cool, rigid shell that geologists call the **lithosphere** ("sphere of rock"). The lithosphere averages about 100 kilometers (more than 60 miles) in thickness but may be more than 250 kilometers (150 miles) thick below the oldest portions of the continents.

Beneath the lithosphere lies a soft and comparatively weak layer known as the **asthenosphere** ("weak sphere"). Within the asthenosphere the temperature/pressure conditions are such that a small amount of melting occurs. The rocks of the asthenosphere are mostly solid; however, they are close enough to their melting temperatures that they are easily deformed. Thus, the asthenosphere is weak because it is near its melting point, just as hot wax is weaker than cold wax.

Below the asthenosphere, increased pressure begins to offset the effects of higher temperature, and the rocks gradually strengthen with depth. Between the depths of 660 kilometers and 2900 kilometers a more rigid layer, called the **lower mantle**, is found. Despite greater strength, the rocks of the lower mantle are very hot and capable of gradual flow.

The core, which is composed mostly of an iron-nickel alloy, is divided into two regions that exhibit different physical properties. The **outer core** is a liquid shell 2260 kilometers thick. The flow of metallic iron within this zone generates Earth's magnetic field. The **inner core** is a sphere having a radius of about 1220 kilometers. Despite its higher temperature, the material in the inner core is compressed into a solid state by the immense pressure.

The Mobile Geosphere

Earth is a dynamic planet! If we could go back in time a few hundred million years, we would find the face of our planet dramatically different from what we see today. There would be no Mount St. Helens, Rocky Mountains, or Gulf of Mexico. Moreover, we would find continents having different sizes and shapes and located in different positions than today's landmasses.

Continental Drift and Plate Tectonics. Within the past several decades a great deal has been learned about the workings of our dynamic planet. This period has seen an unequaled revolution in our understanding of Earth. The revolution began in the early part of the twentieth century with the radical proposal of *continental drift*—the idea that the continents moved about the face of the planet. This proposal contradicted the established view that the continents and ocean basins are permanent and stationary features on the face of Earth. For that reason, the notion of drifting continents was received with great skepticism and even ridicule. More than 50 years passed before enough data were gathered to transform this controversial hypothesis into a sound theory that wove together the basic processes known to operate on Earth. The theory that finally emerged, called **plate tectonics**, provided geologists with the first comprehensive model of Earth's internal workings.

According to the theory of plate tectonics, Earth's rigid outer shell (*lithosphere*) is broken into numerous slabs called **plates**, which are in continual motion. Over a dozen plates exist (Figure 1.12). The largest is the Pacific plate, covering much of the Pacific Ocean basin. Notice that several of the large lithospheric plates include an entire continent plus a large area of the seafloor. Note also that none of the plates are defined entirely by the margins of a continent.

Plate Motion. The lithospheric plates move relative to each other at a very slow but continuous rate that averages about 5 centimeters (2 inches) per year—about as fast as your fingernails grow. Because plates move as coherent units relative to all other plates, they interact along their margins. Where two plates move together, called a *convergent boundary*, one of the plates plunges beneath the other and descends into the mantle (Figure 1.13). It is only those lithospheric plates that are capped with relatively dense oceanic crust that sink into the mantle.

Any portion of a plate that is capped by continental crust is too buoyant to be carried into the mantle. As a result, when two plates carrying continental crust converge, a collision of the two continental margins occurs. The result is the formation of a major mountain belt, as exemplified by the Himalayas.

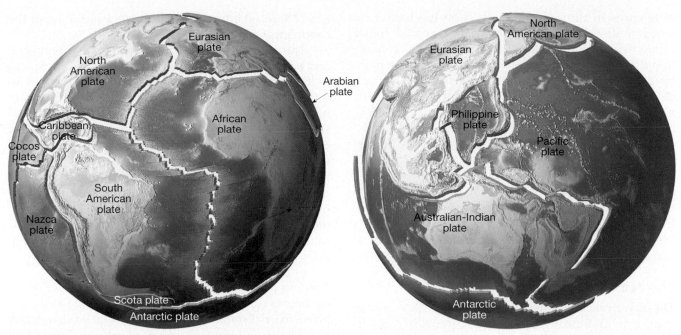

▲ **FIGURE 1.12** Illustration showing some of Earth's lithospheric plates.

Divergent boundaries are located where plates pull apart (Figure 1.13). Here the fractures created as the plates separate are filled with molten rock that wells up from the mantle. This hot material slowly cools to form hard rock, producing new slivers of seafloor. This process occurs along oceanic ridges where over spans of millions of years, hundreds of thousands of square kilometers of new seafloor have been generated (Figure 1.13). Thus, while new seafloor is constantly being added at the oceanic ridges, equal amounts are returned to the mantle along boundaries where two plates converge.

What Drives Plate Motion? The movement of plates is ultimately driven by the unequal distribution of heat within our planet. Temperatures in Earth's interior may exceed 6700°C (more than 12,000°F). By comparison, the surface is relatively cold. These temperature

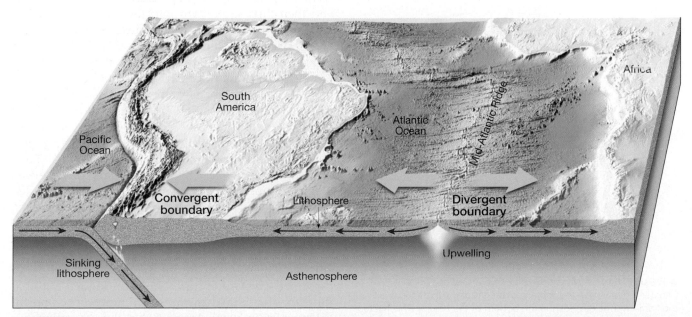

▲ **FIGURE 1.13** *Convergent boundaries* occur where two plates move together, as along the western margin of South America. *Divergent boundaries* are located where adjacent plates move away from one another. The Mid-Atlantic ridge is such a boundary.

differences in turn drive a convection mechanism in which hotter (less dense) material rises and cooler (more dense) material sinks (Figure 1.14). The convection that drives plate motion involves mainly the mantle and the lithospheric plates themselves. Although the mantle is comprised of solid rock, it is very near its melting point and thus quite weak and capable of slow convective motion.

Hot material found deep in the mantle moves slowly upward and serves as one part of our planet's internal convective system. Concurrently, cooler, denser slabs of oceanic lithosphere descend back into the mantle, setting Earth's rigid outer shell in motion. Ultimately, the titanic, grinding movements of Earth's lithospheric plates generate earthquakes, create volcanoes, and deform large masses of rock into mountains.

The Face of Earth

The two principal divisions of Earth's surface are the continents and the ocean basins (Figure 1.15). A significant difference between these two areas is their relative levels. The continents are remarkably flat features that have the appearance of plateaus protruding above sea level. With an average elevation of about 0.8 kilometer (0.5 mile), continents lie close to sea level, except for limited areas of mountainous terrain. By contrast, the average depth of the ocean floor is about 3.8 kilometers (2.4 miles) below sea level, or about 4.5 kilometers (2.8 miles) lower than the average elevation of the continents.

The elevation difference between the continents and ocean basins is primarily the result of differences in their respective densities and thicknesses. Recall that the continents average 35–40 kilometers in thickness and are composed of granitic rocks having a density of about $2.7 \, g/cm^3$. The basaltic rocks that comprise the oceanic crust average only 7 kilometers thick and have an average density of about $3.0 \, g/cm^3$. Thus, the thicker and less dense continental crust is more buoyant than the oceanic crust. As a result, continental crust floats on top of the deformable rocks of the mantle at a higher level than oceanic crust for the same reason that a large, empty (less dense) cargo ship rides higher than a small, loaded (denser) one.

Features of the Continents. The largest features of the continents can be grouped into two distinct categories: extensive, flat stable areas that have been eroded nearly to sea level, and uplifted regions of deformed rocks that make up present-day mountain belts.

The most prominent topographic features of the continents are linear mountain belts. Although the distribution of mountains appears to be random, this is not the case. When the youngest mountains are considered (those less than 100 million years old), we find that they are located principally in two major zones. The circum-Pacific belt (the region surrounding the Pacific Ocean) includes the mountains of the western Americas and continues into the western Pacific in the form of volcanic islands such as the Aleutians, Japan, and the Philippines (Figure 1.15).

The other major mountainous belt extends eastward from the Alps through Iran and the Himalayas and then dips southward into Indonesia. Careful examination of mountainous terrains reveals that most are places where thick sequences of rocks have been squeezed and highly deformed, as if placed in a gigantic vise. Older mountains are also found on the continents. Examples include the Appalachians in the eastern United States and the Urals in Russia. Their once lofty peaks are now worn low, the result of millions of years of erosion.

Unlike the young mountain belts, which have formed within the last 100 million years, the interiors of the continents have been relatively stable (undisturbed) for the last 600 million years or even longer. Typically, these regions were involved in mountain-building episodes much earlier in Earth's history.

Within the stable interiors are areas known as **shields**, which are expansive, flat regions composed of deformed crystalline rock. Notice in Figure 1.16 that the Canadian Shield is exposed in much of the northeastern part of North America. Age determinations for various shields have shown that they are truly ancient

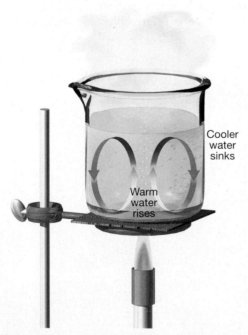

▲ **FIGURE 1.14** A simple example of convection, which is heat transfer that involves the actual movement of a substance. Here the flame warms the water in the bottom of the beaker. This heated water expands, becomes less dense (more buoyant), and rises. Meanwhile, the cooler, denser water near the top sinks.

(Labels in figure: Cooler water sinks; Warm water rises)

regions. All contain Precambrian-age rocks that are over 1 billion years old, with some samples approaching 4 billion years in age (see Figure 1.6 to review the geologic time scale). These oldest-known rocks exhibit evidence of enormous forces that have folded and faulted them and altered them with great heat and pressure. Thus, we conclude that these rocks were once part of an ancient mountain system that has since been eroded away to produce these expansive, flat regions.

Other flat areas of the stable interior exist in which highly deformed rocks, like those found in the shields, are covered by a relatively thin veneer of sedimentary rocks. These areas are called **stable platforms**. The sedimentary rocks in stable platforms are nearly horizontal except where they have been warped to form large basins or domes. In North America a major portion of the stable platforms is located between the Canadian Shield and the Rocky Mountains (Figure 1.16).

Features of the Ocean Basins. If all water were drained from the ocean basins, a great variety of features would be seen, including linear chains of volcanoes, deep canyons, plateaus, and large expanses of monotonously flat plains. In fact, the scenery would be nearly as diverse as that on the continents (see Figure 1.15).

During the past 50 years, oceanographers using modern sonar equipment have slowly mapped significant portions of the ocean floor. From these studies they have defined three major regions: *continental margins, deep-ocean basins,* and *oceanic (mid-ocean) ridges.*

The **continental margin** is that portion of the seafloor adjacent to major landmasses. Although land and sea meet at the shoreline, this is not the boundary between the continents and the ocean basins. Rather, along most coasts a gently sloping platform of material, called the **continental shelf**, extends seaward from the shore. Because it is underlain by continental crust, it is considered a flooded extension of the continents. A glance at Figure 1.15 shows that the width of the continental shelf is variable. For example, it is broad along the East and Gulf coasts of the United States but relatively narrow along the Pacific margin of the continent.

The boundary between the continents and the deep-ocean basins lies along the **continental slope**, which is a relatively steep dropoff that extends from the outer edge of the continental shelf to the floor of the deep ocean (Figure 1.15). Using this as the dividing line, we find that about 60 percent of Earth's surface is represented by ocean basins and the remaining 40 percent by continents.

Beyond the continental margin lies the **deep-ocean basin**. Part of this region consists of incredibly flat features called **abyssal plains**. However, the ocean floor also contains extremely deep depressions that are occasionally more than 11,000 meters (36,000 feet) deep. Although these **deep-ocean trenches** are relatively narrow and represent only a small fraction of the ocean floor, they are nevertheless very significant features. Some trenches are located adjacent to young mountains that flank the continents. For example, in Figure 1.15 the Peru-Chile trench off the west coast of South America parallels the Andes Mountains. Other trenches parallel linear island chains called *volcanic island arcs.* The deep-ocean basins are also dotted with submerged volcanic structures called **seamounts**, which sometimes form long narrow chains.

The most prominent feature on the ocean floor is the **oceanic** or **mid-ocean ridge**. As shown in Figure 1.15, the Mid-Atlantic Ridge and the East Pacific Rise are parts of this system. This broad elevated feature forms a continuous belt that winds for more than 70,000 kilometers (43,000 miles) around the globe in a manner similar to the seam of a baseball. Rather than consisting of highly deformed rock, such as most of the mountains on the continents, the oceanic ridge system consists of layer upon layer of igneous rock that has been fractured and uplifted.

Understanding the topographic features that comprise the face of Earth is critical to our understanding of the mechanisms that have shaped our planet. What is the significance of the enormous ridge system that extends through all the world's oceans? What is the connection, if any, between young, active mountain belts and deep-ocean trenches? What forces crumple rocks to produce majestic mountain ranges? These are questions that will be addressed in some of the coming chapters as we investigate the dynamic processes that shaped our planet in the geologic past and will continue to shape it in the future.

STUDENTS SOMETIMES ASK ...
How do we know about the internal structure of Earth?

You might suspect that the internal structure of Earth has been sampled directly. However, humans have never penetrated beneath the crust!

The internal structure of Earth is determined by using indirect observations. Every time there is an earthquake, waves of energy (called *seismic waves*) penetrate Earth's interior. Seismic waves change their speed and are bent and reflected as they move through zones having different properties. An extensive series of monitoring stations around the world detects and records this energy. The data are analyzed and used to work out the structure of Earth's interior. For more about this technique, see Chapter 7, "Earthquakes and Earth's Interior."

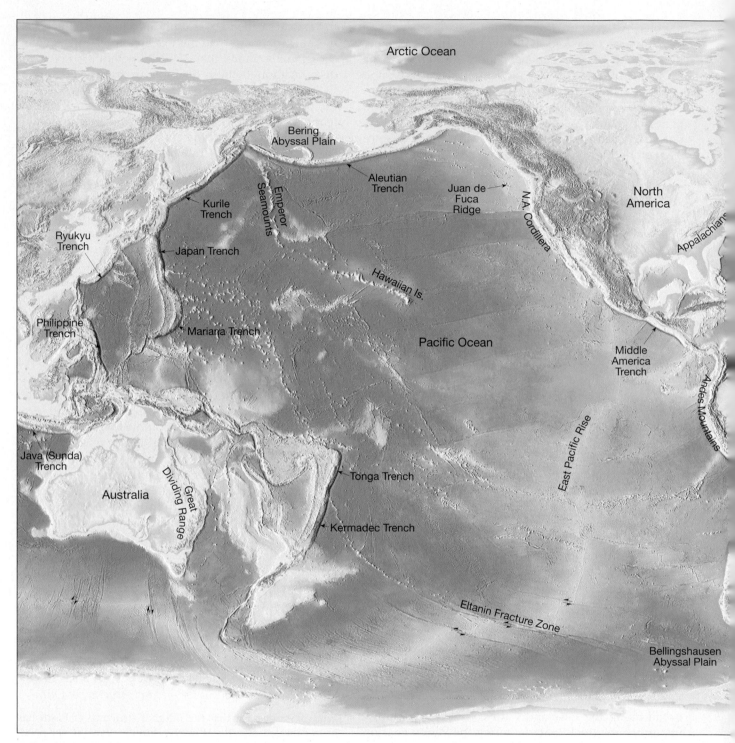

▲ **FIGURE 1.15** Major surface features of the geosphere.

Earth as a System

Anyone who studies Earth soon learns that our planet is a dynamic body with many separate but interacting parts or *spheres*. The hydrosphere, atmosphere, biosphere, and geosphere and all of their components can be studied separately. However, the parts are not isolated. Each is related in some way to the others to produce a complex and continuously interacting whole that we call the *Earth system*.

Earth System Science

A simple example of the interactions among different parts of the Earth system occurs every winter as moisture evaporates from the Pacific Ocean and subsequently falls as rain in the hills of southern California, triggering destructive landslides. The processes that move water from the hydrosphere to the atmosphere and then to the solid Earth have a profound impact on the plants and animals (including humans) that inhabit the affected regions.

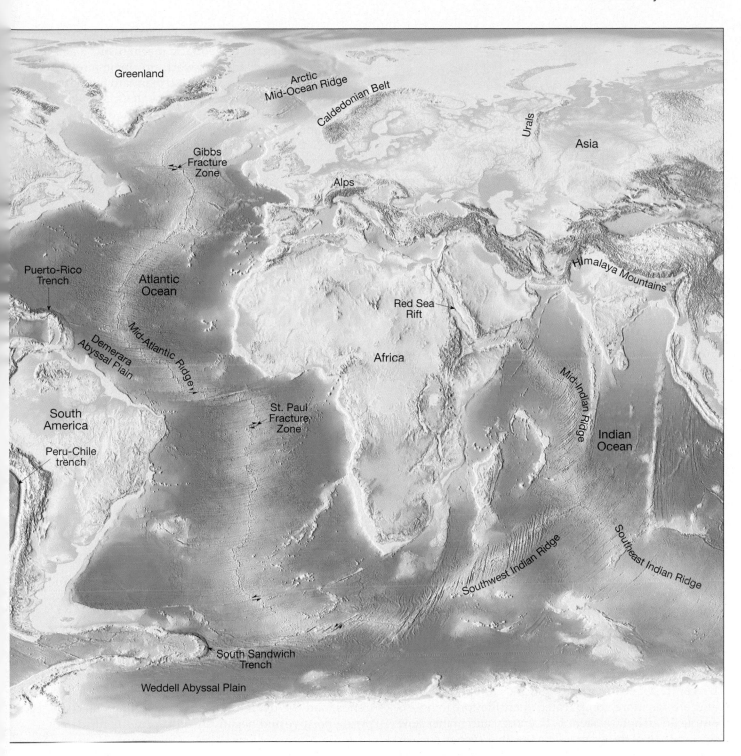

Scientists have recognized that in order to more fully understand our planet, they must learn how its individual components (land, water, air, and life forms) are interconnected. This endeavor, called **Earth system science,** aims to study Earth as a *system* composed of numerous interacting parts, or *subsystems.* Rather than looking through the limited lens of only one of the traditional sciences—geology, atmospheric science, chemistry, biology, etc.—Earth system science attempts to integrate the knowledge of several academic fields. Using this interdisciplinary approach, we hope to achieve the level of understanding necessary to comprehend and solve many of our global environmental problems.

What Is a System? Most of us hear and use the term *system* frequently. We may service our car's cooling *system,* make use of the city's transportation *system,*

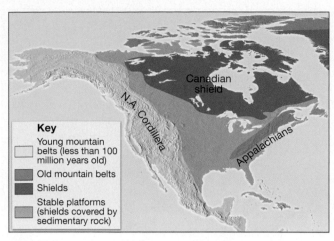

▲ **FIGURE 1.16** Major features of North America.

and participate in the political *system*. A news report might inform us of an approaching weather *system*. Further, we know that Earth is just a small part of a larger system known as the *solar system* which in turn is a *subsystem* of the even larger system called the Milky Way Galaxy.

Loosely defined, a **system** can be any size group of interacting parts that form a complex whole. Most natural systems are driven by sources of energy that move matter and/or energy from one place to another. A simple analogy is a car's cooling system, which contains a liquid (usually water and antifreeze) that is driven from the engine to the radiator and back again. The role of this system is to transfer heat generated by combustion in the engine to the radiator, where moving air removes it from the system. Hence, the term cooling system.

Systems like a car's cooling system are self-contained with regard to matter and are called **closed systems**. Although energy moves freely in and out of a closed system, no matter (liquid in the case of our auto's cooling systems) enters or leaves the system. (This assumes you don't get a leak in your radiator.) By contrast, most natural systems are **open systems** and are far more complicated than the foregoing example. In an open system both energy and matter flow into and out of the system. In a weather system such as a hurricane, factors such as the quantity of water vapor available for cloud formation, the amount of heat released by condensing water vapor, and the flow of air into and out of the storm can fluctuate a great deal. At times the storm may strengthen; at other times it may remain stable or weaken.

Feedback Mechanisms. Most natural systems have mechanisms that tend to enhance change, as well as other mechanisms that tend to resist change and thus stabilize the system. For example, when we get too hot, we perspire to cool down. This cooling phenome-

non works to stabilize our body temperature and is referred to as a **negative feedback mechanism**. Negative feedback mechanisms work to maintain the system as it is or, in other words, to maintain the status quo. By contrast, mechanisms that enhance or drive change are called **positive feedback mechanisms**.

Most of Earth's systems, particularly the climate system, contain a wide variety of negative and positive feedback mechanisms. For example, substantial scientific evidence indicates that Earth has entered a period of global warming. One consequence of global warming is that some of the world's glaciers and ice caps have begun to melt. Highly reflective snow- and ice-covered surfaces are gradually being replaced by brown soils, green trees, or blue oceans, all of which are darker, so they absorb more sunlight. Therefore, as Earth warms and some snow and ice melt, our planet absorbs more sunlight. The result is a positive feedback that contributes to the warming.

On the other hand, an increase in global temperature also causes greater evaporation of water from Earth's land-sea surface. One result of having more water vapor in the air is an increase in cloud cover. Because cloud tops are white and highly reflective, more sunlight is reflected back to space, which diminishes the amount of sunshine reaching Earth's surface and thus reduces global temperatures. Further, warmer temperatures tend to promote the growth of vegetation. Plants in turn remove carbon dioxide (CO_2) from the air. Since carbon dioxide is one of the atmosphere's *greenhouse gases*, its removal has a negative impact on global warming.*

In addition to natural processes, we must also consider the human element. Extensive cutting and clearing of the tropical rain forests and the burning of fossil fuels (oil, natural gas, and coal) result in an increase in atmospheric CO_2. Such activity appears to have contributed to the increase in global temperature that our planet is experiencing. One of the daunting tasks for Earth system scientists is to predict what the climate will be like in the future by taking into account many variables, including technological changes, population trends, and the overall impact of the numerous competing positive and negative feedback mechanisms.

The Earth System

The Earth system has a nearly endless array of subsystems in which matter is recycled over and over again (Figure 1.17). One example that you will learn about in Chapter 3 traces the movements of carbon among Earth's four spheres. It shows us, for example, that the carbon dioxide in the air and the carbon in living

---

*Greenhouse gases absorb heat energy emitted by Earth and thus help keep the atmosphere warm.

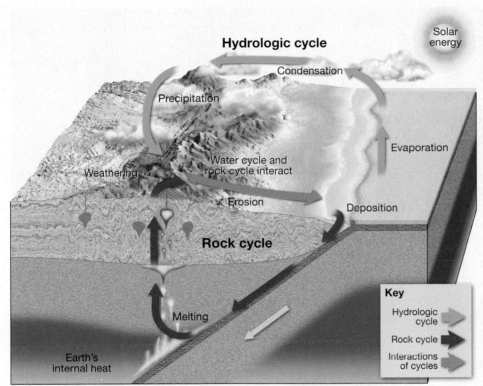

Hydrologic cycle

Solar energy

Condensation

Precipitation

Evaporation

Weathering

Water cycle and rock cycle interact

Erosion

Deposition

Rock cycle

Melting

Earth's internal heat

Key

Hydrologic cycle

Rock cycle

Interactions of cycles

◄ **FIGURE 1.17** Each part of the Earth system is related to every other part to produce a complex interacting whole. The Earth system involves many cycles, including the hydrologic cycle and the rock cycle. Such cycles are not independent of each other. There are many places where they interface.

things and in certain sedimentary rocks is all part of a subsystem described by the *carbon cycle*.

Cycles in the Earth System. A more familiar loop or subsystem is the *hydrologic cycle*. It represents the unending circulation of Earth's water among the hydrosphere, atmosphere, biosphere, and geosphere. Water enters the atmosphere by evaporation from Earth's surface and by transpiration from plants. Water vapor condenses in the atmosphere to form clouds, which in turn produce precipitation that falls back to Earth's surface. Some of the rain that falls onto the land sinks in to be taken up by plants or become groundwater, and some flows across the surface toward the ocean.

Viewed over long time spans, the rocks of the geosphere are constantly forming, changing, and reforming (Figure 1.17). The loop that involves the processes by which one rock changes to another is called the *rock cycle* and will be discussed at some length in Chapter 3. The cycles of the Earth system, such as the hydrologic and rock cycles, are not independent of one another. To the contrary, there are many places where they interface. An **interface** is a common boundary where different parts of a system come in contact and interact. For example, in Figure 1.17, weathering at the surface gradually disintegrates and decomposes solid rock. The work of gravity and running water may eventually move this material to another place and deposit it. Later, groundwater percolating through the debris may leave behind mineral matter that cements the grains together into solid rock (a rock that is often very different from the rock we

started with). This changing of one rock into another could not have occurred without the movement of water through the hydrologic cycle. There are many places where one cycle or loop in the Earth system interfaces with and is a basic part of another.

Energy for the Earth System. The Earth system is powered by energy from two sources. The Sun drives external processes that occur in the atmosphere, hydrosphere, and at Earth's surface. Weather and climate, ocean circulation, and erosional processes are driven by energy from the Sun. Earth's interior is the second source of energy. Heat remaining from when our planet formed, and heat that is continuously generated by decay of radioactive elements, powers the internal processes that produce volcanoes, earthquakes, and mountains.

The Parts are Linked. The parts of the Earth system are linked so that a change in one part can produce changes in any or all of the other parts. For example, when a volcano erupts, lava from Earth's interior may flow out at the surface and block a nearby valley. This new obstruction influences the region's drainage system by creating a lake or causing streams to change course. The large quantities of volcanic ash and gases that can be emitted during an eruption might be blown high into the atmosphere and influence the amount of solar energy that can reach Earth's surface. The result could be a drop in air temperatures over the entire hemisphere.

▲ **FIGURE 1.18** When Mount St. Helens erupted in May 1980, the area shown here was buried by a volcanic mudflow. Now, plants are reestablished and new soil is forming. (Photo by Jack Dykinga)

Where the surface is covered by lava flows or a thick layer of volcanic ash, existing soils are buried. This causes the soil-forming processes to begin anew to transform the new surface material into soil (Figure 1.18). The soil that eventually forms will reflect the interactions among many parts of the Earth system—the volcanic parent material, the type and rate of weathering, and the impact of biological activity. Of course, there would also be significant changes in the biosphere. Some organisms and their habitats would be eliminated by the lava and ash, whereas new settings for life, such as the lake, would be created. The potential climate change could also impact sensitive life forms.

The Earth system is characterized by processes that vary on spatial scales from fractions of millimeters to thousands of kilometers. Time scales for Earth's processes range from milliseconds to billions of years. As we learn about Earth, it becomes increasingly clear that despite significant separations in distance or time, many processes are connected, and a change in one component can influence the entire system.

Humans are *part of* the Earth system, a system in which the living and nonliving components are en-

twined and interconnected. Therefore, our actions produce changes in all of the other parts. When we burn gasoline and coal, build breakwaters along the shoreline, dispose of our wastes, and clear the land, we cause other parts of the system to respond, often in unforeseen ways. Throughout this book you will learn about many of Earth's subsystems: the hydrologic system, the tectonic (mountain-building) system, and the climate system, to name a few. Remember that these components *and we humans* are all part of the complex interacting whole we call the Earth system.

The organization of this text involves traditional groupings of chapters that focus on closely related topics. Nevertheless, the theme of *Earth as a system* keeps recurring through *all* major units of *Earth Science*. It is a thread that weaves through the chapters and helps tie them together. At the end of each chapter there is a section entitled "Examining the Earth System." The questions and problems found there are intended to help you develop an awareness and appreciation for some of the Earth system's important interrelationships.

Chapter Summary

• *Earth science* is the name for all the sciences that collectively seek to understand Earth and its neighbors in space. It includes *geology, oceanography, meteorology,* and *astronomy.* Geology is traditionally divided into two broad areas—*physical* and *historical.*

• *Environment* refers to everything that surrounds and influences an organism. These influences can be biological, social, or physical. When applied to Earth science today, the term *environmental* is usually reserved for those aspects that focus on the relationships between people and the natural environment.

• *Resources* are an important environmental concern. The two broad categories of resources are (1) *renewable*, which means that they can be replenished over relatively short time spans, and (2) *nonrenewable*. As population grows, the demand for resources expands as well.

• Environmental problems can be local, regional, or global. Human-induced problems include urban air pollution, acid rain, ozone depletion, and global warming. Natural hazards include earthquakes, landslides, floods, and hurricanes. As world population grows, pressures on the environment also increase.

• All science is based on the assumption that the natural world behaves in a consistent and predictable manner. The process by which scientists gather facts through observation and careful measurement and formulate scientific *hypotheses* and *theories* is called the *scientific method.* To determine what is occurring in the natural world, scientists often (1) collect facts, (2) develop a scientific hypothesis, (3) construct experiments to validate the hypothesis, and (4) accept, modify, or reject the hypothesis on the basis of extensive testing. Other discoveries represent purely theoretical ideas that have stood up to extensive examination. Still other scientific advancements have been made when a totally unexpected happening occurred during an experiment.

• One of the challenges for those who study Earth is the great variety of space and time scales. The *geologic time scale* subdivides the 4.5 billion years of Earth history into various units.

• The *nebular hypothesis* describes the formation of the solar system. The planets and Sun began forming about 5 billion years ago from a large cloud of dust and gases. As the cloud contracted, it began to rotate and assume a disk shape. Material that was gravitationally pulled toward the center became the *protosun.* Within the rotating disk, small centers, called *protoplanets,* swept up more and more of the cloud's debris. Because of their high temperatures and weak gravitational fields, the inner planets were unable to accumulate and retain many of the lighter components. Because of the very cold temperatures existing far from the Sun, the large outer planets consist of huge amounts of lighter materials. These gaseous substances account for the comparatively large sizes and low densities of the outer planets.

• Earth's physical environment is traditionally divided into three major parts: the solid Earth or *geosphere*; the water portion of our planet, the *hydrosphere*; and Earth's gaseous envelope, the *atmosphere.* In addition, the *biosphere*, the totality of life on Earth, interacts with each of the three physical realms and is an equally integral part of Earth.

• Earth's internal structure is divided into layers based on differences in chemical composition and on the basis of changes in physical properties. Compositionally, Earth is divided into a thin outer *crust*, a solid rocky *mantle*, and a dense *core*. Based on physical properties, the layers of Earth are (1) the *lithosphere*—the cool, rigid outermost layer that averages about 100 kilometers thick, (2) the *asthenosphere*, a relatively weak layer located in the mantle beneath the lithosphere, (3) the more rigid *lower mantle*, where rocks are very hot and capable of very gradual flow, (4) the liquid *outer core*, where Earth's magnetic field is generated, and (5) the solid *inner core.*

• Two principal divisions of Earth's surface are the *continents* and *ocean basins.* A significant difference is their relative levels. The elevation differences between continents and ocean basins is primarily the result of differences in their respective densities and thicknesses.

• The largest features of the continents can be divided into two categories: *mountain belts* and the *stable interior.* The ocean floor is divided into three major topographic units: *continental margins, deep-ocean basins,* and *oceanic ridges.*

• Although each of Earth's four spheres can be studied separately, they are all related in a complex and continuously interacting whole that we call the *Earth system. Earth system science* uses an interdisciplinary approach to integrate the knowledge of several academic fields in the study of our planet and its global environmental problems.

• A *system* is a group of interacting parts that form a complex whole. *Closed systems* are those in which energy moves freely in and out, but matter does not enter or leave the system. In an *open system*, both energy and matter flow into and out of the system.

• The *two sources of energy that power the Earth system are* (1) *the Sun*, which drives the external processes that occur in the atmosphere, hydrosphere, and at Earth's surface, and (2) *heat from Earth's interior*, which powers the internal processes that produce volcanoes, earthquakes, and mountains.

Key Terms

abyssal plain (p. 19)
asthenosphere (p. 16)
astronomy (p. 2)
atmosphere (p. 13)
biosphere (p. 13)
closed system (p. 22)
continental margin (p. 19)
continental shelf (p. 19)
continental slope (p. 19)
core (p. 15)
crust (p. 15)
deep-ocean basin (p. 19)
deep-ocean trench (p. 19)
Earth science (p. 2)
Earth system science (p. 21)
environment (p. 3)

geologic time scale (p. 10)
geology (p. 2)
geosphere (p. 14)
hydrosphere (p. 12)
hypothesis (p. 7)
inner core (p. 16)
interface (p. 23)
lithosphere (p. 16)
lower mantle (p. 16)
mantle (p. 15)
meteorology (p. 2)
model (p. 7)
nebular hypothesis (p. 10)
negative feedback mechanism (p. 22)
nonrenewable resource (p. 3)
oceanic (mid-ocean) ridge (p. 19)

oceanography (p. 2)
open system (p. 22)
outer core (p. 16)
paradigm (p. 7)
physical environment (p. 3)
plate (p. 16)
plate tectonics (p. 16)
positive feedback mechanism (p. 22)
renewable resource (p. 3)
seamount (p. 19)
shield (p. 18)
stable platform (p. 19)
system (p.22)
theory (p. 7)

Review Questions

1. Name the specific Earth science described by each of the following statements.
   (a) The science that deals with the dynamics of the oceans.
   (b) This word literally means "the study of Earth."
   (c) An understanding of the atmosphere is the primary focus of this science.
   (d) This science helps us understand Earth's place in the universe.

2. Contrast renewable and nonrenewable resources. Give one or more examples of each.

3. World population may reach nearly _____ billion by the year 2010. How does this compare to the world population near the beginning of the nineteenth century?

4. List at least four phenomena that could be regarded as natural hazards.

5. How is a scientific hypothesis different from a scientific theory?

6. Briefly describe the events that led to the formation of the solar system.

7. List and briefly define the four "spheres" that constitute our environment.

8. The oceans cover nearly _____ percent of Earth's surface and contain about _____ percent of the planet's total water supply.

9. List and briefly describe the compositional divisions of Earth.

10. Contrast the asthenosphere and the lithosphere.

11. Describe the general distribution of Earth's youngest mountains.

12. Distinguish between shields and stable platforms.

13. List the three major regions of the ocean floor.

14. How is an open system different from a closed system?

15. Contrast positive feedback mechanisms and negative feedback mechanisms.

16. What are the two sources of energy for the Earth system?

Examining the Earth System

1. Examine the chapter-opening photo closely. Selecting among Earth's four major spheres (atmosphere, geosphere, hydrosphere, and biosphere), indicate which sphere is represented by each of the following features: (a) mountain, (b) lake, (c) tree, (d) cloud, (e) ice and snow.

2. Humans are a part of the Earth system. List at least three examples of how you, in particular, influence one or more of Earth's major spheres.

Online Study Guide

The *Earth Science* Website uses the resources and flexibility of the Internet to aid in your study of the topics in this chapter. Written and developed by Earth science instructors, this site will help improve your understanding of Earth science. Visit **http://www.prenhall.com/tarbuck** and click on the cover of *Earth Science* 11e to find:

- **Online review quizzes.**
- **Critical thinking exercises.**
- **Links to chapter-specific Web resources.**
- **Internet-wide key term searches.**

http://www.prenhall.com/tarbuck

CHAPTER 2

Minerals: Building Blocks of Rocks

Sulfur crystals.
(Photo by Gary Retherford/Photo Researchers, Inc.

arth's crust and oceans are the source of a wide variety of useful and essential minerals (Figure 2.1). In fact, practically every manufactured product contains materials obtained from minerals. Most people are familiar with the common uses of many basic metals, including aluminum in beverage cans, copper in electrical wiring, and gold and silver in jewelry. But some people are not aware that pencil lead contains the greasy-feeling mineral graphite and that baby powder comes from the mineral talc. Moreover, many do not know that drill bits impregnated with diamonds are employed by dentists to drill through tooth enamel, or that the common mineral quartz is the source of silicon for computer chips.

As the mineral requirements of modern society grow, the need to locate additional supplies of useful minerals also grows, becoming more challenging as well. In addition to the economic uses of rocks and minerals, all of the processes studied by geologists are in some way related to the properties of these basic Earth materials. Events such as volcanic eruptions, mountain building, weathering and erosion, and even earthquakes involve rocks and minerals. Consequently, a basic knowledge of Earth materials is essential to the understanding of all geologic phenomena.

Minerals: Building Blocks of Rocks

 GEODe **Earth Materials**
EARTH SCIENCE ▼ **Minerals**

The term *mineral* is used in several different ways. For example, those concerned with health and fitness extol the benefits of vitamins and minerals. The mining industry typically uses the word when referring to anything taken out of the ground, such as coal, iron ore, or sand and gravel. The guessing game known as *Twenty*

▼ **FIGURE 2.1** Mineral samples. **A.** Quartz; **B.** Olivine (variety Fosterite); **C.** Fluorite; **D.** Realgar; **E.** Beryl (variety aquamarine); **F.** Bornite and Chalcopyrite; **G.** Native copper; **H.** Gold nugget; **I.** Cut diamond. (Photos A–F by Dennis Tasa; G by E. J. Tarbuck; H and I by Dane Pendland, courtesy of Smithsonian Institution)

A.

B.

C.

D.

E.

F.

G.

H.

I.

Questions usually begins with the question, *Is it Animal, Vegetable or Mineral?* What criteria do geologists use to determine whether something is a mineral?

Geologists define a **mineral** as *any naturally occurring inorganic solid that possesses an orderly crystalline structure and a definite chemical composition.* Thus, those Earth materials that are classified as minerals exhibit the following characteristics:

1. **Naturally occurring.** Minerals form by natural, geologic processes. Consequently, synthetic diamonds and rubies, as well as a variety of other useful materials produced in a laboratory, are not considered minerals.

2. **Solid substance.** Minerals are solids within the temperature ranges normally experienced at Earth's surface. Thus, ice (frozen water) is considered a mineral, whereas liquid water and water vapor are not.

3. **Orderly crystalline structure.** Minerals are crystalline substances, which means their atoms are arranged in an orderly, repetitive manner as shown in Figure 2.2. This orderly packing of atoms is reflected in the regularly shaped objects we call crystals (Figure 2.2D). Some naturally occurring solids, such as volcanic glass (obsidian), lack a repetitive atomic structure and are not considered minerals.

4. **Definite chemical composition.** Most minerals are chemical compounds made up of two or more elements. The common mineral quartz, for example, consists of two oxygen (O) atoms for every silicon (Si) atom, giving it a chemical composition expressed by the formula SiO_2. However, a few minerals, such as gold, sulfur, and silver, consist of only a single element.

5. **Generally inorganic.** Inorganic crystalline solids, as exemplified by ordinary table salt (halite), that are found naturally in the ground are considered minerals. Organic compounds, on the other hand, are generally not. Sugar, a crystalline solid like salt but which comes from sugarcane or sugar beets, is a common example of such an organic compound. However, many marine animals secrete inorganic compounds, such as calcium carbonate (calcite), in the form of shells and coral reefs. These materials are considered minerals by most geologists.

In contrast to minerals, rocks are more loosely defined. Simply, a **rock** is any solid mass of mineral, or mineral-like, matter that occurs naturally as part of our planet. A few rocks are composed almost entirely of one mineral. A common example is the sedimentary rock *limestone*, which consists of impure masses of the mineral calcite. However, most rocks, like the common rock granite shown in Figure 2.3, occur as aggregates of several kinds of minerals. Here, the term *aggregate* implies that the minerals are joined in such a way that the properties of each mineral are retained. Note that

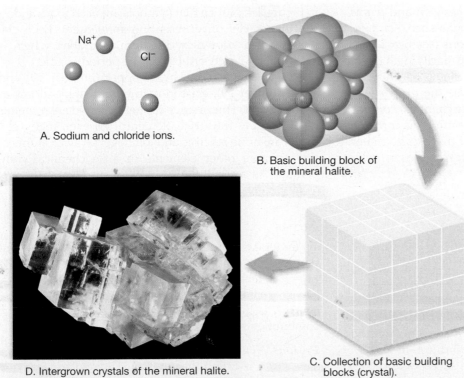

A. Sodium and chloride ions.

B. Basic building block of the mineral halite.

C. Collection of basic building blocks (crystal).

D. Intergrown crystals of the mineral halite.

◀ **FIGURE 2.2** This diagram illustrates the orderly arrangement of sodium and chloride ions in the mineral halite. The arrangement of atoms into basic building blocks having a cubic shape results in regularly shaped cubic crystals. (Photo by M. Claye/Jacana Scientific Control/Photo Researchers, Inc.)

► **FIGURE 2.3** Rocks are aggregates of one or more minerals.

Granite
(Rock)

Quartz
(Mineral)

Hornblende
(Mineral)

Feldspar
(Mineral)

you can easily identify the mineral constituents of the granite in Figure 2.3.

A few rocks are composed of nonmineral matter. These include the volcanic rocks *obsidian* and *pumice*, which are noncrystalline glassy substances, and *coal*, which consists of solid organic debris (see Box 2.1).

Although this chapter deals primarily with the nature of minerals, keep in mind that most rocks are simply aggregates of minerals. Because the properties of rocks are determined largely by the chemical composition and crystalline structure of the minerals contained within them, we will first consider these Earth materials. Then, in Chapter 3, we will take a closer look at Earth's major rock groups.

Elements: Building Blocks of Minerals

You are probably familiar with the names of many elements, including copper, gold, oxygen, and carbon. An **element** is a substance that cannot be broken down into simpler substances by chemical or physical means. There are about 90 different elements found in nature, and consequently about 90 different kinds of atoms. In addition, scientists have succeeded in creating about 23 synthetic elements.

The elements can be organized into rows so that those with similar properties are in the same column. This arrangement, called the **periodic table** is shown in Figure 2.4. Notice that symbols are used to provide a shorthand way of representing an element. Each element is also known by its atomic number, which is shown above each symbol on the periodic table.

Some elements, such as copper (number 29) and gold (number 79), exist in nature with single atoms as the basic unit. Thus, native copper and gold are minerals made entirely from one element. However, many elements are quite reactive and join together with atoms of one or more other elements to form chemical compounds. As a result, most minerals are chemical compounds consisting of two or more different elements.

Atoms

To understand how elements combine to form compounds, we must first consider the atom. The **atom** (*a* = not, *tomos* = cut) is the smallest particle of matter that retains the essential characteristics of an element. It is this extremely small particle that does the combining.

Protons, Neutrons, and Electrons. The central region of an atom is called the **nucleus**. The nucleus contains protons and neutrons. **Protons** are very dense particles

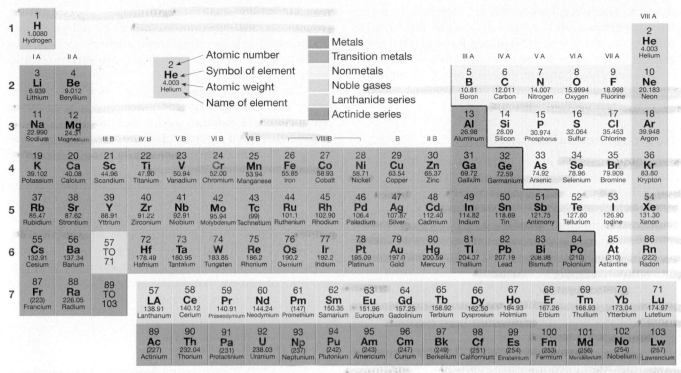

▲ FIGURE 2.4 Periodic Table of the Elements.

with positive electrical charges. **Neutrons** have the same mass as a proton but lack an electrical charge.

Orbiting the nucleus are **electrons**, which have negative electrical charges. For convenience, we sometimes diagram atoms to show the electrons orbiting the nucleus, like the orderly orbiting of the planets around the Sun (Figure 2.5A). However, electrons move in a less predictable way than planets. As a result, they create a sphere-shaped negative zone around the nucleus. A more realistic picture of the positions of electrons can be obtained by envisioning a cloud of negatively charged electrons surrounding the nucleus (Figure 2.5B).

Studies of electron configurations predict that individual electrons move within regions around the nucleus called **principal shells**, or **energy levels**. Furthermore, each of these shells can hold a specific number of electrons, with the outermost principal shell containing the **valence electrons**. Valence electrons are important because, as you will see later, they are involved in chemical bonding.

Atomic Number. Some atoms have only a single proton in their nuclei, while others contain more than 100 protons. The number of protons in the nucleus of an atom is called the **atomic number**. For example, all atoms with six protons are carbon atoms; hence, the atomic number of carbon is 6. Likewise, every atom with eight protons is an oxygen atom.

Atoms in their natural state also have the same number of electrons as protons, so the atomic number also equals the number of electrons surrounding the

▼ FIGURE 2.5 Two models of the atom. **A.** A very simplified view of the atom, which includes a central nucleus, consisting of protons and neutrons, encircled by high-speed electrons. **B.** Another model of the atom showing spherically shaped electron clouds (energy-level shells). Note that these models are not drawn to scale.

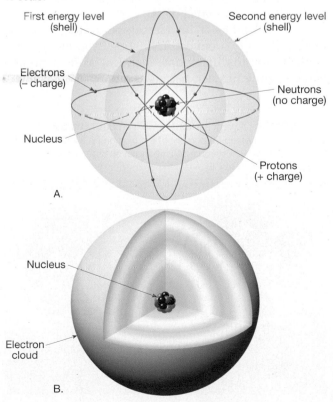

nucleus. Therefore, carbon has six electrons to match its six protons, and oxygen has eight electrons to match its eight protons. Neutrons have no charge, so the positive charge of the protons is exactly balanced by the negative charge of the electrons. Consequently, atoms in the natural state are neutral electrically and have no overall electrical charge.

Isotopes and Radioactive Decay

Subatomic particles are so incredibly small that a special unit, called an *atomic mass unit*, was devised to express their mass. A proton or a neutron has a mass just slightly more than one atomic mass unit, whereas an electron is only about one two-thousandth of an atomic mass unit. Thus, although electrons play an active role in chemical reactions, they do not contribute significantly to the mass of an atom.

The **mass number** of an atom is simply the total of its neutrons and protons. Atoms of the same element always have the same number of protons. But the number of neutrons for atoms of the same element can vary. Atoms with the same number of protons but different numbers of neutrons are **isotopes** of that element. Isotopes of the same element are labeled by placing the mass number after the element's name or symbol.

Carbon has several different isotopes. Models for two of these, carbon-14 and carbon-12, are shown in Figure 2.6. Since all atoms of the same element have the same number of protons, and carbon has six, carbon-12 also has *six neutrons* to give it a mass number of 12. Likewise, carbon-14 has six protons plus *eight neutrons* to give it a mass number of 14.

In chemical behavior, all isotopes of the same element are nearly identical. To distinguish among them is like trying to differentiate identical twins, with one being slightly heavier. Because isotopes of an element react the same chemically, different isotopes can become parts of the same mineral. For example, when the mineral calcite forms from calcium, carbon, and oxygen, some of its carbon atoms are carbon-12 and some are carbon-14.

The nuclei of most atoms are stable. However, many elements do have isotopes in which the nuclei are unstable. "Unstable" means that the isotopes disintegrate through a process called **radioactive decay**. Radioactive decay occurs when the forces that bind the nucleus are not strong enough.

During radioactive decay, unstable atoms radiate energy and emit particles. Some of this energy powers the movements of Earth's crust and upper mantle. The rates at which unstable atoms decay are measurable. Therefore, certain radioactive atoms can be used to determine the ages of fossils, rocks, and minerals. A discussion of radioactive decay and its applications in dating past geologic events is found in Chapter 11.

Why Atoms Bond

Sodium, a soft, silvery metal, is extremely reactive and poisonous. If you were to consume even a small amount of elemental sodium, you would need immediate medical attention. Chlorine, a green poisonous gas, is so toxic it was used as a chemical weapon during World War I. Together, however, these elements produce sodium chloride, a harmless flavor enhancer that we call table salt. When elements combine to form **compounds**, their properties change dramatically.

Why do elements join together to form compounds? From experimentation it has been learned that the forces holding the atoms together are electrical. Further, it is known that chemical bonding results in a change in the electron configuration of the bonded atoms. As we noted earlier, it is the valence electrons (outer-shell electrons) that are generally involved in chemical bonding. Figure 2.7 shows a shorthand way of representing the number of electrons in the outer

▼ **FIGURE 2.6** Two of several known isotopes of carbon. Carbon-12 makes up 99 percent of the carbon on Earth. Although much less common, carbon-14 is found in all organisms and hence is useful for dating fossils.

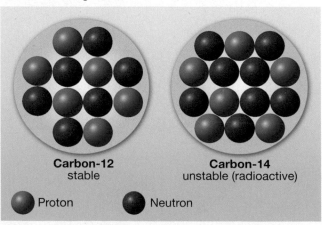

▼ **FIGURE 2.7** Dot diagrams for some representative elements. Each dot represents a valence electron found in the outermost principal shell.

Electron Dot Diagrams for Some Representative Elements							
I	II	III	IV	V	VI	VII	VIII
H ·							He:
Li ·	·Be·	·B·	·Ċ·	·N̈·	:Ö·	:F̈·	:N̈e:
Na ·	·Mg·	·Äl·	·Si·	·P̈·	:S̈·	:C̈l·	:Är:
K ·	·Ca·	·Ga·	·Ge·	·Äs·	:Se·	:Br·	:Kr:

principal shell (valence electrons). Notice that the elements in group I have one valence electron, those in group II have two, and so forth up to group VIII, which has eight valence electrons.

Other than the first shell, which can hold a maximum of two electrons, *a stable configuration occurs when the valence shell contains eight electrons*. Only the noble gases, such as neon and argon, have a complete outermost principal shell. Hence, the noble gases are the least chemically reactive and are designated as "inert."

When an atom's outermost shell does not contain the maximum number of electrons (8), the atom is likely to chemically bond with one or more other atoms. A *chemical bond* is the sharing or transfer of electrons to attain a stable electron configuration among the bonding atoms. If the electrons are transferred, the bond is an *ionic bond*. If the electrons are shared, the bond is called a *covalent bond*. In either case, the bonding atoms get stable electron configurations, which usually consists of eight electrons in the outer shell.

Ionic Bonds: Electrons Transferred

Perhaps the easiest type of bond to visualize is an **ionic bond**. In ionic bonding, one or more valence electrons are transferred from one atom to another. Simply, one atom gives up its valence electrons, and the other uses them to complete its outer shell. A common example of ionic bonding is sodium (Na) and chlorine (Cl) joining to produce sodium chloride (common table salt). This is shown in Figure 2.8A. Notice that sodium gives up its single valence electron to chlorine. As a result, sodium achieves a stable configuration having eight electrons in

its outermost shell. By acquiring the electron that sodium loses, chlorine—which has seven valence electrons—gains the eighth electron needed to complete its outermost shell. Thus, through the transfer of a single electron, both the sodium and chlorine atoms have acquired a stable electron configuration.

Once electron transfer takes place, atoms are no longer electrically neutral. By giving up one electron, a neutral sodium atom becomes *positively charged* (11 protons/10 electrons). Similarly, by acquiring one electron, the neutral chlorine atom becomes *negatively charged* (17 protons/18 electrons). Atoms such as these, which have an electrical charge because of the unequal numbers of electrons and protons, are called **ions**. We know that ions with like charges repel, and those with unlike charges attract. Thus, an *ionic bond* is the attraction of oppositely charged ions to one another, producing an electrically neutral compound. Figure 2.8B illustrates the arrangement of sodium and chlorine ions in the mineral halite.

Covalent Bonds: Electrons Shared

Not all atoms combine by transferring electrons to form ions. Some atoms *share* electrons. For example, the gaseous elements oxygen (O_2), hydrogen (H_2), and chlorine (Cl_2) exist as stable molecules consisting of two atoms bonded together, without a complete transfer of electrons.

Figure 2.9 illustrates the sharing of a pair of electrons between two chlorine atoms to form a molecule of chlorine gas (Cl_2). By overlapping their outer shells, these chlorine atoms share a pair of electrons. Thus, each chlorine atom has acquired, through cooperative action, the needed eight electrons to complete its outer shell. The bond produced by the sharing of electrons is called a **covalent bond**.

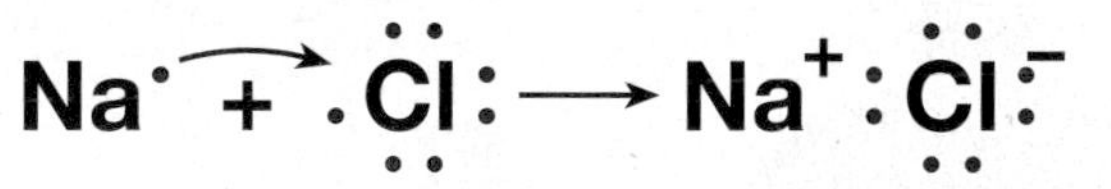

A common analogy may help you visualize a covalent bond. Imagine two people at opposite ends of a dimly lit room, each reading under a separate lamp. By moving the lamps to the center of the room, they are able to combine their light sources so each can see better. Just as the overlapping light beams meld, the shared electrons that provide the "electrical glue" in covalent bonds are indistinguishable

▼ **FIGURE 2.8** Chemical bonding of sodium and chlorine atoms to produce sodium chloride (table salt). **A.** Through the transfer of one electron in the outer shell of a sodium atom to the outer shell of a chlorine atom, the sodium becomes a positive ion and chlorine a negative ion. **B.** Diagram illustrating the arrangement of sodium and chloride ions in table salt.

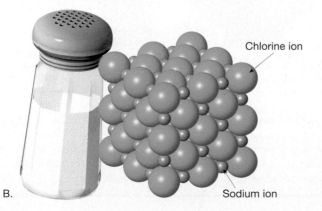

▼ **FIGURE 2.9** Dot diagrams used to illustrate the sharing of a pair of electrons between two chlorine atoms to form a chlorine molecule. Notice that by sharing a pair of electrons, both chlorine atoms achieve a full outer shell (8 electrons).

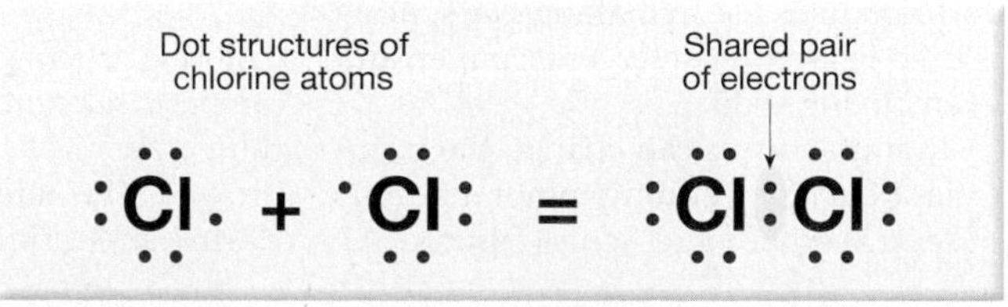

from each other. The most common mineral group, the silicates, contains the element silicon, which readily forms covalent bonds with oxygen.

STUDENTS SOMETIMES ASK...

Are the minerals you talked about in class the same as those found in dietary supplements?

Not ordinarily. From a geologic perspective, a mineral must be a *naturally occurring* crystalline solid. Minerals found in dietary supplements are human-made inorganic compounds that contain *elements* needed to sustain life. These dietary minerals typically contain elements that are metals—calcium, potassium, phosphorus, magnesium, and iron. It should also be noted that vitamins are *organic compounds* produced by living organisms, not *inorganic compounds*, like minerals.

Properties of Minerals

GEODe Earth Materials
▼ Minerals

Minerals are solids formed by inorganic processes. Each mineral has an orderly arrangement of atoms (crystalline structure) and a definite chemical composition, which give it a unique set of physical properties. Because the internal structure and chemical composition of a mineral are difficult to determine without the aid of sophisticated tests and apparatuses, the more easily recognized physical properties are frequently used in identification.

Primary Diagnostic Properties

The diagnostic physical properties of minerals are those that can be determined by observation or by performing a simple test. The primary physical properties

---

BOX 2.1 PEOPLE AND THE ENVIRONMENT
Making Glass from Minerals

Many everyday objects are made of glass, including windowpanes, jars and bottles, and the lenses of some eyeglasses. People have been making glass for at least 2000 years. Glass is manufactured by melting naturally occurring materials and cooling the liquid quickly before the atoms have time to arrange themselves into an orderly crystalline form. (This is the same way that natural glass, called *obsidian*, is generated from lava.)

It is possible to produce glass from a variety of materials, but the primary ingredient (75 percent) of most commercially produced glass is the mineral quartz (SiO_2). Lesser amounts of the minerals calcite (calcium carbonate) and trona (sodium carbonate) are added to the mix. These materials lower the melting temperature and improve the workability of the melt.

In the United States, high-quality quartz (usually quartz sandstone) and calcite (limestone) are readily available in many areas. Trona, on the other hand, is mined almost exclusively in the Green River area of southwestern Wyoming. In addition to its use in making glass, trona is used in making detergents, paper, and even baking soda.

Manufacturers can change the properties of glass by adding minor amounts of several other ingredients (Figure 2.A).

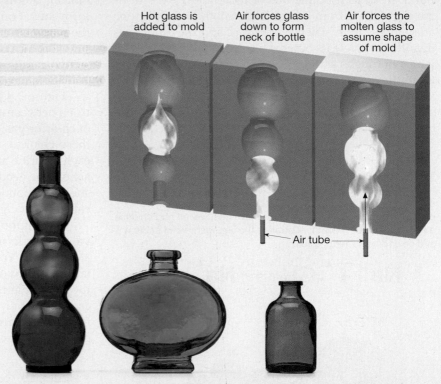

Figure 2.A Glass bottles are made by adding molten glass to a mold and using air to shape the glass. Metallic compounds are mixed with the raw ingredients to color the glass. (Photo by Guy Ryecart)

Coloring agents include iron sulfide (amber), selenium (pink), cobalt oxide (blue), and iron oxides (green, yellow, brown). The addition of lead imparts clarity and brilliance to glass and is therefore used in the manufacture of fine crystal tableware. Ovenware, such as Pyrex®, owes its heat resistance to boron, whereas aluminum makes glass resistant to weathering.

A.

B.

▲ **FIGURE 2.10** Crystal form is the external expression of a mineral's orderly internal structure. **A.** Pyrite, commonly known as "fool's gold," often forms cubic crystals that may contain parallel lines called striations. (Photo by GeoScience/PH) **B.** Quartz sample that exhibits well-developed hexagonal (six-sided) crystals with pyramidal-shaped ends. (Photo by Breck P. Kent)

that are commonly used to identify hand samples of minerals are: crystal form, luster, color, streak, hardness, cleavage or fracture, and specific gravity. Secondary (or "special") properties that are exhibited by a limited number of minerals include: magnetism, taste, feel, smell, elasticity, malleability, double refraction, and chemical reaction to hydrochloric acid.

Crystal Form **Crystal form** is the external expression of a mineral's internal orderly arrangement of atoms. Generally, when a mineral forms without space restrictions, it will develop individual crystals with well-formed crystal faces. Figure 2.10A illustrates this for pyrite, an iron-and-sulfur mineral with cubic crystals. Figure 2.10B shows the distinctive hexagonal crystals of quartz that form when space and time permit. However, most of the time, crystal growth is severely constrained. It is stunted because of competition for space, resulting in an intergrown mass of small, jammed crystals, none of which exhibits its crystal form. This is what happened to the minerals in the granite in Figure 2.3. Thus, most inorganic solid objects are composed of crystals, but they are not clearly visible to the unaided eye.

Luster **Luster** is the appearance or quality of light reflected from the surface of a mineral. Minerals that have the appearance of metals, regardless of color, are said to have a *metallic luster*, like the pyrite crystals in Figure 2.10A. Minerals with a *nonmetallic luster* are described by various adjectives. These include vitreous (glassy, like the quartz crystals in Figure 2.10B), pearly, silky, resinous, and earthy (dull). Some minerals appear somewhat metallic in luster and are said to be *sub-metallic*.

Color Although **color** is an obvious feature of a mineral, it is often an unreliable diagnostic property. Slight

impurities in the common mineral quartz, for example, give it a variety of colors, including pink, purple (amethyst), milky white, and even black.

Streak **Streak** is the color of a mineral in its powdered form, which is a much more reliable indication of color. Streak is obtained by rubbing the mineral across a piece of hard, unglazed porcelain, termed a *streak plate* (Figure 2.11). Whereas the color of a mineral may vary from sample to sample, the streak usually does not and is therefore the more consistent property. Streak can also help to distinguish minerals with metallic lusters from those having nonmetallic lusters. Metallic minerals generally have a dense, dark streak, whereas minerals with nonmetallic lusters do not.

▼ **FIGURE 2.11** Although the color of a mineral may not be very helpful in identification, the streak, which is the color of the powdered mineral, can be very useful. (Photo by Dennis Tasa)

37

Hardness One of the most useful diagnostic properties is **hardness,** a measure of the resistance of a mineral to abrasion or scratching. This property is determined by rubbing the mineral to be identified against another mineral of known hardness. One will scratch the other (unless they have the same hardness).

Geologists use a standard hardness scale, called the **Mohs scale.** It consists of 10 minerals arranged in order from 10 (hardest) to 1 (softest), as shown in Figure 2.12. Any mineral of unknown hardness can be rubbed against these to determine its hardness. In the field, other handy objects work, too. For example, your fingernail has a hardness of 2.5, a copper penny 3.5, and a piece of glass 5.5. The mineral gypsum, which has a hardness of 2, can be easily scratched by your fingernail. Conversely, the similar-looking mineral calcite, which has a hardness of 3, cannot be scratched by your fingernail. Calcite cannot scratch glass, because

▲ **FIGURE 2.13** Sheet-type cleavage common to the micas. (Photo by Chip Clark)

its hardness is less than 5.5. Quartz, the hardest of the common minerals at 7, will scratch a glass plate. Diamonds, hardest of all, scratch anything.

Cleavage In the crystal structure of a mineral, some bonds are weaker than others. These bonds are where a mineral will break when it is stressed. **Cleavage** (*Kleiben* = carve) is the tendency of a mineral to cleave, or break, along planes of weak bonding. Not all minerals have definite planes of weak bonding, but those that possess cleavage can be identified by the distinctive smooth surfaces that are produced when the mineral is broken.

The simplest type of cleavage is exhibited by minerals called micas (Figure 2.13). Because the micas have weak bonds in one direction, they cleave to form thin, flat sheets. Some minerals have several cleavage planes that produce smooth surfaces when broken, whereas others exhibit poor cleavage and still others have no cleavage at all. When minerals break evenly in more than one direction, cleavage is described by the *number of planes* exhibited and the *angles at which they meet* (Figure 2.14).

Do not confuse *cleavage* with *crystal form!* When a mineral exhibits cleavage, it will break into pieces *that have the same geometry as each other.* By contrast, the quartz crystals shown in Figure 2.10B do not have cleavage. If broken, they fracture into pieces that do not resemble each other or the original crystals.

Fracture Minerals that do not exhibit cleavage when broken, such as quartz, are said to **fracture.** Even fracturing has variety: Minerals that break into smooth curved surfaces like those seen in broken glass have a *conchoidal fracture* (Figure 2.15). Others break into splinters or fibers, like asbestos, but most minerals fracture irregularly.

▼ **FIGURE 2.12** Mohs scale of hardness, with the hardness of some common objects.

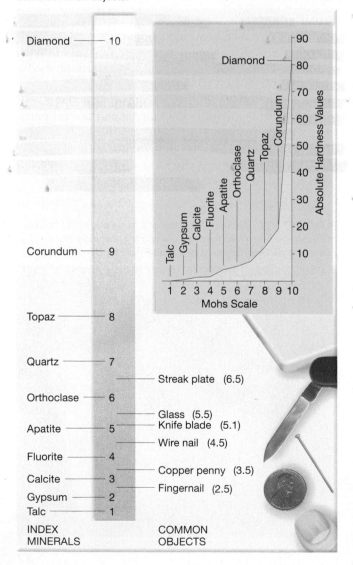

INDEX MINERALS		COMMON OBJECTS
Diamond	10	
Corundum	9	
Topaz	8	
Quartz	7	
		Streak plate (6.5)
Orthoclase	6	
		Glass (5.5)
Apatite	5	Knife blade (5.1)
		Wire nail (4.5)
Fluorite	4	
		Copper penny (3.5)
Calcite	3	Fingernail (2.5)
Gypsum	2	
Talc	1	

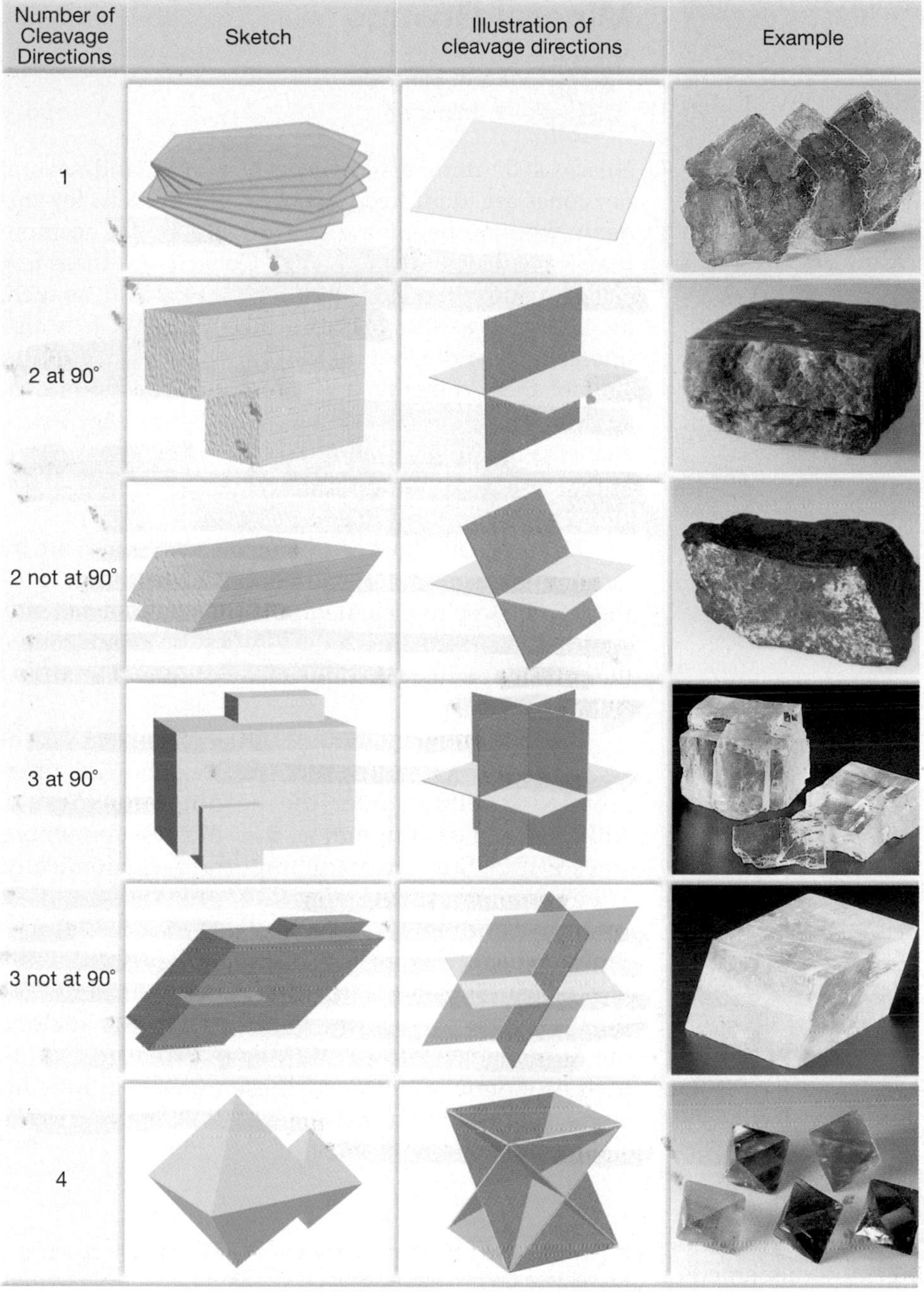

Number of Cleavage Directions	Sketch	Illustration of cleavage directions	Example
1			
2 at 90°			
2 not at 90°			
3 at 90°			
3 not at 90°			
4			

◄ FIGURE 2.14   Common cleavage directions exhibited by minerals.

Specific Gravity **Specific gravity** compares the weight of a mineral to the weight of an equal volume of water. For example, if a cubic centimeter of a mineral weighs three times as much as a cubic centimeter of water, its specific gravity is 3. With a little practice, you can estimate the specific gravity of minerals by hefting them in your hand. For instance, if a mineral feels as heavy as the common rocks you have handled, its specific gravity will probably be somewhere between 2.5 and 3. Some metallic minerals have a specific gravity noticeably greater than that of common rock-forming minerals. Galena, which is an ore of lead, has a specific gravity of roughly 7.5. The specific gravity of pure 24-karat gold is about 20.

STUDENTS SOMETIMES ASK...
Are there any artificial materials harder than diamonds?

Yes, but you won't be seeing them anytime soon. A hard form of carbon nitride (C_3N_4) described in 1989 and synthesized in a laboratory shortly thereafter may be harder than diamond but hasn't been produced in large enough amounts for a proper test. In 1999, researchers discovered that a form of carbon made from fused spheres of 20 and 28 carbon atoms—relatives of the famous "buckyballs"—also could be as hard as a diamond. These materials are expensive to produce, so diamonds continue to be used as abrasives and in certain kinds of cutting tools. Synthetic diamonds, produced since 1955, are now widely used in these industrial applications.

← 2 cm →

▲ **FIGURE 2.15** Conchoidal fracture. The smooth, curved surfaces result when minerals break in a glasslike manner. (Photo by E. J. Tarbuck)

Other Properties of Minerals

In addition, some minerals can be recognized by other distinctive properties. For example, halite is ordinary salt, so it is quickly identified with your tongue. Thin sheets of mica will bend and elastically snap back. Gold is malleable, which means it can be easily hammered or shaped. Talc and graphite both have distinctive feels. Talc feels soapy. Graphite feels greasy (it is a principal ingredient in dry lubricants).

A few minerals, such as magnetite, have a high iron content and can be picked up with a magnet. Some varieties of magnetite (lodestone) are natural magnets and will pick up small iron-based objects such as pins and paper clips. Some minerals exhibit special optical properties. For example, when a transparent piece of calcite is placed over printed material, the letters appear doubled. This optical property is known as *double refraction*. In addition, the streak of many sulfur-bearing minerals smells like rotten eggs.

One very simple chemical test involves placing a drop of dilute hydrochloric acid from a dropper bottle on a freshly broken mineral surface. Certain minerals, called **carbonates**, will effervesce (fizz) with hydrochloric acid. This test is useful in identifying the mineral calcite, which is a common carbonate mineral.

In summary, a number of special physical and chemical properties are useful in identifying particular minerals. These include taste, smell, elasticity, malleability, feel, magnetism, double refraction, and chemical reaction to hydrochloric acid. Remember that every one of these properties depends on the composition (elements) of a mineral and its structure (how the atoms are arranged).

Mineral Groups

Earth Minerals
▼ Minerals

Nearly 4000 minerals have been named, and several new ones are identified each year. Fortunately, for students who are beginning to study minerals, no more than a few dozen are abundant! Collectively, these few make up most of the rocks of Earth's crust and, as such, are classified as the **rock-forming minerals.** It is also interesting to note that *only eight elements* make up the bulk of these minerals and represent over 98 percent (by weight) of the continental crust (Figure 2.16). These elements, in order of abundance, are oxygen (O), silicon (Si), aluminum (Al), iron (Fe), calcium (Ca), sodium (Na), potassium (K), and magnesium (Mg).

As shown in Figure 2.16, silicon and oxygen are by far the most common elements in Earth's crust. Furthermore, these two elements readily combine to form the framework for the most common mineral group, the **silicates**, which account for more than 90 percent of Earth's crust.

Because other mineral groups are far less abundant in Earth's crust than the silicates, they are often grouped together under the heading **nonsilicates**. Although not as common as the silicates, some nonsilicate minerals are very important economically. They provide us with the iron and aluminum to build our automobiles; gypsum for plaster and drywall to construct our homes; and copper for wire to carry electricity and to connect us to the Internet. Some common nonsilicate mineral groups include the carbonates, sulfates, and halides. In addition to their economic importance, these mineral groups include members that are major constituents in sediments and sedimentary rocks.

▼ **FIGURE 2.16** Relative abundance of the eight most abundant elements in the continental crust.

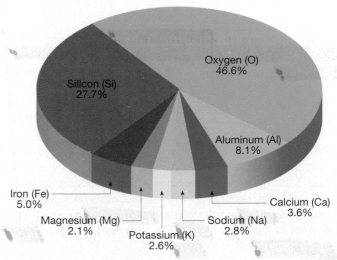

Oxygen (O) 46.6%
Silicon (Si) 27.7%
Aluminum (Al) 8.1%
Iron (Fe) 5.0%
Magnesium (Mg) 2.1%
Potassium (K) 2.6%
Sodium (Na) 2.8%
Calcium (Ca) 3.6%

We will first discuss the most common mineral group, the silicates, and then consider some of the prominent nonsilicate mineral groups.

Silicate Minerals

Each of the silicate minerals contains oxygen and silicon atoms. Except for a few silicate minerals such as quartz, most silicate minerals also contain one or more additional elements in their crystalline structure. These elements give rise to the great variety of silicate minerals and their varied properties.

All silicates have the same fundamental building block, the **silicon-oxygen tetrahedron** (*tetra* = four, *hedra* = a base) This structure consists of four oxygen atoms surrounding a much smaller silicon atom, as shown in Figure 2.17. Thus, a typical hand-size silicate mineral specimen contains millions of these silicon-oxygen tetrahedra, joined together in a variety of ways.

In some minerals, the tetrahedra are joined into chains, sheets, or three-dimensional networks by sharing oxygen atoms (Figure 2.18). These larger silicate structures are then connected to one another by other elements. The primary elements that join silicate structures are iron (Fe), magnesium (Mg), potassium (K), sodium (Na), and calcium (Ca).

Major groups of silicate minerals and common examples are given in Figure 2.18. The **feldspars** are by far the most plentiful group, comprising over 50 percent of Earth's crust. **Quartz**, the second most abundant mineral in the continental crust, is the only common mineral made completely of silicon and oxygen.

Notice in Figure 2.18 that each mineral *group* has a particular silicate *structure*. A relationship exists between this internal structure of a mineral and the *cleavage* it exhibits. Because the silicon-oxygen bonds are strong, silicate minerals tend to cleave between the silicon-oxygen structures rather than across them. For example, the micas have a sheet structure and thus tend to cleave into flat plates (see muscovite in Figure 2.13). Quartz, which has equally strong silicon oxygen bonds in all directions, has no cleavage but fractures instead.

How do silicate minerals form? Most crystallize from molten rock as it cools. This cooling can occur at or near Earth's surface (low temperature and pressure) or at great depths (high temperature and pressure). The *environment* during crystallization and the *chemical composition of the molten rock* mainly determine which minerals are produced. For example, the silicate mineral olivine crystallizes at high temperatures (about 1200°C), whereas quartz crystallizes at much lower temperatures (about 700°C).

In addition, some silicate minerals form at Earth's surface from the weathered products of other silicate minerals. Clay minerals are an example. Still other silicate minerals are formed under the extreme pressures associated with mountain building. Each silicate mineral, therefore, has a structure and a chemical composition that *indicate the conditions under which it formed*. Thus, by carefully examining the mineral makeup of rocks, geologists can often determine the circumstances under which the rocks formed.

▼ **FIGURE 2.17** Two representations of the silicon-oxygen tetrahedron. **A.** The four large spheres represent oxygen ions, and the blue sphere represents a silicon ion. The spheres are drawn in proportion to the radii of the ions. **B.** A model of the tetrahedron using rods to depict the bonds that connect the ions.

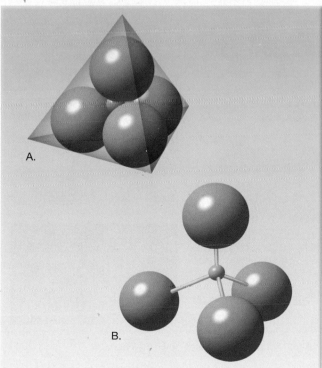

A.

B.

Mineral/Formula	Cleavage	Silicate Structure	Example
Olivine $(Mg, Fe)_2SiO_4$	None	Single tetrahedron 	
Pyroxene group (Augite) $(Mg,Fe)SiO_3$	Two planes at right angles	Single chains 	
Amphibole group (Hornblende) $Ca_2(Fe,Mg)_5Si_8O_{22}(OH)_2$	Two planes at 60° and 120°	Double chains 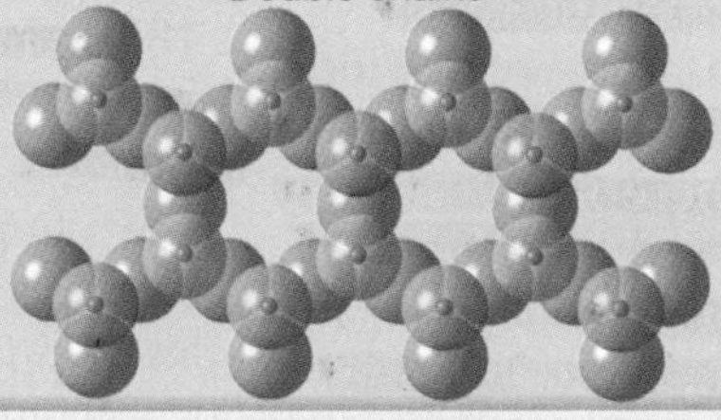	
Micas — Biotite $K(Mg,Fe)_3AlSi_3O_{10}(OH)_2$	One plane	Sheets 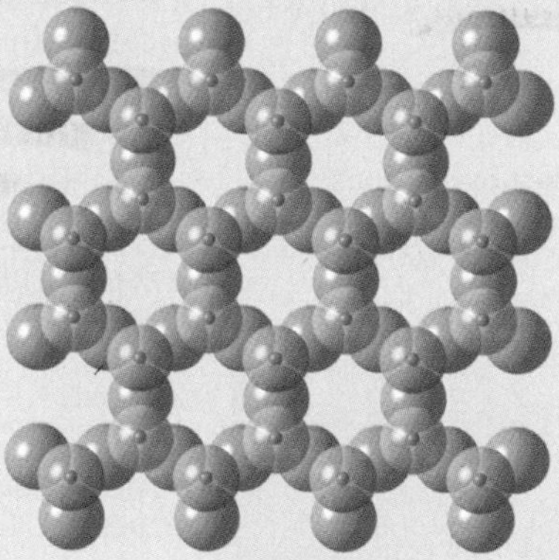	
Micas — Muscovite $KAl_2(AlSi_3O_{10})(OH)_2$	One plane	Sheets	
Feldspars — Potassium feldspar (Orthoclase) $KAlSi_3O_8$	Two planes at 90°	Three-dimensional networks 	
Feldspars — Plagioclase $(Ca,Na)AlSi_3O_8$	Two planes at 90°	Three-dimensional networks	
Quartz SiO_2	None		

▲ FIGURE 2.18 Common silicate minerals. Note that the complexity of silicate structures increases down the chart.

Important Nonsilicate Minerals

Although nonsilicates make up only about 8 percent of Earth's crust, some minerals, such as gypsum, calcite, and halite, are major constituents in sedimentary rocks. Furthermore, many others are important economically. Table 2.1 lists some of the nonsilicate mineral classes and a few examples of each. Some of the most common nonsilicate minerals belong to one of three classes of minerals—the carbonates (CO_3^{2-}), the sulfates (SO_4^{2-}), and the halides (Cl^{1-}, F^{1-}, B^{1-}).

The carbonate minerals are much simpler structurally than the silicates. This mineral group is composed of the carbonate ion (CO_3^{2-}) and one or more kinds of positive ions. The most common carbonate mineral is *calcite*, $CaCO_3$ (calcium carbonate). This mineral is the major constituent in two well-known rocks: limestone and marble. Limestone has many uses, including as road aggregate, as building stone, and as the main ingredient in portland cement. Marble is used decoratively.

Two other nonsilicate minerals frequently found in sedimentary rocks are *halite* and *gypsum*. Both minerals are commonly found in thick layers that are the last vestiges of ancient seas that have long since evaporated (Figure 2.19). Like limestone, both are important nonmetallic resources. Halite is the mineral name for common table salt (NaCl). Gypsum ($CaSO_4 \cdot 2H_2O$), which is calcium sulfate with water

▲ **FIGURE 2.19** Thick bed of halite (salt) at an underground mine in Grand Saline, Texas. (Photo by Tom Bochsler)

bound into the structure, is the mineral of which plaster and other similar building materials are composed.

Most nonsilicate mineral classes contain members that are prized for their economic value. This includes the oxides, whose members hematite and magnetite are important ores of iron (Figure 2.20). Also significant are the sulfides, which are basically compounds

Mineral Groups [key ion(s) or element(s)]	Mineral Name	Chemical Formula	Economic Use
Carbonates (CO_3^{2-})	Calcite	$CaCO_3$	Portland cement, lime
	Dolomite	$CaMg(CO_3)_2$	Portland cement, lime
Halides (Cl^-, F^-, Br^-)	Halite	NaCl	Common salt
	Fluorite (Fluorspar)	CaF_2	Hydrofluoric acid production, steelmaking
	Sylvite	KCl	Fertilizer
Oxides (O^{2-})	Hematite	Fe_2O_3	Ore of iron, pigment
	Magnetite	Fe_3O_4	Ore of iron
	Corundum	Al_2O_3	Gemstone, abrasive
	Ice	H_2O	Solid form of water
Sulfides (S^{2-})	Galena	PbS	Ore of lead
	Sphalerite	ZnS	Ore of zinc
	Pyrite	FeS_2	Sulfuric acid production
	Chalcopyrite	$CuFeS_2$	Ore of copper
	Cinnabar	HgS	Ore of mercury
Sulfates (SO_4^{2-})	Gypsum	$CaSO_4 \cdot 2H_2O$	Plaster
	Anhydrite	$CaSO_4$	Plaster
	Barite	$BaSO_4$	Drilling mud
Native elements (single elements)	Gold	Au	Trade, jewelry
	Copper	Cu	Electrical conductor
	Diamond	C	Gemstone, abrasive
	Sulfur	S	Sulfa drugs, chemicals
	Graphite	C	Pencil lead, dry lubricant
	Silver	Ag	Jewelry, photography
	Platinum	Pt	Catalyst

TABLE 2.1 Common Nonsilicate Mineral Groups

A.

B.

▲ FIGURE 2.20 Magnetite **A.** and hematite **B.** are both oxides and are both important ores of iron. (Photos by E. J. Tarbuck)

of sulfur (S) and one or more metals. Examples of important sulfide minerals include galena (lead), sphalerite (zinc), and chalcopyrite (copper). In addition, native elements, including gold, silver, and carbon (diamonds), plus a host of other nonsilicate minerals—fluorite (flux in making steel), corundum (gemstone, abrasive), and uraninite (a uranium source)—are important economically (see Box 2.2).

Box 2.2 Understanding Earth
Gemstones

Precious stones have been prized since antiquity. But misinformation abounds regarding gems and their mineral makeup. This stems partly from the ancient practice of grouping precious stones by color rather than mineral makeup. For example, *rubies* and red *spinels* are very similar in color, but they are completely different minerals. Classifying by color led to the more common spinels being passed off to royalty as rubies. Even today, with modern identification techniques, common *yellow quartz* is sometimes sold as the more valuable gemstone *topaz*.

Naming Gemstones

Most precious stones are given names that differ from their parent mineral. For example, *sapphire* is one of two gems that are varieties of the same mineral, *corundum*. Trace elements can produce vivid sapphires of nearly every color (Figure 2.B). Tiny amounts of titanium and iron in corundum produce the most prized blue sapphires. When the mineral

corundum contains a sufficient quantity of chromium, it exhibits a brilliant red color, and the gem is called *ruby*. Further, if a specimen is not suitable as a gem, it simply goes by the mineral name *corundum*. Because of its hardness, corundum that is not of gem quality is often crushed and sold as an abrasive.

To summarize, when corundum exhibits a red hue, it is called *ruby*, but if it exhibits any other color, the gem is called *sapphire*. Whereas corundum is the base mineral for two gems, quartz is the parent of more than a dozen gems. Table 2.A lists some well-known gemstones and their parent minerals.

What Constitutes a Gemstone?

When found in their natural state, most gemstones are dull and would be

Figure 2.B Australian sapphires depicting variations in cuts and colors. (Photo by Fred Ward, Black Star)

passed over by most people as "just another rock." Gems must be cut and polished by experienced professionals before their true beauty is displayed (Figure 2.B.). (One of the methods used to shape a gemstone is *cleaving*, the act of splitting the mineral along one of its planes of weakness, or cleavage.) Only those mineral specimens that are of such quality that they can command a price in excess of the cost of processing are considered gemstones.

Gemstones can be divided into two categories: precious and semiprecious. A *precious* gem has beauty, durability, and rarity, whereas a *semiprecious* gem generally has only one or two of these qualities. The gems traditionally held in highest esteem are diamonds, rubies, sapphires, emeralds, and some varieties of opal (Table 2.A). All other gemstones are classified as semiprecious. However, large high-quality specimens of semiprecious stones often command a very high price.

Today translucent stones with evenly tinted colors are preferred. The most favored hues are red, blue, green, purple, rose, and yellow. The most prized stones are pigeon-blood rubies, blue sapphires, grass-green emeralds, and canary-yellow diamonds. Colorless gems are generally less than desirable except for diamonds that display "flashes of color" known as *brilliance*.

TABLE 2.A	Important gemstones	
Gem	**Mineral Name**	**Prized Hues**
Precious		
Diamond	Diamond	Colorless, yellows
Emerald	Beryl	Greens
Opal	Opal	Brilliant hues
Ruby	Corundum	Reds
Sapphire	Corundum	Blues
Semiprecious		
Alexandrite	Chrysoberyl	Variable
Amethyst	Quartz	Purples
Cat's-eye	Chrysoberyl	Yellows
Chalcedony	Quartz (agate)	Banded
Citrine	Quartz	Yellows
Garnet	Garnet	Reds, greens
Jade	Jadeite or nephrite	Greens
Moonstone	Feldspar	Transparent blues
Peridot	Olivine	Olive greens
Smoky quartz	Quartz	Browns
Spinel	Spinel	Reds
Topaz	Topaz	Purples, reds
Tourmaline	Tourmaline	Reds, blue-greens
Turquoise	Turquoise	Blues
Zircon	Zircon	Reds

The durability of a gem depends on its hardness; that is, its resistance to abrasion by objects normally encountered in everyday living. For good durability, gems should be as hard or harder than quartz as defined by the Mohs scale of hardness. One notable exception is opal, which is comparatively soft (hardness 5 to 6.5) and brittle. Opal's esteem comes from its "fire," which is a display of a variety of brilliant colors, including greens, blues, and reds.

It seems to be human nature to treasure that which is rare. In the case of gemstones, large, high-quality specimens are much rarer than smaller stones. Thus, large rubies, diamonds, and emeralds, which are rare in addition to being beautiful and durable, command the very highest prices.

Mineral Resources

Mineral resources are Earth's storehouse of useful minerals that can be recovered for use. Resources include already identified deposits from which minerals can be extracted profitably, called **reserves,** as well as known deposits that are not yet recoverable under present economic conditions or technology. Deposits inferred to exist but not yet discovered are also considered mineral resources.

The term **ore** denotes useful metallic minerals that can be mined at a profit. In common usage, the term *ore* is also applied to some nonmetallic minerals, such as fluorite and sulfur. However, materials used for such purposes as building stone, road paving, abrasives, ceramics, and fertilizers are not usually called ores; rather, they are classified as *industrial rocks* and *minerals*.

Recall that more than 98 percent of Earth's crust is composed of only eight elements. Except for oxygen and silicon, all other elements make up a relatively small fraction of common crustal rocks (see Figure 2.16). Indeed, the natural concentrations of many elements are exceedingly small. A deposit containing only the average crustal percentage of a valuable element like gold is worthless because the cost of extracting it greatly exceeds the value of the material recovered.

To be considered of value, an element must be concentrated above the level of its average crustal abundance. For example, copper makes up about 0.0135 percent of the crust. However, for a material to be considered as copper ore, it must contain a concentration that is about 100 times this amount. Aluminum, in contrast, represents 8.13 percent of the crust and must be concentrated to only about four times its average crustal percentage before it can be extracted profitably.

It is important to realize that economic changes may make a deposit profitable to extract, or lose its profitability. If demand for a metal increases and prices rise sufficiently, the status of a previously unprofitable deposit changes, and it becomes an ore. The status of unprofitable deposits may also change if a technological advance allows the ore to be extracted at a lower cost than before.

Conversely, changing economic factors can turn a once profitable ore deposit into an unprofitable deposit that can no longer be called an ore. This situation is illustrated by a copper mine at Bingham Canyon, Utah, one of the largest open-pit mines on Earth (Figure 2.21). Mining was halted there in 1985 because outmoded equipment had driven the cost of extracting the copper beyond the current selling price. The owners responded by replacing an antiquated 1000-car railroad with conveyor belts and pipelines for transporting the ore and waste. These devices achieved a cost reduction of nearly 30 percent and returned this mining operation to profitability.

STUDENTS SOMETIMES ASK...

According to the textbook, thick beds of halite and gypsum formed when ancient seas evaporated, Has this happened in the recent past?

Yes. During the past 6 million years, the Mediterranean Sea may have dried up and then refilled several times. When 65 percent of seawater evaporates, the mineral gypsum begins to precipitate, meaning it comes out of solution and settles to the bottom. When 90 percent of the water is gone, halite crystals form, followed by salts of potassium and magnesium. Deep-sea drilling in the Mediterranean has encountered the presence of thick beds of gypsum and salt deposits (mostly halite) sitting one atop the other to a maximum depth of 2 kilometers (1.2 miles). These deposits are inferred to have resulted from tectonic events that periodically closed and reopened the connection between the Atlantic Ocean and the Mediterranean Sea (the modern-day Straits of Gibraltar) over the past several million years. During periods when the Mediterranean was cut off from the Atlantic, the warm and dry climate in this region caused the Mediterranean to nearly "dry up". Then, when the connection to the Atlantic was opened, the Mediterranean basin would refill with seawater of normal salinity. This cycle was repeated over and over again, producing the layers of gypsum and salt found on the Mediterranean seafloor.

▼ **FIGURE 2.21** Aerial view of Bingham Canyon copper mine near Salt Lake City, Utah. This huge open-pit mine is about 4 kilometers across and 900 meters deep. Although the amount of copper in the rock is less than 1 percent, the huge volumes of material removed and processed each day (about 200,000 tons) yield significant quantities of metal. (Photo by Michael Collier)

Chapter Summary

- A *mineral* is a naturally occurring inorganic solid possessing a definite chemical structure that gives it a unique set of physical properties. Most *rocks* are aggregates composed of two or more minerals.

- The building blocks of minerals are *elements*. An *atom* is the smallest particle of matter that still retains the characteristics of an element. Each atom has a *nucleus* containing *protons* and *neutrons*. Orbiting the nucleus of an atom are *electrons*. The number of protons in an atom's nucleus determines its *atomic number* and the name of the element. Atoms bond together to form a *compound* by either gaining, losing, or sharing electrons with another atom.

- *Isotopes* are variants of the same element but with a different *mass number* (the total number of neutrons plus protons found in an atom's nucleus). Some isotopes are unstable and disintegrate naturally through a process called *radioactive decay*.

- The properties of minerals include *crystal form, luster, color, streak, hardness, cleavage, fracture,* and *specific gravity*. In addition, a number of special physical and chemical properties (*taste, smell, elasticity, malleability, feel, magnetism, double refraction,* and *chemical reaction to hydrochloric acid*) are useful in identifying certain minerals. Each mineral has a unique set of properties that can be used for identification.

- The eight most abundant elements found in Earth's continental crust (oxygen, silicon, aluminum, iron, calcium, sodium, potassium, and magnesium) also make up the majority of minerals.

- The most common mineral group is the *silicates*. All silicate minerals have the *silicon-oxygen tetrahedron* as their fundamental building block. In some silicate minerals the tetrahedra are joined in chains; in others the tetrahedra are arranged into sheets, or three-dimensional networks. Each silicate mineral has a structure and a chemical composition that indicates the conditions under which it was formed.

- The *nonsilicate* mineral groups include the *oxides* (e.g., magnetite, mined for iron), *sulfides* (e.g., sphalerite, mined for zinc), *sulfates* (e.g., gypsum, used in plaster and frequently found in sedimentary rocks), *native elements* (e.g., graphite, a dry lubricant), *halides* (e.g., halite, common salt and frequently found in sedimentary rocks), and *carbonates* (e.g., calcite, used in portland cement and is a major constituent in two well-known rocks: limestone and marble).

- The term *ore* is used to denote useful metallic minerals, like hematite (mined for iron) and galena (mined for lead), that can be mined for a profit, as well as some nonmetallic minerals, such as fluorite and sulfur, that contain useful substances.

Key Terms

atom (p. 32)
atomic number (p. 33)
carbonates (p. 40)
cleavage (p. 38)
color (p. 37)
compound (p. 34)
covalent bond (p. 35)
crystal form (p. 37)
electron (p. 33)
energy levels (p. 33)
element (p. 32)
feldspar (p. 41)
fracture (p. 38)

hardness (p. 38)
ionic bond (p. 35)
ions (p. 35)
isotope (p. 34)
luster (p. 37)
mass number (p. 34)
mineral (p. 31)
mineral resource (p.45)
Mohs hardness scale (p. 38)
neutron (p. 33)
nonsilicates (40)
nucleus (p. 32)
ore (p. 45)

periodic table (p. 32)
principal shell (p. 33)
proton (p. 32)
quartz (p. 41)
radioactive decay (p. 34)
reserve (p. 45)
rock (p. 31)
rock-forming minerals (p. 40)
silicate (p. 40)
silicon-oxygen tetrahedron (p. 41)
specific gravity (p. 39)
streak (p. 37)
valence electrons (p. 33)

Review Questions

1. List five characteristics an Earth material should have in order to be considered a mineral.

2. Define the term *rock*.

3. List the three main particles of an atom and explain how they differ from one another.

4. If the number of electrons in an atom is 35 and its mass number is 80, calculate the following:
   (a) The number of protons.
   (b) The atomic number.
   (c) The number of neutrons.

5. What is an isotope?

6. What occurs in an atom to produce an ion?

7. Although all minerals have an orderly internal arrangement of atoms (crystalline structure), most mineral samples do not visibly demonstrate their crystal form. Why?

8. Why might it be difficult to identify a mineral by its color?

9. If you found a glassy-appearing mineral while rock hunting and had hopes that it was a diamond, what simple test might help you make a determination?

10. Table 2.1 (p. 43) lists a use for corundum as an abrasive. Explain why it makes a good abrasive in terms of the Mohs hardness scale.

11. Gold has a specific gravity of almost 20. A five-gallon pail of water weighs about 40 pounds, how much would a five-gallon pail of gold weigh?

12. What are the two most common elements in Earth's crust? What is the term used to describe the basic building block of all silicate minerals?

13. What are the two most common silicate minerals in Earth's crust?

14. List three nonsilicate minerals that are commonly found in rocks.

15. Contrast a mineral *resource* and a mineral *reserve*.

16. What might cause a mineral deposit that had not been considered an ore to become reclassified as an ore?

Examining the Earth System

1. Perhaps one of the most significant interrelationships between humans and the Earth system involves the extraction, refinement, and distribution of the planet's mineral wealth. To help you understand these associations, begin by thoroughly researching a mineral commodity that is mined in your local region or state. (You might find useful the information at these United States Geological Survey (USGS) Websites: **http://minerals.er.usgs.gov/minerals/pubs/state/** and **http://minerals.er.usgs.gov:80/minerals/pubs/mcs/** What products are made from this mineral? Do you use any of these products? Describe the mining and refining of the mineral and the local impact these processes have on each of Earth's spheres (atmosphere, hydrosphere, solid Earth, and biosphere). Are any of the effects negative? If so, what, if anything, is being done to end or minimize the damage?

2. Referring to the mineral you described above, in your opinion does the environmental impact of extracting this mineral outweigh the benefits derived from the products produced from this mineral?

Online Study Guide

The *Earth Science* Website uses the resources and flexibility of the Internet to aid in your study of the topics in this chapter. Written and developed by Earth science instructors, this site will help improve your understanding of Earth science. Visit **http://www.prenhall.com/tarbuck** and click on the cover of *Earth Science* 11e to find:

- **Online review quizzes.**
- **Critical thinking exercises.**
- **Links to chapter-specific Web resources.**
- **Internet-wide key term searches.**

http://www.prenhall.com/tarbuck

Climbers scaling the vertical face of
El Capitan in Yosemite National
Park, California.
(Photo by Ron Niebrugge/Mira)

CHAPTER 3

Rocks: Materials of the Solid Earth

hy study rocks? You have already learned that rocks and minerals have great economic value. Furthermore, all Earth processes in some way depend on the properties of these basic materials. Events such as volcanic eruptions, mountain building, weathering, erosion, and even earthquakes involve rocks and minerals. Consequently, a basic knowledge of Earth materials is essential in understanding Earth phenomena.

Every rock contains clues about the environments in which it formed (Figure 3.1). For example, some rocks are composed entirely of small shell fragments. This tells Earth scientists that the particles composing the rock originated in a shallow marine environment. Other rocks contain clues that indicate they formed from a volcanic eruption or deep in Earth during mountain building. Thus, rocks contain a wealth of information about events that have occurred over Earth's long history.

We divide rocks into three groups based on their mode of origin. The groups are igneous, sedimentary, and metamorphic. Before examining each group, we will view the rock cycle, which depicts the interrelationships among these rock groups.

Earth as a System: The Rock Cycle

 Earth Materials
▼ The Rock Cycle

Earth is a system. This means that our planet consists of many interacting parts that form a complex whole. Nowhere is this idea better illustrated than when we examine the rock cycle (Figure 3.2). The **rock cycle** allows us to view many of the interrelationships among different parts of the Earth system. It helps us understand the origin of igneous, sedimentary, and metamorphic rocks

▼ **FIGURE 3.1** Rocks contain information about the processes that produce them. This massive granitic monolith (El Capitan) located in Yosemite National Park, California, was once a molten mass found deep within Earth. (Photo by Tim Fitzharris/Minden Pictures)

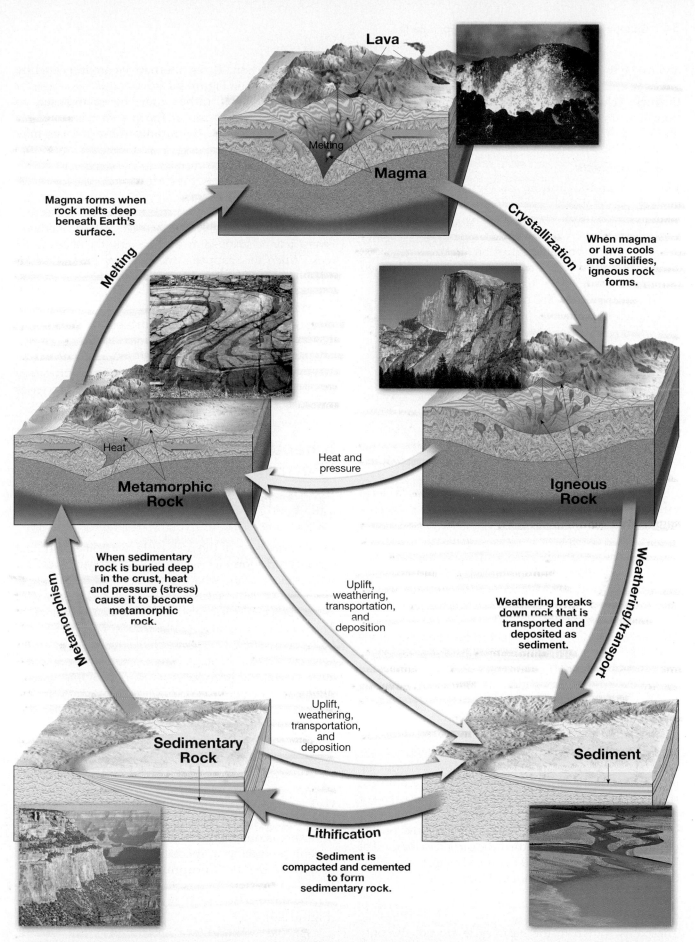

Lava

Magma forms when
rock melts deep
beneath Earth's
surface.

Melting

Crystallization

When magma
or lava cools
and solidifies,
igneous rock
forms.

Melting

Magma

Heat

Heat and
pressure

Metamorphic
Rock

Igneous
Rock

When sedimentary
rock is buried deep
in the crust, heat
and pressure (stress)
cause it to become
metamorphic
rock.

Metamorphism

Uplift,
weathering,
transportation,
and
deposition

Weathering breaks
down rock that is
transported and
deposited as
sediment.

Weathering/transport

Uplift,
weathering,
transportation,
and
deposition

Sedimentary
Rock

Sediment

Lithification

Sediment is
compacted and cemented
to form
sedimentary rock.

▲ FIGURE 3.2 Viewed over long spans, rocks are constantly forming, changing, and reforming. The rock cycle helps us understand the origin of the three basic rock groups. Arrows represent processes that link each group to the others.

and to see that each type is linked to the others by the processes that act upon and within the planet. Learn the rock cycle well; you will be examining its interrelationships in greater detail throughout this chapter and many others.

The Basic Cycle

We begin at the bottom of Figure 3.2. *Magma* is molten material that forms inside Earth. Eventually magma cools and solidifies. This process, called *crystallization*, may occur either beneath the surface or, following a volcanic eruption, at the surface. In either situation, the resulting rocks are called **igneous rocks** (*ignis* = fire).

If igneous rocks are exposed at the surface, they will undergo *weathering*, in which the day-in and day-out influences of the atmosphere slowly disintegrate and decompose rocks. The materials that result are often moved downslope by gravity before being picked up and transported by any of a number of erosional agents, such as running water, glaciers, wind, or waves. Eventually these particles and dissolved substances, called *sediment*, are deposited. Although most sediment ultimately comes to rest in the ocean, other sites of deposition include river floodplains, desert basins, swamps, and sand dunes.

Next the sediments undergo *lithification*, a term meaning "conversion into rock." Sediment is usually lithified into **sedimentary rock** when compacted by the weight of overlying layers or when cemented as percolating groundwater fills the pores with mineral matter.

If the resulting sedimentary rock is buried deep within Earth and involved in the dynamics of mountain building or intruded by a mass of magma, it will be subjected to great pressures and/or intense heat. The sedimentary rock will react to the changing environment and turn into the third rock type, **metamorphic rock**. If metamorphic rock is subjected to still higher temperatures, it will melt, creating magma, which will eventually crystallize into igneous rock, starting the cycle all over again.

Although rocks may seem to be unchanging masses, the rock cycle shows that they are not. The changes, however, take time—great amounts of time. In addition, the rock cycle is operating all over the world, but in different stages. Today new magma is forming under the island of Hawaii, while the Colorado Rockies are slowly being worn down by weathering and erosion. Some of this weathered debris will eventually be carried to the Gulf of Mexico, where it will add to the already substantial mass of sediment that has accumulated there.

Alternative Paths

The paths shown in the basic cycle are not the only ones that are possible. To the contrary, other paths are just as likely to be followed as those described in the preceding section. These alternatives are indicated by the blue arrows in Figure 3.2.

Igneous rocks, rather than being exposed to weathering and erosion at Earth's surface, may remain deeply buried. Eventually these masses may be subjected to the strong compressional forces and high temperatures associated with mountain building. When this occurs, they are transformed directly into metamorphic rocks.

Metamorphic and sedimentary rocks, as well as sediment, do not always remain buried. Rather, overlying layers may be stripped away, exposing the once buried rock. When this happens, the material is attacked by weathering processes and turned into new raw materials for sedimentary rocks.

Where does the energy that drives Earth's rock cycle come from? Processes driven by heat from Earth's interior are responsible for forming igneous and metamorphic rocks. Weathering and the movement of weathered material are external processes powered by energy from the Sun. External processes produce sedimentary rocks.

Igneous Rocks: "Formed by Fire"

 GEODe Earth Materials
▼ Igneous Rocks

In our discussion of the rock cycle, we pointed out that igneous rocks form as *magma* cools and crystallizes. But what is magma, and what is its source? **Magma** is molten rock generated by partial melting of rocks in Earth's mantle and in much smaller amounts, in the lower crust. This molten material consists mainly of the elements found in the silicate minerals. Silicon and oxygen are the main constituents in magma, with lesser amounts of aluminum, iron, calcium, sodium, potassium, magnesium, and others. Magma also contains some gases, particularly water vapor, which are confined within the magma body by the weight of the overlying rocks.

Once formed, a magma body buoyantly rises toward the surface because it is less dense than the surrounding rocks. Occasionally molten rock reaches the surface, where it is called **lava**. Sometimes lava is emitted as fountains that are produced when escaping gases propel molten rock skyward. On other occasions, magma is explosively ejected from a vent producing a spectacular eruption as exemplified by the 1980 eruption of Mount St. Helens. However, most eruptions are not violent; rather, volcanoes more often emit quiet outpourings of lava (Figure 3.3).

Igneous rocks that form when molten rock solidifies *at the surface* are classified as **extrusive** or **volcanic** (after the fire god Vulcan). Extrusive igneous rocks are abundant in western portions of the Americas, including the volcanic cones of the Cascade Range and the extensive lava flows of the Columbia Plateau. In addition, many oceanic islands, typified by the Hawaiian Islands, are composed almost entirely of volcanic igneous rocks.

Most magma, however, loses its mobility before reaching the surface and eventually crystallizes at depth. Igneous rocks that *form at depth* are termed **intrusive** or **plutonic** (after Pluto, the god of the lower world in classical mythology). Intrusive igneous rocks would never be exposed at the surface if portions of the crust were not uplifted and the overlying rocks stripped away by erosion. Exposures of intrusive igneous rocks occur in many places, including Mount Washington, New Hampshire; Stone Mountain, Georgia; the Black Hills of South Dakota; and Yosemite National Park, California.

Magma Crystallizes to Form Igneous Rocks

Magma is basically a very hot, thick fluid, but it also contains solids and gases. The solids are mineral crystals. The liquid portion of a magma body, termed the *melt,* is composed of ions that move about freely.

However, as magma cools, the random movements of the ions slow, and the ions begin to arrange themselves into orderly patterns. This process is called **crystallization**. Usually all of the molten material does not solidify at the same time. Rather, as it cools, numerous small crystals develop. In a systematic fashion, ions are added to these centers of crystal growth. When the crystals grow large enough for their edges to meet, their growth ceases for lack of space, and crystallization continues elsewhere. Eventually, all of the liquid is transformed into a solid mass of interlocking crystals.

The rate of cooling strongly influences crystal size. If magma cools very slowly, relatively few centers of crystal growth develop. Slow cooling also allows ions to migrate over relatively great distances. Consequently, *slow cooling results in the formation of large crystals*. Conversely, if cooling occurs quite rapidly, the ions lose their motion and quickly combine. This results in a large number of tiny crystals that all compete for the available ions. Therefore, the outcome of rapid cooling is the formation of a solid mass of *small intergrown crystals*.

Thus, if a geologist encounters igneous rock containing crystals large enough to be seen with the unaided

▼ **FIGURE 3.3**   Fluid basaltic lava emitted from Hawaii's Kilauea Volcano. (Photo by G. Brad Lewis/Liaison Agency, Inc.)

eye, it means the magma cooled quite slowly. But if the crystals can be seen only with a microscope, the geologist knows that the magma cooled very quickly.

If the molten material is quenched almost instantly, there is not sufficient time for the ions to arrange themselves into a crystalline network. Therefore, solids produced in this manner consist of randomly distributed ions. Such rocks are called *glass* and are quite similar to ordinary manufactured glass. Instant quenching occurs during violent volcanic eruptions that produce tiny shards of glass termed *volcanic ash*.

STUDENTS SOMETIMES ASK...

Are lava and magma the same thing?

No, but their *composition* might be similar. Both are terms that describe molten or liquid rock: Magma exists beneath Earth's surface, and lava is molten rock that has reached the surface. That's the reason why they can be similar in composition: Lava is produced from magma, but it generally has lost materials that escape as a gas, such as water vapor.

In addition to the rate of cooling, the composition of a magma and the amount of dissolved gases influence crystallization. Because magmas differ in each of these aspects, the physical appearance and mineral composition of igneous rocks vary widely. Nevertheless, it is possible to classify igneous rocks based on their texture and mineral constituents. We will now look at both features.

Igneous Textures

Texture describes the overall appearance of an igneous rock based on the *size, shape,* and *arrangement* of its interlocking crystals. Texture is a very important characteristic, because it reveals a great deal about the environment in which the rock formed. You learned that rapid cooling produces small crystals, whereas very slow cooling produces much larger crystals. As you might expect, the rate of cooling is quite slow in magma chambers lying deep within the crust, whereas a thin layer of lava extruded upon Earth's surface may chill rock solid in a matter of hours. Small molten blobs ejected into the air during a violent eruption can solidify almost instantly.

Igneous rocks that form rapidly at the surface or as small masses within the upper crust have a **fine-grained texture**, with the individual crystals too small to be seen with the unaided eye (Figure 3.4A). Common in many fine-grained igneous rocks are voids, called *vesicles*, left by gas bubbles that formed as the lava solidified (Figure 3.5).

▼ **FIGURE 3.4** Igneous rock textures. **A.** Igneous rocks that form at or near Earth's surface cool quickly and often exhibit a fine-grained texture. **B.** Coarse-grained igneous rocks form when magma slowly crystallizes at depth. **C.** During a volcanic eruption in which silica-rich lava is ejected into the atmosphere, a frothy glass called pumice may form. **D.** A porphyritic texture results when magma that already contains some large crystals migrates to a new location where the rate of cooling increases. The resulting rock consists of larger crystals embedded within a matrix of smaller crystals. (Photos courtesy of E. J. Tarbuck)

A. Fine-grained

B. Coarse-grained

Extrusive igneous rocks

Intrusive igneous rocks

C. Glassy (pumice)

D. Porphyritic

▲ **FIGURE 3.5** Scoria is a volcanic rock that exhibits vesicles, which form as gas bubbles escape near the top of a lava flow. (Photo courtesy of E. J. Tarbuck)

▲ **FIGURE 3.6** Obsidian, a glassy volcanic rock. (Photo courtesy of E. J. Tarbuck)

When large masses of magma solidify far below the surface, they form igneous rocks that exhibit a **coarse-grained texture**. These coarse-grained rocks have the appearance of a mass of intergrown crystals, which are roughly equal in size and large enough that the individual minerals can be identified with the unaided eye. Granite is a classic example (Figure 3.4B).

A large mass of magma located at depth may require tens of thousands of years to solidify. Because all materials within a magma do not crystallize at the same rate or at the same time during cooling, it is possible for some crystals to become quite large before others even start to form. If magma that already contains some large crystals suddenly erupts at the surface, the remaining molten portion of the lava would cool quickly. The resulting rock, which has large crystals embedded in a matrix of smaller crystals, is said to have a **porphyritic texture** (Figure 3.4D).

During some volcanic eruptions, molten rock is ejected into the atmosphere, where it is quenched very quickly. Rapid cooling of this type may generate rock with a **glassy texture**. As was indicated, glass results when the ions do not have sufficient time to unite into an orderly crystalline structure. In addition, melts that contain large amounts of silica (SiO_2) are more likely to form rocks that exhibit a glassy texture than melts with a low silica content.

Obsidian, a common type of natural glass, is similar in appearance to a dark chunk of manufactured glass (Figure 3.6). Another volcanic rock that exhibits a glassy texture is *pumice*. Usually found with obsidian, pumice forms when large amounts of gas escape from a melt to generate a gray, frothy mass (Figure 3.4C). In some samples, the vesicles are quite noticeable; in others, the pumice resembles fine shards of intertwined glass. Because of the large volume of air-filled voids, many samples of pumice will float in water.

STUDENTS SOMETIMES ASK...

You mentioned that Native Americans used obsidian for making arrowheads and cutting tools. Is this the only material they used?

No, Native Americans used whatever materials were locally available to make tools, including any hard compact rock material that could be shaped. This includes materials such as the metamorphic rocks slate and quartzite, sedimentary deposits made of silica called jasper, chert, opal, flint, and even jade. Some of these deposits have a limited geographic distribution and can now help anthropologists to reconstruct trade routes between different groups of Indians.

Igneous Compositions

Igneous rocks are mainly composed of silicate minerals. Furthermore, the mineral makeup of a particular igneous rock is ultimately determined by the chemical composition of the magma from which it crystallizes. Recall that magma is composed largely of the eight elements that are the major constituents of the silicate minerals. Chemical analysis shows that silicon and oxygen (usually expressed as the silica [SiO_2] content of magma) are by far the most abundant constituents of igneous rocks. These two elements, plus ions of aluminum (Al), calcium (Ca), sodium (Na), potassium (K), magnesium (Mg), and iron (Fe), make up roughly 98 percent by weight of most magmas.

As magma cools and solidifies, these elements combine to form two major groups of silicate minerals. The *dark silicates* are rich in iron and/or magnesium and are relatively low in silica. *Olivine, pyroxene, amphibole,* and *biotite mica* are the common dark silicate minerals of Earth's crust. By contrast, the *light silicates* contain greater amounts of potassium, sodium, and calcium rather than iron and magnesium. As a group, these minerals are higher in silica than the dark silicates. The light silicates include *quartz, muscovite mica,* and the most abundant mineral group, the *feldspars.* The feldspars make up at least 40 percent of most igneous rocks. Thus, in addition to feldspar, igneous rocks contain some combination of the other light and/or dark silicates listed above.

Classifying Igneous Rocks

Igneous rocks are classified by their texture and mineral composition. Various igneous textures result from different cooling histories, while the mineral compositions are a consequence of the chemical makeup of the parent magma and the environment of crystallization.

Despite their great compositional diversity, igneous rocks can be divided into broad groups according to their proportions of light and dark minerals. A general classification scheme based on texture and mineral composition is provided in Figure 3.7.

Granitic (Felsic) Rocks. Near one end of the continuum are rocks composed almost entirely of light-colored silicates—quartz and potassium feldspar. Igneous rocks in which these are the dominant minerals have a **granitic composition**. Geologists also refer to granitic rocks as being **felsic**, a term derived from *fel*dspar and *si*lica (quartz). In addition to quartz and feldspar, most granitic rocks contain about 10 percent dark silicate minerals, usually biotite mica and amphibole. Granitic rocks are rich in silica (about 70 percent) and are major constituents of the continental crust.

Granite is a coarse-grained igneous rock that forms where large masses of magma slowly solidify at depth. During episodes of mountain building, granite and related crystalline rocks may be uplifted, whereupon the processes of weathering and erosion strip away the overlying crust. Pikes Peak in the Rockies, Mount Rushmore in the Black Hills, Stone Mountain in Georgia, and Yosemite National Park in the Sierra Nevada are all areas where large quantities of granite are exposed at the surface.

Granite is perhaps the best-known igneous rock (Figure 3.8). This is partly because of its natural

▼ **FIGURE 3.7** Classification of the major groups of igneous rocks based on their mineral composition and texture. Coarse-grained rocks are plutonic, solidifying deep underground. Fine-grained rocks are volcanic, or solidify as shallow, thin plutons. Ultramafic rocks are dark, dense rocks, composed almost entirely of minerals containing iron and magnesium. Although relatively rare on Earth's surface, these rocks are believed to be major constituents of the upper mantle.

Chemical Composition		Granitic (Felsic)	Andesitic (Intermediate)	Basaltic (Mafic)	Ultramafic
Dominant Minerals		Quartz Potassium feldspar Sodium-rich plagioclase feldspar	Amphibole Sodium- and calcium-rich plagioclase feldspar	Pyroxene Calcium-rich plagioclase feldspar	Olivine Pyroxene
T E X T U R E	Coarse-grained (phaneritic)	**Granite**	**Diorite**	**Gabbro**	**Peridotite**
	Fine-grained (aphanitic)	**Rhyolite**	**Andesite**	**Basalt**	
	Porphyritic	"Porphyritic" precedes any of the above names whenever there are appreciable phenocrysts			Uncommon
	Glassy	**Obsidian** (compact glass) **Pumice** (frothy glass)			
Rock Color (based on % of dark minerals)		0% to 25%	25% to 45%	45% to 85%	85% to 100%

Granitic (Felsic)	Andesitic (Intermediate)	Basaltic (Mafic)
Intrusive (course-grained) Granite	Diorite	Gabbro
Extrusive (fine-grained) Rhyolite	Andesite	Basalt

▲ **FIGURE 3.8** Common igneous rocks. (Photos by E. J. Tarbuck)

beauty, which is enhanced when polished, and partly because of its abundance. Slabs of polished granite are commonly used for tombstones, monuments, and as building stones.

Rhyolite is the extrusive equivalent of granite and, like granite, is composed essentially of the light-colored silicates (Figure 3.8). This fact accounts for its color, which is usually buff to pink or light gray. Rhyolite is fine-grained and frequently contains glass fragments and voids, indicating rapid cooling in a surface environment. In contrast to granite, which is widely distributed as large plutonic masses, rhyolite deposits are less common and generally less voluminous. Yellowstone Park is one well-known exception. Here rhyolite lava flows and thick ash deposits of similar composition are extensive.

Basaltic (Mafic) Rocks. Rocks that contain substantial amounts of dark colored minerals, mainly olivine and pyroxene, and calcium-rich plagioclase feldspar are said to have a **basaltic composition** (Figure 3.7). Because basaltic rocks contain a high percentage of dark silicate minerals, geologists also refer to them as **mafic** (from *magnesium* and *ferrum*, the Latin name for iron). Because of their iron content, basaltic rocks are typically darker and denser than granitic rocks.

Basalt is a very dark green to black fine-grained volcanic rock composed primarily of pyroxene, olivine, and plagioclase feldspar (Figure 3.8). Basalt is the most common extrusive igneous rock. Many volcanic islands, such as the Hawaiian Islands and Iceland, are composed mainly of basalt. Further, the upper layers of the oceanic crust consist of basalt. In the United States, large portions of central Oregon and Washington were the sites of extensive basaltic outpourings (see Figure 9.19, p. 266).

The coarse-grained, intrusive equivalent of basalt is called *gabbro* (Figure 3.8). Although gabbro is not commonly exposed on the surface, it makes up a significant percentage of the oceanic crust.

Andesitic (Intermediate) Rocks. As you can see in Figure 3.7, rocks with a composition between granitic and basaltic rocks are said to have an **andesitic** or **intermediate composition** after the common volcanic rock *andesite* (Figure 3.8). Andesitic rocks contain a mixture of both light- and dark-colored minerals, mainly amphibole and pyroxene, with the other dominant mineral being plagioclase feldspar. This important category of igneous rocks is associated with volcanic activity that is typically confined to the margins of continents. When magma of intermediate composition crystallizes at depth, it forms the coarse-grained rock called *diorite* (Figure 3.8).

Ultramafic Rocks. Another important igneous rock, *peridotite*, contains mostly the dark-colored minerals olivine and pyroxene and thus falls on the opposite side of the compositional spectrum from granitic rocks (Figure 3.7). Because peridotite is composed almost entirely of dark silicate minerals, its chemical composition is referred to as **ultramafic**. Although ultramafic rocks are rare at Earth's surface, peridotite is believed to be the main constituent of the upper mantle.

How Different Igneous Rocks Form

Because a large variety of igneous rocks exist, it is logical to assume that an equally large variety of magmas must also exist. However, geologists have observed that a single volcano may extrude lavas exhibiting quite different compositions. Data of this type led them to examine the possibility that a single magma body might change (evolve) and thus become the parent to a variety of igneous rocks. To explore this idea, a pioneering investigation into the crystallization of magma was carried out by N. L. Bowen in the first quarter of the twentieth century.

Bowen's Reaction Series. In a laboratory setting, Bowen demonstrated that unlike a pure compound, such as water, which solidifies at a specific temperature, magma with its diverse chemistry crystallizes over a temperature range of at least 200 degrees. Thus as magma cools, certain minerals crystallize first, at relatively high temperatures (top of Figure 3.9). At successively lower temperatures, other minerals crystallize. This arrangement of minerals, shown in Figure 3.9, became known as **Bowen's reaction series**.

Bowen discovered that the first mineral to crystallize from a mass of magma is olivine. Further cooling results in the formation of pyroxene, as well as plagioclase feldspar. At intermediate temperatures the minerals amphibole and biotite begin to crystallize.

During the last stage of crystallization, after most of the magma has solidified, the minerals muscovite and potassium feldspar may form (Figure 3.9). Finally, quartz crystallizes from any remaining liquid. As a result, olivine is not usually found with quartz in the same igneous rock, because quartz crystallizes at much lower temperatures than olivine.

Evidence that this highly idealized crystallization model approximates what can happen in nature comes from the analysis of igneous rocks. In particular, we find that minerals that form in the same general temperature range on Bowen's reaction series are found together in the same igneous rocks. For example, notice in Figure 3.9 that the minerals quartz, potassium feldspar, and muscovite, which are located in the same region of Bowen's diagram, are typically found together as major constituents of the igneous rock *granite*.

Magmatic Differentiation. Bowen demonstrated that different minerals crystallize at different temperatures. But how do Bowen's findings account for the great diversity of igneous rocks? During the crystallization process, the composition of the melt continually changes because it gradually becomes depleted in those elements used to make the earlier formed minerals. This process, coupled with the fact that at one or more stages during crystallization, a separation of the solid and liquid components of magma can occur. One way this happens is called **crystal settling**. This process occurs when the earlier-formed minerals are denser (heavier) than the liquid portion and sink toward the bottom of the magma chamber, as shown in Figure 3.10. When the remaining melt solidifies—either in place or in another location if it migrates into fractures in the surrounding rocks—it will form a rock with a chemical composition much different from the parent magma (Figure 3.10). The formation of one or more secondary magmas from a single parent magma is called **magmatic differentiation.**

At any stage in the evolution of a magma, the solid and liquid components can separate into two chemically distinct units. Further, magmatic differentiation within the secondary melt can generate additional chemically distinct fractions. Consequently, magmatic

▼ **FIGURE 3.9** Bowen's reaction series shows the sequence in which minerals crystallize from a magma. Compare this figure to the mineral composition of the rock groups in Figure 3.7. Note that each rock group consists of minerals that crystallize in the same temperature range.

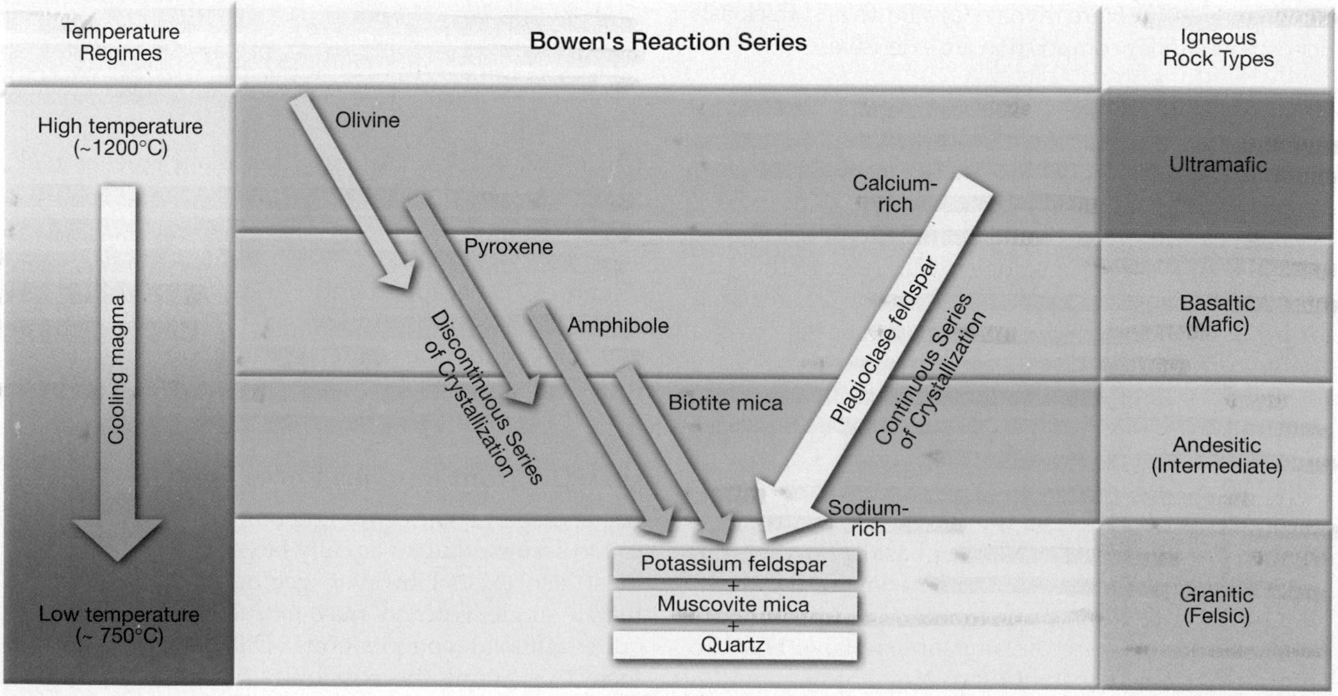

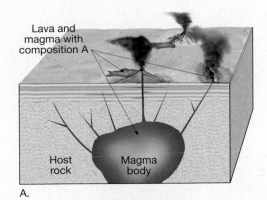

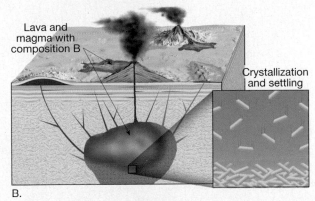

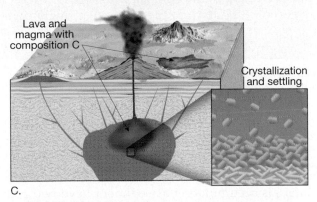

▲ **FIGURE 3.10** Illustration of how a magma evolves as the earliest-formed minerals (those richer in iron, magnesium, and calcium) crystallize and settle to the bottom of the magma chamber, leaving the remaining melt richer in sodium, potassium, and silica (SiO_2) **A.** Emplacement of a magma body and associated igneous activity generates rocks having a composition similar to that of the initial magma. **B.** After a period of time, crystallization and settling changes the composition of the melt, while generating rocks having a composition quite different than the original magma. **C.** Further magmatic differentiation results in another more highly evolved melt with its associated rock types.

differentiation and separation of the solid and liquid components at various stages of crystallization can produce several chemically diverse magmas and ultimately a variety of igneous rocks.

Assimilation and Magma Mixing. Strong evidence suggests that the chemical composition of a magma can change by processes other than magmatic differentiation. For example, as magma migrates upward through the crust, it may incorporate some of the surrounding host rock, a process called *assimilation*. Another means by which the composition of a magma body can be altered is called *magma mixing*. This process occurs whenever one magma body intrudes another. Once combined, convective flow may stir the two magmas to generate a fluid with a different composition.

In summary, N. L. Bowen successfully demonstrated that through magmatic differentiation, a single parent magma can generate several mineralogically different igneous rocks. This process, in concert with magma mixing and contamination by crustal rocks, accounts in part for the great diversity of magmas and igneous rocks.

Sedimentary Rocks: Compacted and Cemented Sediment

GEODe Earth Materials
▼ Sedimentary Rocks

Recall the rock cycle, which shows the origin of sedimentary rocks. Weathering begins the process. Next, gravity and erosional agents (running water, wind, waves, and glacial ice) remove the products of weathering and carry them to a new location where they are deposited. Usually the particles are broken down further during this transport phase. Following deposition, this **sediment** may become lithified, or turned to rock. Commonly, *compaction* and *cementation* transform the sediment into solid sedimentary rock.

The word *sedimentary* indicates the nature of these rocks, for it is derived from the Latin *sedimentum*, which means settling, a reference to a solid material settling out of a fluid. Most sediment is deposited in this fashion. Weathered debris is constantly being swept from bedrock and carried away by water, ice, or wind. Eventually the material is deposited in lakes, river valleys, seas, and countless other places. The particles in a desert sand dune, the mud on the floor of a swamp, the gravels in a stream bed, and even household dust are examples of sediment produced by this never-ending process.

The weathering of bedrock and the transport and deposition of the weathering products are continuous. As piles of sediment accumulate, the materials near the bottom are compacted by the weight of the overlying layers. Over long periods, these sediments are cemented together by mineral matter deposited from water in the spaces between particles. This forms solid sedimentary rock.

Geologists estimate that sedimentary rocks account for only about 5 percent (by volume) of Earth's outer 16 kilometers (10 miles). However, the importance of this group of rocks is far greater than this percentage implies. If you sampled the rocks exposed at Earth's surface, you would find that the great majority are sedimentary

▲ **FIGURE 3.11** Sedimentary rocks exposed in Utah's Capital Reef National Park. Sedimentary rocks occur in layers called strata. About 75 percent of all rock outcrops on the continents are sedimentary rocks. (Photo by Marc Muench/David Muench Photograph)

(Figure 3.11). Indeed, about 75 percent of all rock outcrops on the continents are sedimentary. Therefore, we can think of sedimentary rocks as comprising a relatively thin and somewhat discontinuous layer in the uppermost portion of the crust. This makes sense because sediment accumulates at the surface.

It is from sedimentary rocks that geologists reconstruct many details of Earth's history. Because sediments are deposited in a variety of different settings at the surface, the rock layers that they eventually form hold many clues to past surface environments. They may also exhibit characteristics that allow geologists to decipher information about the method and distance of sediment transport. Furthermore, it is sedimentary rocks that contain fossils, which are vital evidence in the study of the geologic past.

Finally, many sedimentary rocks are important economically. Coal, which is burned to provide a significant portion of U.S. electrical energy, is classified as a sedimentary rock. Other major energy resources (petroleum and natural gas) occur in pores within sedimentary rocks. Other sedimentary rocks are major sources of iron, aluminum, manganese, and fertilizer, plus numerous materials essential to the construction industry.

Classifying Sedimentary Rocks

Materials accumulating as sediment have two principal sources. First, sediments may originate as solid particles from weathered rocks, such as the igneous rocks earlier described. These particles are called *detritus*, and the sedimentary rocks that they form are called **detrital sedimentary rocks** (Figure 3.12).

The second major source of sediment is soluble material produced largely by chemical weathering. When these dissolved substances are precipitated back as solids, they are called chemical sediment, and they form **chemical sedimentary rocks**. We will now look at detrital and chemical sedimentary rocks.

Detrital Sedimentary Rocks. Though a wide variety of minerals and rock fragments may be found in detrital rocks, clay minerals and quartz dominate. Clay minerals are the most abundant product of the chemical weathering of silicate minerals, especially the feldspars. Quartz, in contrast, is abundant because it is extremely durable and very resistant to chemical weathering. Thus, when

Detrital Sedimentary Rocks

Texture (particle size)		Sediment Name	Rock Name
Coarse (over 2 mm)		Gravel (Rounded particles)	**Conglomerate**
		Gravel (Angular particles)	**Breccia**
Medium (1/16 to 2 mm)		Sand (If abundant feldspar is present the rock is called **Arkose**)	**Sandstone**
Fine (1/16 to 1/256 mm)		Mud	**Siltstone**
Very fine (less than 1/256 mm)		Mud	**Shale**

Chemical Sedimentary Rocks

Composition	Texture	Rock Name	
Calcite, $CaCO_3$	Fine to coarse crystalline	**Crystalline Limestone**	
		Travertine	
	Visible shells and shell fragments loosely cemented	**Coquina**	**Biochemical Limestone**
	Various size shells and shell fragments cemented with calcite cement	**Fossiliferous Limestone**	
	Microscopic shells and clay	**Chalk**	
Quartz, SiO_2	Very fine crystalline	**Chert (light colored) Flint (dark colored)**	
Gypsum $CaSO_4 \cdot 2H_2O$	Fine to coarse crystalline	**Rock Gypsum**	
Halite, NaCl	Fine to coarse crystalline	**Rock Salt**	
Altered plant fragments	Fine-grained organic matter	**Bituminous Coal**	

▲ **FIGURE 3.12** Identification of sedimentary rocks. Sedimentary rocks are divided into two major groups, detrital and chemical, based on their source of sediment. The main criterion for naming detrital rocks is particle size, whereas the primary basis for distinguishing among chemical rocks is their mineral composition.

igneous rocks such as granite are weathered, individual quartz grains are set free.

Geologists use particle size to distinguish among detrital sedimentary rocks. Figure 3.12 presents the four size categories for particles making up detrital rocks. When gravel size particles predominate, the rock is called *conglomerate* if the sediment is rounded (Figure 3.13A) and *breccia* if the pieces are angular (Figure 3.13B). Angular fragments indicate that the particles were not transported very far from their source prior to deposition and so have not had corners and rough edges abraded. *Sandstone* is the name given rocks when sand-size grains prevail (Figure 3.13C). *Shale*, the most common sedimentary rock, is made of very fine-grained sediment (Figure 3.13D). *Siltstone*, another rather fine-grained rock, is sometimes difficult to differentiate from rocks such as shale that are composed of even smaller clay-size sediment.

Particle size is not only a convenient method of dividing detrital rocks; the sizes of the component grains also provide useful information about the environment in which the sediment was deposited. Currents of water or air sort the particles by size. The stronger the

current, the larger the particle size carried. Gravels, for example, are moved by swiftly flowing rivers, rockslides, and glaciers. Less energy is required to transport sand; thus, it is common in windblown dunes, river deposits, and beaches. Because silts and clays settle very slowly, accumulations of these materials are generally associated with the quiet waters of a lake, lagoon, swamp, or marine environment.

Although detrital sedimentary rocks are classified by particle size, in certain cases the mineral composition is also part of naming a rock. For example, most sandstones are predominantly quartz-rich and are often referred to as quartz sandstone. In addition, rocks consisting of detrital sediments are rarely composed of grains of just one size. Consequently, a rock containing quantities of both sand and silt can be correctly classified as sandy siltstone or silty sandstone, depending on which particle size dominates.

Chemical Sedimentary Rocks. In contrast to detrital rocks, which form from the solid products of weathering, chemical sediments are derived from material that is carried in solution to lakes and seas. This material

▲ **FIGURE 3.13** Common detrital sedimentary rocks. **A.** Conglomerate (rounded particles). **B.** Breccia (angular particles). **C.** Sandstone **D.** Shale with plant fossil. (Photos by E. J. Tarbuck)

does not remain dissolved in the water indefinitely. When conditions are right, it precipitates to form chemical sediments. This precipitation may occur directly as the result of physical processes, or indirectly through life processes of water-dwelling organisms. Sediment formed in this second way has a *biochemical* origin (see Box 3.1).

An example of a deposit resulting from physical processes is the salt left behind as a body of salt water evaporates. In contrast, many water-dwelling animals and plants extract dissolved mineral matter to form shells and other hard parts. After the organisms die, their skeletons may accumulate on the floor of a lake or ocean.

Limestone is the most abundant chemical sedimentary rock. It is composed chiefly of the mineral calcite ($CaCO_3$). Ninety percent of limestone is biochemical sediment. The rest precipitates directly from water.

One easily identified biochemical limestone is *coquina*, a coarse rock composed of loosely cemented shells and shell fragments (Figure 3.14). Another less obvious but familiar example is *chalk*, a soft, porous rock made up almost entirely of the hard parts of microscopic organisms (Figure 3.15).

Inorganic limestones form when chemical changes or high water temperatures increase the concentration of calcium carbonate to the point that it precipitates. *Travertine*, the type of limestone that decorates caverns, is one example. Groundwater is the source of travertine that is deposited in caves. As water drops reach the air in a cavern, some of the carbon dioxide dissolved in the water escapes, causing calcium carbonate to precipitate.

Dissolved silica (SiO_2) precipitates to form varieties of microcrystalline quartz, including chert (light color), flint (dark), jasper (red), and agate (banded). These chemical sedimentary rocks may have either an inorganic or biochemical origin, but the mode of origin is usually difficult to determine.

Very often, evaporation causes minerals to precipitate from water. Such minerals include halite, the chief component of *rock salt*, and gypsum, the main ingredient of *rock gypsum*. Both materials have significant commercial importance. Halite is familiar to everyone as the common salt used in cooking and seasoning foods. Of course, it has many other uses and has been considered important enough that people have sought, traded, and fought over it for much of human

5 cm

Close up

▲ **FIGURE 3.14** This rock, called coquina, consists of shell fragments; therefore, it has a biochemical origin. (Photo by E. J. Tarbuck)

history. Gypsum is the basic ingredient of plaster of Paris. This material is used most extensively in the construction industry for drywall and plaster.

In the geologic past, many areas that are now dry land were covered by shallow arms of the sea that had only narrow connections to the open ocean. Under these conditions, water continually moved into the bay to replace water lost by evaporation. Eventually

the waters of the bay became saturated and salt deposition began. Today, these arms of the sea are gone, and the remaining deposits are called **evaporites**.

On a smaller scale, evaporite deposits can be seen in such places as Death Valley, California. Here, following rains or periods of snowmelt in the mountains, streams flow from surrounding mountains into an enclosed basin. As the water evaporates, *salt flats* form from dissolved materials left behind as a white crust on the ground (Figure 3.16).

Coal is quite different from other chemical sedimentary rocks. Unlike other rocks in this category, which are calcite- or silica-rich, coal is made mostly of organic matter. Close examination of a piece of coal under a microscope or magnifying glass often reveals plant structures such as leaves, bark, and wood that have been chemically altered but are still identifiable. This supports the conclusion that coal is the end product of the burial of large amounts of plant material over extended periods.

The initial stage in coal formation is the accumulation of large quantities of plant remains. However, special conditions are required for such accumulations, because dead plants normally decompose when exposed to the atmosphere. An ideal environment that allows for the buildup of plant material is a swamp. Because stagnant swamp water is oxygen-deficient, complete decay (oxidation) of the plant material is not possible. At various times during Earth history, such environments have been common. Coal undergoes successive stages of formation. With each successive stage, higher temperatures and pressures drive off impurities and volatiles, as shown in Figure 3.17.

Lignite and bituminous coals are sedimentary rocks, but anthracite is a metamorphic rock. Anthracite forms when sedimentary layers are subjected to the folding and deformation associated with mountain building.

◀ **FIGURE 3.15** The White Chalk Cliffs, Sussex, England. (Photo by Art Wolfe)

▲ **FIGURE 3.16** The Bonneville salt flats in Utah are a well-known example of evaporite deposits. (Photo by Tom & Susan Bean, Inc.)

In summary, we divide sedimentary rocks into two major groups: detrital and chemical. The main criterion for classifying detrital rocks is particle size, whereas chemical rocks are distinguished by their mineral composition. The categories presented here are more rigid than is the actual state of nature. Many detrital sedimentary rocks are a mixture of more than one particle size. Furthermore, many sedimentary rocks classified as chemical also contain at least small quantities of detrital sediment, and practically all detrital rocks are cemented with material that was originally dissolved in water.

? STUDENTS SOMETIMES ASK...

Why are the sedimentary rocks in Figure 3.11 so colorful?

In the western and southwestern United States, steep cliffs and canyon walls made of sedimentary rocks often exhibit a brilliant display of different colors In the walls of Arizona's Grand Canyon we can see layers that may be red, orange, purple, gray, brown, and buff (look at the Chapter 11 opening photo on pages 308–309). The sedimentary rocks in Utah's Bryce Canyon exhibit hues of pink and orange (see Figure 4.10, p. 91). Sedimentary rocks in more humid places are also colorful, but they are usually covered by soil and vegetation.

The most important "pigments" are iron oxides, and only very small amounts are needed to color a rock. Hematite tints rocks red or pink, whereas limonite produces shades of yellow and brown. When sedimentary rocks contain organic matter, it often colors them black or gray.

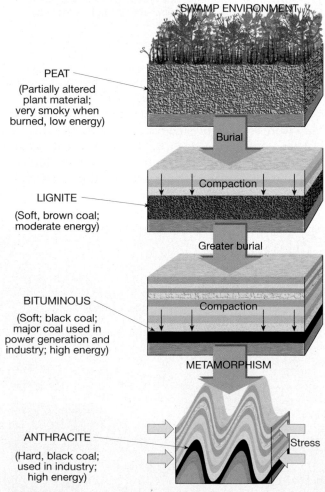

SWAMP ENVIRONMENT

PEAT
(Partially altered plant material; very smoky when burned, low energy)

Burial

Compaction

LIGNITE
(Soft, brown coal; moderate energy)

Greater burial

BITUMINOUS
(Soft; black coal; major coal used in power generation and industry; high energy)

Compaction

METAMORPHISM

ANTHRACITE
(Hard, black coal; used in industry; high energy)

Stress

▲ **FIGURE 3.17** Successive stages in the formation of coal.

Lithification of Sediment

Lithification refers to the processes by which sediments are transformed into solid sedimentary rocks. One of the most common processes is *compaction*. As sediments accumulate through time, the weight of overlying material compresses the deeper sediments. As the grains are pressed closer and closer, pore space is greatly reduced. For example, when clays are buried beneath several thousand meters of material, the volume of the clay might be reduced as much as 40 percent. Compaction is most significant in fine-grained sedimentary rocks such as shale, because sand and other coarse sediments compress little.

Cementation is another important means by which sediments are converted to sedimentary rock. The cementing materials are carried in solution by water percolating through the pore spaces between particles. Through time, the cement precipitates onto the sediment grains, fills the open spaces, and joins the particles. Calcite, silica, and iron oxide are the most common cements. Identification of the cementing material is simple. Calcite cement will effervesce (fizz) with dilute hydrochloric acid. Silica is the hardest cement and thus produces the hardest sedimentary rocks. When a sedimentary rock has an orange or red color, this usually means iron oxide is present.

Features of Sedimentary Rocks

Sedimentary rocks are particularly important evidence of Earth's long history. These rocks form at Earth's surface, and as layer upon layer of sediment accumulates, each records the nature of the environment at the time the sediment was deposited. These layers, called **strata**, or **beds**, are the *single most characteristic feature of sedimentary rocks* (see Figure 3.11).

The thickness of beds range from microscopically thin to dozens of meters thick. Separating the strata are *bedding planes*, flat surfaces along which rocks tend to separate or break. Generally, each bedding plane marks the end of one episode of sedimentation and the beginning of another.

Sedimentary rocks provide geologists with evidence for deciphering past environments. A conglomerate, for example, indicates a high-energy environment, such as a rushing stream, where only the coarse materials can settle out. By contrast, black shale and coal are associated with a low-energy, organic-rich environment such as a swamp or lagoon. Other features found in some sedimentary rocks also give clues to past environments (Figure 3.18).

Fossils, the traces or remains of prehistoric life, are perhaps the most important inclusions found in some sedimentary rock. Knowing the nature of the life forms that existed at a particular time may help to answer many questions about the environment. Was it land or ocean? A lake or swamp? Was the climate

A

B

▲ **FIGURE 3.18** **A**. Ripple marks preserved in sedimentary rocks may indicate a beach or stream channel environment. (Photo by Stephen Trimble) **B**. Mud cracks form when wet mud or clay dries and shrinks, perhaps signifying a tidal flat or desert basin. (Photo by Gary Yeowell/Tony Stone Images)

Box 3.1 Earth as a System
The Carbon Cycle and Sedimentary Rocks

To illustrate the movement of material and energy in the Earth system, let us take a brief look at the *carbon cycle* (Figure 3.A). Pure carbon is relatively rare in nature. It is found predominantly in two minerals: diamond and graphite. Most carbon is bonded chemically to other elements to form compounds such as carbon dioxide, calcium carbonate, and the hydrocarbons found in coal and petroleum. Carbon is also the basic building block of life as it

readily combines with hydrogen and oxygen to form the fundamental organic compounds that compose living things.

In the atmosphere, carbon is found mainly as carbon dioxide (CO_2). Atmospheric carbon dioxide is significant because it is a greenhouse gas, which means it is an efficient absorber of energy emitted by Earth and thus influences the heating of the atmosphere. Because many of the processes that operate on Earth involve carbon dioxide, this gas is constant-

ly moving into and out of the atmosphere (Figure 3.B). For example, through the process of photosynthesis, plants absorb carbon dioxide from the atmosphere to produce the essential organic compounds needed for growth. Animals that consume these plants (or consume other animals that eat plants) use these organic compounds as a source of energy and, through the process of respiration, return carbon dioxide to the atmosphere. (Plants also return some CO_2 to the atmosphere

Figure 3.A Simplified diagram of the carbon cycle, with emphasis on the flow of carbon between the atmosphere and the hydrosphere, geosphere, and biosphere. The colored arrows show whether the flow of carbon is into or out of the atmosphere.

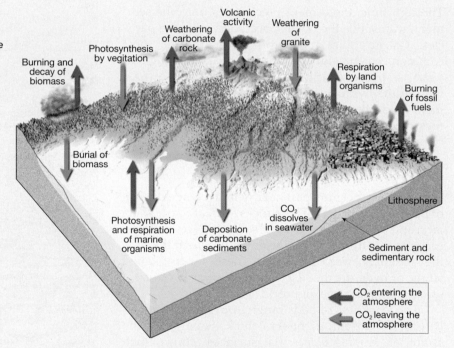

Figure 3.B This map was created using space-based measurements and shows how much CO_2 is taken up by vegetation during photosynthesis minus how much is given off during respiration. The red and yellow areas, which are concentrated near the equator, are regions where CO_2 is being removed from the atmosphere at the highest rate.(NASA image)

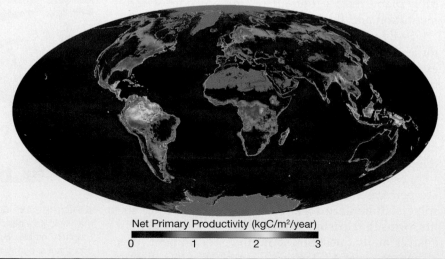

Net Primary Productivity (kgC/m²/year)
0   1   2   3

via respiration.) Further, when plants die and decay or are burned, this biomass is oxidized, and carbon dioxide is returned to the atmosphere.

Not all dead plant material decays immediately back to carbon dioxide. A small percentage is deposited as sediment. Over long spans of geologic time, considerable biomass is buried with sediment. Under the right conditions, some of these carbon-rich deposits are converted to fossil fuels—coal, petroleum, or natural gas. Eventually some of the fuels are recovered (mined or pumped from a well) and burned to run factories and fuel our transportation system. One result of fossil-fuel combustion is the release of huge quantities of CO_2 into the atmosphere. Certainly one of the most active parts of the carbon cycle is the movement of CO_2 from the atmosphere to the biosphere and back again.

Carbon also moves from the geosphere and hydrosphere to the atmosphere and back again. For example, volcanic activity early in Earth's history is thought to be the source of much of the carbon dioxide found in the atmosphere. One way that carbon dioxide makes its way back to the hydrosphere and then to the solid Earth is by first combining with water to form carbonic acid (H_2CO_3), which then attacks the rocks that compose the geosphere. One product of this chemical weathering of solid rock is the soluble bicarbonate ion ($2\,HCO_3{}^-$), which is carried by groundwater and streams to the ocean. Here water-dwelling organisms extract this dissolved material to produce hard parts of calcium carbonate ($CaCO_3$). When the organisms die, these skeletal remains settle to the ocean floor as biochemical sediment and become sedimentary rock. In fact, the

geosphere is by far Earth's largest depository of carbon, where it is a constituent of a variety of rocks, the most abundant being limestone. Eventually the limestone may be exposed at Earth's surface, where chemical weathering will cause the carbon stored in the rock to be released to the atmosphere as CO_2.

In summary, carbon moves among all four of Earth's major spheres. It is essential to every living thing in the biosphere. In the atmosphere carbon dioxide is an important greenhouse gas. In the hydrosphere, carbon dioxide is dissolved in lakes, rivers, and the ocean. In the geosphere, carbon is contained in carbonate sediments and sedimentary rocks and is stored as organic matter dispersed through sedimentary rocks and as deposits of coal and petroleum.

hot or cold, rainy or dry? Was the ocean water shallow or deep, turbid or clear? Furthermore, fossils are important time indicators and play a key role in matching up rocks from different places that are the same age. Fossils are important tools used in interpreting the geologic past and will be examined in some detail in Chapter 11.

Metamorphic Rocks: New Rock from Old

GEODe   Earth Materials
▼ Metamorphic Rocks

Recall from the discussion of the rock cycle that metamorphism is the transformation of one rock type into another. Metamorphic rocks are produced from preexisting igneous, sedimentary, or even from other metamorphic rocks. Thus, every metamorphic rock has a *parent rock*—the rock from which it was formed.

Metamorphism, which means to "change form," is a process that leads to changes in the mineralogy, texture (for example, grain size), and often the chemical composition of rocks. Metamorphism takes place where preexisting rock is subjected to a physical or chemical environment that is significantly different from that in which it initially formed. In response to changes in temperature, pressure (stress), and the introduction of chemically active

fluids, the rock gradually changes until a state of equilibrium with the new environment is reached. Most metamorphic changes occur at the elevated temperatures and pressures that exist in the zone beginning a few kilometers below Earth's surface and extending into the upper mantle.

Metamorphism often progresses incrementally, from slight changes (*low-grade metamorphism*) to substantial changes (*high-grade metamorphism*). For example, under low-grade metamorphism, the common sedimentary rock *shale* becomes the more compact metamorphic rock called *slate*. Hand samples of these rocks are sometimes difficult to distinguish, illustrating that the transition from sedimentary to metamorphic rock is often gradual and the changes subtle.

In more extreme environments, metamorphism causes a transformation so complete that the identity of the parent rock cannot be determined. In high-grade metamorphism, such features as bedding planes, fossils, and vesicles that may have existed in the parent rock are obliterated. Further, when rocks at depth (where temperatures are high) are subjected to directed pressure, they slowly deform to produce a variety of textures as well as large-scale structures such as folds. (Figure 3.19).

In the most extreme metamorphic environments, the temperatures approach those at which rocks melt. However, *during metamorphism the rock must remain essentially solid*, for if complete melting occurs, we have entered the realm of igneous activity.

Most metamorphism occurs in one of two settings:

1. When rock is intruded by a magma body, **contact** or **thermal metamorphism** may take place. Here, change is driven by a rise in temperature within the host rock surrounding a molten igneous body.

2. During mountain building, great quantities of rock are subjected to directed pressures and high temperatures associated with large-scale deformation called **regional metamorphism**.

Extensive areas of metamorphic rocks are exposed on every continent. Metamorphic rocks are an important component of many mountain belts, where they make up a large portion of a mountain's crystalline core. Even the stable continental interiors, which are generally covered by sedimentary rocks, are underlain by metamorphic basement rocks. In all of these settings, the metamorphic rocks are usually highly deformed and intruded by igneous masses. Indeed, significant parts of Earth's continental crust are composed of metamorphic and associated igneous rocks.

Agents of Metamorphism

The agents of metamorphism include *heat, pressure (stress), and chemically active fluids*. During metamorphism, rocks are usually subjected to all three metamorphic agents simultaneously. However, the degree of metamorphism and the contribution of each agent vary greatly from one environment to another.

Heat as a Metamorphic Agent. The most important agent of metamorphism is *heat* because it provides the energy to drive chemical reactions that result in the recrystallization of existing minerals and/or the formation of new minerals. The heat to metamorphose rocks comes mainly from two sources. First, rocks experience a rise in temperature when they are intruded by magma rising from below. This is called *contact meta-*

morphism. Here the adjacent host rock is "baked" by the emplaced magma.

Second, rocks that formed at Earth's surface will experience a gradual increase in temperature as they are transported to greater depths. In the upper crust, this increase in temperature averages between 20°C and 30°C per kilometer. When buried to a depth of about 8 kilometers (5 miles), where temperatures are between 150°C and 200°C, clay minerals tend to become unstable and begin to recrystallize into minerals, such as chlorite and muscovite, that are stable in this environment. (Chlorite is a micalike mineral formed by the metamorphism of iron- and magnesian-rich silicates.) However, many silicate minerals, particularly those found in crystalline igneous rocks—quartz, for example—remain stable at these temperatures. Thus, metamorphic changes in these minerals occur at much higher temperatures.

Pressure (Stress) as a Metamorphic Agent. Pressure, like temperature, also increases with depth as the thickness of the overlying rock increases. Buried rocks are subjected to *confining pressure*, which is analogous to water pressure, where the forces are applied equally in all directions (Figure 3.20A). The deeper you go in the ocean, the greater the confining pressure. The same is true for rock that is buried. Confining pressure causes the spaces between mineral grains to close, producing a more compact rock having a greater density. Further, at great depths, confining pressure may cause minerals to recrystallize into new minerals that display a more compact crystalline form.

During episodes of mountain building, large rock bodies become highly crumpled and metamorphosed (Figure 3.19). The forces that generate mountains are unequal in different directions and are called *differential stress*. Unlike confining pressure, which "squeezes" the rock equally from all directions, differential stresses are greater in one direction than in others. As shown in Figure 3.20B, rocks subjected to differential stress are

▶ **FIGURE 3.19** Deformed strata; north end of Cottonwood Mountains, Death Valley, California. (Photo by Michael Collier)

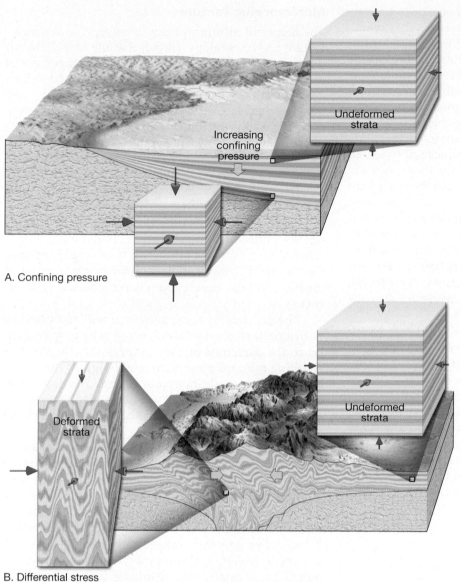

A. Confining pressure

B. Differential stress

shortened in the direction of greatest stress and elongated, or lengthened, in the direction perpendicular to that stress. The deformation caused by differential stresses plays a major role in developing metamorphic textures.

In surface environments where temperatures are comparatively low, rocks are *brittle* and tend to fracture when subjected to differential stress. Continued defor-

mation grinds and pulverizes the mineral grains into small fragments. By contrast, in high-temperature environments rocks are *ductile*. When rocks exhibit ductile behavior, their mineral grains tend to flatten and elongate when subjected to differential stress. This accounts for their ability to deform by flowing (rather than fracturing) to generate intricate folds (Figure 3.21).

◄ **FIGURE 3.21** Deformed metamorphic rocks exposed in a road cut in the Eastern Highland of Connecticut. Imagine the tremendous force required to fold rock in this manner. (Photo by Phil Dombrowski)

Chemically Active Fluids. Fluids composed mainly of water and other volatiles (materials that readily change to a gas at surface conditions), including carbon dioxide, are believed to play an important role in some types of metamorphism. Fluids that surround mineral grains act as catalysts to promote recrystallization by enhancing ion migration. In progressively hotter environments these ion-rich fluids become correspondingly more reactive.

When two mineral grains are squeezed together, the parts of their crystalline structures that touch are the most highly stressed. Ions located at these sites are readily dissolved by the hot fluids and migrate along the surface of the grain to the spaces located between individual grains. Thus, hot fluids aid in the recrystallization of mineral grains by dissolving material from regions of high stress and then precipitating (depositing) this material in areas of low stress. As a result, *minerals tend to recrystallize and grow longer in a direction perpendicular to compressional stresses.*

Where hot fluids circulate freely through rocks, ionic exchange may occur between two adjacent rock layers, or ions may migrate great distances before they are finally deposited. The latter situation is particularly common when we consider hot fluids that escape during the crystallization of an intrusive igneous mass. If the rocks that surround the mass differ markedly in composition from that of the invading fluids, there may be a substantial exchange of ions between the fluids and host rocks. When this occurs, a change in the overall composition of the surrounding rock results.

STUDENTS SOMETIMES ASK...
How hot is it deep in the crust?

The increase in temperature with depth, based on the geothermal gradient, can be expressed as *the deeper one goes, the hotter it gets.* This relationship has been observed by miners in deep mines and in deeply drilled wells. In the deepest mine in the world (the Western Deep Levels mine in South Africa, which is 4 kilometers or 2.5 miles deep), the temperature of the surrounding rock is so hot that it can scorch human skin! In fact, miners often work in groups of two: one to mine the rock, and the other to operate a large fan that keeps the other worker cool.

The temperature is even higher at the bottom of the deepest well in the world, which was completed in the Kola Peninsula of Russia in 1992 and goes down a record 12.3 kilometers (7.7 miles). At this depth it is 245°C (473°F), much greater than the boiling point of water. What keeps water from boiling is the high confining pressure at depth.

Metamorphic Textures

The degree of metamorphism is reflected in the rock's texture and mineralogy. (Recall that the term *texture* is used to describe the size, shape, and arrangement of grains within a rock.) When rocks are subjected to low-grade metamorphism, they become more compact and thus more dense. A common example is the metamorphic rock slate, which forms when shale is subjected to temperatures and pressures only slightly greater than those associated with the compaction that lithifies sediment. In this case, differential stress causes the microscopic clay minerals in shale to align into the more compact arrangement found in slate.

Under more extreme conditions, stress causes certain minerals to recrystallize. In general, recrystallization encourages the growth of larger crystals. Consequently, many metamorphic rocks consist of visible crystals, much like coarse-grained igneous rocks.

The crystals of some minerals will recrystallize with a preferred orientation, essentially perpendicular to the direction of the compressional force. The resulting mineral alignment usually gives the rock a layered or banded appearance termed **foliated texture** (Figure 3.22). Simply, foliation results whenever the minerals of a rock are brought into parallel alignment.

Not all metamorphic rocks have a foliated texture. Such rocks are said to exhibit a **nonfoliated texture**. Metamorphic rocks composed of only one mineral that forms equidimensional crystals are, as a rule, not visibly foliated. For example, limestone, if pure, is composed of only a single mineral, calcite. When a fine-grained limestone is metamorphosed, the small calcite crystals combine to form larger interlocking crystals. The resulting rock resembles a coarse-grained igneous rock. This nonfoliated metamorphic equivalent of limestone is called *marble*.

Common Metamorphic Rocks

To review, metamorphic processes cause many changes in existing rocks, including increased density, growth of larger crystals, foliation (reorientation of the mineral grains into a layered or banded appearance), and the transformation of low-temperature minerals into high-temperature minerals (Figure 3.23). Further, the introduction of ions generates new minerals, some of which are economically important.

Here is a brief look at common rocks produced by metamorphic processes (Figure 3.24).

Foliated Rocks. *Slate* is a very fine-grained foliated rock composed of minute mica flakes. The most noteworthy characteristic of slate is its excellent rock

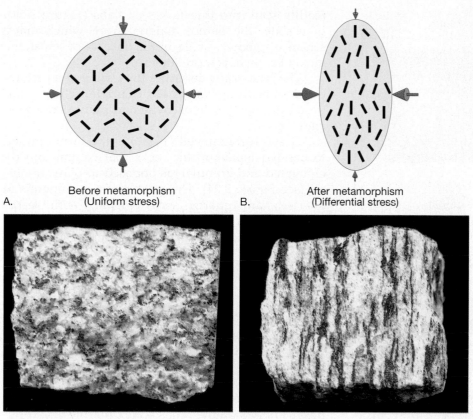

A. Before metamorphism
(Uniform stress)

B. After metamorphism
(Differential stress)

◄ **FIGURE 3.22** Under directed pressure, planar minerals, such as the micas, become reoriented or recrystallized so that their surfaces are aligned at right angles to the stress. The resulting planar orientation of mineral grains gives the rock a foliated texture. If the coarse-grained igneous rock (granite) on the left underwent intense metamorphism, it could end up closely resembling the metamorphic rock on the right (gneiss). (Photos by E. J. Tarbuck)

cleavage, meaning that it splits easily into flat slabs. This property has made slate a most useful rock for roof and floor tile, chalkboards, and billiard tables (Figure 3.25). Slate is most often generated by the low- grade metamorphism of shale, although less frequently it forms from the metamorphism of volcanic ash. Slate can be almost any color, depending on its mineral constituents. Black slate contains organic material;

▼ **FIGURE 3.23**   Classification of common metamorphic rocks.

Rock Name	Texture		Grain Size	Comments	Parent Rock
Slate	Increasing Metamorphism	Foliated	Very fine	Excellent rock cleavage, smooth dull surfaces	Shale, mudstone, or siltstone
Phyllite		Foliated	Fine	Breaks along wavey surfaces, glossy sheen	Slate
Schist		Foliated	Medium to Coarse	Micas dominate, scaly foliation	Phyllite
Gneiss		Foliated	Medium to Coarse	Compositional banding due to segregation of minerals	Schist, granite, or volcanic rocks
Marble		Nonfoliated	Medium to coarse	Interlocking calcite or dolomite grains	Limestone, dolostone
Quartzite		Nonfoliated	Medium to coarse	Fused quartz grains, massive, very hard	Quartz sandstone
Anthracite		Nonfoliated	Fine	Shiny black organic rock that may exhibit conchoidal fracture	Bituminous coal

Foliated metamorphic rocks

Slate

Schist

Gneiss

Nonfoliated metamorphic rocks

Marble

Quartzite

▲ **FIGURE 3.24** Common metamorphic rocks. (Photos by E. J. Tarbuck)

red slate gets its color from iron oxide; and green slate is usually composed of chlorite, a micalike mineral.

Schists are strongly foliated rocks, formed by regional metamorphism. They are platy and can be readily split into thin flakes or slabs (Figure 3.24). Like slate, the parent material from which many schists originate is shale, but in the case of schist, the metamorphism is more intense.

The term *schist* describes the *texture* of a rock regardless of composition. For example, schists composed primarily of muscovite and biotite are called *mica schists*.

Gneiss (pronounced "nice") is the term applied to banded metamorphic rocks that contain mostly elongated and granular, as opposed to platy, minerals (see Figure 3.24). The most common minerals in gneisses are quartz and feldspar, with lesser amounts of muscovite, biotite, and hornblende. Gneisses exhibit strong segregation of light and dark silicates, giving them a characteristic banded texture. While in a plastic state, these banded gneisses can be deformed into intricate folds.

Nonfoliated Rocks. *Marble* is a coarse, crystalline rock whose parent rock is limestone. Marble is composed of large interlocking calcite crystals, which form from the recrystallization of smaller grains in the parent rock.

Because of its color and relative softness (hardness of only 3 on the Mohs scale), marble is a popular building stone. White marble is particularly prized as a stone from which to carve monuments and statues, such as the famous statue of *David* by Michelangelo (Figure 3.26). Often the limestone from which marble forms contains impurities that color the marble. Thus, marble can be pink, gray, green, or even black.

Quartzite is a very hard metamorphic rock most often formed from quartz sandstone. Under moderate- to high-grade metamorphism, the quartz grains in sandstone fuse. Pure quartzite is white, but iron oxide may produce reddish or pinkish stains, and dark minerals may impart a gray color.

Resources from Rocks and Minerals

The outer layer of Earth, which we call the crust, is only as thick when compared to the remainder of the Earth as a peach skin is to a peach, yet it is of supreme importance to us. We depend on it for fossil fuels and as a source of such diverse minerals as the talc for baby powder, salt to flavor food, and gold for world trade. In fact, on occasion, the availability or absence of certain Earth materials has altered the course of history. As the material requirements of modern society grow, the need to locate additional supplies of useful minerals also grows and becomes more challenging (see Box 3.2).

◀ **FIGURE 3.25** Excellent rock cleavage is exhibited by the slate in this quarry near Alta, Norway. (Photo by Fred Bruermmer/DRK Photo) Because slate breaks into flat slabs, it has many uses. In the inset photo, it is used to roof this house in Switzerland. (Photo by E. J. Tarbuck)

Metallic Mineral Resources

Some of the most important accumulations of metals, such as gold, silver, copper, mercury, lead, platinum, and nickel, are produced by igneous and metamorphic processes (Table 3.1). These mineral resources, like most others, result from processes that concentrate desirable materials to the extent that extraction is economically feasible.

The igneous processes that generate some metal deposits are quite straightforward. For example, as a

▼ **FIGURE 3.26** Replica of the statue of *David* by Michelangelo. Like the original, this sculpture was created from a large block of white marble. (Courtesy Getty images/Photo Disk)

TABLE 3.1	Ore minerals of important metals.
Metal	**Ore Mineral**
Aluminum	Bauxite
Chromium	Chromite
Copper	Chalcopyrite
	Bornite
	Chalcocite
Gold	Native gold
Iron	Hematite
	Magnetite
	Limonite
Lead	Galena
Magnesium	Magnesite
	Dolomite
Manganese	Pyrolusite
Mercury	Cinnabar
Molybdenum	Molybdenite
Nickel	Pentlandite
Platinum	Native platinum
Silver	Native silver
	Argentite
Tin	Cassiterite
Titanium	Ilmenite
	Rutile
Tungsten	Wolframite
	Scheelite
Uranium	Uraninite (pitchblende)
Zinc	Sphalerite

Box 3.2 People and the Environment
United States Per Capita Use of Mineral and Energy Resources

Mineral and energy resources from Earth's crust are the raw materials from which the products used by our modern industrial economy are made. Like most people who live in highly industrialized nations, you may not realize the quantity of resources needed to maintain your present standard of living. Figure 3.C shows the approximate annual per capita consumption of several important metallic and nonmetallic mineral resources for the United States. Per capita use of energy resources is also shown. This is each person's proportional share of the materials required by industry to provide the vast array of homes, cars, electronics, cosmetics, packaging, and so on that modern society demands. It represents the energy we need to drive cars, fly airplanes, run machinery, and heat buildings.

Every year about 9000 kilograms (20,000 pounds) of stone, sand, and gravel is mined for each person in the United States. Further, the per capita use of oil, coal, and natural gas exceeds 11,000 kilograms (24,000 pounds). Figures for countries such as Canada, Japan, the United Kingdom, as well as for most other highly industrialized countries are comparable.

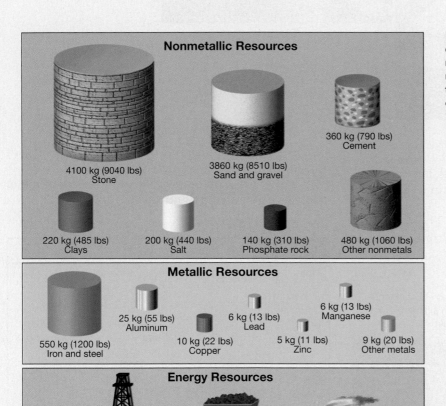

Nonmetallic Resources

4100 kg (9040 lbs)
Stone

3860 kg (8510 lbs)
Sand and gravel

360 kg (790 lbs)
Cement

220 kg (485 lbs)
Clays

200 kg (440 lbs)
Salt

140 kg (310 lbs)
Phosphate rock

480 kg (1060 lbs)
Other nonmetals

Metallic Resources

550 kg (1200 lbs)
Iron and steel

25 kg (55 lbs)
Aluminum

10 kg (22 lbs)
Copper

6 kg (13 lbs)
Lead

5 kg (11 lbs)
Zinc

6 kg (13 lbs)
Manganese

9 kg (20 lbs)
Other metals

Energy Resources

3500 kg (7700 lbs)
Petroleum

3700 kg (8140 lbs)
Coal

3850 kg (8470 lbs)
Natural gas

Figure 3.C The annual per capita consumption of nonmetallic and metallic mineral resources for the United States is nearly 10,000 kilograms! About 94 percent of the materials used are nonmetallic. The per capita use of oil, coal, and natural gas exceeds 11,000 kilograms.

large magma body cools, the heavy minerals that crystallize early tend to settle to the lower portion of the magma chamber. This type of magmatic differentiation is particularly active in large basaltic magmas where chromite (ore of chromium), magnetite, and platinum are occasionally generated. Layers of chromite, interbedded with other heavy minerals, are mined from such deposits in the Bushveld Complex in South Africa, which contains over 70 percent of the world's known reserves of platinum.

Igneous processes are also important in generating other types of mineral deposits. For example, as a granitic magma cools and crystallizes, the residual melt becomes enriched in rare elements and heavy metals, including gold and silver. Further, because water and other volatile substances do not crystallize along with the bulk of the magma body, these fluids make up a high percentage of the melt during the final phase of solidification. Crystallization in a fluid-rich environment enhances the migration of ions and results in the formation of crystals several centimeters, or even a few meters, in length. The resulting rocks, called **pegmatites**, are composed of these unusually large crystals (Figure 3.27).

Most pegmatites are granitic in composition and consist of unusually large crystals of quartz, feldspar, and muscovite. Feldspar is used in the production of ceramics, and muscovite is used for electrical insulation and glitter. In addition to the common silicates, some pegmatites include semiprecious gems such as beryl, topaz, and tourmaline. Moreover, minerals containing the elements lithium, cesium, uranium, and the rare earths are sometimes found. Most pegmatites are located within large igneous masses or as dikes or veins that cut into the host rock surrounding the magma chamber (Figure 3.28).

Among the best-known and most important ore deposits are those generated from **hydrothermal** (*hydra* = water, *therm* = heat) **solutions**. Included in this group are the gold deposits of the Homestake mine in South Dakota; the lead, zinc, and silver ores near Coeur d'Alene, Idaho; the silver deposits of the Comstock Lode in Nevada; and the copper ores of the Keweenaw Peninsula in Michigan.

The majority of hydrothermal deposits are thought to originate from hot metal-rich fluids that are associated with cooling magma bodies. During solidification, liquids plus various metallic ions accumulate near the top of the magma chamber. Be-

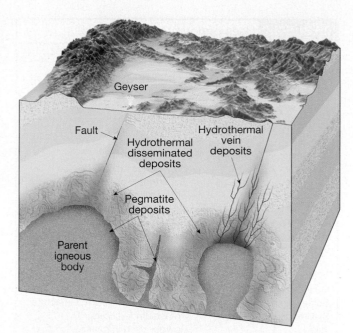

▲ **FIGURE 3.28** Illustration of the relationship between a parent igneous body and the associated pegmatite and hydrothermal mineral deposits.

cause these hot fluids are very mobile, they can migrate great distances through the surrounding rock before they are eventually deposited. Some of this fluid moves along fractures or bedding planes, where it cools and precipitates the metallic ions to produce **vein deposits** (Figure 3.28). Many of the most productive deposits of gold, silver, and mercury occur as hydrothermal vein deposits.

Another important type of accumulation generated by hydrothermal activity is called a **disseminated deposit**. Rather than being concentrated in narrow veins and dikes, these ores are distributed as minute masses throughout the entire rock mass (Figure 3.28). Much of the world's copper is extracted from disseminated deposits, including the huge Bingham Canyon copper mine in Utah. Because these accumulations contain only 0.4 to 0.8 percent copper, between 125 and 250 metric tons of ore must be mined for every ton of metal recovered. The environmental impact of these large excavations, including the problems of waste disposal, is significant.

Nonmetallic Mineral Resources

Mineral resources not used as fuels or processed for the metals they contain are referred to as *nonmetallic mineral resources*. These materials are extracted and processed either to make use of the nonmetallic

▼ **FIGURE 3.27** A granite pegmatite composed mainly of quartz and feldspar (salmon color). The elongated quartz crystal on the right is about the size of a person's index finger. (Photo by Colin Keates)

TABLE 3.2	Uses of nonmetallic minerals.
Mineral	**Uses**
Apatite	Phosphorous fertilizers
Asbestos (chrysotile)	Incombustible fibers
Calcite	Aggregate; steelmaking; soil conditioning; chemicals; cement; building stone
Clay minerals (kaolinite)	Ceramics; china
Corundum	Gemstones; abrasives
Diamond	Gemstones; abrasives
Fluorite	Steelmaking; aluminum refining; glass; chemicals
Garnet	Abrasives; gemstones
Graphite	Pencil lead; lubricant; refractories
Gypsum	Plaster of Paris
Halite	Table salt; chemicals; ice control
Muscovite	Insulator in electrical applications
Quartz	Primary ingredient in glass
Sulfur	Chemicals; fertilizer manufacture
Sylvite	Potassium fertilizers
Talc	Powder used in paints, cosmetics, etc.

▲ **FIGURE 3.29** Rock of Ages granite quarry, Barre, Vermont. Granite is used as a building stone as well as for making monuments. (Photo by Paul Rocheleau/Index Stock Imagery)

elements they contain or for the physical and chemical properties they possess (Table 3.2). Nonmetallic mineral resources are commonly divided into two broad groups: *building materials* and *industrial minerals* (Figure 3.29). As some substances have many different uses, they are found in both categories. Limestone, perhaps the most versatile and widely used rock of all, is the best example. As a building material, it is used not only as crushed rock and building stone but also in the making of cement. Moreover, as an industrial mineral, it is an ingredient in the manufacture of steel and is used in agriculture to neutralize acidic soils.

Besides aggregate (sand, gravel, and crushed rock) and cut stone, the other important building materials include gypsum for plaster and wallboard, clay for tile and bricks, and cement, which is made from limestone and shale. Cement and aggregate go into the making of concrete, a material that is essential to practically all construction.

Many and various nonmetallic resources are classified as industrial minerals. People often do not realize the importance of industrial minerals because they see only the products that resulted from their use and not the minerals themselves. That is, many nonmetallics are used up in the process of creating other products. Examples include fluorite and limestone, which are part of the steelmaking process; corundum and garnet, which are used as abrasives to make machinery parts; and sylvite, which is employed in the production of the fertilizers used to grow a food crop.

Chapter Summary

• *Igneous rock* forms from *magma* that cools and solidifies in a process called *crystallization*. *Sedimentary rock* forms from the *lithification of sediment*. *Metamorphic rock* forms from rock that has been subjected to great pressure and heat in a process called *metamorphism*.

• The rate of cooling of magma greatly influences the size of mineral crystals in igneous rock. The four basic igneous rock textures are (1) *fine-grained*, (2) *coarse-grained*, (3) *porphyritic*, and (4) *glassy*.

• Igneous rocks are classified by their *texture* and *mineral composition*. Igneous rocks are divided into broad compositional groups based on the percentage of dark and light silicate minerals they contain. *Felsic rocks* (e.g., granite and rhyolite) are composed mostly of the light-colored silicate minerals potassium feldspar and quartz. Rocks of *intermediate* composition (e.g., andesite) contain plagioclase feldspar and amphibole. *Mafic rocks* (e.g., basalt) contain abundant olivine, pyroxene, and calcium feldspar.

- The mineral makeup of an igneous rock is ultimately determined by the chemical composition of the magma from which it crystallized. N. L. Bowen showed that as magma cools, minerals crystallize in an orderly fashion. *Magmatic differentiation* changes the composition of magma and causes more than one rock type to form from a common parent magma.

- *Detrital sediments* are materials that originate and are transported as solid particles derived from weathering. *Chemical sediments* are soluble materials produced largely by chemical weathering that are precipitated by either inorganic or organic processes. *Detrital sedimentary rocks*, which are classified by particle size, contain a variety of mineral and rock fragments, with clay minerals and quartz the chief constituents. *Chemical sedimentary rocks* often contain the products of biological processes such as shells or mineral crystals that form as water evaporates and minerals precipitate. *Lithification* refers to the processes by which sediments are transformed into solid sedimentary rocks.

- Common detrital sedimentary rocks include *shale* (the most common sedimentary rock), *sandstone*, and *conglomerate*. The most abundant chemical sedimentary rock is *limestone*, composed chiefly of the mineral calcite. *Rock gypsum* and *rock salt* are chemical rocks that form as water evaporates and triggers the deposition of chemical precipates.

- Some of the features of sedimentary rocks that are often used in the interpretation of Earth history and past environments include *strata*, or *beds* (the single most characteristic feature), *fossils, ripple marks*, and *mud cracks*.

- Two types of metamorphism are (1) *regional metamorphism* and (2) *contact or thermal metamorphism*. The agents of metamorphism include *heat, pressure (stress)*, and *chemically active fluids*. Heat is perhaps the most important because it provides the energy to drive the reactions that result in the *recrystallization* of minerals. Metamorphic processes cause many changes in rocks, including *increased density*, growth of *larger mineral crystals, reorientation of the mineral grains* into a layered or banded appearance known as *foliation*, and the formation of *new minerals*.

- Some common metamorphic rocks with a *foliated texture* include *slate, schist*, and *gneiss*. Metamorphic rocks with a *nonfoliated texture* include *marble* and *quartzite*.

- Some of the most important accumulations of *metallic mineral resources* are produced by igneous and metamorphic processes. *Vein deposits* (deposits in fractures or bedding planes) and *disseminated deposits* (deposits distributed throughout the entire rock mass) are produced from *hydrothermal solutions*—hot metal-rich fluids associated with cooling magma bodies.

- *Nonmetallic mineral resources* are mined for the nonmetallic elements they contain or for the physical and chemical properties they possess. The two groups of nonmetallic mineral resources are (1) *building materials* (e.g., limestone and gypsum) and (2) *industrial minerals* (e.g., fluorite and corundum).

Key Terms

andesitic composition (p. 59)
basaltic composition (p. 59)
Bowen's reaction series (p. 60)
chemical sedimentary rock (p. 62)
coarse-grained texture (p. 56)
contact metamorphism (p. 70)
crystallization (p. 55)
crystal settling (p. 60)
detrital sedimentary rock (p. 62)
disseminated deposit (p. 77)
evaporite (p. 65)
extrusive (volcanic) (p. 54)
felsic (p. 58)
fine-grained texture (p. 56)

fossils (p. 67)
foliated texture (p. 72)
glassy texture (p. 57)
granitic composition (p. 58)
hydrothermal solution (p. 77)
igneous rock (p. 54)
intermediate composition (p. 59)
intrusive (plutonic) (p. 55)
lava (p. 54)
lithification (p. 67)
mafic (p. 59)
magma (p. 54)
magmatic differentiation (p. 60)
metamorphic rock (p. 54)

metamorphism (p. 69)
nonfoliated texture (p. 72)
pegmatite (p. 77)
porphyritic texture (p. 57)
regional metamorphism (p. 70)
rock cycle (p. 52)
sediment (p. 61)
sedimentary rock (p. 54)
strata (beds) (p. 67)
texture (p. 56)
thermal metamorphism (p. 70)
ultramafic composition (p. 59)
vein deposit (p. 77)

Review Questions

1. Using the rock cycle, explain the statement "One rock is the raw material for another."

2. If a lava flow at Earth's surface had a basaltic composition, what rock type would the flow likely be (see Figure 3.7)? What igneous rock would form from the same magma if it did not reach the surface but instead crystallized at great depth?

3. What does a porphyritic texture indicate about the history of an igneous rock?

4. How are granite and rhyolite different? The same (see Figure 3.7)?

5. Relate the classification of igneous rocks to Bowen's reaction series.

6. What minerals are most common in detrital sedimentary rocks? Why are these minerals so abundant?

7. What is the primary basis for distinguishing among various detrital sedimentary rocks?

8. Distinguish between the two categories of chemical sedimentary rocks.

9. What are evaporite deposits? Name a rock that is an evaporite.

10. Compaction is an important lithification process with which sediment size?

11. What is probably the single most characteristic feature of sedimentary rocks?

12. What is metamorphism?

13. List the three agents of metamorphism and describe the role of each.

14. Distinguish between regional and contact metamorphism.

15. What feature would easily distinguish schist and gneiss from quartzite and marble?

16. In what ways do metamorphic rocks differ from the igneous and sedimentary rocks from which they formed?

17. List two general types of hydrothermal deposits.

18. Nonmetallic resources are commonly divided into two broad groups. List the two groups and some examples of materials that belong to each.

Examining the Earth System

1. The sedimentary rocks coquina and shale have each formed in response to interactions among two or more of Earth's spheres. List the spheres associated with the formation of each of the rocks and write a short explanation for each of your choices.

2. Of the two main sources of energy that drive the rock cycle—Earth's internal heat and solar energy—which is primarily responsible for each of the three groups of rocks found on and within Earth? Explain your reasoning.

3. Every year about 20,000 pounds of stone, sand, and gravel are mined for each person in the United States. Calculate how many pounds of stone, sand, and gravel will be needed for an individual during a 75-year lifespan. Since a cubic yard of rock weighs roughly 1700 pounds, how big a hole must be dug to supply an individual with 75 years' worth of stone, sand, and gravel? (Note that a typical pickup truck will carry about a half of a cubic yard of rock.)

Online Study Guide

The *Earth Science* Website uses the resources and flexibility of the Internet to aid in your study of the topics in this chapter. Written and developed by Earth science instructors, this site will help improve your understanding of Earth science. Visit **http://www.prenhall.com/tarbuck** and click on the cover of *Earth Science* 11e to find:

- **Online review quizzes.**
- **Critical thinking exercises.**
- **Links to chapter-specific Web resources.**
- **Internet-wide key term searches.**

http://www.prenhall.com/tarbuck

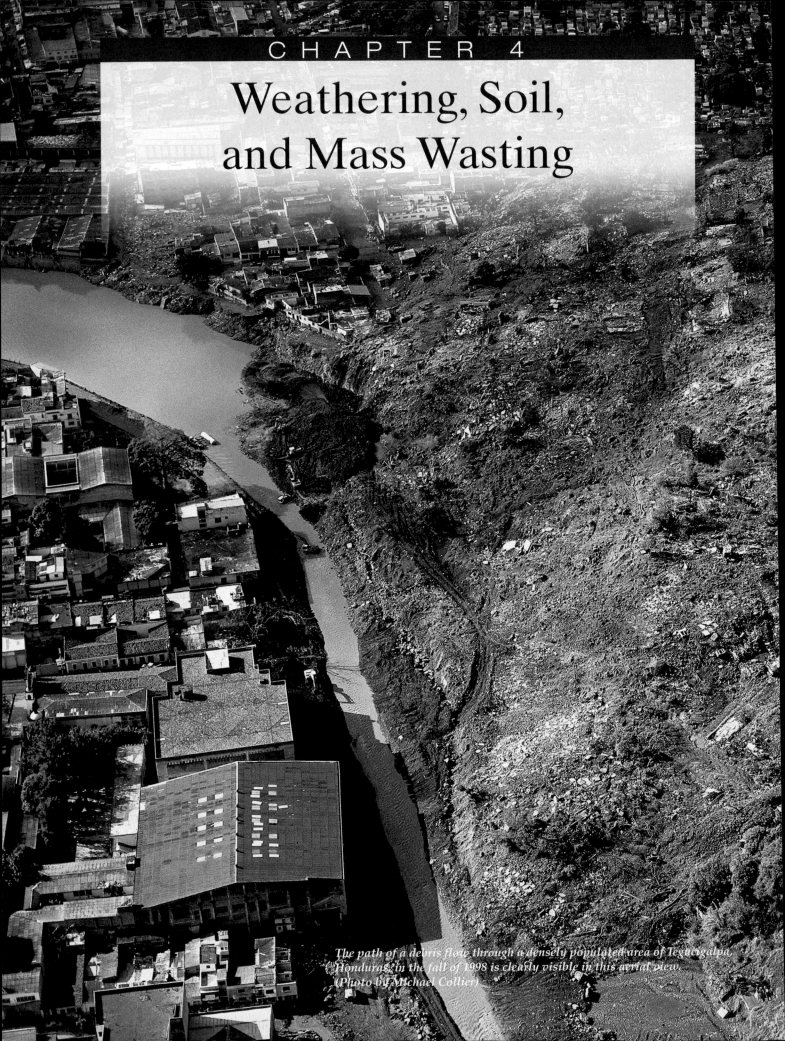

CHAPTER 4

Weathering, Soil, and Mass Wasting

The path of a debris flow through a densely populated area of Tegucigalpa, Honduras, in the fall of 1998 is clearly visible in this aerial view. (Photo by Michael Collier)

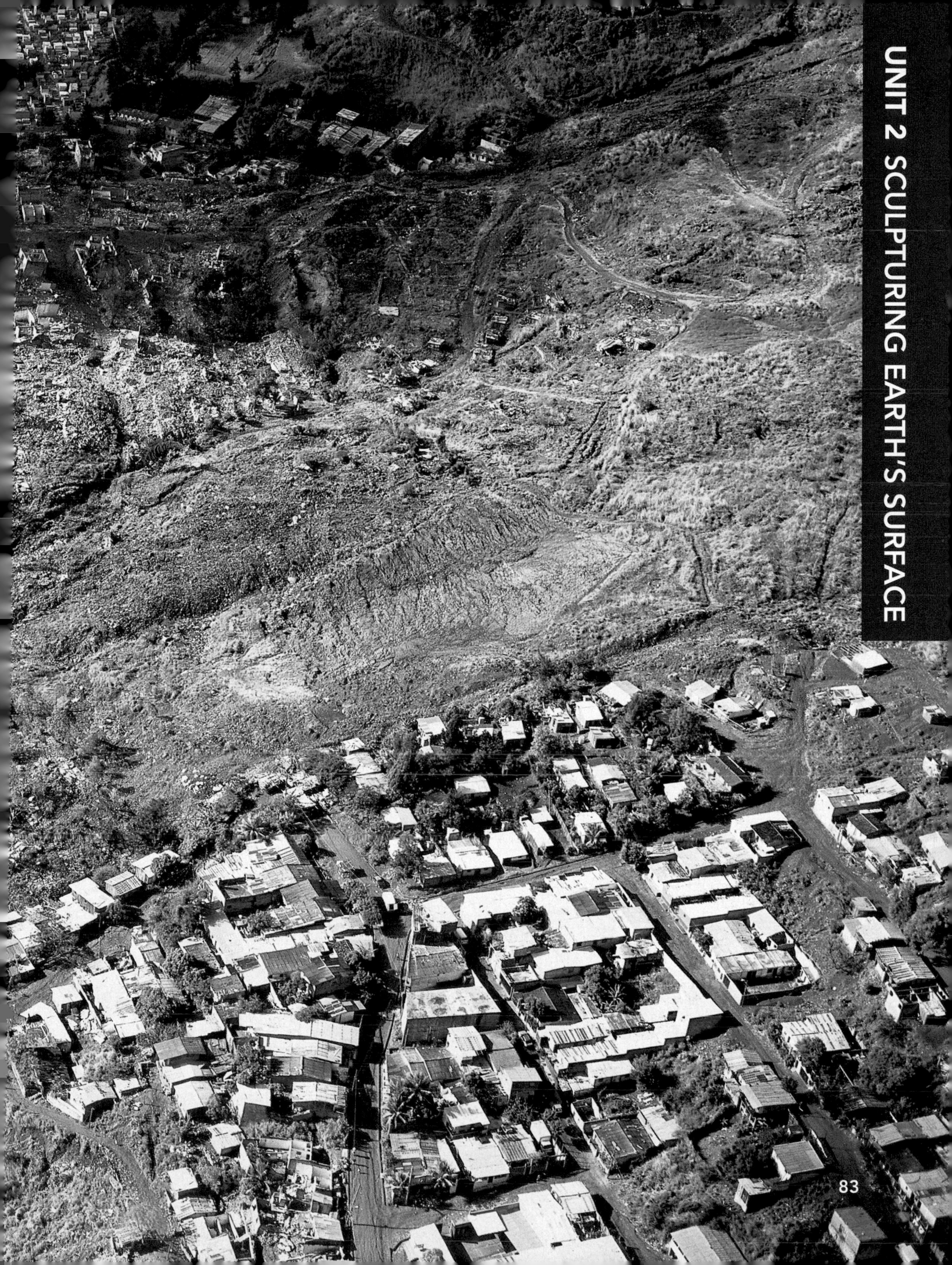

UNIT 2 SCULPTURING EARTH'S SURFACE

arth's surface is constantly changing. Rock is disintegrated and decomposed, moved to lower elevations by gravity, and carried away by water, wind, or ice. In this manner Earth's physical landscape is sculptured. This chapter focuses on the first two steps of this never-ending process—weathering and mass wasting. What causes solid rock to crumble, and why does the type and rate of weathering vary from place to place? What mechanisms act to move weathered debris downslope? Soil, an important product of the weathering process and a vital resource, is also examined.

Earth's External Processes

GEODe
Sculpturing Earth's Surface
▼ External vs. Internal Processes

Weathering, mass wasting, and erosion are called **external processes** because they occur at or near Earth's surface and are powered by energy from the Sun. External processes are a basic part of the rock cycle because they are responsible for transforming solid rock into sediment.

To the casual observer, the face of Earth may appear to be without change, unaffected by time. In fact, 200 years ago most people believed that mountains, lakes, and deserts were permanent features of an Earth that was thought to be no more than a few thousand years old. Today we know that Earth is 4.5 billion years old and that mountains eventually succumb to weathering and erosion, lakes fill with sediment or are drained by streams, and deserts come and go with changes in climate.

Earth is a dynamic body. Some parts of Earth's surface are gradually elevated by mountain building and volcanic activity. These **internal processes** derive their energy from Earth's interior. Meanwhile, opposing external processes are continually breaking rock apart and moving the debris to lower elevations (Figure 4.1). The latter processes include:

1. **Weathering**—the physical breakdown (disintegration) and chemical alteration (decomposition) of rocks at or near Earth's surface.
2. **Mass wasting**—the transfer of rock and soil downslope under the influence of gravity.
3. **Erosion**—the physical removal of material by mobile agents such as water, wind, or ice.

We will first turn our attention to the process of weathering and the products generated by this activity. However, weathering cannot be easily separated from the other two processes because, as weathering breaks rocks apart, it facilitates the movement of rock debris by mass wasting and erosion. Conversely, the transport of material by mass wasting and erosion further disintegrates and decomposes the rock.

Weathering

Weathering goes on all around us, but it seems like such a slow and subtle process that it is easy to underestimate its importance. However, it is worth remembering that weathering is a basic part of the rock cycle and thus a key process in the Earth system.

All materials are susceptible to weathering. Consider, for example, the fabricated product we call concrete, which closely resembles the sedimentary rock called conglomerate. A newly poured concrete sidewalk has a smooth, unweathered look. However, not many years later, the same sidewalk will appear

► **FIGURE 4.1** Slopes are places where materials are continually moving from higher to lower elevations. Weathering begins the process by attacking the solid rock exposed at the surface. Next, gravity moves the weathered debris downslope. This step, termed mass wasting, may range from a slow and gradual creep to a thundering landslide. Eventually, the material that was once high up the slope reaches the stream at the bottom. The moving water then transports the debris away. The walls of the Grand Canyon extend far from the Colorado River. This results primarily from the transfer of weathered debris downslope to the river and its tributaries by mass-wasting processes. (Photo by Tom and Susan Bean, Inc.)

chipped, cracked, and rough, with pebbles exposed at the surface. If a tree is nearby, its roots may grow under the concrete, heaving and buckling it. The same natural processes that eventually break apart a concrete sidewalk also act to disintegrate rocks, regardless of their type or strength.

Weathering occurs when rock is mechanically fragmented (disintegrated) and/or chemically altered (decomposed). **Mechanical weathering** is accomplished by physical forces that break rock into smaller and smaller pieces without changing the rock's mineral composition. **Chemical weathering** involves a chemical transformation of rock into one or more new compounds. These two concepts can be illustrated with a piece of paper. The paper can be disintegrated by tearing it into smaller and smaller pieces, whereas decomposition occurs when the paper is set afire and burned.

In the following sections we will discuss the various modes of mechanical and chemical weathering. Although we will consider these two processes separately, keep in mind that they usually work simultaneously in nature.

Mechanical Weathering

GEODe
Earth Materials
▼ Sedimentary Rocks

When a rock undergoes *mechanical weathering*, it is broken into smaller and smaller pieces, each retaining the characteristics of the original material. The end result is many small pieces from a single large one. Figure 4.2 shows that breaking a rock into smaller pieces increases the surface area available for chemical attack. Hence, by breaking rocks into smaller pieces, mechanical weathering increases the amount of surface area available for chemical weathering.

In nature, three physical processes are especially important in breaking rocks into smaller fragments: frost wedging, expansion resulting from unloading, and biological activity. In addition, although the work of erosional agents such as waves, wind, glacial ice, and running water is usually considered separately from mechanical weathering, it is nevertheless important. As these mobile agents move rock debris, they relentlessly disintegrate these materials.

Frost Wedging

Repeated cycles of freezing and thawing of water represent an important process of mechanical weathering. Liquid water has the unique property of expanding about 9 percent when it freezes. This increase in volume occurs because, as ice forms, the water molecules arrange themselves into a very open crystalline structure. As a result, when water freezes, it expands and exerts a tremendous outward force. This can be verified by completely filling a container with water and freezing it. The formation of ice will rupture the container.

In nature, water works its way into every crack or void in rock and, upon freezing, expands and enlarges these openings. After many freeze-thaw cycles, the rock is broken into pieces. This process is appropriately called **frost wedging** (Figure 4.3). Frost wedging is most pronounced in mountainous regions in the middle latitudes where a daily freeze-thaw cycle often exists (see Box 4.1). Here, sections of rock are wedged loose and may tumble into large piles called **talus slopes**, which often form at the base of steep rocky cliffs (Figure 4.3).

Frost wedging also causes great destruction to highways in the northern United States and Canada, particularly in the early spring when the freeze-thaw cycle is well established. Roadways acquire numerous potholes and are occasionally heaved and buckled by this destructive force.

Unloading

When large masses of igneous rock, particularly those composed of granite, are exposed by erosion, concentric slabs begin to break loose. The process generating these onion-like layers is called **sheeting** and probably

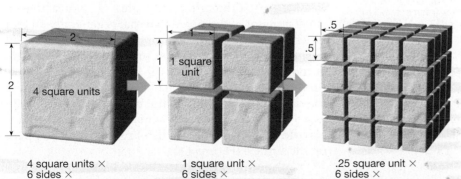

◀ **FIGURE 4.2** Chemical weathering can occur only to those portions of a rock that are exposed to the elements. Mechanical weathering breaks rock into smaller and smaller pieces, thereby increasing the surface area available for chemical attack.

4 square units ×
6 sides ×
1 cube =
_____
24 square units

1 square unit ×
6 sides ×
8 cubes =
_____
48 square units

.25 square unit ×
6 sides ×
64 cubes =
_____
96 square units

Frost wedging

▲ **FIGURE 4.3** Frost wedging. As water freezes, it expands, exerting a force great enough to break rock. When frost wedging occurs in a setting such as this, the broken rock fragments fall to the base of the cliff and create a cone-shaped accumulation known as talus. (Photo by Tom & Susan Bean, Inc.)

takes place in response to the great reduction in pressure that occurs when the overlying rock is eroded away. Accompanying this *unloading*, the outer layers expand more than the rock below and thus separate from the rock body (Figure 4.4A,B). Continued weathering eventually causes the slabs to separate and spall off, creating **exfoliation domes** (*ex* = off, *folium* = leaf). Excellent examples of exfoliation domes include Stone Mountain, Georgia, and Half

Dome (Figure 4.4C) and Liberty Cap in Yosemite National Park.

Deep underground mining provides us with another example of how rocks behave once the confining pressure is removed. Large rock slabs sometimes explode off the walls of newly cut mine tunnels because of the abruptly reduced pressure. Evidence of this type, plus the fact that fracturing occurs parallel to the floor of a quarry when large blocks of rock are removed, strongly supports the process of unloading as the cause of sheeting.

Biological Activity

Weathering is also accomplished by the activities of organisms, including plants, burrowing animals, and humans. Plant roots in search of minerals and water grow into fractures, and as the roots grow, they wedge the rock apart (Figure 4.5). Burrowing animals further break down the rock by moving fresh material to the surface, where physical and chemical processes can more effectively attack it. Decaying organisms also produce acids, which contribute to chemical weathering. Where rock has been blasted in search of minerals or for road construction, the impact of humans is particularly noticeable.

Chemical Weathering

 GEODe Earth Materials ▼ Sedimentary Rocks

Chemical weathering involves the complex processes that alter the internal structures of minerals by removing and/or adding elements. During this transformation, the original rock decomposes into substances that are stable in the surface environment. Consequently, the products of chemical weathering will remain essentially unchanged as long as they remain in an environment similar to the one in which they formed.

Water and Carbonic Acid

Water is by far the most important agent of chemical weathering. Although pure water is nonreactive, a small amount of dissolved material is generally all that is needed to activate it. Oxygen dissolved in water will *oxidize* some materials. For example, when an iron nail is found in moist soil, it will have a coating of rust (iron oxide), and if the time of exposure has been long, the nail will be so weak that it can be broken as easily as a toothpick. When rocks containing iron-rich minerals oxidize, a yellow to reddish-brown rust will appear on the surface (Figure 4.6).

Carbon dioxide (CO_2) dissolved in water (H_2O) forms carbonic acid (H_2CO_3), the same weak acid produced when soft drinks are carbonated. Rain dis-

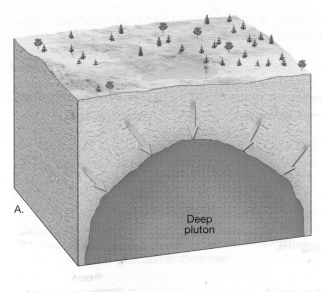

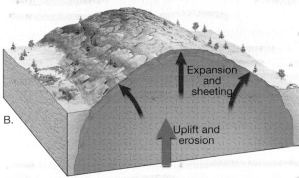

▲ FIGURE 4.4 Sheeting is caused by the expansion of crystalline rock as erosion removes the overlying material. When the deeply buried pluton in **A** is exposed at the surface following uplift and erosion in **B,** the igneous mass fractures into thin slabs. The photo in **C** is of the summit of Half Dome in Yosemite National Park, California. It is an exfoliation dome and illustrates the onionlike layers created by sheeting. (Photo by Breck P. Kent)

solves some carbon dioxide as it falls through the atmosphere, and additional amounts released by decaying organic matter are acquired as the water percolates through the soil. Carbonic acid ionizes to form the very reactive hydrogen ion (H^+) and the bicarbonate ion (HCO_3^-).

How Granite Weathers

To illustrate how rock chemically weathers when attacked by carbonic acid, we will consider the weathering of granite, an abundant continental rock. Recall that granite consists mainly of quartz and potassium feldspar. The weathering of the potassium feldspar component of granite takes place as follows:

$$2\,KAlSi_3O_8 + 2(H^+ + HCO_3^-) + H_2O \rightarrow$$

<table>
<tr><td>potassium feldspar</td><td>carbonic acid</td><td>water</td></tr>
</table>

$$Al_2Si_2O_5(OH)_4 + 2\,K^+ + 2\,HCO_3^- + 4\,SiO_2$$

<table>
<tr><td></td><td>potassium
ion</td><td>bicarbonite
ion</td><td>silica</td></tr>
<tr><td>clay mineral</td><td colspan="3" align="center">in solution</td></tr>
</table>

▼ FIGURE 4.5 Root wedging widens fractures in rock and aids the process of mechanical weathering. (Photo by Tom and Susan Bean, Inc).

▲ FIGURE 4.6 Iron reacts with oxygen to form iron oxide as seen on these rusted barrels. (Photo by Stephen J. Krasemann/DRK Photo)

In this reaction, the hydrogen ions (H^+) attack and replace potassium ions (K^+) in the feldspar structure, thereby disrupting the crystalline network. Once removed, the potassium is available as a nutrient for plants or becomes the soluble salt potassium bicarbonate ($KHCO_3$), which may be incorporated into other minerals or carried to the ocean in dissolved form by streams.

The most abundant products of the chemical breakdown of feldspar are residual clay minerals. Clay minerals are the end product of weathering and are very stable under surface conditions. Consequently, clay minerals make up a high percentage of the inorganic material in soils. Moreover, the most abundant sedimentary rock, shale, contains a high proportion of clay minerals.

In addition to the formation of clay minerals during this reaction, some silica is removed from the feldspar structure and is carried away by groundwater (water beneath Earth's surface). This dissolved silica will eventually precipitate to produce nodules of chert or flint, fill in the pore spaces between sediment grains, or be carried to the ocean, where microscopic animals will remove it to build hard silica shells.

To summarize, the weathering of potassium feldspar generates a residual clay mineral, a soluble salt (potassium bicarbonate), and some silica that enters into solution.

Quartz, the other main component of granite, is *very resistant* to chemical weathering; it remains substantially unaltered when attacked by weakly acidic solutions. As a result, when granite weathers, the feldspar crystals dull

and slowly turn to clay, releasing the once interlocked quartz grains, which still retain their fresh, glassy appearance. Although some quartz remains in the soil, much is transported to the sea or to other sites of deposition, where it becomes the main constituent of such features as sandy beaches and sand dunes. In time it may become lithified to form the sedimentary rock *sandstone*.

Weathering of Silicate Minerals

Table 4.1 lists the weathered products of some of the most common silicate minerals. Remember that silicate minerals make up most of Earth's crust and that these minerals are composed essentially of only eight elements. When chemically weathered, silicate minerals yield sodium, calcium, potassium, and magnesium ions, which form soluble products that may be removed by groundwater. The element iron combines with oxygen, producing relatively insoluble iron oxides, which give soil a reddish-brown or yellowish color. Under most conditions the three remaining elements—aluminum, silicon, and oxygen—join with water to produce residual clay minerals. However, even the highly insoluble clay minerals are very slowly removed by subsurface water.

Spheroidal Weathering

In addition to altering the internal structure of minerals, chemical weathering causes physical changes as well. For instance, when angular rock masses are attacked by water that enters along joints, the rocks tend to take on a spherical shape. Gradually the corners and edges of the angular blocks become more rounded. The corners are attacked most readily because of their greater surface area, as compared to the edges and faces. This process, called **spheroidal weathering**, gives the weathered rock a more rounded or spherical shape (Figure 4.7A).

Sometimes during the formation of spheroidal boulders, successive shells separate from the rock's main body (Figure 4.7B). Eventually the outer shells spall off, allowing the chemical weathering activity to penetrate deeper into the boulder. This spherical scaling results because, as the minerals in the rock weather to clay, they increase in size through the addition of water

TABLE 4.1	Products of weathering.	
Mineral	**Residual Products**	**Material in Solution**
Quartz	Quartz grains	Silica
Feldspars	Clay minerals	Silica
		K^+, Na^+, Ca^{2+}
Amphibole	Clay minerals	Silica
(hornblende)	Limonite	Ca^{2+}, Mg^{2+}
	Hematite	
Olivine	Limonite	Silica
	Hematite	Mg^{2+}

▲ **FIGURE 4.7** **A.** Spheroidal weathering is evident in this exposure of granite in California's Joshua Tree National Monument. Because the rocks are attacked more vigorously on the corners and edges, they take on a spherical shape. The lines visible in the rock are called *joints*. Joints are important rock structures that allow water to penetrate and start the weathering process long before the rock is exposed. (Photo by E. J. Tarbuck) **B.** Sometimes successive shells are loosened as the weathering process continues to penetrate ever deeper into the rock. (Photo by Martin Schmidt, Jr.)

to their structure. This increased bulk exerts an outward force that causes concentric layers of rock to break loose and fall off. Hence, chemical weathering does produce forces great enough to cause mechanical weathering.

This type of spheroidal weathering, in which shells spall off, should not be confused with the phenomenon of sheeting discussed earlier. In sheeting, the fracturing occurs as a result of unloading, and the rock layers that separate from the main body are largely unaltered at the time of separation.

Rates of Weathering

Several factors influence the type and rate of rock weathering. We have already seen how mechanical weathering affects the rate of weathering. By breaking rock into smaller pieces, the amount of surface area exposed to chemical weathering is increased. Other important factors examined here include rock characteristics and climate.

Rock Characteristics

Rock characteristics encompass all of the chemical traits of rocks, including mineral composition and solubility. In addition, any physical features, such as joints (cracks), can be important because they influence the ability of water to penetrate rock.

The variations in weathering rates, due to the mineral constituents, can be demonstrated by comparing old headstones made from different rock types. Headstones of granite, which is composed of silicate minerals, are relatively resistant to chemical weathering. We can see this by examining the inscriptions on the headstones shown in Figure 4.8. This is not true of the marble headstone, which shows signs

► **FIGURE 4.8** An examination of headstones reveals the rate of chemical weathering on diverse rock types. The granite headstone (left) was erected four years before the marble headstone (right). The inscription date of 1872 on the marble monument is nearly illegible. (Photos by E. J. Tarbuck)

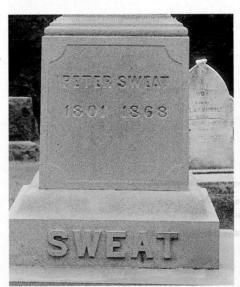

of extensive chemical alteration over a relatively short period. Marble is composed of calcite (calcium carbonate), which readily dissolves even in a weakly acidic solution.

The silicates, the most abundant mineral group, weather in essentially the same sequence as their order of crystallization. By examining Bowen's reaction series (see Figure 3.9, p. 60), you can see that olivine crystallizes first and is therefore the least resistant to chemical weathering, whereas quartz, which crystallizes last, is the most resistant.

Climate

Climatic factors, particularly temperature and moisture, are crucial to the rate of rock weathering. One important example from mechanical weathering is that the frequency of freeze-thaw cycles greatly affects the amount of frost wedging. Temperature and moisture also exert a strong influence on the rates of chemical weathering and on the kind and amount of vegetation present. Regions with lush vegetation generally have a thick mantle of soil rich in decayed organic matter from which chemically active fluids such as carbonic and humic acids are derived.

The optimum environment for chemical weathering is a combination of warm temperatures and abundant moisture. In polar regions chemical weathering is ineffective because frigid temperatures keep the available moisture locked up as ice, whereas in arid regions there is insufficient moisture to foster rapid chemical weathering.

Human activities can influence the composition of the atmosphere, which in turn can impact the rate of chemical weathering. One well-known example is acid rain (Figure 4.9).

Differential Weathering

Masses of rock do not weather uniformly. Take a moment to look at the photo of Shiprock, New Mexico, in Figure 9.21 (p. 267). The durable volcanic neck protrudes high above the surrounding terrain. A glance at the chapter-opening photo shows an additional example of this phenomenon, called **differential weathering**. The results vary in scale from the rough, uneven surface of the marble headstone in Figure 4.8 to the boldly sculpted exposures in Bryce Canyon (Figure 4.10).

Many factors influence the rate of rock weathering. Among the most important are variations in the composition of the rock. More resistant rock protrudes as ridges or pinnacles, or as steeper cliffs on a canyon wall (see Figure 11.2 p. 313). The number and spacing of joints can also be a significant factor (see Figure 4.7A and Figure 10.15, p. 294). Differential weathering and subsequent erosion are responsible for creating many

▲ **FIGURE 4.9** As a consequence of burning large quantities of coal and petroleum, more than 40 million tons of sulfur and nitrogen oxides are released into the atmosphere each year in the United States. Through a series of complex chemical reactions, some of these pollutants are converted into acids that then fall to Earth's surface as rain or snow. Among its many environmental effects, acid rain accelerates the chemical weathering of stone monuments and structures, including this building façade in Leipzig, Germany. (Photo by Doug Plummer/Photo Researchers, Inc.)

unusual and sometimes spectacular rock formations and landforms.

Soil

Soil covers most land surfaces. Along with air and water, it is one of our most indispensable resources (Figure 4.11). Also like air and water, soil is taken for granted by many of us. The following quote helps put this vital layer in perspective.

> Science, in recent years, has focused more and more on the Earth as a planet, one that for all we know is unique—where a thin blanket of air, a thinner film of water, and the thinnest veneer of soil combine to support a web of life of wondrous diversity in continuous change.*

Soil has accurately been called "the bridge between life and the inanimate world." All life—the entire biosphere—owes its existence to a dozen or so elements that must ultimately come from Earth's crust. Once weathering and other processes create soil, plants carry out the intermediary role of assimilating the necessary elements and making them available to animals, including humans.

---

*Jack Eddy. "A Fragile Seam of Dark Blue Light," in *Proceedings of the Global Change Research Forum.* U.S. Geological Survey Circular 1086, 1993, p. 15.

▲ **FIGURE 4.10** Differential weathering is illustrated by these sculpted rock pinnacles in Bryce National Park, Utah. (Photo by Barbara Gerlach/DRK Photo)

An Interface in the Earth System

When Earth is viewed as a system, soil is referred to as an *interface*—a common boundary where different parts of a system interact. This is an appropriate designation because soil forms where the solid Earth, the atmosphere, the hydrosphere, and the biosphere meet. Soil is a material that develops in response to complex environmental interactions among different parts of the Earth system. Over time, soil gradually evolves to a state of equilibrium or balance with the environment. Soil is dynamic and sensitive to almost every aspect of its surroundings.

▼ **FIGURE 4.11** Soil is an essential resource that we often take for granted. Soil is not a living entity, but it contains a great deal of life. Moreover, this complex medium supports nearly all plant life, which in turn supports animal life. (Photo by Tom Bean/Corbis/Stock Market)

Thus, when environmental changes occur, such as climate, vegetative cover, and animal (including human) activity, the soil responds. Any such change produces a gradual alteration of soil characteristics until a new balance is reached. Although thinly distributed over the land surface, soil functions as a fundamental interface, providing an excellent example of the integration among many parts of the Earth system.

STUDENTS SOMETIMES ASK...

I've seen photos of footprints left on the lunar surface by astronauts. Does this mean the Moon has soil?

Not exactly. The Moon has no atmosphere, no water, and lacks biological activity. Therefore, the chemical weathering, frost wedging, and other weathering processes we are familiar with on Earth are lacking on the Moon. However, all lunar terrains are mantled with a soil-like layer of gray debris, called *lunar regolith,* derived from a few billion years of bombardment by meteorites. The rate of surface change is so slow that the footprints left by the *Apollo* astronauts (see Figure 22.7 on p. 618), will likely remain fresh-looking for millions of years to come.

What Is Soil?

With few exceptions, Earth's land surface is covered by **regolith** (*rhegos* = blanket, *lithos* = stone), the layer of rock and mineral fragments produced by weathering. Some would call this material soil, but soil is more than an accumulation of weathered debris. **Soil** is a combination of mineral and organic matter,

91

water, and air—that portion of the regolith that supports the growth of plants. Although the proportions of the major components in soil vary, the same four components always are present to some extent (Figure 4.12). About one half of the total volume of a good-quality surface soil is a mixture of disintegrated and decomposed rock (mineral matter) and *humus*, the decayed remains of animal and plant life (organic matter). The remaining half consists of pore spaces among the solid particles where air and water circulate.

Although the mineral portion of the soil is usually much greater than the organic portion, humus is an essential component. In addition to being an important source of plant nutrients, humus enhances the soil's ability to retain water. Because plants require air and water to live and grow, the portion of the soil consisting of pore spaces that allow for the circulation of these fluids is as vital as the solid soil constituents.

Soil water is far from "pure" water; instead, it is a complex solution containing many soluble nutrients. Soil water not only provides the necessary moisture for the chemical reactions that sustain life; it also supplies plants with nutrients in a form they can use. The pore spaces not filled with water contain air. This air is the source of necessary oxygen and carbon dioxide for most microorganisms and plants that live in the soil.

Soil Texture and Structure

Most soils are far from uniform and contain particles of different sizes. **Soil texture** refers to the proportions of different particle sizes. Texture is a very basic soil property because it strongly influences the soil's ability to

retain and transmit water and air, both of which are essential to plant growth. Sandy soils may drain too rapidly and dry out quickly. At the opposite extreme, the pore spaces of clay-rich soils may be so small that they inhibit drainage, and long-lasting puddles result. Moreover, when the clay and silt content is very high, plant roots may have difficulty penetrating the soil.

Because soils rarely consist of particles of only one size, *textural categories* have been established based upon the varying proportions of clay, silt, and sand. The standard system of classes used by the U.S. Department of Agriculture is shown in Figure 4.13. For example, point *A* on this triangular diagram (left center) represents a soil composed of 10 percent silt, 40 percent clay, and 50 percent sand. Such a soil is called a *sandy clay*. The soils called *loam*, which occupy the central portion of the diagram, are those in which no single particle size predominates over the other two. Loam soils are best suited to support plant life because they generally have better moisture characteristics and nutrient storage ability than do soils composed predominantly of clay or coarse sand.

Soil particles are seldom completely independent of one another. Rather, they usually form clumps called *peds* that give soils a particular structure. Four basic soil structures are recognized: platy, prismatic, blocky, and spheroidal. Soil structure is important because it influences the ease of a soil's cultivation as well as the susceptibility of a soil to erosion. In addition, soil structure affects the porosity and permeability of soil (that is, the ease with which water can penetrate). This in turn influences the movement of

▼ **FIGURE 4.12** Composition (by volume) of a soil in good condition for plant growth. Although the percentages vary, each soil is composed of mineral and organic matter, water, and air.

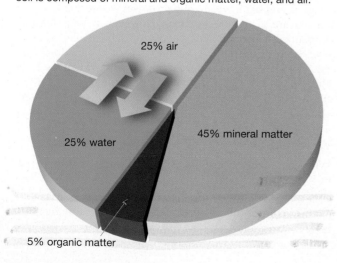

▼ **FIGURE 4.13** The texture of any soil can be represented by a point on this soil-texture diagram. Soil texture is one of the most significant factors used to estimate agricultural potential and engineering characteristics. (After U.S. Department of Agriculture)

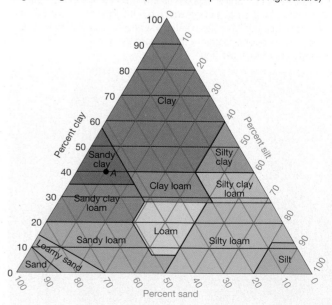

nutrients to plant roots. Prismatic and blocky peds usually allow for moderate water infiltration, whereas platy and spheroidal structures are characterized by slower infiltration rates.

Controls of Soil Formation

Soil is the product of the complex interplay of several factors. The most important of these are parent material, time, climate, plants and animals, and topography. Although all of these factors are interdependent, their roles will be examined separately.

Parent Material

The source of the weathered mineral matter from which soils develop is called the **parent material**, and it is a major factor influencing a newly forming soil. Gradually it undergoes physical and chemical changes as the processes of soil formation progress. Parent material may be the underlying bedrock, or it can be a layer of unconsolidated deposits, as in a stream valley. When the parent material is bedrock, the soils are termed *residual soils*. By contrast, those developed on unconsolidated sediment are called *transported soils* (Figure 4.14). Note that transported soils form *in place* on parent materials that have been carried from elsewhere and deposited by gravity, water, wind, or ice.

The nature of the parent material influences soils in two ways. First, the type of parent material affects the rate of weathering and thus the rate of soil formation. (Consider the weathering rates of granite versus limestone.) Also, because unconsolidated deposits are already partly weathered and provide more surface area for chemical weathering, soil development on such material usually progresses more rapidly. Second, the chemical makeup of the parent material affects the soil's fertility. This influences the character of the natural vegetation the soil can support.

At one time the parent material was believed to be the primary factor causing differences among soils. Today soil scientists realize that other factors, especially climate, are more important. In fact, it has been found that similar soils are often produced from different parent materials and that dissimilar soils have developed from the same parent material. Such discoveries reinforce the importance of the other soil-forming factors.

Time

Time is an important component of every geological process, and soil formation is no exception. The nature of soil is strongly influenced by the length of time that

▼ **FIGURE 4.14** The parent material for residual soils is the underlying bedrock, whereas transported soils form on unconsolidated deposits. Also note that as slopes become steeper, soil becomes thinner.

No soil development because of very steep slope

Residual soil is developed on bedrock

Transported soil is developed on unconsolidated deposits

Thinner soil on slope because of erosion

Unconsolidated deposits

Bedrock

processes have been operating. If weathering has been going on for a comparatively short time, the parent material determines to a large extent the characteristics of the soil. As weathering processes continue, the influence of parent material on soil is overshadowed by the other soil-forming factors, especially climate. The amount of time required for various soils to evolve cannot be specified because the soil-forming processes act at varying rates under different circumstances. However, as a rule, the longer a soil has been forming, the thicker it becomes and the less it resembles the parent material.

Climate

Climate is the most influential control of soil formation. Just as temperature and precipitation are the climatic elements that influence people the most, so too are they the elements that exert the strongest impact on soil formation. Variations in temperature and precipitation determine whether chemical or mechanical weathering predominates. They also greatly influence the rate and depth of weathering. For instance, a hot, wet climate may produce a thick layer of chemically weathered soil in the same amount of time that a cold, dry climate produces a thin mantle of mechanically weathered debris. Also, the amount of precipitation influences the degree to which various materials are removed (leached) from the soil, thereby affecting soil fertility. Finally, climatic conditions are important factors controlling the type of plant and animal life present.

Plants and Animals

The biosphere plays a vital role in soil formation. The types and abundance of organisms present have a strong influence on the physical and chemical properties of a soil (Figure 4.15). In fact, for well-developed soils in many regions, the significance of natural vegetation is frequently implied in the description used by soil scientists. Such phrases as *prairie soil, forest soil,* and *tundra soil* are common.

Plants and animals furnish organic matter to the soil. Certain bog soils are composed almost entirely of organic matter, whereas desert soils may contain only a tiny percentage. Although the quantity of organic matter varies substantially among soils, it is the rare soil that completely lacks it.

The primary source of organic matter is plants, although animals and the uncountable microorganisms also contribute. When organic matter decomposes, important nutrients are supplied to plants, as well as to animals and microorganisms living in the soil. Consequently, soil fertility depends in part on the amount of organic matter present. Furthermore, the decay of plant and animal remains causes the formation of various organic acids. These complex acids

▼ **FIGURE 4.15** The northern coniferous forest in Alaska's Denali National Park. The type of vegetation strongly influences soil formation. The organic litter received by the soil from the conifers is high in acid resins, which contribute to an accumulation of acid in the soil. As a result, intensive acid leaching is an important soil-forming process here. (Photo by Yva Momatiuk and John Eastcott/Photo Researchers, Inc).

hasten the weathering process. Organic matter also has a high water-holding ability and thus aids water retention in a soil.

Microorganisms, including fungi, bacteria, and single-celled protozoa, play an active role in the decay of plant and animal remains. The end product is *humus*, a material that no longer resembles the plants and animals from which it is formed. In addition, certain microorganisms aid soil fertility because they have the ability to convert atmospheric nitrogen gas into soil nitrogen compounds.

Earthworms and other burrowing animals act to mix the mineral and organic portions of a soil. Earthworms, for example, feed on organic matter and thoroughly mix soils in which they live, often moving and enriching many tons per acre each year. Burrows and holes also aid the passage of water and air through the soil.

Topography

The lay of the land can vary greatly over short distances. Such variations in topography can lead to the development of a variety of localized soil types. Many of the differences exist because the length and steepness of slopes have a significant impact on the amount of erosion and the water content of soil.

On steep slopes, soils are often poorly developed. In such situations little water can soak in, and as a result, soil moisture may be insufficient for vigorous plant growth. Further, because of accelerated erosion on steep slopes, the soils are thin or nonexistent (Figure 4.14).

In contrast, waterlogged soils in poorly drained bottomlands have a much different character. Such soils are usually thick and dark. The dark color results from the large quantity of organic matter that accumulates because saturated conditions retard the decay of vegetation. The optimum terrain for soil development is a flat-to-undulating upland surface. Here we find good drainage, minimum erosion, and sufficient infiltration of water into the soil.

Slope orientation, the direction a slope is facing, also is significant. In the midlatitudes of the Northern Hemisphere, a south-facing slope receives a great deal more sunlight than does a north-facing slope. In fact, a steep north-facing slope may receive no direct sunlight at all. The difference in the amount of solar radiation received causes substantial differences in soil temperature and moisture, which in turn may influence the nature of the vegetation and the character of the soil.

Although we have dealt separately with each of the soil-forming factors, remember that *all of them work together* to form soil. No single factor is responsible for a soil being as it is. Rather, it is the combined influence of parent material, time, climate, plants and animals, and topography that determines a soil's character.

The Soil Profile

It is important to realize that soil-forming processes *operate from the surface downward*. Thus, variations in composition, texture, structure, and color gradually evolve at varying depths. These vertical differences, which usually become more pronounced as time passes, divide the soil into zones or layers known as *horizons*. If you were to dig a trench in soil, you would see that its walls are layered. Such a vertical section through all of the soil horizons constitutes the **soil profile** (Figure 4.16).

Figure 4.17 presents an idealized view of a well-developed soil profile in which five horizons are identified. From the surface downward, they are designated as *O, A, E, B,* and *C,* respectively. These five horizons are common to soils in temperate regions. The characteristics and extent of development of horizons vary in different environments. Thus, different localities exhibit soil profiles that can contrast greatly with one another.

The *O* horizon consists largely of organic material. This is in contrast to the layers beneath it, which consist mainly of mineral matter. The upper portion of the *O* horizon is primarily plant litter such as loose leaves and other organic debris that are still recognizable. By contrast, the lower portion of the *O* horizon is made up of partly decomposed organic matter (humus) in which plant structures can no longer be identified. In addition to plants, the *O* horizon is teeming with microscopic life, including bacteria, fungi, algae, and insects. All of these organisms contribute oxygen, carbon dioxide, and organic acids to the developing soil.

Underlying the organic-rich *O* horizon is the *A* horizon. This zone is largely mineral matter, yet biological activity is high and humus is generally present—up to 30 percent in some instances. Together the *O* and *A* horizons make up what is commonly called *topsoil*. Below the *A* horizon, the *E* horizon is a light-colored layer that contains little organic material. As water percolates downward through this zone, finer particles are carried away. This washing out of the fine soil components is termed **eluviation** (*elu* = get away from, *via* = a way). Water percolating downward also dissolves soluble inorganic soil components and carries them to deeper zones. This depletion of soluble materials from the upper soil is termed **leaching**.

Immediately below the *E* horizon is the *B* horizon, or *subsoil*. Much of the material removed from the *E* horizon by eluviation is deposited in the *B* horizon, which is often referred to as the *zone of accumulation*. The accumulation of the fine clay particles enhances water retention in the subsoil. However, in extreme cases clay accumulation can form a very compact and impermeable layer called *hardpan*. The *O, A, E,* and *B* horizons together constitute the **solum, or "true soil."** It is in the solum that the soil-forming processes are

A.

B.

▲ **FIGURE 4.16**   We all know that a book should not be judged by its cover. Neither should a soil be evaluated only by examining its surface. A soil profile is a vertical cross-section from the surface through all of the soil's horizons and into the parent material. **A.** This profile shows a well-developed soil in southeastern South Dakota. (Photo by E. J. Tarbuck) **B.** The boundaries between horizons in this soil in Puerto Rico are indistinct, giving it a relatively uniform appearance. (Photo courtesy of Soil Science Society of America)

active and that living roots and other plant and animal life are largely confined.

Below the solum and above the unaltered parent material is the C horizon, a layer characterized by partially altered parent material. Whereas the O, A, E, and B horizons bear little resemblance to the parent material, it is easily identifiable in the C horizon. Although this material is undergoing changes that will eventually transform it into soil, it has not yet crossed the threshold that separates regolith from soil.

The characteristics and extent of development can vary greatly among soils in different environments. The boundaries between soil horizons can be very distinct or the horizons may blend gradually from one to another. A well-developed soil profile indicates that environmental conditions have been relatively stable over an extended time span and that the soil is *mature*. By contrast, some soils lack horizons altogether. Such soils are called *immature* because soil building has been going on for only a short time. Immature soils are also characteristic of steep slopes, where erosion continually strips away the soil, preventing full development.

Classifying Soils

There are many variations from place to place and from time to time among the factors that control soil formation. These differences lead to a bewildering variety of soil types. To cope with such variety, it is essential to devise some means of classifying the vast array of data to be studied. By establishing groups consisting of items that have certain important characteristics in common, order and simplicity are introduced. Bringing order to large quantities of information not only aids comprehension and understanding but also facilitates analysis and explanation.

In the United States, soil scientists have devised a system for classifying soils known as the **Soil Taxonomy**. It emphasizes the physical and chemical properties of the soil profile and is organized on the basis of observable soil characteristics. There are six hierarchical categories of classification, ranging from *order*, the broadest category, *to series*, the most specific category. The system recognizes 12 soil orders and more than 19,000 soil series.

The names of the classification units are combinations of syllables, most of which are derived from

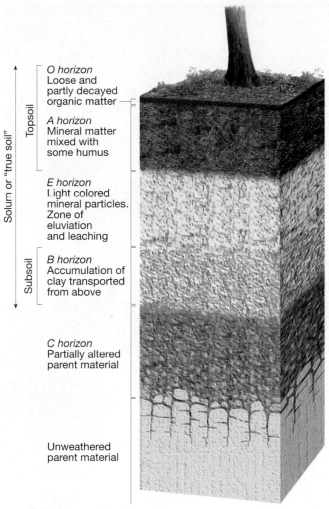

O horizon
Loose and
partly decayed
organic matter

A horizon
Mineral matter
mixed with
some humus

E horizon
Light colored
mineral particles.
Zone of
eluviation
and leaching

B horizon
Accumulation of
clay transported
from above

C horizon
Partially altered
parent material

Unweathered
parent material

Solum or "true soil"

Topsoil

Subsoil

▲ **FIGURE 4.17** Idealized soil profile from a humid climate in the middle latitudes.

TABLE 4.2	World Soil Orders
Alfisols	Moderately weathered soils that form under boreal forests or broadleaf deciduous forests, rich in iron and aluminum. Clay particles accumulate in a subsurface layer in response to leaching in moist environments. Fertile, productive soils, because they are neither too wet nor too dry.
Andisols	Young soils in which the parent material is volcanic ash and cinders, deposited by recent volcanic activity.
Aridosols	Soils that develop in dry places; insufficient water to remove soluble minerals, may have an accumulation of calcium carbonate, gypsum, or salt in subsoil; low organic content.
Entisols	Young soils having limited development and exhibiting properties of the parent material. Productivity ranges from very high for some formed on recent river deposits to very low for those forming on shifting sand or rocky slopes.
Gelisols	Young soils with little profile development that occur in regions with permafrost. Low temperatures and frozen conditions for much of the year slow soil-forming processes.
Histosols	Organic soils with little or no climatic implications. Can be found in any climate where organic debris can accumulate to form a bog soil. Dark, partially decomposed organic material commonly referred to as *peat*.
Inceptisols	Weakly developed young soils in which the beginning (inception) of profile development is evident. Most common in humid climates, they exist from the Arctic to the tropics. Native vegetation is most often forest.
Mollisols	Dark, soft soils that have developed under grass vegetation, generally found in prairie areas. Humus-rich surface horizon that is rich in calcium and magnesium. Soil fertility is excellent. Also found in hardwood forests with significant earthworm activity. Climatic range is boreal or alpine to tropical. Dry seasons are normal (see Figure 4.16A).
Oxisols	Soils that occur on old land surfaces unless parent materials were strongly weathered before they were deposited. Generally found in the tropics and subtropical regions. Rich in iron and aluminum oxides, oxisols are heavily leached, hence are poor soils for agricultural activity (see Figure 4.16B).
Spodosols	Soils found only in humid regions on sandy material. Common in northern coniferous forests (see Figure 4.15) and cool humid forests. Beneath the dark upper horizon of weathered organic material lies a light-colored horizon of leached material, the distinctive property of this soil.
Ultisols	Soils that represent the products of long periods of weathering. Water percolating through the soil concentrates clay particles in the lower horizons (argillic horizons). Restricted to humid climates in the temperate regions and the tropics, where the growing season is long. Abundant water and a long frost-free period contribute to extensive leaching, hence poorer soil quality.
Vertisols	Soils containing large amounts of clay, which shrink upon drying and swell with the addition of water. Found in subhumid to arid climates, provided that adequate supplies of water are available to saturate the soil after periods of drought. Soil expansion and contraction exert stresses on human structures.

Latin or Greek. The names are descriptive. For example, soils of the order Aridosol (from the Latin *aridus* = dry, and *solum* = soil) are characteristically dry soils in arid regions. Soils in the order Inceptisols (from Latin *inceptum* = beginning, and *solum* = soil) are soils with only the beginning or inception of profile development.

Brief descriptions of the 12 basic soil orders are provided in Table 4.2. Figure 4.18 shows the complex worldwide distribution pattern of the Soil Taxonomy's 12 soil orders. Like many classification systems, the Soil Taxonomy is not suitable for every purpose. It is especially useful for agricultural and related land-use purposes, but it is not a useful system for engineers who are preparing evaluations of potential construction sites.

Soil Erosion

Soils are just a tiny fraction of all Earth materials, yet they are a vital resource. Because soils are necessary for the growth of rooted plants, they are the very

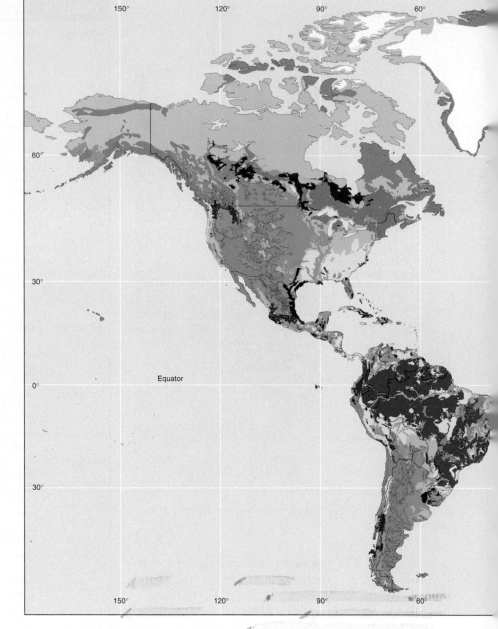

Alfisols (High-Nutrient Soils)

Andisols (Volcanic Soils)

Aridisols (Desert Soils)

Entisols (New Soils)

Gelisols (Permafrost Soils)

Histosols (Organic Soils)

Inceptisols (Young Soils)

Mollisols (Prairie Soils)

Oxisols (Tropical Forest Soils)

Spodosols (Conifer Forest Soils)

Ultisols (Low-Nutrient Soils)

Vertisols (Swelling Clay Soils)

Rock Land

Shifting Sands

Ice/Glacier

▲ **FIGURE 4.18** Global soil regions. Worldwide distribution of the Soil Taxonomy's 12 soil orders. (After U.S. Department of Agriculture, Natural Resources Conservation Service, World Soil Resources Staff)

foundation of the human life-support system. Just as human ingenuity can increase the agricultural productivity of soils through fertilization and irrigation, soils can be damaged or destroyed by carelessness. Despite their basic role in providing food, fiber, and other basic materials, soils are among our most abused resources.

Perhaps this neglect and indifference has occurred because a substantial amount of soil seems to remain even where soil erosion is serious. Nevertheless, although the loss of fertile topsoil may not be obvious to the untrained eye, it is a growing problem as human activities expand and disturb more and more of Earth's surface.

How Soil Is Eroded

Soil erosion is a natural process; it is part of the constant recycling of Earth materials that we call the *rock cycle*. Once soil forms, erosional forces, especially water and wind, move soil components from one place to another. Every time it rains, raindrops strike the land with surprising force (Figure 4.19). Each drop acts like a tiny bomb, blasting movable soil particles out of their positions in the soil mass. Then, water flowing across the surface carries away the dislodged soil particles. Because the soil is moved by thin sheets of water, this process is termed *sheet erosion*.

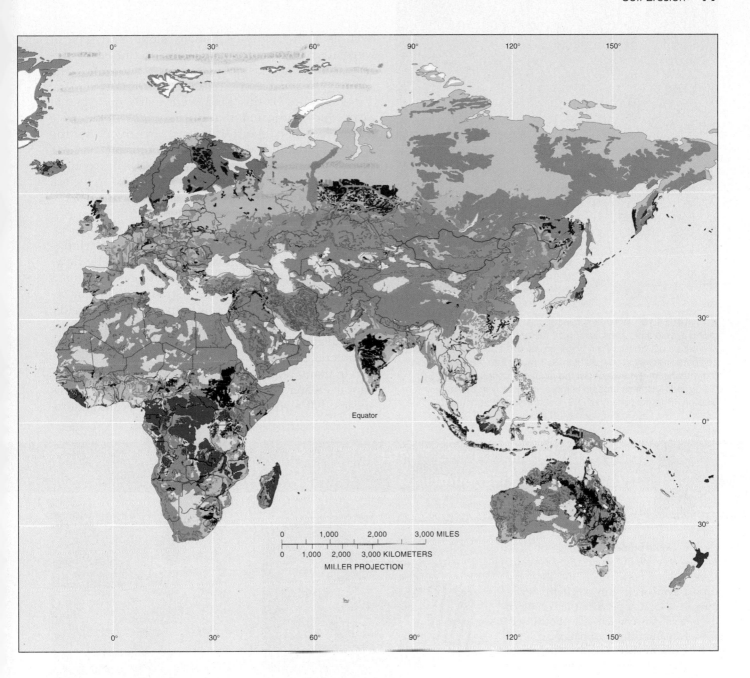

0 1,000 2,000 3,000 MILES

0 1,000 2,000 3,000 KILOMETERS

MILLER PROJECTION

Equator

After flowing as a thin, unconfined sheet for a relatively short distance, threads of current typically develop and tiny channels called *rills* begin to form. Still deeper cuts in the soil, known as *gullies*, are created as rills enlarge (Figure 4.20). When normal farm cultivation cannot eliminate the channels, we know the rills have grown large enough to be called gullies. Although most dislodged soil particles move only a short distance during each rainfall, substantial quantities eventually leave the fields and make their way downslope to a stream. Once in the stream channel, these soil particles, which can now be called *sediment*, are transported downstream and eventually deposited.

Rates of Erosion

We know that soil erosion is the ultimate fate of practically all soils. In the past, erosion occurred at slower rates than it does today because more of the land surface was covered and protected by trees, shrubs, grasses, and other plants. However, human activities such as farming, logging, and construction, which remove or disrupt the natural vegetation, have greatly accelerated the rate of soil erosion. Without the stabilizing effect of plants, the soil is more easily swept away by the wind or carried downslope by sheet wash.

Natural rates of soil erosion vary greatly from one place to another and depend on soil characteristics as

▲ **FIGURE 4.19** When it is raining, millions of water drops are falling at velocities approaching 10 meters per second (35 kilometers per hour). When water drops strike an exposed surface, soil particles may splash as high as 1 meter into the air and land more than a meter away from the point of raindrop impact. Soil dislodged by splash erosion is more easily moved by sheet erosion. (Photo courtesy of U.S. Department of Agriculture)

well as such factors as climate, slope, and type of vegetation. Over a broad area, erosion caused by surface runoff may be estimated by determining the sediment loads of the streams that drain the region. When studies of this kind were made on a global scale, they indicated that prior to the appearance of humans, sediment transport by rivers to the oceans amounted to just over 9 billion metric tons per year (1 metric ton = 1000 kilograms). By contrast, the amount of material currently transported to the sea by rivers is about 24 billion metric tons per year, or more than two and a half times the earlier rate.

It is more difficult to measure the loss of soil due to wind erosion. However, the removal of soil by wind is generally much less significant than erosion by flowing water except during periods of prolonged drought. When dry conditions prevail, strong winds can remove large quantities of soil from unprotected fields. Such was the case in the 1930s in the portions of the Great Plains that came to be called the *Dust Bowl* (see Chapter 6, page 179).

In many regions the rate of soil erosion is significantly greater than the rate of soil formation. This means that a renewable resource has become nonre-

BOX 4.1 UNDERSTANDING EARTH
The Old Man of the Mountain

The Old Man of the Mountain, also known as The Great Stone Face or simply The Profile, was one of New Hampshire's (*the Granite State*) best-known and most enduring symbols (Figure 4.A, left). Beginning in 1945, it appeared at the center of the official state emblem (see photo inset). It was a natural rock formation sculpted from Conway Red Granite that, when viewed from the proper location, gave the appearance of an old man. Each year hundreds of thousands of people traveled to view the Old Man, which protruded from high on Cannon Mountain, 360 meters (1200 feet) above Profile Lake in northern New Hampshire's Franconia Notch State Park.

On Saturday morning, May 3, 2003, the people of New Hampshire learned that the famous landmark had succumbed to nature and collapsed (Figure 4.A, right). The collapse ended decades of efforts to protect the state symbol from the same natural processes that created it in the first place. Ultimately, frost wedging and other weathering processes prevailed.

Figure 4.A (left)—The Old Man of the Mountain, high above Franconia Notch in New Hampshire's White Mountains, as it appeared prior to May 3, 2003.(Associated Press Photo) The inset shows the state emblem of New Hampshire.
Figure 4.A (right)—The famous granite outcrop after it collapsed on May 3, 2003. The natural processes that sculpted the Old Man ultimately destroyed it.(Associated Press Photo)

A.   B.

▲ FIGURE 4.20 **A.** Soil erosion from this field in northeastern Wisconsin is obvious. Just 1 millimeter of soil lost from a single acre of land amounts to about 5 tons. (Photo by D. P. Burnside/Photo Researchers, Inc.) **B.** Gully erosion in poorly protected soil, southern Colombia. (Photo by Carl Purcell/Photo Researchers, Inc.)

newable in these places. At present, it is estimated that topsoil is eroding faster than it forms on more than one-third of the world's croplands. The result is lower productivity, poorer crop quality, reduced agricultural income, and an ominous future.

STUDENTS SOMETIMES ASK...

Is soil erosion actually causing the amount of farmland to diminish?

Yes, indeed. It's been estimated that between 3 and 5 million acres of prime U.S. farmland are lost each year through mismanagement (including soil erosion) and conversion to nonagricultural uses. According to the United Nations, since 1950, more than one-third of the world's farmable land has been lost to soil erosion.

Sedimentation and Chemical Pollution

Another problem related to excessive soil erosion involves the deposition of sediment. Each year in the United States hundreds of millions of tons of eroded soil are deposited in lakes, reservoirs, and streams. The detrimental impact of this process can be significant. For example, as more and more sediment is deposited in a reservoir, the capacity of the reservoir is reduced, limiting its usefulness for flood control, water supply, and/or hydroelectric power generation. In addition, sedimentation in streams and other waterways can restrict navigation and lead to costly dredging operations.

In some cases soil particles are contaminated with pesticides used in farming. When these chemicals are introduced into a lake or reservoir, the quality of the water supply is threatened and aquatic organisms can be endangered. In addition to pesticides, nutrients found naturally in soils as well as those added by agricultural fertilizers make their way into streams and lakes, where they stimulate the growth of plants. Over a period of time, excessive nutrients accelerate the process by which plant growth leads to the depletion of oxygen and an early death of the lake.

The availability of good soils is critical if the world's rapidly growing population is to be fed. On every continent, unnecessary soil loss is occurring because appropriate conservation measures are not being used. Although it is a recognized fact that soil erosion can never be completely eliminated, soil conservation programs can substantially reduce the loss of this basic resource. Windbreaks (rows of trees), terracing, and plowing along the contours of hills are some of the effective measures, as are special tillage practices and crop rotation.

Weathering Creates Ore Deposits

Weathering creates many important mineral deposits by concentrating minor amounts of metals that are scattered through unweathered rock into economically valuable concentrations. Such a transformation is often termed **secondary enrichment** and takes place in one of two ways. In one situation, chemical weathering coupled with downward-percolating water removes undesired materials from decomposing rock, leaving the desired elements enriched in the upper zones of the soil. The second way is basically the reverse of the first. That is, the desirable elements that are found in low concentrations near the surface are removed and carried to lower zones, where they are redeposited and become more concentrated.

Bauxite

The formation of *bauxite*, the principal ore of aluminum, is one important example of an ore created as a result of enrichment by weathering processes (Figure 4.21). Although aluminum is the third most abundant element in Earth's crust, economically valuable concentrations of this important metal are not common, because most aluminum is tied up in silicate minerals from which it is extremely difficult to extract.

Bauxite forms in rainy tropical climates. When aluminum-rich source rocks are subjected to the intense and prolonged chemical weathering of the tropics, most of the common elements, including calcium, sodium, and potassium, are removed by leaching. Because aluminum is extremely insoluble, it becomes concentrated in the soil (as bauxite, a hydrated aluminum oxide). Thus, the formation of bauxite depends on climatic conditions in which chemical weathering and leaching are pronounced, plus, of course, the presence of aluminum-rich source rock. In a similar manner, important deposits of nickel and cobalt develop from igneous rocks rich in silicate minerals such as olivine.

There is significant concern regarding the mining of bauxite and other residual deposits because they tend to occur in environmentally sensitive areas of the tropics. Mining is preceded by the removal of tropical vegetation, thus destroying rain forest ecosystems. Moreover, the thin moisture-retaining layer of organic matter is also disturbed. When the soil dries out in the

hot sun, it becomes bricklike and loses its moisture-retaining qualities. Such soil cannot be productively farmed nor can it support significant forest growth. The long-term consequences of bauxite mining are clearly of concern for developing countries in the tropics where this important ore is mined.

Other Deposits

Many copper and silver deposits result when weathering processes concentrate metals that are dispersed through a low-grade primary ore. Usually such enrichment occurs in deposits containing pyrite (FeS_2), the most common and widespread sulfide mineral. Pyrite is important because when it chemically weathers, sulfuric acid forms, which enables percolating waters to dissolve the ore metals. Once dissolved, the metals gradually migrate downward through the primary ore body until they are precipitated. Deposition takes place because of changes that occur in the chemistry of the solution when it reaches the groundwater zone (the zone beneath the surface where all pore spaces are filled with water). In this manner, the small percentage of dispersed metal can be removed from a large volume of rock and redeposited as a higher-grade ore in a smaller volume of rock.

Mass Wasting: The Work of Gravity

Earth's surface is never perfectly flat but instead consists of slopes. Some are steep and precipitous; others are moderate or gentle. Some are long and gradual; others are short and abrupt. Some slopes are mantled with soil and covered by vegetation; others consist of barren rock and rubble. Their form and variety are great. Although most slopes appear to be stable and unchanging, they are not static features because the force of gravity causes material to move downslope. At one extreme the movement may be gradual and practically imperceptible. At the other extreme it may consist of a roaring debris flow or a thundering rock avalanche. Landslides are a worldwide natural hazard. When these natural processes lead to loss of life and property, they become natural disasters.

Most mass wasting, whether spectacular or subtle, is the result of circumstances that are completely independent of human activities. Very few landslides occur where they cannot be anticipated. In places where mass wasting is a recognized threat, steps can often be taken to control downslope movements or limit the damages that such movements can cause. If the potential for mass wasting goes unrecognized or is ignored, the results can be costly and dangerous. We should also note that although most downslope movements occur whether people are present or not, many occurrences are aggravated or even triggered by human actions.

▼ **FIGURE 4.21** Bauxite is the ore of aluminum and forms as a result of weathering processes under tropical conditions. Its color varies from red or brown to nearly white. (Photo by E. J. Tarbuck)

Mass Wasting and Landform Development

Landslides are spectacular examples of a basic geologic process called *mass wasting*. Mass wasting refers to the downslope movement of rock, regolith, and soil under the direct influence of gravity. It is distinct from the erosional processes that are examined in subsequent chapters because mass wasting does not require a transporting medium such as water, wind, or glacial ice.

The Role of Mass Wasting

In the evolution of most landforms, mass wasting is the step that follows weathering. By itself, weathering does not produce significant landforms. Rather, landforms develop as the products of weathering are removed from the places where they originate. Once weathering weakens and breaks rock apart, mass wasting transfers the debris downslope, where a stream, acting as a conveyor belt, usually carries it away (see Figure 4.1). Although there may be many intermediate stops along the way, the sediment is eventually transported to its ultimate destination: the sea.

The combined effects of mass wasting and running water produce stream valleys, which are the most common and conspicuous of Earth's landforms. If streams alone were responsible for creating the valleys in which they flow, the valleys would be very narrow features. However, the fact that most river valleys are much wider than they are deep is a strong indication of the significance of mass-wasting processes in supplying material to streams. The walls of a canyon extend far from the river because of the transfer of weathered debris downslope to the river and its tributaries by mass-wasting processes. In this manner, streams and mass wasting combine to modify and sculpture the surface. Of course, glaciers, groundwater, waves, and wind are also important agents in shaping landforms and developing landscapes.

Slopes Change Through Time

It is clear that if mass wasting is to occur, there must be slopes that rock, soil, and regolith can move down. It is Earth's mountain-building and volcanic processes that produce these slopes through sporadic changes in the elevations of landmasses and the ocean floor. If dynamic internal processes did not continually produce regions having higher elevations, the system that moves debris to lower elevations would gradually slow and eventually cease.

Most rapid and spectacular mass-wasting events occur in areas of rugged, geologically young mountains. Newly formed mountains are rapidly eroded by rivers and glaciers into regions characterized by steep and unstable slopes. It is in such settings that massive destructive landslides occur. As mountain building subsides, mass wasting and erosional processes lower the land. Through time, steep and rugged mountain slopes give way to gentler, more subdued terrain. Thus, as a landscape ages, massive and rapid mass-wasting processes give way to smaller, less dramatic downslope movements.

Controls and Triggers of Mass Wasting

Gravity is the controlling force of mass wasting, but several factors play an important role in overcoming inertia and creating downslope movements. Long before a landslide occurs, various processes work to weaken slope material, gradually making it more and more susceptible to the pull of gravity. During this span, the slope remains stable but gets closer and closer to being unstable. Eventually, the strength of the slope is weakened to the point that something causes it to cross the threshold from stability to instability. Such an event that initiates downslope movement is called a *trigger*. Remember that the trigger is not the sole cause of the mass-wasting event, but just the last of many causes. Among the common factors that trigger mass-wasting processes are saturation of material with water, oversteepening of slopes, removal of anchoring vegetation, and ground vibrations from earthquakes.

The Role of Water

Mass wasting is sometimes triggered when heavy rains or periods of snowmelt saturate surface materials. This was the case in October 1998 when torrential downpours associated with Hurricane Mitch triggered devastating mudflows in Central America (Figure 4.22).

When the pores in sediment become filled with water, the cohesion among particles is destroyed, allowing them to slide past one another with relative ease. For example, when sand is slightly moist, it sticks together quite well. However, if enough water is added to fill the openings between the grains, the sand will ooze out in all directions (Figure 4.23). Thus, saturation reduces the internal resistance of materials, which are then easily set in motion by the force of gravity. When clay is wetted, it becomes very slick—another example of the "lubricating" effect of water. Water also adds considerable weight to a mass of material. The added weight in itself may be enough to cause the material to slide or flow downslope.

Oversteepened Slopes

Oversteepening of slopes is another trigger of many mass movements. There are many situations in nature where this takes place. A stream undercutting a valley

▶ FIGURE 4.22 Heavy rains from Hurricane Mitch in November 1998 created devastating mudflows in Central America. Water plays an important role in many mass-wasting processes. (Associated Press Photo)

wall and waves pounding against the base of a cliff are two familiar examples. Furthermore, through their activities, people often create oversteepened and unstable slopes that become prime sites for mass wasting.

Unconsolidated, granular (sand-size or coarser) particles assume a stable slope called the **angle of repose** (*reposen* = to be at rest). This is the steepest angle at which material remains stable. Depending on the size and shape of the particles, the angle varies from 25 to 40 degrees. The larger, more angular particles maintain the steepest slopes. If the angle is increased, the rock debris will adjust by moving downslope.

Oversteepening is important not only because it triggers movements of unconsolidated granular materials, but it also produces unstable slopes and mass movements in cohesive soils, regolith, and bedrock. The response will not be immediate, as with loose, granular material, but sooner or later one or more mass-wasting processes will eliminate the oversteepening and restore stability to the slope.

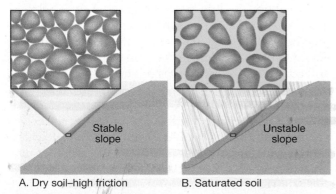

A. Dry soil–high friction   B. Saturated soil

▲ FIGURE 4.23 The effect of water on mass wasting can be great. **A.** When little or no water is present, friction among the closely packed soil particles on the slope holds them in place. **B.** When the soil is saturated, the grains are forced apart and friction is reduced, allowing the soil to move downslope.

Removal of Vegetation

Plants protect against erosion and contribute to the stability of slopes because their root systems bind soil and regolith together. Where plants are lacking, mass wasting is enhanced, especially if slopes are steep and water is plentiful. When anchoring vegetation is removed by forest fires or by people (for timber, farming, or development), surface materials frequently move downslope.

An unusual example illustrating the anchoring effect of plants occurred several decades ago on steep slopes near Menton, France. Farmers replaced olive trees, which have deep roots, with a more profitable but shallow-rooted crop: carnations. When the less stable slope failed, the landslide took 11 lives.

In July 1994 a severe wildfire swept Storm King Mountain west of Glenwood Springs, Colorado, denuding the slopes of vegetation. Two months later heavy rains resulted in numerous debris flows, one of which blocked Interstate 70 and threatened to dam the Colorado River. A 5-kilometer (3-mile) length of the highway was inundated with tons of rock, mud, and burned trees. The closure of Interstate 70 imposed costly delays on this major highway. Events such as this are relatively common in the American West where slopes are often steep and summer wildfires may affect millions of acres each year (Figure 4.24).

Earthquakes as Triggers

Conditions favoring mass wasting can exist in an area for a long time without movement occurring. An additional factor is sometimes necessary to trigger the movement. Among the more important and dramatic triggers are earthquakes. An earthquake and its aftershocks can dislodge enormous volumes of rock and unconsolidated material. The mass-wasting events shown in Figure 1.2 (p. 6) and Figure 4.27 were both

▲ **FIGURE 4.24** During the summer, wildfires are common occurrences in many parts of the west. Millions of acres are burned each year. The loss of anchoring vegetation sets the stage for accelerated mass wasting. (Photo by Raymond Gehman)

triggered by earthquakes. In many areas that are jolted by earthquakes, it is not ground vibrations directly but landslides and ground subsidence triggered by the vibrations that cause the greatest damage.

Classifying Mass-Wasting Processes

Geologists include several processes under the name of mass wasting. Four are illustrated in Figure 4.25. Generally, each process is defined by the type of material involved, the kind of motion, and the velocity of the movement.

If soil and regolith dominate, terms such as *debris*, *mud*, or *earth* are used. In contrast, when a mass of bedrock breaks loose and moves downslope, the term *rock* may be part of the description. Generally, the kind of motion is described as either a fall, a slide, or a flow.

Type of Motion

When the movement involves the freefall of detached individual pieces of any size, it is termed a **fall**. Falls are common on slopes that are too steep for loose material to remain on the surface (Figure 4.26). Many falls result when freeze and thaw cycles or the action of plant roots loosen rock to the point that gravity takes over. Rockfall is the primary way in which talus slopes are built and maintained (see Figure 4.3). Sometimes falls may trigger other forms of downslope movement.

Many mass-wasting processes are **slides**, which occur whenever material remains fairly coherent and moves along a well-defined surface. Sometimes the surface is a joint, a fault, or a bedding plane that is roughly parallel to the slope. However, in the movement called *slump*, the descending material moves en masse along a curved surface of rupture.

A third type of movement common to mass wasting is termed **flow**. Flow occurs when material moves downslope as a viscous fluid. Most flows are saturated with water and typically move as lobes or tongues.

Rate of Movement

When mass-wasting events make the news, a large quantity of material has in all likelihood moved rapidly downslope and has had a disastrous effect upon people and property. Indeed, during events called *rock avalanches*, rock and debris can hurtle downslope at speeds exceeding 200 kilometers (125 miles) per hour. Many researchers believe that rock avalanches, such as the one that produced the scene in Figure 4.27, must literally "float on air" as they move downslope. That is, high velocities result when air becomes trapped and compressed beneath the falling mass of debris, allowing it to move as a buoyant, flexible sheet across the surface.

Most mass movements, however, do not move with the speed of a rock avalanche. In fact, a great deal of mass wasting is imperceptibly slow. One process that we will examine later, termed *creep*, results in

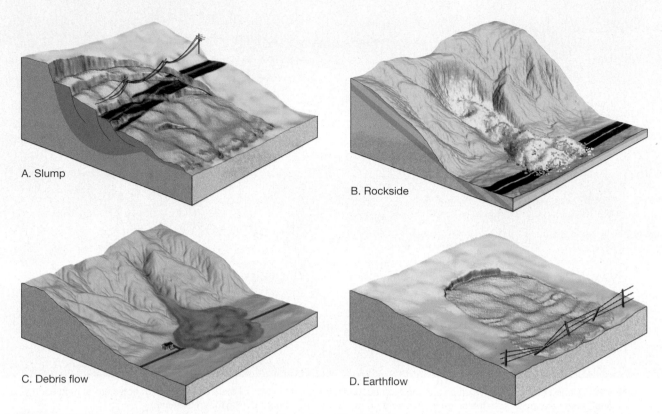

A. Slump

B. Rockside

C. Debris flow

D. Earthflow

▲ **FIGURE 4.25** The four processes illustrated here are all considered to be relatively rapid forms of mass wasting. Because material in slumps **A.** and rockslides **B.** move along well-defined surfaces, they are said to move by sliding. By contrast, when material moves downslope as a viscous fluid, the movement is described as a flow. Debris flow **C.** and earthflow **D.** advance downslope in this manner.

particle movements that are usually measured in millimeters or centimeters per year. Thus, as you can see, rates of movement can be spectacularly sudden or exceptionally gradual. Although various types of mass wasting are often classified as either rapid or slow, such a distinction is highly subjective because a wide range of rates exists between the two extremes. Even the velocity of a single process at a particular site can vary considerably from one time to another.

▼ **FIGURE 4.26** Rockfall blocking the Karakoram Highway in Pakistan's Hunza Valley. (Photo by Robert Holmes/CORBIS)

STUDENTS SOMETIMES ASK...

*Are snow avalanches considered
a type of mass wasting?*

Sure. Sometimes these thundering downslope movements of snow and ice move large quantities of rock, soil, and trees. Of course, snow avalanches are very dangerous, especially to skiers on high mountain slopes and to buildings and roads at the bottom of slopes in avalanche-prone regions.

About 10,000 snow avalanches occur each year in the mountainous western United States. In an average year they claim between 15 and 25 lives in the U.S. and Canada. They are a growing problem as more people become involved in winter sports and recreation.

Slump

Slump refers to the downward sliding of a mass of rock or unconsolidated material moving as a unit along a *curved* surface (Figure 4.25A). Usually the slumped material does not travel spectacularly fast nor very far. This is a common form of mass wasting, especially in thick accumulations of cohesive materials such as clay. As the movement occurs, a crescent-shaped scarp (cliff) is created at the head, and the block's upper surface is sometimes tilted backward.

Rock avalanche debris

▲ **FIGURE 4.27** This 4-kilometer-long tongue of rubble was deposited atop Alaska's Sherman Glacier by a rock avalanche. The event was triggered by a tremendous earthquake in March 1964. (Photo by Austin Post, U.S. Geological Survey)

Slump commonly occurs because a slope has been oversteepened. The material on the upper portion of a slope is held in place by the material at the bottom of the slope. As this anchoring material at the base is removed, the material above is made unstable and reacts to the pull of gravity. A common example is a valley wall that becomes oversteepened by a meandering river. Another is a coastal area that has been undercut by wave activity at its base.

Rockslide

Rockslides frequently occur in high mountain areas such as the Andes, Alps, and Canadian Rockies. They are sudden and rapid movements that happen when detached segments of bedrock break loose and slide downslope (Figure 4.25B). As the moving mass thunders along the surface, it breaks into many smaller pieces. Such events are among the fastest and most destructive mass movements.

Rockslides usually take place where there is an inclined surface of weakness. Such surfaces tend to form where strata are tilted or where joints and fractures exist parallel to the slope. When rock in such a setting is undercut at the base of the slope, it loses support and eventually gives way. Sometimes an earthquake is the trigger. On other occasions the rockslide is triggered when rain or melting snow lubricates the underlying surface to the point that friction is no longer sufficient to hold the rock unit in place. As a result, rockslides tend to be more common during the spring, when heavy rains and melting snow are most prevalent. The massive Gros Ventre slide shown in Figure 4.28 is a classic example.

Debris Flow

Debris flow is a relatively rapid type of mass wasting that involves a flow of soil and regolith containing a large amount of water (Figure 4.25C). Debris flows, which are also called *mudflows*, are most characteristic of semiarid mountainous regions and are also common on the slopes of some volcanoes. Because of their fluid properties, debris flows follow canyons and stream channels. As Figure 4.29 illustrates, in populated areas debris flows can pose a significant hazard to life and property (see Box 4.2).

Debris Flows in Semiarid Regions

When a cloudburst or rapidly melting mountain snows create a sudden flood in a semiarid region, large quantities of soil and regolith are washed into nearby stream channels because there is usually little vegetation to anchor the surface material. The end product is a flowing tongue of well-mixed mud, soil, rock, and water. Its consistency may range from that of wet concrete to a soupy mixture not much thicker than muddy water. The rate of flow therefore

▶ **FIGURE 4.28** On June 23, 1925, a massive rockslide took place in the valley of the Gros Ventre River in northwestern Wyoming following heavy spring rains and snowmelt. The volume of debris, estimated at 38 million cubic meters, created a 70-meter-high dam. Later, the lake created by the debris dam overflowed, resulting in a devastating flood downstream. **A.** Cross-sectional view. The slide occurred when the tilted and undercut sandstone bed could no longer maintain its position atop the saturated bed of clay. **B.** Even though the Gros Ventre rockslide occurred in 1925, the scar left on the side of Sheep Mountain is still a prominent feature. (Part A after W. C. Alden, "Landslide and Flood at Gros Ventre, Wyoming," *Transactions* (AIME) 76 (1928); 348. Part B photo by Stephen Trimble)

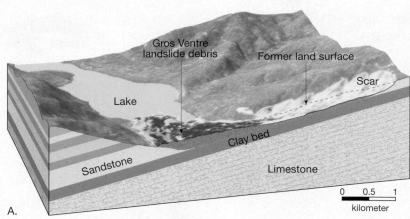

A.

B.

depends not only on the slope but also on the water content. When dense, debris flows are capable of carrying or pushing large boulders, trees, and even houses with relative ease.

Debris flows pose a serious hazard to development in dry mountainous areas such as southern California. The construction of homes on canyon hillsides and the removal of anchoring vegetation by brush fires

▶ **FIGURE 4.29** In February 1998, a debris flow literally buried this pickup truck in Rio Nido, California. Heavy rains triggered the event. (Photo by Michael Collier)

and other means have increased the frequency of these destructive events.

Lahars

Debris flows composed mostly of volcanic materials on the flanks of volcanoes are called **lahars**. The word originated in Indonesia, a volcanic region that has experienced many of these often destructive events. Historically, lahars have been one of the deadliest volcano hazards. They can occur either during an eruption or when a volcano is quiet. They take place when highly unstable layers of ash and debris become saturated with water and flow down steep volcanic slopes, generally following existing stream channels. Heavy rainfalls often trigger these flows. Others are triggered when large volumes of ice and snow are suddenly melted by heat flowing to the surface from within the volcano or by the hot gases and near-molten debris emitted during a violent eruption.

In November 1985 lahars were produced when Nevado del Ruiz, a 5300-meter (17,400-foot) volcano in the Andes Mountains of Colombia, erupted. The eruption melted much of the snow and ice that capped the uppermost 600 meters (2000 feet) of the peak, producing torrents of hot, thick mud, ash, and debris. The lahars moved outward from the volcano, following the valleys of three rain-swollen rivers that radiate from the peak. The flow that moved down the valley of the Lagunilla River was the most destructive, devastating the town of Armero, 48 kilometers (30 miles) from the mountain. Most of the more than 25,000 deaths caused by the event occurred in this once-thriving agricultural community.

Earthflow

We have seen that debris flows are frequently confined to channels in semiarid regions. In contrast, **earthflows** most often form on hillsides in humid areas during times of heavy precipitation or snowmelt (see Figure 4.25D). When water saturates the soil and regolith on a hillside, the material may break away, leaving a scar on the slope and forming a tongue- or teardrop-shaped mass that flows downslope (Figure 4.30). The materials most commonly involved are rich in clay and silt and contain only small proportions of sand and coarser particles. Earthflows range in size from bodies a few meters long, a few meters wide, and less than 1 meter deep to masses more than 1 kilometer long, several hundred meters wide, and more than 10 meters deep.

Because earthflows are quite viscous, they generally move at slower rates than the more fluid debris flows described in the preceding section. They move

▲ **FIGURE 4.30** This small, tongue-shaped earthflow occurred on a newly formed slope along a recently constructed highway. It formed in clay-rich material following a period of heavy rain. Notice the small slump at the head of the earthflow. (Photo by E. J. Tarbuck)

slowly and persistently for periods ranging from days to years. Depending on slope steepness and the material's consistency, velocities range from less than 1 millimeter per day up to several meters per day. Movement is typically faster during wet periods. In addition to occurring as isolated hillside phenomena, earthflows commonly take place in association with large slumps. In this situation, they may be seen as tonguelike flows at the base of the slump.

A special type of earthflow, known as *liquefaction*, sometimes occurs in association with earthquakes. Porous clay- to sand-size sediments that are saturated with water are most vulnerable. When shaken suddenly, the grains lose cohesion and the ground flows. Liquefaction can cause buildings to sink or tip on their sides and underground storage tanks and sewer lines to float upward. To say the least, damage can be substantial. You will learn more about this in Chapter 7.

Slow Movements

Movements such as rockslides, rock avalanches, and lahars are certainly the most spectacular and catastrophic forms of mass wasting. These dangerous events deserve intensive study to enable more effective prediction, timely warnings, and better controls to save lives. However, because of their large size and spectacular nature, they give us a false impression of their importance as a mass-wasting process. Indeed, sudden movements transport less material than the slow, subtle action of creep. Whereas rapid types of mass wasting are characteristic of mountains and steep hillsides, creep takes place on both steep and gentle slopes and is thus much more widespread.

Box 4.2 People and the Environment
Debris Flows on Alluvial Fans: A Case Study from Venezuela*

In December 1999, heavy rains triggered thousands of landslides along the coast of Venezuela (Figure 4.B). Debris flows and flash floods caused severe property damage and the tragic loss of an estimated 19,000 lives. The sites of most of the death and destruction were *alluvial fans*. These landforms are gently sloping, cone- to fan-shaped accumulations of sediment that are commonly found where high-gradient streams leave narrow valleys in mountainous areas and abruptly meet flat terrain.**

Several hundred thousand people live in the narrow coastal zone north of Caracas, Venezuela. They occupy alluvial fans located at the base of steep mountains that rise to elevations of more than 2000 meters (6600 feet) because these sites are the only areas that are not too steep to build on (Figure 4.C). Such settings are highly vulnerable to rainfall-induced landslides.

An unusually wet period in December 1999 included rains of 20 centimeters (8 inches) on December 2 and 3, followed by an additional 91 centimeters (36 inches) between December 14 and 16. The heavy rains triggered thousands of debris flows and other types of mass wasting. Once created, these moving masses of mud and rock coalesced to form giant debris flows that moved rapidly through steep, narrow canyons before exiting onto the alluvial fans.

On virtually every alluvial fan in the area, debris flows and flash floods brought massive amounts of sediment,

*Based on material prepared by the U.S. Geological Survey.

**For more about alluvial fans, see p. 130 in Chapter 5 and p. 177–78 in Chapter 6.

Figure 4.B The red arrow near the top of the map points to the area of Venezuela affected by disastrous debris flows and flash floods in 1999.

Figure 4.C Aerial view of the highly developed alluvial fan at Caraballeda, Venezuela, covered by debris-flow material.(Associated Press Photo)

including boulders as large as 10 meters (33 feet) in diameter. Hundreds of houses and other structures were damaged or destroyed (Figure 4.D). Total damages approached $2 billion.

This example from Venezuela shows the potential for extreme loss of life and property damage where large numbers of people occupy alluvial fans. The possibility for similar events of comparable magnitude exists in other parts of the world.

Building communities on alluvial fans can transform natural processes into major lethal events. Kofi Annan, Secretary General of the United Nations, put it this way: "The term 'natural disaster' has become an increasingly anachronistic misnomer. In reality, human behavior transforms natural hazards into what should really be called unnatural disasters.***

***Matthew C. Larsen, et al. *Natural Hazards on Alluvial Fans: The Venezuela Delvis Flow and Flash Flood Disaster*, U.S. Geological Survey Fact Sheet PS 103, p. 4.

Figure 4.D Debris-flow damage. Huge boulders (in excess of 300 tons) were transported by some flows.(AP/Wide World Photo)

Creep

Creep is a type of mass wasting that involves the gradual downhill movement of soil and regolith. One factor that contributes to creep is the alternate expansion and contraction of surface material caused by freezing and thawing or wetting and drying. As shown in Figure 4.31, freezing or wetting lifts particles at right angles to the slope, and thawing or drying allows the particles to fall back to a slightly lower level. Each cycle therefore moves the material a short distance downhill.

Creep is aided by anything that disturbs the soil. For example, raindrop impact and disturbance by plant roots and burrowing animals may contribute. Creep is also promoted when the ground becomes saturated with water. Following a heavy rain or snowmelt, a waterlogged soil may lose its internal cohesion, allowing gravity to pull the material downslope. Because creep is imperceptibly slow, the process cannot be observed in action. What can be observed, however, are the effects of creep. Creep causes fences and utility poles to tilt and retaining walls to be displaced.

STUDENTS SOMETIMES ASK...

How many deaths and how much damage is caused by mass wasting each year?

The U.S. Geological Survey estimates that between 25 and 50 people are killed by landslides annually in the United States. The worldwide death toll, of course, is much higher. The annual cost of damage by all types of mass wasting in the U.S. is about $4 billion. However, this figure is probably conservative because slow movements like creep do a substantial amount of damage that goes unreported.

Solifluction

When soil is saturated with water, the soggy mass may flow downslope at a rate of a few millimeters or a few centimeters per day or per year. Such a

▲ **FIGURE 4.32** Solifluction lobes northeast of Fairbanks, Alaska. Solifluction occurs in permafrost regions when the active layer thaws in summer. (Photo by James E. Patterson)

process is called **solifluction** (literally, "soil flow"). It is a type of mass wasting that is common wherever water cannot escape from the saturated surface layer by infiltrating to deeper levels. A dense clay hardpan in soil or an impermeable bedrock layer can promote solifluction.

Solifluction is also common in regions underlain by **permafrost**. Permafrost refers to the permanently frozen ground that occurs in Earth's harsh tundra and ice cap climates. It occurs in a zone above the permafrost called the *active layer*, which thaws in summer and refreezes in winter. During summer, water is unable to percolate into the impervious permafrost layer below. As a result, the active layer becomes saturated and slowly flows. The process can occur on slopes as gentle as 2 to 3 degrees. Where there is a well-developed mat of vegetation, a solifluction sheet may move in a series of well-defined lobes or overriding folds (Figure 4.32).

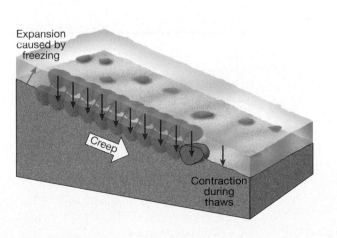

Expansion caused by freezing

Creep

Contraction during thaws

◀ **FIGURE 4.31** The repeated expansion and contraction of the surface material causes a net downslope migration of rock particles—a process called *creep*.

Chapter Summary

- External processes include (1) *weathering*—the disintegration and decomposition of rock at or near the surface, (2) *mass wasting*—the transfer of rock material downslope under the influence of gravity, and (3) *erosion*—the incorporation and transportation of material by a mobile agent, usually water, wind, or ice. They are called *external processes* because they occur at or near Earth's surface and are powered by energy from the Sun. By contrast, *internal processes*, such as volcanism and mountain building, derive their energy from Earth's interior.

- *Mechanical weathering* is the physical breaking up of rock into smaller pieces. *Chemical weathering* alters a rock's chemistry, changing it into different substances. Rocks can be broken into smaller fragments by *frost wedging, unloading*, and *biological activity*. Water is by far the most important agent of chemical weathering. Oxygen in water can *oxidize* some materials, while carbon dioxide (CO_2) dissolved in water forms *carbonic acid*. The chemical weathering of silicate minerals frequently produces (1) soluble products containing sodium, calcium, potassium, and magnesium, (2) insoluble iron oxides, and (3) clay minerals.

- The rate at which rock weathers depends on such factors as (1) *particle size*—small pieces generally weather faster than large pieces; (2) *mineral makeup*—calcite readily dissolves in mildly acidic solutions, and silicate minerals that form first from magma are least resistant to chemical weathering; and (3) *climatic factors*, particularly temperature and moisture. Frequently, rocks exposed at Earth's surface do not weather at the same rate. This *differential weathering* of rocks is influenced by such factors as mineral makeup and degree of jointing.

- *Soil* is a combination of mineral and organic matter, water, and air—that portion of the *regolith* (the layer of rock and mineral fragments produced by weathering) that supports the growth of plants. *Soil texture* refers to the proportions of different particle sizes (clay, silt, and sand) found in soil. The most important *factors that control soil formation are parent material, time, climate, plants and animals, and topography*.

- Soil-forming processes operate from the surface downward and produce zones or layers in the soil called *horizons*. From the surface downward the horizons are designated as O, A, E, B, and C, respectively.

- In the United States, soils are classified using a system known as the *Soil Taxonomy*. It is based on physical and chemical properties of the soil profile and includes six hierarchical categories. The system is especially useful for agricultural and related land-use purposes.

- Soil erosion is a natural process; it is part of the constant recycling of Earth materials that we call the *rock cycle. Rates of soil erosion vary* from one place to another and depend on the soil's characteristics as well as such factors as climate, slope, and type of vegetation. Human activities have greatly accelerated the rate of soil erosion in many areas.

- Weathering creates mineral deposits by concentrating metals into economically valuable deposits. The process, called *secondary enrichment*, is accomplished by either (1) removing undesirable materials and leaving the desired elements enriched in the upper zones of the soil or (2) removing and carrying the desirable elements to lower soil zones where they are redeposited and thus become more concentrated. *Bauxite*, the principal ore of aluminum, is one important ore created by secondary enrichment.

- In the evolution of most landforms, mass wasting is the step that follows weathering. The combined effect of mass wasting and erosion by running water produce stream valleys. *Gravity is the controlling force of mass wasting*. Other factors that influence or trigger downslope movements are saturation of the material with water, oversteepening of slopes beyond the *angle of repose*, removal of anchoring vegetation, and ground vibrations from earthquakes.

- The various processes included under the name of mass wasting are classified and described on the basis of (1) the type of material involved (debris, mud, earth, or rock), (2) the kind of motion (fall, slide, or flow), and (3) the rate of movement (fast, slow). The various kinds of mass wasting include the more rapid forms called *slump, rockslide, debris flow*, and *earthflow*, as well as the slow movements referred to as *creep* and *solifluction*.

Key Terms

angle of repose (p. 104)
chemical weathering (p. 85)
creep (p. 11)
debris flow (p. 107)
differential weathering (p. 90)
earthflow (p. 109)
eluviation (p. 95)
erosion (p. 84)
exfoliation dome (p. 86)
external processes (p. 84)
fall (p. 105)
flow (p. 105)

frost wedging (p. 85)
horizon (p. 95)
internal processes (p. 84)
lahar (p. 109)
leaching (p. 95
mass wasting (p. 84)
mechanical weathering (p. 85)
parent material (p. 93)
permafrost (p. 111)
regolith (p. 91)
rockslide (p. 107)
secondary enrichment (p. 101)

sheeting (p. 85)
slide (p. 105)
slump (p. 106)
soil (p. 91)
soil profile (p. 95)
Soil Taxonomy (p. 96)
soil texture (p. 92)
solifluction (p. 111)
solum (p. 95)
spheroidal weathering (p. 88)
talus slope (p. 85)
weathering (p. 84)

Review Questions

1. Describe the role of external processes in the rock cycle.

2. If two identical rocks were weathered—one mechanically and the other chemically—how would the products of weathering for the two rocks differ?

3. How does mechanical weathering add to the effectiveness of chemical weathering?

4. Describe the formation of an exfoliation dome. Give an example of such a feature.

5. Granite and basalt are exposed at the surface in a hot, wet region.
   (a) Which type of weathering (mechanical or chemical) will predominate?
   (b) Which of the rocks will weather most rapidly? Why?

6. Heat speeds up a chemical reaction. Why then does chemical weathering proceed slowly in a hot desert?

7. How is carbonic acid (H_2CO_3) formed in nature? What results when this acid reacts with potassium feldspar?

8. Using the soil-texture diagram (Figure 4.13), name the soil that consists of 60 percent sand, 30 percent silt, and 10 percent clay.

9. What factors might cause different soils to develop from the same parent material, or similar soils to form from different parent materials?

10. Which of the controls of soil formation is most important? Explain.

11. How can slope affect the development of soil? What is meant by the term *slope orientation?*

12. List the characteristics associated with each of the horizons in a well-developed soil profile. Which of the horizons constitute the solum? Under what circumstances do soils lack horizons?

13. Is soil erosion a natural process or primarily the result of inappropriate land use by people? Explain.

14. List three detrimental effects of soil erosion other than the loss of topsoil from croplands.

15. Name the primary ore of aluminum and describe its formation.

16. Describe how mass-wasting processes contribute to the development of stream valleys.

17. What is the controlling force of mass wasting? What other factors can influence or trigger mass wasting?

18. List the criteria that are used to classify mass-wasting processes.

19. What factors led to the massive rockslide at Gros Ventre, Wyoming?

20. What type of mass-wasting event killed thousands of people in the vicinity of the Colombian volcano Nevado del Ruiz in 1985?

21. Describe the mechanisms or factors that contribute to the slow downslope movement called creep.

22. During what season does solifluction occur in permafrost regions?

Examining the Earth System

1. Which of Earth's spheres are associated with the production of carbonic acid?

2. The level of carbon dioxide (CO_2) in the atmosphere has been increasing for more than a century. Should this increase tend to accelerate or slow down the rate of chemical weathering of Earth's surface rocks? Explain how you arrived at your conclusion.

3. Many gases are emitted into the atmosphere as a result of human activities. List some gases other than carbon dioxide that create acids and therefore accelerate the rate of chemical weathering. Do these acids have other effects on the Earth system? You will find the information about acid rain at this Web site helpful: **http://epa.gov/airmarkets/acidrain/**

4. Discuss the interaction of the atmosphere, geosphere, biosphere, and hydrosphere in the formation of soil.

Online Study Guide

The *Earth Science* Website uses the resources and flexibility of the Internet to aid in your study of the topics in this chapter. Written and developed by Earth science instructors, this site will help improve your understanding of Earth science. Visit **http://www.prenhall.com/tarbuck** and click on the cover of *Earth Science* 11e to find:

- **On-line review quizzes.**
- **Critical thinking exercises.**
- **Links to chapter-specific Web resources.**
- **Internet-wide key term searches.**
http://www.prenhall.com/tarbuck

CHAPTER 5

Running Water and Groundwater

Muddy flow of the Little Colorado River in semiarid northern Arizona shortly after a heavy rain.
(Photo by Tom Brownold Photography)

All the rivers run into the sea; yet the sea is not full; unto the place from whence the rivers come, thither they return again. (Ecclesiastes 1:7)

As the perceptive writer of Ecclesiastes indicated, water is continually on the move, from the ocean to the land and back again in an endless cycle. This chapter deals with that part of the hydrologic cycle that returns water to the sea. Some water travels quickly via a rushing stream, and some moves more slowly below the surface. When viewed as part of the Earth system, streams and groundwater represent basic links in the constant cycling of the planet's water. In Chapter 5, we will examine the factors that influence the distribution and movement of water, as well as look at how water sculptures the landscape. To a great extent, the Grand Canyon, Niagara Falls, Old Faithful, and Mammoth Cave all owe their existence to the action of water on its way to the sea.

Earth as a System: The Hydrologic Cycle

Sculpturing Earth's Surface
▼ Hydrologic Cycle

Water is just about everywhere on Earth—in the oceans, glaciers, rivers, lakes, the air, soil, and in living tissue. All of these reservoirs constitute Earth's hydrosphere. In all, the water content of the hydrosphere comprises about 1.36 billion cubic kilometers (326 million cubic miles). The vast bulk of it, about 97.2 percent, is stored in the global oceans (Figure 5.1). Ice sheets and glaciers account for another 2.15 percent, leaving only 0.65 percent to be divided among lakes, streams, groundwater, and the atmosphere (Figure 5.1). Although the percentages of Earth's total water found in each of the latter sources is but a small fraction of the total inventory, the absolute quantities are great.

The water found in each of the reservoirs depicted in Figure 5.1 does not remain in these places indefinitely. Water can readily change from one state of matter (solid, liquid, or gas) to another at the temperatures and pressures occurring at Earth's surface. Therefore, water is constantly moving among the oceans, the atmosphere, the solid Earth, and the biosphere. This unending circulation of Earth's water supply is called the **hydrologic cycle**. The cycle shows us many critical interrelationships among different parts of the Earth system.

The hydrologic cycle is a gigantic worldwide system powered by energy from the Sun in which the atmosphere provides the vital link between the oceans and continents (Figure 5.2). Water evaporates into the atmosphere from the ocean and to a much lesser extent from the continents. Winds transport this moisture-laden air, often great distances, until conditions cause

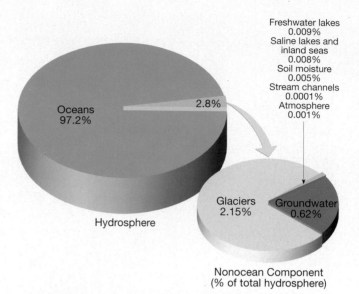

▲ **FIGURE 5.1** Distribution of Earth's water.

the moisture to condense into clouds and to precipitate and fall. The precipitation that falls into the ocean has completed its cycle and is ready to begin another. The water that falls on the continents, however, must make its way back to the ocean.

What happens to precipitation once it has fallen on land? A portion of the water soaks into the ground (called **infiltration**), slowly moving downward, then laterally, finally seeping into lakes, streams, or directly into the ocean. When the rate of rainfall exceeds Earth's ability to absorb it, the surplus water flows over the surface into lakes and streams, a process called **runoff.** Much of the water that infiltrates or runs off eventually returns to the atmosphere because of evaporation from the soil, lakes, and streams. Also, some of the water that infiltrates the ground surface is absorbed by plants, which then release it into the atmosphere. This process is called **transpiration** (*trans* = across, *spiro* = to breathe).

Each year a field of crops may transpire the equivalent of a water layer 60 centimeters (2 feet) deep over the entire field. The same area of trees may pump twice this amount into the atmosphere. Because we cannot clearly distinguish between the amount of water that is evaporated and the amount that is transpired by plants, the term **evapotranspiration** is often used for the combined effect.

When precipitation falls in very cold areas—at high elevations or high latitudes—the water may not immediately soak in, run off, or evaporate. Instead, it may become part of a snowfield or a glacier. In this way, glaciers store large quantities of water on land. If present-day glaciers were to melt and release all their water, sea level would rise by several dozen meters. This would submerge many heavily populated coastal areas. As you will see in Chapter 6, over the past 2 million years, huge ice sheets have formed and melted on several occasions, each time changing the balance of the hydrologic cycle.

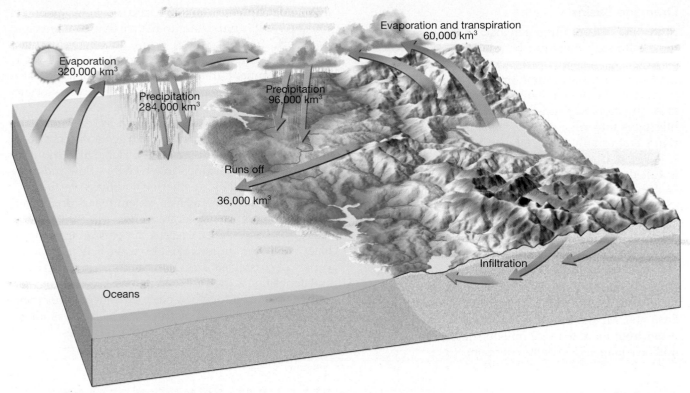

▲ **FIGURE 5.2** Earth's water balance. About 320,000 cubic kilometers of water evaporate each year from the oceans, while evaporation from the land (including lakes and streams) contributes 60,000 cubic kilometers of water. Of this total of 380,000 cubic kilometers of water, about 284,000 cubic kilometers fall back to the ocean, and the remaining 96,000 cubic kilometers fall on Earth's land surface. Since 60,000 cubic kilometers of water evaporate from the land, 36,000 cubic kilometers of water remain to erode the land during the journey back to the ocean.

Figure 5.2 also shows Earth's overall *water balance,* or the volume of water that passes through each part of the cycle annually. The amount of water vapor in the air at any one time is just a tiny fraction of Earth's total water supply. But the *absolute* quantities that are cycled through the atmosphere over a one-year period are immense—some 380,000 cubic kilometers— enough to cover Earth's entire surface to a depth of about 1 meter (39 inches). Estimates show that over North America almost six times more water is carried by moving currents of air than is transported by all the continents' rivers.

It is important to know that the hydrologic cycle is *balanced.* Because the total amount of water vapor in the atmosphere remains about the same, the average annual precipitation over Earth must be equal to the quantity of water evaporated. However, for all of the continents taken together, precipitation exceeds evaporation. Conversely, over the oceans, evaporation exceeds precipitation. Because the level of the world ocean is not dropping, the system must be in balance. In Figure 5.2, the 36,000 cubic kilometers of water that annually runs off from the land to the ocean causes enormous erosion. In fact, this immense volume of moving water is the *single most important agent sculpturing Earth's land surface.*

To summarize, the hydrologic cycle is the continuous movement of water from the oceans to the atmosphere, from the atmosphere to the land, and from the land back to the sea. The land-back-to-the-sea step is the primary action that wears down Earth's land surface. In this chapter, we will first observe the work of water running over the surface, including floods, erosion, and the formation of valleys. Then we will look underground at the slow labors of groundwater as it forms springs and caverns and provides drinking water on its long migration to the sea.

Running Water

 GEODe Sculpturing Earth's Surface
▼ Running Water

Running water is of great importance to people. We depend on rivers for energy, transportation, and irrigation. Their fertile floodplains have been favored sites for agriculture and industry since the dawn of civilization. Furthermore, as the dominant agent of erosion, running water has shaped much of our physical environment.

Although we have always depended on running water, its source eluded us for centuries. Not until the 1500s did we realize that streams were supplied by surface runoff and the migration of underground water.

Drainage Basins

All water in rivers ultimately had its source as rain and snow, although there can be considerable lag time before this water actually enters a river system. Upon reaching the surface, some precipitation soaks into the ground, while the remaining water flows over the surface. This *overland flow* is routed across the landscape by hillslopes into rills, gullies, streams, and finally large rivers that usually flow to the ocean.

The land area that contributes water to a river system is called a **drainage basin** (Figure 5.3). The drainage basin of one stream is separated from the drainage basin of another by an imaginary line called a **divide** (Figure 5.3). Divides range in scale from a ridge separating two small gullies on a hillside to a *continental divide* which splits whole continents into enormous drainage basins. The Mississippi River has the largest drainage basin in North America. Extending between the Rocky Mountains in the West and the Appalachian Mountains in the East, the Mississippi River and its tributaries collect water from more than 3.2 million square kilometers (1.2 million square miles) of the continent.

River Systems

Rivers and streams can be simply defined as water flowing in a channel. They have three important roles in the formation of a landscape: They erode the channels in which they flow, they transport sediments provided by weathering and slope processes, and they produce a wide variety of erosional and depositional landforms. In fact, in most areas, including many arid regions, river systems have shaped the varied landscape that we humans inhabit.

A river system consists of three main parts in which different processes dominate: a zone of erosion, a zone of sediment transport, and a zone of sediment deposition. It is important to realize that some erosion, transport, and deposition occur in all three zones; however, within each zone, one of these processes is usually dominant. In addition, the parts of a river system are interdependent, so that the processes occurring in one part influence the others.

In large river systems erosion is the dominant process in the upstream area, which generally consists of mountainous or hilly topography. Here small tributary streams erode the channels in which they flow and carry material provided by weathering and mass wasting.

The region within a river system that is dominated by deposition is usually located where the stream enters a large body of water. Here sediments accumulate to form a delta, or are reworked by wave action to form a variety of coastal features. Between the zones of erosion and deposition is the *trunk stream* that serves to transport sediments. Taken together, erosion, transportation, and deposition are the processes by which rivers move Earth's surface materials and sculpt landscapes.

STUDENTS SOMETIMES ASK...
What's the difference between a stream and a river?

In common usage, these terms imply relative size (a river is larger than a stream, both of which are larger than a creek or a brook). However, in geology this is not the case: The word stream is used to denote channelized flow of any size, from a small creek to the mightiest river. It is important to note that although the terms river and stream are sometimes used interchangeably, the term river is often preferred when describing a main stream into which several tributaries flow.

▶ **FIGURE 5.3** A drainage basin is the land area drained by a stream and its tributaries. The drainage basin of the Mississippi River, North America's largest river, covers about 3 million square kilometers. Divides are the boundaries that separate drainage basins from each other. Drainage basins and divides exist for all streams regardless of size.

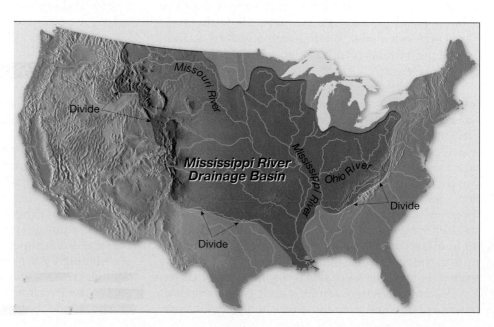

Streamflow

Water may flow in one of two ways, either as **laminar flow** or **turbulent flow**. In very slow moving streams the flow is often laminar and the water particles move in roughly straight-line paths that parallel the stream channel. However, streamflow is usually turbulent, with the water moving in an erratic fashion that can be characterized as a swirling motion. Strong turbulent flow may exhibit whirlpools and eddies, as well as roiling whitewater rapids. Even streams that appear smooth on the surface often exhibit turbulent flow near the bottom and sides of the channel. Turbulence contributes to the stream's ability to erode its channel because it acts to lift sediment from the streambed.

Water makes its way to the sea under the influence of gravity. The time required for the journey depends on the velocity of the stream. Velocity is the distance that water travels in a unit of time. Some sluggish streams flow at less than 1 kilometer per hour, whereas a few rapid ones may exceed 30 kilometers per hour. Velocities are measured at gaging stations (Figure 5.4A). Along straight stretches, the highest velocities are near the center of the channel just below the surface, where friction is lowest (Figure 5.4B). But when a stream curves, its zone of maximum speed shifts toward its outer bank (Figure 5.4C).

The ability of a stream to erode and transport materials depends on its velocity. Even slight changes in velocity can lead to significant changes in the load of sediment that water can transport. Several factors determine the velocity of a stream, including (1) gradient; (2) shape, size, and roughness of the channel; and (3) discharge.

Gradient and Channel Characteristics

The slope of a stream channel expressed as the vertical drop of a stream over a specified distance is **gradient**. Portions of the lower Mississippi River, for example, have very low gradients of 10 centimeters per kilometer or less. By contrast, some mountain stream channels decrease in elevation at a rate of more than 40 meters per kilometer, or a gradient 400 times steeper than the lower Mississippi (Figure 5.5). Gradient varies not only among different streams but also over a particular stream's length. The steeper the gradient, the more energy available for streamflow. If two streams were identical in every respect except gradient, the stream with the higher gradient would obviously have the greater velocity.

A stream's channel is a conduit that guides the flow of water, but the water encounters friction as it flows. The shape, size, and roughness of the channel affect the amount of friction. Larger channels have more efficient flow because a smaller proportion of

A. Gaging station

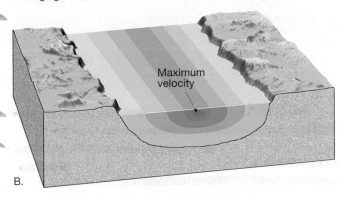

B.

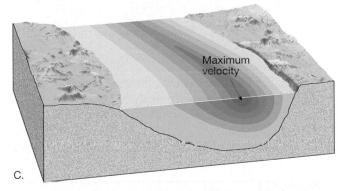

C.

▲ **FIGURE 5.4   A.** Continuous records of stage and discharge are collected by the U.S. Geological Survey at more than 7000 gaging stations in the United States. Average velocities are determined by using measurements from several spots across the stream. This station is on the Rio Grande south of Taos, New Mexico. (Photo by E. J. Tarbuck) **B.** Along straight stretches, stream velocity is highest at the center of the channel. **C.** When a stream curves, its zone of maximum speed shifts toward the outer bank.

water is in contact with the channel. A smooth channel promotes a more uniform flow, whereas an irregular channel filled with boulders creates enough turbulence to slow the stream significantly.

▲ FIGURE 5.5 Rapids are common in mountain streams where the gradient is steep and the channel is rough and irregular. (Photo by Bruce Gaylord/Visuals Unlimited)

Discharge

The **discharge** of a stream is the volume of water flowing past a certain point in a given unit of time. This is usually measured in cubic meters per second or cubic feet per second. Discharge is determined by multiplying a stream's cross-sectional area by its velocity:

$$\text{discharge } (\text{m}^3/\text{second}) = \text{channel width (meters)}$$
$$\times \text{ channel depth (meters)}$$
$$\times \text{ velocity (meters/second)}$$

Table 5.1 lists the world's largest rivers in terms of discharge. The largest river in North America, the Mississippi, discharges an average of 17,300 cubic meters (611,000 cubic feet) per second. Although this is a huge quantity of water, it is nevertheless dwarfed

by the mighty Amazon in South America, the world's largest river. Fed by a vast rainy region that is nearly three-fourths the size of the conterminous United States, the Amazon discharges 12 times more water than the Mississippi.

The discharges of most rivers are far from constant. This is true because of such variables as rainfall and snowmelt. In areas with seasonal variations in precipitation, streamflow will tend to be highest during the wet season, or during spring snowmelt, and lowest during the dry season or during periods when high temperature increases the water losses through evapotranspiration. However, not all channels maintain a continuous flow of water. Streams that exhibit flow only during "wet" periods are referred to as *intermittent streams*. In arid climates many streams carry water only occasionally after a heavy rainstorm and are called *ephemeral streams*.

Changes from Upstream to Downstream

One useful way of studying a stream is to examine its *profile*. A profile is simply a cross-sectional view of a stream from its source area (called the *head* or *headwaters*) to its *mouth*, the point downstream where the river empties into another water body. By examining Figure 5.6, you can see that the most obvious feature of a typical profile is a constantly decreasing gradient from the head to the mouth. Although many local irregularities may exist, the overall profile is a smooth concave upward curve.

The profile shows that the gradient decreases downstream. To see how other factors change in a downstream direction, observations and measurements must be made. When data are collected from several gaging stations along a river, they show that in a humid region discharge increases from the head toward the mouth. This should come as no surprise because, as we move downstream, more and more

			Drainage Area		Average Discharge	
Rank	River	Country	Square kilometers	Square miles	Cubic meters	Cubic feet
1	Amazon	Brazil	5,778,000	2,231,000	212,400	7,500,000
2	Congo	Zaire	4,014,500	1,550,000	39,650	1,400,000
3	Yangtze	China	1,942,500	750,000	21,800	770,000
4	Brahmaputra	Bangladesh	935,000	361,000	19,800	700,000
5	Ganges	India	1,059,300	409,000	18,700	660,000
6	Yenisei	Russia	2,590,000	1,000,000	17,400	614,000
7	Mississippi	United States	3,222,000	1,244,000	17,300	611,000
8	Orinoco	Venezuela	880,600	340,000	17,000	600,000
9	Lena	Russia	2,424,000	936,000	15,500	547,000
10	Parana	Argentina	2,305,000	890,000	14,900	526,000

TABLE 5.1 World's largest rivers ranked by discharge

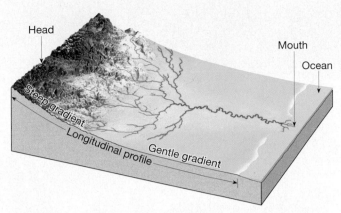

Head

Steep gradient

Longitudinal profile

Gentle gradient

Mouth

Ocean

▲ **FIGURE 5.6**   A longitudinal profile is a cross-section along the length of a stream. Note the concave-upward curve of the profile, with a steeper gradient upstream and a gentler gradient downstream. Moving downstream from the head, the discharge of most streams increases because tributaries and groundwater contribute water to the main channel.

tributaries contribute water to the main channel (Figure 5.6). Furthermore, in most humid regions, additional water is added from the groundwater supply. Thus, as you move downstream, the stream's width, depth, and velocity change in response to the increased volume of water carried by the stream.

Streams that begin in mountainous areas where precipitation is abundant and then flow through arid regions may experience the opposite situation. Here discharge may actually decrease downstream because of water loss due to evaporation, infiltration into the streambed, and removal by irrigation. The Colorado River in the southwestern United States is such an example.

The Work of Running Water

Streams are Earth's most important erosional agent. Not only do they have the ability to downcut and widen their channels but streams also have the capacity to transport the enormous quantities of sediment that are delivered to the stream by sheet flow, mass wasting, and groundwater. Eventually much of this material is dropped by the water to create a variety of depositional features.

Erosion

A stream's ability to accumulate and transport soil and weathered rock is aided by the work of raindrops, which knock sediment particles loose (see Figure 4.19 on p. 100). When the ground is saturated, rainwater begins to flow downslope, transporting some of the material it has dislodged. On barren slopes the flow of muddy water, called *sheet flow,* will often erode small channels, or *rills,* which in time may evolve into larger *gullies* (see Figure 4.20 on p. 101).

Once the surface flow reaches a stream, its ability to erode is greatly enhanced by the increase in water volume. When the flow of water is sufficiently strong, it can dislodge particles from the channel and lift them into the moving water. In this manner, the force of running water swiftly erodes poorly consolidated materials on the bed and sides of a stream channel. On occasion, the banks of the channel may be undercut, dumping even more loose debris into the water to be carried downstream.

In addition to eroding unconsolidated materials, the hydraulic force of streamflow can also cut a channel into solid bedrock. A stream's ability to erode bedrock is greatly enhanced by the particles it carries. These particles can be any size, from large boulders in very fast-flowing waters to sand and gravel-size particles in somewhat slower flow. Just as the particles of grit on sandpaper can wear away a piece of wood, so too can the sand and gravel carried by a stream abrade a bedrock channel. Moreover, pebbles caught in swirling eddies can act like "drills" and bore circular *potholes* into the channel floor.

Transportation

All streams regardless of size transport some rock material. Streams also sort the solid sediment they transport because finer, lighter material is carried more rapidly than coarser, heavier rock debris. Depending on the nature of the rock material, the stream load consists of material (1) in solution (**dissolved load**), (2) in suspension (**suspended load**), and (3) sliding or rolling along the bottom (**bed load**).

Dissolved Load. Most of the *dissolved load* is brought to a stream by groundwater and is dispersed throughout the flow. The quantity of material carried in solution is highly variable and is most abundant in humid areas where limestone and other soluble rock forms the bedrock. Usually the amount of dissolved load is small and therefore is expressed as parts of dissolved material per million parts of water (parts per million, or ppm). Although some rivers may have a dissolved load of 1000 ppm or more, the average figure for the world's rivers is estimated at 115 to 120 ppm.

Suspended Load. Most large rivers carry the largest part of their load in *suspension.* Indeed, the visible cloud of sediment suspended in the water is the most obvious portion of a stream's load. Usually only fine particles consisting of silt and clay can be carried this way, but during a flood, sand and even gravel-size particles are transported as well. Also, during a flood, the total quantity of material carried in suspension increases dramatically, as can be verified by anyone whose home has been a site for the deposition of this material.

► **FIGURE 5.7** The suspended load is clearly visible because it gives this flooding river a "muddy" appearance. During floods both capacity and competency increase. Therefore, the greatest erosion and sediment transport occur during these high-water periods. The flooding Ohio River at Cincinnati, Ohio, March 1997. This regional flood caused more than 50 deaths and approximately $500 million in damages. (Photo by Mark Lyons/Liaison Agency, Inc.)

Bed Load. A portion of a stream's load consists of sand, gravel, and occasionally large boulders. These coarser particles, which are too large to be carried in suspension, move along the bottom (bed) of the stream channel and constitute the *bed load*. Unlike the suspended and dissolved loads, which are constantly in motion, the bed load is in motion only intermittently, when the force of the water is sufficient to move the larger particles. The smaller particles, mainly sand and gravel, are moved along the stream by *saltation*, which resembles a series of jumps or skips. Larger particles either roll or slide along the bottom, depending on their shape.

Each year, the Mississippi River transports about 750 million tons of material to the Gulf of Mexico. Of this total, it is estimated that approximately 500 million tons are carried in suspension, 200 million tons in solution, and the remaining 50 million tons as bed load; however, such proportions vary widely from one stream to another.

Competence and Capacity. Streams vary in their ability to carry a load. Their ability is determined by two criteria. First, the **competence** of a stream measures the maximum size of particles it is capable of transporting. The stream's velocity determines its competence. If the velocity of a stream doubles, its competence increases four times; if the velocity triples, its competence increases nine times; and so forth. This explains how large boulders that seem immovable can be transported during a flood, which greatly increases a stream's velocity.

Second, the **capacity** of a stream is the maximum load it can carry. The capacity of a stream is directly related to its discharge. The greater the volume of water flowing in a stream, the greater is its capacity for hauling sediment.

By now it should be clear why the greatest erosion and transportation of sediment occur during floods (Figure 5.7). The increase in discharge results in a greater capacity, and the increase in velocity results in greater competence. With rising velocity the water becomes more turbulent, and larger and larger particles are set in motion. In just a few days or perhaps a few hours a stream in flood stage can erode and transport more sediment than it does during months of normal flow.

Deposition

Whenever a stream slows down, the situation reverses. As its velocity decreases, its competence is reduced and sediment begins to drop out, largest particles first. Each particle size has a *critical settling velocity*. As streamflow drops below the critical setting velocity of a certain particle size, sediment in that category begins to settle out. Thus, stream transport provides a mechanism by which solid particles of various sizes are separated. This process, called **sorting,** explains why particles of similar size are deposited together.

The material deposited by a stream is called **alluvium,** the general term for any stream-deposited sediment. Many different depositional features are composed of alluvium. Some occur within stream channels, some occur on the valley floor adjacent to the channel, and some exist at the mouth of the stream. We will consider the nature of these features later.

Stream Channels

A basic characteristic of streamflow that distinguishes it from overland flow is that it is usually confined to a channel. A stream channel can be thought of as an open conduit that consists of the streambed and banks that act to confine the flow except during floods.

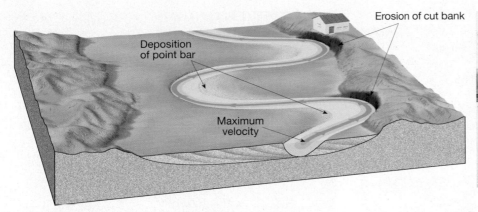

▲ **FIGURE 5.8** When a stream meanders, its zone of maximum speed shifts toward the outer bank. Because the outside of a meander is a zone of active erosion, it is called the *cut bank*. The house in the photo is located next to a cut bank. A point bar is deposited when water on the inside of a meander slows.

Although somewhat oversimplified, we can divide stream channels into two types. *Bedrock channels* are those in which the streams are actively cutting into solid rock. In contrast, when the bed and banks are composed mainly of unconsolidated sediment, the channel is called an *alluvial channel*.

Bedrock Channels

In their headwaters, where the gradient is steep, most rivers cut into bedrock. These mountain streams typically transport coarse particles that actively abrade the bedrock channel. Potholes are often visible evidence of the erosional forces at work.

Bedrock channels typically alternate between relatively gently sloping segments where alluvium tends to accumulate, and steeper segments where bedrock is exposed. These steeper areas may contain rapids or occasionally a waterfall. The channel pattern exhibited by streams cutting into bedrock is controlled by the underlying geologic structure. Even when flowing over rather uniform bedrock, streams tend to exhibit a winding or irregular pattern rather than flowing in a straight channel. Anyone who has gone on a white-water rafting trip has observed the steep, winding nature of a stream flowing in a bedrock channel.

Alluvial Channels

Many stream channels are composed of loosely consolidated sediment (alluvium) and therefore can undergo major changes in shape because the sediments are continually being eroded, transported, and redeposited. The major factors affecting the shapes of these channels is the average size of the sediment being transported, the channel gradient, and the discharge.

Alluvial channel patterns reflect a stream's ability to transport its load at a uniform rate, while expending the least amount of energy. Thus, the size and type of sediment being carried help determine the nature of the stream channel. Two common types of alluvial channels are *meandering channels* and *braided channels*.

Meandering Streams. Streams that transport much of their load in suspension generally move in sweeping bends called **meanders**. These streams flow in relatively deep, smooth channels and transport mainly mud (silt and clay). The lower Mississippi River exhibits a channel of this type.

Because of the cohesiveness of consolidated mud, the banks of stream channels carrying fine particles tend to resist erosion. As a consequence, most of the erosion in such channels occurs on the outside of the meander, where velocity and turbulence are greatest. In time, the outside bank is undermined, especially during periods of high water. Because the outside of a meander is a zone of active erosion, it is often referred to as the *cut bank* (Figure 5.8). The debris acquired by the stream at the cut bank moves downstream with the coarser material generally being deposited as *point bars* in zones of decreased velocity on the insides of meanders. In this manner, meanders migrate laterally by eroding the outside of the bends and depositing on the inside.

In addition to migrating laterally, the bends in a channel also migrate down the valley. This occurs because erosion is more effective on the downstream (downslope) side of the meander. Sometimes the downstream migration of a meander is slowed when it reaches a more resistant material. This allows the next meander upstream to "catch up" and overtake it as shown in Figure 5.9. Gradually the neck of land between the meanders is narrowed. Eventually the river may erode through the narrow neck of land to the next loop (Figure 5.9). The new, shorter channel segment is called a **cutoff** and, because of its shape, the abandoned bend is called an **oxbow lake** (Figure 5.10).

Braided Streams. Some streams consist of a complex network of converging and diverging channels that thread their way among numerous islands or gravel

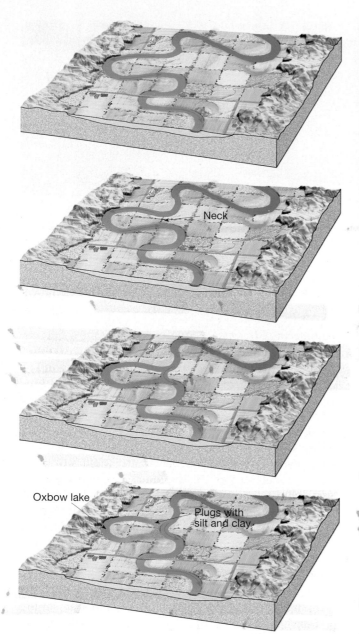

▲ **FIGURE 5.9**   Formation of a cutoff and oxbow lake.

▲ **FIGURE 5.10**   Oxbow lakes occupy abandoned meanders. As they fill with sediment, oxbow lakes gradually become swampy meander scars. Aerial view of an oxbow lake created by the meandering Green River near Bronx, Wyoming. (Photo by Michael Collier)

bars (Figure 5.11). Because these channels have an interwoven appearance, these streams are said to be **braided**. Braided channels form where a large proportion of the stream's load consists of coarse material (sand and gravel) and the stream has a highly variable discharge. Because the bank material is readily erodable, braided channels are wide and shallow.

One circumstance in which braided streams form is at the end of a glacier where there is a large seasonal variation in discharge. Here, large amounts of ice-eroded sediment are dumped into the meltwater streams flowing away from the glacier. When flow is sluggish, the stream is unable to move all of the sediment and therefore deposits the coarsest material as bars that force the flow to split and follow several

paths. Usually the laterally shifting channels completely rework most of the surface sediments each year, thereby transforming the entire streambed. In some braided streams, however, the bars have built up to form islands that are anchored by vegetation.

In summary, meandering channels develop where the load consists largely of fine-grained particles that are transported as suspended load in a deep, smooth channel. By contrast, wide, shallow braided channels develop where coarse-grained alluvium is transported as bedload.

Base Level and Stream Erosion

Streams cannot endlessly erode their channels deeper and deeper. There is a lower limit to how deep a stream can erode, and that limit is called **base level**. Most often a stream's base level occurs where a stream enters the ocean, a lake, or another stream.

Two general types of base level are recognized. Sea level is considered the *ultimate base level*, because it is the lowest level to which stream erosion could lower the land. *Temporary*, or *local, base levels* include lakes, resistant layers of rock, and main streams that act as base level for their tributaries. For example, when a stream enters a lake, its velocity quickly approaches zero and its ability to erode ceases. Thus, the lake prevents the stream from eroding below its level at any

▲ **FIGURE 5.11** Braided stream choked with sediment near the terminus of a melting glacier. (Photo by Bradford Washburn)

point upstream from the lake. However, because the outlet of the lake can cut downward and drain the lake, the lake is only a temporary hindrance to the stream's ability to downcut its channel. In a similar manner, the layer of resistant rock at the lip of the waterfall in Figure 5.12 acts as a temporary base level. Until the ledge of hard rock is eliminated, it will limit the amount of downcutting upstream.

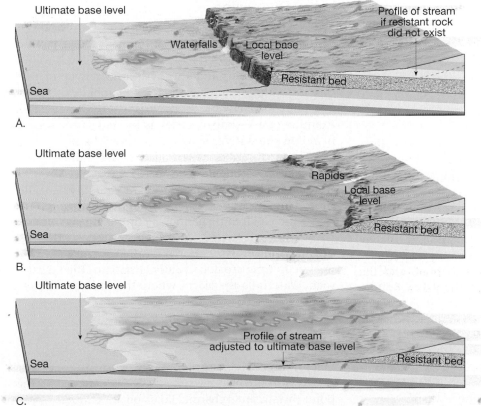

◄ **FIGURE 5.12** A resistant layer of rock can act as a local (temporary) base level. Because the durable layer is eroded more slowly, it limits the amount of downcutting upstream.

▶ **FIGURE 5.13** When a dam is built and a reservoir forms, the stream's base level is raised. This reduces the stream's velocity and leads to deposition and a reduction of the gradient upstream from the reservoir.

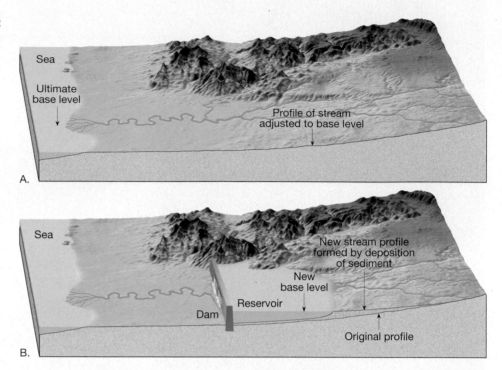

Any change in base level will cause a corresponding readjustment of stream activities. When a dam is built along a stream, the reservoir that forms behind it raises the base level of the stream (Figure 5.13). Upstream from the dam the gradient is reduced, lowering the stream's velocity and, hence, its sediment-transporting ability. The stream, now having too little energy to transport its entire load, will deposit sediment. This builds up its channel. Deposition will be the dominant process until the stream's gradient increases sufficiently to transport its load.

Shaping Stream Valleys

Streams, with the aid of weathering and mass wasting, shape the landscape through which they flow. As a result, streams continuously modify the valleys that they occupy.

A **stream valley** consists not only of the channel but also the surrounding terrain that directly contributes water to the stream. Thus it includes the valley bottom, which is the lower, flatter area that is partially or totally occupied by the stream channel, and the sloping valley walls that rise above the valley bottom on both sides. Most stream valleys are much broader at the top than is the width of their channel at the bottom. This would not be the case if the only agent responsible for eroding valleys were the streams flowing through them. The sides of most valleys are shaped by a combination of weathering, overland flow, and mass wasting. In some arid regions, where weathering is

slow and where rock is particularly resistant, narrow valleys having nearly vertical walls are common.

Stream valleys can be divided into two general types—narrow V-shaped valleys and wide valleys with flat floors—with many gradations between.

Valley Deepening

When a stream's gradient is steep and the channel is well above base level, downcutting is the dominant activity. Abrasion caused by bed load sliding and rolling along the bottom, and the hydraulic power of fast-moving water, slowly lowers the streambed. The result is usually a V-shaped valley with steep sides. A classic example of a V-shaped valley is located in the section of Yellowstone River shown in Figure 5.14.

The most prominent features of a V-shaped valley are *rapids* and *waterfalls*. Both occur where the stream's gradient increases significantly, a situation usually caused by variations in the erodability of the bedrock into which a stream channel is cutting. Resistant beds create rapids by acting as a temporary base level upstream while allowing downcutting to continue downstream. In time erosion usually eliminates the resistant rock. Waterfalls are places where the stream makes an abrupt vertical drop.

Valley Widening

Once a stream has cut its channel closer to base level, downward erosion becomes less dominant. At this point the stream's channel takes on a meandering pat-

▲ **FIGURE 5.14** V-shaped valley of the Yellowstone River. The rapids and waterfalls indicate that the river is vigorously downcutting. (Photo by Art Wolfe)

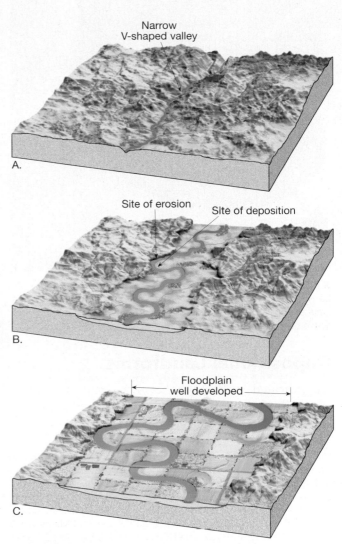

▲ **FIGURE 5.15** Stream eroding its floodplain.

tern, and more of the stream's energy is directed from side to side. The result is a widening of the valley as the river cuts away first at one bank and then at the other (Figure 5.15). The continuous lateral erosion caused by shifting of the stream's meanders produces an increasingly broader, flat valley floor covered with alluvium. This feature, called a **floodplain**, is appropriately named because when a river overflows its banks during flood stage, it inundates the floodplain.

Over time the floodplain will widen to the point that the stream is only actively eroding the valley walls in a few places. In fact, in large rivers such as the lower Mississippi River valley, the distance from one valley wall to another can exceed 100 miles.

Changing Base Level and Incised Meanders

We usually expect a stream with a highly meandering course to be on a floodplain in a wide valley. However, certain rivers exhibit meandering channels that flow in steep, narrow valleys. Such meanders are called

incised (*incisum* = to cut into) **meanders** (Figure 5.16). How do such features form?

Originally, the meanders probably developed on the floodplain of a stream that was relatively near base level. Then, a change in base level caused the stream to begin downcutting. One of two events could have occurred. Either base level dropped or the land upon which the river was flowing was uplifted.

An example of the first circumstance happened during the Ice Age when large quantities of water were withdrawn from the ocean and locked up in glaciers on land. The result was that sea level (ultimate base level) dropped, causing rivers flowing into the ocean to begin to downcut.

Regional uplift of the land, the second cause for incised meanders, is exemplified by the Colorado Plateau in the southwestern United States. Here, as the plateau was gradually uplifted, numerous meandering rivers adjusted to being higher above base level by downcutting (Figure 5.16).

▲ **FIGURE 5.16** Incised meanders of the Colorado River in Canyonlands National Park, Utah. Here, as the Colorado Plateau was gradually uplifted, the meandering river adjusted to being higher above base level by downcutting. (Photos by Michael Collier)

Depositional Landforms

As indicated earlier, whenever a stream's velocity slows, it begins to deposit some of the sediment it is carrying. Also recall that streams continually pick up sediment in one part of their channel and redeposit it downstream. These channel deposits are most often composed of sand and gravel and are commonly referred to as **bars**. Such features, however, are only temporary, for the material will be picked up again and eventually carried to the ocean. In addition to sand and gravel bars, streams also create other depositional features that have a somewhat longer life span. These include deltas, natural levees, and alluvial fans.

Deltas

When a stream enters the relatively still waters of an ocean or lake, its velocity drops abruptly, and the resulting deposits form a **delta** (Figure 5.17). As the delta grows outward, the stream's gradient continually lessens. This circumstance eventually causes the channel to become choked with sediment deposited from the slowing water. As a consequence, the river seeks a shorter, higher-gradient route to base level, as illustrated in Figure 5.17 B. This illustration shows the main channel dividing into several smaller ones, called **distributaries**. Most deltas are characterized by these shifting channels that act in an opposite way to that of tributaries.

▼ **FIGURE 5.17** **A.** Structure of a simple delta that forms in the relatively quiet waters of a lake. **B.** Growth of a simple delta. As a stream extends its channel, the gradient is reduced. Frequently, during flood stage the river is diverted to a higher-gradient route, forming a new distributary. Old, abandoned distributaries are gradually invaded by aquatic vegetation and fill with sediment. (After Ward's Natural Science Establishment, Inc., Rochester, N.Y.) **C.** Satellite image of a portion of the delta of the Mississippi River in May 2001. For the past 500 years or so, the main flow of the river has been along its present course, extending southeast from New Orleans. During that span, the delta advanced into the Gulf of Mexico at a rate of about 10 kilometers (6 miles) per century. Notice the numerous distributaries. (NASA/GSFC/METI/ERSDAC/JAROS, and U.S./Japan ASTER Science Team)

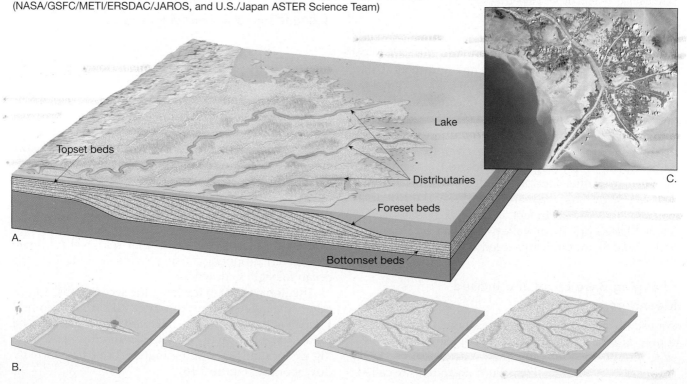

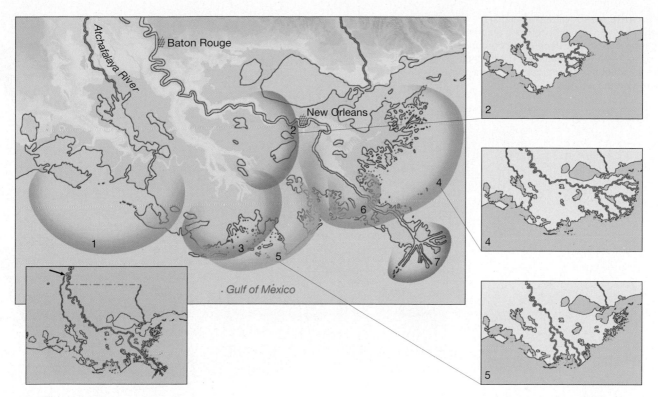

▲ **FIGURE 5.18** During the past 5000 to 6000 years, the Mississippi River has built a series of seven coalescing subdeltas. The numbers indicate the order in which the subdeltas were deposited. The present birdfoot delta (number 7) represents the activity of the past 500 years. Without ongoing human efforts, the present course will shift and follow the path of the Atchafalaya River. The inset on left shows the point where the Mississippi may someday break through (arrow) and the shorter path it would take to the Gulf of Mexico. (After C. R. Kolb and J. R. Van Lopik)

Rather than carrying water into the main channel, distributaries carry water away from the main channel. After numerous shifts of the channel, a delta may grow into a rough triangular shape like the Greek letter delta (Δ), for which it is named. Note, however, that many deltas do not exhibit the idealized shape. Differences in the configurations of shorelines and variations in the nature and strength of wave activity result in many shapes. Many large rivers have deltas extending over thousands of square kilometers. The delta of the Mississippi River is one example. It resulted from the accumulation of huge quantities of sediment derived from the vast region drained by the river and its tributaries. Today, New Orleans rests where there was ocean less than 5000 years ago. Figure 5.18 shows that portion of the Mississippi delta that has been built over the past 5000 to 6000 years. As you can see, the delta is actually a series of seven coalescing subdeltas. Each formed when the river left its existing channel in favor of a shorter, more direct path to the Gulf of Mexico. The individual subdeltas interfinger and partially cover one another to produce a very complex structure. The present subdelta, called a *bird-foot delta* because of the configuration of its distribu-

taries, has been built by the Mississippi in the last 500 years.

Natural Levees

Some rivers occupy valleys with broad floodplains and build **natural levees** that parallel their channels on both banks (Figure 5.19). Natural levees are built by successive floods over many years. When a stream overflows its banks, its velocity immediately diminishes, leaving coarse sediment deposited in strips bordering the channel. As the water spreads out over the valley, a lesser amount of fine sediment is deposited over the valley floor. This uneven distribution of material produces the very gentle slope of the natural levee.

The natural levees of the lower Mississippi rise 6 meters (20 feet) above the floodplain. The area behind the levee is characteristically poorly drained for the obvious reason that water cannot flow up the levee and into the river. Marshes called **backswamps** result. A tributary stream that cannot enter a river because levees block the way often has to flow parallel to the river until it can breach the levee. Such streams are called **yazoo tributaries** after the Yazoo River, which parallels the Mississippi for over 300 kilometers (about 190 miles).

▶ **FIGURE 5.19** Natural levees are gently sloping deposits that are created by repeated floods. Because the ground next to the stream channel is higher than the adjacent floodplain, back swamps and yazoo tributaries may develop.

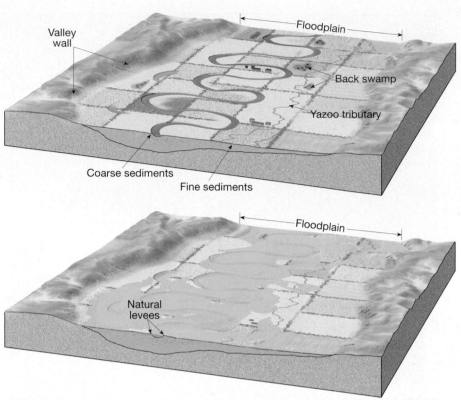

Alluvial Fans

Alluvial fans typically develop where a high-gradient stream leaves a narrow valley in mountainous terrain and comes out suddenly onto a broad, flat plain or valley floor (see Figure 6.28, p. 178). Alluvial fans form in response to the abrupt drop in gradient combined with the change from a narrow channel of a mountain stream to less confined channels at the base of the mountains. The sudden drop in velocity causes the stream to dump its load of sediment quickly in a distinctive cone- or fan-shaped accumulation. As illustrated by Figure 6.28, the surface of the fan slopes outward in a broad arc from an apex at the mouth of the steep valley. Usually, coarse material is dropped near the apex of the fan, while fine material is carried toward the base of the deposit.

Drainage Patterns

GEODe

Sculpturing Earth's Surface
▼ Running Water

Drainage systems are networks of streams that together form distinctive patterns. The nature of a drainage pattern can vary greatly from one type of ter-

rain to another, primarily in response to the kinds of rock on which the streams developed or the structural pattern of faults and folds.

The most commonly encountered drainage pattern is the **dendritic pattern** (Figure 5.20A). This pattern of irregularly branching tributary streams resembles the branching pattern of a deciduous tree. In fact, the word *dendritic* means "treelike." The dendritic pattern forms where the underlying material is relatively uniform. Because the surface material is essentially uniform in its resistance to erosion, it does not control the pattern of streamflow. Rather, the pattern is determined chiefly by the direction of slope of the land.

When streams diverge from a central area like spokes from the hub of a wheel, the pattern is said to be **radial** (Figure 5.20B). This pattern typically develops on isolated volcanic cones and domal uplifts.

Figure 5.20C illustrates a **rectangular** pattern, in which many right-angle bends can be seen. This pattern develops when the bedrock is crisscrossed by a series of joints and/or faults. Because these structures are eroded more easily than unbroken rock, their geometric pattern guides the directions of valleys.

Figure 5.20D illustrates a **trellis** drainage pattern, a rectangular pattern in which tributary streams are nearly parallel to one another and have the appearance of a gar-

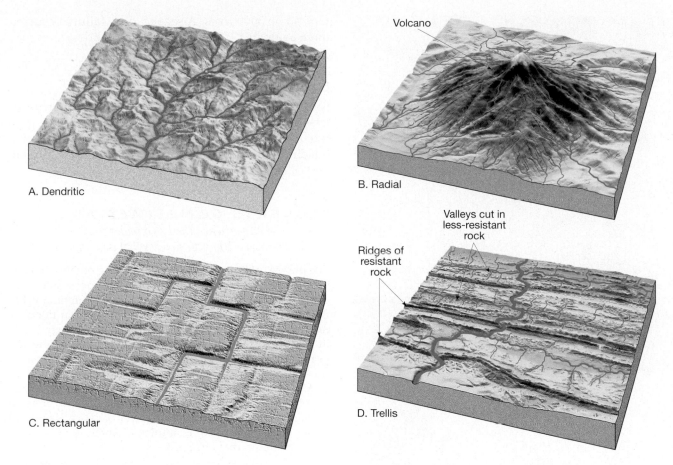

▲ **FIGURE 5.20** Drainage patterns. **A.** Dendritic. **B.** Radial. **C.** Rectangular. **D.** Trellis.

den trellis. This pattern forms in areas underlain by alternating bands of resistant and less-resistant rock.

Floods and Flood Control

When the discharge of a stream becomes so great that it exceeds the capacity of its channel, it overflows its banks as a flood. Floods are the most common and most destructive of all geologic hazards. They are, nevertheless, simply part of the *natural* behavior of streams.

Causes of Floods

Rivers flood because of the weather. Rapid melting of snow in the spring and/or major storms that bring heavy rains over a large region cause most floods. The extensive 1997 flood along the Red River of the North is a recent example of an event triggered by rapid snowmelt. Exceptional rains caused the devastating floods in the upper Mississippi River Valley during the summer of 1993 (Figure 5.21). Heavy rains were also responsible for the Ohio River Flood shown in Figure 5.7 on page 122.

Unlike the extensive regional floods just mentioned, *flash floods* are more limited in extent. Flash floods occur with little warning and can be deadly because they produce a rapid rise in water levels and can have a devastating flow velocity (see Box 5.1). Several factors influence flash flooding. Among them are rainfall intensity and duration, topography, and surface conditions. Mountainous areas are susceptible because steep slopes can quickly funnel runoff into narrow canyons. Urban areas are susceptible to flash floods because a high percentage of the surface area is composed of impervious surfaces such as roofs, streets, and parking lots where runoff is very rapid. In fact, a recent study indicated that the area of impervious surfaces in the United States (excluding Alaska and Hawaii) amounts to more than 112,600 square kilometers (nearly 44,000 square miles), which is slightly less than the area of the state of Ohio.*

Human interference with the stream system can worsen or even cause floods. A prime example is the

---

*C. D. Elvidge, et al. "U.S. Constructed Area Approaches the Size of Ohio," in *EOS, Transactions, American Geophysical Union*, Vol. 85, No. 24, 15 June 2004, p. 233.

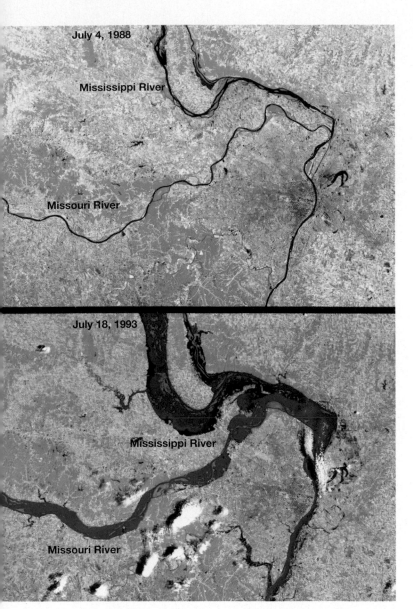

▲ **FIGURE 5.21** Satellite views of the Missouri River flowing into the Mississippi River. St. Louis is just south of their confluence. The upper image shows the rivers during a drought that occurred in summer 1988. The lower image depicts the peak of the record-breaking 1993 flood. Exceptional rains produced the wettest spring and early summer of the twentieth century in the upper Mississippi River basin. In all, nearly 14 million acres were inundated, displacing at least 50,000 people. (Photos courtesy of Spaceimaging.com)

failure of a dam or an artificial levee. These structures are built for flood protection. They are designed to contain floods of a certain magnitude. If a larger flood occurs, the dam or levee is overtopped. If the dam or levee fails or is washed out, the water behind it is released to become a flash flood. The bursting of a dam in 1889 on the Little Conemaugh River caused the devastating Johnstown, Pennsylvania, flood that

took some 3000 lives. A second dam failure occurred there again in 1977 and caused 77 fatalities.

Flood Control

Several strategies have been devised to eliminate or lessen the catastrophic effects of floods. Engineering efforts include the construction of artificial levees, the building of flood-control dams, and river channelization.

STUDENTS SOMETIMES ASK...
Sometimes when there is a major flood, it is described as a 100-year flood. What does that mean?

The phrase "100-year flood" is misleading because it leads people to believe that such an event happens only once every 100 years. The truth is that an uncommonly big flood can happen any year. The phrase "100-year flood" is really a statistical designation, indicating that there is a 1-in-100 chance that a flood this size will happen during any year. Perhaps a better term would be the "1-in-100-chance flood."

Many flood designations are reevaluated and changed over time as more data are collected or when a river basin is altered in a way that affects the flow of water. Dams and urban development are examples of some human influences in a basin that affect floods.

Artificial Levees. *Artificial levees* are earthen mounds built on the banks of a river to increase the volume of water the channel can hold. These most common of stream-containment structures have been used since ancient times and continue to be used today.

Artificial levees are usually easy to distinguish from natural levees because their slopes are much steeper. When a river is confined by levees during periods of high water, it frequently deposits material in its channel as the discharge diminishes. This is sediment that otherwise would have been dropped on the floodplain. Thus, each time there is a high flow, deposits are left on the river bed, and the bottom of the channel is built up. With the buildup of the bed, less water is required to overflow the original levee. As a result, the height of the levee may have to be raised periodically to protect the floodplain. Moreover, many artificial levees are not built to withstand periods of extreme flooding. For example, levee failures were numerous in the Midwest during the summer of 1993, when the upper Mississippi and many of its tributaries experienced record floods (Figure 5.22).

Flood-Control Dams. *Flood-control dams* are built to store floodwater and then let it out slowly. This lowers the flood crest by spreading it out over a longer time span. Since the 1920s, thousands of dams have been built on nearly every major river in the United States. Many dams have significant non-flood-related functions, such as providing water for irrigated agriculture and for hydroelectric power generation. Many reservoirs are also major regional recreational facilities.

Although dams may reduce flooding and provide other benefits, building these structures also has significant costs and consequences. For example, reservoirs created by dams may cover fertile farmland, useful forests, historic sites, and scenic valleys. Of course, dams trap sediment. Therefore, deltas and floodplains downstream erode because they are no longer replenished with silt during floods. Large dams can also cause significant ecological damage to river environments that took thousands of years to establish.

Building a dam is not a permanent solution to flooding. Sedimentation behind a dam means that the volume of its reservoir will gradually diminish, reducing the effectiveness of this flood-control measure.

Channelization. *Channelization* involves altering a stream channel in order to speed the flow of water to prevent it from reaching flood height. This may simply involve clearing a channel of obstructions or dredging a channel to make it wider and deeper.

A more radical alteration involves straightening a channel by creating *artificial cutoffs.* The idea is that by shortening the stream, the gradient and hence the velocity are both increased. By increasing velocity, the larger discharge associated with flooding can be dispersed more rapidly.

Since the early 1930s, the U.S. Army Corps of Engineers has created many artificial cutoffs on the Mississippi for the purpose of increasing the efficiency of the channel and reducing the threat of flooding. In all, the river has been shortened more

than 240 kilometers (150 miles). The program has been somewhat successful in reducing the height of the river in flood stage. However, because the river's tendency toward meandering still exists, preventing the river from returning to its previous course has been difficult.

A Nonstructural Approach. All of the flood-control measures described so far have involved structural solutions aimed at "controlling" a river. These solutions are expensive and often give people residing on the floodplain a false sense of security.

Today many scientists and engineers advocate a nonstructural approach to flood control. They suggest that an alternative to artificial levees, dams, and channelization is sound floodplain management. By identifying high-risk areas, appropriate zoning regulations can be implemented to minimize development and promote more appropriate land use.

Groundwater: Water Beneath the Surface

GEODe Sculpturing Earth's Surface
▼ Groundwater

Groundwater is one of our most important and widely available resources. Yet people's perceptions of groundwater are often unclear and incorrect. The reason is that groundwater is hidden from view except in caves and mines, and the impressions people gain from these subsurface openings are often misleading. Observations on the land surface give an impression that Earth is solid. This view is not changed very much when we enter a cave and see water flowing in a channel that appears to have been cut into solid rock. Because of such observations, many people believe that groundwater occurs

◄ **FIGURE 5.22** Water rushes through a break in an artificial levee in Monroe County, Illinois. During the record-breaking 1993 Midwest floods, many artificial levees could not withstand the force of the floodwaters. Sections of many weakened structures were overtopped or simply collapsed. (Photo by James A. Finley/AP/Wide World Photos)

Box 5.1 People and the Environment:
Flash Floods

Tornadoes and hurricanes are nature's most awesome storms. Yet surprisingly, these dreaded events are not responsible for the greatest number of storm-related deaths. That distinction is reserved for flash floods. For a recent 10-year period (1992–2001), the number of storm-related deaths in the United States from flooding averaged 127 per year. By contrast, tornado fatalities averaged 71 annually and hurricanes 16.

Flash floods are local floods of great volume and short duration. The rapidly rising surge of water usually occurs with little advance warning and can destroy roads, bridges, homes, and other substantial structures (Figure 5.A). Discharges quickly reach a maximum and diminish almost as rapidly. Flood flows often contain large quantities of sediment and debris as they sweep channels clean.

Frequently flash floods result from the torrential rains associated with a slow-moving severe thunderstorm or take place when a series of thunderstorms repeatedly pass over the same location. Occasionally floating debris or ice can accumulate at a natural or artificial obstruction and restrict the flow of water. When such temporary dams fail, torrents of water can be released as a flash flood.

Figure 5.A Flash flooding in Las Vegas, Nevada, in August 2003. Parts of the city received nearly half the average annual rainfall in a matter of hours. Here firefighters are rescued from a firetruck that was caught in a torrent of water.(Photo by John Locher/*Las Vegas Review-Journal*)

only in underground "rivers." But actual rivers underground are extremely rare.

In reality, most of the subsurface environment is not solid at all. It includes countless tiny *pore spaces* between grains of soil and sediment, plus narrow joints and fractures in bedrock. Together these spaces add up to an immense volume. It is in these tiny openings that groundwater collects and moves.

The Importance of Groundwater

Considering the entire hydrosphere, or all of Earth's water, only about six-tenths of 1 percent occurs underground. Nevertheless, this small percentage, stored in the rocks and sediments beneath Earth's surface, is a vast quantity. When the oceans are excluded and only sources of freshwater are considered, the significance of groundwater becomes more apparent. Figure 5.23

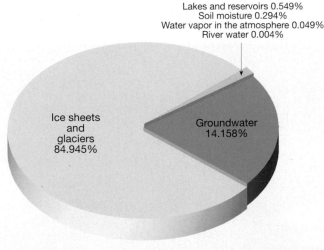

Lakes and reservoirs 0.549%
Soil moisture 0.294%
Water vapor in the atmosphere 0.049%
River water 0.004%

Ice sheets and glaciers 84.945%

Groundwater 14.158%

▲ **FIGURE 5.23** Estimated distribution of freshwater in the hydrosphere according to the U.S. Geological Survey.

Flash floods can take place in almost any area of the country. They are particularly common in mountainous terrain, where steep slopes can quickly channel runoff into narrow valleys. The hazard is most acute when the soil is already nearly saturated from earlier rains or consists of impermeable materials.

Why do so many people perish in flash floods? Aside from the factor of surprise (many are caught sleeping), people do not appreciate the power of moving water. A glance at Figure 5.B helps illustrate the force of a flood wave. Just 15 centimeters (6 inches) of fast-moving floodwater can knock a person down. Most automobiles will float and be swept away in only 0.6 meter (2 feet) of water. *More than half of all U.S. flash-flood fatalities are auto related!* Clearly, people should never attempt to drive over a flooded road. The depth of water is not always obvious. Also, the road bed may have been washed out under water. Present-day flash floods are calamities with potential for very high death tolls and huge property losses. Although efforts are being made to improve observations and warnings, flash floods remain elusive natural killers.

Figure 5.B Debris piled around the sign for a campground in Spain's Pyrenees Mountains on August 8, 1996, after a flash flood flowed through the campground killing at least 67 people and injuring another 180. (AP Photo/Christophe Ena)

contains estimates of the distribution of freshwater in the hydrosphere. Clearly, the largest volume occurs as glacial ice. Second in rank is groundwater, with slightly more than 14 percent of the total. However, when ice is excluded and just liquid water is considered, more than 94 percent is groundwater. Without question, *groundwater represents the largest reservoir of freshwater that is readily available to humans.* Its value in terms of economics and human well-being is incalculable.

Worldwide, wells and springs provide water for cities, crops, livestock, and industry. In the United States, groundwater is the source of about 40 percent of the water used for all purposes (except hydroelectric power generation and power-plant cooling). Groundwater is the drinking water for more than 50 percent of the population, is 40 percent of the water for irrigation, and provides more than 25 percent of industry's needs. In some areas, however, overuse of this basic resource has caused serious problems, including streamflow depletion, land subsidence, and increased pumping costs. In addition, groundwater contamination due to human activities is a real and growing threat in many places.

Groundwater's Geological Roles

Geologically, groundwater is important as an erosional agent. The dissolving action of groundwater slowly removes soluble rock, allowing surface depressions known as sinkholes to form as well as creating subterranean caverns (Figure 5.24). Groundwater is also an equalizer of streamflow. Much of the water that flows in rivers is not direct runoff from rain and snowmelt. Rather, a large percentage of precipitation soaks in and then moves slowly underground to stream channels.

► **FIGURE 5.24** A view of the interior of Carlsbad Caverns in southeastern New Mexico. The dissolving action of groundwater created the caverns. Later, groundwater deposited the limestone decorations. (Photo by David Muench/David Muench Photography, Inc./CORBIS)

Groundwater is thus a form of storage that sustains streams during periods when rain does not fall. When we observe water flowing in a river during a dry period, it is water from rain that fell at some earlier time and was stored underground.

Distribution and Movement of Groundwater

 GEODe Sculpturing Earth's Surface
▼ Groundwater

When rain falls, some of the water runs off, some evaporates, and the remainder soaks into the ground. This last path is the primary source of practically all subsurface water. The amount of water that takes each of these paths, however, varies considerably both in time and space. Influential factors include the steepness of the slopes, the nature of the surface materials, the intensity of the rainfall, and the type and amount of vegetation. Heavy rains falling on steep slopes underlain by impervious materials will obviously result in a high percentage of the water running off. Conversely, if rain falls steadily and gently on more gradual slopes composed of materials that are easily penetrated by water, a much larger percentage of the water soaks into the ground.

Distribution

Some of the water that soaks in does not travel far, because it is held by molecular attraction as a surface film on soil particles. This near-surface zone is called the *belt of soil moisture*. It is crisscrossed by roots, voids left by decayed roots, and animal and worm burrows that enhance the infiltration of rainwater into the soil. Soil water is used by plants in life functions and tran-

spiration. Some water also evaporates directly back into the atmosphere.

Water that is not held as soil moisture percolates downward until it reaches a zone where all of the open spaces in sediment and rock are completely filled with water. This is the **zone of saturation**. Water within it is called **groundwater**. The upper limit of this zone is known as the **water table**. The area above the water table where the soil, sediment, and rock are not saturated is called the **zone of aeration** (Figure 5.25). Although a considerable amount of water can be present in the zone of aeration, this water cannot be pumped by wells, because it clings too tightly to rock and soil particles. By contrast, below the water table, the water pressure is great enough to allow water to enter wells, thus permitting groundwater to be withdrawn for use. We will examine wells more closely later in the chapter.

The water table is rarely level as we might expect a table to be. Instead, its shape is usually a subdued replica of the surface, reaching its highest elevations beneath hills and decreasing in height toward valleys (Figure 5.25). When you see a wetland (swamp), it indicates that the water table is right at the surface. Lakes and streams generally occupy areas low enough that the water table is above the land surface.

Several factors contribute to the irregular surface of the water table. One important influence is the fact that groundwater moves very slowly. Because of this, water tends to "pile up" beneath high areas between stream valleys. If rainfall were to cease completely, these water "hills" would slowly subside and gradually approach the level of the valleys. However, new supplies of rainwater are usually added often enough to prevent this. Nevertheless, in times of extended drought, the water table may drop enough to dry up shallow wells. Other causes for the uneven water table are variations in rainfall and permeability from place to place.

Factors Influencing the Storage and Movement of Groundwater

The nature of subsurface materials strongly influences the rate of groundwater movement and the amount of groundwater that can be stored. Two factors are especially important—porosity and permeability.

Porosity. Water soaks into the ground because bedrock, sediment, and soil contain countless voids or openings. These openings are similar to those of a sponge and are often called pore spaces. The quantity of groundwater that can be stored depends on the **porosity** of the material, which is the percentage of the total volume of rock or sediment that consists of pore spaces. Voids most often are spaces between sedimentary particles, but also common are joints, faults, cavities formed by the dissolving of soluble rock such as limestone, and vesicles (voids left by gases escaping from lava).

Variations in porosity can be great. Sediment is commonly quite porous, and open spaces may occupy 10 to 50 percent of the sediment's total volume. Pore space depends on the size and shape of the grains, how they are packed together, the degree of sorting, and in sedimentary rocks, the amount of cementing material. Most igneous and metamorphic rocks, as well as some sedimentary rocks, are composed of tightly interlocking crystals so the voids between grains may be negligible. In these rocks, fractures must provide the voids.

Permeability. Porosity alone cannot measure a material's capacity to yield groundwater. Rock or sediment may be very porous and still prohibit water from moving through it. The **permeability** of a material indicates its ability to *transmit* a fluid. Groundwater moves by twisting and turning through interconnected small openings. The smaller the pore spaces, the slower the groundwater moves. If the spaces between particles are too small, water cannot move at all. For example, clay's ability to store water can be great, owing to its high porosity, but its pore spaces are so small that water is unable to move through it. Thus, we say that clay is *impermeable.*

Aquitards and Aquifers. Impermeable layers such as clay that hinder or prevent water movement are termed **aquitards** (*aqua* = water, *tard* = slow). In contrast, larger particles, such as sand or gravel, have larger pore spaces. Therefore, the water moves with relative ease. Permeable rock strata or sediments that transmit groundwater freely are called **aquifers** ("water carriers"). Aquifers are important because they are the water-bearing layers sought after by well drillers.

Groundwater Movement

The movement of most groundwater is exceedingly slow, from pore to pore. A typical rate is a few centimeters per day. The energy that makes the water move is

▼ **FIGURE 5.25** This diagram illustrates the relative positions of many features associated with subsurface water.

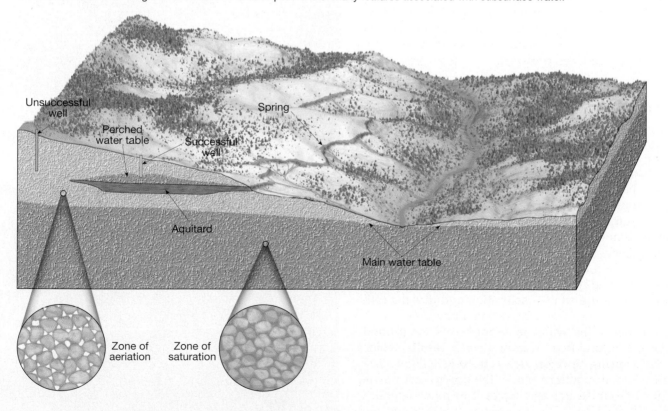

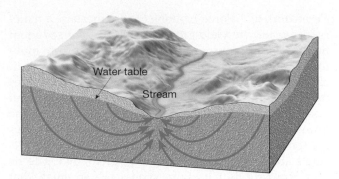

▲ **FIGURE 5.26** Arrows indicate groundwater movement through uniformly permeable material. The looping curves may be thought of as a compromise between the downward pull of gravity and the tendency of water to move toward areas of reduced pressure.

provided by the force of gravity. In response to gravity, water moves from areas where the water table is high to zones where the water table is lower (see Box 5.2). This means that water usually gravitates toward a stream channel, lake, or spring. Although some water takes the most direct path down the slope of the water table, much of the water follows long, curving paths toward the zone of discharge.

Figure 5.26 shows how water percolates into a stream from all possible directions. Some paths clearly turn upward, apparently against the force of gravity, and enter through the bottom of the channel. This is easily explained: The deeper you go into the zone of saturation, the greater the water pressure. Thus, the looping curves followed by water in the saturated zone may be thought of as a compromise between the downward pull of gravity and the tendency of water to move toward areas of reduced pressure.

Springs

GEODe   Sculpturing Earth's Surface
▼ Groundwater

Springs have aroused the curiosity and wonder of people for thousands of years. The fact that springs were (and to some people still are) rather mysterious phenomena is not difficult to understand, for here is water flowing freely from the ground in all kinds of weather in seemingly inexhaustible supply but with no obvious source. Today we know that the source of springs is water from the zone of saturation and that the ultimate source of this water is precipitation.

Whenever the water table intersects the ground surface, a natural flow of groundwater results, which we call a **spring**. Springs such as the one in Figure 5.27 form when an aquitard blocks the downward movement of groundwater and forces it to move laterally.

When the permeable bed (aquifer) outcrops in a valley, a spring or series of springs results.

Another situation that can produce a spring is illustrated in Figure 5.25. Here an aquitard is situated above the main water table. As water percolates downward, a portion accumulates above the aquitard to create a localized zone of saturation and a *perched water table*. Springs, however, are not confined to places where a perched water table creates a flow at the surface. Many geological situations lead to the formation of springs because subsurface conditions vary greatly from place to place.

Hot Springs

By definition, the water in **hot springs** is 6–9°C (10–15°F) warmer than the mean annual air temperature for the localities where they occur. In the United States alone, there are well over 1000 such springs.

Temperatures in deep mines and oil wells usually rise with an increase in depth averaging about 2°C per 100 meters (1°F per 100 feet). Therefore, when ground-

▼ **FIGURE 5.27**   Spring flowing from a valley wall in Arizona's Marble Canyon. (Photo by Michael Collier)

Box 5.2 Understanding Earth
Measuring Groundwater Movement

The foundations of our modern understanding of groundwater movement began in the mid nineteenth century with the work of the French scientist-engineer Henri Darcy. Among the experiments carried out by Darcy was one that showed that the velocity of groundwater flow is proportional to the slope of the water table—the steeper the slope, the faster the water moves (because the steeper the slope, the greater the pressure difference between two points). The water-table slope is known as the hydraulic gradient and can be expressed as follows:

$$\text{hydraulic gradient} = \frac{h_1 - h_2}{d}.$$

Where h_1 is the elevation of one point on the water table, h_2 is the elevation of a second point, and d is the horizontal distance between the two points (Figure 5.C).

Darcy also discovered that the flow velocity varied with the permeability of the sediment—groundwater flows more rapidly through sediments having greater permeability than through materials having lower permeability. This factor is known as *hydraulic conductivity* and is a coefficient that takes into account the permeability of the aquifer and the viscosity of the fluid.

To determine discharge (Q)—that is, the actual volume of water that flows through an aquifer in a specified time—the following equation is used:

$$Q = \frac{K\,A(h_1 - h_2)}{d}.$$

Where $\frac{h_1 - h_2}{d}$ is the hydraulic gradient, K is the coefficient that represents hydraulic conductivity and A is the cross-sectional area of the aquifer. This expression has come to be called *Darcy's law*.

Figure 5.C The hydraulic gradient is determined by measuring the difference in elevation between two points on the water table (h_1-h_2) divided by the distance between them, d. Wells are used to determine the height of the water table.

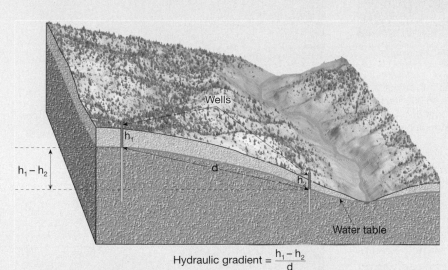

Wells

h_1

$h_1 - h_2$

d

h_2

Water table

Hydraulic gradient $= \dfrac{h_1 - h_2}{d}$

water circulates at great depths, it becomes heated, and if it rises to the surface, the water may emerge as a hot spring. The water of some hot springs in the United States, particularly in the East, is heated in this manner. However, the great majority (over 95 percent) of the hot springs (and geysers) in the United States are found in the West. The reason for such a distribution is that the source of heat for most hot springs is cooling igneous rock, and it is in the West that igneous activity has been most recent.

Geysers

Geysers are intermittent hot springs or fountains in which columns of water are ejected with great force at various intervals, often rising 30 to 60 meters (100 to 200 feet). After the jet of water ceases, a column of steam rushes out, usually with a thundering roar. Figure 5.28 shows a wintertime eruption of Old Faithful in Yellowstone National Park, perhaps the most famous geyser in the world, which erupts about once each hour. Geysers are also found in other parts of the world, including New Zealand and Iceland. In fact, the Icelandic word *geysa*—to gush—gives us the name *geyser*.

Geysers occur where extensive underground chambers exist within hot igneous rocks. How they operate is shown in Figure 5.29. As relatively cool groundwater enters the chambers, it is heated by the surrounding rock. At the bottom of the chamber, the water is under great pressure because of the weight of the overlying water. This great pressure prevents the water from boiling at the normal surface temperature of 100°C (212°F). For example, at the bottom of

▲ **FIGURE 5.28** A wintertime eruption of Old Faithful, one of the world's most famous geysers. It emits as much as 45,000 liters (almost 12,000 gallons) of hot water and steam about once each hour. (Photo by Marc Muench/David Muench Photography, Inc.)

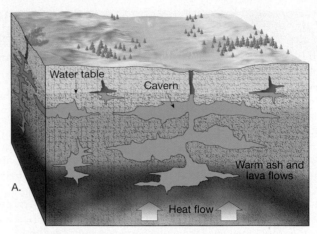

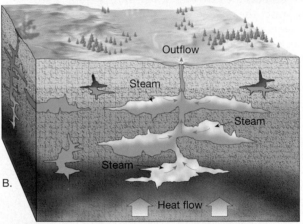

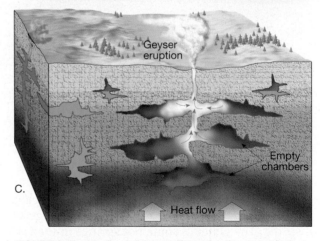

▲ **FIGURE 5.29** Idealized diagrams illustrating the stages in the eruption cycle of a geyser. **A.** Groundwater enters underground caverns and fractures in hot igneous rock, where it is heated to near its boiling point. **B.** Heating causes the water to expand, with some being forced out at the surface. The loss of water reduces the pressure on the remaining water, thus reducing its boiling temperature. Some of the water flashes to steam. **C.** The rapidly expanding steam forces the hot water out of the chambers to produce a geyser. The empty chambers fill again, and the cycle starts anew.

a 300-meter (1000-foot) water-filled chamber, water must attain a temperature of nearly 230°C (450°F) before it will boil. The heating causes the water to expand, with the result that some is forced out at the surface. This loss of water reduces the pressure on the remaining water in the chamber, which lowers the boiling point. A portion of the water deep within the chamber quickly turns to steam and causes the geyser to erupt. Following the eruption, cool groundwater again seeps into the chamber, and the cycle begins anew.

Wells

Sculpturing Earth's Surface
▼ Groundwater

The most common method for removing groundwater is the **well**, a hole bored into the zone of saturation (see Figure 5.25). Wells serve as small reservoirs into which groundwater moves and from which it can be pumped to the surface. The use of wells dates back many centuries and continues to be an important method of obtaining water today. By far the single greatest use of this water in the United

States is irrigation for agriculture. More than 65 percent of the groundwater used each year is for this purpose. Industrial uses rank a distant second, followed by the amount used by homes in cities and rural areas.

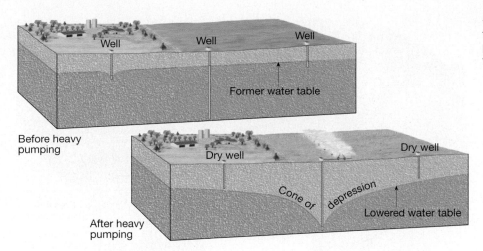

The level of the water table may fluctuate considerably during the course of a year, dropping during dry seasons and rising following periods of rain. Therefore, to ensure a continuous supply of water, a well must penetrate far below the water table. Whenever a substantial amount of water is withdrawn from a well, the water table around the well is lowered. This effect, termed **drawdown**, decreases with increasing distance from the well. The result is a depression in the water table, roughly conical in shape, known as a **cone of depression** (Figure 5.30). For most small domestic wells, the cone of depression is negligible. However, when wells are used for irrigation or for industrial purposes, the withdrawal of water can be great enough to create a very wide and steep cone of depression that may substantially lower the water table in an area and cause nearby shallow wells to become dry. Figure 5.30 illustrates this situation.

STUDENTS SOMETIMES ASK...

I have heard people say that supplies of groundwater can be located using a forked stick. Can this actually be done?

What you describe is a practice called "water dowsing." In the classic method, a person holding a forked stick walks back and forth over an area. When water is detected, the bottom of the "Y" is supposed to be attracted downward.

Geologists and engineers are dubious, to say the least. Case histories and demonstrations may seem convincing, but when dowsing is exposed to scientific scrutiny, it fails. Most "successful" examples of water dowsing occur in places where water would be hard to miss. In a region of adequate rainfall and favorable geology, it is difficult to drill and *not* find water!

Artesian Wells

GEODe Sculpturing Earth's Surface
▼ Groundwater

In most wells, water does not rise on its own. If water is first encountered at 30 meters of depth, it remains at that level, fluctuating perhaps a meter or two with seasonal wet and dry periods. However, in some wells, water rises, sometimes overflowing at the surface.

The term **artesian** is applied to any situation in which groundwater rises in a well above the level where it was initially encountered. For such a situation to occur, two conditions must exist (Figure 5.31): (1) Water must be confined to an aquifer that is inclined so that one end is exposed at the surface, where it can receive water; and (2) aquitards both above and below the aquifer must be present to prevent the water from escaping. When such a layer is tapped, the pressure created by the weight of the water above will force the water to rise. If there were no friction, the water in the well would rise to the level of the water at the top of the aquifer. However, friction reduces the height of this pressure surface. The greater the distance from the recharge area (area where water enters the inclined aquifer), the greater the friction and the less the rise of water.

In Figure 5.31, Well 1 is a *nonflowing artesian well*, because at this location the pressure surface is below ground level. When the pressure surface is above the ground and a well is drilled into the aquifer, a *flowing artesian well* is created (Well 2, Figure 5.31).

Artesian systems act as conduits, transmitting water from remote areas of recharge great distances to the points of discharge. In this manner, water that fell in central Wisconsin years ago is now taken from the ground and used by communities many kilometers to

▶ **FIGURE 5.31** Artesian systems occur when an inclined aquifer is surrounded by impermeable beds.

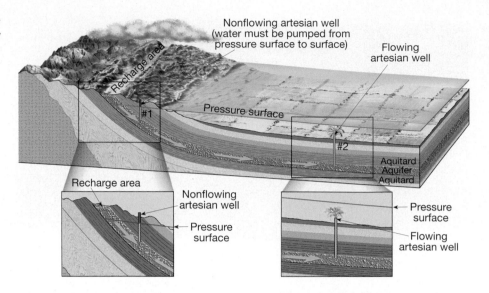

Nonflowing artesian well (water must be pumped from pressure surface to surface)

Flowing artesian well

Recharge area

#1

Pressure surface

#2

Aquitard
Aquifer
Aquitard

Recharge area

Nonflowing artesian well

Pressure surface

Pressure surface

Flowing artesian well

the south in Illinois. In South Dakota, such a system brings water eastward across the state from the Black Hills in the west.

On a different scale, city water systems may be considered examples of artificial artesian systems. The water tower, into which water is pumped, may be considered the area of recharge, the pipes the confined aquifer, and the faucets in homes the flowing artesian wells.

Environmental Problems Associated with Groundwater

As with many of our valuable natural resources, groundwater is being exploited at an increasing rate. In some areas, overuse threatens the groundwater supply. In other places, groundwater withdrawal has caused the ground and everything resting upon it to sink. Still other localities are concerned with the possible contamination of their groundwater supply.

Treating Groundwater as a Nonrenewable Resource

For many, groundwater appears to be an endlessly renewable resource, for it is continually replenished by rainfall and melting snow. But in some regions, groundwater has been and continues to be treated as a *nonrenewable* resource. Where this occurs, the amount of water available to recharge the aquifer is significantly less than the amount being withdrawn.

The High Plains, a relatively dry region that extends from South Dakota to western Texas, provides one example (Figure 5.32A). Here an extensive agricultural economy is largely dependent on irrigation.

Today nearly 170,000 wells are being used to irrigate more than 65,000 square kilometers (16 million acres) of land (Figure 5.32B). In the southern part of this region, which includes the Texas panhandle, the natural recharge of the aquifer is very slow and the problem of declining groundwater levels is acute. In fact, in years of average or below-average precipitation, recharge is negligible because nearly all of the meager rainfall is returned to the atmosphere by evaporation and transpiration.

Therefore, where intense irrigation has been practiced for an extended period, depletion of groundwater can be severe. Declines in the water table as great as 1 meter per year have led to an overall drop of between 15 and 60 meters (50 and 200 feet) in some areas. Under these circumstances, it can be said that the groundwater is literally being "mined." Even if pumping were to cease immediately, it would take thousands of years for the groundwater to be fully replenished.

Land Subsidence Caused by Groundwater Withdrawal

As you will see later in this chapter, surface subsidence can result from natural processes related to groundwater. However, the ground may also sink when water is pumped from wells faster than natural recharge processes can replace it. This effect is particularly pronounced in areas underlain by thick layers of loose sediments. As water is withdrawn, the ground subsides because the weight of the overburden packs the sediment grains more tightly together.

Many areas can be used to illustrate land subsidence caused by excessive pumping of groundwater from relatively loose sediment. A classic example in the United States occurred in the San Joaquin Valley of

▲ **FIGURE 5.32** **A.** The High Plains extend from the western Dakotas south to Texas. Despite being a land of little rain, this is an important agricultural region. The reason is a vast endowment of groundwater that makes irrigation possible through most of the region. The source of most of this water is the Ogallala formation, the largest aquifer in the United States. **B.** In some agricultural regions, water is pumped from the ground faster than it is replenished. In such instances, groundwater is being treated as a nonrenewable resource. This aerial view shows circular crop fields irrigated by center pivot irrigation systems in semiarid eastern Colorado. (Photo by James L. Amos/Corbis/Bettmann)

California (Figure 5.33). This important agricultural region relies heavily on irrigation. Land subsidence due to groundwater withdrawal began in the valley in the mid-1920s and locally exceeded 8 meters (28 feet) by 1970. Then, because of the importation of surface water and a decrease in groundwater pumping, water levels in the aquifer recovered and subsidence ceased. However, during a drought in 1976 and 1977, heavy groundwater pumping led to renewed subsidence. This time, water levels dropped at a much faster rate than during the previous period, because of the reduced storage capacity caused by earlier compaction of material in the aquifer. In all, more than 13,400 square kilometers (5200 square miles) of irrigable land—one half the entire valley—were affected by subsidence. Damage to structures, including highways, bridges, water lines, and wells, was extensive. Many other examples of land subsidence due to groundwater pumping occur in the United States and elsewhere in the world.

Groundwater Contamination

The pollution of groundwater is a serious matter, particularly in areas where aquifers provide a large part of the water supply. One common source of groundwater pollution is sewage. Its sources include an ever increasing number of septic tanks, as well as farm wastes and inadequate or broken sewer systems.

If sewage water that is contaminated with bacteria enters the groundwater system, it may become purified through natural processes. The harmful bacteria can be mechanically filtered by the sediment through which the water percolates, destroyed by chemical oxidation, and/or assimilated by other organisms. For purification to occur, however, the aquifer must be of the correct composition.

For example, extremely permeable aquifers (such as highly fractured crystalline rock, coarse gravel, or cavernous limestone) have such large openings that contaminated groundwater may travel long distances without being cleansed. In this case, the water flows too rapidly and is not in contact with the surrounding material long enough for purification to occur. This is the problem at Well 1 in Figure 5.34A.

Conversely, when the aquifer is composed of sand or permeable sandstone, the water can sometimes be purified after traveling only a few dozen meters through it. The openings between sand grains are large enough to permit water movement, yet the movement of the water is slow enough to allow ample time for its purification (Well 2, Figure 5.34B).

▲ **FIGURE 5.33** The shaded area on the map shows California's San Joaquin Valley. The marks on the utility pole in the photo indicate the level of the surrounding land in preceding years. Between 1925 and 1975 this part of the San Joaquin Valley subsided almost 9 meters because of the withdrawal of groundwater and the resulting compaction of sediments. (Photo courtesy of U.S. Geological Survey)

Other sources and types of contamination also threaten groundwater supplies (Figure 5.35). These include widely used substances such as highway salt, fertilizers that are spread across the land surface, and pesticides. In addition, a wide array of chemicals and industrial materials may leak from pipelines, storage tanks, landfills, and holding ponds. Some of these pollutants are classified as *hazardous,* meaning that they are either flammable, corrosive, explosive, or toxic. As rainwater oozes through the refuse, it may dissolve a variety of potential contaminants. If the leached material reaches the water table, it will mix with the groundwater and contaminate the supply.

Because groundwater movement is usually slow, polluted water might go undetected for a long time. In fact, most contamination is discovered only after drinking water has been affected and people become ill. By this time, the volume of polluted water might be very large, and even if the source of contamination is removed immediately, the problem is not solved. Although the sources of groundwater contamination are numerous, the solutions are relatively few.

Once the source of the problem has been identified and eliminated, the most common practice is simply to abandon the water supply and allow the pollutants to be flushed away gradually. This is the least costly and easiest solution, but the aquifer must remain unused for many years. To accelerate this process, polluted water is sometimes pumped out and

▶ **FIGURE 5.34** **A.** Although the contaminated water has traveled more than 100 meters before reaching Well 1, the water moves too rapidly through the cavernous limestone to be purified. **B.** As the discharge from the septic tank percolates through the permeable sandstone, it is purified in a relatively short distance.

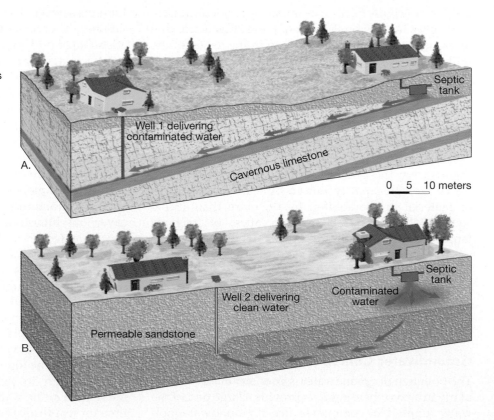

A.

B.

▲ FIGURE 5.35 Sometimes agricultural chemicals **A.** and materials leached from landfills **B.** find their way into the groundwater. These are two of the potential sources of groundwater contamination. (Photo A by Roy Morsch/The Stock Market; Photo B by F. Rossotto/The Stock Market)

treated. Following removal of the tainted water, the aquifer is allowed to recharge naturally or, in some cases, the treated water or other freshwater is pumped back in. This process is costly, time consuming, and it may be risky because there is no way to be certain that all of the contamination has been removed. Clearly, the most effective solution to groundwater contamination is prevention.

The Geologic Work of Groundwater

Groundwater dissolves rock. This fact is key to understanding how caverns and sinkholes form. Because soluble rocks, especially limestone, underlie millions of square kilometers of Earth's surface, it is here that groundwater carries on its important role as an erosional agent. Limestone is nearly insoluble in pure water but is quite easily dissolved by water containing small quantities of carbonic acid. Most natural water contains this weak acid because rainwater readily dissolves carbon dioxide from the air and from decaying plants. Therefore, when groundwater comes in contact with limestone, the carbonic acid reacts with calcite in the rocks to form calcium bicarbonate, a soluble material that is then carried away in solution.

Caverns

The most spectacular results of groundwater's erosional handiwork are limestone **caverns**. In the United States alone about 17,000 caves have been discovered. Although most are relatively small, some have spectacular dimensions. Carlsbad Caverns in southeastern New Mexico and Mammoth Cave in Kentucky are famous examples. One chamber in Carlsbad Caverns has an area equivalent to 14 football fields and enough

height to accommodate the U.S. Capitol Building. At Mammoth Cave, the total length of interconnected caverns extends for more than 540 kilometers (340 miles).

Most caverns are created at or below the water table in the zone of saturation. Here acidic groundwater follows lines of weakness in the rock, such as joints and bedding planes. As time passes, the dissolving process slowly creates cavities and gradually enlarges them into caverns. Material that is dissolved by the groundwater is eventually discharged into streams and carried to the ocean.

Certainly the features that arouse the greatest curiosity for most cavern visitors are the stone formations that give some caverns a wonderland appearance (see Figure 5.24). These are not erosional features, like the caverns in which they reside, but depositional features. They are created by the seemingly endless dripping of water over great spans of time. The calcium carbonate that is left behind produces the limestone we call *travertine*. These cave deposits, however, are also commonly called *dripstone*, an obvious reference to their mode of origin.

Although the formation of caverns takes place in the zone of saturation, the deposition of dripstone is not possible until the caverns are above the water table in the zone of aeration. This commonly occurs as nearby streams cut their valleys deeper, lowering the water table as the elevation of the rivers drops. As soon as the chamber is filled with air, the conditions are right for the decoration phase of cavern building to begin.

Of the various dripstone features found in caverns, perhaps the most familiar are **stalactites**. These icicle-like pendants hang from the ceiling of the cavern and form where water seeps through cracks above. When water reaches air in the cave, some of the dissolved carbon dioxide escapes from the drop and calcite begins to precipitate. Deposition occurs as a ring around the

▲ FIGURE 5.36 "Live" solitary soda-straw stalactites. Lehman Caves. Great Basin National Park, Nevada. (Photo by Tom Bean)

edge of the water drop. As drop after drop follows, each leaves an infinitesimal trace of calcite behind, and a hollow limestone tube is created. Water then moves through the tube, remains suspended momentarily at the end, contributes a tiny ring of calcite, and falls to the cavern floor. The stalactite just described is appropriately called a *soda straw* (Figure 5.36). Often the hollow tube of the soda straw becomes plugged or its supply of water increases. In either case, the water is forced to flow and deposit along the outside of the tube. As deposition continues, the stalactite takes on the more common conical shape.

Formations that develop on the floor of a cavern and reach upward toward the ceiling are called **stalagmites**. The water supplying the calcite for stalagmite growth falls from the ceiling and splatters over the surface. As a result, stalagmites do not have a central tube and are usually more massive in appearance and more rounded on their upper ends than stalactites. Given enough time, a downward-growing stalactite and an upward-growing stalagmite may join to form a *column*.

Karst Topography

Many areas of the world have landscapes that to a large extent have been shaped by the dissolving power of groundwater. Such areas are said to exhibit **karst topography**, named for the *Krs* region in Slovenia (formerly a part of Yugoslavia) where such topography is strikingly developed. In the United States, karst landscapes occur in many areas that are underlain by limestone, including portions of Kentucky, Tennessee, Alabama, southern Indiana, and central and northern Florida (Figure 5.37). Generally, arid and semiarid

▶ FIGURE 5.37 This diagram shows a hypothetical region with well-developed karst features. Sinkholes are plentiful, and surface streams are funneled below ground. With the passage of time, caverns grow larger and the number and size of sinkholes increase. Collapse of caverns and coalescence of sinkholes may form large, flat-floored depressions. Eventually, the dissolving power of acidic groundwater may remove most of the limestone from the area, leaving only isolated remnants.

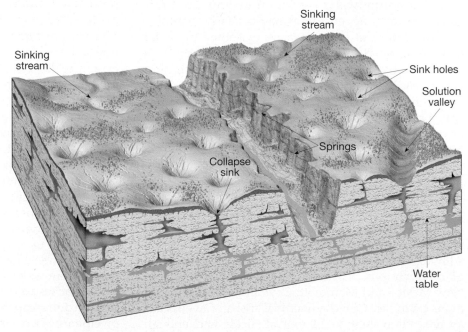

146

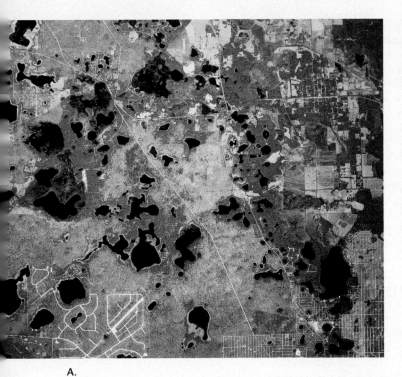

A.

B.

▲ **FIGURE 5.38** **A.** This high-altitude infrared image shows an area of karst topography in central Florida. The numerous lakes occupy sinkholes. (Courtesy of USDA-ASCS) **B.** This small sinkhole formed suddenly in 1991 when the roof of a cavern collapsed, destroying this home in Frostproof, Florida. (Photo by *St. Petersburg Times*/Liaison Agency, Inc.)

areas do not develop karst topography because there is insufficient groundwater. When solution features exist in such regions, they are likely to be remnants of a time when rainier conditions prevailed.

Karst areas typically have irregular terrain punctuated with many depressions called **sinkholes** or, simply, **sinks**. In the limestone areas of Florida, Kentucky, and southern Indiana, there are literally tens of thousands of these depressions varying in depth from just a meter or two to a maximum of more than 50 meters (Figure 5.38A).

Sinkholes commonly form in one of two ways. Some develop gradually over many years without any physical disturbance to the rock. In these situations, the limestone immediately below the soil is dissolved by downward-seeping rainwater that is freshly charged with carbon dioxide. These depressions are usually not deep and are characterized by relatively gentle slopes. By contrast, sinkholes can also form suddenly and without warning when the roof of a cavern collapses under its own weight. Typically, the depressions created in this manner are steep-sided and deep. When they form in populous areas, they may represent a serious geologic hazard. Such a situation is clearly the case in Figure 5.38B and in Box 5.3.

In addition to a surface pockmarked by sinkholes, karst regions characteristically show a striking lack of surface drainage (streams). Following a rainfall, runoff is quickly funneled below ground through sinks. It then flows through caverns until it finally reaches the water table. Where streams do exist at the surface, their paths are usually short. The names of such streams often give a clue to their fate. In the Mammoth Cave area of Kentucky, for example, there is Sinking Creek, Little Sinking Creek, and Sinking Branch. Some sinkholes become plugged with clay and debris, creating small lakes or ponds.

? STUDENTS SOMETIMES ASK...
Is limestone the only rock type that develops karst features?

No. For example, karst development can occur in other carbonate rocks such as marble and dolostone. In addition, evaporates such as gypsum and salt (halite) are highly soluble and are readily dissolved to form karst features, including sinkholes, caves, and disappearing streams. This latter situation is termed *evaporite karst*.

147

Lake Chesterfield was a pleasant 9.3 hectare (23-acre) man-made lake in a quiet suburb of St. Louis where people living along its shore could fish from their small paddleboats—until it disappeared! In June 2004 residents witnessed the entire lake drain in less than three days (Figure 5.D).

"It was like someone pulled the plug," said Donna Ripp, who lives across the street from the lake, which is now a giant mud hole. Ripp said she began to notice the water level sinking and that by the second day, the lake was half empty. A day later it was completely gone.

What happened? The culprit is clear: At the north end of the lake there is a gaping sinkhole estimated to be about 20 meters (70 feet) in diameter. What geologists are now investigating is what the larger subterranean network looks like. This part of Missouri has many caves, including many that are large enough for humans to explore.

Geologist David Taylor, who inspected the lake shortly after the water drained into the ground, said the sinkhole itself "is really not that big." But it doesn't take a very large sinkhole to knock out an entire lake. Taylor said a hole 0.3 meter (1 foot) in diameter can drain at least 3800 liters (about 1000 gallons) a minute. Taylor is the head of a St. Charles–based company called Strata Services, Inc., that specializes in repairing lakes that are draining into subterranean cavities. "In my business I have fixed hundreds of leaky lakes," he said.

But before Taylor can consider repairing Lake Chesterfield, he and his colleagues first must get a sense of the network of cavities under the lake—a task that he said is exceedingly difficult. "There's all kinds of crazy stuff going on down there," he said. "This is all subsurface work. It's very unpredictable and very difficult."

Taylor found that the subsurface cavity responsible for the sinkhole under Lake Chesterfield runs laterally underground for several kilometers. A tracing dye placed near the sinkhole reemerges in a spring about 5.5 kilometers (3.5 miles) from the lake. In order to develop a better picture of the subterranean cavities, Taylor drilled five test holes at 12-meter (40-foot) intervals, finding that two revealed empty cavities below. But he estimated that it would require 600 holes in a 12-meter grid to even begin to understand the region completely.

Once that picture emerges, Taylor's company then fills the cavities with a cementlike substance so that other sinkholes don't open and create a similar problem. "If we just put a Band-Aid over the hole and fill the lake back up, the same thing will happen again," he said.

In the meantime, nearby residents are getting a crash course on karst topography. "I didn't even know there were underground caves here until all this happened," Donna Ripp said.

*Prepared with the assistance of Chris Wilson.

Figure 5.D Residents examine emptied Lake Chesterfield, a 23-acre reservoir that drained in three days when a sinkhole opened beneath it. (Photo by Hillary Levin/*St. Louis Post-Dispatch*)

Chapter Summary

- The *hydrologic cycle* describes the continuous interchange of water among the oceans, atmosphere, and continents. Powered by energy from the Sun, it is a global system in which the atmosphere provides the link between the oceans and continents. The processes involved in the hydrologic cycle include *precipitation, evaporation, infiltration* (the movement of water into rocks or soil through cracks and pore spaces), *runoff* (water that flows over the land rather than infiltrating into the ground), and *transpiration* (the release of water vapor to the atmosphere by plants). *Running water is the single most important agent sculpturing Earth's land surface.*

- The land area that contributes water to a stream is its *drainage basin*. Drainage basins are separated by imaginary lines called *divides*.

- River systems consist of three main parts: the zones of erosion, transportation, and deposition.

- The factors that determine a stream's *velocity* are *gradient* (slope of the stream channel), *shape, size,* and *roughness* of the channel, and the stream's *discharge* (amount of water passing a given point per unit of time, frequently measured in cubic feet per second). Most often, the gradient and roughness of a stream decrease downstream, while width, depth, discharge, and velocity increase.

- Streams transport their load of sediment in solution (*dissolved load*), in suspension (*suspended load*), and along the bottom of the channel (*bed load*). Much of the dissolved load is contributed by groundwater. Most streams carry the greatest part of their load in suspension. The bed load moves only intermittently and is usually the smallest portion of a stream's load.

- A stream's ability to transport solid particles is described using two criteria: *capacity* (the maximum load of solid particles a stream can carry) and *competence* (the maximum particle size a stream can transport). Competence increases as the square of stream velocity, so if velocity doubles, water's force increases fourfold.

- Streams deposit sediment when velocity slows and competence is reduced. This results in *sorting*, the process by which like-sized particles are deposited together. Stream deposits are called *alluvium* and may occur as channel deposits called *bars*, as floodplain deposits, which include *natural levees*, and as *deltas* or *alluvial fans* at the mouths of streams.

- Stream channels are of two basic types: *bedrock channels* and *alluvial channels*. Bedrock channels are most common in headwaters regions where gradients are steep. Rapids and waterfalls are common features. Two types of alluvial channels are *meandering channels* and *braided channels*.

- The two general types of *base level* (the lowest point to which a stream may erode its channel) are (1) *ultimate base level* and (2) *temporary*, or *local base level*. Any change in base level will cause a stream to adjust and establish a new balance. Lowering base level will cause a stream to downcut, whereas raising base level results in deposition of material in the channel.

- When a stream has cut its channel closer to base level, its energy is directed from side to side, and erosion produces a flat valley floor, or *floodplain*. Streams that flow upon floodplains often move in sweeping bends called *meanders*. Widespread meandering may result in shorter channel segments, called *cutoffs*, and/or abandoned bends, called *oxbow lakes*.

- *Floods* are triggered by heavy rains and/or snowmelt. Sometimes human interference can worsen or even cause floods. Flood-control measures include the building of *artificial levees* and dams, as well as *channelization*, which could involve creating *artificial cutoffs*. Many scientists and engineers advocate a nonstructural approach to flood control that involves more appropriate land use.

- Common *drainage patterns* produced by streams include (1) *dendritic*, (2) *radial*, (3) *rectangular*, and (4) *trellis*.

- As a resource, *groundwater* represents the largest reservoir of freshwater that is readily available to humans. Geologically, the dissolving action of groundwater produces *caves* and *sinkholes*. Groundwater is also an equalizer of streamflow.

- Groundwater is water that occupies the pore spaces in sediment and rock in a zone beneath the surface called the *zone of saturation*. The upper limit of this zone is the *water table*. The *zone of aeration* is above the water table where the soil, sediment, and rock are not saturated.

- The quantity of water that can be stored depends on the *porosity* (the volume of open spaces) of the material. The *permeability* (the ability to transmit a fluid through interconnected pore spaces) of a material is a very important factor controlling the movement of groundwater.

- Materials with very small pore spaces (such as clay) hinder or prevent groundwater movement and are called *aquitards*. *Aquifers* consist of materials with larger pore spaces (such as sand) that are permeable and transmit groundwater freely.

- *Springs* occur whenever the water table intersects the land surface and a natural flow of groundwater results. *Wells*, openings drilled into the zone of saturation, withdraw groundwater and create roughly conical depressions in the water table known as *cones of depression*. *Artesian wells* occur when water rises above the level at which it was initially encountered.

• When groundwater circulates at great depths, it becomes heated. If it rises, the water may emerge as a *hot spring*. *Geysers* occur when groundwater is heated in underground chambers, expands, and some water quickly changes to steam, causing the geyser to erupt. The source of heat for most hot springs and geysers is hot igneous rock.

• Some of the current environmental problems involving groundwater include (1) *overuse* by intense irrigation, (2) *land subsidence* caused by groundwater withdrawal, and (3) *contamination* by pollutants.

• Most *caverns* form in limestone at or below the water table when acidic groundwater dissolves rock along lines of weakness, such as joints and bedding planes. *Karst topography* exhibits an irregular terrain punctuated with many depressions, called *sinkholes*.

Key Terms

alluvial fan (p. 130)
alluvium (p. 122)
aquifer (p. 137)
aquitard (p. 137)
artesian well (p. 141)
backswamp (p. 129)
bar (p. 128)
base level (p. 124)
bed load (p. 124)
braided stream (p. 124)
capacity (p. 122)
cavern (p. 145)
competence (p. 122)
cone of depression (p. 141)
cutoff (p. 123)
delta (p. 128)
dendritic pattern (p. 130)
discharge (p. 120)
dissolved load (p. 121)
distributary (p. 128)

divide (p. 118)
drainage basin (p. 118)
drawdown (p. 141)
evapotranspiration (p. 116)
flood (p. 131)
floodplain (p. 127)
geyser (p. 139)
gradient (p. 119)
groundwater (p. 136)
hot spring (p. 138)
hydrologic cycle (p. 116)
incised meander (p. 127)
infiltration (p. 116)
karst topography (p. 146)
laminar flow (p. 119)
meander (p. 123)
natural levee (p. 129)
oxbow lake (p. 123)
permeability (p. 137)
porosity (p. 137)

radial pattern (p. 130)
rectangular pattern (p. 130)
runoff (p. 116)
sinkhole (sink) (p. 147)
sorting (p. 122)
spring (p. 138)
stalactite (p. 145)
stalagmite (p. 145)
stream valley (p. 126)
suspended load (p. 121)
transpiration (p. 116)
trellis pattern (p. 130)
turbulent flow (p. 119)
water table (p. 136)
well (p. 140)
yazoo tributary (p. 129)
zone of aeration (p. 136)
zone of saturation (p. 136)

Review Questions

1. Describe the movement of water through the hydrologic cycle. Once precipitation has fallen on land, what paths are available to it?

2. What are the three main parts (zones) of a river system?

3. A stream starts out 2000 meters above sea level and travels 250 kilometers to the ocean. What is its average gradient in meters per kilometer?

4. Suppose that the stream mentioned in Question 3 developed extensive meanders so that its course was lengthened to 500 kilometers. Calculate its new gradient. How does meandering affect gradient?

5. When the discharge of a stream increases, what happens to the stream's velocity?

6. In what three ways does a stream transport its load?

7. If you collect a jar of water from a stream, what part of its load will settle to the bottom of the jar? What portion will remain in the water? What part of a stream's load would probably *not* be present in your sample?

8. Differentiate between *competency* and *capacity*.

9. Are bedrock channels more likely to be found near the head or near the mouth of a stream?

10. Describe a situation that might cause a stream channel to become braided.

11. Define base level. Name the main river in your area. For what streams does it act as base level? What is the base level for the Mississippi River? The Missouri River?

12. Describe two situations that would trigger the formation of incised meanders.

13. Briefly describe the formation of a natural levee. How is this feature related to backswamps and yazoo tributaries?

14. List two major depositional features, other than natural levees, that are associated with streams. Under what circumstances does each form?

15. List and briefly describe three basic flood-control strategies. What are some drawbacks of each?

16. Each of the following statements refers to a particular drainage pattern. Identify the pattern. (a) Streams diverge from a central high area such as a volcano. (b) Streams form a branching, treelike pattern. (c) A pattern that develops when bedrock is crisscrossed by joints and faults.

17. What percentage of freshwater is groundwater (see Figure 5.23)? If glacial ice is excluded and only liquid freshwater is considered, about what percentage is groundwater?

18. Geologically, groundwater is important as an erosional agent. Name another significant geological role of groundwater.

19. Define *groundwater* and relate it to the water table.

20. How do *porosity* and *permeability* differ?

21. Distinguish between an *aquifer* and an *aquitard*.

22. What is the source of heat for most hot springs and geysers? How is this reflected in the distribution of these features?

23. What is meant by the term *artesian*? In order for artesian wells to exist, two conditions must be present. List these conditions.

24. What problem is associated with the pumping of groundwater for irrigation in the southern part of the High Plains?

25. Briefly explain what happened in the San Joaquin Valley of California as the result of excessive groundwater withdrawal.

26. Which would be most effective in purifying polluted groundwater: an aquifer composed mainly of coarse gravel, sand, or cavernous limestone?

27. Differentiate between stalactites and stalagmites. How do these features form?

28. If you were to explore an area that exhibited karst topography, what features might you find? This area would probably be underlain by what rock type? Name a region that exhibits such features.

29. Describe two ways in which sinkholes are created.

Examining the Earth System

1. List the process(es) involved in moving water through the hydrologic cycle from the (a) hydrosphere to the atmosphere, (b) atmosphere to the solid Earth, (c) biosphere to the atmosphere, and (d) hydrosphere (land) to the hydrosphere (ocean).

2. Over the oceans, evaporation exceeds precipitation, yet sea level does not drop. Why?

3. List at least three specific examples of interactions between humans and the hydrologic cycle (for example, the construction of a dam). Briefly describe the consequence(s) of each of these interactions.

4. Describe the role of the atmosphere, geosphere and biosphere in determining the quantity and quality of groundwater available for human consumption.

Online Study Guide

The *Earth Science* Website uses the resources and flexibility of the Internet to aid in your study of the topics in this chapter. Written and developed by Earth science instructors, this site will help improve your understanding of Earth science. Visit **http://www.prenhall.com/tarbuck** and click on the cover of *Earth Science* 11e to find:

- **On-line review quizzes.**
- **Critical thinking exercises.**
- **Links to chapter-specific Web resources.**
- **Internet-wide key term searches.**
http://www.prenhall.com/tarbuck

Franz Josef Glacier in New Zealand's Southern Alps.
(Photo by Colin Monteath)

CHAPTER 6

Glaciers, Deserts, and Wind

Like the running water and groundwater that were the focus of Chapter 5, glaciers and wind are significant erosional processes. They are responsible for creating many different landforms and are part of an important link in the rock cycle in which the products of weathering are transported and deposited as sediment.

Climate has a strong influence on the nature and intensity of Earth's external processes. This fact is dramatically illustrated in this chapter. The existence and extent of glaciers is largely controlled by Earth's changing climate. Another excellent example of the strong link between climate and geology is seen when we examine the development of arid landscapes.

Today glaciers cover nearly 10 percent of Earth's land surface; however, in the recent geologic past, ice sheets were three times more extensive, covering vast areas with ice thousands of meters thick. Many regions still bear the mark of these glaciers. The first part of this chapter examines glaciers and the erosional and depositional features they create. The second part is devoted to dry lands and the geologic work of wind. Because desert and near-desert conditions prevail over an area as large as that affected by the massive glaciers of the Ice Age, the nature of such landscapes is indeed worth investigating.

Glaciers: A Part of Two Basic Cycles in the Earth System

 GEODe Sculpturing Earth's Surface
▼ Glaciers

Many present-day landscapes were modified by the widespread glaciers of the most recent Ice Age and still strongly reflect the handiwork of ice. The basic character of such diverse places as the Alps, Cape Cod, and Yosemite Valley was fashioned by now vanished masses of glacial ice. Moreover, Long Island, the Great Lakes, and the fiords of Norway and Alaska all owe their existence to glaciers. Glaciers, of course, are not just a phenomenon of the geologic past. As you will see, they are still sculpting and depositing debris in many regions today.

Glaciers are a part of two fundamental cycles in the Earth system—the hydrologic cycle and the rock cycle. Earlier you learned that the water of the hydrosphere is constantly cycled through the atmosphere, biosphere, and geosphere. Time and time again the same water is evaporated from the oceans into the atmosphere, precipitated upon the land, and carried by rivers and underground flow back to the sea. However, when precipitation falls at high elevations or high latitudes, the water may not immediately make its way toward the sea. Instead, it may become part of a glacier. Although the ice will eventually melt, allowing the water to continue its path to the sea, water can be stored as glacial ice for many tens, hundreds, or even thousands of years.

A **glacier** is a thick ice mass that forms over hundreds or thousands of years. It originates on land from the accumulation, compaction, and recrystallization of snow. A glacier appears to be motionless, but it is not—glaciers move very slowly. Like running water, groundwater, wind, and waves, glaciers are dynamic erosional agents that accumulate, transport, and deposit sediment. As such, glaciers are among the processes that perform a very basic function in the rock cycle. Although glaciers are found in many parts of the world, most are located in remote areas, either near Earth's poles or in high mountains.

Valley (Alpine) Glaciers

Literally thousands of relatively small glaciers exist in lofty mountain areas, where they usually follow valleys originally occupied by streams. Unlike the rivers that previously flowed in these valleys, the glaciers advance slowly, perhaps only a few centimeters each day. Because of their setting, these moving ice masses are termed **valley glaciers** or **alpine glaciers** (See chapter-opening photo). Each glacier is a stream of ice, bounded by precipitous rock walls, that flows downvalley from a snow accumulation center near its head. Like rivers, valley glaciers can be long or short, wide or narrow, single or with branching tributaries. Generally, the widths of alpine glaciers are small compared to their lengths; some extend for just a fraction of a kilometer, whereas others go on for many dozens of kilometers. The west branch of the Hubbard Glacier, for example, runs through 112 kilometers (nearly 70 miles) of mountainous terrain in Alaska and the Yukon Territory.

Ice Sheets

In contrast to valley glaciers, **ice sheets** exist on a *much* larger scale. In fact, they are often referred to as *continental ice sheets*. These enormous masses flow out in all directions from one or more centers and completely obscure all but the highest areas of underlying terrain. Although many ice sheets have existed in the past, just two achieve this status at present (Figure 6.1). Nevertheless, their combined areas represent almost 10 percent of Earth's land area. In the Northern Hemisphere, Greenland is covered by an imposing ice sheet that occupies 1.7 million square kilometers (0.7 million square miles), or about 80 percent of this large island. Averaging nearly 1500 meters (5000 feet) thick, in places the ice extends 3000 meters (10,000 feet) above the island's bedrock floor.

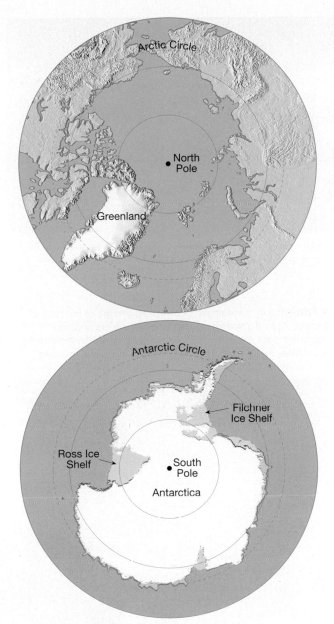

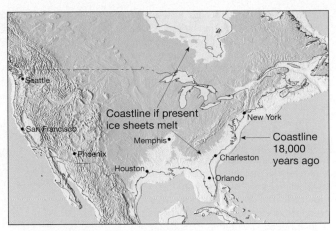

▲ **FIGURE 6.2** This map of a portion of North America shows the present-day coastline compared to the coastline that existed during the last ice-age maximum (18,000 years ago) and the coastline that would exist if present ice sheets in Greenland and Antarctica melted. (After R. H. Dott, Jr. and R. L. Battan, *Evolution of the Earth*, New York: McGraw-Hill, 1971.)

▲ **FIGURE 6.1** The only present-day continental ice sheets are those covering Greenland and Antarctica. Their combined areas represent almost 10 percent of Earth's land area. Greenland's ice sheet occupies 1.7 million square kilometers, or about 80 percent of the island. The area of the Antarctic Ice Sheet is almost 14 million square kilometers. Along portions of the Antarctic coast, glacial ice flows into bays, creating features called *ice shelves*. These large masses of floating ice remain attached to the land on one or more sides. Ice shelves occupy an additional 1.4 million square kilometers adjacent to the Antarctic Ice Sheet.

In the Southern Hemisphere, the huge Antarctic Ice Sheet attains a maximum thickness of nearly 4300 meters (14,000 feet) and covers an area of more than 13.9 million square kilometers (5.4 million square miles). The Antarctic Ice Sheet represents 80 percent of the world's ice and nearly two-thirds of Earth's fresh water. If this ice melted, sea level would rise an esti-

mated 60 to 70 meters (200 to 230 feet) and the ocean would inundate many densely populated coastal areas (Figure 6.2).

Other Types of Glaciers

In addition to valley glaciers and ice sheets, other types of glaciers are also identified. Covering some uplands and plateaus are masses of glacial ice called **ice caps**. Like ice sheets, ice caps completely bury the underlying landscape but are much smaller than the continental-scale features. Ice caps occur in many places, including Iceland and several of the large islands in the Arctic Ocean. Another type, known as **piedmont glaciers**, occupy broad lowlands at the bases of steep mountains and form when one or more valley glaciers emerge from the confining walls of mountain valleys. Here the advancing ice spreads out to form a broad sheet. The size of individual piedmont glaciers varies greatly. Among the largest is the broad Malaspina Glacier along the coast of southern Alaska. It covers more than 5000 square kilometers (2000 square miles) of the flat coastal plain at the foot of the lofty St. Elias range (Figure 6.3).

STUDENTS SOMETIMES ASK...
How many glaciers are there on Earth?

Approximately 160,000 glaciers occupy Earth's polar regions and high mountain environments. Since most glaciers are in remote and often inaccessible places, satellite images are indispensable in identifying and monitoring glaciers worldwide.

▶ **FIGURE 6.3** Malaspina glacier in southeastern Alaska is considered a classic example of a piedmont glacier. Piedmont glaciers occur where valley glaciers exit a mountain range onto broad lowlands, are no longer laterally confined, and spread to become wide lobes. Malaspina Glacier is actually a compound glacier, formed by the merger of several valley glaciers, the most prominent of which seen here are Agassiz Glacier (left) and Seward Glacier (right). In total, Malaspina Glacier is up to 65 kilometers (40 miles) wide and extends up to 45 kilometers (28 miles) from the mountain front nearly to the sea. This perspective view looking north covers an area about 55 kilometers × 55 kilometers

(34 × 34 miles). It was created from a Landsat satellite image and an elevation model generated by the Shuttle Radar Topography Mission (SRTM). Such images are excellent tools for mapping the geographic extent of glaciers and for determining whether such glaciers are thinning or thickening. (Image from NASA/JPL)

How Glaciers Move

GEODe
Sculpturing Earth's Surface
▼ Glaciers

The movement of glacial ice is generally referred to as *flow*. The fact that glacial movement is described in this way seems paradoxical—how can a solid flow? The way in which ice flows is complex and is of two basic types. The first of these, *plastic flow*, involves movement *within* the ice. Ice behaves as a brittle solid until the pressure upon it is equivalent to the weight of about 50 meters (165 feet) of ice. Once that load is surpassed, ice behaves as a plastic material and flow begins. A second and often equally important mechanism of glacial movement consists of the entire ice mass slipping along the ground. The lowest portions of most glaciers are thought to move by this sliding process.

The upper 50 meters or so of a glacier is not under sufficient pressure to exhibit plastic flow. Rather, the ice in this uppermost zone is brittle and is appropriately referred to as the *zone of fracture*. The ice in this zone is carried along "piggyback" style by the ice below. When the glacier moves over irregular terrain, the zone of fracture is subjected to tension, resulting in cracks called **crevasses** (Figure 6.4). These gaping cracks can make travel across glaciers dangerous and can extend to depths of 50 meters (165 feet). Beyond this depth, plastic flow seals them off.

Observing and Measuring Movement

Unlike streamflow, glacial movement is not obvious. If we could watch a valley glacier move, we would see that like the water in a river, all of the ice does not move downstream at the same rate. Flow is greatest in the center of the glacier because of the drag created by the walls and floor of the valley.

▼ **FIGURE 6.4** Crevasses form in the brittle ice of the zone of fracture. They can extend to depths of 50 meters and can obviously make travel across glaciers dangerous. Gasherbrum II expedition, Pakistan. (Photo by Galen Rowell/Mountain Light Photography, Inc.)

Early in the nineteenth century, the first experiments involving the movement of glaciers were designed and carried out in the Alps. Markers were placed in a straight line across an alpine glacier. The position of the line was marked on the valley walls so that if the ice moved, the change in position could be detected. Periodically the positions of the markers were noted, revealing the movement just described. Although most glaciers move too slowly for direct visual detection, the experiments succeeded in demonstrating that movement nevertheless occurs. The experiment illustrated in Figure 6.5 was carried out at Switzerland's Rhone Glacier later in the nineteenth century. It not only traced the movement of markers within the ice but also mapped the position of the glacier's terminus.

How rapidly does glacial ice move? Average rates vary considerably from one glacier to another. Some move so slowly that trees and other vegetation may become well established in the debris that accumulates on the glacier's surface. Others advance up to several meters per day. The movement of some glaciers is characterized by periods of extremely rapid advance followed by periods during which movement is practically nonexistent.

Budget of a Glacier

Snow is the raw material from which glacial ice originates. Therefore, glaciers form in areas where more snow falls in winter than can melt during the summer (See Box 6.1). Glaciers are constantly gaining and losing ice. Snow accumulation and ice formation occur in the **zone of accumulation** (Figure 6.6). Here the addition of snow thickens the glacier and promotes move-

ment. Beyond this area of ice formation is the **zone of wastage**. Here there is a net loss to the glacier when the snow from the previous winter melts, as does some of the glacial ice (Figure 6.6).

In addition to melting, glaciers also waste as large pieces of ice break off the front of a glacier in a process called *calving*. Where glaciers reach the sea, calving creates *icebergs* (Figure 6.7). Because icebergs are just slightly less dense than seawater, they float very low in the water, with more than 80 percent of their mass submerged. The margins of the Greenland Ice Sheet produce thousands of icebergs each year. Many drift southward and find their way into the North Atlantic, where they are a hazard to navigation.

Whether the margin of a glacier is advancing, retreating, or remaining stationary depends on the *budget* of the glacier. The glacial budget is the balance or lack of balance between accumulation at the upper end of a glacier and loss at the lower end. This loss is termed **ablation**. If ice accumulation exceeds ablation, the glacial front advances until the two factors balance. At this point, the terminus of the glacier becomes stationary.

If a warming trend increases ablation and/or if a drop in snowfall decreases accumulation, the ice front will retreat. As the terminus of the glacier retreats, the extent of the zone of wastage diminishes. Therefore, in time a new balance will be reached between accumulation and ablation, and the ice front will again become stationary.

Whether the margin of a glacier is advancing, retreating, or stationary, the ice within the glacier continues to flow forward. In the case of a receding glacier, the ice still flows forward, but not rapidly enough to offset wastage. This point is illustrated in

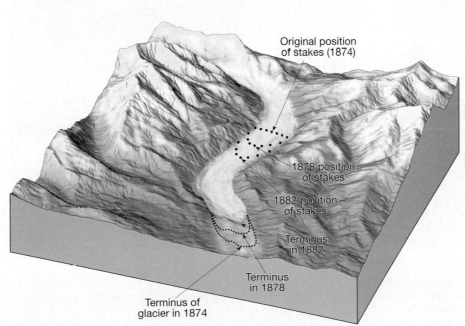

Original position of stakes (1874)
1878 position of stakes
1882 position of stakes
Terminus in 1882
Terminus in 1878
Terminus of glacier in 1874

◀ **FIGURE 6.5** Ice movement and changes in the terminus at Rhone Glacier, Switzerland. In this classic study of a valley glacier, the movement of stakes clearly showed that ice along the sides of the glacier moves slowest. Also notice that even though the ice front was retreating, the ice within the glacier was advancing.

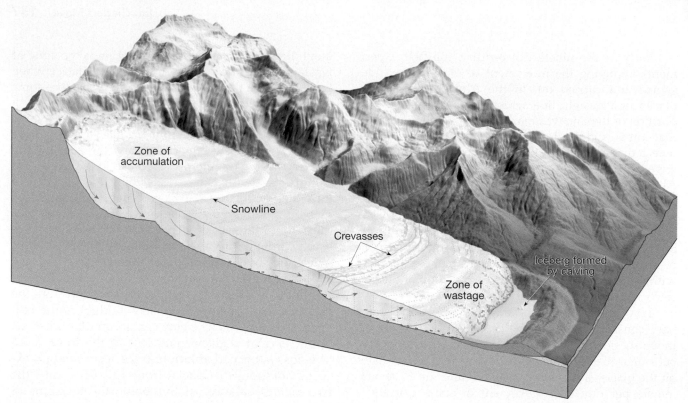

▲ **FIGURE 6.6** The snowline separates the zone of accumulation and the zone of wastage. Above the snowline, more snow falls each winter than melts each summer. Below the snowline, the snow from the previous winter completely melts, as does some of the underlying ice. Whether the margin of a glacier advances, retreats, or remains stationary depends on the balance or lack of balance between accumulation and wastage (ablation). When a glacier moves across irregular terrain, *crevasses* form in the brittle portion.

▲ **FIGURE 6.7** Icebergs are created when large pieces calve from the front of a glacier after it reaches a water body. Here ice is calving from the terminus of Alaska's Hubbard Glacier in Wrangell–St. Elias National Park, Alaska. Only about 20 percent of an iceberg protrudes above the water line. (Photo by Tom & Susan Bean, Inc.)

Figure 6.5. As the line of stakes within the Rhone Glacier continued to move downvalley, the terminus of the glacier slowly retreated upvalley.

Glacial Erosion

Sculpturing Earth's Surface
▼ Glaciers

Glaciers are capable of great erosion. For anyone who has observed the terminus of an alpine glacier, the evidence of its erosive force is clear. You can witness firsthand the release of rock fragments of various sizes from the ice as it melts. All signs lead to the conclusion that the ice has scraped, scoured, and torn rock debris from the floor and walls of the valley and carried it downvalley. Indeed, as a transporter of sediment, ice has no equal.

Once rock debris is acquired by the glacier, it cannot settle out as does the load carried by a stream or by the wind. Consequently, glaciers can carry huge blocks that no other erosional agent could possibly budge. Although today's glaciers are of limited importance as erosional agents, many landscapes that were modified by the widespread glaciers of the recent Ice Age still reflect to a high degree the work of ice.

How Glaciers Erode

Glaciers erode land primarily in two ways—plucking and abrasion. First, as a glacier flows over a fractured bedrock surface, it loosens and lifts blocks of rock and incorporates them into the ice. This process, known as **plucking**, occurs when meltwater penetrates the cracks and joints along the rock floor of the glacier and freezes. As the water expands, it exerts tremendous leverage that pries the rock loose. In this manner, sediment of all sizes becomes part of the glacier's load.

The second major erosional process is **abrasion** (Figure 6.8). As the ice and its load of rock fragments slide over bedrock, they function like sandpaper to

A.

B.

▲ **FIGURE 6.8   A.** Glacial abrasion created the scratches and grooves in this bedrock. Glacier Bay National Park, Alaska. (Photo © by Carr Clifton) **B.** Glacially polished granite in California's Yosemite National Park. (Photo by E. J. Tarbuck)

smooth and polish the surface below. The pulverized rock produced by the glacial gristmill is appropriately called **rock flour**. So much rock flour may be produced that meltwater streams leaving a glacier often have the grayish appearance of skim milk—visible evidence of the grinding power of the ice.

When the ice at the bottom of a glacier contains large rock fragments, long scratches and grooves called **glacial striations** may be gouged into the bedrock (Figure 6.8A). These linear scratches on the bedrock surface provide clues to the direction of glacial movement. By mapping the striations over large areas, glacial flow patterns can often be reconstructed. Conversely, not all abrasive action produces striations. The rock surface over which the glacier moves may also become highly polished by the ice and its load of finer particles. The broad expanses of smoothly polished granite in California's Yosemite National Park provide an excellent example (Figure 6.8B).

As is the case with other agents of erosion, the rate of glacial erosion is highly variable. This differential erosion by ice is largely controlled by four factors: (1) rate of glacial movement; (2) thickness of the ice; (3) shape,

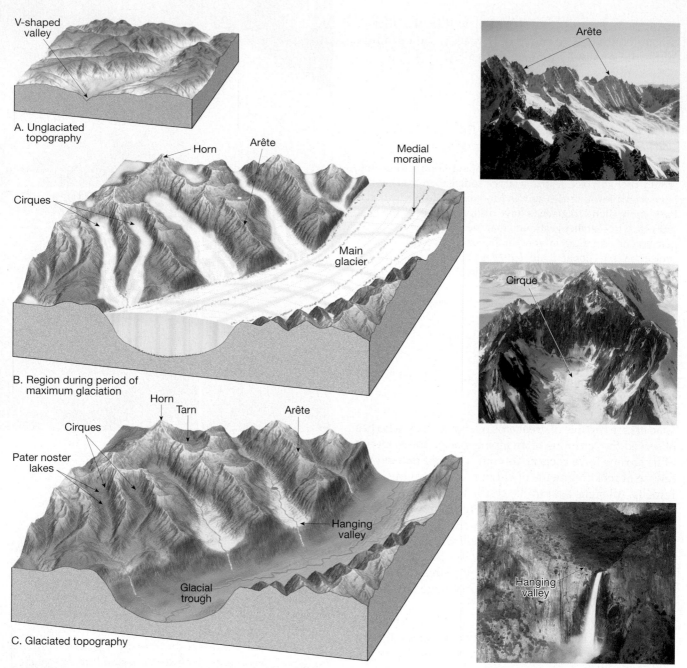

▲ **FIGURE 6.9** Erosional landforms created by alpine glaciers. The unglaciated landscape in part **A** is modified by valley glaciers in part **B.** After the ice recedes, in part **C,** the terrain looks very different than before glaciation. (Arête photo from James E. Patterson Collection, Cirque photo by Martin G. Miller, Hanging Valley photo by Marc Muench)

abundance, and hardness of the rock fragments contained in the ice at the base of the glacier; and (4) the erodibility of the surface beneath the glacier. Variations in any or all of these factors from time to time and/or from place to place mean that the features, effects, and degree of landscape modification in glaciated regions can vary greatly.

Landforms Created by Glacial Erosion

Although the erosional accomplishments of ice sheets can be tremendous, landforms carved by these huge ice masses usually do not inspire the same awe as do

the erosional features created by valley glaciers. In regions where the erosional effects of ice sheets are significant, glacially scoured surfaces and subdued terrain are the rule. By contrast, in mountainous areas, erosion by valley glaciers produces many truly spectacular features. Much of the rugged mountain scenery so celebrated for its majestic beauty is the product of erosion by valley glaciers.

Take a moment to study Figure 6.9, which shows a mountain setting before, during, and after glaciation. You will refer to this often in the following discussion.

◄ **FIGURE 6.10** Prior to glaciation, a mountain valley is typically narrow and V-shaped. During glaciation, an alpine glacier widens, deepens, and straightens the valley, creating the U-shaped glacial trough seen here. The string of lakes is called pater noster lakes. This valley is in Glacier National Park, Montana. (Photo by John Montagne)

Glaciated Valleys. Prior to glaciation, alpine valleys are characteristically V-shaped because streams are well above base level and are therefore downcutting (Figure 6.9A). However, in mountainous regions that have been glaciated, the valleys are no longer narrow. As a glacier moves down a valley once occupied by a stream, the ice modifies it in three ways: The glacier widens, deepens, and straightens the valley, so that what was once a narrow V-shaped valley is transformed into a U-shaped **glacial trough** (Figures 6.9C and 6.10).

The amount of glacial erosion depends in part on the thickness of the ice. Consequently, main glaciers, also called *trunk glaciers*, cut their valleys deeper than do their smaller tributary glaciers. Thus, after the ice has receded, the valleys of tributary glaciers are left standing above the main glacial trough and are termed **hanging valleys** (Figure 6.9C). Rivers flowing through hanging valleys can produce spectacular waterfalls, such as those in Yosemite National Park, California.

Cirques. At the head of a glacial valley is a characteristic and often imposing feature associated with an alpine glacier—a **cirque**. As Figure 6.9 illustrates, these bowl-shaped depressions have precipitous walls on three sides but are open on the downvalley side. The cirque is the focal point of the glacier's growth because it is the area of snow accumulation and ice formation. Cirques begin as irregularities in the mountainside that are subsequently enlarged by frost wedging and plucking along the sides and bottom of the glacier. The glacier in turn acts as a conveyor belt that carries away the debris. After the glacier has melted away, the cirque basin is sometimes occupied by a small lake.

Arêtes and Horns. The Alps, Northern Rockies, and many other mountain landscapes carved by valley glaciers reveal more than glacial troughs and cirques. In addition, sinuous, sharp-edged ridges called **arêtes** and sharp, pyramid-like peaks termed **horns** project above the surroundings (Figure 6.9C). Both features can originate from the same basic process: the enlargement of cirques produced by plucking and frost action. Several cirques around a single high mountain create the spires of rock called *horns*. As the cirques enlarge and converge, an isolated horn is produced. A famous example is the Matterhorn in the Swiss Alps (Figure 6.11).

▼ **FIGURE 6.11** Horns are sharp, pyramid-like peaks that are fashioned by alpine glaciers. This example is the famous Matterhorn in the Swiss Alps. (Photo by Gavriel Jecan)

BOX 6.1 UNDERSTANDING EARTH

Glacial Ice—A Storehouse of Climate Data

Climatology operates under a handicap when compared to many other sciences. In other fields of study, hypotheses can be tested by direct experimentation in the laboratory. However, this is usually not possible in the study of climate. Rather, scientists must construct computer models of how our planet's climate system works. If we understand the climate system correctly and construct the model appropriately, then the behavior of the model climate system should mimic the behavior of Earth's climate system.

One of the best ways to test such a model is to see if it can reproduce climate changes that have already occurred. To do this requires detailed climate records that extend back hundreds of thousands of years. Ice cores are an indispensable source of data for reconstructing past climates. Research based on vertical cores taken from the Greenland and Antarctic ice sheets has changed our basic understanding of how the climate system works.

Scientists collect samples with a drilling rig, like a small version of an oil drill. A hollow shaft follows the drill head into the ice, and an ice core is extracted. In this way, cores that sometimes exceed 2000 meters (6500 feet) in length and may represent more than 200,000 years of climate history are acquired for study (Figure 6.A).

The ice provides a detailed record of changing air temperatures and snowfall. Air bubbles trapped in the ice record variations in atmospheric composition. Changes in carbon dioxide and methane are linked to fluctuating temperatures. The cores also include atmospheric fallout such as windblown dust, volcanic ash, pollen, and modern-day pollution.

Past temperatures are determined by *oxygen isotope analysis*. This technique is based on precise measurement of the ratio between two isotopes of oxygen: O^{16}, which is the most common, and the heavier O^{18}. More O^{18} is evaporated from the oceans when temperatures are

Figure 6.A The National Ice Core Laboratory is a physical plant for storing and studying cores of ice taken from glaciers around the world. These cores represent a long-term record of material deposited from the atmosphere. The lab provides scientists with the capability to conduct examinations of ice cores, and it preserves the integrity of these samples in a repository for the study of global climate change and past environmental conditions. (Photo by USGS/National Ice Core Laboratory)

high, and less is evaporated when temperatures are low. Therefore, the heavier isotope is more abundant in the precipitation of warm eras and less abundant during colder periods. Using this principle, scientists are able to produce a record of past temperature changes. A portion of such a record is shown in Figure 6.B.

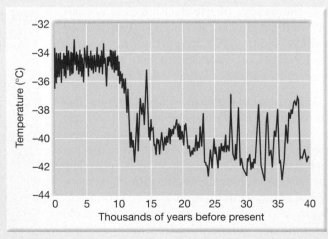

Figure 6.B This graph showing temperature variations over the past 40,000 years is derived from oxygen isotope analysis of ice cores recovered from the Greenland ice sheet. (After U. S. Geological Survey)

Arêtes can form in a similar manner except that the cirques are not clustered around a point but rather exist on opposite sides of a divide. As the cirques grow, the divide separating them is reduced to a very narrow, knifelike partition. An arête can also be created in another way. When two glaciers occupy parallel valleys, an arête can form when the land separating the moving tongues of ice is progressively narrowed as the glaciers scour and widen their valleys.

Fiords. **Fiords** are deep, often spectacular, steep-sided inlets of the sea that exist in many high-latitude areas of the world where mountains are adjacent to the ocean (Figure 6.12). Norway, British Columbia, Greenland, New Zealand, Chile, and Alaska all have coastlines characterized by fiords. They are glacial troughs that became submerged as the ice left the valley and sea level rose following the Ice Age.

The depths of some fiords can exceed 1000 meters (3300 feet). However, the great depths of these flooded troughs are only partly explained by the post–Ice Age rise in sea level. Unlike the situation governing the downward erosional work of rivers, sea level does not act as a base level for glaciers. As a consequence, glaciers are capable of eroding their beds far below the surface of the sea. For example, a valley glacier 300 meters (1000 feet) thick can carve its valley floor more than 250 meters (800 feet) below sea level before downward erosion ceases and the ice begins to float.

Glacial Deposits

Sculpturing Earth's Surface
▼ Glaciers

Glaciers pick up and transport a huge load of debris as they slowly advance across the land. Ultimately these materials are deposited when the ice melts. In regions where glacial sediment is deposited, it can play a truly significant role in forming the physical landscape. For example, in many areas once covered by the ice sheets of the recent Ice Age, the bedrock is rarely exposed, because glacial deposits that are dozens or even hundreds of meters thick completely mantle the terrain. The general effect of these deposits is to reduce the local relief and thus level the topography. Indeed, rural country scenes that are familiar to many of us—rocky pastures in New England, wheat fields in the Dakotas, rolling farmland in the Midwest—result directly from glacial deposition.

Types of Glacial Drift

Long before the theory of an extensive Ice Age was proposed, much of the soil and rock debris covering portions of Europe was recognized as coming from elsewhere. At the time, these foreign materials were believed to have been "drifted" into their present positions by floating ice during an ancient flood. As a consequence, the term *drift* was applied to this sediment. Although rooted in a concept that was not correct, this term was so well established by the time the

▼ **FIGURE 6.12** Like other fiords, this one at Tracy Arm, Alaska is a drowned glacial trough. (Photo by Tom & Susan Bean, Inc.)

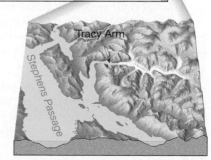

▲ FIGURE 6.14 **A.** A large, glacial erratic, central Wyoming. (Photo by Yva Momatiuk and John Eastcott/Photo Researchers, Inc.) **B.** Stone wall made of glacial erratics near West Bend, Wisconsin. (Photo by Tom Bean)

Close up
of cobble

▲ FIGURE 6.13 Glacial till is an unsorted mixture of many different sediment sizes. A close examination often reveals cobbles that have been scratched as they were dragged along by the glacier. (Photos by E. J. Tarbuck)

cording to the size and weight of the fragments. As ice is not capable of such sorting activity, these sediments are not deposited directly by the glacier. Rather, they reflect the sorting action of glacial meltwater.

Some deposits of stratified drift are made by streams issuing directly from the glacier. Other stratified deposits involve sediment that was originally laid down as till and later picked up, transported, and redeposited by meltwater beyond the margin of the ice. Accumulations of stratified drift often consist largely of sand and gravel, because the meltwater is not capable of moving larger material and because the finer rock flour remains suspended and is commonly carried far from the glacier. An indication that stratified drift consists primarily of sand and gravel can be seen in many areas where these deposits are actively mined as aggregate for road work and other construction projects.

Boulders found in the till or lying free on the surface are called **glacial erratics** if they are different from the bedrock below (Figure 6.14). Of course, this means that they must have been derived from a source outside the area where they are found. Although the locality of origin for most erratics is unknown, the origin of some can be determined. Therefore, by studying glacial erratics as well as the mineral composition of the till, geologists can sometimes trace the path of a lobe of ice. In portions of New England as well as other areas, erratics can be seen dotting pastures and farm fields. In some places, these rocks were cleared from fields and piled to make fences and walls (Figure 6.14 inset).

true glacial origin of the debris became widely recognized that it remained in the glacial vocabulary. Today, **glacial drift** is an all-embracing term for sediments of glacial origin, no matter how, where, or in what form they were deposited.

Glacial drift is divided into two distinct types: (1) materials deposited directly by the glacier, which are known as *till*, and (2) sediments laid down by glacial meltwater, called *stratified drift*. Here is the difference: **Till** is deposited as glacial ice melts and drops its load of rock debris. Unlike moving water and wind, ice cannot sort the sediment it carries; therefore, deposits of till are characteristically unsorted mixtures of many particle sizes (Figure 6.13). **Stratified drift** is sorted ac-

Moraines, Outwash Plains, and Kettles

Perhaps the most widespread features created by glacial deposition are *moraines*, which are simply layers or ridges of till. Several types of moraines are identified; some are common only to mountain valleys, and others are associated with areas affected by either ice sheets or valley glaciers. Lateral and medial moraines fall in the first category, whereas end moraines and ground moraines are in the second.

Lateral and Medial Moraines. The sides of a valley glacier accumulate large quantities of debris from the valley walls. When the glacier wastes away, these materials are left as ridges, called **lateral moraines**, along the sides of the valley (Figure 6.15). **Medial moraines** are formed when two valley glaciers coalesce to form a single ice stream. The till that was once carried along the edges of each glacier joins to form a single dark stripe of debris within the newly enlarged glacier. Creation of these dark stripes within the ice stream is one obvious proof that glacial ice moves, because the medial moraine could not form if the ice did not flow downvalley (Figure 6.15). It is common to see several medial moraines within a large alpine glacier, because a streak will form whenever a tributary glacier joins the main valley.

End Moraines and Ground Moraines. An **end moraine** is a ridge of till that forms at the terminus of a glacier. These relatively common landforms are deposited when a state of equilibrium is attained between ablation and ice accumulation. That is, the end moraine forms when the ice is melting near the end of the glacier at a rate equal to the forward advance of the glacier from its region of nourishment. Although the terminus of the glacier is stationary, the ice continues to flow forward, delivering a continuous supply of sediment in the same manner a conveyor belt delivers goods to the end of a production line. As the ice melts, the till is dropped and the end moraine grows. The longer the ice front remains stable, the larger the ridge of till will become.

Eventually the time comes when ablation exceeds nourishment. At this point, the front of the glacier begins to recede in the direction from which it originally advanced. However, as the ice front retreats, the conveyor-belt action of the glacier continues to provide fresh supplies of sediment to the terminus. In this manner a large quantity of till is deposited as the ice melts away, creating a rock-strewn, undulating plain. This gently rolling layer of till deposited as the ice front recedes is termed **ground moraine**. Ground moraine has a leveling effect, filling in low spots and clogging old stream channels, often leading to a derangement of the existing drainage system. In areas where this layer of till is still relatively fresh, such as the northern Great Lakes region, poorly drained swampy lands are quite common.

Periodically, a glacier will retreat to a point where ablation and nourishment once again balance. When this happens, the ice front stabilizes and a new end moraine forms.

The pattern of end moraine formation and ground moraine deposition may be repeated many times before the glacier has completely vanished. Such a pattern is illustrated by Figure 6.16. The very first end moraine to form marks the farthest advance of the glacier and is called the *terminal end moraine*. Those end moraines that form as the ice front occasionally stabilizes during retreat are termed *recessional end moraines*. Terminal and recessional moraines are essentially alike; the only difference between them is their relative positions.

◄ **FIGURE 6.15** Lateral moraines form from the accumulation of debris along the sides of a valley glacier. Medial moraines form when the lateral moraines of merging valley glaciers join. Medial moraines could not form if the ice did not advance downvalley. Therefore, these dark stripes are proof that glacial ice moves. Saint Elias National Park, Alaska. (Photo by Tom Bean)

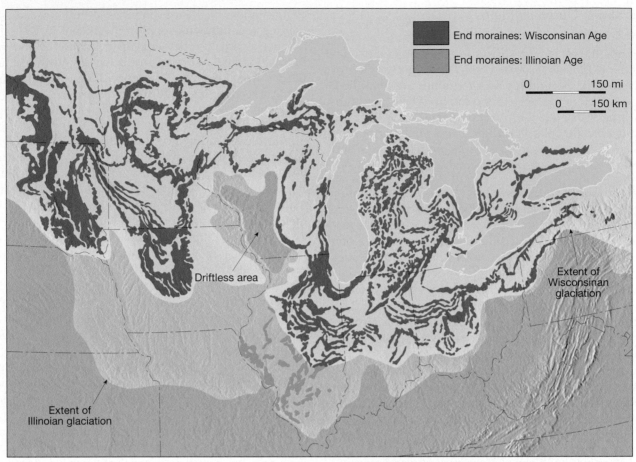

▲ **FIGURE 6.16** End moraines of the Great Lakes region. Those deposited during the most recent (Wisconsinan) stage are most prominent.

End moraines deposited by the most recent major stage of Ice Age glaciation are prominent features in many parts of the Midwest and Northeast. In Wisconsin, the wooded, hilly terrain of the Kettle Moraine near Milwaukee is a particularly picturesque example. A well-known example in the Northeast is Long Island. This linear strip of glacial sediment that extends northeastward from New York City is part of an end moraine complex that stretches from eastern Pennsylvania to Cape Cod, Massachusetts (Figure 6.17).

Figure 6.18 represents a hypothetical area during and following glaciation. It shows the end moraines that were just described as well as the depositional features that are discussed in the sections that follow. This figure depicts landscape features similar to what might be encountered if you were traveling in the upper Midwest or New England. As you read about

▶ **FIGURE 6.17** End moraines make up substantial parts of Long Island, Cape Cod, Martha's Vineyard, and Nantucket. Although portions are submerged, the Ronkonkoma moraine (a terminal end moraine) extends through central Long Island, Martha's Vineyard, and Nantucket. It was deposited about 20,000 years ago. The recessional Harbor Hill moraine, which formed about 14,000 years ago, extends along the north shore of Long Island, through southern Rhode Island and Cape Cod.

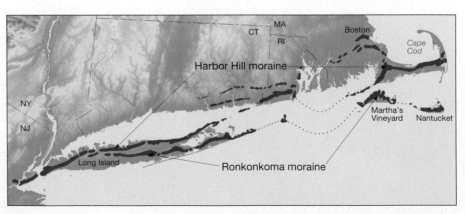

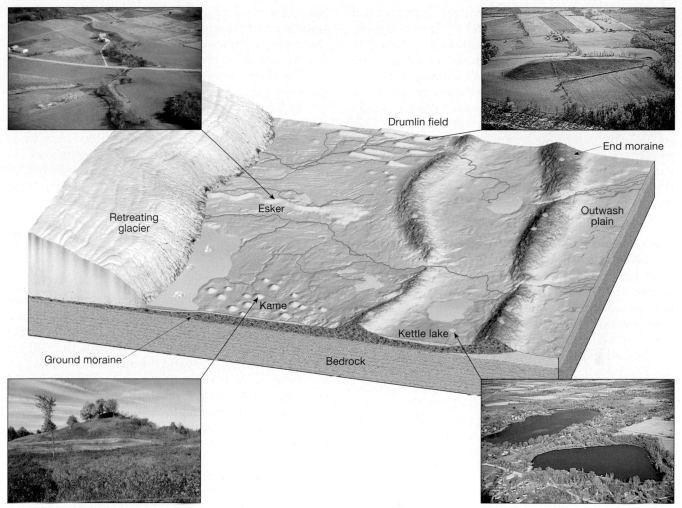

▲ **FIGURE 6.18** This hypothetical area illustrates many common depositional landforms. The outermost end moraine marks the limit of glacial advance and is called the *terminal end moraine*. End moraines that form as the ice front occasionally becomes stationary during retreat are called *recessional end moraines*. (Drumlin photo courtesy by Ward's Natural Science Establishment; Kame, Esker, and Kettle photos by Richard P. Jacobs/JLM Visuals)

other glacial deposits, you will be referred to this figure again.

Outwash Plains and Valley Trains. At the same time that an end moraine is forming, meltwater emerges from the ice in rapidly moving streams. Often they are choked with suspended material and carry a substantial bed load. As the water leaves the glacier, it rapidly loses velocity and much of its bed load is dropped. In this way a broad, ramplike accumulation of stratified drift is built adjacent to the downstream edge of most end moraines. When the feature is formed in association with an ice sheet, it is termed an **outwash plain**, and when it is confined to a mountain valley, it is usually referred to as a **valley train** (Figure 6.18).

Kettles. Often end moraines, outwash plains, and valley trains are pockmarked with basins or depressions known as **kettles** (Figure 6.18). Kettles form when blocks of stagnant ice become buried in drift and eventually melt, leaving pits in the glacial sediment. Most kettles do not exceed 2 kilometers in diameter, and the typical depth of most kettles is less than 10 meters (33 feet). Water often fills the depression and forms a pond or lake. One well-known example is Walden Pond near Concord, Massachusetts. It is here that Henry David Thoreau lived alone for two years in the 1840s and about which he wrote *Walden*, his classic of American literature.

Drumlins, Eskers, and Kames

Moraines are not the only landforms deposited by glaciers. Some landscapes are characterized by numerous elongate parallel hills made of till. Other areas exhibit conical hills and relatively narrow winding ridges composed mainly of stratified drift.

Drumlins.   **Drumlins** are streamlined asymmetrical hills composed of till (Figure 6.18). They range in height from 15 to 60 meters (50 to 200 feet) and average 0.4 to 0.8 kilometer (0.25–0.50 mile) in length. The steep side of the hill faces the direction from which the ice advanced, whereas the gentler slope points in the direction the ice moved. Drumlins are not found singly but rather occur in clusters, called *drumlin fields.* One such cluster, east of Rochester, New York, is estimated to contain about 10,000 drumlins. Their streamlined shape indicates that they were molded in the zone of flow within an active glacier. It is thought that drumlins originate when glaciers advance over previously deposited drift and reshape the material.

Eskers and Kames.   In some areas that were once occupied by glaciers, sinuous ridges composed largely of sand and gravel might be found. These ridges, called **eskers,** are deposits made by streams flowing in tunnels beneath the ice, near the terminus of a glacier (Figure 6.18). They may be several meters high and extend for many kilometers. In some areas they are mined for sand and gravel, and for this reason, eskers are disappearing in some localities.

Kames are steep-sided hills that, like eskers, are composed of sand and gravel (Figure 6.18). Kames originate when glacial meltwater washes sediment into openings and depressions in the stagnant wasting terminus of a glacier. When the ice eventually melts away, the stratified drift is left behind as mounds or hills.

Glaciers of the Ice Age

In the preceding pages, we mentioned the Ice Age, a time when ice sheets and alpine glaciers were far more extensive than they are today. There was a time when the most popular explanation for what we now know to be glacial deposits was that the material had been drifted in by means of icebergs or perhaps simply swept across the landscape by a catastrophic flood. However, during the nineteenth century, field investigations by many scientists provided convincing evidence that an extensive Ice Age was responsible for these deposits and for many other features.

By the beginning of the twentieth century, geologists had largely determined the extent of Ice Age glaciation. Further, they discovered that many glaciated regions had not one layer of drift but several. Close examination of these older deposits showed well-developed zones of chemical weathering and soil formation as well as the remains of plants that require warm temperatures. The evidence was clear: There had not been just one glacial advance but many, each separated by extended periods when climates were as warm or warmer than the present. The Ice Age had not simply been a time when the ice advanced over the land, lingered for a while, and then receded. Rather, the period was a very complex event characterized by a number of advances and withdrawals of glacial ice.

The glacial record on land is punctuated by many erosional gaps. This makes it difficult to reconstruct the episodes of the Ice Age. But sediment on the ocean floor provides an uninterrupted record of climate cycles for this period. Studies of cores drilled from these seafloor sediments show that glacial/interglacial cycles have occurred about every 100,000 years. Approximately 20 such cycles of cooling and warming were identified for the span we call the Ice Age.

During the glacial age, ice left its imprint on almost 30 percent of Earth's land area, including about 10 million square kilometers of North America, 5 million square kilometers of Europe, and 4 million square kilometers of Siberia (Figure 6.19). The amount of glacial ice in the Northern Hemisphere was roughly twice that of the Southern Hemisphere. The primary reason is that the Southern Hemisphere has little land in the middle latitudes, and therefore the southern polar ice could not spread far beyond the margins of Antarctica. By contrast, North America and Eurasia provided great expanses of land for the spread of ice sheets.

Today we know that the Ice Age began between 2 million and 3 million years ago. This means that most of the major glacial episodes occurred during a division of the geologic time scale called the **Pleistocene epoch**. Although the Pleistocene is commonly used as a synonym for the Ice Age, this epoch does not encom-

▼ **FIGURE 6.19**   Maximum extent of ice sheets in the Northern Hemisphere during the Ice Age.

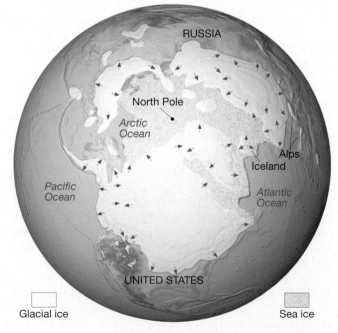

Glacial ice   Sea ice

pass it all. The Antarctic Ice Sheet, for example, formed at least 14 million years ago and, in fact, might be much older.

Some Indirect Effects of Ice Age Glaciers

In addition to the massive erosional and depositional work carried on by Pleistocene glaciers, the ice sheets had other, sometimes profound, effects on the landscape. For example, as the ice advanced and retreated, animals and plants were forced to migrate. This led to stresses that some organisms could not tolerate. Furthermore, many present-day stream courses bear little resemblance to their preglacial routes. The Missouri River once flowed northward toward Canada's Hudson Bay. The Mississippi River followed a path through central Illinois, and the head of the Ohio River reached only as far as Indiana (Figure 6.20). Other rivers that today carry only a trickle of water but nevertheless occupy broad channels are a testimony to the fact that they once carried torrents of glacial meltwater.

In areas that were centers of ice accumulation, such as Scandinavia and the Canadian Shield, the land has been slowly rising for the past several thousand years. Uplifting of almost 300 meters (1000 feet) has occurred in the Hudson Bay region. This, too, is the result of continental ice sheets. But how does glacial ice cause such vertical movement? We now understand that the land is rising because the added weight of the 3-kilometer-thick (2-mile-thick) mass of ice caused downwarping of Earth's crust. Following removal of this immense load, the crust has been adjusting by gradually rebounding upward ever since.*

A far-reaching effect of the Ice Age was the worldwide change in sea level that accompanied each advance and retreat of the ice sheets. The snow that nourishes glaciers ultimately comes from moisture evaporated from the oceans. Therefore, when the ice sheets increased in size, sea level fell and the shoreline shifted seaward. Estimates suggest that sea level was as much as 100 meters (330 feet) lower than today. Thus, land that is presently flooded by the oceans was dry. The Atlantic Coast of the United States lay more than 100 kilometers (60 miles) to the east of New York City (see Figure 6.2, p. 155). Moreover, France and Britain were joined where the English Channel is today, Alaska and Siberia were connected across the

*For a more complete discussion of this concept, termed *isostatic adjustment*, see the section on "Isostasy" in Chapter 10, pp. 303.

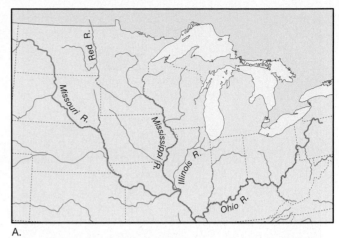

A.

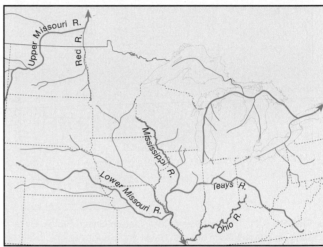

B.

▲ **FIGURE 6.20** **A.** This map shows the Great Lakes and the familiar present-day pattern of rivers in the central United States. Pleistocene ice sheets played a major role in creating this pattern. **B.** Reconstruction of drainage systems in the central United States prior to the Ice Age. The pattern was very different from today, and there were no Great Lakes.

Bering Strait, and Southeast Asia was tied by dry land to the islands of Indonesia.

The formation and growth of ice sheets was an obvious response to significant changes in climate. But the existence of the glaciers themselves triggered climatic changes in the regions beyond their margins. In arid and semiarid areas on all continents, temperatures were lower, which meant evaporation rates were also lower. At the same time, precipitation was moderate. This cooler, wetter climate resulted in the formation of many lakes called **pluvial lakes** (from the Latin term *pluvia*, meaning rain). In North America, pluvial lakes were concentrated in the vast Basin and Range region of Nevada and Utah (Figure 6.21). Although most are now gone, a few remnants remain, the largest being Utah's Great Salt Lake.

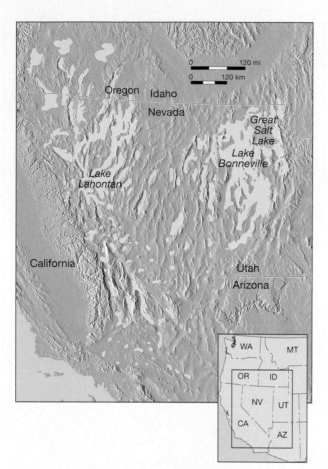

▲ **FIGURE 6.21**   Pluvial lakes of the Western United States. By far the largest pluvial lake in the vast Basin and Range region of Nevada and Utah was Lake Bonneville. With maximum depths exceeding 300 meters and an area of 50,000 square kilometers, Lake Bonneville was nearly the same size as present-day Lake Michigan. The Great Salt Lake is a remnant of this huge pluvial lake. (After R. F. Flint, *Glacial and Quaternary Geology*, New York: John Wiley & Sons)

Causes of Glaciation

A great deal is known about glaciers and glaciation. Much has been learned about glacier formation and movement, the extent of glaciers past and present, and the features created by glaciers, both erosional and depositional. However, the causes of glacial ages are not completely understood.

Although widespread glaciation has been rare in Earth's history, the Pleistocene Ice Age is not the only glacial period for which a record exists. Two Precambrian glacial episodes have been identified in the geologic record, the first approximately 2 billion years ago and the second about 600 million years ago. Further, a well-documented record of an earlier glacial age is found in late Paleozoic rocks that are about 250 million years old and which exist on several landmasses.*

---

*The terms *Precambrian* and *Paleozoic* refer to time spans on the geologic time scale of Earth history. For more on the geologic time scale, see Chapter 11 and Figure 11.16.

Any theory that attempts to explain the causes of glacial ages must successfully answer two basic questions: (1) *What causes the onset of glacial conditions?* For continental ice sheets to have formed, average temperatures must have been somewhat lower than at present and perhaps substantially lower than throughout much of geologic time. Thus, a successful theory would have to account for the cooling that finally leads to glacial conditions. (2) *What caused the alternation of glacial and interglacial stages that have been documented for the Pleistocene epoch?* The first question deals with long-term trends in temperature on a scale of millions of years, but this second question relates to much shorter-term changes.

Although the literature of science contains many hypotheses relating to the possible causes of glacial periods, we will discuss only a few major ideas to summarize current thought.

Plate Tectonics

Probably the most attractive proposal for explaining the fact that extensive glaciations have occurred only a few times in the geologic past comes from the theory of plate tectonics.† Because glaciers can form only on land, we know that landmasses must exist somewhere in the higher latitudes before an ice age can commence. Many scientists suggest that ice ages have occurred only when Earth's shifting crustal plates have carried the continents from tropical latitudes to more poleward positions.

Glacial features in present-day Africa, Australia, South America, and India indicate that these regions, which are now tropical or subtropical, experienced an ice age near the end of the Paleozoic era, about 250 million years ago. However, there is no evidence that ice sheets existed during this same period in what are today the higher latitudes of North America and Eurasia. For many years this puzzled scientists. Was the climate in these relatively tropical latitudes once like it is today in Greenland and Antarctica? Why did glaciers not form in North America and Eurasia? Until the plate tectonics theory was formulated, there had been no reasonable explanation.

Today scientists realize that the areas containing these ancient glacial features were joined together as a single supercontinent (Pangaea) located at latitudes far to the south of their present positions. Later this landmass broke apart, and its pieces, each moving on a different plate, migrated toward their present locations (Figure 6.22). It is now understood that during the geologic past, plate movements accounted for many dramatic climatic changes as landmasses shifted in relation to one another and moved to different latitudinal positions. Changes in

---

†A complete discussion of plate tectonics is presented in Chapter 8.

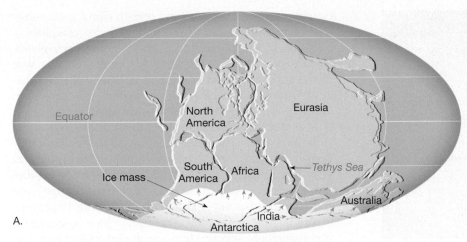

A.

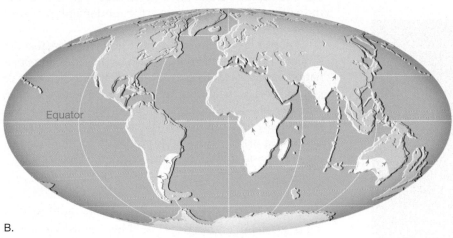

B.

◀ **FIGURE 6.22** **A.** The supercontinent Pangaea showing the area covered by glacial ice 300 million years ago. **B.** The continents as they are today. The white areas indicate regions where evidence of the old ice sheets exists.

oceanic circulation also must have occurred, altering the transport of heat and moisture and consequently the climate as well. Because the rate of plate movement is very slow—a few centimeters per year—appreciable changes in the positions of the continents occur only over great spans of geologic time. Thus, climatic changes brought about by shifting plates are extremely gradual and occur on a scale of millions of years.

Variations in Earth's Orbit

Because climatic changes brought about by moving plates are extremely gradual, the plate tectonics theory cannot be used to explain the alternation between glacial and interglacial climates that occurred during the Pleistocene epoch. Therefore, we must look to some other triggering mechanism that may cause climatic change on a scale of thousands rather than millions of years. Today many scientists strongly suspect that the climatic oscillations that characterized the Pleistocene may be linked to variations in Earth's orbit. This hypothesis was first developed and strongly advocated by the Yugoslavian scientist Milutin Milankovitch and is based on the premise that variations in incoming solar radiation are a principal factor in controlling Earth's climate.

Milankovitch formulated a comprehensive mathematical model based on the following elements (Figure 6.23):

1. Variations in the shape (*eccentricity*) of Earth's orbit about the Sun;
2. Changes in *obliquity*—that is, changes in the angle that the axis makes with the plane of Earth's orbit; and
3. The wobbling of Earth's axis, called *precession*.

Using these factors, Milankovitch calculated variations in the receipt of solar energy and the corresponding surface temperature of Earth back into time in an attempt to correlate these changes with the climatic fluctuations of the Pleistocene. In explaining climatic changes that result from these three variables, note that they cause little or no variation in the total solar energy reaching the ground. Instead, their impact is felt because they change the degree of contrast between the seasons. Somewhat milder winters in the middle to high latitudes means greater snowfall totals, whereas cooler summers would bring a reduction in snowmelt.

Among the studies that added considerable credibility to this astronomical hypothesis is one in which

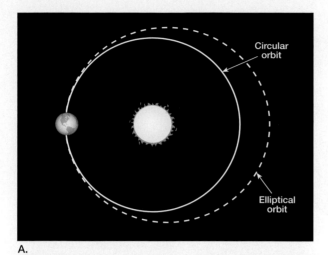

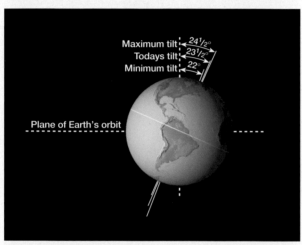

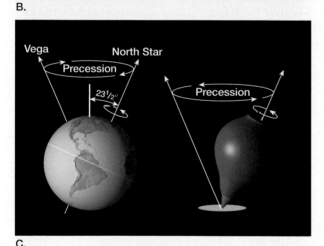

▲ **FIGURE 6.23** Orbital variations. **A.** The shape of Earth's orbit changes during a cycle that spans about 100,000 years. It gradually changes from nearly circular to one that is more elliptical and then back again. This diagram greatly exaggerates the amount of change. **B.** Today the axis of rotation is tilted about 23.5° to the plane of Earth's orbit. During a cycle of 41,000 years, this angle varies from 22° to 24.5°. **C.** Precession. Earth's axis wobbles like that of a spinning top. Consequently, the axis points to different spots in the sky during a cycle of about 26,000 years.

deep-sea sediments containing certain climatically sensitive microorganisms were analyzed to establish a chronology of temperature changes going back nearly 500,000 years.* This time scale of climatic change was then compared to astronomical calculations of eccentricity, obliquity, and precession to determine if a correlation did indeed exist. Although the study was very involved and mathematically complex, the conclusions were straightforward. The researchers found that major variations in climate over the past several hundred thousand years were closely associated with changes in the geometry of Earth's orbit; that is, cycles of climatic change were shown to correspond closely with the periods of obliquity, precession, and orbital eccentricity. More specifically, the researchers stated: "It is concluded that changes in the earth's orbital geometry are the fundamental cause of the succession of Quaternary ice ages."[†]

Let us briefly summarize the ideas that were just described. The theory of plate tectonics provides an explanation for the widely spaced and nonperiodic onset of glacial conditions at various times in the geologic past, whereas the astronomical model proposed by Milankovitch and supported by the work of J. D. Hays and his colleagues furnishes an explanation for the alternating glacial and interglacial episodes of the Pleistocene.

Other Factors

Variations in Earth's orbit correlate closely with the timing of glacial-interglacial cycles. However, the variations in solar energy reaching Earth's surface caused by these orbital changes do not adequately explain the magnitude of the temperature changes that occurred during the most recent ice age. Other factors must also have contributed. One factor involves variations in the chemical composition of the atmosphere. Other influences involve changes in the reflectivity of Earth's surface and in ocean circulation. Let's take a brief look at these factors.

Chemical analyses of air bubbles that become trapped in glacial ice at the time of ice formation indicate that the ice-age atmosphere contained less of the gases carbon dioxide and methane than the post ice-age atmosphere (see Box 6.1). Carbon dioxide and methane are important "greenhouse" gases, which means that they trap radiation emitted by Earth and contribute to the heating of the atmosphere.[‡] When the amount of carbon dioxide and methane in the atmosphere increase, global temperatures rise, and when there is a reduction in these gases, as occurred during the Ice Age, temperatures fall. Therefore, reductions in the concentrations of greenhouse gases help explain

*J. D. Hays, John Imbrie, and N. J. Shackelton, "Variations in the Earth's Orbit: Pacemaker of the Ice Ages", *Science* 194 (1976): 1121–32.

†J. D. Hays et al., p. 1131. The term *Quaternary* refers to the period on the geologic time scale that encompasses the last 1.6 million years.

‡For more on this idea, see the section on "Heating the Atmosphere: The Greenhouse Effect" in Chapter 16 and the discussion on "Carbon Dioxide, Trace Gases, and Global Warming" in Chapter 20.

the magnitude of the temperature drop that occurred during glacial times. Although scientists know that concentrations of carbon dioxide and methane dropped, they do not know what caused the drop. As often occurs in science, observations gathered during one investigation yield information and raise questions that require further analysis and explanation.

Obviously, whenever Earth enters an ice age, extensive areas of land that were once ice free are covered with ice and snow. In addition, a colder climate causes the area covered by sea ice (frozen surface sea water) to expand as well. Ice and snow reflect a large portion of incoming solar energy back to space. Thus, energy that would have warmed Earth's surface and the air above is lost and global cooling is reinforced.*

Yet another factor that influences climate during glacial times related to ocean currents, which, as you will learn in Chapter 15, are a complex matter. Research has shown that ocean circulation changes during ice ages. For example, studies suggest that the warm current that transports large amounts of heat from the tropics toward higher latitudes in the North Atlantic was significantly weaker during the ice age. This would lead to a colder climate in Europe, amplifying the cooling attributable to orbital variations.

In conclusion, it should be noted that our understanding of the causes of glacial ages is not complete. The ideas that were just discussed do not represent all of the possible explanations. Additional factors may be, and probably are, involved.

Deserts

Sculpturing Earth's Surface
▼ **Deserts**

The dry regions of the world encompass about 42 million square kilometers—a surprising 30 percent of Earth's land surface (Figure 6.24). No other climatic group covers so large a land area.[†] The word *desert* literally means deserted or unoccupied. For many dry regions this is a very appropriate description. Yet where water is available in deserts, plants and animals thrive. Nevertheless, the world's dry regions are among the least familiar land areas on Earth outside of the polar realm.

Desert landscapes frequently appear stark. Their profiles are not softened by a carpet of soil and abundant plant life. Instead, barren rocky outcrops with steep, angular slopes are common. At some places the rocks are tinted orange and red. At others they are gray and brown and streaked with black. For many

---

*Recall from Chapter 1 that something that reinforces (adds to) the initial change is called a *positive feedback mechanism*. To review this idea, see the discussion on feedback mechanisms in the section on "Earth as a System" in Chapter 1.

[†]An examination of dry climates is found in Chapter 20.

▲ **FIGURE 6.24** In this view of Earth from space, North Africa's Sahara Desert, the adjacent Arabian Desert, and the Kalahari and Namib deserts in southern Africa are clearly visible as tan-colored, cloud-free zones. These low-latitude deserts are dominated by the dry, subsiding air associated with pressure belts known as the *subtropical highs*. By contrast, the band of clouds that extends across central Africa and the adjacent oceans coincides with the equatorial low-pressure belt, the rainiest region on Earth. (Image courtesy of NASA)

visitors desert scenery exhibits a striking beauty; to others the terrain seems bleak. No matter which feeling is elicited, it is clear that deserts are very different from the more humid places where most people live.

As you will see, arid regions are not dominated by a single geologic process. Rather, the effects of tectonic forces, running water, and wind are all apparent. Because these processes combine in different ways from place to place, the appearance of desert landscapes varies a great deal as well (Figure 6.25).

STUDENTS SOMETIMES ASK...

I heard somewhere that deserts are expanding. Is that actually occurring?

Yes. The problem is called *desertification*, and it refers to the alteration of land to desertlike conditions as the result of human activities. It commonly takes place on the margins of deserts and results primarily from inappropriate land use. It is triggered when the modest natural vegetation in marginal areas is removed by plowing or grazing. When drought occurs, as it inevitably does in these regions, and the vegetative cover has been destroyed beyond the minimum to hold the soil against erosion, the destruction becomes irreversible. Desertification is occurring in many places but is particularly serious in the region south of the Sahara Desert known as the Sahel.

▲ **FIGURE 6.25** Scene in Nevada's Great Basin Desert. The appearance of desert landscapes varies a great deal from place to place. (Photo by Charlie Ott/Photo Researchers, Inc.)

Geologic Processes in Arid Climates

GEODe

Sculpturing Earth's Surface
▼ Deserts

The angular rock exposures, the sheer canyon walls, and the pebble- or sand-covered surface of the desert contrast sharply with the rounded hills and curving slopes of more humid places. To a visitor from a humid region, a desert landscape may seem to have been shaped by forces different from those operating in wetter areas. However, although the contrasts might be striking, they do not reflect different processes. They merely disclose the differing effects of the same processes that operate under contrasting climatic conditions.

Weathering

In humid regions, relatively well-developed soils support an almost continuous cover of vegetation. Here the slopes and rock edges are rounded. Such a landscape reflects the strong influence of chemical weathering in a humid climate. By contrast, much of the weathered debris in deserts consists of unaltered rock and mineral fragments—the result of mechanical weathering processes. In dry lands rock weathering of any type is greatly reduced because of the lack of moisture and the scarcity of organic acids from decaying plants. Chemical weathering, however, is not com-

pletely lacking in deserts. Over long spans of time, clays and thin soils do form, and many iron-bearing silicate minerals oxidize, producing the rust-colored stain found tinting some desert landscapes.

The Role of Water

Permanent streams are normal in humid regions, but practically all desert streams are dry most of the time (Figure 6.26A). Desert streams are said to be **ephemeral**, which means that they carry water only in response to specific episodes of rainfall. A typical ephemeral stream might flow only a few days or perhaps just a few hours during the year. In some years the channel may carry no water at all.

This fact is obvious even to the casual observer who, while traveling in a dry region, notices the number of bridges with no streams beneath them or the number of dips in the road where dry channels cross. However, when the rare heavy showers do occur, so much rain falls in such a short time that all of it cannot soak in. Because the vegetative cover is sparse, runoff is largely unhindered and consequently rapid, often creating flash floods along valley floors (Figure 6.26B). Such floods, however, are quite unlike floods in humid regions. A flood on a river such as the Mississippi may take many days to reach its crest and then subside. But desert floods arrive suddenly and subside quickly. Because much of the surface material is not anchored by vegetation, the amount of erosional work that occurs during a single short-lived rain event is impressive.

A.   B.

▲ **FIGURE 6.26** **A.** Most of the time, desert stream channels are dry. **B.** An ephemeral stream shortly after a heavy shower. Although such floods are short-lived, large amounts of erosion occur. (Photos by E. J. Tarbuck)

In the dry western United States a number of different names are used for ephemeral streams. Two of the most common are *wash* and *arroyo*. In other parts of the world, a dry desert stream may be called a *wadi* (Arabia and North Africa), a *donga* (South America), or a *nullah* (India).

Humid regions are notable for their integrated drainage systems. But in arid regions streams usually lack an extensive system of tributaries. In fact, a basic characteristic of desert streams are that they are small and die out before reaching the sea. Because the water table is usually far below the surface, few desert streams can draw upon it as streams do in humid regions. Without a steady supply of water, the combination of evaporation and infiltration soon depletes the stream.

The few permanent streams that do cross arid regions, such as the Colorado and Nile rivers, originate *outside* the desert, often in well-watered mountains. Here the water supply must be great to compensate for the losses occurring as the stream crosses the desert (Box 6.2). For example, after the Nile leaves the lakes and mountains of central Africa that are its source, it traverses almost 3000 kilometers (nearly 1900 miles) of the Sahara *without a single tributary*. By contrast, in humid regions the discharge of a river increases as it flows downstream because tributaries and groundwater contribute additional water along the way.

It should be emphasized that *running water, although infrequent, nevertheless does most of the erosional work in deserts*. This is contrary to a common belief that wind is the most important erosional agent sculpturing desert landscapes. Although wind erosion is indeed more significant in dry areas than elsewhere, most desert landforms are nevertheless carved by run-

ning water. As you will see shortly, the main role of wind is in the transportation and deposition of sediment, which creates and shapes the ridges and mounds we call dunes.

Basin and Range: The Evolution of a Mountainous Desert Landscape

Because arid regions typically lack permanent streams, they are characterized as having **interior drainage**. This means that they have a discontinuous pattern of intermittent streams that do not flow out of the desert to the ocean. In the United States, the dry Basin and Range region provides an excellent example. The region includes southern Oregon, all of Nevada, western Utah, southeastern California, southern Arizona, and southern New Mexico. The name Basin and Range is an apt description for this almost 800,000-square-kilometer (more than 300,000-square-mile) region, because it is characterized by more than 200 relatively small mountain ranges that rise 900 to 1500 meters (3000 to 5000 feet) above the basins that separate them.

In this region, as in others like it around the world, most erosion occurs without reference to the ocean (ultimate base level), because the interior drainage never reaches the sea. Even where permanent streams flow to the ocean, few tributaries exist, and thus only a narrow strip of land adjacent to the stream has sea level as its ultimate level of land reduction.

The block models in Figure 6.27 depict how the landscape has evolved in the Basin and Range region. During and following uplift of the mountains, running water begins carving the elevated mass and depositing large quantities of debris in the basin. In this early stage, relief is greatest, and as erosion lowers

175

BOX 6.2 PEOPLE AND THE ENVIRONMENT:
The Disappearing Aral Sea

The Aral Sea lies on the border between Uzbekistan and Kazakhstan in central Asia (Figure 6.C). The setting is the Turkestan desert, a middle-latitude desert in the rainshadow of Afghanistan's high mountains. In this region of interior drainage, two large rivers, the Amu Darya and the Syr Darya, carry water from the mountains of northern Afghanistan across the desert to the Aral Sea. Water leaves the sea by evaporation. Thus, the size of the water body depends upon the balance between river inflow aned evaporation.

In 1960 the Aral Sea was one of the world's largest inland water bodies, with an area of about 67,000 square kilometers (26,000 square miles). Only the Caspian Sea, Lake Superior, and Lake Victoria were larger. By the year 2000 the area of the Aral Sea was less than 50 percent of its 1960 size, and its volume was reduced by 80 percent. The shrinking of this water body is depicted in Figure 6.D. By about 2010 all that will remain will be three shallow remnants.

What caused the Aral Sea to dry up over the past 40 years? The answer is that the flow of water from the mountains that supplied the sea was significantly reduced and then all but eliminated. As recently as 1965, the Aral Sea received about 50 cubic kilometers (12 cubic miles) of fresh water per year. By the early 1980s this number fell to nearly zero. The reason was that the waters of the Amu Darya and Syr Darya were diverted to supply a major expansion of irrigated agriculture in this dry realm.

The intensive irrigation greatly increased agricultural productivity, but not without significant costs. The deltas of the two major rivers have lost their wetlands, and wildlife has dissappeared. The once thriving fishing industry is dead, and the 24 species of fish that once lived in the Aral Sea are no longer there. The shoreline is now tens of kilometers from the towns that were once fishing centers (Figure 6.E).

The shrinking sea has exposed millions of acres of former seabed to sun and wind. The surface is encrusted with salt and with agrochemicals brought by the rivers. Strong winds routinely pick up and deposit thousands of tons of newly exposed material every year. This process has not only contributed to a significant reduction in air quality for people living in the region but has also appreciably affected crop yields due to the deposition of salt-rich sediments on arable land.

The shrinking Aral Sea has had a noticeable impact on the region's climate. Without the moderating effect of a large water body, there are greater extremes of temperature, a shorter growing season, and reduced local precipitation. These changes have caused many farms to switch from growing cotton to growing rice, which demands even more diverted water.

Environmental experts agree that the current situation cannot be sustained. Could this crisis be reversed if enough fresh water were to once again flow into the Aral Sea?

Prospects appear grim. Experts estimate that restoring the Aral Sea to about twice its present size would require stopping all irrigation from the two major rivers for 50 years. This could not be done without ruining the economies of the countries that rely on that water.*

The decline of the Aral Sea is a major environmental disaster that sadly is of human making.

*For more on this, see "Coming to Grips with the Aral Sea's Grim Legacy," in *Science*, vol. 284, 2 April 1999, pp. 30–31.

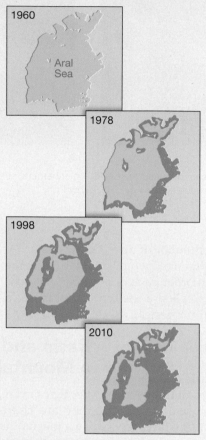

Figure 6.D The shrinking Aral Sea. By the year 2010 all that will remain are three small remnants.

Figure 6.C The Aral Sea lies east of the Caspian Sea in the Turkestan Desert. Two rivers, the Amu Darya and Syr Darya, bring water from the mountains to the sea.

Figure 6.E In the town of Jamboul, Kazakhstan, boats now lie in the sand because the Aral Sea has dried up. (Photo by Ergun Cagatay/Liaison Agency, Inc.)

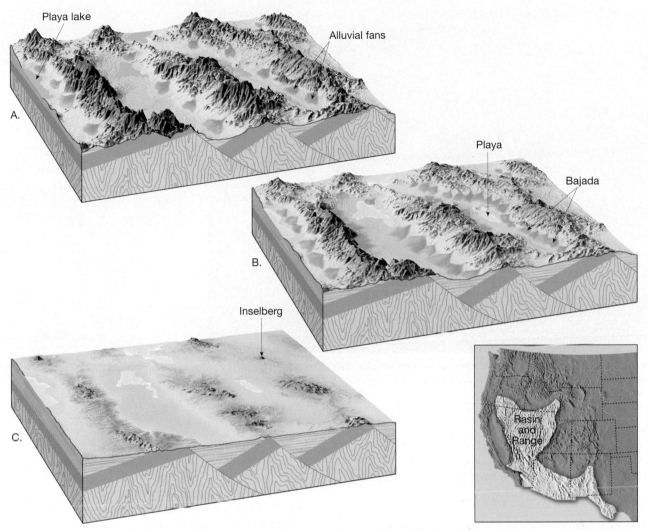

▲ **FIGURE 6.27**   Stages of landscape evolution in a mountainous desert such as the Basin and Range region of the West. As erosion of the mountains and deposition in the basins continue, relief diminishes. **A.** Early stage. **B.** Middle stage. **C.** Late stage.

the mountains and sediment fills the basins, elevation differences diminish.

When the occasional torrents of water produced by sporadic rains move down the mountain canyons, they are heavily loaded with sediment. Emerging from the confines of the canyon, the runoff spreads over the gentler slopes at the base of the mountains and quickly loses velocity. Consequently, most of its load is dumped within a short distance. The result is a cone of debris known as an **alluvial fan** at the mouth of a canyon (Figure 6.28). Over the years, a fan enlarges, eventually coalescing with fans from adjacent canyons to produce an apron of sediment (*bajada*) along the mountain front.

On the rare occasions of abundant rainfall, or snowmelt in the mountains, streams may flow across the alluvial fans to the center of the basin, converting the basin floor into a shallow **playa lake**. Playa lakes last only a few days or weeks, before evaporation and

infiltration remove the water. The dry, flat lake bed that remains is termed a *playa*.

Playas occasionally become encrusted with salts left behind by evaporation. These precipitated salts may be unusual. A case in point is the sodium borate (better known as borax) mined from ancient playa lake deposits in Death Valley, California.

With the ongoing erosion of the mountain mass and the accompanying sedimentation, the local relief continues to diminish. Eventually nearly the entire mountain mass is gone. Thus, by the late stages of erosion, the mountain areas are reduced to a few large bedrock knobs (called *inselbergs*) projecting above the sediment-filled basin.

Each of the stages of landscape evolution in an arid climate depicted in Figure 6.27 can be observed in the Basin and Range region. Recently uplifted mountains in an early stage of erosion are found in southern Oregon and northern Nevada. Death Valley,

▲ **FIGURE 6.28** Alluvial fans develop where the gradient of a stream changes abruptly from steep to flat. Such a situation exists in Death Valley, California, where streams emerge from the mountains into a flat basin. As a result, Death Valley has many large alluvial fans. (Photo by Michael Collier)

California, and southern Nevada fit into the more advanced middle stage, whereas the late stage, with its inselbergs, can be seen in southern Arizona.

STUDENTS SOMETIMES ASK...

Do the terms "arid" and "drought" mean about the same thing?

The concept of drought differs from that of aridity. Drought is a *temporary* happening that is considered an atmospheric hazard. By contrast, aridity describes regions where low rainfall is a *permanent* feature of climate. Although natural disasters such as floods and hurricanes generate more attention, droughts can be just as devastating and carry a bigger price tag. On the average, droughts cost the United States $6 to $8 billion annually compared to $2.4 billion for floods and $1.2 to $4.8 billion for hurricanes.

Wind Erosion

GEODe
Sculpturing Earth's Surface
▼ Deserts

Moving air, like moving water, is turbulent and able to pick up loose debris and transport it to other locations. Just as in a stream, the velocity of wind increases with height above the surface. Also like a stream, wind transports fine particles in suspension while heavier ones are carried as bed load (Figure 6.29). However,

the transport of sediment by wind differs from that of running water in two significant ways. First, wind's lower density compared to water renders it less capable of picking up and transporting coarse materials. Second, because wind is not confined to channels, it can spread sediment over large areas, as well as high into the atmosphere.

Compared to running water and glaciers, wind is a relatively insignificant erosional agent. Recall that even in deserts, most erosion is performed by intermittent running water, not by the wind. Wind erosion is more effective in arid lands than in humid areas because in humid places moisture binds particles together and vegetation anchors the soil. For wind to be an effective erosional force, dryness and scanty vegetation are essential. When such circumstances exist, wind may pick up, transport, and deposit great quantities of fine sediment. During the 1930s, parts of the Great Plains experienced vast dust storms. The plowing under of the natural vegetative cover for farming, followed by severe drought, exposed the land to wind erosion and led to the area being labeled the Dust Bowl (Figure 6.30).

One way that wind erodes is by **deflation** (*de* = out, *flat* = blow), the lifting and removal of loose material. Because the competence (ability to transport different-size particles) of moving air is low, it can suspend only fine sediment, such as clay and silt. Larger grains of sand are rolled or skipped along the surface (a process called *saltation*) and comprise the

A.

B.

▲ **FIGURE 6.29** **A.** Fine dust particles can be swept high into the atmosphere and carried in suspension. This satellite image shows thick plumes of dust from the Sahara Desert blowing across the Mediterranean Sea toward Italy on July 16, 2003. Such dust storms are common in arid North Africa. In fact, this region is the largest dust source in the world. Satellites are excellent tools for studying the transport of dust on a global scale. They show that dust storms can cover huge areas and that dust can be transported great distances. (NASA Image) **B.** The bed load carried by winds consists of sand grains, many of which move by bouncing along the surface. Sand never travels far from the surface, even when winds are very strong. (Photo by Stephen Trimble)

bed load. Particles larger than sand are usually not transported by wind. Deflation sometimes is difficult to notice because the entire surface is being lowered at the same time, but it can be significant. In portions of the 1930s Dust Bowl, the land was lowered by a meter or more in only a few years (Figure 6.31).

The most noticeable results of deflation in some places are shallow depressions called **blowouts.** In the Great Plains region, from Texas north to Montana, thousands of blowouts can be seen. They range from small dimples less than 1 meter deep and 3 meters wide to depressions that are over 45 meters deep and several kilometers across.

In portions of many deserts, the surface is characterized by a layer of coarse pebbles and cobbles that are too large to be moved by the wind. This stony veneer, called **desert pavement,** is created as deflation lowers the surface by removing sand and silt until eventually only a continuous cover of coarse particles remains (Figure 6.32). Once desert pavement becomes established, a process that might take hundreds of years, the surface is effectively protected from further deflation if left undisturbed. However, as the layer is only one or two stones thick, the passage of vehicles or animals can dislodge

▼ **FIGURE 6.30** Dust blackens the sky on May 21, 1937, near Elkhart, Kansas. It was because of storms like this that portions of the Great Plains were called the Dust Bowl in the 1930s. (Photo reproduced from the collection of the Library of Congress)

▼ **FIGURE 6.31** This photo was taken north of Granville, North Dakota, in July 1936, during a prolonged drought. Strong winds removed the soil that was not anchored by vegetation. The mounds are 1.2 meters (4 feet) high and show the level of the land prior to deflation. (Photo courtesy of the State Historical Society of North Dakota, 0278-01)

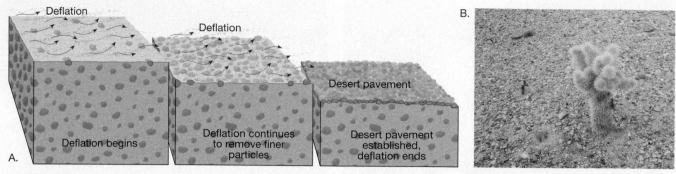

▲ **FIGURE 6.32** **A.** Formation of desert pavement. As these cross-sections illustrate, coarse particles gradually become concentrated into a tightly packed layer as deflation lowers the surface by removing sand and silt. **B.** If left undisturbed, desert pavement such as this in Arizona's Sonoran Desert will protect the surface from further deflation. (Photo by David Muench)

the pavement and expose the fine-grained material below. If this happens, the surface is once again subject to deflation.

Like glaciers and streams, wind erodes in part by *abrasion.* In dry regions as well as along some beaches, windblown sand will cut and polish exposed rock surfaces. Abrasion is often credited for accomplishments beyond its actual capabilities. Such features as balanced rocks that stand high atop narrow pedestals, and intricate detailing on tall pinnacles, are not the results of abrasion. Sand seldom travels more than a meter above the surface, so the wind's sandblasting effect is obviously limited in vertical extent. However, in areas prone to such activity, telephone poles have actually been cut through near their bases. For this reason, collars are often fitted on the poles to protect them from being sawed down.

Wind Deposits

Sculpturing Earth's Surface
▼ Deserts

Although wind is relatively unimportant in producing *erosional* landforms, significant *depositional* landforms are created by the wind in some regions. Accumulations of windblown sediment are particularly conspicuous in the world's dry lands and along many sandy coasts. Wind deposits are of two distinctive types: (1) extensive blankets of silt, called *loess,* which once were carried in suspension, and (2) mounds and ridges of sand from the wind's bed load, which we call *dunes.*

Loess

In some parts of the world the surface topography is mantled with deposits of windblown silt termed **loess.** Over thousands of years dust storms deposited this material. When loess is breached by streams or road cuts, it tends to maintain vertical cliffs and lacks any visible layers, as you can see in Figure 6.33.

The distribution of loess worldwide indicates that there are two primary sources for this sediment: deserts and glacial deposits of stratified drift. The thickest and most extensive deposits of loess on Earth occur in western and northern China. They were blown there from the extensive desert basins of central Asia. Accumulations of 30 meters are not uncommon, and thicknesses of more than 100 meters have been measured. It is this fine, buff-colored sediment that gives the Yellow River (Hwang Ho) its name.

In the United States, deposits of loess are significant in many areas, including South Dakota, Nebraska, Iowa, Missouri, and Illinois, as well as portions of the Columbia Plateau in the Pacific Northwest. Unlike the deposits in China, the loess in the United States, as well as in Europe, is an indirect product of glaciation. Its source is deposits of stratified drift. During the retreat of the ice sheets, many river valleys were choked with sediment deposited by meltwater. Strong westerly winds sweeping across the barren floodplains picked up the finer sediment and dropped it as a blanket on the eastern sides of the valleys.

Sand Dunes

Like running water, wind releases its load of sediment when its velocity falls and the energy available for transport diminishes. Thus, sand begins to accumulate wherever an obstruction across the path of the wind

▼ **FIGURE 6.33** This vertical loess bluff near the Mississippi River in southern Illinois is about 3 meters high. (Photo by James E. Patterson)

A.

B.

▲ **FIGURE 6.34** **A.** Dunes composed of gypsum sand at White Sands National Monument in southeastern New Mexico. These dunes are gradually migrating from the background (top) toward the foreground (bottom) of the photo. Strong winds move sand up the more gentle windward slopes. As sand accumulates near the dune crest, the slope steepens and some of the sand slides down the steeper *slip face*. **B.** Sand sliding down the steep slip face of a dune in White Sands National Monument, New Mexico. (Photos by Michael Collier)

slows its movement. Unlike deposits of loess, which form blanketlike layers over broad areas, winds commonly deposit sand in mounds or ridges called *dunes* (Figure 6.34).

As moving air encounters an object, such as a clump of vegetation or a rock, the wind sweeps around and over it, leaving a shadow of more slowly moving air behind the obstacle as well as a smaller zone of quieter air just in front of the obstacle. Some of the saltating sand grains moving with the wind come to rest in these wind shadows. As the accumulation of sand continues, it forms an increasingly efficient wind barrier to trap even more sand. If there is a sufficient supply of sand and the wind blows steadily long enough, the mound of sand grows into a dune.

Many dunes have an asymmetrical profile, with the leeward (sheltered) slope being steep and the windward slope more gently inclined (Figure 6.34). Sand moves up the gentler slope on the windward side by saltation. Just beyond the crest of the dune, where wind velocity is reduced, the sand accumulates. As more sand collects, the slope steepens, and eventually some of it slides or slumps under the pull of gravity (Figure 6.34B). In this way, the leeward slope of the dune, called the **slip face,** maintains an angle of about 34 degrees. Continued sand accumulation, coupled with periodic slides down the slip face, results in the slow migration of the dune in the direction of air movement.

As sand is deposited on the slip face, it forms layers inclined in the direction the wind is blowing. These slop-

ing layers are called **cross beds** (Figure 6.35). When the dunes are eventually buried under other layers of sediment and become part of the sedimentary rock record, their asymmetrical shape is destroyed, but the cross beds remain as a testimony to their origin. Nowhere is cross-bedding more prominent than in the sandstone walls of Zion Canyon in Utah (Figure 6.35D).

STUDENTS SOMETIMES ASK...
Aren't deserts mostly covered with sand dunes?

A common misconception about deserts is that they consist of mile after mile of drifting sand dunes. It is true that sand accumulations do exist in some areas and may be striking features. But, perhaps surprisingly, sand accumulations worldwide represent only a small percentage of the total desert area. For example, in the Sahara—the world's largest desert—accumulations of sand cover only *one-tenth* of its area. The sandiest of all deserts is the Arabian, one-third of which consists of sand.

Types of Sand Dunes

Dunes are not just random heaps of windblown sediment. Rather, they are accumulations that usually assume patterns that are surprisingly consistent (Figure 6.36). A broad assortment of dune forms exist, generally simplified to a few major types for discussion. Of course, gradations exist among different forms as well as irregularly shaped dunes that do not fit easily into any

181

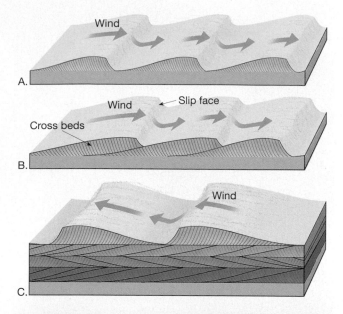

D.

▲ **FIGURE 6.35** As parts **A** and **B** illustrate, dunes commonly have an asymmetrical shape. The steeper leeward side is called the *slip face*. Sand grains deposited on the slip face create the cross-bedding of the dunes. **C.** A complex pattern develops in response to changes in wind direction. Also notice that when dunes are buried and become part of the sedimentary record, the cross-bedded structure is preserved. **D.** Cross beds are an obvious characteristic of the Navajo Sandstone in Zion National Park, Utah. (Photo by David Muench)

category. Several factors influence the form and size that dunes ultimately assume. These include wind direction and velocity, availability of sand, and the amount of vegetation present. Six basic dune types are shown in Figure 6.36, with arrows indicating wind directions.

Barchan Dunes. Solitary sand dunes shaped like crescents and with their tips pointing downwind are called **barchan dunes** (Figure 6.36A). These dunes form where

supplies of sand are limited and the surface is relatively flat, hard, and lacking vegetation. They migrate slowly with the wind at a rate of up to 15 meters annually. Their size is usually modest, with the largest barchans reaching heights of about 30 meters while the maximum spread between their horns approaches 300 meters. When the wind direction is nearly constant, the crescent form of these dunes is nearly symmetrical. However, when the wind direction is not perfectly fixed, one tip becomes larger than the other.

Transverse Dunes. In regions where the prevailing winds are steady, sand is plentiful, and vegetation is sparse or absent, the dunes form a series of long ridges that are separated by troughs and oriented at right angles to the prevailing wind. Because of this orientation, they are termed **transverse dunes** (Figure 6.36B). Typically, many coastal dunes are of this type. In addition, transverse dunes are common in many arid regions where the extensive surface of wavy sand is sometimes called a *sand sea*. In some parts of the Sahara and Arabian deserts, transverse dunes reach heights of 200 meters, are 1 to 3 kilometers across, and can extend for distances of 100 kilometers or more.

There is a relatively common dune form that is intermediate between isolated barchans and extensive waves of transverse dunes. Such dunes, called **barchanoid dunes,** form scalloped rows of sand oriented at right angles to the wind (Figure 6.36C). The rows resemble a series of barchans that have been positioned side by side. Visitors exploring the gypsum dunes at White Sands National Monument, New Mexico, will recognize this form.

Longitudinal Dunes. **Longitudinal dunes** are long ridges of sand that form more or less parallel to the prevailing wind and where sand supplies are moderate (Figure 6.36D). Apparently the prevailing wind direction must vary somewhat but still remain in the same quadrant of the compass. Although the smaller types are only 3 or 4 meters high and several tens of meters long, in some large deserts longitudinal dunes can reach great size. For example, in portions of North Africa, Arabia, and central Australia, these dunes may approach a height of 100 meters and extend for distances of more than 100 kilometers (62 miles).

Parabolic Dunes. Unlike the other dunes that have been described thus far, **parabolic dunes** form where vegetation partially covers the sand. The shape of these dunes resembles the shape of barchans except that their tips point into the wind rather than downwind (Figure 6.36E). Parabolic dunes often form along coasts where there are strong onshore winds and abundant sand. If the sand's sparse vegetative cover is disturbed at some spot, deflation creates a blowout. Sand is then transported out of the depression and deposited as a curved rim that grows higher as deflation enlarges the blowout.

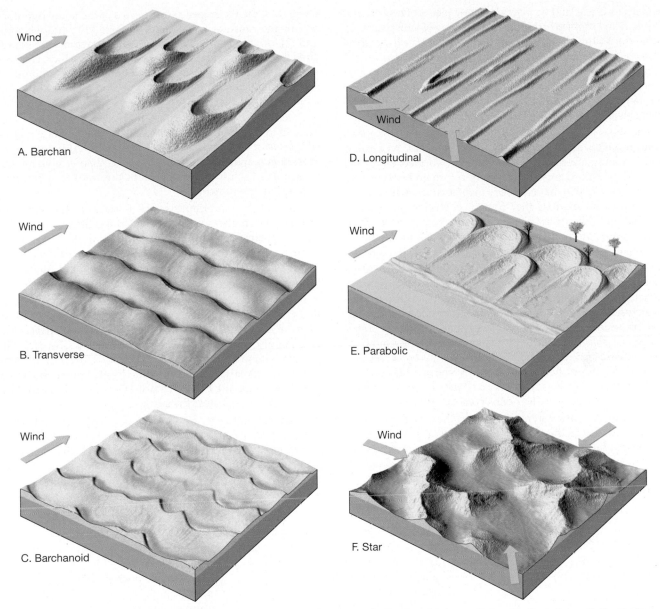

▲ **FIGURE 6.36** Sand dune types. **A.** Barchan dunes. **B.** Transverse dunes. **C.** Barchanoid dunes. **D.** Longitudinal dunes. **E.** Parabolic dunes. **F.** Star dunes.

Star Dunes. Confined largely to parts of the Sahara and Arabian deserts, **star dunes** are isolated hills of sand that exhibit a complex form (Figure 6.36F). Their name is derived from the fact that the bases of these dunes resemble multipointed stars. Usually three or four sharp-crested ridges diverge from a central high point that in some cases may approach a height of 90 meters. As their form suggests, star dunes develop where wind directions are variable.

Chapter Summary

• A *glacier* is a thick mass of ice originating on land from the compaction and recrystallization of snow, and it shows evidence of past or present flow. Today, *valley* or *alpine glaciers* are found in mountain areas where they usually follow valleys that were originally occupied by streams. *Ice sheets* exist on a much larger scale, covering most of Greenland and Antarctica.

• Near the surface of a glacier, in the *zone of fracture,* ice is brittle. However, below about 50 meters, pressure is great, causing ice to *flow* like a *plastic material.* A second important mechanism of glacial movement consists of the whole ice mass *slipping* along the ground.

• Glaciers form in areas where more snow falls in winter than melts during summer. Snow accumulation and ice formation occur in the *zone of accumulation.* Beyond this area is the *zone of wastage,* where there is a net loss to the glacier. The *glacial budget* is the balance, or lack of balance, between accumulation at the upper end of the glacier, and loss at the lower end.

• Glaciers erode land by *plucking* (lifting pieces of bedrock out of place) and *abrasion* (grinding and scraping of a rock surface). Erosional features produced by valley glaciers include *glacial troughs, hanging valleys, cirques, arêtes, horns,* and *fiords.*

• Any sediment of glacial origin is called *drift.* The two distinct types of glacial drift are (1) *till,* which is unsorted sediment deposited directly by the ice; and (2) *stratified drift,* which is relatively well-sorted sediment laid down by glacial meltwater.

• The most widespread features created by glacial deposition are layers or ridges of till, called *moraines.* Associated with valley glaciers are *lateral moraines,* formed along the sides of the valley, and *medial moraines,* formed between two valley glaciers that have joined. *End moraines,* which mark the former position of the front of a glacier, and *ground moraine,* an undulating layer of till deposited as the ice front retreats, are common to both valley glaciers and ice sheets.

• Perhaps the most convincing evidence for the occurrence of several glacial advances during the *Ice Age* is the widespread existence of *multiple layers of drift* and an uninterrupted record of climate cycles preserved in *seafloor sediments.* In addition to massive erosional and depositional work, other effects of Ice Age glaciers included the *migration of organisms, changes in stream courses, adjustment of the crust* by rebounding after the removal of the immense load of ice, and *climate changes* caused by the existence of the glaciers themselves. In the sea, the most far-reaching effect of the Ice Age was the *worldwide change in sea level* that accompanied each advance and retreat of the ice sheets.

• Any theory that attempts to explain the causes of glacial ages must answer the two basic questions: (1) What causes the onset of glacial conditions? and (2) What caused the alternating glacial and interglacial stages that have been documented for the Pleistocene epoch? Two of the many hypotheses for the cause of glacial ages involve (1) plate tectonics and (2) variations in Earth's orbit. Other factors that are related to climate change during glacial ages include changes in atmospheric composition, variations in the amount of sunlight reflected by Earth's surface, and changes in ocean circulation.

• Practically all desert streams are dry most of the time and are said to be *ephemeral.* Nevertheless, *running water is responsible for most of the erosional work in a desert.* Although wind erosion is more significant in dry areas than elsewhere, the main role of wind in a desert is in the transportation and deposition of sediment.

• Many of the landscapes of the Basin and Range region of the western and southwestern United States are the result of streams eroding uplifted mountain blocks and depositing the sediment in interior basins. *Alluvial fans, playas,* and *playa lakes* are features often associated with these landscapes.

• For wind erosion to be effective, dryness and scant vegetation are essential. *Deflation,* the lifting and removal of loose material, often produces shallow depressions called *blowouts* and can also lower the surface by removing sand and silt, leaving behind a stony veneer called *desert pavement. Abrasion,* the sandblasting effect of wind, is often given too much credit for producing desert features. However, abrasion does cut and polish rock near the surface.

• Wind deposits are of two distinct types: (1) extensive *blankets of silt,* called *loess,* carried by wind in *suspension,* and (2) *mounds and ridges of sand,* called *dunes,* which are formed from sediment that is carried as part of the wind's *bed load.*

Key Terms

ablation (p. 157)
abrasion (p. 159)
alluvial fan (p. 177)
alpine glacier (p. 154)
arête (p. 161)
barchan dune (p. 182)
barchanoid dune (p. 182)
blowout (p. 179)
cirque (p. 161)
crevasse (p. 156)
cross beds (p. 181)
deflation (p. 178)
desert pavement (p. 179)
drumlin (p. 168)
end moraine (p. 165)
ephemeral (p. 174)
esker (p. 168)
fiord (p. 163)

glacial drift (p. 164)
glacial erratic (p. 164)
glacial striations (p. 159)
glacial trough (p. 161)
glacier (p. 154)
ground moraine (p. 165)
hanging valley (p. 161)
horn (p. 161)
ice cap (p. 155)
ice sheet (p. 154)
interior drainage (p. 175)
kame (p. 168)
kettle (p. 167)
lateral moraine (p. 165)
loess (p. 180)
longitudinal dune (p. 182)
medial moraine (p. 165)
outwash plain (p. 167)

parabolic dune (p. 182)
piedmont glacier (p. 155)
playa lake (p. 177)
Pleistocene epoch (p. 168)
plucking (p. 159)
pluvial lake (p. 169)
rock flour (p. 159)
slip face (p. 181)
star dune (p. 183)
stratified drift (p. 164)
till (p. 164)
transverse dune (p. 182)
valley glacier (p. 154)
valley train (p. 167)
zone of accumulation (p. 157)
zone of wastage (p. 157)

Review Questions

1. What is a glacier? What percentage of Earth's land area do glaciers cover? Where are glaciers found today?

2. Each of the following statements refers to a particular type of glacier. Name the type of glacier.

   (a) The term *continental* is often used to describe this type of glacier.

   (b) This type of glacier is also called an *alpine glacier.*

(c) This is a glacier formed when one or more valley glaciers spreads out at the base of a steep mountain front.

(d) Greenland is the only example in the Northern Hemisphere.

3. Describe how glaciers fit into the hydrologic cycle. What role do they play in the rock cycle?

4. Describe the two components of glacial flow. At what rates do glaciers move? In a valley glacier does all of the ice move at the same rate? Explain.

5. Why do crevasses form in the upper portion of a glacier but not below a depth of about 50 meters?

6. Under what circumstances will the front of a glacier advance? Retreat? Remain stationary?

7. Describe two basic processes of glacial erosion.

8. How does the shape of a glaciated mountain valley differ from a mountain valley that was not glaciated?

9. List and describe the erosional features you might expect to see in an area where valley glaciers exist or have recently existed.

10. What is glacial drift? What is the difference between till and stratified drift? What general effect do glacial deposits have on the landscape?

11. List and briefly describe the four basic moraine types. What do all moraines have in common?

12. List and briefly describe four depositional features other than moraines.

13. About what percentage of Earth's land surface was covered at some time by Pleistocene glaciers? How does this compare to the area presently covered by ice sheets and glaciers? (Check your answer with Question 1.)

14. List three indirect effects of Ice Age glaciers.

15. How might plate tectonics help us understand the cause of ice ages? Can plate tectonics explain the alternation between glacial and interglacial climates during the Pleistocene?

16. How extensive are Earth's dry regions?

17. What is the most important erosional process in deserts?

18. Describe the characteristics associated with each of the stages in the evolution of a mountainous desert region such as the Basin and Range region in the western United States.

19. Why is wind erosion relatively more important in dry regions than in humid areas?

20. Although sand dunes are the best-known wind deposits, accumulations of loess are very significant in some parts of the world. What is loess? Where are such deposits found? What are the origins of this sediment?

21. How do sand dunes migrate?

22. Identify each of the dunes described in the following statements.

    (a) Long ridges of sand oriented parallel to the prevailing wind.

    (b) Solitary, crescent-shaped dunes oriented with their tips pointing downwind.

    (c) Ridges of sand oriented at right angles to the prevailing wind.

    (d) Dunes whose tips point into the wind.

    (e) Scalloped rows of sand oriented at right angles to the wind.

    (f) Isolated dunes consisting of three or four sharp-crested ridges diverging from a central high point.

Examining the Earth System

1. Assume you are teaching an introductory Earth science class and that you have just assigned Chapter 6 of this text. A student in the class asks why glaciers, deserts, and wind are treated in the same chapter. Formulate a response that connects these topics.

2. Glaciers are solid, but they are a basic part of the hydrologic cycle. Should glaciers be considered a part of the sphere we call the geosphere, or do they belong to the hydrosphere?

3. Some scientists think that ice should be a separate sphere of the Earth system, called the *cryosphere*. Does such an idea have merit?

4. Wind erosion occurs at the interface of the atmosphere, geosphere, and biosphere and is influenced by the hydrosphere and human activity. With this in mind, describe how human activity contributed to the Dust Bowl, the period of intense wind erosion in the Great Plains in the 1930s. You might find it helpful to research Dust Bowl on the Internet using a search engine such as Google **(http://www.google.com/)** or Yahoo **(http://www. yahoo.com/)**. You may also find the Wind Erosion Research Unit site to be informative: **(http://www.weru.ksu.edu/)**.

Online Study Guide

The *Earth Science* Website uses the resources and flexibility of the Internet to aid in your study of the topics in this chapter. Written and developed by Earth science instructors, this site will help improve your understanding of Earth science. Visit **http://www.prenhall.com/tarbuck** and click on the cover of *Earth Science* 11e to find:

- **Online review quizzes.**
- **Critical thinking exercises.**
- **Links to chapter-specific Web resources.**
- **Internet-wide key term searches.**

http://www.prenhall.com/tarbuck

CHAPTER 7

Earthquakes and Earth's Interior

In January 2001, a magnitude 7.9 earthquake struck Bhuj, India. It resulted in an estimated 20,000 deaths and left one million people homeless. (Photo by DPA/RAJ/The Image Works)

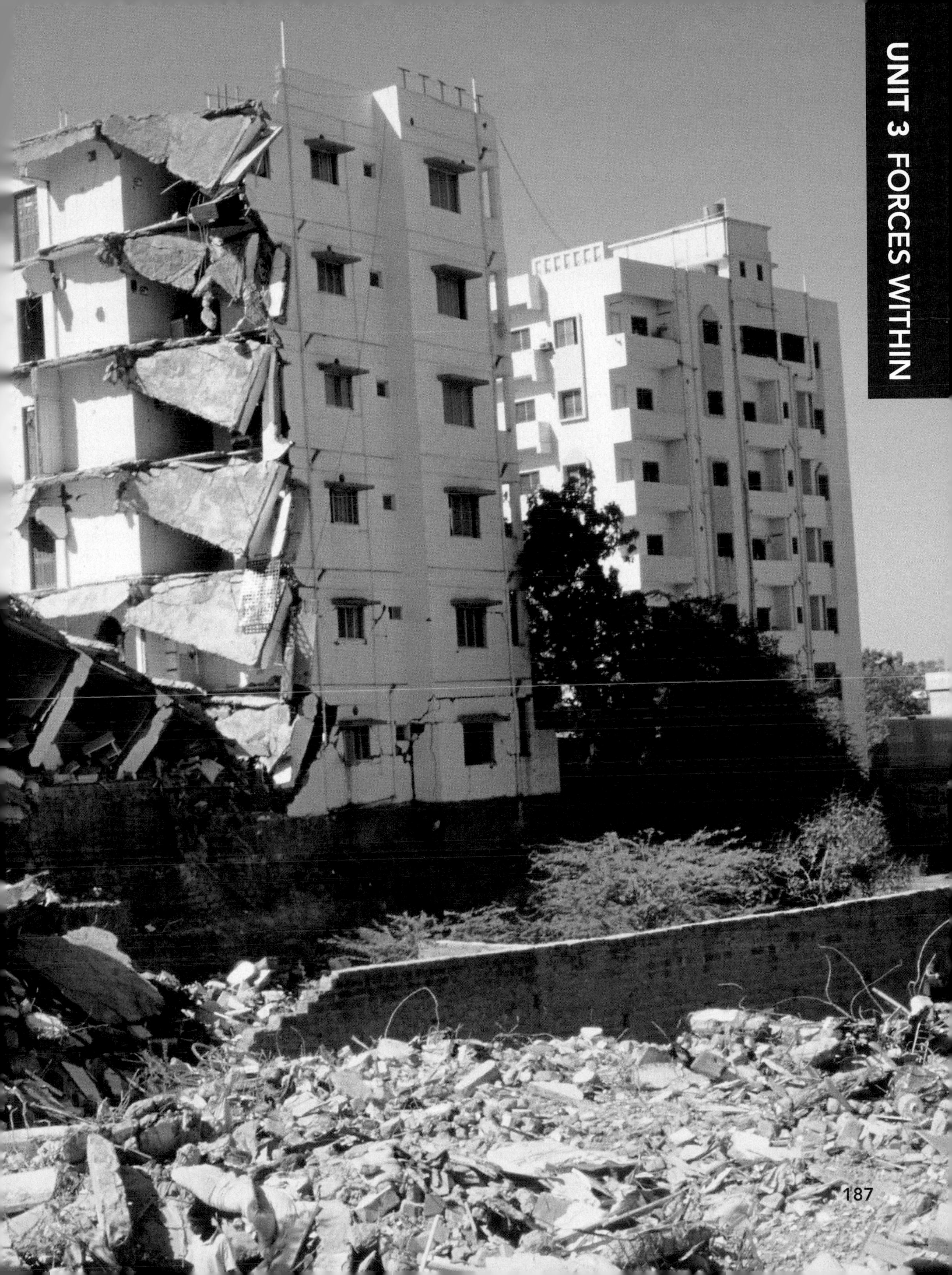

On October 17, 1989, at 5:04 p.m. Pacific daylight time, millions of television viewers around the world were settling in to watch the third game of the World Series. Instead, they saw their television sets go black as tremors hit San Francisco's Candlestick Park. Although the earthquake was centered in a remote section of the Santa Cruz Mountains, 100 kilometers to the south, major damage occurred in the Marina District of San Francisco (Figure 7.1).

The most tragic result of the violent shaking was the collapse of some double-decked sections of Interstate 880, also known as the Nimitz Freeway. The ground motions caused the upper deck to sway, shattering the concrete support columns along a mile-long section of the freeway. The upper deck then collapsed onto the lower roadway, flattening cars as if they were aluminum beverage cans. This earthquake, named the Loma Prieta quake for its point of origin, claimed 67 lives.

In mid-January 1994, less than five years after the Loma Prieta earthquake devastated portions of the San Francisco Bay Area, a major earthquake struck the Northridge area of Los Angeles. Although not the fabled "Big One," this moderate 6.7-magnitude earthquake left 57 dead, over 5000 injured, and tens of thousands of households without water and electricity. The damage exceeded $40 billion and was attributed to an apparently unknown fault that ruptured 18 kilometers (11 miles) beneath Northridge.

The Northridge earthquake began at 4:31 a.m. and lasted roughly 40 seconds. During this brief period, the quake terrorized the entire Los Angeles area. In the three-story Northridge Meadows apartment complex, 16 people died when sections of the upper floors collapsed onto the first-floor units. Nearly 300 schools were seriously damaged, and a dozen major roadways buckled. Among these were two of California's major arteries—the Golden State Freeway (Interstate 5), where an overpass collapsed completely and blocked the roadway, and the Santa Monica Freeway. Fortunately, these roadways had practically no traffic at this early morning hour.

In nearby Granada Hills, broken gas lines were set ablaze while the streets flooded from broken water mains. Seventy homes burned in the Sylmar area. A 64-car freight train derailed, including some cars carrying hazardous cargo. But it is remarkable that the destruction was not greater. Unquestionably, the upgrading of structures to meet the requirements of building codes developed for this earthquake-prone

▲ **FIGURE 7.1** Damage in San Francisco's Marina District following the 1989 Loma Prieta earthquake. (Photo by David Weintraub/Photo Researchers, Inc.)

area helped minimize what could have been a much greater human tragedy.

STUDENTS SOMETIMES ASK...
How often do earthquakes occur?

All the time—in fact, there are literally thousands of earthquakes daily! Fortunately, the majority of them are too small to be felt by people, and many of them occur in remote regions. Their existence is known only because of sensitive seismographs.

What Is an Earthquake?

GEODe Forces Within
▼ Earthquakes

An **earthquake** is the vibration of Earth produced by the rapid release of energy. Most often earthquakes are caused by slippage along a fault in Earth's crust. The energy released radiates in all directions from its source, the **focus** (*foci* = a point), in the form of waves. These waves are analogous to those produced when a stone is dropped into a calm pond (Figure 7.2). Just as the impact of the stone sets water waves in motion, an earthquake generates seismic waves that radiate throughout the Earth. Even though the energy dissipates rapidly with increasing distance from the focus, sensitive instruments located throughout the world record the event.

Over 30,000 earthquakes that are strong enough to be felt occur worldwide annually. Fortunately, most are minor tremors and do very little damage. Generally, only about 75 significant earthquakes take place each year, and many of these occur in remote regions.

▼ **FIGURE 7.2** The focus of an earthquake is located at depth. The surface location directly above it is called the epicenter.

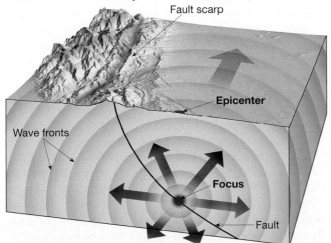

However, occasionally a large earthquake occurs near a large population center. Under these conditions, an earthquake is among the most destructive natural forces on Earth.

The shaking of the ground, coupled with the liquefaction of some soils, wreaks havoc on buildings and other structures. In addition, when a quake occurs in a populated area, power and gas lines are often ruptured, causing numerous fires. In the famous 1906 San Francisco earthquake, much of the damage was caused by fires (Figure 7.3), which quickly became uncontrollable when broken water mains left firefighters with only trickles of water.

Earthquakes and Faults

The tremendous energy released by atomic explosions or by volcanic eruptions can produce an earthquake, but these events are relatively weak and infrequent. What mechanism produces a destructive earthquake? Ample evidence exists that Earth is not a static planet. We know that Earth's crust has been uplifted at times, because we have found numerous ancient wave-cut benches many meters above the level of the highest tides. Other regions exhibit evidence of extensive subsidence. In addition to these vertical displacements, offsets in fence lines, roads, and other structures indicate that horizontal movement is common (Figure 7.4). These movements are associated with large fractures in Earth's crust called **faults**.

Typically, earthquakes occur along preexisting faults that formed in the distant past along zones of weakness in Earth's crust. Some are very large and can generate major earthquakes. One example is the San Andreas Fault, which is a transform fault boundary that separates two great sections of Earth's lithosphere: the North American plate and the Pacific plate. Other faults are small and produce only minor earthquakes.

Most faults are not perfectly straight or continuous; instead, they consist of numerous branches and smaller fractures that display kinks and offsets. Such a pattern is displayed in Figure 10.A (p. 292), which shows that the San Andreas Fault is actually a system that consists of several large faults and innumerable small fractures (not shown).

It is also clear that most faults are locked, except for brief, abrupt movements that accompany an earthquake rupture. The primary reason faults are locked is that the confining pressure exerted by the overlying crust is enormous. Because of this, the fractures in the crust are essentially squeezed shut. Nevertheless, even faults that have been inactive for thousands of years can rupture again if the stresses acting on the region increase sufficiently.

Discovering the Cause of Earthquakes

The actual mechanism of earthquake generation eluded geologists until H.F. Reid of Johns Hopkins University conducted a study following the great 1906 San Francisco earthquake. The earthquake was

▶ **FIGURE 7.3** San Francisco in flames after the 1906 earthquake. (Reproduced from the collection of the Library of Congress)

accompanied by horizontal surface displacements of several meters along the northern portion of the San Andreas Fault. Field investigations determined that during this single earthquake, the Pacific plate lurched as much as 4.7 meters (15 feet) northward past the adjacent North American plate.

The mechanism for earthquake formation that Reid deduced from this information is illustrated in Figure 7.5. In part A of the figure, you see an existing fault, or break in the rock. In part B, tectonic forces ever so slowly deform the crustal rocks on both sides of the fault, as demonstrated by the bent features. Under these conditions, rocks are bending

▼ **FIGURE 7.4** Slippage along a fault produced an offset in this orange grove east of Calexico, California. (Photo by John S. Shelton)

and storing elastic energy, much like a wooden stick does if bent. Eventually, the frictional resistance holding the rocks in place is overcome. As slippage occurs at the weakest point (the focus), displacement will exert stress farther along the fault, where additional slippage will occur, releasing the built-up strain (Figure 7.5C). This slippage allows the deformed rock to "snap back." The vibrations we know as an earthquake occur as the rock elastically returns to its original shape. The "springing back" of the rock was termed **elastic rebound** by Reid, because the rock behaves elastically, much like a stretched rubber band does when it is released.

Most of the motion along faults can be satisfactorily explained by the plate tectonics theory, which states that large slabs of Earth's lithosphere are in continual slow motion. These mobile plates interact with neighboring plates, straining and deforming the rocks at their margins. In fact, it is along faults associated with plate boundaries that most earthquakes occur. Furthermore, earthquakes are repetitive: As soon as one is over, the continuous motion of the plates adds strain to the rocks until they eventually fail again.

In summary, most earthquakes are produced by the rapid release of elastic energy stored in rock that has been subjected to great stress. Once the strength of the rock is exceeded, it suddenly ruptures, causing the vibrations of an earthquake. Earthquakes most often occur along existing faults whenever the frictional forces on the fault surfaces are overcome (see Box 7.1).

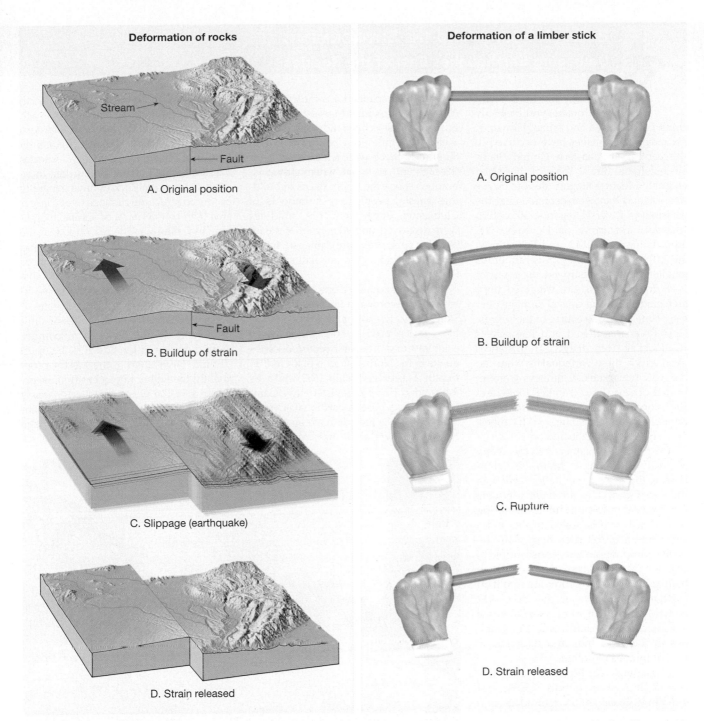

Deformation of rocks

A. Original position

B. Buildup of strain

C. Slippage (earthquake)

D. Strain released

Deformation of a limber stick

A. Original position

B. Buildup of strain

C. Rupture

D. Strain released

▲ **FIGURE 7.5** Elastic rebound. As rock is deformed it bends, storing elastic energy. Once the rock is strained beyond its breaking point it ruptures, releasing the stored-up energy in the form of earthquake waves.

Foreshocks and Aftershocks

The intense vibrations of the 1906 San Francisco earthquake lasted about 40 seconds. Although most of the displacement along the fault occurred in this rather short period, additional movements along this and other nearby faults continued for several days following the main quake. The adjustments that follow a major earthquake often generate smaller earthquakes called **aftershocks**. Although

these aftershocks are usually much weaker than the main earthquake, they can sometimes destroy already badly weakened structures. This occurred, for example, during a 1988 earthquake in Armenia. A large aftershock of magnitude 5.8 collapsed many structures that had been weakened by the main tremor.

In addition, small earthquakes called **fore-shocks** often precede a major earthquake by days or,

BOX 7.1 PEOPLE AND THE ENVIRONMENT:
Damaging Earthquakes East of the Rockies

When you think earthquake, you probably think of California and Japan. However, six major earthquakes have occurred in the central and eastern United States since colonial times. Three of these had estimated Richter magnitudes of 7.5, 7.3, and 7.8, and they were centered near the Mississippi River Valley in southeastern Missouri. Occurring on December 16, 1811, January 23, 1812, and February 7, 1812, these earthquakes, plus numerous smaller tremors, destroyed the town of New Madrid, Missouri, triggered massive landslides, and caused damage over a six-state area. The course of the Mississippi River was altered, and Tennessee's Reelfoot Lake was enlarged. The distances over which these earthquakes were felt are truly remarkable. Chimneys were reported downed in Cincinnati, Ohio, and Richmond, Virginia, while Boston residents, located 1770 kilometers (1100 miles) to the northeast, felt the tremor.

Despite the history of the New Madrid earthquake, Memphis, Tennessee, the largest population center in the area today, does not have adequate earthquake provisions in its building code. Further, because Memphis is located on unconsolidated floodplain deposits, buildings are more susceptible to damage than similar structures built on bedrock. It has been estimated that if an earthquake the size of the 1811–1812 New Madrid event were to strike in the next decade, it would result in casualties in the thousands and damages in tens of billions of dollars.

Damaging earthquakes that occurred in Aurora, Illinois (1909), and Valentine, Texas (1931), remind us that other areas in the central United States are vulnerable.

The greatest historical earthquake in the eastern states occurred August 31, 1886, in Charleston, South Carolina. The event, which spanned 1 minute, caused 60 deaths, numerous injuries, and great economic loss within a radius of 200 kilometers (120 miles) of Charleston. Within 8 minutes, effects were felt as far away as Chicago and St. Louis, where strong vibrations shook the upper floors of buildings, causing people to rush outdoors. In Charleston alone, over 100 buildings were destroyed, and 90 percent of the remaining structures were damaged. It was difficult to find a chimney still standing (Figure 7.A).

Numerous other strong earthquakes have been recorded in the eastern United States. New England and adjacent areas have experienced sizable shocks since colonial times. The first reported earthquake in the Northeast took place in Plymouth, Massachusetts, in 1683, and was followed in 1755 by the destructive Cambridge, Massachusetts, earthquake. Moreover, ever since records have been kept, New York State alone has experienced over 300 earthquakes large enough to be felt.

Earthquakes in the central and eastern United States occur far less frequently than in California. Yet history indicates that the East is vulnerable. Further, these shocks east of the Rockies have generally produced structural damage over a larger area than counterparts of similar magnitude in California. The reason is that the underlying bedrock in the central and eastern United States is older and more rigid. As a result, seismic waves are able to travel greater distances with less weakening than in the western United States. It is estimated that for earthquakes of similar magnitude, the region of maximum ground motion in the East may be up to 10 times larger than in the West. Consequently, the higher rate of earthquake occurrence in the western United States is balanced somewhat by the fact that central and eastern U.S. quakes can damage larger areas.

Figure 7.A   Damage to Charleston, South Carolina, caused by the August 31, 1886, earthquake. Damage ranged from toppled chimneys and broken plaster to total collapse. (Photo courtesy of U.S. Geological Survey)

---

in some cases, by as much as several years. Monitoring of these foreshocks has been used as a means of predicting forthcoming major earthquakes, with mixed success. We will consider the topic of earthquake prediction in a later section of this chapter.

San Andreas Fault: An Active Earthquake Zone

The San Andreas is undoubtedly the most studied fault system in the world. Over the years, investigations have shown that displacement occurs along discrete segments that are 100 to 200 kilometers long. Further, each fault segment behaves somewhat differently from the others. Some portions of the San Andreas exhibit a slow, gradual displacement known as **fault creep**, which occurs relatively smoothly and therefore with little noticable seismic activity. Other segments regularly slip, producing small earthquakes.

Still other segments remain locked and store elastic energy for hundreds of years before rupturing in great earthquakes. The latter process is described as *stick-slip* motion, because the fault exhibits alternating periods of locked behavior followed by sudden slippage. It is estimated that great earthquakes should occur about every 50 to 200 years along those sections of the San Andreas Fault that exhibit stick-slip motion. This knowledge is useful when assigning a potential earthquake risk to a given segment of the fault zone.

STUDENTS SOMETIMES ASK...

Do moderate earthquakes decrease the chances of a major quake in the same region?

No. This is due to the vast increase in release of energy associated with higher-magnitude earthquakes. For instance, an earthquake with a magnitude of 8.5 releases millions of times more energy than the smallest earthquakes felt by humans. Similarly, thousands of moderate tremors would be needed to release the huge amount of energy equal to one "great" earthquake.

The tectonic forces along the San Andreas Fault zone that were responsible for the 1906 San Francisco earthquake are still active. Currently, laser beams are used to measure the relative motion between the opposite sides of this fault. These measurements reveal a displacement of 2 to 5 centimeters (1 to 2 inches) per year. Although this seems slow, it produces substantial movement over millions of years.

To illustrate, in 30 million years this rate of displacement would slide the western portion of California northward so that Los Angeles, on the Pacific plate, would be adjacent to San Francisco on the North American plate! More important in the short term, a displacement of just 2 centimeters each year produces 2 meters of offset every 100 years. Consequently, the 4 meters of displacement produced during the 1906 San Francisco earthquake should occur at least every 200 years along this segment of the fault zone. This fact lies behind California's concern for making buildings earthquake-resistant in anticipation of the inevitable "Big One."

Seismology: The Study of Earthquake Waves

GEODe
Forces Within
▼ Earthquakes

The study of earthquake waves, **seismology** (*seismos* = shake, *ology* = the study of), dates back to attempts by the Chinese almost 2000 years ago to determine the direction of the source of each earth-quake. Modern **seismographs** (*seismos* = shake, *graph* = write) are instruments that record earthquake waves. Their principle is simple: A weight is freely suspended from a support that is attached to bedrock (Figure 7.6). When waves from an earthquake reach the instrument, the inertia of the weight keeps it stationary, while Earth and the support vibrate. The movement of Earth in relation to the stationary weight is recorded on a rotating drum. (*Inertia* is the tendency of a stationary object to remain still, or a moving object to stay in motion.)

Modern seismographs amplify and record ground motion, producing a trace as shown in Figure 7.7. These records, called **seismograms** (*seismos* = shake, *gramma* = what is written), reveal that seismic waves are elastic energy. This energy radiates outward in all directions from the focus, as you saw in Figure 7.2. The transmission of this energy can be compared to the shaking of gelatin in a bowl that is jarred. Seismograms reveal that two main types of seismic waves are generated by the slippage of a rock mass. Some travel along Earth's outer layer and are called **surface waves**. Others travel through Earth's interior and are called **body waves**. Body waves are further divided into **primary waves (P waves)** and **secondary waves (S waves)**.

Body waves are divided into P and S waves by their mode of travel through intervening materials. P waves are push-pull waves—they push (compress) and pull (expand) rocks in the direction the wave is traveling (Figure 7.8A). Imagine holding someone by the shoulders and shaking the person. This push-pull movement is how P waves move through the Earth. This wave motion is analogous to that generated by human vocal cords as they move air to create sound. Solids, liquids, and gases resist a change in volume when compressed and will elastically spring back once the force is removed (Figure 7.8B). Therefore, P waves, which are compressional waves, can travel through all these materials.

In contrast, S waves shake the particles at right angles to their direction of travel. This can be illustrated by fastening one end of a rope and shaking the other end, as shown in Figure 7.8C. Unlike P waves, which temporarily change the *volume* of the

▶ **FIGURE 7.6** Principle of the seismograph. The inertia of the suspended mass tends to keep it motionless, while the recording drum, which is anchored to bedrock, vibrates in response to seismic waves. Thus, the stationary mass provides a reference point from which to measure the amount of displacement occurring as the seismic wave passes through the ground below.

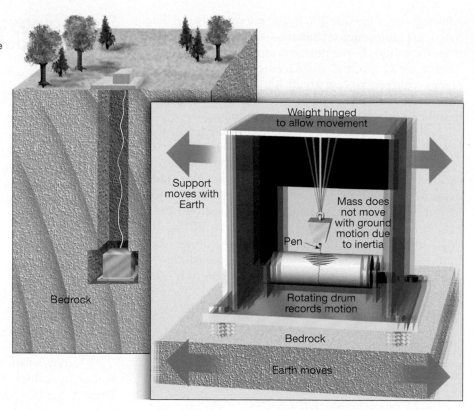

intervening material by alternately compressing and expanding it, S waves temporarily change the *shape* of the material that transmits them (Figure 7.8D). Because fluids (gases and liquids) do not respond elastically to changes in shape, they will not transmit S waves.

The motion of surface waves is somewhat more complex. As surface waves travel along the ground, they cause the ground and anything resting upon it to move, much like ocean swells toss a ship. In addition to their up-and-down motion, surface waves have a side-to-side motion similar to an S wave oriented in a horizontal plane. This latter motion is particularly damaging to the foundations of structures.

By observing a typical seismic record, as shown in Figure 7.7, you can see a major difference among these seismic waves: P waves arrive at the recording station first, then S waves, and then surface waves. This is a consequence of their speeds. To illustrate, the velocity of P waves through granite within the crust is about 6 kilometers per second. Under the same conditions, S waves travel at 3.5 kilometers per second. Differences in density and elastic properties of the rock greatly influence the velocities of these waves. Generally, in any solid material, P waves travel about 1.7 times faster than S waves, and surface waves can be expected to travel at 90 percent of the velocity of the S waves.

As you shall see, seismic waves allow us to determine the location and magnitude of earthquakes. In addition, seismic waves provide us with a tool for probing Earth's interior.

Locating an Earthquake

Forces Within
▼ **Earthquakes**

Recall that the *focus* is the place within Earth where earthquake waves originate. The **epicenter** (*epi* = upon, *center* = a point) is the location on the surface directly above the focus (see Figure 7.2).

▼ **FIGURE 7.7** Typical seismic record. Note the time interval (about 5 minutes) between the arrival of the first P wave and the arrival of the first S wave.

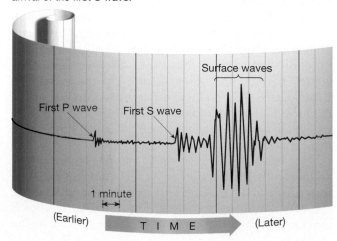

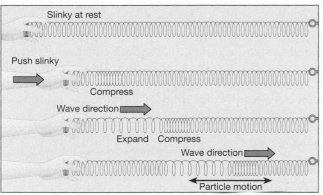

A. P waves generated using a slinky

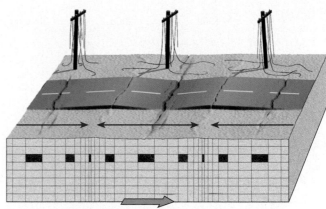

B. P waves traveling along the surface

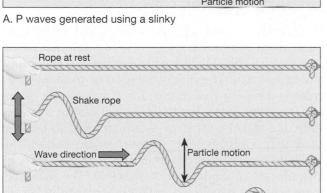

C. S waves generated using a rope

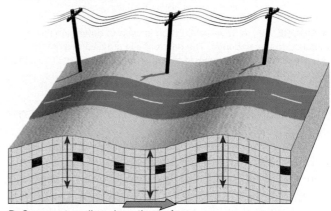

D. S waves traveling along the surface

▲ **FIGURE 7.8** Types of seismic waves and their characteristic motion. (Note that during a strong earthquake, ground shaking consists of a combination of various kinds of seismic waves.) **A.** As illustrated by a slinky, P waves are compressional waves that alternately compress and expand the material through which they pass. **B.** The back-and-forth motion produced as compressional waves travel along the surface can cause the ground to buckle and fracture, and may cause power lines to break. **C.** S waves cause material to oscillate at right angles to the direction of wave motion. **D.** Because S waves can travel in any plane, they produce up-and-down and sideways shaking of the ground.

The difference in velocities of P and S waves provides a method for locating the epicenter. The principle used is analogous to a race between two autos, one faster than the other. The P wave always wins the race, arriving ahead of the S wave. But the greater the length of the race, the greater will be the difference in the arrival times at the finish line (the seismic station). Therefore, the greater the interval measured on a seismogram between the arrival of the first P wave and the first S wave, the greater the distance to the earthquake source.

A system for locating earthquake epicenters was developed by using seismograms from earthquakes whose epicenters could be easily pinpointed from physical evidence. From these seismograms, travel-time graphs were constructed (Figure 7.9). The first travel-time graphs were greatly improved when seismograms became available from nuclear explosions, because the precise location and time of detonation were known.

Using the sample seismogram in Figure 7.7 and the travel-time curve in Figure 7.9, we can determine the distance separating the recording station from the earthquake in two steps: (1) using the seismogram, determine the time interval between the arrival of the first P wave and the first S wave, and (2) using the travel-time graph, find the P-S interval on the vertical axis and use that information to determine the distance to the epicenter on the horizontal axis. From this information, we can determine that this earthquake occurred 3400 kilometers (2100 miles) from the recording instrument.

Now we know the *distance*, but what *direction*? The epicenter could be in any direction from the seismic station. As shown in Figure 7.10, the precise location can be found when the distance is known from three or more different seismic stations. On a globe, we draw a circle around each seismic station. Each circle represents the epicenter distance for each station. The point where the three circles intersect is

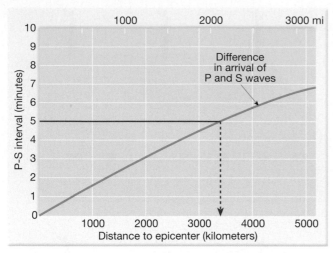

▲ **FIGURE 7.9** A travel-time graph is used to determine the distance to the epicenter. The difference in arrival times of the first P wave and the first S wave in the example is 5 minutes. Thus, the epicenter is roughly 3400 kilometers (2100 miles) away.

the epicenter of the quake. This method is called triangulation.

About 95 percent of the energy released by earthquakes originates in a few relatively narrow zones (Figure 7.11). The greatest energy is released along a path around the outer edge of the Pacific Ocean known as the *circum-Pacific belt*. Included in this zone are regions of great seismic activity, such as Japan, the Philippines, Chile, and numerous volcanic island chains, as exemplified by Alaska's Aleutian Islands.

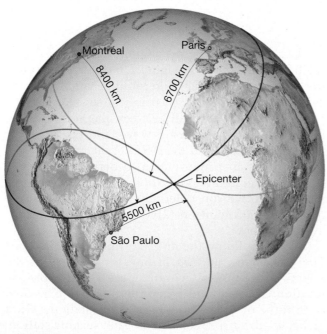

▲ **FIGURE 7.10** Earthquake epicenter is located using the distance obtained from three seismic stations.

Figure 7.11 reveals another continuous belt that extends for thousands of kilometers through the world's oceans. This zone coincides with the oceanic ridge system, an area of frequent but low-intensity seismic activity. By comparing this figure with Figure 8.9 (pp. 224-225), you can see a close correlation between the location of earthquake epicenters and plate boundaries.

Measuring the Size of Earthquakes

Historically, seismologists have employed a variety of methods to obtain two fundamentally different measures that describe the size of an earthquake—intensity and magnitude. The first of these to be used was **intensity**—a measure of the degree of earthquake shaking at a given locale based on the amount of damage. With the development of seismographs, it became clear that a quantitative measure of an earthquake based on seismic records rather than uncertain personal estimates of damage was desirable. The measurement that was developed, called **magnitude**, relies on calculations that use data provided by seismic records (and other techniques) to estimate the amount of energy released at the source of the earthquake.

Intensity Scales

Until a little over a century ago, historical records provided the only accounts of the severity of earthquake shaking and destruction. Using these descriptions—which were compiled without any established standards for reporting—made accurate comparisons of earthquake sizes difficult, at best.

In order to standardize the study of earthquake severity, workers developed various intensity scales that considered damage done to buildings, as well as individual descriptions of the event, and secondary effects such as landslides and the extent of ground rupture. By 1902, Guiseppe Mercalli had developed a relatively reliable intensity scale, which in a modified form is still used today. The **Modified Mercalli Intensity Scale** shown in Table 7.1 was developed using California buildings as its standard, but it is appropriate for use throughout most of the United States and Canada to estimate the strength of an earthquake. For example, if some well-built wood structures and most masonry buildings are destroyed by an earthquake, a region would be assigned an intensity of X on the Mercalli scale (Table 7.1).

Despite their usefulness in providing seismologists with a tool to compare earthquake severity, particularly in regions where there are no seismographs,

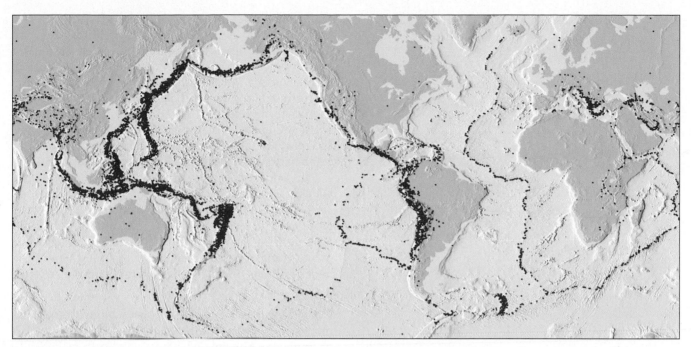

▲ **FIGURE 7.11** Distribution of the 14,229 earthquakes with magnitudes equal to or greater than 5 for a 10-year period. (Data from National Geophysical Data Center/NOAA)

intensity scales have severe drawbacks. In particular, intensity scales are based on effects (largely destruction) of earthquakes that depend not only on the severity of ground shaking but also on factors such as population density, building design, and the nature of surface materials.

Magnitude Scales

In order to compare earthquakes across the globe, a measure was needed that does not rely on parameters that vary considerably from one part of the world to another, such as construction practices. As a consequence, a number of magnitude scales were developed.

Richter Magnitude. In 1935 Charles Richter of the California Institute of Technology developed the first magnitude scale using seismic records to estimate the relative sizes of earthquakes. As shown in Figure 7.12 (top), the **Richter scale** is based on the amplitude of the largest seismic wave (P, S, or surface wave) recorded on a seismogram. Because seismic waves weaken as the distance between the earthquake focus and the seismograph increases (in a manner similar to light), Richter developed a method that accounted for the decrease in wave amplitude with increased distance. Theoretically, as long as the same, or equivalent, instruments were used, monitoring stations at various locations would obtain the same Richter magnitude

TABLE 7.1	Modified Mercalli Intensity Scale.
I	Not felt except by a very few under especially favorable circumstances.
II	Felt only by a few persons at rest, especially on upper floors of buildings.
III	Felt quite noticeably indoors, especially on upper floors of buildings, but many people do not recognize it as an earthquake.
IV	During the day felt indoors by many, outdoors by few. Sensation like heavy truck striking building.
V	Felt by nearly everyone, many awakened. Disturbances of trees, poles, and other tall objects sometimes noticed.
VI	Felt by all; many frightened and run outdoors. Some heavy furniture moved; few instances of fallen plaster or damaged chimneys. Damage slight.
VII	Everybody runs outdoors. Damage negligible in buildings of good design and construction; slight-to-moderate in well-built ordinary structures; considerable in poorly built or badly designed structures.
VIII	Damage slight in specially designed structures; considerable in ordinary substantial buildings with partial collapse; great in poorly built structures. (Fall of chimneys, factory stacks, columns, monuments, walls.)
IX	Damage considerable in specially designed structures. Buildings shifted off foundations. Ground cracked conspicuously.
X	Some well-built wooden structures destroyed. Most masonry and frame structures destroyed. Ground badly cracked.
XI	Few, if any, (masonry) structures remain standing. Bridges destroyed. Broad fissures in ground.
XII	Damage total. Waves seen on ground surfaces. Objects thrown upward into air.

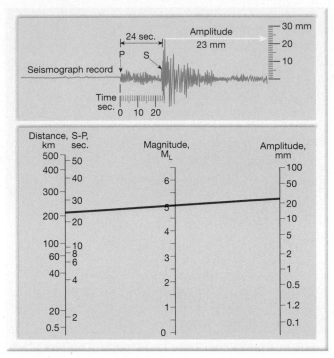

▲ FIGURE 7.12 Illustration showing how the Richter magnitude of an earthquake can be determined graphically using a seismograph record from a Wood-Anderson instrument. First, measure the height (amplitude) of the largest wave on the seismogram (23 mm) and then the distance to the focus using the time interval between S and P waves (24 seconds). Next, draw a line between the distance scale (left) and the wave amplitude scale (right). By doing this, you should obtain the Richter magnitude (M_L) of 5. (Data from California Institute of Technology)

for every recorded earthquake. (Richter selected the Wood-Anderson seismograph as the standard recording device.)

Although the Richter scale has no upper limit, the largest magnitude recorded on a Wood-Anderson seismograph was 8.9. These great shocks release approximately 10^{26} ergs of energy—roughly equivalent to the detonation of 1 billion tons of TNT. Conversely, earthquakes with a Richter magnitude of less than 2.0 are not felt by humans. With the development of more sensitive instruments, tremors of a magnitude of minus 2 were recorded. Table 7.2 shows how Richter magnitudes and their effects are related.

Earthquakes vary enormously in strength, and great earthquakes produce wave amplitudes that are thousands of times larger than those generated by weak tremors. To accommodate this wide variation, Richter used a *logarithmic scale* to express magnitude, where a *tenfold* increase in wave amplitude corresponds to an increase of 1 on the magnitude scale. Thus, the amount of ground shaking for a 5-magnitude earthquake is 10 times greater than that produced by an earthquake having a Richter magnitude of 4.

In addition, each unit of Richter magnitude equates to roughly a *32-fold energy increase*. Thus, an earthquake with a magnitude of 6.5 releases 32 times more energy than one with a magnitude of 5.5, and roughly 1000 times more energy than a 4.5-magnitude quake. A major earthquake with a magnitude of 8.5 releases millions of times more energy than the smallest earthquakes felt by humans.

Other Magnitude Scales. Richter's original goal was modest in that he only attempted to rank the earthquakes of southern California (shallow-focus earthquakes) into groups of large, medium, and small magnitude. Hence, Richter magnitude was designed to study nearby (or local) earthquakes and is denoted by the symbol M_L— where M is for *magnitude* and L is for *local*.

The convenience of describing the size of an earthquake by a single number that could be calculated quickly from seismograms makes the Richter scale a powerful tool. Further, unlike intensity scales that could only be applied to populated areas of the globe, Richter magnitudes could be assigned to earthquakes in more remote regions and even to events that occurred in the ocean basins. As a result, the method devised by Richter was adapted to a number of different seismographs located throughout the world. In time, seismologists modified Richter's work and developed new magnitude scales.

However, despite their usefulness, none of these "Richter-like" magnitude scales are adequate for describing very large earthquakes. For example, the 1906 San Francisco earthquake and the 1964 Alaskan earthquake have roughly the same Richter magnitudes of about 8.3. However, the Alaskan earthquake released considerably more energy than the San Francisco quake,

TABLE 7.2	Earthquake Magnitudes and Expected World Incidence.	
Richter Magnitudes	**Effects Near Epicenter**	**Estimated Number per Year**
<2.0	Generally not felt, but recorded.	600,000
2.0–2.9	Potentially perceptible.	300,000
3.0–3.9	Felt by some.	49,000
4.0–4.9	Felt by most.	6200
5.0–5.9	Damaging shocks.	800
6.0–6.9	Destructive in populous regions.	266
7.0–7.9	Major earthquakes. Inflict serious damage.	18
8.0 and above	Great earthquakes. Destroy communities near epicenter.	1.4
Source: *Earthquake Information Bulletin* and others.		

based on the size of the fault zone and the amount of displacement observed. Thus, the Richter scale (as well as the other related magnitude scales) are said to be *saturated* for large earthquakes because they cannot distinguish between the size of very large events.

Moment Magnitude. In recent years seismologists have been employing a more precise measure called **moment magnitude** (M_W), which can be calculated using several techniques. In one method the moment magnitude is calculated from field studies using a combination of factors that include the average amount of displacement along the fault, the area of the rupture surface, and the shear strength of the faulted rock—a measure of how much energy a rock can store before it suddenly slips and releases this energy in the form of an earthquake (and heat).

The moment magnitude can also be readily calculated from seismograms by examining very long period seismic waves. The values obtained have been calibrated so that small- and moderate-sized earthquakes have moment magnitudes that are roughly equivalent to Richter magnitudes. However, moment magnitudes are much better for describing very large earthquakes. For example, on the moment magnitude scale, the 1906 San Francisco earthquake, which had a Richter magnitude of 8.3, would be demoted to 7.9 on the moment magnitude scale, whereas the 1964 Alaskan earthquake with an 8.3 Richter magnitude would be increased to 9.2. The strongest earthquake on record is the 1960 Chilean earthquake with a moment magnitude of 9.5.

Moment magnitude has gained wide acceptance among seismologists and engineers because: (1) it is the only magnitude scale that estimates adequately the size of very large earthquakes; (2) it is a measure that can be derived mathematically from the size of the rupture surface and the amount of displacement, thus it better reflects the total energy released during an earthquake; and (3) it can be verified by two independent methods—field studies that are based on measurements of fault displacement and by seismographic methods using long-period waves.

Destruction from Earthquakes

The most violent earthquake to jar North America this century—the Good Friday Alaskan Earthquake—occurred in 1964. Felt throughout that state, the earthquake had a moment magnitude of 9.2 and reportedly lasted 3 to 4 minutes. This event left 131 people dead, thousands homeless, and the economy of the state badly disrupted because it occurred near major towns and seaports (Figure 7.13). Had the schools and business districts been open on this holiday, the toll surely would have been higher. Within 24 hours of the initial shock, 28 aftershocks were recorded, 10 of which exceeded a Richter magnitude of 6.

STUDENTS SOMETIMES ASK...

I've heard that the safest place to be in a house during an earthquake is in a doorframe. Is that really the best place?

No! An enduring earthquake image of California is a collapsed adobe home with the doorframe as the only standing part. From this came the belief that a doorway is the safest place to be during an earthquake. In modern homes, doorways are no stronger than any other part of the house and usually have doors that will swing and can injure you.

If you're inside, the best advice is to *duck, cover,* and *hold.* When you feel an earthquake, *duck* under a desk or sturdy table. Stay away from windows, bookcases, file cabinets, heavy mirrors, hanging plants, and other heavy objects that could fall. Stay under *cover* until the shaking stops. And, *hold* on to the desk or table: If it moves, move with it.

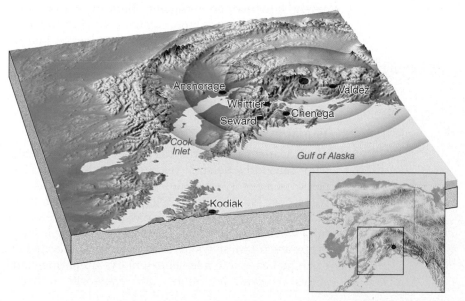

◄ **FIGURE 7.13** Region most affected by the Good Friday earthquake of 1964. Note the location of the epicenter (red dot). (After U.S. Geological Survey)

Destruction from Seismic Vibrations

The 1964 Alaskan earthquake provided geologists with new insights into the role of *ground shaking* as a destructive force. As the energy released by an earthquake travels along Earth's surface, it causes the ground to vibrate in a complex manner by moving up and down as well as from side to side. The amount of structural damage attributable to the vibrations depends on several factors, including (1) the intensity and (2) duration of the vibrations, (3) the nature of the material upon which the structure rests, and (4) the design of the structure.

All multistory structures in Anchorage were damaged by the vibrations, but the more flexible wood-frame residential buildings fared best. Figure 7.14 offers a striking example of how construction variations affect earthquake damage; you can see that the steel-frame building on the left withstood the vibrations, whereas the poorly designed J.C. Penney building was badly damaged. Engineers have learned that unreinforced masonry buildings are the most serious safety threat in earthquakes.

Most large structures in Anchorage were damaged, even though they were built according to the earthquake provisions of the Uniform Building Code. Perhaps some of that destruction can be attributed to the unusually long duration of this earthquake. Most quakes consist of tremors that last less than a minute. For example, the 1994 Northridge earthquake was felt for about 40 seconds, and the strong vibrations of the 1989 Loma Prieta earthquake lasted less than 15 seconds. But the Alaska quake reverberated for 3 to 4 minutes.

▼ **FIGURE 7.14** Damage to the five-story JC Penney Co. building, Anchorage, Alaska. Very little structural damage was incurred by the adjacent building. (Courtesy of NOAA)

Amplification of Seismic Waves Although the region within 20 to 50 kilometers of the epicenter will experience about the same intensity of ground shaking, the destruction varies considerably within this area. This difference is mainly attributable to the nature of the ground on which the structures are built. Soft sediments, for example, generally amplify the vibrations more than solid bedrock. Thus, the buildings located in Anchorage, which were situated on unconsolidated sediments, experienced heavy structural damage. By contrast, most of the town of Whittier, although much nearer the epicenter, rests on a firm foundation of granite and hence suffered much less damage. However, Whittier was damaged by a seismic sea wave (described in the next section).

The 1985 Mexican earthquake gave seismologists and engineers a vivid reminder of what had been learned from the 1964 Alaskan earthquake. The Mexican coast, where the earthquake was centered, experienced unusually mild tremors despite the strength of the quake. As expected, the seismic waves became progressively weaker with increasing distance from the epicenter. However, in the central section of Mexico City, nearly 400 kilometers from the source, the vibrations intensified to five times that experienced in outlying districts (Figure 7.15). Much of this amplified ground motion can be attributed to soft sediments, remnants of an ancient lakebed, that underlie portions of the city.

Liquefaction In areas where unconsolidated materials are saturated with water, earthquake vibrations can generate a phenomenon known as **liquefaction** (*liqueo* = to be fluid, *facio* = to make). Under these conditions, what had been a stable soil turns into a mobile fluid that is not capable of supporting buildings or other structures (Figure 7.16). As a result, underground objects such as storage tanks and sewer lines may literally float toward the surface of their newly liquefied environment. Buildings and other structures may settle and collapse. During the 1989 Loma Prieta earthquake, in San Francisco's Marina District, foundations failed and geysers of sand and water shot from the ground, indicating that liquefaction had occurred (Figure 7.17).

Tsunami

Most deaths associated with the 1964 Alaskan quake were caused by **seismic sea waves,** or **tsunami.** (The name *tsunami* is a Japanese word meaning "harbor wave," for Japanese harbors have suffered from them many times.) These destructive waves often are called tidal waves by the media. However, this name is inappropriate, for these waves are most often generated by earthquakes and to a lesser extent by submarine landslides, volcanic eruptions, and meteorite impacts.

◄ **FIGURE 7.15** During the 1985 Mexican earthquake, multistory buildings swayed back and forth as much as 1 meter. (Photo by James L. Beck)

They are *not* produced by the tidal effect of the Moon or Sun.

Tsunami triggered by an earthquake occur where a slab of oceanic crust is displaced vertically along a fault (Figure 7.18), or where the vibration of a quake sets an underwater landslide into motion.

Once formed, a tsunami resembles the ripples created when a pebble is dropped into a pond. In contrast to ripples, tsunami advance across the ocean at speeds between 500 and 950 kilometers (300 and 600 miles) per hour. Despite this, a tsunami in the open ocean can pass undetected because its height is usually less than 1 meter and the distance between wave crests is great, ranging from 100 to 700 kilometers. However, upon entering shallower coastal water, these destructive waves are slowed and the water begins to pile up to heights that occasionally exceed 30 meters (100 feet), as shown in Figure 7.18. As the crest of a tsunami approaches shore, it appears as a rapid rise in sea level with a turbulent and chaotic surface.

Usually the first warning of an approaching tsunami is a rather rapid withdrawal of water from beaches. Coastal residents have learned to heed this warning and move to higher ground, because about 5 to 30 minutes following the retreat of water a surge capable of extending hundreds of meters inland often occurs. In a successive fashion, each surge is followed by a rapid oceanward retreat of the water. These waves are separated by intervals of between 10 and 60 minutes. They are able to traverse thousands of kilometers of the ocean before their energy is dissipated.

Tsunami Damage from the 2004 Indonesian Earthquake
A massive undersea earthquake of moment magnitude 9.0 occurred near the island of Sumatra on December 26, 2004, and sent waves of water racing across the Indian

◄ **FIGURE 7.16** Effects of liquefaction. This tilted building rests on unconsolidated sediment that imitated quicksand during the 1985 Mexican earthquake. (Photo by James L. Beck)

▲ FIGURE 7.17 These "mud volcanoes" were produced by the Loma Prieta earthquake of 1989. They formed when geysers of sand and water shot from the ground, an indication that liquefaction occurred. (Photo by Richard Hilton, courtesy of Dennis Fox)

Ocean and Bay of Bengal. This tsunami was one of the deadliest natural disasters of any kind in modern times, killing more than 200,000 people. The killer waves generated by this massive quake achieved heights as great as 10 meters (33 feet) and struck many unprepared areas within three hours of the event.

The tsunami ravaged coastal regions throughout the northern Indian Ocean basin (Figure 7.19). Devastation was severe along the southeast coast of Sri Lanka, in the Indonesian province of Aceh, in the Indian state of Tamil Nada, and on Thailand's resort island of Phuket. Damages were reported as far away as the Somalia coast of Africa, 4100 kilometers (25,000 miles) west of the earthquake epicenter.

Tsunami Warning System. In 1946, a large tsunami struck the Hawaiian Islands without warning. A wave more than 15 meters (50 feet) high left several coastal villages in shambles. This destruction motivated the U.S. Coast and Geodetic Survey to establish a tsunami warning system for coastal areas of the Pacific. From seismic observatories throughout the region, large earthquakes are reported to the Tsunami Warning Center in Honolulu. Scientists at the Center use tidal gauges to determine whether a tsunami has formed. Within an hour a warning is issued. Although tsunamis travel very rapidly, there is sufficient time to evacuate all but the region nearest the epicenter. For example, a tsunami generated near the Aleutian Islands would take five hours to reach Hawaii, and one generated near the coast of Chile would travel 15 hours before reaching Hawaii (Figure 7.20).

Fortunately, most earthquakes do not generate tsunamis. On the average, only about one or two de-

▼ FIGURE 7.18 Schematic drawing of a tsunami generated by displacement of the ocean floor. The speed of a wave moving across the ocean correlates with ocean depth. As shown, waves moving in deep water advance at speeds in excess of 800 kilometers per hour. Speed gradually slows to 50 kilometers per hour at depths of 20 meters. Decreasing depth slows the movement of the wave column. As waves slow in shallow water, they grow in height until they topple and rush onto shore with tremendous force. The size and spacing of these swells are not to scale.

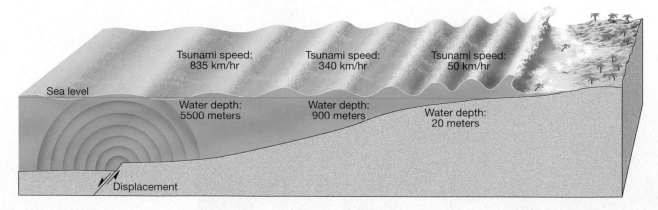

Tsunami speed: 835 km/hr

Tsunami speed: 340 km/hr

Tsunami speed: 50 km/hr

Sea level

Water depth: 5500 meters

Water depth: 900 meters

Water depth: 20 meters

Displacement

◀ FIGURE 7.19 A massive earthquake of magnitude 9.0 off the Indonesian island of Sumatra sent tsunami racing across the Indian Ocean and Bay of Bengal on December 26, 2004. Here, unsuspecting foreign tourists who at first walked out on the sand after the water receded now rush toward shore as the first of six tsunami start to roll towards Hat Rai Lay Beach near Krabi in southern Thailand. (AFP/Getty Images Inc.)

structive tsunamis are generated worldwide annually. Of these, only about one every 10 years is catastrophic.

STUDENTS SOMETIMES ASK...

What is the largest wave triggered by an earthquake?

The largest wave ever recorded occurred in Lituya Bay, about 200 kilometers (125 miles) west of Juneau, Alaska's capital. On July 9, 1958, an earthquake triggered an enormous rockslide that dumped 90 million tons of rock into the upper part of the bay. The rockslide created a huge *splash wave* (different than a tsunami, these waves are produced when an object splashes into water) that swept over the ridge facing the rockslide and uprooted or snapped off trees 1740 feet above the bay. Even larger splash waves have occurred in prehistoric times, including an estimated 3000-foot wave that is thought to have resulted from a meteorite impact in the Gulf of Mexico about 65 million years ago.

Landslides and Ground Subsidence

In the 1964 Alaskan earthquake, the greatest damage to structures was from landslides and ground subsidence triggered by the vibrations. At Valdez and Seward, the violent shaking caused river-delta materials to experience liquefaction; the subsequent slumping carried both waterfronts away. Because the disaster could happen again, the entire town of Valdez was relocated about 7 kilometers away on more stable ground. In Valdez, 31 people on a dock died when it slid into the sea.

Most of the damage in Anchorage was attributed to landslides. Many homes were destroyed in Turnagain Heights when a layer of clay lost its strength and over 200 acres of land slid toward the ocean (Figure 7.21). A portion of this landslide has been left in its natural condition as a reminder of this destructive event. The site was named "Earthquake Park." Downtown Anchorage was also disrupted as sections of the main business district dropped by as much as 3 meters (10 feet).

▼ FIGURE 7.20 Tsunami travel times to Honolulu, Hawaii, from selected locations throughout the Pacific. (Data from NOAA)

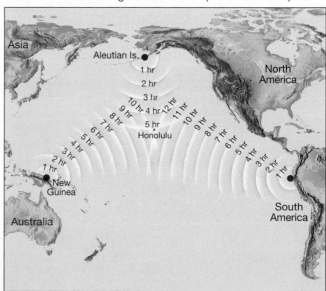

▲ **FIGURE 7.21** Photo of a small portion of the Turnagain Heights slide. (Photo courtesy of U.S. Geological Survey)

Fire

The 1906 San Francisco earthquake reminds us of the formidable threat of fire. The central city contained mostly large, older wooden structures and brick buildings. The greatest destruction was caused by fires that started when gas and electrical lines were severed. The fires raged uncontrolled for three days and devastated over 500 city blocks (see Figure 7.3). The problem was compounded by the initial ground shaking, which broke the city's water lines into hundreds of unconnected pieces.

The fire was finally contained when buildings were dynamited along a wide boulevard to create a *fire break,* the same strategy used in fighting a forest fire. Although only a few deaths were attributed to the fires, such is not always the case. A 1923 earthquake in Japan (their worst quake prior to the 1995 Kobe tremor) triggered an estimated 250 fires, which devastated the city of Yokohama and destroyed more than half the homes in Tokyo. Over 100,000 deaths were attributed to the fires, which were driven by unusually high winds.

Can Earthquakes Be Predicted?

The vibrations that shook Northridge, California, in 1994 inflicted 57 deaths and about $40 billion in damage (Figure 7.22). This was from a brief earthquake (about 40 seconds) of moderate rating (M_W 6.7). Seismologists warn that earthquakes of comparable or greater strength will occur along the San Andreas Fault, which cuts a 1300-kilometer (800-mile) path through the state. The obvious question is, can earthquakes be predicted?

Short-Range Predictions

The goal of short-range earthquake prediction is to provide a warning of the location and magnitude of a large earthquake within a narrow time frame. Substan-

▼ **FIGURE 7.22** Damage to Interstate 5 caused by the January 17, 1994, Northridge earthquake. (Photo by Tom McHugh/Photo Researchers, Inc.)

TABLE 7.3 Some Notable Earthquakes.

Year	Location	Deaths (est.)	Magnitude[†]	Comments
1556	Shensi, China	830,000		Possibly the greatest natural disaster.
1755	Lisbon, Portugal	70,000		Tsunami damage extensive.
1811–1812*	New Madrid, Missouri	Few	7.9	Three major earthquakes.
1886*	Charleston, South Carolina	60		Greatest historical earthquake in the eastern United States.
1906*	San Francisco, California	1500	7.8	Fires caused extensive damage.
1908	Messina, Italy	120,000		
1923	Tokyo, Japan	143,000	7.9	Fire caused extensive destruction.
1960	Southern Chile	5700	9.6	Possibly the largest-magnitude earthquake ever recorded.
1964*	Alaska	131	9.2	Greatest North American earthquake.
1970	Peru	66,000	7.8	Great rockslide.
1964*	San Fernando, California	65	6.5	Damage exceeded $1 billion.
1975	Liaoning Province, China	1328	7.5	First major earthquake to be predicted.
1976	Tangshan, China	240,000	7.6	Not predicted.
1985	Mexico City	9500	8.1	Major damage occurred 400 km from epicenter.
1988	Armenia	25,000	6.9	Poor construction practices.
1989*	Loma Prieta, California	62	6.9	Damages exceeded $6 billion.
1990	Iran	50,000	7.3	Landslides and poor construction practices caused great damage.
1993	Latur, India	10,000	6.4	Located in stable continental interior.
1994*	Northridge, California	57	6.7	Damages in excess of $40 billion.
1995	Kobe, Japan	5472	6.9	Damages estimated to exceed $100 billion.
1999	Izmit, Turkey	17,127	7.4	Nearly 44,000 injured and more than 250,000 displaced
1999	Chi-Chi, Taiwan	2300	7.6	Severe destruction; 8700 injuries.
2001	El Salvador	1000	7.6	Triggered many landslides.
2001	Bhuj, India	20,000[†]	7.9	1 million or more homeless.
2003	Bam, Iran	41,000[†]	6.6	Ancient city with poor construction.

*U.S. earthquakes.

[†]Widely differing magnitudes have been estimated for some of these earthquakes. When available, moment magnitudes are used.

Source: U.S. Geological Survey

tial efforts to achieve this objective are being put forth in Japan, the United States, China, and Russia—countries where earthquake risks are high (Table 7.3). This research has concentrated on monitoring possible *precursors*—phenomena that precede and thus provide a warning of a forthcoming earthquake. In California, for example, some seismologists are measuring uplift, subsidence, and strain in the rocks near active faults. Some Japanese scientists are studying peculiar anomalous behavior that may precede a quake.

One claim of a successful short-range prediction was made by Chinese seismologists after the February 4, 1975, earthquake in Liaoning Province. According to reports, very few people were killed, although more than 1 million lived near the epicenter, because the earthquake was predicted and the population was evacuated. Recently, some Western seismologists have questioned this claim and suggest instead that an intense swarm of foreshocks that began 24 hours before the main earthquake may have caused many people to evacuate spontaneously. Further, an official Chinese government report issued 10 years later stated that 1328 people died and 16,980 injuries resulted from this earthquake.

One year after the Liaoning earthquake at least 240,000 people died in the Tangshan, China, earthquake, which was not predicted. The Chinese have also issued false alarms. In a province near Hong Kong, people reportedly left their dwellings for over a month, but no earthquake followed. Clearly, whatever method the Chinese employ for short-range predictions, it is *not* reliable.

For a short-range prediction scheme to warrant general acceptance, it must be both accurate and reliable. Thus, *it must have a small range of uncertainty as regards to location and timing, and it must produce few failures, or false alarms.* Can you imagine the debate that would precede an order to evacuate a large city in the United States, such as Los Angeles or San Francisco? The cost of evacuating millions of people, arranging for living accommodations, and providing for their lost work time and wages would be staggering.

Long-Range Forecasts

In contrast to short-range predictions, which aim to predict earthquakes within a time frame of hours or at most days, long-range forecasts give the probability of a certain magnitude earthquake occurring on a time scale of 30 to 100 years or more. Stated another way, these forecasts give statistical estimates of the expected intensity of ground motion for a given area over a specified time frame. Although long-range forecasts may not be as informative as we might like, the data are important for updating the Uniform Building Code, which contains nationwide standards for designing earthquake-resistant structures.

Long-range forecasts are based on the premise that earthquakes are repetitive or cyclical, like the weather. In other words, as soon as one earthquake is over, the continuing motions of Earth's plates begin to build strain in the rocks again, until they fail once more. This has led seismologists to study historical records of earthquakes to see if there are any discernible patterns so that the probability of recurrence might be established.

One study conducted by the U.S. Geological Survey gives the probability of a rupture occurring along various segments of the San Andreas Fault for the 30 years between 1988 and 2018 (Figure 7.23). From this investigation, the Santa Cruz Mountains area was given a 30 percent probability of producing a 6.5-magnitude earthquake during this time period. In fact, it produced the Loma Prieta quake in 1989, of 7.1 magnitude.

The region along the San Andreas Fault given the highest probability (90 percent) of generating a quake is the Parkfield section. This area has been called the "Old Faithful" of earthquake zones because activity here has been very regular since record keeping began in 1857. In late September 2004, a magnitude 6.0 earthquake again struck this area. Although the event was more than a decade overdue, it did demonstrate the potential usefulness of long-range forcasts. Another section between Parkfield and the Santa Cruz Mountains is given a very low probability of generating an earthquake. This area has experienced very little seismic activity in historical times; rather, it exhibits a slow, continual movement known as *fault creep*. Such movement is beneficial because it prevents strain from building to high levels in the rocks.

In summary, it appears that the best prospects for making useful earthquake predictions involve forecasting magnitudes and locations in time scales of years, or perhaps even decades. These forecasts are important because they provide information used to develop the Uniform Building Code and assist in land-use planning.

Earth's Layered Structure

Forces Within
▼ Earthquakes

Earth's interior lies just below us; however, access is very limited. The deepest well yet drilled has penetrated Earth's crust only 12 kilometers (7.5 miles), less than 0.2 percent of the distance to the planet's center. Consequently, most knowledge of our planet's interior comes from the study of earthquake waves that travel through Earth and vibrate the surface at some distant point.

▶ **FIGURE 7.23** Probability of a major earthquake from 1988 to 2018 along the San Andreas Fault.

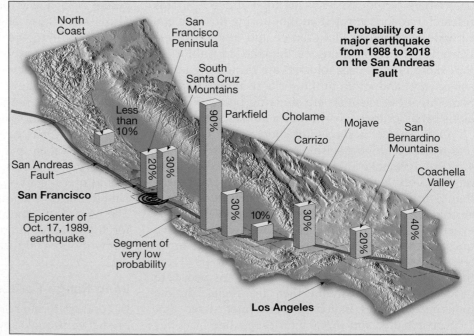

If Earth were a perfectly homogeneous body, seismic waves would spread through it in a straight line at a constant speed. However, this is not the case. It so happens that the seismic waves reaching seismographs located farther from an earthquake travel at faster average speeds than those recorded at locations closer to the event. This general increase in speed with depth is a consequence of increased pressure, which enhances the elastic properties of deeply buried rock. As a result, the paths of seismic rays through Earth are refracted (bent) as they travel (Figure 7.24).

As more sensitive seismographs were developed, it became apparent that in addition to gradual changes in seismic-wave velocities, rather abrupt velocity changes also occur at particular depths. Because these changes were detected worldwide, seismologists concluded that Earth must be composed of distinct shells having varying compositions and/or mechanical properties (Figure 7.24).

Layers Defined by Composition

Compositional layering occurred early in Earth's history as meteorite impacts and decay of radioactive elements caused the temperature of our planet to steadily increase. Eventually Earth became hot enough so that at least some melting occurred. During this period of partial melting the heavier elements, principally iron and nickel, sank as the lighter rocky components floated upward. This segregation of material is still occurring, but at a much reduced rate. Because of this chemical differentiation, Earth's interior is not homogeneous. Rather, it consists of three major zones each defined by its chemical composition—the crust, mantle, and core (Figure 7.25).

Crust. The **crust,** Earth's comparatively thin, rocky outer skin, is generally divided into oceanic and continental crust. The oceanic crust is roughly 7 kilometers (5 miles) thick and composed of the dark igneous rock called *basalt.* By contrast, the continental crust averages 35–40 kilometers (25 miles) thick but may exceed 70 kilometers (40 miles) in some mountainous regions. Unlike the oceanic crust, which has a relatively homogeneous chemical composition, the continental crust consists of many rock types. The upper crust has an average composition of a *granitic rock* called *granodiorite,* whereas the composition of the lowermost continental crust is more akin to basalt. Continental rocks have an average density of about 2.7 g/cm^3, and some have been discovered that are 4 billion years old. The rocks of the oceanic crust are younger (180 million years or less) and more dense (about 3.0 g/cm^3) than continental rocks.

Mantle. Over 82 percent of Earth's volume is contained in the **mantle,** a solid, rocky shell that extends to a depth of 2900 kilometers (1800 miles). The boundary between the crust and mantle represents a marked change in chemical composition. The dominant rock type in the uppermost mantle is *peridotite,* which has a density of 3.3 g/cm^3. At greater depths peridotite changes by assuming a more compact crystalline structure and hence a greater density.

Core. The **core** is a sphere composed of an iron-nickel alloy having a radius of 3486 kilometers (2161 miles). At the extreme pressures found in the core, the iron-rich material has an average density of nearly 11 g/cm^3 and approaches 14 times the density of water at Earth's center.

Layers Defined by Physical Properties

Earth's interior is characterized by a gradual increase in temperature, pressure, and density with depth. Estimates put the temperature at a depth of 100 kilometers at between 1200° and 1400°C, whereas the temperature at Earth's center may exceed 6700°C. Clearly, Earth's interior has retained much of the energy acquired during its formative years, despite the fact that heat is continuously flowing toward the surface, where it is lost to space.

The gradual increase in temperature and pressure with depth affects the physical properties and hence the mechanical behavior of Earth materials. When a substance is heated, its chemical bonds weaken and its mechanical strength (resistance to deformation) is reduced. If the temperature exceeds the melting point of

▼ **FIGURE 7.24** A few of the many possible paths that seismic rays take through Earth.

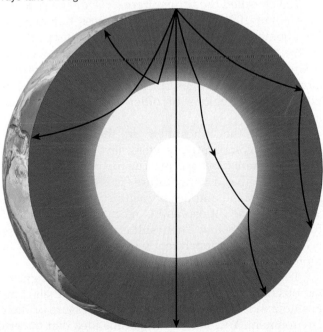

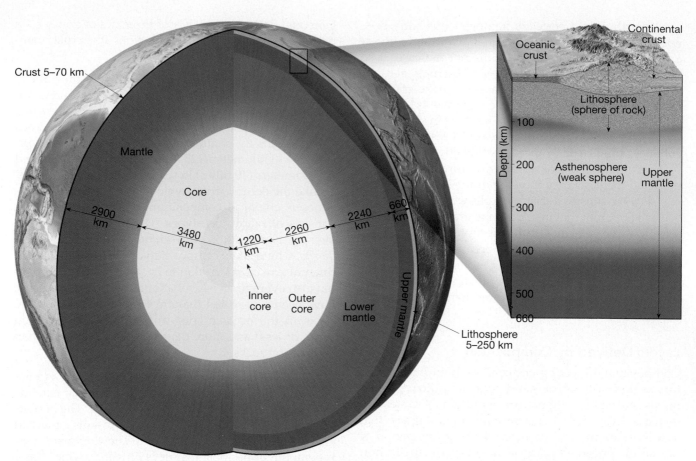

▲ **FIGURE 7.25**   Views of Earth's layered structure. The left side of the large cross section shows that Earth's interior is divided into three different layers based on compositional differences—the crust, mantle, and core. The right side of the large cross section depicts the five main layers of Earth's interior based on physical properties and hence mechanical strength—the lithosphere, asthenosphere, lower mantle, outer core, and inner core. The block diagram to the right of the large cross section shows enlarged views of the upper portion of Earth's interior.

an Earth material, the material's chemical bonds break and melting ensues. If temperature were the only factor that determined whether a substance melted, our planet would be a molten ball covered with a thin, solid outer shell. However, pressure also increases with depth and tends to increase rock strength. Furthermore, because melting is accompanied by an increase in volume, it occurs at higher temperatures at depth because of greater confining pressure. Thus, depending on the physical environment (temperature and pressure), a particular Earth material may behave like a brittle solid, deform in a puttylike manner, or even melt and become liquid.

Earth can be divided into five main layers based on physical properties and hence mechanical strength— the *lithosphere, asthenosphere, mesosphere (lower mantle), outer core,* and *inner core* (Figure 7.25).

Lithosphere and Asthenosphere.   Based on physical properties, Earth's outermost layer consists of the crust and uppermost mantle and forms a relatively cool, rigid shell. Although this layer is composed of materials with markedly different chemical composi-

tions, it tends to act as a unit that exhibits rigid behavior—mainly because it is cool and thus strong. This layer, called the **lithosphere** (*sphere of rock*), averages about 100 kilometers in thickness but may be 250 kilometers thick or more below the older portions of the continents (Figure 7.25). Within the ocean basins, the lithosphere is only a few kilometers thick along the oceanic ridges but increases to perhaps 100 kilometers in regions of older and cooler oceanic crust.

Beneath the lithosphere in the upper mantle lies a soft, comparatively weak layer known as the **asthenosphere** (*weak sphere*). The top portion of the asthenosphere has a temperature/pressure regime that results in a small amount of melting. Within this very weak zone, the lithosphere is mechanically detached from the layer below. The result is that the lithosphere is able to move independently of the asthenosphere, a topic we will consider in the next chapter.

It is important to emphasize that the strength of various Earth materials is a function of both their composition and of the temperature and pressure of their environment. You should not get the idea that the entire lithosphere behaves like a brittle solid similar to rocks found on the surface. Rather, the rocks of the lithosphere get progressively hotter and weaker (more easily deformed) with increasing depth. At the depth of the uppermost asthenosphere, the rocks are close enough to their melting temperatures (some melting may actually occur) that they are very easily deformed. Thus, the uppermost asthenosphere is weak because it is near its melting point, just as hot wax is weaker than cold wax.

Lower Mantle. Below the zone of weakness in the uppermost asthenosphere, increased pressure counteracts the effects of higher temperature, and the rocks gradually strengthen with depth. Between the depths of 660 kilometers and 2900 kilometers a more rigid layer called the **lower mantle** is found (Figure 7.25). Despite their strength, the rocks of the mesosphere are still very hot and capable of very gradual flow.

Inner and Outer Core. The core, which is composed mostly of an iron-nickel alloy, is divided into two regions that exhibit very different mechanical strengths (Figure 7.25). The **outer core** is a *liquid layer* 2270 kilometers (1410 miles) thick. It is the convective flow of metallic iron within this zone that generates Earth's magnetic field. The **inner core** is a sphere having a radius of 1216 kilometers (754 miles). Despite its higher temperature, the material in the inner core is stronger (because of immense pressure) than the outer core and behaves like a *solid.*

Discovering Earth's Major Layers

The story of how seismologists discovered Earth's core and layers is interesting. In 1909, a pioneering Yugoslavian seismologist, Andrija Mohorovičić presented the first convincing evidence for layering within Earth. By studying seismic records, he found that the velocity of seismic waves increases abruptly below about 50 kilometers of depth. This boundary separates the crust from the underlying mantle and is known as the **Mohorovičić discontinuity** in his honor. For reasons that are obvious, the name for this boundary was quickly shortened to **Moho.**

A few years later another boundary was discovered by the German seismologist Beno Gutenberg. Generally, seismic waves from even small earthquakes are strong enough to travel around the world. This is why a seismograph in Antarctica can record earthquakes in California or Italy. But Gutenberg observed that P waves diminish and eventually die out about 105 degrees around the globe from an earthquake. Then, about 140 degrees away, the P waves reappear, but about 2 minutes later than would be expected, based on the distance traveled. This belt, where direct P waves are absent, is about 35 degrees

wide and has been named the **shadow zone.** Figure 7.26 illustrates how this works.

Gutenberg realized that the shadow zone could be explained if Earth contained a core composed of material unlike the overlying mantle. The core must somehow hinder the transmission of P waves in a manner similar to the light rays being blocked by an opaque object that casts a shadow. However, rather than actually stopping the P waves, the shadow zone bends them, as shown in Figure 7.26. It was further learned that S waves could not travel through the core. Therefore, geologists concluded that at least a portion of this region is liquid.

In 1936 the last major subdivision of Earth's interior was predicted by Inge Lehmann, a Danish seismologist (see Box 7.2). Lehmann discovered a new region of seismic reflection within the core. Hence, Earth has a core within a core. The size of the inner core was not accurately calculated until the early 1960s, when underground nuclear tests were conducted in Nevada. Because the precise locations and times of the explosions were known, echoes from seismic waves that bounced off the inner core provided an accurate means of determining its size.

Over the past few decades, advances in seismology have allowed for much refinement of the gross view of Earth's interior presented so far. Of major importance was the discovery of the lithosphere and asthenosphere.

STUDENTS SOMETIMES ASK...

As compared to continental crust, the ocean crust is quite thin. Has there ever been an attempt to drill through it to obtain a sample of the mantle?

Yes. Project Mohole was initiated in 1958 to retrieve a sample of material from Earth's mantle by drilling a hole through Earth's crust to the *Mohorovičić discontinuity,* or *Moho.* The plan was to drill to the Moho to gain valuable information on Earth's age, makeup, and internal processes. Despite a successful test phase, drilling was halted as control of the project was shifted from one governmental organization to another. Although Project Mohole failed in its intended purpose, it did show that deep-ocean drilling is a viable means of obtaining geological samples.

Discovering Earth's Composition

We have examined Earth's structure, so let us now look at the composition of each layer. Composition tells us much about how our planet has developed over its estimated age of 4.5 billion years.

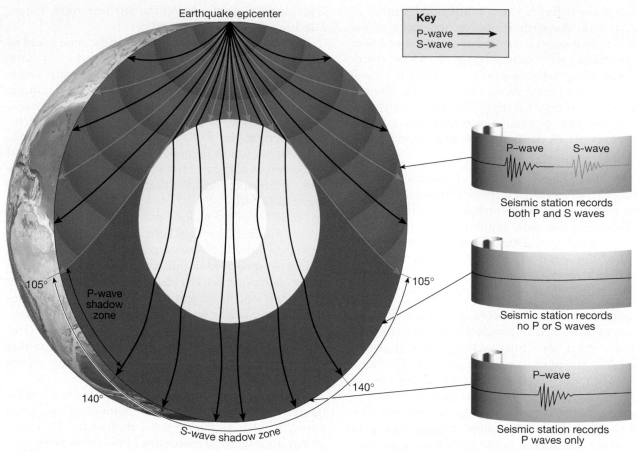

▲ **FIGURE 7.26** View of Earth's interior showing P and S wave paths. The abrupt change in physical properties at the mantle-core boundary causes the wave paths to bend sharply. This change in wave direction results in a shadow zone for P waves between about 105 and 140 degrees. Further, any location more than 105 degrees from an earthquake epicenter will not receive direct S waves since the outer core will not transmit them.

Earth's crust varies in thickness, exceeding 70 kilometers (40 miles) in some mountainous regions and being thinner than 3 kilometers (2 miles) in some oceanic areas (see Figure 7.25). Early seismic data indicated that the continental crust, which is mostly made of lighter, granitic rocks, is quite different in composition from denser oceanic crust. Until the late 1960s, however, scientists had only seismic evidence from which to determine the composition of oceanic crust, because it lies beneath an average of 3 kilometers of water as well as hundreds of meters of sediment. With the development of deep-sea drilling technology the recovery of ocean-floor samples was made possible. The samples were of basaltic composition very different from the rocks that compose the continents.

Our knowledge of the rocks of the mantle and core is much more speculative. However, we do have some clues. Recall that some of the lava that reaches Earth's surface originates in the partially melted asthenosphere within the mantle. In the laboratory, experiments have shown that partial melting of a rock called *peridotite* results in a melt that has a basaltic composition similar to lava that emerges during volcanic activity of oceanic islands. Denser rocks like peridotite are thought to make up the mantle and provide the lava for oceanic eruptions.

Surprisingly, meteorites—or shooting stars—that collide with Earth provide evidence of Earth's inner composition. Because meteorites are part of the solar system, they are assumed to be representative samples. Their composition ranges from metallic meteorites made of iron and nickel to stony meteorites composed of dense rock similar to peridotite.

Because Earth's crust contains a much smaller percentage of iron than do meteorites, geologists believe that the dense iron, and other dense metals, literally sank toward Earth's center during the planet's early history. By the same token, lighter substances may have floated to the surface, creating the less-dense crust. Thus, Earth's core is thought to be mainly dense iron and nickel, similar to metallic meteorites, whereas the surrounding mantle is believed to be composed of rocks similar to stony meteorites.

The concept of a molten iron outer core is further supported by Earth's magnetic field. Our planet acts as a large magnet. The most widely accepted mechanism explaining why Earth has a magnetic field requires that the core be made of a material that

BOX 7.1 UNDERSTANDING EARTH:
Inge Lehmann: A Pioneering Seismologist*

Inge Lehmann was a pioneering scientist at a time when few women had careers in science and mathematics (Figure 7.B). Born in Denmark in 1888, Lehmann lived a long and productive life that included important contributions to our understanding of Earth's interior. She died in 1993 at the age of 105.

Inge Lehmann's career in seismology started in 1925 when she helped establish seismic networks in Denmark and Greenland. It was a paper that she published in 1936 that established her place in the history of geophysics. Known simply as *P* (*P-prime*), the paper identified a new region of seismic reflection and refraction in Earth's interior, now known as the *Lehmann discontinuity*. Because of her exacting scrutiny of seismic records, Inge Lehmann had discovered the boundary that divides Earth's core into inner and outer parts.

Many honors came to Lehmann in recognition of her extraordinary accomplishments. In 1997 the Lehmann Medal was created by the American Geophysical Union (AGU) to recognize outstanding research on the structure, composition, and dynamics of Earth's mantle and core. It was the first medal awarded by the AGU that is named for a woman and the first to be named for someone who worked outside the United States.

Figure 7.B Inge Lehmann, 1888–1993. (Photo courtesy of Susan M. Landon)

*This box was prepared by Nancy L. Lutgens.

---

conducts electricity, such as iron, and which is mobile enough to allow circulation. Both of these conditions are met by the model of Earth's core that was established on the basis of seismic data.

Not only does an iron core explain Earth's magnetic field but it also explains the high density of inner Earth, about 14 times that of water at Earth's center. Even under the extreme pressure at those depths, average crustal rocks with densities 2.8 times that of water would not have the density calculated for the core. But iron, which is 3 times more dense than crustal rocks, has the required density.

In summary, although earthquakes can be very destructive, much of our knowledge about Earth's interior comes from the study of these phenomena. As our knowledge of earthquakes and their causes improve, we learn more of our planet's internal workings and possibly how to reduce the consequences of tremors. In the next chapter, you will see that most earthquakes originate at the boundaries of Earth's great lithospheric plates.

Chapter Summary

- *Earthquakes* are vibrations of Earth produced by the rapid release of energy from rocks that rupture because they have been subjected to stresses beyond their limit. This energy, which takes the form of waves, radiates in all directions from the earthquake's source, called the *focus*. The movements that produce most earthquakes occur along large fractures, called *faults*, that are associated with plate boundaries.

- Two main groups of *seismic waves* are generated during an earthquake: (1) *surface waves*, which travel along the outer layer of Earth; and (2) *body waves*, which travel through Earth's interior. Body waves are further divided into *primary*, or *P*, waves, which push (compress) and pull (expand) rocks in the direction the wave is traveling, and *secondary*, or *S*, waves, which shake the particles in rock at right angles to their direction of travel. P waves can travel through solids, liquids, and gases. Fluids (gases and liquids) will not transmit S waves. In any solid material, P waves travel about 1.7 times faster than do S waves.

- The location on Earth's surface directly above the focus of an earthquake is the *epicenter*. An epicenter is determined using the difference in velocities of P and S waves.

- *There is a close correlation between earthquake epicenters and plate boundaries.* The principal earthquake epicenter zones are along the outer margin of the Pacific Ocean, known as the *circum-Pacific belt* and through the world's oceans along the *oceanic ridge system*.

- Seismologists use two fundamentally different measures to describe the size of an earthquake—intensity and magnitude. *Intensity* is a measure of the degree of ground shaking at a given locale based on the amount of damage. The *Modified Mercalli Intensity Scale* uses damages to buildings in California to estimate the intensity of ground shaking for a local earthquake. *Magnitude* is calculated from seismic records and estimates the amount of energy released at the source of an earthquake. Using the *Richter scale*, the magnitude of an earthquake is estimated by measuring the *amplitude* (maximum displacement) of the largest seismic wave recorded. A logarithmic scale is used to express magnitude, in which a tenfold increase in ground shaking corresponds to an increase of 1 on the magnitude scale. *Moment magnitude* is currently used to estimate the size of moderate and large earthquakes. It is calculated using the average displacement of the fault, the area of the fault surface, and the sheer strength of the faulted rock.

• The most obvious factors that determine the amount of destruction accompanying an earthquake are the *magnitude* of the earthquake and the *proximity* of the quake to a populated area. *Structural damage* attributable to earthquake vibrations depends on several factors, including (1) *intensity*, (2) *duration* of the vibrations, (3) *nature of the material* upon which the structure rests, and (4) the *design* of the structure. Secondary effects of earthquakes include *tsunamis, landslides, ground subsidence,* and *fire*.

• Substantial research to predict earthquakes is under way in Japan, the United States, China, and Russia—countries where earthquake risk is high. No consistent method of short-range prediction has yet been devised. Long-range forecasts are based on the premise that earthquakes are repetitive or cyclical. Seismologists study the history of earthquakes for patterns, so their occurrences might be predicted.

• As indicated by the behavior of P and S waves as they travel through Earth, the four major zones of Earth's interior are the (1) *crust* (the very thin outer layer), (2) *mantle* (a rocky layer located below the crust with a thickness of 2885 kilometers), (3) *outer core* (a layer about 2270 kilometers thick, which exhibits the characteristics of a mobile liquid), and (4) *inner core* (a solid metallic sphere with a radius of about 1216 kilometers).

• The *continental crust* is primarily made of *granitic* rocks, while the *oceanic crust* is of *basaltic* composition. *Ultramafic rocks,* such as *peridotite,* are thought to make up the *mantle.* The *core* is composed mainly of *iron* and *nickel.*

• The crust and uppermost mantle form Earth's cool rigid outer shell called the *lithosphere.* Beneath the lithosphere lies a soft, relatively weak layer of the mantle known as the *asthenosphere.*

Key Terms

aftershock (p. 191)
asthenosphere (p. 208)
body wave (p. 193)
core (p. 207)
crust (p. 207)
earthquake (p. 189)
elastic rebound (p. 190)
epicenter (p. 194)
fault (p. 189)
fault creep (p. 193)
focus (p. 189)
foreshock (p. 191)

inner core (p. 209)
intensity (p. 196)
liquefaction (p. 200)
lithosphere (p. 208)
magnitude (p. 196)
mantle (p. 207)
mesosphere (lower mantle) (p. 209)
Modified Mercalli Intensity Scale
    (p. 196)
Mohorovičić discontinuity (Moho)
    (p. 209)
moment magnitude (p. 199)

outer core (p. 209)
primary (P) wave (p. 193)
Richter scale (p. 197)
secondary (S) wave (p. 193)
seismic sea wave (tsunami) (p. 200)
seismogram (p. 193)
seismograph (p. 193)
seismology (p. 193)
shadow zone (p. 209)
surface wave (p. 193)

Review Questions

1. What is an earthquake? Under what circumstances do earthquakes occur?

2. How are faults, foci, and epicenters related?

3. Who was first to explain the actual mechanism by which earthquakes are generated?

4. Explain what is meant by elastic rebound.

5. Faults that are experiencing no active creep may be considered safe. Rebut or defend this statement.

6. Describe the principle of a seismograph.

7. Using Figure 7.9, determine the distance between an earthquake and a seismic station if the first S wave arrives 3 minutes after the first P wave.

8. List the major differences between P and S waves.

9. Which type of seismic wave causes the greatest destruction to buildings?

10. Most strong earthquakes occur in a zone on the globe known as the _____.

11. What factor contributed most to the extensive damage that occurred in the central portion of Mexico City during the 1985 earthquake?

12. In 1988 an earthquake in Armenia caused approximately 25,000 deaths. A year later an earthquake in the Bay Area of California, having the same Richter magnitude of 6.9, had only 62 deaths. Suggest a reason for the greater loss of life in the Armenian event.

13. An earthquake measuring 7 on the Richter scale releases about _____ times more energy than an earthquake with a magnitude of 6.

14. List three reasons why the moment magnitude scale has gained wide acceptance among seismologists.

15. List four factors that affect the amount of destruction caused by seismic vibrations.

16. In addition to the destruction created directly by seismic vibrations, list three other types of destruction associated with earthquakes.

17. Distinguish between the Mercalli scale and the Richter scale.

18. What is a tsunami? In what two ways do earthquakes generate tsunami?

19. Cite some reasons why an earthquake with a moderate magnitude might cause more extensive damage than a quake with a high magnitude.

20. What evidence do we have that Earth's outer core is molten?

21. Contrast the physical makeup of the asthenosphere and the lithosphere.

22. Why are meteorites considered important clues to the composition of Earth's interior?

23. Describe the composition of the following:
    (a) continental crust
    (b) oceanic crust
    (c) mantle
    (d) core

Examining the Earth System

1. What potentially disastrous phenomenon often occurs when the energy of an earthquake is tranferred from the solid Earth to the hydrosphere (ocean) at their interface on the floor of the ocean? When the energy from this event is expended along a coast, how might coastal lands and the biosphere be altered?

2. Several research projects are considering ways that humans could interact with the earthquake-generating process and prevent, or significantly reduce, the release of destructive energy. Begin your research of earthquake prevention at this University of Colorado at Boulder Website, **http://www.colorado.edu/hazards/ resources/sites.html.** Following your investigation, describe some of the current earthquake-prevention proposals being considered by researchers. Do you believe that the proposals are feasible? What are some other ways to lessen the devastating impact of earthquakes?

Online Study Guide

The *Earth Science* Website uses the resources and flexibility of the Internet to aid in your study of the topics in this chapter. Written and developed by Earth science instructors, this site will help improve your understanding of Earth science. Visit **http://www.prenhall.com/tarbuck** and click on the cover of *Earth Science 11e* to find:

- **On-line review quizzes.**
- **Critical thinking exercises.**
- **Links to chapter-specific Web resources.**
- **Internet-wide key term searches.**
http://www.prenhall.com/tarbuck

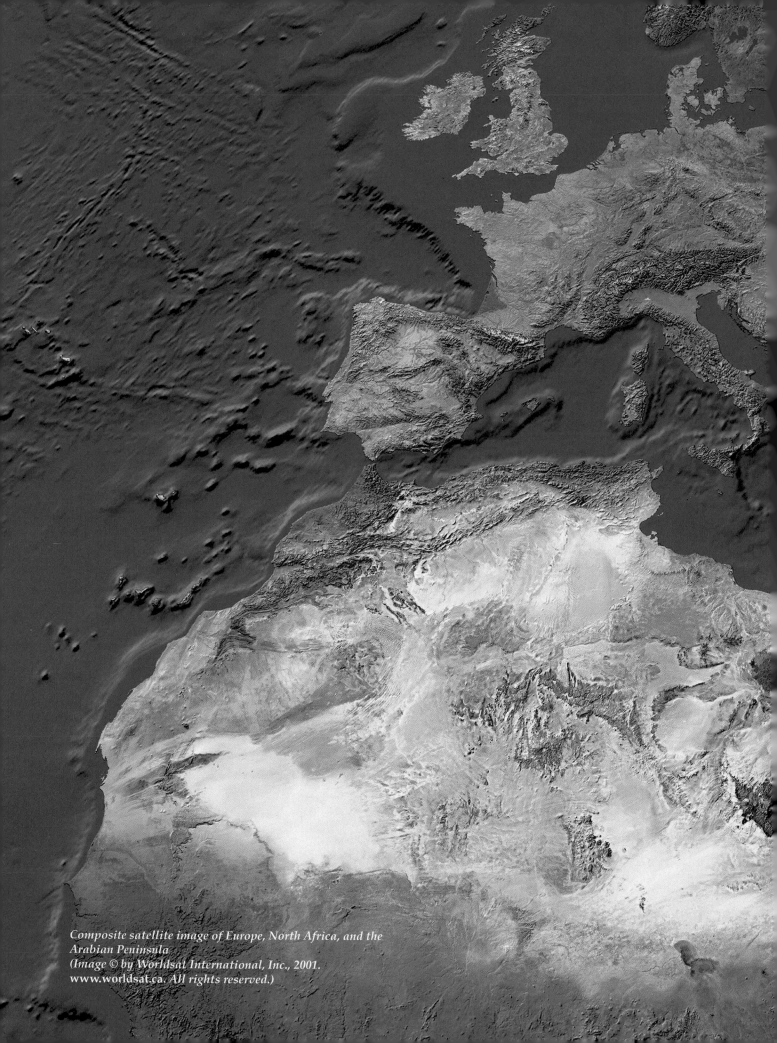

Composite satellite image of Europe, North Africa, and the
Arabian Peninsula
(Image © by Worldsat International, Inc., 2001.
www.worldsat.ca. All rights reserved.)

CHAPTER 8

Plate Tectonics: A Scientific Theory Unfolds

Early in the Twentieth Century, most geologists believed that the geographic positions of the ocean basins and continents were fixed. During the last few decades, however, vast amounts of new data have dramatically changed our understanding of the nature and workings of our planet. Earth scientists now realize that the continents gradually migrate across the globe. Where landmasses split apart, new ocean basins are created between the diverging blocks. Meanwhile, older portions of the seafloor are carried back into the mantle in regions where trenches occur in the deep ocean floor. Because of these movements, blocks of continental crust eventually collide and form Earth's great mountain ranges (Figure 8.1). In short, a revolutionary new model of Earth's tectonic* processes has emerged.

This profound reversal of scientific understanding has been appropriately described as a *scientific revolution*. The revolution began as a relatively straightforward proposal by Alfred Wegener, called *continental drift*. After many years of heated debate, Wegener's hypothesis of drifting continents was rejected by the vast majority of the scientific community. The concept of a mobile Earth was particularly distasteful to North American geologists, perhaps because much of the supporting evidence had been gathered from the southern continents, with which most North American geologists were unfamiliar.

During the 1950s and 1960s, new kinds of evidence began to rekindle interest in this nearly abandoned proposal. By 1968 these new developments led to the unfolding of a far more encompassing explanation, which incorporated aspects of continental drift and seafloor spreading—a theory known as *plate tectonics*.

In this chapter, we will examine the events that led to this dramatic reversal of scientific opinion in an attempt to provide some insight into how science works. We will also briefly trace the developments of the concept of continental drift, examine why it was first rejected, and consider the evidence that finally led to the acceptance of the theory of plate tectonics.

*Tectonics refers to the deformation of Earth's crust and results in the formation of structural features such as mountains.

▼ **FIGURE 8.1** Climbers camping on a shear rock face of a mountain known as K7 in Pakistan's Karakoram, a part of the Himalayas. These mountains formed as India collided with Eurasia. (Photo by Jimmy Chin/National Geographic/Getty)

Continental Drift: An Idea Before Its Time

The idea that continents, particularly South America and Africa, fit together like pieces of a jigsaw puzzle originated with improved world maps. However, little significance was given this idea until 1915, when Alfred Wegener, a German meteorologist and geophysicist, published *The Origin of Continents and Oceans*. In this book, Wegener set forth his radical hypothesis of **continental drift**.*

Wegener suggested that a supercontinent he called **Pangaea** (meaning "all land") once existed (Figure 8.2). He further hypothesized that, about 200 million years ago, this supercontinent began breaking into smaller continents, which then "drifted" to their present positions. (see Box 8.1)

Wegener and others collected substantial evidence to support these claims. The fit of South America and Africa, and the geographic distribution of fossils, rock structures, and ancient climates all seemed to support the idea that these now separate landmasses were once joined. Let us examine their evidence.

Evidence: The Continental Jigsaw Puzzle

Like a few others before him, Wegener first suspected that the continents might have been joined when he noticed the remarkable similarity between the coastlines on opposite sides of the South Atlantic. However, his use of present-day shorelines to make a fit of the continents was challenged immediately by other Earth scientists. These opponents correctly argued that shorelines are continually modified by erosional processes, and even if continental displacement had taken place, a good fit today would be unlikely. Wegener appeared to be aware of this problem, and, in fact, his original jigsaw fit of the continents was only very crude.

Scientists have determined that a much better approximation of the true outer boundary of the continents is the seaward edge of the continental shelf, which lies submerged several hundred meters below sea level. In the early 1960s, scientists produced a map that attempted to fit the edges of the continental shelves at a depth of 900 meters (3000 feet). The

---

*Wegener's ideas were actually preceded by those of an American geologist, F.B. Taylor, who in 1910 published a paper on continental drift. Taylor's paper provided little supporting evidence for continental drift, which may have been the reason that it had a relatively small impact on the scientific community.

▼ **FIGURE 8.2** Reconstruction of Pangaea as it is thought to have appeared 200 million years ago. **A.** Modern reconstruction. **B.** Reconstruction done by Wegener in 1915.

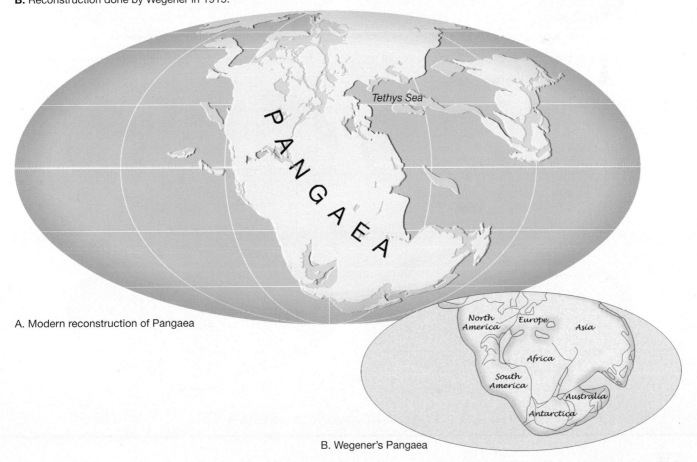

A. Modern reconstruction of Pangaea

B. Wegener's Pangaea

BOX 8.1 UNDERSTANDING EARTH:
The Breakup of Pangaea

Wegener used evidence from fossils, rock types, and ancient climates to create a jigsaw-puzzle fit of the continents—thereby creating his supercontinent of Pangaea. In a similar manner, but employing modern tools not available to Wegener, geologists have re-created the steps in the breakup of this supercontinent, an event that began nearly 200 million years ago. From this work, the dates when individual crustal fragments separated from one another and their relative motions have been well established (Figure 8.A).

An important consequence of the breakup of Pangaea was the creation of a "new" ocean basin: the Atlantic. As you can see in part B of Figure 8.A, splitting of the supercontinent did not occur simultaneously along the margins of the Atlantic. The first split developed between North America and Africa.

Here, the continental crust was highly fractured, providing pathways for huge quantities of fluid lavas to reach the surface. Today these lavas are represented by weathered igneous rocks found along the Eastern Seaboard of the United States—primarily buried beneath the sedimentary rocks that form the continental shelf. Radiometric dating of these solidified lavas indicate that rifting began in various stages between 180 million and 165 million years ago. This time span can be used as the "birth date" for this section of the North Atlantic.

By 130 million years ago, the South Atlantic began to open near the tip of what is now South Africa. As this zone of rifting migrated northward, it gradually opened the South Atlantic (compare Figure 8.A, parts B and C). Continued breakup of the southern landmass led to the separation of Africa and Antarctica and sent India on a northward journey. By the early Cenozoic, about 50 million years ago, Australia had separated from Antarctica, and the South Atlantic had emerged as a full-fledged ocean (Figure 8.A, part D).

A modern map (Figure 8.A, part F) shows that India eventually collided with Asia, an event that began about 45 million years ago and created the Himalayas as well as the Tibetan Highlands. About the same time, the separation of Greenland from Eurasia completed the breakup of the northern landmass. During the last 20 million years or so of Earth history, Arabia rifted from Africa to form the Red Sea, and Baja California separated from Mexico to form the Gulf of California (Figure 8.A, part E). Meanwhile, the Panama Arc joined North America and South America to produce our globe's familiar, modern appearance.

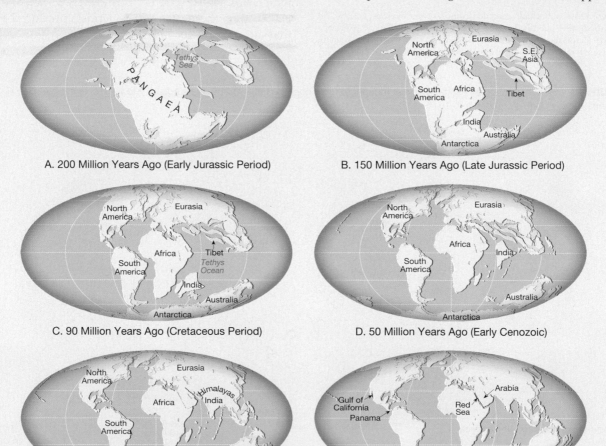

A. 200 Million Years Ago (Early Jurassic Period)

B. 150 Million Years Ago (Late Jurassic Period)

C. 90 Million Years Ago (Cretaceous Period)

D. 50 Million Years Ago (Early Cenozoic)

E. 20 Million Years Ago (Late Cenozoic)

F. Present

Figure 8.A Several views of the breakup of Pangaea over a period of 200 million years.

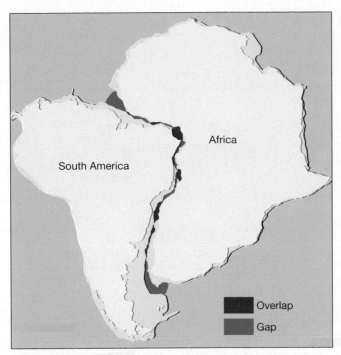

▲ **FIGURE 8.3** This shows the best fit of South America and Africa along the continental slope at a depth of 500 fathoms (about 900 meters). The areas where continental blocks overlap appear in brown. (After A. G. Smith. "Continental Drift," in *Understanding the Earth*, edited by I. G. Gass)

remarkable fit that was obtained is shown in Figure 8.3. Although the continents overlap in a few places, these are regions where streams have deposited large quantities of sediment, thus enlarging the continental shelves. The overall fit was even better than researchers suspected it would be.

❓ STUDENTS SOMETIMES ASK...

If all the continents were joined during the time of Pangaea, what did the rest of Earth look like?

When all the continents were together, there must also have been one huge ocean surrounding them. This ocean is called *Panthalassa* (*pan* = all, *thalassa* = sea). About 200 million years ago the supercontinent of Pangaea began to split apart, and the various continental masses we know today started to move toward their present geographic positions. Today all that remains of Panthalassa is the Pacific Ocean, which has been decreasing in size since the breakup of Pangaea.

Evidence: Fossils Match Across the Seas

Although the seed for Wegener's hypothesis came from the remarkable similarities of the continental margins on opposite sides of the Atlantic, he thought the idea of a mobile Earth was improbable. Not until

he learned that identical fossil organisms were known from rocks in both South America and Africa did he begin to seriously pursue this idea. Through a review of the literature, Wegener learned that most paleontologists (scientists who study the fossilized remains of organisms) were in agreement that some type of land connection was needed to explain the existence of identical fossils of Mesozoic life forms on widely separated landmasses. (Just as modern life forms native to North America are quite different from those of Africa, one would expect that during the Mesozoic era, organisms on widely separated continents would be quite distinct.)

Mesosaurus To add credibility to his argument for the existence of a supercontinent, Wegener cited documented cases of several fossil organisms that were found on different landmasses despite the unlikely possibility that their living forms could have crossed the vast ocean presently separating these continents. The classic example is *Mesosaurus*, an aquatic fish-catching reptile whose fossil remains are found only in black shales of Permian age (about 260 million years ago) in eastern South America and southern Africa (Figure 8.4). If *Mesosaurus* had been able to make the long journey across the vast South Atlantic Ocean, its remains should be more widely distributed. As this is not the case, Wegener argued that South America and Africa must have been joined during that period of Earth history.

How did scientists during Wegener's era explain the existence of identical fossil organisms in places separated by thousands of kilometers of open ocean? Transoceanic land bridges were the most widely accepted explanations of such migrations (Figure 8.5). We know, for example, that during the recent Ice Age, the lowering of sea level allowed animals to cross the narrow Bering Strait between Asia and North America. Was it possible that land bridges once connected Africa and South America but later subsided below sea level? Modern maps of the seafloor substantiate Wegener's contention that land bridges of this magnitude had not existed. If they had, remnants would still lie below sea level.

Present-Day Organisms. In a later edition of his book, Wegener also cited the distribution of present-day organisms as evidence to support the drifting of continents. For example, modern organisms with similar ancestries clearly had to evolve in isolation during the last few tens of millions of years. Most obvious of these are the Australian marsupials (such as kangaroos), which have a direct fossil link to the marsupial opossums found in the Americas. After the breakup of Pangaea, the Australian marsupials followed a different evolutionary path than related life forms in the Americas.

▶ **FIGURE 8.4** Fossils of *Mesosaurus* have been found on both sides of the South Atlantic and nowhere else in the world. Fossil remains of this and other organisms on the continents of Africa and South America appear to link these landmasses during the late Paleozoic and early Mesozoic eras.

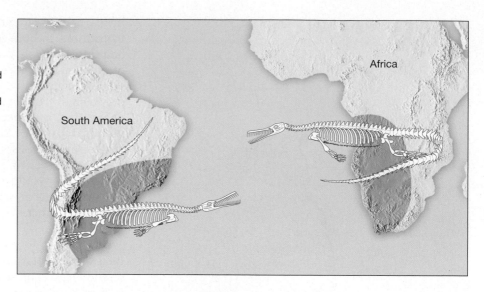

Evidence: Rock Types and Structures Match

Anyone who has worked a picture puzzle knows that in addition to the pieces fitting together, the picture must be continuous as well. The "picture" that must match in the "continental drift puzzle" is one of rock types and mountain belts found on the continents. If the continents were once together, the rocks found in a particular region on one continent should closely match in age and type those found in adjacent positions on the adjoining continent. Wegener found evidence of 2.2-billion-year-old igneous rocks in Brazil that closely resembled similarly aged rocks in Africa.

Similar evidence exists in the form of mountain belts that terminate at one coastline, only to reappear on landmasses across the ocean. For instance, the mountain belt that includes the Appalachians trends northeastward through the eastern United States and disappears off the coast of Newfoundland. Mountains of comparable age and structure are found in Greenland, the British Isles, and Scandinavia. When these landmasses are reassembled, as in Figure 8.6, the mountain chains form a nearly continuous belt.

Wegener must have been convinced that the similarities in rock structure on both sides of the Atlantic linked these landmasses when he said, "It is just as if

▶ **FIGURE 8.5** These sketches by John Holden illustrate various explanations for the occurrence of similar species on landmasses that are presently separated by vast oceans. (Reprinted with permission of John Holden)

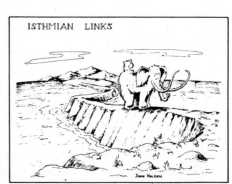

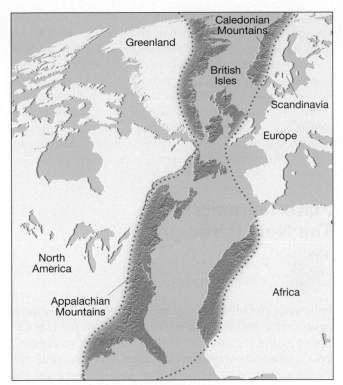

▲ **FIGURE 8.6** Matching mountain ranges across the North Atlantic. The Appalachian Mountains trend along the eastern flank of North America and disappear off the coast of Newfoundland. Mountains of comparable age and structure are found in Greenland, the British Isles, and Scandinavia. When these landmasses are placed in their predrift locations, these ancient mountain chains form a nearly continuous belt. These folded mountain belts formed roughly 300 million years ago as the landmasses collided during the formation of the supercontinent of Pangaea.

we were to refit the torn pieces of a newspaper by matching their edges and then check whether the lines of print run smoothly across. If they do, there is nothing left but to conclude that the pieces were in fact joined in this way.*

Evidence: Ancient Climates

Because Alfred Wegener was a meteorologist by profession, he was keenly interested in obtaining paleoclimatic (*paleo* = ancient, *climatic* = climate) data to support continental drift. His efforts were rewarded when he found evidence for apparently dramatic global climatic changes during the geologic past. In particular, he learned of ancient glacial deposits that indicated that near the end of the Paleozoic era (about 300 million years ago), ice sheets covered extensive areas of the Southern Hemisphere and India (Figure 8.7A). Layers of glacially transported sediments of the same age were found in southern Africa and South America, as well as in India and Australia. Much of the land area containing evidence

_____
*Alfred Wegener, *The Origin of Continents and Oceans*. Translated from the 4th revised German edition of 1929 by J. Birman (London: Methuen, 1966).

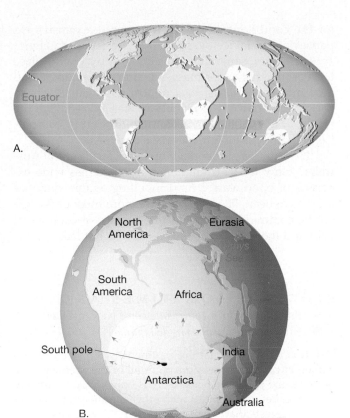

▲ **FIGURE 8.7** Paleoclimatic evidence for continental drift. **A.** Near the end of the Paleozoic era (about 300 million years ago) ice sheets covered extensive areas of the Southern Hemisphere and India. Arrows show the direction of ice movement that can be inferred from glacial grooves in the bedrock. **B.** Shown are the continents restored to their former position with the South Pole located roughly between Antarctica and Africa. This configuration accounts for the conditions necessary to generate a vast ice sheet and also explains the directions of ice movement that radiated away from the South Pole.

of this late Paleozoic glaciation presently lies within 30 degrees of the equator in subtropical or tropical climates.

Could Earth have gone through a period of sufficient cooling to have generated extensive ice sheets in areas that are presently tropical? Wegener rejected this explanation because during the late Paleozoic, large tropical swamps existed in the Northern Hemisphere. These swamps, with their lush vegetation, eventually became the major coal fields of the eastern United States, Europe, and Siberia.

Fossils from these coal fields indicate that the tree ferns that produced the coal deposits had large fronds, which are indicative of tropical settings. Furthermore, unlike trees in colder climates, these trees lacked growth rings, a characteristic of tropical plants that grow in regions having minimal fluctuations in temperature.

Wegener suggested that a more plausible explanation for the late Paleozoic glaciation was provided by the supercontinent of Pangaea. In this configuration the southern continents are joined together and located near the South Pole (Figure 8.7B). This would account for the conditions necessary to generate extensive expanses of glacial ice over much of the Southern Hemisphere. At the same time, this geography would place today's northern landmasses nearer the equator and account for their vast coal deposits. Wegener was so convinced that his explanation was correct that he wrote, "This evidence is so compelling that by comparison all other criteria must take a back seat."

How does a glacier develop in hot, arid central Australia? How do land animals migrate across wide expanses of open water? As compelling as this evidence may have been, 50 years passed before most of the scientific community accepted the concept of continental drift and the logical conclusions to which it led.

? STUDENTS SOMETIMES ASK...

Someday will the continents come back together and form a single landmass?

Yes. It is very likely that the continents will come back together, but not anytime soon. Since all of the continents are on the same planetary body, there is only so far a continent can travel before it collides with other continents. Recent research suggests that the continents may form a supercontinent about once every 500 million years or so. Since it has been about 200 million years since Pangaea broke up, we have only about 300 million years before the next supercontinent is completed.

The Great Debate

Wegener's proposal did not attract much open criticism until 1924, when his book was translated into English. From this time on, until his death in 1930, his drift hypothesis encountered a great deal of hostile criticism. To quote the respected American geologist T. C. Chamberlin, "Wegener's hypothesis . . . takes considerable liberty with our globe, and is less bound by restrictions or tied down by awkward, ugly facts than most of its rival theories. Its appeal seems to lie in the fact that it plays a game in which there are few restrictive rules and no sharply drawn code of conduct."

One of the main objections to Wegener's hypothesis stemmed from his inability to provide a mechanism that was capable of moving the continents across the globe. Wegener proposed that the tidal influence of the Moon was strong enough to give the continents a westward motion. However, the prominent physicist Harold Jeffreys quickly countered with the argument that tidal friction of the magnitude needed to displace the continents would bring Earth's rotation to a halt in a matter of a few years.

Wegener also proposed that the larger and sturdier continents broke through the oceanic crust, much like ice breakers cut through ice. However, no evidence existed to suggest that the ocean floor was weak enough to permit passage of the continents without themselves being appreciably deformed in the process.

Although most of Wegener's contemporaries opposed his views, even to the point of open ridicule, a few considered his ideas plausible. For those geologists who continued the search for additional evidence, the exciting concept of continents in motion held their interest. Others viewed continental drift as a solution to previously unexplainable observations.

Plate Tectonics: The New Paradigm

GEODe
Forces Within
▼ Plate Tectonics

Following World War II, oceanographers equipped with new marine tools and ample funding from the U.S. Office of Naval Research embarked on an unprecedented period of oceanographic exploration. Over the next two decades a much better picture of large expanses of the seafloor slowly and painstakingly began to emerge. From this work came the discovery of a global **oceanic ridge system** that winds through all of the major oceans in a manner similar to the seams on a baseball.

In other parts of the ocean, new discoveries were also being made. Earthquake studies conducted in the western Pacific demonstrated that tectonic activity was occuring at great depths beneath deep-ocean trenches. Of equal importance was the fact that dredging of the seafloor did not bring up any oceanic crust that was older than 180 million years. Further, sediment accumulations in the deep-ocean basins were found to be thin, not the thousands of meters that were predicted.

By 1968, these developments, among others, led to the unfolding of a far more encompassing theory than continental drift, known as **plate tectonics**. The implications of plate tectonics are so far-reaching that this theory is today the framework within which to view most geologic processes.

Earth's Major Plates

According to the plate tectonics model, the uppermost mantle, along with the overlying crust, behave as a strong, rigid layer, known as the **lithosphere**, which is broken into pieces called **plates** (Figure 8.8). Lithospheric plates are thinnest in the oceans where their thickness may vary from as little as a few kilometers at the oceanic ridges to 100 kilometers in the deep-ocean basins. By contrast, continental lithosphere is generally 100–150 kilometers thick but may be more than 250 kilometers thick below older portions of landmasses. The lithosphere overlies a weaker region in the mantle known as the **asthenosphere** (weak sphere). The

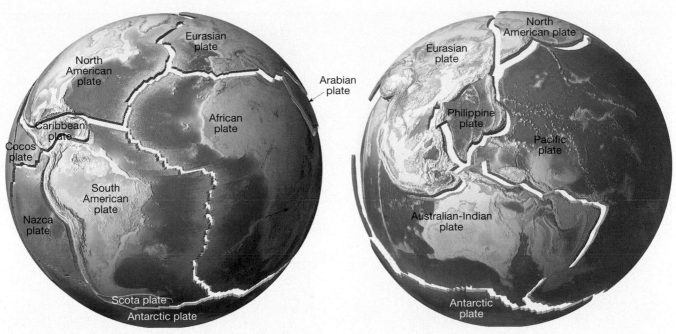

▲ **FIGURE 8.8** Illustration of some of Earth's lithospheric plates.

temperature/pressure regime in the upper asthenosphere is such that the rocks there are near their melting temperatures. This results in a very weak zone that permits the lithosphere to be effectively detached from the layers below. Thus, the weak rock within the upper asthenosphere allows Earth's rigid outer shell to move.

As shown in Figure 8.9, seven major lithospheric plates are recognized. They are the North American, South American, Pacific, African, Eurasian, Australian-Indian, and Antarctic plates. The largest is the Pacific plate, which encompasses a significant portion of the Pacific Ocean basin. Notice from Figure 8.9 that most of the large plates include an entire continent plus a large area of ocean floor (for example, the South American plate). This is a major departure from Wegener's continental drift hypothesis, which proposed that the continents moved through the ocean floor, not with it. Note also that none of the plates are defined entirely by the margins of a continent.

Intermediate-sized plates include the Caribbean, Nazca, Philippine, Arabian, Cocos, Scotia, and Juan de Fuca plates. In addition, there are over a dozen smaller plates that have been identified but are not shown in Figure 8.9.

One of the main tenets of the plate tectonic theory is that plates move as coherent units relative to all other plates. As plates move, the distance between two locations on the same plate—New York and Denver, for example—remains relatively constant, whereas the distance between sites on different plates, such as New York and London, gradually changes. (Recently it has been shown that plates can suffer *some* internal deformation, particularly oceanic lithosphere.)

Lithospheric plates move relative to each other at a very slow but continuous rate that averages about 5 centimeters (2 inches) per year. This movement is ultimately driven by the unequal distribution of heat within Earth. Hot material found deep in the mantle moves slowly upward and serves as one part of our planet's internal convection system. Concurrently, cooler, denser slabs of oceanic lithosphere descend into the mantle, setting Earth's rigid outer shell into motion. Ultimately, the titanic, grinding movements of Earth's lithospheric plates generate earthquakes, create volcanoes, and deform large masses of rock into mountains.

Plate Boundaries

Lithospheric plates move as coherent units relative to all other plates. Although the interiors of plates may experience some deformation, all major interactions among individual plates (and therefore most deformation) occur along their *boundaries*. In fact, plate boundaries were first established by plotting the locations of earthquakes. Moreover, plates are bounded by three distinct types of boundaries, which are differentiated by the type of movement they exhibit. These boundaries are depicted at the bottom of Figure 8.9 and are briefly described here:

1. **Divergent boundaries** (*constructive margins*)— where two plates move apart, resulting in upwelling of material from the mantle to create new seafloor (Figure 8.9A).

2. **Convergent boundaries** (*destructive margins*)— where two plates move together, resulting in oceanic lithosphere descending beneath an

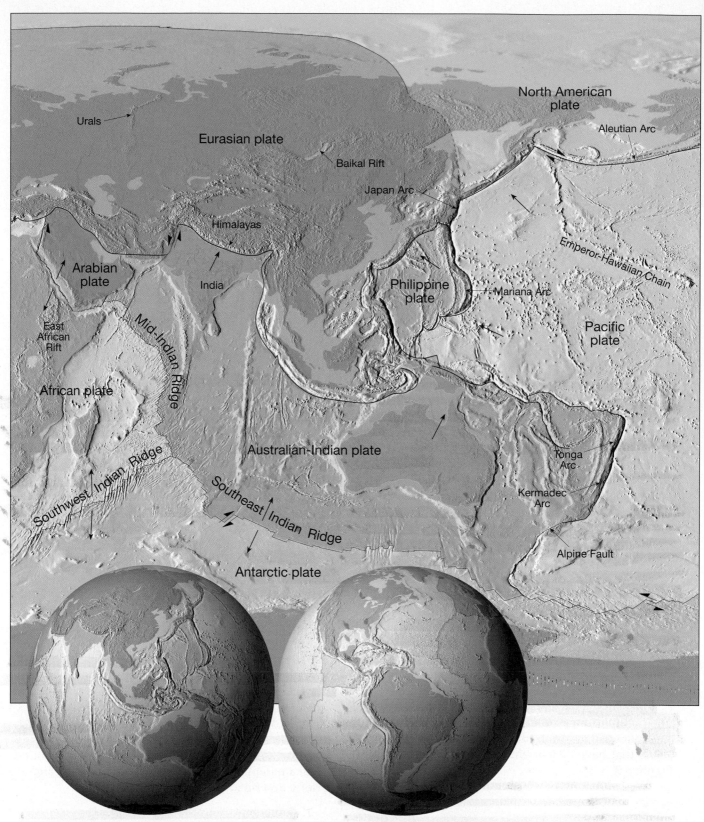

▲ **FIGURE 8.9** A mosaic of rigid plates constitutes Earth's outer shell. (After W. B. Hamilton, U.S. Geological Survey)

overriding plate, eventually to be reabsorbed into the mantle, or possibly in the collision of two continental blocks to create a mountain system (Figure 8.9B).

3. **Transform fault boundaries** (*conservative margins*)—where two plates grind past each other without the production or destruction of lithosphere (Figure 8.9C).

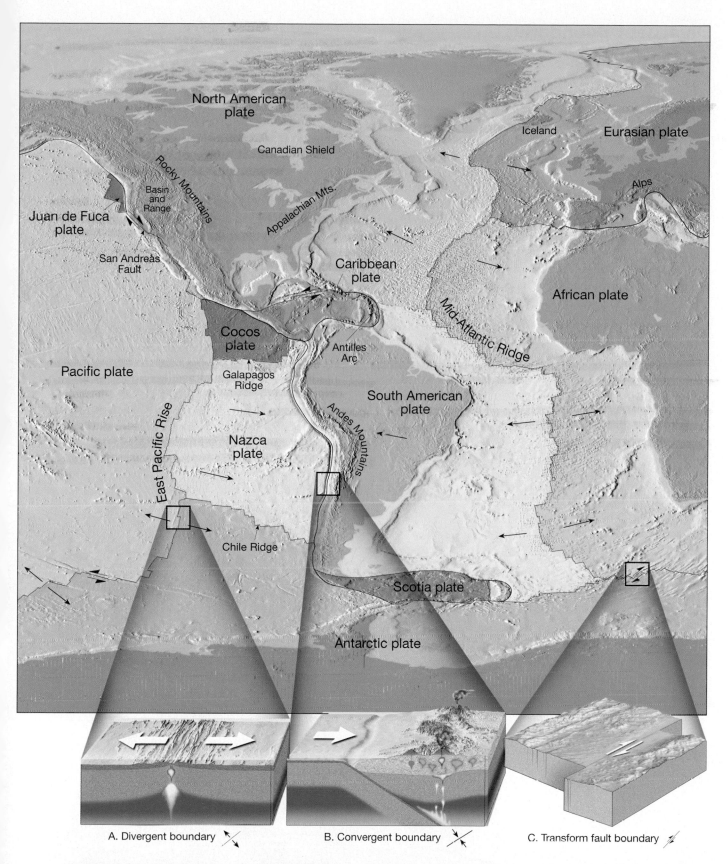

North American plate

Canadian Shield

Iceland

Eurasian plate

Rocky Mountains

Basin and Range

Alps

Juan de Fuca plate

San Andreas Fault

Appalachian Mts.

Caribbean plate

African plate

Mid-Atlantic Ridge

Cocos plate

Antilles Arc

Pacific plate

Galapagos Ridge

South American plate

East Pacific Rise

Andes Mountains

Nazca plate

Chile Ridge

Scotia plate

Antarctic plate

A. Divergent boundary

B. Convergent boundary

C. Transform fault boundary

Each plate is bounded by a combination of these three types of plate margins. For example, the Juan de Fuca plate has a divergent zone on the west, a convergent boundary on the east, and numerous transform faults, which offset segments of the oceanic ridge (Figure 8.9).

In the following sections we will briefly summarize the nature of the three types of plate boundaries.

Divergent Boundaries

Forces Within
▼ Plate Tectonics

Most **divergent boundaries** (*di* = apart, *vergere* = to move) are located along the crests of oceanic ridges and can be thought of as *constructive plate margins* since this is where new oceanic lithosphere is generated (Figure 8.10). Here, as the plates move away from the ridge axis, the fractures that form are filled with molten rock that wells up from the hot mantle below. Gradually, this magma cools to produce new slivers of seafloor. In a continuous manner, adjacent plates spread apart and new oceanic lithosphere forms between them. As we shall see later, divergent boundaries are not confined to the ocean floor but can also form on the continents.

Oceanic Ridges and Seafloor Spreading

Along well-developed divergent plate boundaries, the seafloor is elevated, forming the *oceanic ridge*. The interconnected oceanic ridge system is the longest topographic feature on Earth's surface, exceeding 70,000 kilometers (43,000 miles) in length. Representing 20 percent of Earth's surface, the oceanic ridge system winds through all major ocean basins like the seam on a base-ball. Although the crest of the oceanic ridge is commonly 2 to 3 kilometers higher than the adjacent ocean basins, the term "ridge" may be misleading as this feature is not narrow but has widths from 1000 to 4000 kilometers. Further, along the axis of some ridge segments is a deep downfaulted structure called a **rift valley.**

The mechanism that operates along the oceanic ridge system to create new seafloor is appropriately called **seafloor spreading**. Typical rates of spreading average around 5 centimeters (2 inches) per year. This is roughly the same rate at which human fingernails grow. Comparatively slow spreading rates of 2 centimeters per year are found along the Mid-Atlantic Ridge, whereas spreading rates exceeding 15 centimeters (6 inches) have been measured along sections of the East Pacific Rise. Although these rates of lithospheric production are slow on a human time scale, they are nevertheless rapid enough to have generated all of Earth's ocean basins within the last 200 million years. In fact, none of the ocean floor that has been dated exceeds 180 million years in age.

The primary reason for the elevated position of the oceanic ridge is that newly created oceanic crust is hot, and occupies more volume, which makes it less dense than cooler rocks. As new lithosphere is formed along the oceanic ridge, it is slowly yet continually displaced away from the zone of upwelling along the ridge axis. Thus, it begins to cool and contract, thereby increasing in density.

▶ **FIGURE 8.10** Most divergent plate boundaries are situated along the crests of oceanic ridges.

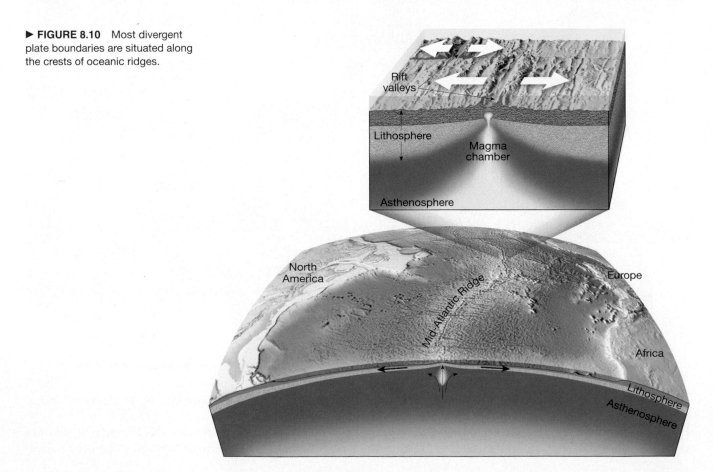

This thermal contraction accounts for the greater ocean depths that exist away from the ridge crest.

It takes about 80 million years before the cooling and contracting cease completely. By this time, rock that was once part of the elevated oceanic ridge system is located in the deep-ocean basin, where it is buried by substantial accumulations of sediment. In addition, cooling causes the mantle rocks below the oceanic crust to strengthen, thereby adding to the plate's thickness. Stated another way, the thickness of oceanic lithosphere is age-dependent. The older (cooler) it is, the greater its thickness.

Continental Rifting

Divergent plate boundaries can also develop within a continent, in which case the landmass may split into two or more smaller segments, as Alfred Wegener had pro-posed for the breakup of Pangaea (Figure 8.11). The splitting of a continent is thought to begin with the formation of an elongated depression called a *continental rift*. A modern example of a continental rift is the East African Rift (Figure 8.12). Whether this rift will develop into a full-fledged spreading center and eventually split the continent of Africa is a matter of much speculation.

Nevertheless, the East African Rift represents the initial stage in the breakup of a continent. Here tensional forces have stretched and thinned the continental crust. As a result, molten rock ascends from the asthenosphere and initiates volcanic activity at the surface (Figure 8.11A). Large volcanic mountains such as Mount Kilimanjaro and Mount Kenya exemplify the extensive volcanic activity that accompanies continental rifting. Research suggests that if tensional forces are maintained, the rift valley will lengthen and deepen, eventually extending out to the margin of the continent, splitting it in

▶ **FIGURE 8.11** Continental rifting and the formation of a new ocean basin. **A.** Continental rifting is thought to occur where tensional forces stretch and thin the crust. As a result, molten rock ascends from the asthenosphere and initiates volcanic activity at the surface. **B.** As the crust is pulled apart, large slabs of rock sink, generating a rift valley. **C.** Further spreading generates a narrow sea. **D.** Eventually, an expansive ocean basin and ridge system are created.

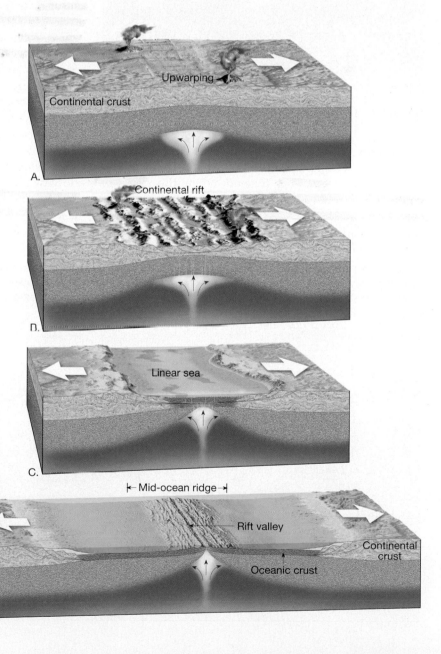

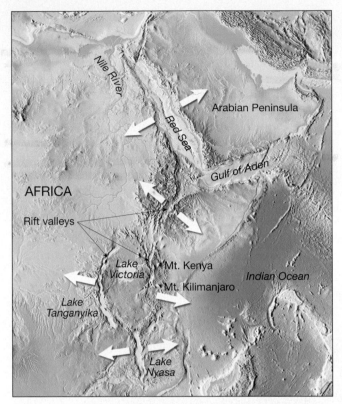

▲ **FIGURE 8.12** East African rift valleys and associated features.

two (Figure 8.11C). At this point, the rift becomes a narrow sea with an outlet to the ocean, similar to the Red Sea. The Red Sea formed when the Arabian Peninsula rifted from Africa, an event that began about 20 million years ago. If spreading continues, the Red Sea will grow wider and develop an elevated oceanic ridge similar to the Mid-Atlantic Ridge (Figure 8.11D).

Not all rift valleys develop into full-fledged spreading centers. Running through the central United States is an aborted rift zone that extends from Lake Superior into central Kansas. This once active rift valley is filled with rock that was extruded onto the crust more than a billion years ago. Why one rift valley develops into a full-fledged oceanic spreading center while others are abandoned is not yet known.

Convergent Boundaries

 GEODe

Forces Within

▼ Plate Tectonics

Although new lithosphere is constantly being produced at the oceanic ridges, our planet is not growing larger—its total surface area remains constant. To balance the addition of newly created lithosphere, older portions of oceanic lithosphere descend into the mantle along **convergent boundaries** (*con* = together, *vergere* = to move). Because lithosphere is "destroyed" at convergent boundaries, they are also called *destructive plate margins*.

Convergent plate margins occur where two plates move toward each other and the motion is accommodated by one plate sliding beneath the other. As two plates slowly converge, the leading edge of one is bent downward, allowing it to slide beneath the other. The surface expression produced by the descending plate is a **deep-ocean trench**, such as the Peru–Chile trench (Figure 8.13). Trenches formed in this manner may be

▼ **FIGURE 8.13** Distribution of the world's oceanic trenches, ridge system, and transform faults. Where transform faults offset ridge segments, they permit the ridge to change direction (curve) as can be seen in the Atlantic Ocean.

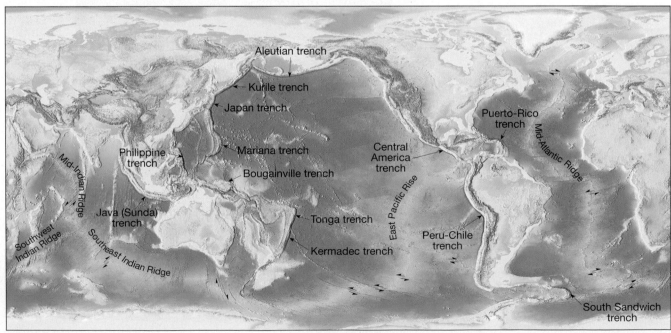

thousands of kilometers long, 8 to 12 kilometers deep, and between 50 and 100 kilometers wide.

Convergent boundaries are also called **subduction zones**, (*sub* = under, *duct* = lead), because they are sites where lithosphere is descending (being subducted) into the asthenosphere. Subduction occurs because the density of the descending lithospheric plate is greater than that of the underlying asthenosphere. In general, oceanic lithosphere is more dense than the underlying asthenosphere, whereas continental lithosphere is less dense and resists subduction. As a consequence, it is always the lithosphere that is capped with oceanic crust that is subducted.

Slabs of oceanic lithosphere descend into the asthenosphere at angles that vary from a few degrees to nearly vertical (90 degrees), but average about 45 degrees. The angle at which oceanic lithosphere descends into the asthenosphere depends on its density. For example, when a spreading center is located near a subduction zone, the lithosphere is young and, therefore, warm and buoyant. Hence, the angle of descent is small. This is the situation along parts of the Peru–Chile trench. Low dip angles usually result in considerable interaction between the descending slab and the overriding plate. Consequently, these regions experience great earthquakes.

As oceanic lithosphere ages (gets farther from the spreading center), it gradually cools, which causes it to thicken and increase in density. Once oceanic lithosphere is about 15 million years old, it becomes more dense than the supporting asthenosphere and will sink when given the opportunity. In parts of the western Pacific, some oceanic lithosphere is over 180 million years old. This is the thickest and most dense in today's oceans. The subducting slabs in this region typically descend at angles approaching 90 degrees. Examples can be found in the subduction zones associated with the Tonga, Mariana, and Kurile trenches (Figure 8.13).

Although all convergent zones have the same basic characteristics, they are highly variable features. Each is controlled by the type of crustal material involved and the tectonic setting. Convergent boundaries can form between two oceanic plates, one oceanic and one continental plate, or two continental plates. All three situations are illustrated in Figure 8.14.

Oceanic–Continental Convergence

Whenever the leading edge of a plate capped with continental crust converges with a slab of oceanic lithosphere, the buoyant continental block remains "floating," while the denser oceanic slab sinks into the mantle (Figure 8.14A). When a descending oceanic slab reaches a depth of about 100 kilometers, melting is triggered within the wedge of hot asthenosphere that lies above it. But how does the subduction of a cool slab of oceanic lithosphere cause mantle rock to melt? The answer lies in the fact that volatiles (mainly water) act like salt does to melt ice. That is, "wet" rock, in a high-pressure environment, melts at substantially lower temperatures than "dry" rock of the same composition.

Sediments and oceanic crust contain a large amount of water which is carried to great depths by a subducting plate. As the plate plunges downward, water is "squeezed" from the pore spaces as confining pressure increases. At even greater depths, heat and pressure drive water from hydrated (water-rich) minerals such as the *amphiboles*. At a depth of roughly 100 kilometers and several kilometers from the upper boundary of the cool subducting oceanic slab, the mantle is sufficiently hot that the introduction of water leads to some melting. This process, called **partial melting**, generates as little as 10 percent molten material, which is intermixed with unmelted mantle rock. Being less dense than the surrounding mantle, this hot mobile mixture (magma) gradually rises toward the surface as a teardrop-shaped structure. Depending on the environment, these mantle-derived magmas may ascend through the crust and give rise to a volcanic eruption. However, much of this molten rock never reaches the surface; rather, it solidifies at depth where it acts to thicken the crust.

Partial melting of mantle rock generates molten rock that has a *basaltic composition* similar to what erupts on the Island of Hawaii. In a continental setting, however, basaltic magma typically melts and assimilates some of the crustal rocks through which it ascends. The result is the formation of a silica-rich (SiO_2) magma having an *intermediate (andesitic) composition*. On occasions when andesitic magmas reach the surface, they often erupt explosively, generating large columns of volcanic ash and gases. A classic example of such an eruption is the 1980 eruption of Mount St. Helens.

The volcanoes of the towering Andes are the product of magma generated by the subduction of the Nazca plate beneath the South American continent. Mountains such as the Andes, which are produced in part by volcanic activity associated with the subduction of oceanic lithosphere, are called **continental volcanic arcs**. Another active continental volcanic arc is located in the western United States. The Cascade Range of Washington, Oregon, and California consists of several well-known volcanic mountains, including Mount Rainier, Mount Shasta, and Mount St. Helens. This active volcanic arc also extends into Canada, where it includes Mount Garibaldi, Mount Silverthrone and others. As the continuing activity of Mount St. Helens testifies, the Cascade Range is still active.

▶ **FIGURE 8.14** Three types of convergent plate boundaries.
A. Oceanic–continental.
B. Oceanic–oceanic.
C. Continental–continental.

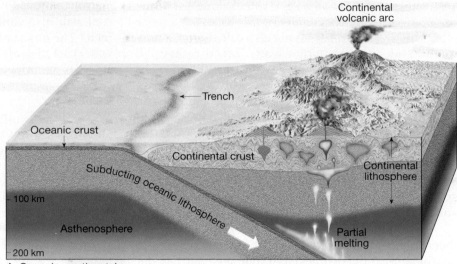

A. Oceanic-continental

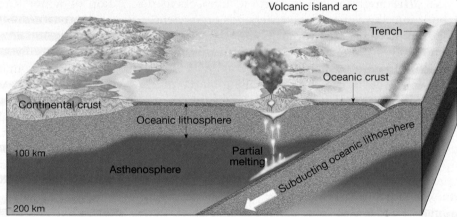

B. Oceanic-oceanic

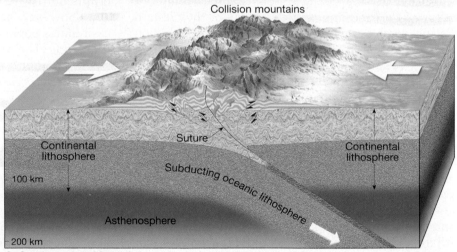

C. Continental-continental

Oceanic–Oceanic Convergence

An oceanic–oceanic convergent boundary has many features in common with oceanic–continental plate margins. The differences are mainly attributable to the nature of the crust capping the overriding plate. Where two oceanic slabs converge, one descends beneath the other, initiating volcanic activity by the same mechanism that operates at oceanic–continental plate boundaries. Water "squeezed" from the subducting slab of oceanic lithosphere triggers melting in the hot wedge of mantle rock that lies above. In this setting, volcanoes grow up from the ocean floor, rather than upon a continental platform. When subduction is sustained, it will eventually build a chain of volcanic structures that emerge as islands. The volcanic islands are spaced about 80 kilometers apart and are built upon submerged ridges of volcanic material a few hundred kilometers wide. This newly formed land consisting of an arc-shaped chain of small volcanic islands is called a **volcanic island arc**, or simply an **island arc** (Figure 8.14B).

The Aleutian, Mariana, and Tonga islands are examples of volcanic island arcs. Island arcs such as these are generally located 100 to 300 kilometers (60 to 200 miles) from a deep-ocean trench. Located adjacent to the island arcs just mentioned are the Aleutian trench, the Mariana trench, and the Tonga trench (see Figure 8.13).

Most volcanic island arcs are located in the western Pacific. At these sites the subducting Pacific crust is relatively old and dense and therefore will readily sink into the mantle. This accounts for the steep angle of descent (often approaching 90 degrees) common in the trenches of this region. Further, many of these subduction zones lack the large earthquakes that are associated with some other convergent zones, such as the Peru–Chile trench.

Only two volcanic island arcs are located in the Atlantic—the Lesser Antilles adjacent to the Caribbean Sea, and the Sandwich Islands in the South Atlantic. The Lesser Antilles are a product of the subduction of the Atlantic beneath the Caribbean plate. Located within this arc is the island of Martinique, where Mount Pelée erupted in 1902, destroying the town of St. Pierre and killing an estimated 28,000 people; and the island of Montserrat, where volcanic activity has occurred very recently.

Relatively young island arcs are fairly simple structures that are underlain by deformed oceanic crust that is generally less than 20 kilometers (12 miles) thick. Examples include the arcs of Tonga, the Aleutians, and the Lesser Antilles. By contrast, older island arcs are more complex and are underlain by crust that ranges in thickness from 20 to 35 kilometers. Examples include the Japanese and Indonesian arcs, which are built upon material generated by earlier episodes of subduction or sometimes on a small piece of continental crust (see Box 8.2).

Continental–Continental Convergence

As you saw earlier, when an oceanic plate is subducted beneath continental lithosphere, an Andean-type volcanic arc develops along the margin of the continent. However, if the subducting plate also contains continental lithosphere, continued subduction eventually brings the two continental blocks together (Figure 8.15A). Whereas oceanic lithosphere is relatively dense and sinks into the asthenosphere, continental lithosphere is buoyant, which prevents it from being subducted to any great depth. The result is a collision between the two continental fragments (Figure 8.15C).

Such a collision occurred when the subcontinent of India "rammed" into Asia and produced the Himalayas—the most spectacular mountain range on Earth (see Figure 8.1). During this collision, the continental crust buckled, fractured, and was generally shortened and thickened. In addition to the Himalayas, several other major mountain systems, including the Alps, Appalachians, and Urals, formed during continental collisions.

Prior to a continental collision, the landmasses involved are separated by an ocean basin. As the continental blocks converge, the intervening seafloor is subducted beneath one of the plates. Subduction initiates partial melting in the overlying mantle, which in turn results in the growth of a volcanic arc. Depending on the location of the subduction zone, the volcanic arc could develop on either of the converging landmasses, or if the subduction zone developed several hundred kilometers seaward from the coast, a volcanic island arc would form. Eventually, as the intervening seafloor is consumed, these continental masses collide. This folds and deforms the accumulation of sediments and sedimentary rocks along the continental margin as if they had been placed in a gigantic vise. The result is the formation of a new mountain range composed of deformed and metamorphosed sedimentary rocks, fragments of the volcanic arc, and often slivers of oceanic crust.

Transform Fault Boundaries

GEODe
Forces Within
▼ Plate Tectonics

The third type of plate boundary is the **transform fault**, where plates slide horizontally past one another without the production or destruction of lithosphere (*conservative plate margins*). Transform faults were first identified where they join offset segments of an oceanic ridge (Figure 8.16). At first it was erroneously assumed that the ridge system originally formed a long and continuous chain that was later offset by horizontal displacement along these large faults. However, the displacement along these faults was found to be in the exact opposite direction required to produce the offset ridge segments.

The true nature of transform faults was discovered in 1965 by J. Tuzo Wilson of the University of Toronto. Wilson suggested that these large faults connect the global active belts (convergent boundaries, divergent boundaries, and other transform faults) into a continuous network that divides Earth's outer shell into

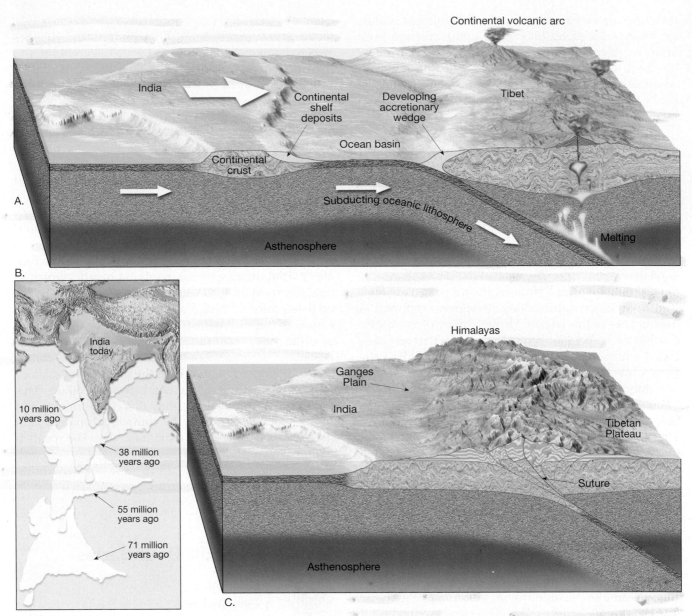

▲ **FIGURE 8.15** The ongoing collision of India and Asia, starting about 45 million years ago, produced the majestic Himalayas.
A. Converging plates generated a subduction zone, while partial melting triggered by the subducting oceanic slab produced a continental volcanic arc. Sediments scraped from the subducting plate were added to the accretionary wedge. **B.** Position of India in relation to Eurasia at various times. (Modified after Peter Molnar) **C.** Eventually the two landmasses collided, deforming and elevating the sediments that had been deposited along their continental margins.

several rigid plates. Thus, Wilson became the first to suggest that Earth was made of individual plates, while at the same time identifying the faults along which relative motion between the plates is made possible.

Most transform faults join two segments of an oceanic ridge (Figure 8.16). Here, they are part of prominent linear breaks in the oceanic crust known as **fracture zones**, which include both the active transform faults as well as their inactive extentions into the plate interior. These fracture zones are present approximately every 100 kilometers along the trend of a ridge axis. As shown in Figure 8.16, active transform faults lie *only between* the two offset ridge segments. Here seafloor produced at one ridge axis moves in the opposite direction as seafloor produced at an opposing ridge segment. Thus, between the ridge segments these adjacent slabs of oceanic crust are grinding past each other along the fault. Beyond the ridge crests are the inactive zones, where the fractures are preserved as linear topographic scars. The trend of these fracture zones roughly parallels the direction of plate motion at the time of their formation. Thus, these structures can be used to map the direction of plate motion in the geologic past.

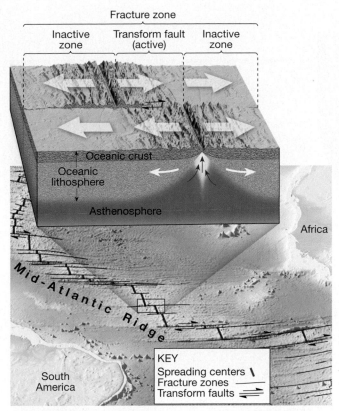

Fracture zone

Inactive zone | Transform fault (active) | Inactive zone

Oceanic crust
Oceanic lithosphere
Asthenosphere

Africa

Mid-Atlantic Ridge

South America

KEY
Spreading centers ⟍
Fracture zones ——
Transform faults ⇄

▲ **FIGURE 8.16** Illustration of a transform fault joining segments of the Mid-Atlantic Ridge.

In another role, transform faults provide the means by which the oceanic crust created at ridge crests can be transported to a site of destruction: the deep-ocean trenches. Figure 8.17 illustrates this situation. Notice that the Juan de Fuca plate moves in a southeasterly direction, eventually being subducted under the West Coast of the United States. The southern end of this plate is bounded by the Mendocino fault. This transform fault boundary connects the Juan de Fuca ridge to the Cascadia subduction zone (Figure 8.17). Therefore, it facilitates the movement of the crustal material created at the ridge crest to its destination beneath the North American continent.

Although most transform faults are located within the ocean basins, a few cut through continental crust. Two examples are the earthquake-prone San Andreas Fault of California and the Alpine Fault of New Zealand. Notice in Figure 8.17 that the San Andreas Fault connects a spreading center located in the Gulf of California to the Cascadia subduction zone and the Mendocino fault located along the northwest coast of the United States. Along the San Andreas Fault, the Pacific plate is moving toward the northwest, past the North American plate. If this movement continues, that part of California west of the fault zone, including the Baja Peninsula, will eventually become an island off the West Coast of the United States and Canada. It

could eventually reach Alaska. However, a more immediate concern is the earthquake activity triggered by movements along this fault system.

? STUDENTS SOMETIMES ASK...
If the continents move, do other features like segments of the oceanic ridge also move?

That's a good observation, and yes, they do! It is interesting to note that very little is really fixed in place on Earth's surface. When we talk about movement of features on Earth, we must consider the question, "Moving relative to what?" Certainly, the oceanic ridge does move relative to the continents (which sometimes causes segments of the oceanic ridges to be subducted beneath the continents). In addition, the oceanic ridge is moving relative to a fixed location outside Earth. This means that an observer orbiting above Earth would notice, after only a few million years, that all continental and seafloor features—as well as plate boundaries—are indeed moving.

Testing the Plate Tectonics Model

With the development of the theory of plate tectonics, researchers from all of the Earth sciences began testing this new model of how Earth works. Some of the evidence supporting continental drift and seafloor spreading has already been presented. In addition, some of the evidence that was instrumental in solidifying the support for this new idea follows. Note that much of this evidence was not new; rather, it was new interpretations of already existing data that swayed the tide of opinion.

Evidence: Ocean Drilling

Some of the most convincing evidence confirming seafloor spreading has come from drilling into the ocean floor. From 1968 until 1983, the source of these important data was the Deep Sea Drilling Project, an international program sponsored by several major oceanographic institutions and the National Science Foundation. The primary goal was to gather firsthand information about the age of the ocean basins and the processes that formed them. To accomplish this, a new drilling ship, the *Glomar Challenger*, was built.

Operations began in August 1968, in the South Atlantic. At several sites holes were drilled through the entire thickness of sediments to the basaltic rock below. An important objective was to gather samples of sediment from just above the igneous crust as a means of dating the seafloor at each site.* Because

---
*Radiometric dates of the ocean crust itself are unreliable because of the alteration of basalt by seawater.

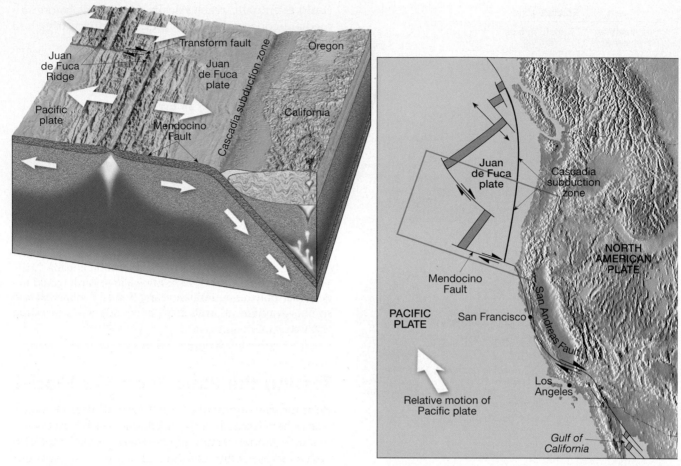

▲ **FIGURE 8.17**   The Mendocino transform fault permits seafloor generated at the Juan de Fuca ridge to move southeastward past the Pacific plate and beneath the North American plate. Thus, this transform fault connects a divergent boundary to a subduction zone. Furthermore, the San Andreas Fault, also a transform fault, connects two spreading centers: the Juan de Fuca ridge and a divergent zone located in the Gulf of California.

sedimentation begins immediately after the oceanic crust forms, remains of microorganisms found in the oldest sediments—those resting directly on the crust—can be used to date the ocean floor at that site.

When the oldest sediment from each drill site was plotted against its distance from the ridge crest, the plot demonstrated that the age of the sediment increased with increasing distance from the ridge. This finding supported the seafloor-spreading hypothesis, which predicted the youngest oceanic crust would be found at the ridge crest and the oldest oceanic crust would be adjacent to the continental margins.

The data from the Deep Sea Drilling Project also reinforced the idea that the ocean basins are geologically youthful because no sediment with an age in excess of 180 million years was found. By comparison, continental crust that exceeds 4 billion years in age has been dated.

The thickness of ocean-floor sediments provided additional verification of seafloor spreading. Drill cores from the *Glomar Challenger* revealed that sedi-

ments are almost entirely absent on the ridge crest and the sediment thickens with increasing distance from the ridge. Because the ridge crest is younger than the areas farther away from it, this pattern of sediment distribution should be expected if the seafloor-spreading hypothesis is correct.

The Ocean Drilling Program succeeded the Deep Sea Drilling Project and, like its predecessor, was a major international program. The more technologically advanced drilling ship, the *JOIDES Resolution*, continued the work of the *Glomar Challenger*.* The *JOIDES Resolution* can drill in water depths as great as 8200 meters (27,000 feet) and contains onboard laboratories equipped with a large and sophisticated array of seagoing scientific research equipment (Figure 8.18)

In October 2003, the *JOIDES Resolution* became part of a new program, the Integrated Ocean Drilling Program (IODP). This new international effort will not

---

*JOIDES stands for Joint Oceanographic Institutions for Deep Earth Sampling. For more on the *JOIDES Resolution*, see Box 13.3, p. 377.

◀ **FIGURE 8.18** The *JOIDES Resolution* is the drilling ship that replaced the *Glomar Challenger* in the important work of sampling the floor of the world's oceans. (Photo courtesy of Ocean Drilling Program)

rely on just one drilling ship, but will use multiple vessels for exploration. One of the new additions is the massive 210-meter-long *Chikyu*, which is scheduled to begin operations in 2006.

Evidence: Hot Spots

Mapping of seamounts (submarine volcanoes) in the Pacific Ocean revealed several linear chains of volcanic structures. One of the most studied chains extends from the Hawaiian Islands to Midway Island and continues northward toward the Aleutian trench (Figure 8.19). This nearly continuous string of volcanic islands and seamounts is called the Hawaiian Island–Emperor Seamount chain. Radiometric dating of these structures showed that the volcanoes increase in age with increasing distance from Hawaii. Hawaii, the youngest island in the chain, began rising from the seafloor less than a million years ago, whereas Midway Island is 27 million years old, and Suiko Seamount, near the Aleutian trench, is 65 million years old (Figure 8.19).

Taking a closer look at the Hawaiian Islands, we see a similar increase in age from the volcanically active island of Hawaii, at the southeastern end of the chain, to the inactive volcanoes that make up the island of Kauai in the northwest (Figure 8.19).

Researchers are in agreement that a rising plume of mantle material is located beneath the island of

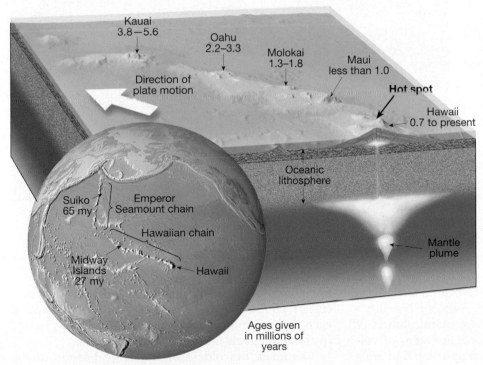

◀ **FIGURE 8.19** The chain of islands and seamounts that extends from Hawaii to the Aleutian trench results from the movement of the Pacific plate over an apparently stationary hot spot. Radiometric dating of the Hawaiian Islands shows that the volcanic activity decreases in age toward the island of Hawaii.

Kauai
3.8–5.6

Oahu
2.2–3.3

Molokai
1.3–1.8

Maui
less than 1.0

Direction of plate motion

Hot spot

Hawaii
0.7 to present

Oceanic lithosphere

Suiko
65 my

Emperor Seamount chain

Hawaiian chain

Mantle plume

Midway Islands
27 my

Hawaii

Ages given in millions of years

BOX 8.2 UNDERSTANDING EARTH:
Susan DeBari—A Career in Geology

I discovered geology the summer that I worked doing trail maintenance in the North Cascades mountains of Washington State. I had just finished my freshman year in college and had never before studied Earth Science. But a coworker (now my best friend) began to describe the geological features of the mountains that we were hiking in—the classic cone shape of Mount Baker volcano, the U-shaped glacial valleys, the advance of active glaciers, and other wonders. I was hooked and went back to college that fall with a geology passion that hasn't abated. As an undergraduate, I worked as a field assistant to a graduate student and did a senior thesis project on rocks from the Aleutian island arc. From that initial spark, island arcs have remained my top research interest, on through Ph.D. research at Stanford University, postdoctoral work at the University of Hawaii, and as a faculty member at San Jose State University and Western Washington University.

My special interest is the deep crust of arcs, close to the discontinuity (fondly known as the Moho). What processes are occurring down there at the base of the crust in island arcs? What is the source of magmas that make their way to the surface (The mantle? The deep crust itself?). How do these magmas interact with the crust as they make their way upward? What do these early magmas look like chemically? Are they very different from what is erupted at the surface?

Obviously, geologists cannot go down to the base of the crust (typically 20 to 40 kilometers beneath Earth's surface). So what they do is play a bit of a detective game. They must use rocks that are *now exposed at the surface* that were originally formed in the deep crust of an island arc. The rocks must have been brought to the surface rapidly along fault zones to preserve their original features. Thus, I can

walk on rocks of the deep crust without really leaving the Earth's surface! There are a few places around the world where these rare rocks are exposed. Some of the places that I have worked include the Chugach Mountains of Alaska, the Sierras Pampeanas of Argentina, the Karakorum range in Pakistan, Vancouver Island's west coast, and the North Cascades of Washington. Fieldwork has involved hiking most commonly, but also extensive use of mules and trucks.

I also went looking for exposed pieces of the deep crust of island arcs in a less obvious place, in one of the deepest oceanic trenches of the world, the Izu Bonin trench (Figure 8.B). Here I dove into the ocean in a submersible called the *Shinkai 6500* (pictured to my right in the background). The *Shinkai 6500* is a Japanese submersible that has the capability to dive to 6500 meters below the surface of the ocean (approximately 4 miles). My plan was to take rock samples from the

wall of the trench at its deepest levels using the submersible's mechanical arm. Because preliminary data suggested that vast amounts of rock were exposed for several kilometers in a vertical sense, this could be a great way to sample the deep arc basement. I dove in the submersible three times, reaching a maximum depth of 6497 meters. Each dive lasted nine hours, spent in a space no bigger than the front seat of a Honda, shared with two of the Japanese pilots that controlled the submersible's movements. It was an exhilarating experience!

I am now on the faculty at Western Washington University, where I continue to do research on the deep roots of volcanic arcs, and encourage students to become involved as well. I am also involved in science education training for K–12 teachers, hoping to motivate young people to ask questions about the fascinating world that surrounds them!

Figure 8.B Susan DeBari photographed with the Japanese submersible, *Shinkai 6500*, which she used to collect rock samples from the Izu Bonin trench. (Photo courtesy of Susan DeBari)

Hawaii. As the ascending **mantle plume** enters the low-pressure environment at the base of the lithosphere, melting occurs. The surface manifestation of this activity is a **hot spot**. Hot spots are areas of volcanism, high heat flow, and crustal uplifting that are a few hundred kilometers across. As the Pacific plate

moved over this hot spot, successive volcanic structures were built. As shown in Figure 8.19, the age of each volcano indicates the time when it was situated over the relatively stationary mantle plume.

Kauai is the oldest of the large islands in the Hawaiian chain. Five million years ago, when it was positioned over the hot spot, Kauai was the only Hawaiian Island (Figure 8.19). Visible evidence of the age of Kauai can be seen by examining its extinct volcanoes, which have been eroded into jagged peaks and vast canyons. By contrast, the relatively young island of Hawaii exhibits fresh lava flows, and two of Hawaii's volcanoes, Mauna Loa and Kilauea, remain active.

Two island groups parallel the Hawaiian Island–Emperor Seamount chain. One chain consists of the Tuamotu and Line islands, and the other includes the Austral, Gilbert, and Marshall islands. In each case, the most recent volcanic activity has occurred at the southeastern end of the chain, and the islands get progressively older to the northwest. Thus, like the Hawaiian Island–Emperor Seamount chain, these volcanic structures apparently formed by the same motion of the Pacific plate over fixed mantle plumes. Not only does this evidence support the fact that the plates do indeed move relative to Earth's interior but the hot spot "tracks" trace the direction of plate motion. For example, hot spots found on the floor of the Atlantic have increased our understanding of the migration of landmasses following the breakup of Pangaea.

Research suggests that at least some mantle plumes originate at great depth, perhaps at the mantle-core boundary. Others, however, may have a much shallower origin. Of the 40 or so hot spots that have been identified, over a dozen are located near spreading centers. For example, the mantle plume located beneath Iceland is responsible for the large accumulation of volcanic rocks found along the northern section of the Mid-Atlantic Ridge.

The existence of mantle plumes and their association with hot spots is well documented. Most mantle plumes are long-lived features that appear to maintain relatively fixed positions within the mantle. However, recent evidence has shown that some hot spots may slowly migrate. If this is the case, models of past plate motion that are based on a fixed hot-spot frame of reference will need to be reevaluated.

Evidence: Paleomagnetism

Anyone who has used a compass to find direction knows that Earth's magnetic field has a north and a south magnetic pole. Today these magnetic poles align closely, but not exactly, with the geographic poles. (The geographic poles, or true north and south poles, are where Earth's rotational axis intersects the surface.) Earth's magnetic field is similar to that produced by a simple bar magnet. Invisible lines of force pass through the planet and extend from one magnetic pole to the other as shown in Figure 8.20. A compass needle, itself a small magnet free to rotate on an axis, becomes aligned with the magnetic lines of force and points to the magnetic poles.

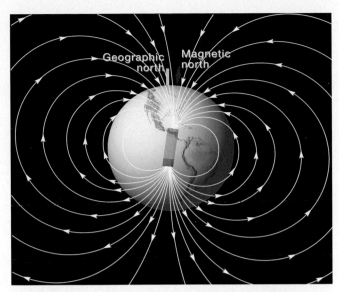

▲ **FIGURE 8.20**   Earth's magnetic field consists of lines of force much like those a giant bar magnet would produce if placed at the center of Earth.

Unlike the pull of gravity, we cannot feel Earth's magnetic field, yet its presence is revealed because it deflects a compass needle. In a similar manner, certain rocks contain minerals that serve as "fossil compasses." These iron-rich minerals, such as *magnetite*, are abundant in lava flows of basaltic composition.* When heated above a temperature known as the **Curie point**, these magnetic minerals lose their magnetism. However, when these iron-rich grains cool below their Curie point (about 585°C for magnetite), they gradually become magnetized in the direction of the existing magnetic lines of force. Once the minerals solidify, the magnetism they possess will usually remain "frozen" in this position. In this regard, they behave much like a compass needle; they "point" toward the position of the magnetic poles at the time of their formation. Then, if the rock is moved, the rock magnetism will retain its original alignment. Rocks that formed thousands or millions of years ago and contain a "record" of the direction of the magnetic poles at the time of their formation are said to possess **fossil magnetism**, or **paleomagnetism**.

Apparent Polar Wandering   A study of rock magnetism conducted during the 1950s in Europe led to an interesting discovery. The magnetic alignment in the iron-rich minerals in lava flows of different ages indicated that many different paleomagnetic poles once existed. A plot of the apparent positions of the magnetic north pole with respect to Europe revealed that during the past 500 million years, the location of the pole

---

*Some sediments and sedimentary rocks contain enough iron-bearing mineral grains to acquire a measurable amount of magnetization.

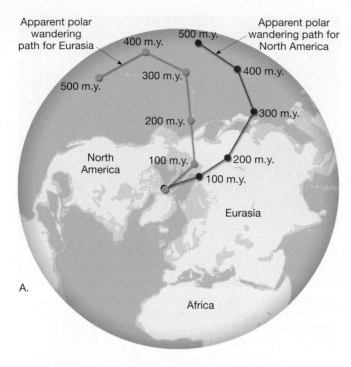

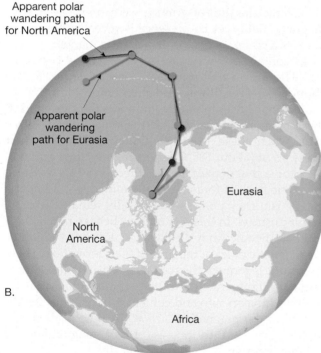

▲ **FIGURE 8.21** Simplified apparent polar wandering paths as established from North American and Eurasian paleomagnetic data. **A.** The more westerly path determined from North American data is thought to have been caused by the westward drift of North America by about 24 degrees from Eurasia. **B.** The positions of the wandering paths when the landmasses are reassembled in their predrift locations.

had gradually wandered from a location near Hawaii northward through eastern Siberia and finally to its present location (Figure 8.21A). This was strong evidence that either the magnetic poles had migrated through time, an idea known as *polar wandering*, or

that the lava flows moved—in other words, Europe had drifted in relation to the poles.

Although the magnetic poles are known to move in an erratic path around the geographic poles, studies of paleomagnetism from numerous locations show that the positions of the magnetic poles, averaged over thousands of years, correspond closely to the positions of the geographic poles. Therefore, a more acceptable explanation for the apparent polar wandering paths was provided by Wegener's hypothesis. If the magnetic poles remain stationary, their *apparent movement* is produced by continental drift.

The latter idea was further supported by comparing the latitude of Europe as determined from fossil magnetism with evidence obtained from paleoclimatic studies. Recall that during the Pennsylvanian period (about 300 million years ago) coal-producing swamps covered much of Europe. During this same time period, paleomagnetic evidence places Europe near the equator—a fact consistent with the tropical environment indicated by these coal deposits.

Further evidence for continental drift came a few years later when a polar-wandering path was constructed for North America (Figure 8.21A). It turned out that paths for North America and Europe had similar shapes but were separated by about 30° of longitude. At the time these rocks crystallized, could there have been two magnetic north poles that migrated parallel to each other? Investigators found no evidence to support this possibility. The differences in these migration paths, however, can be reconciled if the two presently separated continents are placed next to one another, as we now believe they were prior to the opening of the Atlantic Ocean. Notice in Figure 8.21B that these apparent wandering paths nearly coincided during the period from about 400 to 160 million years ago. This is evidence that North America and Europe were joined during this period and moved relative to the poles as part of the same continent.

Magnetic Reversals and Seafloor Spreading Another discovery came when geophysicists learned that Earth's magnetic field periodically reverses polarity; that is, the north magnetic pole becomes the south magnetic pole, and vice versa. A rock solidifying during one of the periods of reverse polarity will be magnetized with the polarity opposite that of rocks being formed today. When rocks exhibit the same magnetism as the present magnetic field, they are said to possess **normal polarity**, whereas rocks exhibiting the opposite magnetism are said to have **reverse polarity.**

Evidence for **magnetic reversals** was obtained when investigators measured the magnetism of lavas and sediments of various ages around the world. They found that normally and reversely magnetized rocks of a given age in one location matched the magnetism of rocks of the same age found in all other locations.

This was convincing evidence that Earth's magnetic field had indeed reversed.

Once the concept of magnetic reversals was confirmed, researchers set out to establish a time scale for magnetic reversals. The task was to measure the magnetic polarity of hundreds of lava flows and use radiometric dating techniques to establish their ages. Figure 8.22 shows the **magnetic time scale** established for the last few million years. The major divisions of the magnetic time scale are called *chrons* and last for roughly 1 million years. As more measurements became available, researchers realized that several, short-lived reversals (less than 200,000 years long) occur during any one chron.

Meanwhile, oceanographers had begun to do magnetic surveys of the ocean floor in conjunction with their efforts to construct detailed maps of seafloor topography. These magnetic surveys were accomplished by towing very sensitive instruments called **magnetometers** behind research vessels. The goal of these geophysical surveys was to map variations in the strength of Earth's magnetic field that arise from differences in the magnetic properties of the underlying crustal rocks.

The first comprehensive study of this type was carried out off the Pacific Coast of North America and had an unexpected outcome. Researchers discovered

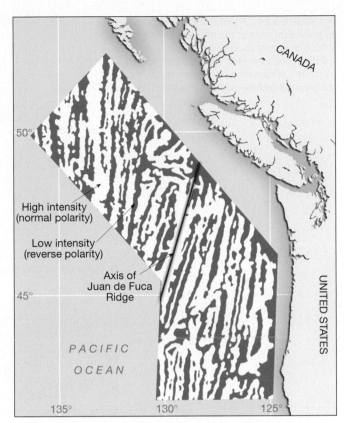

▲ **FIGURE 8.23** Pattern of alternating stripes of high- and low-intensity magnetism discovered off the Pacific Coast of North America.

▼ **FIGURE 8.22** Time scale of Earth's magnetic field in the recent past. This time scale was developed by establishing the magnetic polarity for lava flows of known age. (Data from Allen Cox and G. B. Dalrymple)

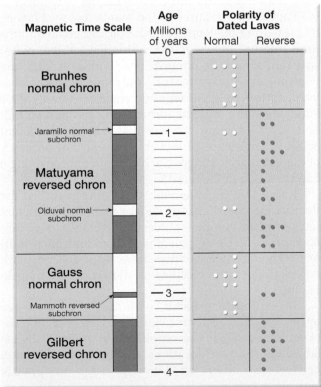

alternating strips of high- and low-intensity magnetism as shown in Figure 8.23.

This relatively simple pattern of magnetic variation defied explanation until 1963, when Fred Vine and D. H. Matthews demonstrated that the high- and low-intensity stripes supported the concept of seafloor spreading. Vine and Matthews suggested that the stripes of high-intensity magnetism are regions where the paleomagnetism of the ocean crust exhibits normal polarity (Figure 8.23). Consequently, these rocks *enhance* (reinforce) Earth's magnetic field. Conversely, the low-intensity stripes are regions where the ocean crust is polarized in the reverse direction and therefore *weakens* the existing magnetic field. But how do parallel stripes of normally and reversely magnetized rock become distributed across the ocean floor?

Vine and Matthews reasoned that as magma solidifies along narrow rifts at the crest of an oceanic ridge, it is magnetized with the polarity of the existing magnetic field (Figure 8.24). Because of seafloor spreading, this stripe of magnetized crust would gradually increase in width. When Earth's magnetic field reverses polarity, any newly formed seafloor (having the opposite polarity) would form in the middle of the old stripe. Gradually, the two parts of the old stripe are carried in opposite directions away from the ridge crest. Subsequent reversals would build a pattern of

▶ **FIGURE 8.24** As new basalt is added to the ocean floor at the mid-ocean ridges, it is magnetized according to Earth's existing magnetic field. Hence, it behaves much like a tape recorder as it records each reversal of the planet's magnetic field.

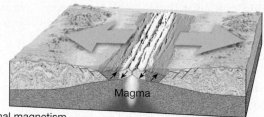

A. Period of normal magnetism

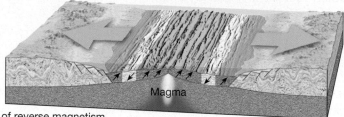

B. Period of reverse magnetism

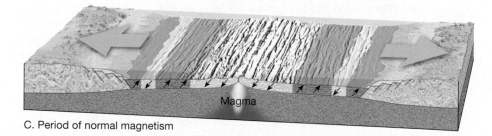

C. Period of normal magnetism

normal and reverse stripes as shown in Figure 8.24. Because new rock is added in equal amounts to both trailing edges of the spreading ocean floor, we should expect that the pattern of stripes (size and polarity) found on one side of an oceanic ridge to be a mirror image of the other side. A few years later a survey across the Mid-Atlantic Ridge just south of Iceland revealed a pattern of magnetic stripes exhibiting a remarkable degree of symmetry to the ridge axis (Figure 8.25).

Because the dates of magnetic reversals going back nearly 200 million years have been established, the rate at which spreading occurs at the various ridges can be determined accurately. In the Pacific Ocean, for example, the magnetic stripes are much wider for corresponding time intervals than those of the Atlantic Ocean. Hence, we conclude that a faster spreading rate exists for the spreading center of the Pacific as compared to the Atlantic. When we apply numerical dates to these magnetic events, we find that the spreading rate for the North Atlantic Ridge is only 2 centimeters (1 inch) per year. The rate is somewhat faster for the South Atlantic. The spreading rates for the East Pacific Rise generally range between 6 and 12 centimeters (2.5 to 5 inches) per year, with a maximum rate of about 20 centimeters (8 inches) a year. Thus, we have a magnetic tape recorder that records changes in Earth's magnetic field. This recorder also permits us to determine the rate of seafloor spreading.

Measuring Plate Motion

A number of methods have been employed to establish the direction and rate of plate motion. As noted earlier, hot spot "tracks" like those of the Hawaiian Island–Emperor Seamount chain trace the movement of the Pacific Plate relative to the mantle below. Further, by measuring the length of this volcanic chain and the time interval between the formation of the oldest structure (Suiko Seamount) and youngest structure (Hawaii), an average rate of plate motion can be calculated. In this case the volcanic chain is roughly 3000 kilometers long and has formed over the past 65 million years—making the average rate of movement about 9 centimeters (4 inches) per year. The accuracy of this calculation hinges on the hot spot maintaining a fixed position in the mantle.

Recall that the magnetic stripes measured on the floor of the ocean also provide a method to measure rates of plate motion—at least as averaged over millions of years. Using paleomagnetism and other indirect techniques, researchers have been able to work out relative plate velocities as shown on the map in Figure 8.26.

It is currently possible, with the use of space-age technology, to directly measure the relative motion between plates. This is accomplished by periodically establishing the exact locations, and hence the distance, between two observing stations situated on opposite

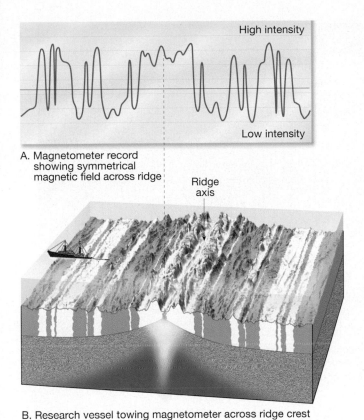

High intensity

Low intensity

A. Magnetometer record
showing symmetrical
magnetic field across ridge

Ridge
axis

B. Research vessel towing magnetometer across ridge crest

▲ **FIGURE 8.25**   The ocean floor as a magnetic tape recorder.
A. Schematic representation of magnetic intensities recorded as a
magnetometer is towed across a segment of the oceanic ridge.
B. Notice the symmetrical stripes of low- and high-intensity
magnetism that parallel the ridge crest. Vine and Matthews suggested
that the stripes of high-intensity magnetism occur where normally
magnetized oceanic basalts enhance the existing magnetic field.
Conversely, the low-intensity stripes are regions where the crust is
polarized in the reverse direction, which weakens the existing
magnetic field.

sides of a plate boundary. Two of the methods used for
this calculation are *Very Long Baseline Interferometry*
(VLBI) and a satellite positioning technique that employs the *Global Positioning System* (GPS). The Very
Long Baseline Interferometry system utilizes large
radio telescopes to record signals from very distant
quasars (quasi-stellar objects). Quasars lie billions of
light-years from Earth, so they act as stationary reference points. The millisecond differences in the arrival
times of the same signal at different Earth-bound observatories provide a means of establishing the precise
distance between receivers. A typical survey may take
a day to perform and involves two widely spaced
radio telescopes observing perhaps a dozen quasars, 5
to 10 times each. This scheme provides an estimate of
the distance between these observatories, which is accurate to about 2 centimeters. By repeating this experiment at a later date, researchers can establish the
relative motion of these sites. This method has been
particularly useful in establishing large-scale plate mo-

tions, such as the separation that is occurring between
the United States and Europe.

You may be familiar with GPS, which uses 21
satellites to accurately locate any individual who is
equipped with a handheld receiver. By using two
spaced receivers, signals obtained by these instruments can be used to calculate their relative positions with considerable accuracy. Techniques using
GPS receivers have been shown to be useful in establishing small-scale crustal movements such as those
that occur along local faults in regions known to be
tectonically active.

Confirming data obtained from these and other
techniques leave little doubt that real plate motion has
been detected (Figure 8.26). Calculations show that
Hawaii is moving in a northwesterly direction and is
approaching Japan at 8.3 centimeters per year. A site
located in Maryland is retreating from one in England
at a rate of about 1.7 centimeters per year—a rate that
is close to the 2.3-centimeters-per-year spreading rate
that was established from paleomagnetic evidence.

What Drives Plate Motion?

The plate tectonics theory *describes* plate motion and
the role that this motion plays in generating and/or
modifying the major features of Earth's crust. Therefore, acceptance of plate tectonics does not rely on
knowing exactly what drives plate motion. This is fortunate, because none of the models yet proposed can
account for all major facets of plate tectonics. Nevertheless, researchers generally agree on the following:

1. Convective flow in the rocky 2900-kilometer-
   thick mantle—in which warm, buoyant rock rises
   and cooler, dense material sinks—is the underlying driving force for plate movement.

2. Mantle convection and plate tectonics are part of
   the same system. Subducting oceanic plates
   drive the cold downward-moving portion of
   convective flow while shallow upwelling of hot
   rock along the oceanic ridge and buoyant mantle
   plumes are the upward-flowing arm of the convective mechanism.

3. The slow movements of Earth's plates and mantle are ultimately driven by the unequal distribution of heat within Earth's interior.

What is not known with any high degree of certainty is
the precise nature of this convective flow.

Forces that Drive Plate Motion

Several mechanisms contribute to plate motion; these
include *slab pull*, *ridge push*, and *slab suction*. There is
general agreement that the subduction of cold, dense

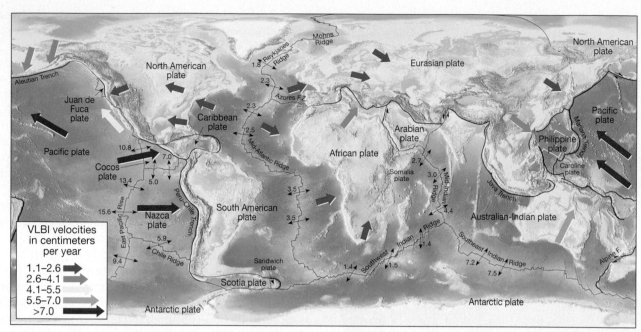

▲ **FIGURE 8.26** This map illustrates directions and rates of plate motion in centimeters per year. Seafloor-spreading velocities (as shown with black arrows and labels) are based on the spacing of dated magnetic stripes (anomalies). The colored arrows show Very Long Baseline Interferometry (VLBI) data of plate motion at selected locations. The data obtained by these methods are typically consistent. (Seafloor data from DeMets and others, VLBI data from Ryan and others)

slabs of oceanic lithosphere is the main driving force of plate motion (Figure 8.27). As these slabs sink into the asthenosphere, they "pull" the trailing plate along. This phenomenon, called **slab pull**, results because old slabs of oceanic lithosphere are more dense than the underlying asthenosphere and hence "sink like a rock."

Another important driving force is called **ridge push** (Figure 8.27). This gravity-driven mechanism results from the elevated position of the oceanic ridge, which causes slabs of lithosphere to "slide" down the flanks of the ridge. Ridge push appears to contribute far less to plate motions than slab pull. Note that de-

spite its greater average height above the seafloor, spreading rates along the Mid-Atlantic Ridge are considerably less than spreading rates along the less steep East Pacific Rise. Also supporting the notion that slab pull is more important than ridge push is the fact that when more than 20 percent of the perimeter of a plate consists of subduction zones, rates of plate movement are relatively rapid. Examples include the Pacific, Nazca, and Cocos plates, all of which have spreading rates that exceed 10 centimeters per year.

Yet another driving force arises from the drag of a subducting slab on the adjacent mantle. The result is

▶ **FIGURE 8.27** Illustration of some of the forces that act on plates.

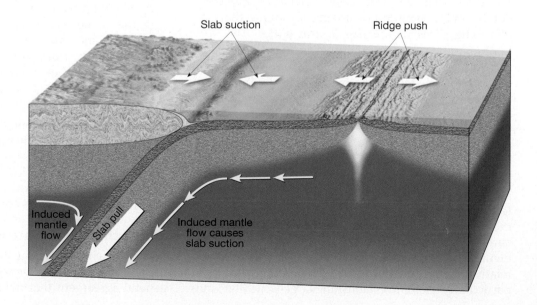

an induced mantle circulation that pulls both the subducting and overriding plates toward the trench. Because this mantle flow tends to "suck" in nearby plates (similar to pulling the plug on a bathtub), it is called **slab suction** (Figure 8.27). Even if a subducting slab becomes detached from the overlying plate, its descent will continue to create flow in the mantle and hence will continue to drive plate motion.

Models of Plate-Mantle Convection

Any model of mantle-plate convection must be consistent with observed physical and chemical properties of the mantle. When seafloor spreading was first proposed, geologists suggested that convection in the mantle consisted of upcurrents coming from deep in the mantle beneath oceanic ridges. Upon reaching the base of the lithosphere, these currents were thought to spread laterally and pull the plates apart. Thus, plates were viewed as being carried passively by the flow in the mantle. However, based on physical evidence, it became clear that upwelling beneath oceanic ridges is shallow and not related to deep convection in the mantle. It is the horizontal movement of lithospheric plates away from the ridge that causes mantle upwelling, not the other way around. We have also learned that plate motion is the dominant source of convective flow in the mantle. As the plates move, they drag the adjacent material along, thereby inducing flow in the mantle. Thus, modern models have plates being an integral part of mantle convection and perhaps even its most active component.

Layering at 660 Kilometers. One of the earliest proposals regarding mantle convection came to be called the "layer cake" model. As shown in Figure 8.28A, this model has two zones of convection—a thin convective layer above 660 kilometers and a thick one located below. This model successfully explains why the basaltic lavas that erupt along oceanic ridges have a somewhat different composition than those that erupt in Hawaii as a result of hot-spot activity. The mid-ocean ridge basalts come from the upper convective layer, which is well mixed, whereas the mantle plume that feeds the Hawaiian volcanoes taps a deeper source of magma that resides in the lower convective layer.

Despite evidence that supports this model, other research suggests that subducting slabs of cold oceanic lithosphere penetrate the 660-kilometer boundary. If true, the subducting lithosphere should serve to mix the upper and lower layers. As a result, the layered mantle structure proposed in this model would not exist.

Whole-Mantle Convection. Other researchers favor some type of *whole-mantle convection*. In a whole-mantle convection model, slabs of cold oceanic lithosphere descend to great depths and stir the entire mantle (Figure 8.28B). Simultaneously, hot mantle plumes originating near the mantle-core boundary transport heat and material toward the surface.

One whole-mantle model suggests that the subducting slabs of oceanic lithosphere accumulate at the mantle-core boundary. Over time this material is thought to melt and buoyantly rise toward the surface in the form of a mantle plume.

Deep-Layer Model. A remaining possibility is layering deeper in the mantle. One deep-layer model has been described as analogous to a lava lamp on a low setting. As shown in Figure 8.28C, the lower, perhaps one-third, of the mantle is like the colored fluid in the bottom layer of a lava lamp. Like a lava lamp on low, heat from Earth's interior causes the two layers to slowly swell and shrink in complex patterns without substantial mixing. A small amount of material from the lower layer flows upward as mantle plumes to generate hot-spot volcanism at the surface.

This model provides the two chemically different mantle sources for basalt that are required by observational data. Further, it is compatible with seismic images that show cold lithospheric plates sinking deep into the mantle. Despite its attractiveness, there is little seismic evidence to suggest that a deep mantle layer of this nature exists, except for the very thin layer located at the mantle-core boundary.

Although there is still much to be learned about the mechanisms that cause plates to move, some facts are clear. The unequal distribution of heat in Earth generates some type of thermal convection that ultimately drives plate-mantle motion. Furthermore, the descending lithospheric plates are active components of downwelling, and they serve to transport cold material into the mantle. Exactly how this convective flow operates is yet to be determined.

Plate Tectonics into the Future

Geologists have also extrapolated present-day plate movements into the future. Figure 8.29 illustrates where Earth's landmasses may be 50 million years from now if present plate movements persist during this time span.

In North America we see that the Baja Peninsula and the portion of southern California that lies west of the San Andreas Fault will have slid past the North American plate. If this northward migration takes place, Los Angeles and San Francisco will pass each other in about 10 million years, and in about 60 million years Los Angeles will begin to descend into the Aleutian Trench.

If Africa continues on a northward path, it will collide with Eurasia, closing the Mediterranean and

▶ **FIGURE 8.28** Proposed models for mantle convection. **A.** The model shown in this illustration consists of two convection layers—a thin, convective layer above 660 kilometers and a thick one below. **B.** In this whole-mantle convection model, cold oceanic lithosphere descends into the lowermost mantle, while hot mantle plumes transport heat toward the surface. **C.** This deep-layer model suggests that the mantle operates similar to a lava lamp on a low setting. Earth's heat causes these layers of convection to slowly swell and shrink in complex patterns without substantial mixing. Some material from the lower layer flows upward as mantle plumes.

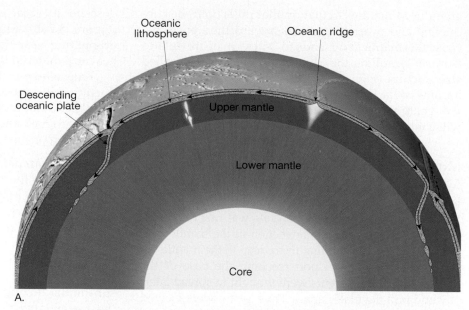

A.

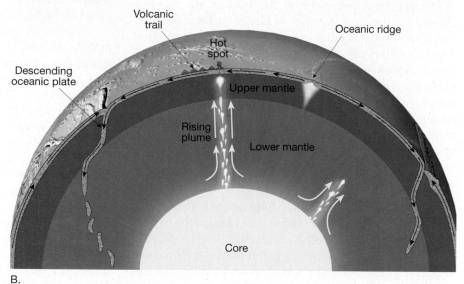

B.

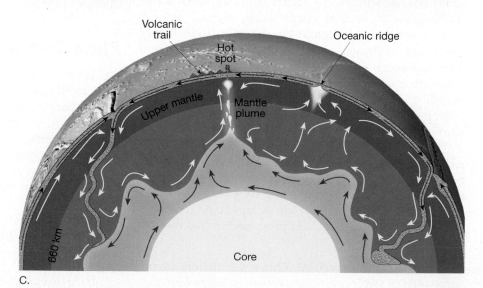

C.

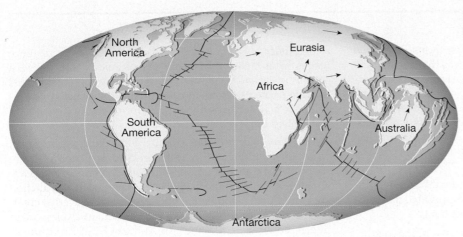

◄ **FIGURE 8.29** The world as it may look 50 million years from now. (Modified after Robert S. Dietz, John C. Holden, C. Scotese, and others)

initiating a major mountain-building episode (Figure 8.29). In other parts of the world, Australia will be astride the equator and, along with New Guinea, will be on a collision course with Asia. Meanwhile, North and South America will begin to separate, while the Atlantic and Indian oceans continue to grow at the expense of the Pacific Ocean.

A few geologists have even speculated on the nature of the globe 250 million years into the future. As shown in Figure 8.30, the next supercontinent may form as a result of subduction of the floor of the Atlantic Ocean, resulting in the collision of the Americas with the Eurasian–African landmass. Support for the possible closing of the Atlantic comes from a similar event when the proto-Atlantic closed to form the Appalachian and Caledonian mountains. During the next 250 million years, Australia is also projected to collide with Southeast Asia. If this scenario is accurate, the dispersal of Pangaea will end when the continents reorganize into the next supercontinent.

Such projections, although interesting, must be viewed with considerable skepticism because many assumptions must be correct for these events to unfold as just described. Nevertheless, changes in the shapes and positions of continents that are equally profound will undoubtedly occur for many hundreds of millions of years to come. Only after much more of Earth's internal heat has been lost will the engine that drives plate motions cease.

? STUDENTS SOMETIMES ASK...
Will plate tectonics eventually "turn off" and cease to operate on Earth?

Because plate tectonic processes are powered by heat from within Earth (which is of a finite amount), the forces will slow sometime in the distant future to the point that the plates will cease to move. The work of external forces (wind, water, and ice), however, will continue to erode Earth's features, most of which will be eventually eroded flat. What a different world it will be—an Earth with no earthquakes, no volcanoes, and no mountains. Flatness will prevail!

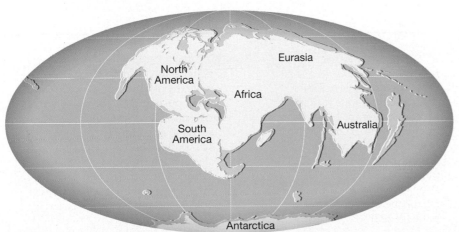

◄ **FIGURE 8.30** Reconstruction of Earth as it may appear 250 million years into the future. (Modified after C. Scotese and others)

Chapter Summary

• In the early 1900s *Alfred Wegener* set forth his *continental drift* hypothesis. One of its major tenets was that a supercontinent called *Pangaea* began breaking apart into smaller continents about 200 million years ago. The smaller continental fragments then "drifted" to their present positions. To support the claim that the now-separate continents were once joined, Wegener and others used the *fit of South America and Africa, the distribution of ancient climates, fossil evidence,* and *rock structures.*

• One of the main objections to the continental drift hypothesis was the inability of its supporters to provide an acceptable mechanism for the movement of continents.

• The theory of *plate tectonics,* a far more encompassing theory than continental drift, holds that Earth's rigid outer shell, called the *lithosphere,* consists of seven large and numerous smaller segments called *plates* that are in motion relative to each other. Most of Earth's *seismic activity, volcanism,* and *mountain building* occur along the dynamic margins of these plates.

• A major departure of the plate tectonics theory from the continental drift hypothesis is that large plates contain both continental and ocean crust and the entire plate moves. By contrast, in continental drift, Wegener proposed that the sturdier continents "drifted" by breaking through the oceanic crust, much like ice breakers cut through ice.

• *Divergent plate boundaries* occur where plates move apart, resulting in upwelling of material from the mantle to create new seafloor. Most divergent boundaries occur along the axis of the oceanic ridge system and are associated with seafloor spreading, which occurs at rates between about 2 and 15 centimeters per year. New divergent boundaries may form within a continent (for example, the East African rift valleys), where they may fragment a landmass and develop a new ocean basin.

• *Convergent plate boundaries* occur where plates move together, resulting in the subduction of oceanic lithosphere into the mantle along a deep oceanic trench. Convergence between an oceanic and continental block results in subduction of the oceanic slab and the formation of a *continental volcanic arc* such as the Andes of South America. Oceanic–oceanic convergence results in an arc-shaped chain of volcanic islands called a *volcanic island arc.* When two plates carrying continental crust converge, both plates are too buoyant to be subducted. The result is a "collision" resulting in the formation of a mountain belt such as the Himalayas.

• *Transform fault boundaries* occur where plates grind past each other without the production or destruction of lithosphere. Most transform faults join two segments of an oceanic ridge. Others connect spreading centers to subduction zones and thus facilitate the transport of oceanic crust created at a ridge crest to its site of destruction, at a deep-ocean trench. Still others, like the San Andreas Fault, cut through continental crust.

• The theory of plate tectonics is supported by (1) *paleomagnetism,* the direction and intensity of Earth's magnetism in the geologic past; (2) the global distribution of *earthquakes* and their close association with plate boundaries; (3) the ages of *sediments* from the floors of the deep-ocean basins; and (4) the existence of island groups that formed over *hot spots* and that provide a frame of reference for tracing the direction of plate motion.

• Three basic models for mantle convection are currently being evaluated. Mechanisms that contribute to this convective flow are slab-pull, ridge-push, and mantle plumes. *Slab-pull* occurs where cold, dense oceanic lithosphere is subducted and pulls the trailing lithosphere along. *Ridge-push* results when gravity sets the elevated slabs astride oceanic ridges in motion. Hot, buoyant *mantle plumes* are considered the upward flowing arms of mantle convection. One model suggests that mantle convection occurs in two layers separated at a depth of 660 kilometers. Another model proposes whole-mantle convection that stirs the entire 2900-kilometer-thick rocky mantle. Yet another model suggests that the bottom third of the mantle gradually bulges upward in some areas and sinks in others without appreciable mixing. Yet another model suggests that the bottom third of the mantle gradually bulges upward in some areas and sinks in others without appreciable mixing.

Key Terms

asthenosphere (p. 222)
continental drift (p. 217)
continental volcanic arc (p. 229)
convergent boundary (p. 228)
Curie point (p. 227)
deep-ocean trench (p. 228)
divergent boundary (p. 226)
fossil magnetism (p. 237)
fracture zone (p. 232)
hot spot (p. 236)
island arc (p. 231)

lithosphere (p. 222)
magnetic reversal (p. 238)
magnetic time scale (p. 239)
magnetometer (p. 239)
mantle plume (p. 236)
normal polarity (p. 238)
oceanic ridge system (p. 222)
paleomagnetism (p. 237)
Pangaea (p. 217)
partial melting (p. 229)
plate (p. 222)

plate tectonics (p. 232)
reverse polarity (p. 238)
ridge push (p. 242)
rift (rift valley) (p. 226)
seafloor spreading (p. 226)
slab pull (p. 242)
slab suction (p. 243)
subduction zone (p. 229)
transform fault boundary (p. 231)
volcanic island arc (p. 231)

Review Questions

1. Who is credited with developing the continental drift hypothesis?

2. What was probably the first evidence that led some to suspect the continents were once connected?

3. What was Pangaea?

4. List the evidence that Wegener and his supporters gathered to substantiate the continental drift hypothesis.

5. Explain why the discovery of the fossil remains of *Mesosaurus* in both South America and Africa, but nowhere else, supports the continental drift hypothesis.

6. Early in this century, what was the prevailing view of how land animals migrated across vast expanses of ocean?

7. How did Wegener account for the existence of glaciers in the southern landmasses, while at the same time areas in North America, Europe, and Siberia supported lush tropical swamps?

8. On what basis were plate boundaries first established?

9. What are the three major types of plate boundaries? Describe the relative plate motion at each of these boundaries.

10. What is seafloor spreading? Where is active seafloor spreading occurring today?

11. What is a subduction zone? With what type of plate boundary is it associated?

12. Where is lithosphere being consumed? Why must the production and destruction of lithosphere be going on at approximately the same rate?

13. Briefly describe how the Himalaya Mountains formed.

14. Differentiate between transform faults and the other two types of plate boundaries.

15. Some predict that California will sink into the ocean. Is this idea consistent with the theory of plate tectonics?

16. Define the term *paleomagnetism*.

17. How does the continental drift hypothesis account for the apparent wandering of Earth's magnetic poles?

18. What is the age of the oldest sediments recovered by deep-ocean drilling? How do the ages of these sediments compare to the ages of the oldest continental rocks?

19. How do hot spots and the plate tectonics theory account for the fact that the Hawaiian Islands vary in age?

20. Briefly describe the three mechanisms that drive plate motion.

21. With what type of plate boundary are the following places or features associated: Himalayas, Aleutian Islands, Red Sea, Andes Mountains, San Andreas Fault, Iceland, Japan, Mount St. Helens?

Examining the Earth System

1. As an integral subsystem of the Earth system, plate tectonics has played a major role in determining the events that have taken place on Earth since its formation about 4.5 billion years ago. Briefly comment on the general effect that the changing positions of the continents and the redistribution of land and water over Earth's surface have had on the atmosphere, hydrosphere, and biosphere through time. You may want to review plate tectonics by investigating the topic on the Internet using a search engine such as HotBot (**http://www.hotbot.com/**) or Yahoo (**http://www.yahoo.com/**) and visiting the United States Geological Survey's (USGS) This Dynamic Earth: The Story of Plate Tectonics Website at **http://pubs.usgs.gov/publications/text/dynamic.html.**

2. Assume that plate tectonics did not cause the breakup of the supercontinent Pangaea. Using Figure 8.2, describe how the climate (atmosphere), vegetation and animal life (biosphere), and geological features (geosphere) of those locations currently occupied by the cities of Seattle, WA; Chicago, IL; New York, NY; and your college campus location would be different from the conditions that exist today.

Online Study Guide

The *Earth Science* Website uses the resources and flexibility of the Internet to aid in your study of the topics in this chapter. Written and developed by Earth science instructors, this site will help improve your understanding of Earth science. Visit **http://www.prenhall.com/tarbuck** and click on the cover of *Earth Science 11e* to find:

- **On-line review quizzes.**
- **Critical thinking exercises.**
- **Links to chapter-specific Web resources.**
- **Internet-wide key term searches.**

http://www.prenhall.com/tarbuck

A recent eruption of Italy's Mount Etna.
(Photo by Art Wolfe)

Volcanoes
and Other Igneous Activity

The significance of igneous activity may not be obvious at first glance. However, because volcanoes extrude molten rock that formed at great depth, they provide the only windows we have for direct observation of processes that occur many kilometers below Earth's surface. Furthermore, the atmosphere and oceans are thought to have evolved from gases emitted during volcanic eruptions. Either of these facts is reason enough for igneous activity to warrant our attention. All volcanic eruptions are spectacular, but why are some destructive and others quiescent? Is the entire island of Hawaii a volcano as high as Mount Everest resting on the ocean floor? Do the large volcanoes of Washington and Oregon present a threat to human lives? This chapter considers these and other questions as we explore the formation and movement of magma.

On Sunday, May 18, 1980, one of the largest volcanic eruptions to occur in North America in historic times transformed a picturesque volcano into a decapitated remnant (Figure 9.1). On this date in southwestern Washington State, Mount St. Helens erupted with tremendous force. The blast blew out the entire north flank of the volcano, leaving a gaping hole. In one brief moment, a prominent volcano whose summit had been more than 2900 meters (9500 feet) above sea level was lowered by more than 400 meters (1350 feet).

The event devastated a wide swath of timber-rich land on the north side of the mountain (Figure 9.2). Trees within a 400-square-kilometer (160-square-mile) area lay intertwined and flattened, stripped of their branches and appearing from the air like toothpicks strewn about. The accompanying mudflows carried ash, trees, and water-saturated rock debris 29 kilometers (18 miles) down the Toutle River. The eruption claimed 59 lives, some dying from the intense heat and the suffocating cloud of ash and gases, others from being hurled by the blast, and still others from entrapment in the mudflows.

The eruption ejected nearly a cubic kilometer of ash and rock debris. Following the devastating explosion, Mount St. Helens continued to emit great quantities of hot gases and ash, some of which were propelled more than 18,000 meters (over 11 miles) into

▼ FIGURE 9.1 Before-and-after photographs show the transformation of Mount St. Helens caused by the May 18, 1980, eruption. The dark area in the "after" photo is debris-filled Spirit Lake, partially visible in the "before" photo. (Photos courtesy of U.S. Geological Survey)

the stratosphere. During the next few days, this very fine-grained material was carried around Earth by strong upper-air winds. Measurable deposits were reported in Oklahoma and Minnesota, with crop damage into central Montana. Meanwhile, ash fallout in the immediate vicinity exceeded 2 meters (6 feet) in depth. The air over Yakima, Washington (130 kilometers to the east), was so filled with ash that residents experienced midnightlike darkness at noon.

Mount St. Helens is one of 15 large volcanoes and innumerable smaller ones that comprise the Cascade Range, which extends from British Columbia to northern California. Eight of the largest volcanoes have been active in the past few hundred years. Of the remaining seven active volcanoes, the most likely to erupt again are Mount Baker and Mount Rainier in Washington, Mount Shasta and Lassen Peak in California, and Mount Hood in Oregon.

Not all volcanic eruptions are as violent as the 1980 Mount St. Helens event. Some volcanoes, such as Hawaii's Kilauea volcano, generate relatively quiet outpourings of fluid lavas. These "gentle" eruptions are not without some fiery displays; occasionally, fountains of incandescent lava spray hundreds of meters into the air. Such events, however, typically pose minimal threat to human life and property.

Testimony to the quiet nature of Kilauea's eruptions is the fact that the Hawaiian Volcanoes Observatory has operated on its summit since 1912. This, despite the fact that Kilhauea has had more than 50 eruptive phases since record keeping began in 1823. Further, the longest and largest of Kilauea's eruptions began in 1983 and remains active, although it has received only modest media attention.

Why do volcanoes like Mount St. Helens erupt explosively, whereas others like Kilauea are relatively quiet? Why do volcanoes occur in chains like the Aleutian Islands or the Cascade Range? Why do some volcanoes form on the ocean floor, while others occur on the continents? This chapter will deal with these and other questions as we explore the nature and movement of magma and lava (Figure 9.3).

STUDENTS SOMETIMES ASK...
After all the destruction during the eruption of Mount St. Helens, what does the area look like today?

The area continues to make a slow recovery. Surprisingly, many organisms survived the blast, including animals that live underground and plants (particularly those protected by snow or near streams, where erosion quickly removed the ash). Others have adaptations that allow them to quickly repopulate devastated areas. Fully 20 years after the blast, plants have revegetated the area, first-growth forests are beginning to be established, and many animals have returned. Once the old-growth forest is complete (in a few

Spirit Lake

► **FIGURE 9.2** Douglas fir trees were snapped off or uprooted by the lateral blast of Mount St. Helens on May 18, 1980. (Photo by Krafft-Explorer/Photo Researchers, Inc.)

hundred years), it might be hard to find much evidence of the destruction, other than a thick weathered ash layer in the soil.

The volcano itself is rebuilding, too. A large lava dome is forming inside the summit crater, suggesting that the mountain will build up again. Many volcanoes similar to Mount St. Helens exhibit this behavior: rapid destruction followed by slow rebuilding. If you really want to see what it looks like, go to the Mount St. Helens National Volcanic Home Page at **http://www.fs.fed.us/gpnf/mshnvm/**, where they have a "volcanocam" with real-time images of the mountain.

The Nature of Volcanic Eruptions

Forces Within

▼ **Igneous Activity**

Volcanic activity is commonly perceived as a process that produces a picturesque, cone-shaped structure that periodically erupts in a violent manner. Although some eruptions are very explosive, many others are not. What determines whether a volcano extrudes magma violently or gently? The primary factors include the magma's *composition*, its *temperature*, and the

▼ **FIGURE 9.3** A river of basaltic lava from the January 17, 2002, erruption of Mount Nyiragongo destroyed many homes in Goma, Congo. (Photo by Marco Longari/Agence France Presse/Getty Images)

amount of *dissolved gases* it contains. To varying degrees these factors affect the magma's **viscosity** (*viscos* = sticky). The more viscous the material, the greater its resistance to flow. For example, syrup is more viscous than water. Magma associated with an explosive eruption may be thousands of times more viscous than magma that is extruded in a quiescent manner.

Factors Affecting Viscosity

The effect of temperature on viscosity is easily seen. Just as heating syrup makes it more fluid (less viscous), the mobility of lava is strongly influenced by temperature. As a lava flow cools and begins to congeal, its mobility decreases and eventually the flow halts.

A more significant influence on volcanic behavior is the chemical composition of magmas. This was discussed in Chapter 3 along with the classification of igneous rocks. Recall that a major difference among various igneous rocks is their silica (SiO_2) content. The same is true of the magmas from which rocks form. Magmas that produce basaltic rocks contain about 50 percent silica, whereas magmas that produce granitic rocks contain over 70 percent silica (Table 9.1).

A magma's viscosity is directly related to its silica content. In general, the more silica in magma, the greater its viscosity. The flow of magma is impeded because silicate structures link together into long chains even before crystallization begins. Consequently, because of their high silica content, rhyolitic (felsic) lavas are very viscous and tend to form comparatively short, thick flows. By contrast, basaltic (mafic) lavas, which contain less silica, tend to be more fluid and have been known to travel distances of 150 kilometers (90 miles) or more before congealing (Figure 9.3).

In Hawaiian eruptions, the magmas are hot and basaltic, so they are extruded with ease. By contrast, highly viscous magmas are more difficult to force through a vent. On occasion, the vent may become plugged with viscous magma, which can result in a buildup of gases that produce an explosive eruption. However, a viscous magma is not explosive by itself. It is the gas content that puts the bang into a violent eruption.

Importance of Dissolved Gases in Magma

During explosive eruptions, it is the **volatiles** (the gaseous components of magma) that provide the force that extrudes molten rock from the vent. These gases are mostly water vapor and carbon dioxide. As magma moves into a near-surface environment, such as within a volcano, the confining pressure in the uppermost portion of the magma body is greatly reduced. The reduction in confining pressure allows dissolved gases to be released suddenly, just as opening a soda bottle allows dissolved carbon dioxide gas bubbles to escape.

At high temperatures and low near-surface pressures, these gases will expand to occupy hundreds of times their original volume. Very fluid basaltic magmas allow the expanding gases to bubble upward and escape from the vent with relative ease. As they escape, the gases will often propel incandescent lava hundreds of meters into the air, producing lava fountains. Although spectacular, such fountains are mostly harmless and not generally associated with major explosive events that cause great loss of life and property. Rather, eruptions of fluid basaltic lavas, such as those that occur in Hawaii, are relatively quiescent.

At the other extreme, highly viscous magmas impede the upward migration of expanding gases. The gases collect in bubbles and pockets that increase in size until they explosively eject the semimolten rock from the volcano (Figure 9.4). The result is a Mount St. Helens or Mt. Pinatubo.

To summarize, the viscosity of magma, plus the quantity of dissolved gases and the ease with which they can escape, determines the nature of a volcanic eruption. We can now understand the "gentle" volcanic eruptions of hot, fluid lavas in Hawaii and the explosive, violent eruptions of viscous lavas from volcanoes such as Mount St. Helens.

TABLE 9.1	Magmas have different compositions, which cause their properties to vary.				
Composition	**Silica Content**	**Viscosity**	**Gas Content**	**Tendency to Form Pyroclastics**	**Volcanic Landform**
Basaltic (Mafic)	Least (~50%)	Least	Least (1–2%)	Least	Shield Volcanoes Basalt Plateaus Cinder Cones
Andesitic (Intermediate)	Intermediate (~60%)	Intermediate	Intermediate (3–4%)	Intermediate	Composite Cones
Rhyolitic (Felsic)	Most (~70%)	Greatest	Most (4–6%)	Greatest	Pyroclastic Flows Volcanic Domes

► **FIGURE 9.4** Mount St. Helens emitting volcanic ash on July 22, 1980, two months after the huge May erruption. (Photo by David Weintraub/Photo Researchers, Inc.)

What Is Extruded During Eruptions?

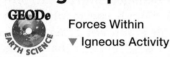

Forces Within
▼ Igneous Activity

Lava may appear to be the primary material extruded from a volcano, but this is not always the case. Just as often, explosive eruptions eject huge quantities of broken rock, lava bombs, fine ash, and dust. Moreover, all volcanic eruptions emit large amounts of gas. In this section we will examine each of these materials associated with a volcanic eruption.

Lava Flows

Because of their low silica content, hot basaltic lavas are usually very fluid. They flow in thin, broad sheets or streamlike ribbons. On the island of Hawaii, such lavas have been clocked at speeds of 30 kilometers (20 miles) per hour down steep slopes. These velocities are rare, however, and flow rates of 10 to 300 meters per hour are more common. In contrast, the movement of silica-rich (rhyolitic) lava is on occasion too slow to be perceptible.

When fluid basaltic lavas of the Hawaiian type congeal, they commonly form a relatively smooth skin that wrinkles as the still-molten subsurface lava continues to advance (Figure 9.5A). These are known as **pahoehoe flows** (pronounced *pah-hoy-hoy*) and resemble the twisted braids in ropes.

Another common type of basaltic lava has a surface of rough, jagged blocks with dangerously sharp edges and spiny projections (Figure 9.5B). The name **aa** (pronounced *ah-ah*) is given to these flows. In general,

Hawaiian aa flows are relatively cool and thick and, depending on the slope, move more slowly than pahoehoe flows. Further, gases escaping from the surface produce numerous voids and sharp spines in the congealing lava. As the molten interior advances, the outer crust is broken further, giving the flow the appearance of an advancing mass of lava rubble.

Gases

Magmas contain varied amounts of dissolved gases (*volatiles*) held in the molten rock by confining pressure, just as carbon dioxide is held in soft drinks. As with soft drinks, as soon as the pressure is reduced, the gases begin to escape. Obtaining gas samples from an erupting volcano is often difficult and dangerous, so geologists can often only estimate the amount of gas originally contained within the magma.

The gaseous portion of most magmas is believed to make up from 1 to 6 percent of the total weight, with most of this being water vapor. Although the percentage may be small, the actual quantity of emitted gas can exceed thousands of tons each day.

The composition of volcanic gases is important to scientists because they have contributed significantly to gases that make up the atmosphere. Analysis of samples taken during Hawaiian eruptions indicates that the gases are about 70 percent water vapor, 15 percent carbon dioxide, 5 percent nitrogen, 5 percent sulfur, and lesser amounts of chlorine, hydrogen, and argon. Sulfur compounds are easily recognized by their pungent odor and because they readily form sulfuric acid—a natural source of air pollution.

A.

B.

▲ **FIGURE 9.5** **A.** Typical pahoehoe (ropy) lava flow, Kilauea, Hawaii. (Photo by Doug Perrine/DRK) **B.** Typical slow-moving aa flow. (Photo by J. D. Griggs/U.S. Geological Survey)

Pyroclastic Materials

When basaltic lava is extruded, dissolved gases escape quite freely and continually. These gases propel incandescent blobs of lava to great heights. Some of this ejected material may land near the vent and build a cone-shaped structure, whereas smaller particles will be carried great distances by the wind. By contrast, viscous (rhyolitic) magmas are highly charged with gases, and upon release they expand greatly as they blow pulverized rock, lava, and glass fragments from the vent. The particles produced in both of these situations are referred to as **pyroclastic materials** (*pyro* = fire, *clast* = fragment). These ejected fragments range in size from very fine dust and sand-size volcanic ash (less than 2 millimeters) to pieces that weigh several tons.

Ash and *dust* particles are produced from gas-laden viscous magma during an explosive eruption (see Figure 9.4). As magma moves up in the vent, the gases rapidly expand, generating a froth of melt that might resemble froth that flows from a just opened bottle of champagne. As the hot gases expand explosively, the froth is blown into very fine glassy fragments. When the hot ash falls, the glassy shards often fuse to form a rock called *welded tuff*. Sheets of this material, as well as ash deposits that later consolidate, cover vast portions of the western United States.

Also common are pyroclasts that range in size from small beads to walnuts termed *lapilli* ("little stones"). These ejecta are commonly called *cinders* (2–64 millimeters). Particles larger than 64 millimeters (2.5 inches) in diameter are called *blocks* when they are made of hardened lava and *bombs* when they are ejected as incandescent lava. Because bombs are semi-molten upon ejection, they often take on a streamlined shape as they hurtle through the air (Figure 9.6). Because of their size, bombs and blocks usually fall on the slopes of a cone; however, they are occasionally propelled far from the volcano by the force of escaping gases. For instance, bombs 6 meters (20 feet) long and weighing about 200 tons were blown 600 meters (2000 feet) from the vent during an eruption of the Japanese volcano Asama.

Volcanic Structures and Eruptive Styles

 GEODe

Forces Within
▼ Igneous Activity

The popular image of a volcano is that of a solitary, graceful, snowcapped cone, such as Mount Hood in Oregon or Japan's Fujiyama. These picturesque, conical mountains are produced by volcanic activity that occurred intermittently over thousands, or even hundreds of thousands, of years. However, many volcanoes do not fit this image. Some volcanoes are only 30 meters (100 feet) high and formed during a single eruptive phase that may have lasted only a few days.

Volcanic landforms come in a wide variety of shapes and sizes, and each structure has a unique eruptive history. Nevertheless, volcanologists have been able to classify volcanic landforms and determine their eruptive patterns. In this section we will consider the general anatomy of a volcano and look at three major volcanic types: *shield volcanoes, cinder cones*, and *composite cones*. This discussion will be followed by an overview of other significant volcanic landforms.

Anatomy of a Volcano

Volcanic activity frequently begins when a **fissure** (crack) develops in the crust as magma moves forcefully toward the surface. As the gas-rich magma moves up this linear fissure, its path is usually localized into a

▲ **FIGURE 9.6** Volcanic bombs forming during an eruption of Hawaii's Kilauea volcano. Ejected lava fragments take on a streamlined shape as they sail through the air. The bomb in the insert is about 10 centimeters (4 inches) long. (Photo by Arthur Roy/National Audubon Society/Photo Researchers, Inc.; inset photo by E. J. Tarbuck)

circular **conduit**, or **pipe**, that terminates at a surface opening called a **vent** (Figure 9.7). Successive eruptions of lava, pyroclastic material, or frequently a combination of both often separated by long periods of inactivity eventually build the structure we call a **volcano**.

Located at the summit of many volcanoes is a steep-walled depression called a **crater** (*crater* = a bowl). Craters are constructional features that were built upward as ejected fragments collected around the vent to form a doughnut-like structure. Some volcanoes have multiple summit craters, whereas others have very large, more or less circular depressions called **calderas**. Calderas are large collapse structures that may or may not form in association with a volcano. (We will consider the formation of various types of calderas later.)

During early stages of growth most volcanic discharges come from a central summit vent. As a volcano matures, material also tends to be emitted from fissures that develop along the flanks, or base, of the volcano. Continued activity from a flank eruption may produce a small **parasitic cone** (*parasitus* = one who eats at the table of another). Mount Etna in Italy, for example, has more than 200 secondary vents, some of which have built cones. Many of these vents, however, emit only gases and are appropriately called **fumaroles** (*fumus* = smoke).

▶ **FIGURE 9.7** Anatomy of a "typical" composite cone (see also Figures 9.8 and 9.11 for a comparison with a shield and cinder cone, respectively).

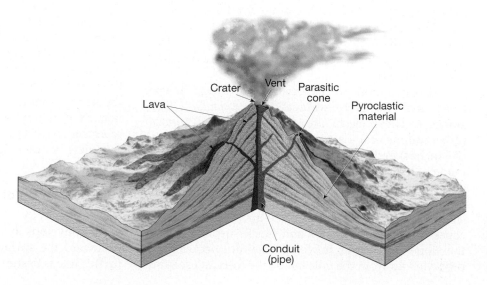

Crater Vent Parasitic cone

Lava

Pyroclastic material

Conduit (pipe)

The form of a particular volcano is largely determined by the composition of the contributing magma. As you will see, fluid Hawaiian-type lavas tend to produce broad structures with gentle slopes, whereas more viscous silica-rich lavas (and some gas-rich basaltic lavas) tend to generate cones with moderate to steep slopes.

Shield Volcanoes

Shield volcanoes are produced by the accumulation of fluid basaltic lavas and exhibit the shape of a broad, slightly domed structure that resembles a warrior's shield (Figure 9.8). Most (but not all) shield volcanoes have grown up from the deep-ocean floor to form islands or seamounts. For example, the islands of the Hawaiian chain, Iceland, and the Galápagos are either a single shield volcano or the coalescence of several shields. Extensive study of the Hawaiian Islands confirms that each shield was built from a myriad of basaltic lava flows averaging a few meters thick. Also, these islands consist of only about 1 percent pyroclastic ejecta.

Mauna Loa: Earth's Largest Volcano. Mauna Loa is one of five overlapping shield volcanoes that together comprise the Big Island of Hawaii (Figure 9.8). From its base on the floor of the Pacific Ocean to its summit, Mauna Loa is over 9 kilometers (6 miles) high, exceeding that of Mount Everest. This massive pile of basaltic rock has a volume of 40,000 cubic kilometers that was extruded over a period of nearly a million years. For comparison, the volume of material composing Mauna Loa is roughly 200 times greater than the amount composing a large composite cone such as Mount Rainier (Figure 9.9). Most shields, however, are more modest in size. For example, the classic Icelandic shield, Skjalbreidur, rises to a height of only about 600 meters (2000 feet) and is 10 kilometers (6 miles) across its base.

Young shields, particularly those located in Iceland, emit very fluid lava from a central summit vent and have sides with gentle slopes that vary from 1 to 5 degrees. Mature shields, as exemplified by Mauna Loa, have steeper slopes in the middle sections (about 10 degrees), while their summits and flanks are comparatively flat. During the mature stage, lavas are discharged from the summit vents, as well as from rift zones that develop along the slopes. Most lava is discharged as the fluid pahoehoe type, but as these flows cool downslope, many change into a clinkery aa flow. Once an eruption is well established, a large fraction of the lava (perhaps 80 percent) flows through a well-developed system of lava tubes. This greatly increases the distance lava can travel before it solidifies. Thus, lava emitted near the summit often reaches the sea, thereby adding to the width of the cone at the expense of its height.

▼ **FIGURE 9.8** Shield volcanoes are built primarily of fluid basaltic lava flows and contain only a small percentage of pyroclastic materials. These broad, slightly domed structures, exemplified by the Hawaiian Islands, are the largest volcanoes on Earth. (Photo by Greg Vaughn)

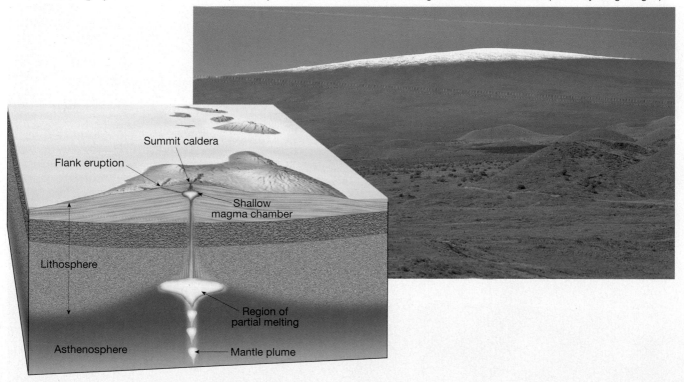

Summit caldera

Flank eruption

Shallow magma chamber

Lithosphere

Region of partial melting

Asthenosphere — Mantle plume

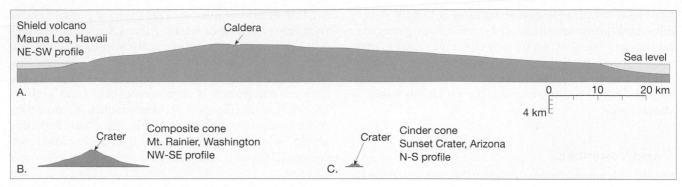

▲ **FIGURE 9.9** Profiles of volcanic landforms. **A.** Profile of Mauna Loa, Hawaii, the largest shield volcano in the Hawaiian chain. Note size comparison with Mount Rainier, Washington, a large composite cone. **B.** Profile of Mount Rainier, Washington. Note how it dwarfs a typical cinder cone. **C.** Profile of Sunset Crater, Arizona, a typical steep-sided cinder cone.

Another feature common to a mature, active shield volcano is a large steep-walled caldera that occupies its summit. Calderas form when the roof of the volcano collapses as magma from the central magma reservoir migrates to the flanks, often feeding fissure eruptions. Mauna Loa's summit caldera measures 2.6 by 4.5 kilometers (1.6 by 2.8 miles) and has a depth that averages about 150 meters (500 feet).

Kilauea, Hawaii: Eruption of a Shield Volcano. Kilauea, the most active and intensely studied shield volcano in the world, is located on the island of Hawaii in the shadow of Mauna Loa. More than 50 eruptions have been witnessed here since record keeping began in 1823. Several months before each eruptive phase, Kilauea in-

flates as magma gradually migrates upward and accumulates in a central reservoir located a few kilometers below the summit. For up to 24 hours in advance of an eruption, swarms of small earthquakes warn of the impending activity.

Most of the activity on Kilauea during the past 50 years occurred along the flanks of the volcano in a region called the East Rift Zone. The longest and largest rift eruption ever recorded on Kilauea began in 1983 and continues to this day, with no signs of abating. The first discharge began along a 6-kilometer (4-mile) fissure where a 100-meter (300-foot) high "curtain of fire" formed as red-hot lava was ejected skyward (Figure 9.10). When the activity became localized, a cinder and spatter cone given the Hawaiian name *Puu Oo*

▼ **FIGURE 9.10** Lava extruded along the East Rift Zone, Kilauea, Hawaii. (Photo by Greg Vaughn)

was built. Over the next three years the general eruptive pattern consisted of short periods (hours to days) when fountains of gas-rich lava sprayed skyward. Each event was followed by nearly a month of inactivity.

By the summer of 1986 a new vent opened up 3 kilometers downrift. Here smooth-surfaced pahoehoe lava formed a lava lake. Occasionally the lake overflowed, but more often lava escaped through tunnels to feed pahoehoe flows that moved down the southeastern flank of the volcano toward the sea. These flows destroyed nearly a hundred rural homes, covered a major roadway, and eventually reached the sea. Lava has been intermittently pouring into the ocean ever since, adding new land to the island of Hawaii.

Cinder Cones

As the name suggests, **cinder cones** (also called **scoria cones**) are built from ejected lava fragments that take on the appearance of cinders or clinkers as they begin to harden while in flight. These fragments range in size from fine ash to bombs but consist most-ly of pea- to walnut-size lapilli. Usually a product of relatively gas-rich basaltic magma, cinder cones consist of rounded to irregular fragments that are markedly vesicular (containing voids) and have a black to reddish-brown color. Occasionally an eruption of silica-rich magma will generate a light-colored cinder cone composed of ash and pumice fragments. Although cinder cones are composed mostly of loose pyroclastic material, they sometimes extrude lava. On such occasions the discharges come from vents located at or near the base rather than from the summit crater.

Cinder cones have a very simple, distinctive shape determined by the slope that loose pyroclastic material maintains as it comes to rest (Figure 9.11). Because cinders have a high angle of repose (the steepest angle at which material remains stable), young cinder cones are steep-sided, having *stopes* between 30 and 40 degrees. In addition, cinder cones have large, deep craters in relation to the overall size of the structure. Although relatively symmetrical, many cinder cones are elongated, and higher on the side that was downwind during the eruptions.

▼ **FIGURE 9.11** SP Crater, a cinder cone north of Flagstaff, Arizona. (Photo by Michael Collier)

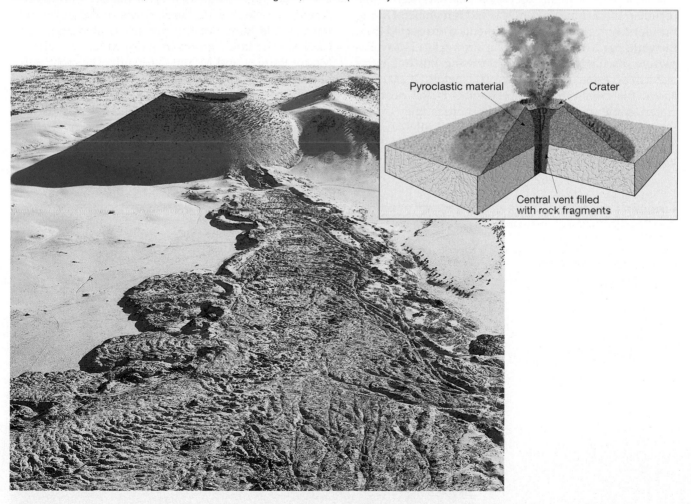

Cinder cones are usually the product of a single eruptive episode that sometimes lasts only a few weeks and rarely exceeds a few years. Once this event ceases, the magma in the pipe connecting the vent to the magma chamber solidifies, and the volcano never erupts again. As a consequence of this short life span, cinder cones are small, usually between 30 meters (100 feet) and 300 meters (1000 feet) and rarely exceed 700 meters (2100 feet) in height (see Figure 9.9).

Cinder cones are found by the thousands all around the globe. Some are located in volcanic fields like the one located near Flagstaff, Arizona, which consists of about 600 cones. Others are parasitic cones of larger volcanoes. Mount Etna, for example, has dozens of cinder cones dotting its flanks.

Parícutin: Life of a Garden-Variety Cinder Cone. One of the very few volcanoes studied by geologists from beginning to end is the cinder cone called Parícutin, located about 320 kilometers (200 miles) west of Mexico City. In 1943 its eruptive phase began in a cornfield owned by Dionisio Pulido, who witnessed the event as he prepared the field for planting.

For two weeks prior to the first eruption, numerous Earth tremors caused apprehension in the nearby village of Parícutin. Then on February 20 sulfurous gases began billowing from a small depression that had been in the cornfield for as long as people could remember. During the night, hot, glowing rock fragments were ejected from the vent, producing a spectacular fireworks display. Explosive discharges continued, throwing hot fragments and ash occasionally as high as 6000 meters (20,000 feet) above the crater rim. Larger fragments fell near the crater, some remaining incandescent as they rolled down the slope. These built an aesthetically pleasing cone, while finer ash fell over a much larger area, burning and eventually covering the village of Parícutin. In the first day the cone grew to 40 meters (130 feet), and by the fifth day it was over 100 meters (330 feet) high. Within the first year more than 90 percent of the total ejecta had been discharged.

The first lava flow came from a fissure that opened just north of the cone, but after a few months flows began to emerge from the base of the cone itself. In June 1944 a clinkery aa flow 10 meters (30 feet) thick moved over much of the village of San Juan Parangaricutiro, leaving only the church steeple exposed (Figure 9.12). After nine years of intermittent pyroclastic explosions and nearly continuous discharge of lava from vents at its base, the activity ceased almost as quickly as it had begun. Today, Parícutin is just another one of the scores of cinder cones dotting the landscape in this region of Mexico. Like the others, it will not erupt again.

Composite Cones

Earth's most picturesque yet potentially dangerous volcanoes are **composite cones** or **stratovolcanoes** (Figure 9.13). Most are located in a relatively narrow zone that rims the Pacific Ocean, appropriately called the *Ring of Fire* (see Figure 9.28, p. 273). This active zone includes a chain of continental volcanoes that are distributed along the west coast of South and North America, including the large cones of the Andes and the Cascade Range of the western United States and Canada. The latter group includes Mount St. Helens, Mount Rainier, and Mount Garibaldi. The most active regions in the Ring of Fire are located along curved belts of volcanic islands situated adjacent to the deep ocean trenches of the northern and western Pacific. This nearly continuous chain of volcanoes stretches from the Aleutian Islands to Japan and the Philippines and ends on the North Island of New Zealand.

▶ **FIGURE 9.12** The village of San Juan Parangaricutiro engulfed by lava from Parícutin, shown in the background. Only the church towers remain. (Photo by Tad Nichols)

▲ **FIGURE 9.13** Mount Shasta, California, one of the largest composite cones in the Cascade Range. Shastina is the smaller parasitic cone on the left. (Photo by David Muench)

The classic composite cone is a large, nearly symmetrical structure composed of both lava and pyroclastic deposits. Just as shield volcanoes owe their shape to fluid basaltic lavas, composite cones reflect the nature of the erupted material. For the most part, composite cones are the product of gas-rich magma having an andesitic composition. Relative to shields, the silica-rich magmas typical of composite cones generate thick viscous lavas that travel short distances. In addition, composite cones may generate explosive eruptions that eject huge quantities of pyroclastic material.

The growth of a "typical" composite cone begins with both pyroclastic material and lava being emitted from a central vent. As the structure matures, lavas tend to flow from fissures that develop on the lower flanks of the cone. This activity may alternate with explosive eruptions that eject pyroclastic material from the summit crater. Sometimes both activities occur simultaneously.

A conical shape, with a steep summit area and more gradually sloping flanks, is typical of many large composite cones. This classic profile, which adorns calendars and postcards, is partially a consequence of the way viscous lavas and pyroclastic ejecta contribute to the growth of the cone. Coarse fragments ejected from the summit crater tend to accumulate near their source. Because of their high angle of repose, coarse materials contribute to the steep slopes of the summit area. Finer ejecta, on the other hand, are deposited as a thin layer over a large area. This acts to flatten the flank of the cone. In addition, during the early stages of growth, lavas tend to be more abundant and flow greater distances from the vent than do later lavas. This contributes to the cone's broad base. As the vol-

cano matures, the short flows that come from the central vent serve to armor and strengthen the summit area. Consequently, steep summit slopes exceeding 40 degrees are sometimes possible. Two of the most perfect cones—Mount Mayon in the Philippines and Fujiyama in Japan—exhibit the classic form we expect of a composite cone, with its steep summit and gently sloping flanks.

Despite their symmetrical form, most composite cones have a complex history. Huge mounds of volcanic debris surrounding many cones provide evidence that, in the distant past, a large section of the volcano slid downslope as a massive landslide. Others develop horseshoe-shaped depressions at their summits as a result of explosive eruptions, or as occurred during the 1980 eruption of Mount St. Helens, a combination of a landslide and the eruption of 0.6 cubic kilometer of magma left a gaping void on the north side of the cone. Often, so much rebuilding has occurred since these eruptions that no trace of the amphitheater-shaped scar remains.

Living in the Shadow of a Composite Cone

More than 50 volcanoes have erupted in the United States in the past 200 years (see Figure 9.30, p. 278). Fortunately, the most explosive of these eruptions, except for Mount St. Helens in 1980, occurred in sparsely inhabited regions of Alaska. On a global scale numerous destructive eruptions have occurred during the past few thousand years, a few of which may have influenced the course of human civilization.

Nuée Ardente: A Deadly Pyroclastic Flow

One of the most devastating phenomena associated with composite cones are **pyroclastic flows**, which consist of hot gases infused with incandescent ash and larger rock fragments. The most destructive of these fiery flows, called **nuée ardentes** (also referred to as *glowing avalanches*), are capable of racing down steep volcanic slopes at speeds that can approach 200 kilometers (125 miles) per hour (Figure 9.14).

The ground-hugging portion of a glowing avalanche is rich in particulate matter, which is suspended by jets of buoyant gases passing upward through the flow. Some of these gases have escaped from newly erupted volcanic fragments. In addition, air that is overtaken and trapped by an advancing flow may be heated sufficiently to provide buoyancy to the particulate matter of the nuée ardente. Thus, these flows, which can include large rock fragments in addition to ash, travel downslope in a nearly frictionless environment. This helps to explain why some nuée ardente deposits are found more than 100 kilometers (60 miles) from their source.

The pull of gravity is the force that causes these heavier-than-air flows to sweep downslope much like a snow avalanche. Some pyroclastic flows result when a powerful eruption blasts pyroclastic material laterally out the side of a volcano. Probably more often, nuée ardentes form from the collapse of tall eruption columns that form over a volcano during an explosive event. Once gravity overcomes the initial upward thrust provided by

the escaping gases, the ejecta begin to fall. Massive amounts of incandescent blocks, ash, and pumice fragments that fall onto the summit area begin to cascade downslope under the influence of gravity. The largest fragments have been observed bouncing down the flanks of a cone, while the finer materials travel rapidly as an expanding tongue-shaped cloud.

The Destruction of St. Pierre. In 1902 an infamous nuée ardente from Mount Pelée, a small volcano on the Caribbean island of Martinique, destroyed the port town of St. Pierre. The destruction happened in moments and was so devastating that almost all of St. Pierre's 28,000 inhabitants were killed. Only one person on the outskirts of town—a prisoner protected in a dungeon—and a few people on ships in the harbor were spared (Figure 9.15).

Shortly after this calamitous eruption, scientists arrived on the scene. Although St. Pierre was mantled by only a thin layer of volcanic debris, they discovered that masonry walls nearly a meter thick were knocked over like dominoes; large trees were uprooted and cannons were torn from their mounts. A further reminder of the destructive force of this nuée ardente is preserved in the ruins of the mental hospital. One of the immense steel chairs that had been used to confine alcoholic patients can be seen today, contorted, as though it were made of plastic.

Lahars: Mudflows on Active and Inactive Cones

In addition to violent eruptions, large composite cones may generate a type of mudflow referred to by its Indonesian name **lahar**. The destructive mudflows occur when volcanic debris becomes saturated with water and rapidly moves down steep volcanic slopes, generally following gullies and stream valleys. Some lahars are triggered when large volumes of ice and snow melt during an eruption. Others are generated when heavy rainfall saturates weathered volcanic deposits. Thus, lahars can occur even when a volcano is *not* erupting.

When Mount St. Helens erupted in 1980, several lahars formed. These flows and accompanying floodwaters raced down the valleys of the north and south forks of the Toutle River at speeds exceeding 30 kilometers per hour. Water levels in the river rose to 4 meters (13 feet) above flood stage, destroying or severely damaging nearly all the homes and bridges along the impacted area (Figure 9.16). Fortunately, the area was not densely populated.

In 1985 deadly lahars were produced during a small eruption of Nevado del Ruiz, a 5300-meter (17,400-foot) volcano in the Andes Mountains of Colombia. Hot pyroclastic material melted ice and snow that capped the mountain (Nevado means *snow* in Spanish) and sent torrents of ash and debris down three major river valleys that flank the volcano. Reach-

▼ **FIGURE 9.14** Nuée ardente races down the slope of Mount St. Helens on August 7, 1980, at speeds in excess of 100 kilometers (60 miles) per hour. (Photo by Peter W. Lipman, U.S. Geological Survey)

▲ **FIGURE 9.15**   St. Pierre as it appeared shortly after the eruption of Mount Pelée, 1902. (Reproduced from the collection of the Library of Congress)

ing speeds of 100 kilometers (60 miles) per hour, these mudflows tragically took 25,000 lives.

Mount Rainier, Washington, is considered by many to be America's most dangerous volcano because, like Nevado del Ruiz, it has a thick year-round mantle of snow and ice. Adding to the risk is the fact that 100,000 people live in the valleys around Rainier, and many homes are built on lahars that flowed down the volcano hundreds or thousands of years ago. A future eruption, or perhaps just a period of heavy rainfall, may produce lahars that will likely take similar paths.

STUDENTS SOMETIMES ASK...
If volcanoes are so dangerous, why do people live on or near them?

Realize that many who live near volcanoes did not choose the location; they were simply born there. Their ancestors may have lived in the region for generations. Historically, many have been drawn to volcanic regions because of their fertile soils. Not all volcanoes have explosive eruptions, but all active volcanoes are dangerous. Certainly, choosing to live close to an active composite cone like Mount St. Helens or Italy's Mount Vesuvius has a high inherent risk. However, the time interval between successive eruptions might be several decades or more—plenty of time for generations of people to forget the last eruption and consider the volcano to be dormant (*dormin* = to sleep) and therefore safe. Many people that choose to live near an active volcano have the belief that the *relative* risk is no higher than in other hazard-prone places. In essence, they are gambling that they will be able to live out their lives before the next major eruption.

Other Volcanic Landforms

The most obvious volcanic structure is a cone. But other distinctive and important landforms are also associated with volcanic activity.

Calderas

Calderas (*caldaria* = a cooking pot) are large collapse depressions having a more or less circular form. Their diameters exceed 1 kilometer, and many are tens of kilometers across. (Those less than a kilometer across are called *collapse pits*.) Most calderas are formed by one of the following processes: 1) the collapse of the summit

▼ **FIGURE 9.16**   A house damaged by a lahar along the Toutle River, west-northwest of Mount St. Helens. The end section of the house was torn free and lodged against trees. (Photo by D. R. Crandell, U.S. Geological Survey)

BOX 9.1 PEOPLE AND THE ENVIRONMENT:
Eruption of Vesuvius A.D. 79

In addition to producing some of the most violent volcanic activity, composite cones can erupt unexpectedly. One of the best documented of these events was the A.D. 79 eruption of the Italian volcano we now call Vesuvius. Prior to this eruption, Vesuvius had been dormant for centuries and had vineyards adorning its sunny slopes. On August 24, however, the tranquility ended, and in less than 24 hours the city of Pompeii (near Naples) and more than 2000 of its 20,000 residents perished. Most were entombed beneath a layer of pumice nearly 3 meters (10 feet) thick. They remained this way for nearly 17 centuries, until the city was partially excavated, giving archaeologists a superbly detailed picture of ancient Roman life.

By reconciling historical records with detailed scientific studies of the region, volcanologists have pieced together the chronology of the destruction of Pompeii. The eruption most likely began as steam discharges on the morning of August 24. By early afternoon fine ash and pumice fragments formed a tall eruptive cloud emanating from Vesuvius. Shortly thereafter, debris from this cloud began to shower Pompeii, located 9 kilometers (6 miles) downwind of the volcano. Undoubtedly, many people fled during this early phase of the eruption. For the next several hours pumice fragments as large as 5 centimeters (2 inches) fell on Pompeii. One historical record of this eruption states that people located more distant than Pompeii tied pillows to their heads in order to fend off the flying fragments.

The pumice fall continued for several hours, accumulating at the rate of 12 to 15 centimeters (5 to 6 inches) per hour. Most of the roofs in Pompeii eventually gave way. Despite the accumulation of more than 2 meters of pumice, many of the people that had not evacuated Pompeii were probably still alive the morning of August 25. Then suddenly and unexpectedly a surge of searing hot dust and gas swept rapidly down the flanks of Vesuvius. This blast killed an estimated 2000 people who had somehow managed to survive the pumice fall. Some may have been killed by flying debris, but most died of suffocation as a result of inhaling ash-laden gases. Their remains were quickly buried by the falling ash, which rain cemented into a hard mass before their bodies had time to decay. The subsequent decomposition of the bodies produced cavities in the hardened ash that replicated the form of the entombed bodies, preserving in some cases even facial expressions. Nine-teenth-century excavators found these cavities and created casts of the corpses by pouring plaster of Paris into the voids (Figure 9.A). Some of the plaster casts show victims trying to cover their mouth in an effort to draw what would be their last breath.

Volcanologists now realize that several destructive flows of hot, asphyxiating ash-laden gas swept over the countryside surrounding Vesuvius. Skeletons excavated from the nearby town of Herculaneum indicate that most of its inhabitants were probably killed by these flows. Further, many of those who evacuated Pompeii probably met a similar fate. An estimated 16,000 people may have perished in this tragic and unexpected event.

Figure 9.A Plaster casts of several victims of the A.D. 79 eruption of Mount Vesuvius that destroyed the Italian city of Pompeii.(Photo by Leonard von Matt/Photo Researchers, Inc.)

of a large composite volcano following an explosive eruption of silica-rich pumice and ash fragments; 2) the collapse of the top of a shield volcano caused by subterranean drainage from a central magma chamber; and 3) the collapse of a large area caused by the discharge of colossal volumes of silica-rich pumice and ash along ring fractures.

Crater Lake–Type Calderas. Crater Lake, Oregon, is located in a caldera that has a maximum diameter of 10 kilometers (6 miles) and is 1175 meters (over 3800 feet) deep. This caldera formed about 7000 years ago when a composite cone, later named Mount Mazama, violently extruded 50 to 70 cubic kilometers of pyroclastic material (Figure 9.17). With the loss of support, 1500 meters (near-

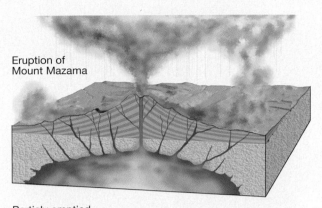

Eruption of
Mount Mazama

Partialy emptied
magma chamber

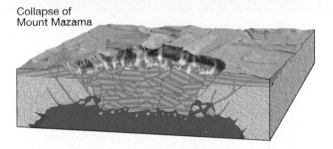

Collapse of
Mount Mazama

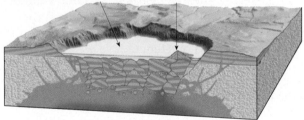

Formation of Crater Lake and Wizard Island

▲ **FIGURE 9.17** Sequence of events that formed Crater Lake, Oregon. About 7000 years ago, the summit of former Mount Mazama collapsed following a violent eruption that partly emptied the magma chamber. Subsequent eruptions produced the cinder cone called Wizard Island. Rainfall and groundwater contributed to form the lake. (After H. Williams, *The Ancient Volcanoes of Oregon*, p. 47. Courtesy of the University of Oregon)

ly a mile) of the summit of this once prominent cone collapsed. After the collapse, rainwater filled the caldera (Figure 9.18). Later volcanic activity built a small cinder cone in the lake. Today this cone, called Wizard Island, provides a mute reminder of past activity.

Hawaiian-Type Calderas. Although most calderas are produced by *collapse following an explosive eruption*, some are not. For example, Hawaii's active shield vol-

canoes, Mauna Loa and Kilauea, both have large calderas at their summits. Kilauea's measures 3.3 by 4.4 kilometers (about 2 by 3 miles) and is 150 meters (500 feet) deep. Each caldera formed by gradual subsidence of the summit as magma slowly drained laterally from the central magma chamber to a rift zone, often producing flank eruptions.

Yellowstone-Type Calderas. Although the 1980 eruption of Mount St. Helens was spectacular, it pales by comparison to what happened 630,000 years ago in the region now occupied by Yellowstone National Park. Here approximately 1000 cubic kilometers of pyroclastic material erupted, eventually producing a caldera 70 kilometers (43 miles) across. This event produced showers of ash as far away as the Gulf of Mexico. Vestiges of this activity are the many hot springs and geysers in the region.

The formation of a large Yellowstone-type caldera begins when a silica-rich (rhyolitic) magma body is emplaced near the surface, upwarping the overlying rocks. Next, ring fractures develop in the roof, providing a pathway to the surface for the highly gas-rich magma. This initiates an explosive eruption of colossal proportions, ejecting huge volumes (usually exceeding 100 cubic kilometers) of pyroclastic materials, mainly in the form of ash and pumice fragments. Typically, these materials form a pyroclastic flow that spreads across the landscape at speeds that may exceed 100 kilometers (60 miles) per hour, destroying most living things in its path. After coming to rest, the hot fragments of ash and pumice fuse together, forming a welded tuff that closely resembles a solidified lava flow. Finally, with the loss of support, the roof of the magma chamber collapses, generating a large caldera.

Calderas of the Yellowstone type are the largest volcanic structures on Earth. Some geologists have compared their destructive force with that of the impact of a small asteroid. Fortunately, no eruption of this type has occurred in historic times. Other examples of large calderas located in the United States are California's Long Valley Caldera and Valles Caldera located west of Los Alamos, New Mexico.

Fissure Eruptions and Lava Plateaus

We think of volcanic eruptions as building a cone or shield from a central vent. But by far the greatest volume of volcanic material is extruded from fractures in the crust called **fissures** (*fissura* = to split). Rather than building a cone, these long, narrow cracks may emit a low-viscosity basaltic lava, blanketing a wide area.

The extensive Columbia Plateau in the northwestern United States was formed this way (Figure 9.19).

▲ **FIGURE 9.18** Crater Lake in Oregon occupies a caldera about 10 kilometers (6 miles) in diameter. Wizard Island is a cinder cone located in the caldera. (Photo by Greg Vaughn/Tom Stack and Associates)

Here, numerous **fissure eruptions** extruded very fluid basaltic lava (Figure 9.20). Successive flows, some 50 meters (160 feet) thick, buried the existing landscape as they built a lava plateau nearly a mile thick. The fluid nature of the lava is evident, because some remained molten long enough to flow 150 kilometers (90 miles) from its source. The term **flood basalts** appropriately describes these flows. Massive accumulations of basaltic lava, similar to those of the Columbia Plateau, occur worldwide. One of the largest is the Deccan Traps, a thick sequence of flat-lying basalt flows covering nearly 500,000 square kilometers (195,000 square miles) of west central India. When the Deccan Traps formed about 66 million years ago, nearly 2 million cubic kilometers of lava were extruded in less than 1 million years. Another huge deposit of flood basalts, called the Ontong Java Plateau, is found on the floor of the Pacific Ocean.

▼ **FIGURE 9.19** Volcanic areas in the northwestern United States. The Columbia River basalts cover an area of nearly 200,000 square kilometers (80,000 square miles). Activity here began about 17 million years ago as lava poured out of large fissures, eventually producing a basalt plateau with an average thickness of more than 1 kilometer. (After U.S. Geological Survey)

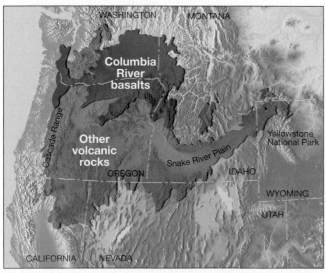

Volcanic Pipes and Necks

Most volcanoes are fed magma through short conduits, called *pipes*, that connect a magma chamber to the surface. In rare circumstances, pipes may extend tubelike to depths exceeding 200 kilometers (125 miles). When this occurs, the ultramafic magmas that migrate up these structures produce rocks that are thought to be samples of the mantle that have undergone very little alteration during their ascent. Geologists consider these unusually deep conduits to be "windows" into Earth, for they allow us to view rock normally found only at great depth.

The best-known volcanic pipes are the diamond-bearing structures of South Africa. Here, the rocks filling the pipes originated at depths of at least 150 kilometers

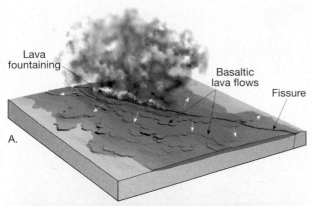

▲ FIGURE 9.20 Basaltic fissure eruption. **A.** Lava fountaining from a fissure forming fluid lava flows called flood basalts. **B.** Photo of basalt flows (dark) near Idaho Falls. (Photo by John S. Shelton)

rock is exceptional. This fact accounts for the scarcity of natural diamonds.

Volcanoes on land are continually being lowered by weathering and erosion. Cinder cones are easily eroded, because they are composed of unconsolidated materials. However, all volcanoes will eventually succumb to relentless erosion over geologic time. As erosion progresses, the rock occupying the volcanic pipe is often more resistant and may remain standing above the surrounding terrain long after most of the cone has vanished. Ship Rock, New Mexico, is such a feature and is called a **volcanic neck** (Figure 9.21). This structure, higher than many skyscrapers, is but one of many such landforms that protrude conspicuously from the red desert landscapes of the American Southwest.

Intrusive Igneous Activity

GEODe
Forces Within
▼ Igneous Activity

Although volcanic eruptions can be among the most violent and spectacular events in nature and therefore worthy of detailed study, most magma is emplaced at depth. Thus, an understanding of intrusive igneous activity is as important to geologists as the study of volcanic events.

The structures that result from the emplacement of igneous material at depth are called **plutons**, named for Pluto, the god of the lower world in classical mythology. Because all plutons form out of view beneath Earth's surface, they can be studied only after uplifting and erosion have exposed them. The challenge lies in reconstructing the events that generated these structures millions or even hundreds of millions of years ago.

Plutons are known to occur in a great variety of sizes and shapes. Some of the most common types are illustrated in Figure 9.22. Notice that some of these structures have a tabular (tabletop) shape,

(90 miles), where pressure is high enough to generate diamonds and other high-pressure minerals. The task of transporting essentially unaltered magma (along with diamond inclusions) through 150 kilometers of solid

▶ FIGURE 9.21 Ship Rock, New Mexico, is a volcanic neck. This structure, which stands over 420 meters (1380 feet) high, consists of igneous rock that crystallized in the vent of a volcano that has long since been eroded away. The tabular structure in the background is a dike that served to feed lava flows along the flanks of the once active volcano. (Photo by Tom Bean/DRK Photo)

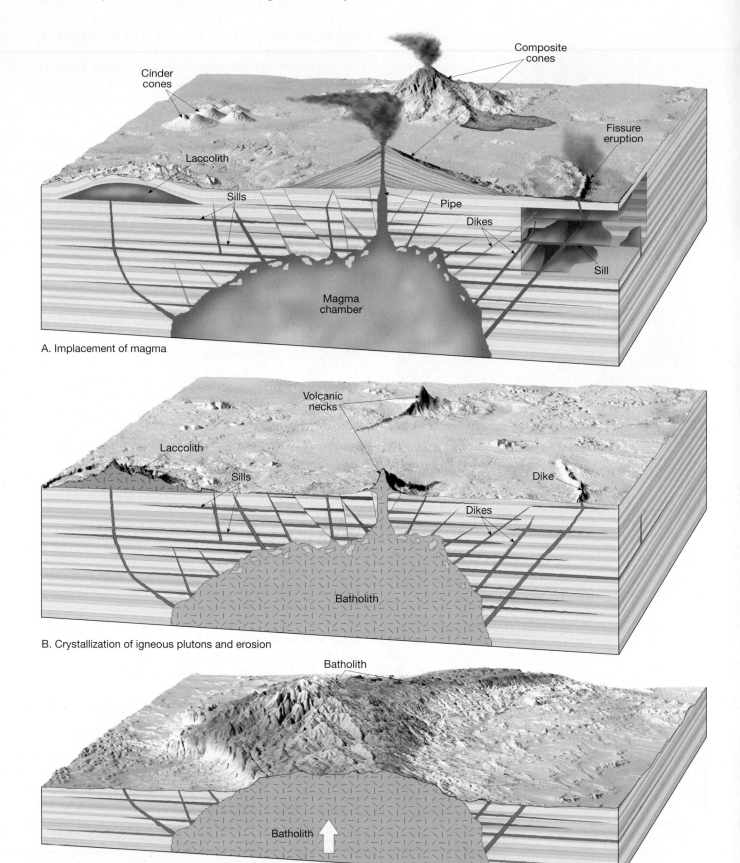

A. Implacement of magma

B. Crystallization of igneous plutons and erosion

C. Extensive uplift and erosion exposes batholith

▲ **FIGURE 9.22**   Illustrations showing basic igneous plutons. **A.** This block diagram shows the relationship between volcanism and intrusive igneous activity. **B.** This view illustrates the basic intrusive igneous structures, some of which have been exposed by erosion long after their formation. **C.** After millions of years of uplifting and erosion a batholith is exposed at the surface.

whereas others are quite massive. Also, observe that some of these bodies cut across existing structures, such as layers of sedimentary rock; others form when magma is injected between sedimentary layers. Because of these differences, intrusive igneous bodies are generally classified according to their shape as either *tabular* (*tabula* = table) or *massive* and by their orientation with respect to the host rock. Plutons are said to be *discordant* (*discordare* = to *disagree*) if they cut across existing structures and *concordant* (*concordare* = to agree) if they form parallel to features such as sedimentary strata. As you can see in Figure 9.22A, plutons are closely associated with volcanic activity.

Dikes

Dikes are tabular discordant bodies that are produced when magma is injected into fractures. The force exerted by the emplaced magma can be great enough to separate the walls of the fracture further. Once crystallized, these sheetlike structures have thicknesses ranging from less than a centimeter to more than a kilometer. The largest have lengths of hundreds of kilometers. Most dikes, however, are a few meters thick and extend laterally for no more than a few kilometers.

Dikes are often found in groups that once served as vertically oriented pathways followed by molten rock that fed ancient lava flows. The parent pluton is generally not observable. Some dikes are found radiating, like spokes on a wheel, from an eroded volcanic neck. In these situations the active ascent of magma is thought to have generated fissures in the volcanic cone out of which lava flowed.

Sills and Laccoliths

Sills and laccoliths are concordant plutons that form when magma is intruded in a near-surface environment. They differ in shape and usually differ in composition.

Sills. **Sills** are tabular plutons formed when magma is injected along sedimentary bedding surfaces (Figure 9.23). Horizontal sills are the most common, although all orientations, even vertical, are known to exist. Because of their relatively uniform thickness and large areal extent, sills are likely the product of very fluid magmas. Magmas having a low silica content are more fluid, so most sills are composed of the rock basalt.

The emplacement of a sill requires that the overlying sedimentary rock be lifted to a height equal to the thickness of the sill. Although this is a formidable task, in shallow environments it often requires less energy than forcing the magma up the remaining

▲ **FIGURE 9.23** Salt River Canyon, Arizona. The dark, essentially horizontal band is a sill of basaltic composition that intruded into horizontal layers of sedimentary rock. (Photo by E. J. Tarbuck)

distance to the surface. Consequently, sills form only at shallow depths, where the pressure exerted by the weight of overlying rock layers is low. Although sills are intruded between layers, they can be locally discordant. Large sills frequently cut across sedimentary layers and resume their concordant nature at a higher level.

One of the largest and most studied of all sills in the United States is the Palisades Sill. Exposed for 80 kilometers (50 miles) along the west bank of the Hudson River in southeastern New York and northeastern New Jersey, this sill is about 300 meters (nearly 1000 feet) thick. Because of its resistant nature, the Palisades Sill forms an imposing cliff that can be seen easily from the opposite side of the Hudson.

In many respects, sills closely resemble buried lava flows. Both are tabular and often exhibit columnar jointing (Figure 9.24). **Columnar joints** form as igneous rocks cool and develop shrinkage fractures that produce elongated, pillarlike columns. Further, because sills generally form in near-surface environments and may be only a few meters thick, the emplaced magma often cools quickly enough to generate a fine-grained texture.

Laccoliths. **Laccoliths** are similar to sills because they form when magma is intruded between sedimentary layers in a near-surface environment. However, the magma that generates laccoliths is more viscous. This less fluid magma collects as a lens-shaped mass that arches the overlying strata upward (Figure 9.22). Consequently, a laccolith can occasionally be detected because of the dome-shaped bulge it creates at the surface.

Most large laccoliths are probably not much wider than a few kilometers. The Henry Mountains in southeastern Utah are largely composed of several laccoliths believed to have been fed by a much larger magma body emplaced nearby.

269

▲ **FIGURE 9.24** Columnar jointing in basalt, Giants Causeway National Park, Northern Ireland. These five-to-seven-sided columns are produced by contraction and fracturing that results as a lava flow or sill gradually cools. (Photo by Tom Till)

Batholiths

By far the largest intrusive igneous bodies are **batholiths** (*bathos* = depth, *lithos* = *stone*). Most often, batholiths occur in groups that form linear structures several hundreds of kilometers long and up to 100 kilometers (62 miles) wide. The Idaho batholith, for example, encompasses an area of more than 40,000 square kilometers (15,360 square miles) and consists of many plutons. Indirect evidence gathered from gravitational studies indicates that batholiths are also very thick, possibly extending dozens of kilometers into the crust. Based on the amount exposed by erosion, some batholiths are at least several kilometers thick.

By definition, a plutonic body must have a surface exposure greater than 100 square kilometers (40 square miles) to be considered a batholith. Smaller plutons of this type are termed *stocks*. Many stocks appear to be portions of batholiths that are not yet fully exposed.

Batholiths may compose the core of mountain systems. Here uplifting and erosion have removed the surrounding rock, thereby exposing the resistant igneous body (Figure 9.25). Some of the highest peaks in the Sierra Nevada, such as Mount Whitney, are carved from such a granitic mass.

Large expanses of granitic rock also occur in the stable interiors of the continents, such as the Canadian Shield of North America. These relatively flat exposures are the remains of ancient mountains that have long since been leveled by erosion. Thus, the rocks that make up the batholiths of youthful mountain ranges, such as the Sierra Nevada, were generated near the top of a magma chamber, whereas in shield areas, the roots of former mountains and, thus, the lower portions of batholiths, are exposed. In Chapter 10 we will further consider the role of igneous activity as it relates to mountain building.

Origin of Magma

The origin of magma has been controversial in geology, almost from the beginning of the science. How do magmas of different compositions form? Why do volcanoes in the deep-ocean basins primarily extrude basaltic lava, whereas those adjacent to oceanic trenches extrude mainly andesitic lava? These are some of the questions we will address in the following sections.

◀ **FIGURE 9.25** Half Dome in Yosemite National Park, California, is a part of the Sierra Nevada Batholith. (Photo by Charles Gurche)

Generating Magma from Solid Rock

Based on available scientific evidence, *Earth's crust and mantle are composed primarily of solid, not molten, rock*. Although the outer core is a fluid, its iron-rich material is very dense and remains deep within Earth. So what is the source of magma that produces igneous activity?

Geologists conclude that magma originates when essentially solid rock, located in the crust and upper mantle, melts. The most obvious way to generate magma from solid rock is to raise the temperature above the level at which the rock begins to melt.

Role of Heat. What source of heat is sufficient to melt rock? Workers in underground mines know that temperatures get higher as they go deeper. Although the rate of temperature change varies from place to place, it *averages* between 20°C and 30°C per kilometer in the *upper* crust. This change in temperature with depth is known as the **geothermal gradient**. Estimates indicate that the temperature at a depth of 100 kilometers (62 miles) ranges between 1200°C and 1400°C. At these high temperatures, rocks in the lower crust and upper mantle are near but somewhat cooler than their melting point temperatures. Thus, they are very hot but still essentially solid.

One source of heat to melt crustal rocks is basaltic magma that originates in the mantle. As basaltic magma buoyantly rises, it often "ponds" beneath crustal rocks, which have a lower density and are already near their melting temperatures. This results in the partial melting of crustal rocks and the formation of a secondary, silica-rich magma. But what is the source of heat to melt mantle rock? As you will learn, the vast bulk of mantle-derived magma forms without the aid of an additional heat source.

Role of Pressure. If temperature were the only factor that determined whether or not rock melts, our planet would be a molten ball covered with a thin, solid outer shell. This, of course, is not the case. The reason is that pressure also increases with depth.

Melting, which is accompanied by an increase in volume, *occurs at higher temperatures at depth* because of greater confining pressure. Consequently, an increase in confining pressure causes an increase in the rock's melting temperature. Conversely, reducing confining pressure lowers a rock's melting temperature. When confining pressure drops enough, **decompression melting** is triggered. This may occur when mantle rock *ascends* as a result of convective upwelling, thereby moving into zones of lower pressure. Decompression melting is responsible for generating magma

271

▶ **FIGURE 9.26** As hot mantle rock ascends, it continually moves into zones of lower pressure. This drop in confining pressure can trigger melting, even without additional heat.

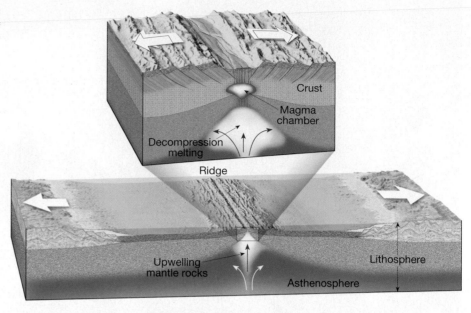

along oceanic ridges where plates are rifting apart (Figure 9.26). Here, below the ridge crest, hot mantle rock rises upward to replace the material that has shifted horizontally.

Role of Volatiles. Another important factor affecting the melting temperature of rock is its water content. Water and other volatiles act as salt does to melt ice. That is, volatiles cause rock to melt at lower temperatures. Further, the effect of volatiles is magnified by increased pressure. Consequently, "wet" rock buried at depth has a much lower melting temperature than does "dry" rock of the same composition and under the same confining pressure. Therefore, in addition to a rock's composition, its temperature, depth (confining pressure), and water content determine whether it exists as a solid or liquid.

Volatiles play an important role in generating magma in regions where cool slabs of oceanic lithosphere descend into the mantle (Figure 9.27). As an oceanic plate sinks, both heat and pressure drive water from the subducting crustal rocks. These volatiles, which are very mobile, migrate into the wedge of hot mantle that lies above. This process is believed to lower the melting temperature of mantle rock sufficiently to generate some melt. Laboratory studies have shown that the melting point of basalt can be lowered by as much as 100°C by the addition of only 0.1 percent water. When enough mantle-derived basaltic magma forms, it will buoyantly rise toward the surface.

In summary, magma can be generated under three sets of conditions: (1) *heat* may be added; for example, a magma body from a deeper source intrudes and melts crustal rock; (2) *a decrease in pressure* (without the

▶ **FIGURE 9.27** As an oceanic plate descends into the mantle, water and other volatiles are driven from the subducting crustal rocks. These volatiles lower the melting temperature of mantle rock sufficiently to generate magma.

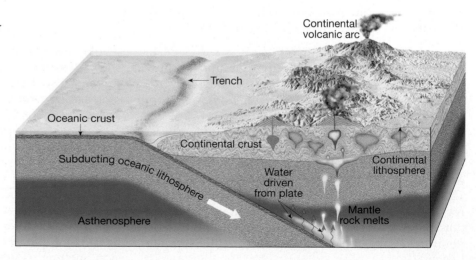

addition of heat) can result in *decompression melting*; and (3) the *introduction of volatiles* (principally water) can lower the melting temperature of mantle rock sufficiently to generate magma.

Partial Melting and Magma Compositions

An important difference exists between the melting of a substance that consists of a single compound, such as ice, and melting igneous rocks, which are mixtures of several different minerals. Ice melts at a specific temperature, whereas igneous rocks melt over a temperature range of about 200°C. As rock is heated, minerals with the lowest melting points tend to start melting first. Should melting continue, minerals with higher melting points begin to melt, and the composition of the magma steadily approaches the overall composition of the rock from which it was derived. Most often, melting is not complete. This process, known as **partial melting**, produces most, if not all, magma.

An important consequence of partial melting is the *production of a magma with a higher silica content than the original rock*. Recall from the discussion of Bowen's reaction series that basaltic (mafic) rocks contain mostly high-melting-temperature minerals that are comparatively low in silica, whereas granitic (felsic) rocks are composed primarily of low-melting-temperature silicates that are enriched in silica (see Chapter 3). Because silica-rich minerals melt first, magmas generated by partial melting are nearer to the granitic end of the compositional spectrum than are the rocks from which they formed (see Figure 3.7, p. 58).

Plate Tectonics and Igneous Activity

Geologists have known for decades that the global distribution of volcanism is not random. Of the more than 800 active volcanoes that have been identified, most are located along the margins of the ocean basins—most notably within the circum-Pacific belt known as the *Ring of Fire* (Figure 9.28). This group of volcanoes consists mainly of composite cones that emit volatile-rich magma having an intermediate (andesitic) composition that occasionally produce awe-inspiring eruptions.

The volcanoes comprising a second group emit very fluid basaltic lavas and are confined to the deep-ocean basins, including well-known examples on Hawaii and Iceland. In addition, this group contains many active submarine volcanoes that dot the ocean floor; particularly notable are the innumerable small seamounts that lie along the axis of the oceanic ridge.

A third group includes those volcanic structures that are irregularly distributed in the interiors of the continents. Volcanism on continents is the most diverse, ranging from eruptions of very fluid basaltic lavas, like those that generated the Columbia River basalts to explosive eruptions of silica-rich rhyolitic magma as occurred in Yellowstone.

▼ **FIGURE 9.28** Locations of some of Earth's major volcanoes. Note the concentration of volcanoes encircling the Pacific basin, known as the "Ring of Fire."

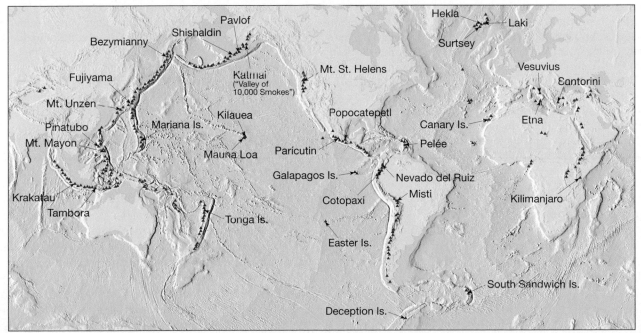

Box 9.2 Earth as a System:
Can Volcanoes Change Earth's Climate?

One example of the interplay between different parts of the Earth system is the relationship between volcanic activity and changes in climate. We know that changes in the composition of the atmosphere can have a significant impact on climate. Moreover, we know that volcanic eruptions can emit large quantities of gases and particles into the atmosphere, thus altering its composition. So do volcanic eruptions actually influence Earth's climate?

The idea that explosive volcanic eruptions might alter Earth's climate was first proposed many years ago. It is still regarded as a plausible explanation for some aspects of climatic variability. Explosive eruptions emit huge quantities of gases and fine-grained debris into the atmosphere (Figure 9.B). The greatest eruptions are sufficiently powerful to inject material high into the stratosphere (an atmospheric layer that extends between the heights of about 10 and 50 kilometers), where it spreads around the globe and remains for many months or even years.

The Basic Premise

The basic premise is that this suspended volcanic material will filter out a portion of the incoming solar radiation, which in turn will drop temperatures in the lowest layer of the atmosphere (this layer, called the *troposphere*, extends from Earth's surface to a height of about 10 kilometers).

More than 200 years ago Benjamin Franklin used this idea to argue that material from the eruption of a large Icelandic volcano could have reflected sunlight back to space and therefore might have been responsible for the unusually cold winter of 1783–1784.

Perhaps the most notable cool period linked to a volcanic event is the "year without a summer" that followed the 1815 eruption of Mount Tambora in Indonesia. The eruption of Tambora is the largest of modern times. During April 7–12, 1815, this nearly 4000-meter-high (13,000-foot) volcano violently expelled more than 100 cubic kilometers (24 cubic miles) of volcanic debris. The impact of the volcanic aerosols on climate is believed to have been widespread in the Northern Hemisphere. From May through September 1816 an unprecedented series of cold spells affected the northeastern United States and adjacent portions of Canada. There was heavy snow in June and frost in July and August. Abnormal cold was also experienced in much of Western Europe. Similar, although apparently less dramatic, effects were associated with other great explosive volcanoes, including Indonesia's Krakatau in 1883.

Three Modern Examples

Three major volcanic events have provided considerable data and insight regarding the impact of volcanoes on global temperatures. The eruptions of Washington State's Mount St. Helens in 1980, the Mexican volcano El Chichón in 1982, and the Philippines' Mount Pinatubo in 1991 have given scientists an opportunity to study the atmospheric effects of volcanic eruptions with the aid of more sophisticated technology than had been available in the past. Satellite images and remote-sensing instruments allowed scientists to monitor closely the effects of the clouds of gases and ash that these volcanoes emitted.

Mount St. Helens When Mount St. Helens erupted, there was immediate speculation about the possible effects on our climate. Could such an eruption cause our climate to change? There is no doubt that the large quantity of volcanic ash emitted by the explosive eruption had significant local and regional effects for a short period. Still, studies indicated that any longer-term lowering of hemispheric temperatures was negligible. The cooling was so slight, probably less than 0.1°C (0.2°F), that it could not be distinguished from other natural temperature fluctuations.

El Chichón Two years of monitoring and studies following the 1982 El Chichón eruption indicated that its cooling effect on global mean temperature was greater than that of Mount St. Helens, on the order of 0.3 to 0.5°C (0.5 to 0.9°F). The eruption of El Chichón was *less explosive* than the Mount St. Helens blast, so why did it have a greater impact on global temperatures? The reason is that the material emitted by Mount St. Helens was largely fine ash that settled out in a relatively short time. El Chichón, on the other hand, emitted far greater quantities of sulfur dioxide gas (an estimated 40 times more) than Mount St. Helens. This gas combines with water vapor in the stratosphere to produce a dense cloud of tiny sulfuric-acid particles. The particles, called *aerosols*, take several years to settle out completely. They lower the tropo-

---

Until the late 1960s, geologists had no explanation for the apparently haphazard distribution of continental volcanoes, nor were they able to account for the almost continuous chain of volcanoes that circles the margin of the Pacific basin. With the development of the theory of plate tectonics, the picture was greatly clarified. Recall that most magma originates in the upper mantle and that the mantle is essentially solid, *not molten* rock. The basic connection between plate tectonics and volcanism is that *plate motions provide the mechanisms by which mantle rocks melt to generate magma.*

We will examine three zones of igneous activity and their relationship to plate boundaries (Figure 9.29). These active areas are located 1) along convergent plate boundaries, where plates move toward each other and one sinks beneath the other; 2) along divergent plate boundaries, where plates move away from each other and new seafloor is created; and 3) areas within the plates proper that are not associated with any plate boundary.

sphere's mean temperature because they reflect solar radiation back to space.

We now understand that volcanic clouds that remain in the stratosphere for a year or more are composed largely of sulfuric-acid droplets and not of dust, as was once thought. Thus, the volume of fine debris emitted during an explosive event is not an accurate criterion for predicting the global atmosphere effects of an eruption.

Mount Pinatubo The Philippines volcano, Mount Pinatubo, erupted explosively in June 1991, injecting 25 million to 30 million tons of sulfur dioxide into the stratosphere. The event provided scientists with an opportunity to study the climatic impact of a major explosive volcanic eruption using NASA's spaceborne Earth Radiation Budget Experiment. During the next year the haze of tiny aerosols increased the percentage of light reflected by the atmosphere and thus lowered global temperatures by 0.5° C (0.9°F).

It may be true that the impact on global temperature of eruptions like El Chichón and Mount Pinatubo is relatively minor, but many scientists agree that the cooling produced could alter the general pattern of atmospheric circulation for a limited period. Such a change, in turn, could influence the weather in some regions. Predicting or even identifying specific regional effects still presents a considerable challenge to atmospheric scientists.

The preceding examples illustrate that the impact on climate of a single volcanic eruption, no matter how great, is relatively small and short-lived. Therefore, if volcanism is to have a pronounced impact over an extended period, many great eruptions, closely spaced in time, need to occur. If this happens, the stratosphere could become loaded with enough sulfur dioxide and volcanic dust to seriously diminish the amount of solar radiation reaching the surface.

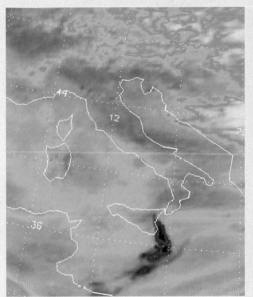 

Figure 9.B Mount Etna, a volcano on the island of Sicily, erupting in late October, 2002. Mount Etna is Eurorpe's largest and most active volcano. **Left.** This image from the Atmospheric Infrared Sounder on NASA's *Aqua* satellite shows the sulfur dioxide (SO$_2$) plume in shades of purple and black. **Right.** This photo of Mount Etna looking southeast was taken by a crew member aboard the International Space Station. It shows a plume of volcanic ash streaming southeastward from the volcano. (Images courtesy of NASA)

Igneous Activity at Convergent Plate Boundaries

Recall that at convergent plate boundaries, slabs of occanic crust are bent as they descend into the mantle, generating an oceanic trench. As a slab sinks deeper into the mantle, the increase in temperature and pressure drives volatiles (mostly H$_2$O) from the oceanic crust. These mobile fluids migrate upward into the wedge-shaped piece of mantle located between the subducting slab and overriding plate (see Figure 9.27). Once the sinking slab reaches a depth of about 100 to 150 kilometers, these water-rich fluids reduce the melting point of hot mantle rock sufficiently to trigger some melting. The partial melting of mantle rock generates magma with a basaltic composition. After a sufficient quantity of magma has accumulated, it slowly migrates upward.

Volcanism at a convergent plate margin results in the development of a linear or slightly curved chain of

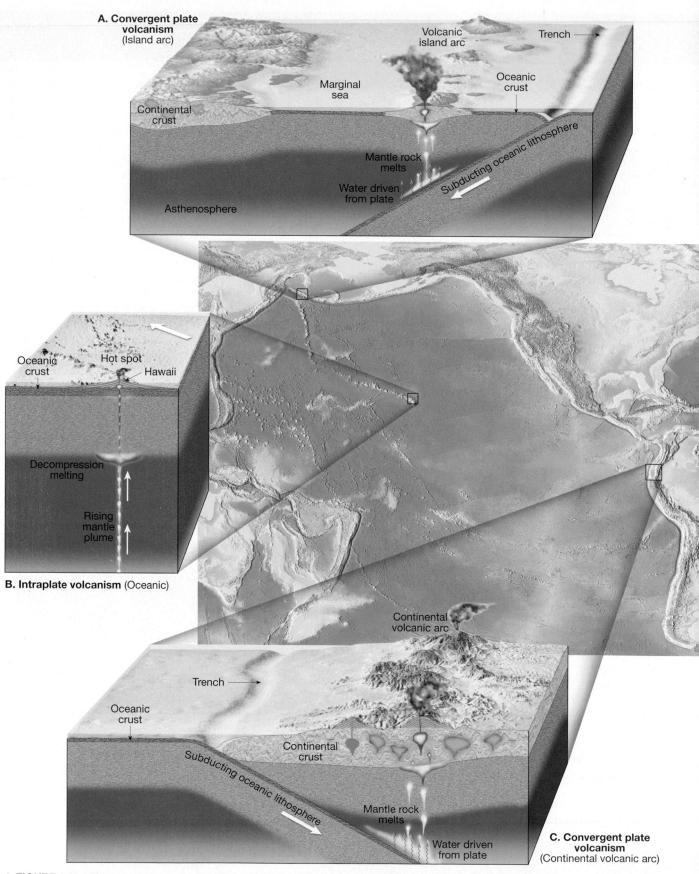

A. Convergent plate volcanism (Island arc)

Trench

Volcanic island arc

Marginal sea

Oceanic crust

Continental crust

Mantle rock melts

Water driven from plate

Subducting oceanic lithosphere

Asthenosphere

B. Intraplate volcanism (Oceanic)

Oceanic crust

Hot spot

Hawaii

Decompression melting

Rising mantle plume

Continental volcanic arc

Trench

Oceanic crust

Continental crust

Subducting oceanic lithosphere

Mantle rock melts

Water driven from plate

C. Convergent plate volcanism (Continental volcanic arc)

▲ **FIGURE 9.29** Three zones of volcanism. Two of these zones are plate boundaries, and the third is the interior area of the plates.

D. Divergent plate volcanism
(Oceanic ridge)

Oceanic crust

Magma chamber

Decompression melting

Asthenosphere

E. Intraplate volcanism
(Continental)

Flood basalts

Hot spot

Continental crust

Decompression melting

Rising mantle plume

Rift valley

Continental crust

Decompression melting

F. Divergent plate volcanism
(Continental rifting)

volcanoes called a *volcanic arc*. These volcanic chains develop roughly parallel to the associated trench—at distances of 200 to 300 kilometers (100 to 200 miles). Volcanic arcs can be constructed on oceanic, or continental, lithosphere. Those that develop within the ocean and grow large enough for their tops to rise above the surface are labeled *island archipelagos* in most atlases. Geologists prefer the more descriptive term **volcanic island arcs**, or simply **island arcs** (Figure 9.29, upper left). Several young volcanic island arcs of this type border the western Pacific basin, including the Aleutians, the Tongas, and the Marianas.

Volcanism associated with convergent plate boundaries may also develop where slabs at oceanic lithosphere are subducted under continental lithosphere to produce a **continental volcanic arc** (Figure 9.29, lower left). The mechanisms that generate these magmas are essentially the same as those operating at island arcs. The major difference is that continental crust is much thicker and is composed of rocks having a higher silica content than oceanic crust. Hence through the assimilation of silica-rich crustal rocks a magma body may change composition as it rises through continental crust. Stated another way, magmas generated in the mantle may change from a comparatively dry, fluid basaltic magma to a viscous andesitic or rhyolitic magma having a high concentration of volatiles as it moves up through the continental crust. The volcanic chain of the Andes Mountains along the western edge of South America is perhaps the best example of a mature continental volcanic arc.

Since the Pacific basin is essentially bordered by convergent plate boundaries (and associated subduction zones), it is easy to see why the irregular belt of explosive volcanoes we call the Ring of Fire formed in this region. The volcanoes of the Cascade Range in the northwestern United States, including Mount Hood, Mount Rainier, and Mount Shasta, are included in this group (Figure 9.30).

Igneous Activity at Divergent Plate Boundaries

The greatest volume of magma (perhaps 60 percent of Earth's total yearly output) is produced along the oceanic ridge system in association with seafloor spreading (Figure 9.29, upper right). Here, below the ridge axis where the lithospheric plates are being continually pulled apart, the solid yet mobile mantle responds to the decrease in overburden and rises upward to fill in the rift. Recall that as rock rises, it experiences a decrease in confining pressure and undergoes melting without the addition of heat. This process, called *decompression melting*, generates large quantities of magma.

Partial melting of mantle rock at spreading centers produces basaltic magma having a composition that is surprisingly similar to that generated along convergent plate boundaries. Because this newly formed

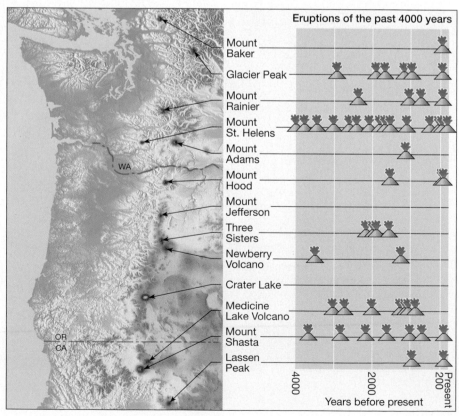

► **FIGURE 9.30** Of the 13 potentially active volcanoes in the Cascade Range, 11 have erupted in the past 4000 years and 7 in just the past 200 years. More than 100 eruptions, most of which were explosive, have occurred in the past 4000 years. Mount St. Helens is the most active volcano in the Cascades. Its eruptions have ranged from relatively quiet outflows of lava to explosive events much larger than that of May 18, 1980. Each eruption symbol in the diagram represents from one to several dozen eruptions closely spaced in time. (After U.S. Geological Survey)

basaltic magma is less dense than the mantle rock from which it was derived, it buoyantly rises.

Collecting in reservoirs located just beneath the ridge crest, about 10 percent of this magma eventually migrates upward along fissures to erupt as flows on the ocean floor. This activity continuously adds new basaltic rock to the plate margins, temporarily welding them together, only to break again as spreading continues. Along some ridges, outpouring of bulbous pillow lavas build numerous small seamounts. At other locations erupted lavas produce fluid flows that create more subdued topography.

Although most spreading centers are located along the axis of an oceanic ridge, some are not. In particular, the East African Rift is a site where continental crust is being ripped apart (Figure 9.29, lower right). Vast outpourings of fluid basaltic lavas are common in this region. The East African Rift zone also contains numerous small volcanoes and even a few large composite cones, as exemplified by Mount Kilimanjaro.

Intraplate Igneous Activity

We know why igneous activity is initiated along plate boundaries, but why do eruptions occur in the interiors of plates? Hawaii's Kilauea is considered the world's most active volcano, yet it is situated thousands of kilometers from the nearest plate boundary in the middle of the vast Pacific plate. Other sites of **intraplate volcanism** (meaning "within the plate") include the Canary Islands, Yellowstone, and several volcanic centers that you may be surprised to learn are located in the Sahara Desert of northern Africa.

We now recognize that most intraplate volcanism occurs where a mass of hotter than normal mantle material called a **mantle plume** ascends toward the surface (Figure 9.29, left). Although the depth at which (at least some) mantle plumes originate is still hotly debated, many appear to form deep within Earth at the core–mantle boundary. These plumes of solid yet mobile mantle rock rise toward the surface in a manner similar to the blobs that form within a lava lamp. (These are the gaudy lamps that contain two immiscible liquids in a glass container. As the base of the lamp is heated, the denser liquid at the bottom becomes buoyant and forms blobs that rise to the top.) Like the blobs in a lava lamp, a mantle plume has a bulbous head that draws out a narrow stalk beneath it as it rises. Once the plume head nears the top of the mantle, decompression melting generates basaltic magma that may eventually trigger volcanism at the surface. The result is a localized volcanic region a few hundred kilometers across called a **hot spot** (Figure 9.29 right). More than 100 hot spots have been identified, and most have persisted for millions of years. The land surface around hot spots is often elevated, showing that it is buoyed up by a plume of warm low-density material. Furthermore, by measuring the heat flow in these regions, geologists have determined that the mantle beneath hot spots must be 100–150°C hotter than normal.

The volcanic activity on the island of Hawaii, with its outpourings of basaltic lava, is certainly the result of hot-spot volcanism (Figure 9.29, left). Where a mantle plume has persisted for long periods of time, a chain of volcanic structures may form as the overlying plate moves over it. In the Hawaiian Islands, hot-spot activity is currently centered on Kilauea. However, over the past 80 million years the same mantle plume generated a chain of volcanic islands (and seamounts) that extend thousands of kilometers from the Big Island in a northwestward direction across the Pacific.

Mantle plumes are also thought to be responsible for the vast outpourings of basaltic lava that create large basalt plateaus such as the Columbia Plateau in the northwestern United States, India's Deccan Plateau, and the Ontong Java Plateau in the western Pacific (Figure 9.29, right).

Although the plate tectonics theory has answered many questions regarding the distribution of igneous activity, many new questions have arisen: Why does seafloor spreading occur in some areas but not others? How do mantle plumes and associated hot spots originate? These and other questions are the subject of continuing geologic research.

STUDENTS SOMETIMES ASK...

Some of the larger volcanic eruptions, like the eruption of Krakatau, must have been impressive. What was it like?

On August 27, 1883, in what is now Indonesia, the volcanic island of Krakatau exploded and was nearly obliterated. The sound of the explosion was heard an incredible 4800 kilometers (3000 miles) away at Rodriguez Island in the western Indian Ocean. Dust from the explosion was propelled into the atmosphere and circled Earth on high-altitude winds. This dust produced unusual and beautiful sunsets for nearly a year.

Not many were killed directly by the explosion, because the island was uninhabited. However, the displacement of water from the energy released during the explosion was enormous. The resulting *seismic sea wave* or *tsunami* exceeded 35 meters (116 feet) in height. It devastated the coastal region of the Sunda Strait between the nearby islands of Sumatra and Java, drowning over 1000 villages and taking more than 36,000 lives. The energy carried by this wave reached every ocean basin and was detected by tide-recording stations as far away as London and San Francisco.

Chapter Summary

• The primary factors that determine the nature of volcanic eruptions include the magma's *temperature,* its *composition,* and the *amount of dissolved gases* it contains. As lava cools, it begins to congeal, and as *viscosity* increases, its mobility decreases. *The viscosity of magma is directly related to its silica content.* Rhyolitic lava, with its high silica content, is very viscous and forms short, thick flows. *Basaltic* lava, with a lower silica content, is more fluid and may travel a long distance before congealing. Dissolved gases provide the force that propels molten rock from the vent of a volcano.

• The materials associated with a volcanic eruption include *lava flows* (*pahoehoe* and *aa* flows for basaltic lavas), *gases* (primarily in the form of *water vapor*), and *pyroclastic material* (pulverized rock and lava fragments blown from the volcano's vent, which include *ash, pumice, lapilli, cinders, blocks,* and *bombs*).

• Successive eruptions of lava from a central vent result in a mountainous accumulation of material known as a *volcano.* Located at the summit of many volcanoes is a steep-walled depression called a *crater. Shield cones* are broad, slightly domed volcanoes built primarily of fluid, basaltic lava. *Cinder cones* have steep slopes composed of pyroclastic material. *Composite cones,* or *stratovolcanoes,* are large, nearly symmetrical structures built of interbedded lavas and pyroclastic deposits. Composite cones produce some of the most violent volcanic activity. Often associated with a violent eruption is a *nuée ardente,* a fiery cloud of hot gases infused with incandescent ash that races down steep volcanic slopes. Large composite cones may also generate a type of mudflow known as a *lahar.*

• Most volcanoes are fed by *conduits* or *pipes.* As erosion progresses, the rock occupying the pipe is often more resistant and may remain standing above the surrounding terrain as a *volcanic neck.* The summits of some volcanoes have large, nearly circular depressions called *calderas* that result from collapse following an explosive eruption. Calderas also form on shield volcanos by subterranean drainage from a central magma chamber, and the largest calderas form by the discharge of colossal volumes of silica-rich pumice along ring fractures. Although volcanic eruptions from a central vent are the most familiar, by far the largest amounts of volcanic material are extruded from cracks in the crust called *fissures.* The term *flood basalts* describes the fluid, waterlike, basaltic lava flows that cover an extensive region in the northwestern United States known as the Columbia Plateau. When silica-rich magma is extruded, *pyroclastic flows* consisting largely of ash and pumice fragments usually result.

• Igneous intrusive bodies are classified according to their *shape* and by their *orientation with respect to the host rock,* generally sedimentary rock. The two general shapes are *tabular* (tablelike) and *massive.* Intrusive igneous bodies that cut across existing sedimentary beds are said to be *discordant,* whereas those that form parallel to existing sedimentary beds are *concordant.*

• *Dikes* are tabular, discordant igneous bodies produced when magma is injected into fractures that cut across rock layers. Tabular, concordant bodies called *sills* form when magma is injected along the bedding surfaces of sedimentary rocks. *Laccoliths* are similar to sills but form from less-fluid magma that collects as a lens-shaped mass that arches the overlying strata upward. *Batholiths,* the largest intrusive igneous bodies with surface exposures of more than 100 square kilometers (40 square miles), frequently make up the cores of mountains.

• Magma originates from essentially solid rock of the crust and mantle. In addition to a rock's composition, its temperature, depth (confining pressure), and water content determine whether it exists as a solid or liquid. Thus, magma can be generated by *raising a rock's temperature,* as occurs when a hot mantle plume "ponds" beneath crustal rocks. A *decrease in pressure* can cause *decompression melting.* Further, the *introduction of volatiles* (water) can lower a rock's melting point sufficiently to generate magma. Because melting is generally not complete, a process called *partial melting* produces a melt made of the lowest-melting-temperature minerals, which are higher in silica than the original rock. Thus, magmas generated by partial melting are nearer to the granitic (felsic) end of the compositional spectrum than are the rocks from which they formed.

• *Most active volcanoes are associated with plate boundaries.* Active areas of volcanism are found along oceanic ridges where seafloor spreading is occurring (*divergent plate boundaries*), in the vicinity of ocean trenches where one plate is being subducted beneath another (*convergent plate boundaries*), and in the interiors of plates themselves (*intraplate volcanism*). Rising plumes of hot mantle rock are the source of most intraplate volcanism.

Key Terms

aa flow (p. 254)
batholith (p. 270)
caldera (p. 263)
cinder cone (p. 259)
columnar joint (p. 269)
composite cone (p. 260)
conduit (p. 256)
continental volcanic arc (p. 278)
crater (p. 256)
decompression melting (p. 271)
dike (p. 269)
fissure (p. 255, 265)
fissure eruption (p. 266)
flood basalt (p. 266)

furmarole (p. 256)
geothermal gradient (p. 271)
hot spot (p. 279)
intraplate volcanism (p. 279)
island arc (p. 278)
laccolith (p. 269)
lahar (p. 262)
mantle plume (p. 279)
nuée ardente (p. 262)
pahoehoe flow (p. 254)
parasitic cone (p. 256)
partial melting (p. 273)
pipe (p. 256)
pluton (p. 267)

pyroclastic flow (p. 255)
pyroclastic materials (p. 262)
scoria cone (p. 259)
shield volcano (p. 257)
sill (p. 269)
stratovolcanoes (p. 260)
vent (p. 256)
viscosity (p. 253)
volatiles (p. 253)
volcanic island arc (p. 278)
volcanic neck (p. 267)
volcano (p. 256)

Review Questions

1. What is the difference between *magma* and *lava*?

2. What three factors determine the nature of a volcanic eruption? What role does each play?

3. Why is a volcano fed by highly viscous magma likely to be a greater threat than a volcano supplied with very fluid magma?

4. Describe *pahoehoe* and *aa* lava.

5. List the main gases released during a volcanic eruption.

6. Analysis of samples taken during Hawaiian eruptions on the island of Hawaii indicate that _____ was the most abundant gas released.

7. How do volcanic bombs differ from blocks of pyroclastic debris?

8. Compare a *volcanic crater* to a *caldera*.

9. Compare and contrast the three main types of volcanoes (size, shape, eruptive style, and so forth).

10. Name one example of each of the three types of volcanoes.

11. Compare the formation of Hawaii with that of Parícutin.

12. Describe the formation of Crater Lake. Compare it to the caldera formed during the eruption of Kilauea.

13. What are the largest volcanic structures on Earth?

14. What is Ship Rock, New Mexico, and how did it form?

15. How do the eruptions that created the Columbia Plateau differ from eruptions that create volcanic peaks?

16. Describe each of the four intrusive igneous features discussed in the text (*dike, sill, laccolith,* and *batholith*).

17. Why might a laccolith be detected at Earth's surface before being exposed by erosion?

18. What is the largest of all intrusive igneous bodies? Is it tabular or massive? Concordant or discordant?

19. Explain how most magma is thought to originate.

20. Spreading-center volcanism is associated with which rock type? What causes rocks to melt in regions of spreading-center volcanism?

21. What is the Ring of Fire?

22. Are volcanic eruptions in the Ring of Fire generally quiet or violent? Name a volcano that would support your answer.

23. The Hawaiian Islands and Yellowstone Park are thought to be associated with which of the three zones of volcanism?

24. Volcanic islands in the deep ocean are composed primarily of what igneous rock type?

25. Briefly describe the mechanism by which explosive volcanic eruptions are thought to influence Earth's climate (See Box 9.2).

Examining the Earth System

1. Speculate about some of the possible consequences that a great and prolonged increase in explosive volcanic activity might have on each of Earth's four spheres.

2. Despite the potential for devastating destruction, humans live, work, and play on or near many active volcanoes. What are some of the benefits that a volcano or volcanic region might offer? (List some volcanoes and the assets they provide.)

Online Study Guide

The *Earth Science* Website uses the resources and flexibility of the Internet to aid in your study of the topics in this chapter. Written and developed by Earth science instructors, this site will help improve your understanding of Earth science. Visit **http://www.prenhall.com/tarbuck** and click on the cover of *Earth Science* 11e to find:

- **On-line review quizzes.**
- **Critical thinking exercises.**
- **Links to chapter-specific Web resources.**
- **Internet-wide key term searches.**

http://www.prenhall.com/tarbuck

Hikers camping in Pakistan's Charakusa Valley with the bold peaks of the Karakoram Range in the background. (Photo by Jimmy Chin/National Geographic/Getty)

CHAPTER 10

Mountain Building

Mountains provide some of the most spectacular scenery on our planet (Figure 10.1). This splendor has been captured by poets, painters, and songwriters alike. Geologists believe that at some time all continental regions were mountainous masses and have concluded that the continents grow by the addition of mountains to their flanks. Consequently, as geologists unravel the secrets of mountain formation, they also gain a deeper understanding of the evolution of Earth's continents. If continents do indeed grow by adding mountains to their flanks, how do geologists explain the existence of mountains (the Urals, for example) that are located in the interior of a landmass? To answer this and related questions, this chapter attempts to piece together the sequence of events believed to generate these lofty structures. We begin our look at mountain building by examining the process of rock deformation and the structures that result.

Rock Deformation

Every body of rock, no matter how strong, has a point at which it will fracture or flow. **Deformation** (*de* = out, *forma* = form) is a general term that refers to all changes in the original shape and/or size of a rock body. Most crustal deformation occurs along plate margins. Recall from Chapter 8 that the lithosphere consists of large segments (plates) that move relative to one another. Plate motions and the interactions along plate boundaries generate forces that cause rock to deform.

When rocks are subjected to forces (stresses) greater than their own strength, they begin to deform, usually by folding, flowing, or fracturing (Figure 10.2). It is easy to visualize how rocks break, because we normally think of them as being brittle. But how can rock masses be bent into intricate folds without being broken during the process? To answer this question, geologists performed laboratory experiments in which rocks were subjected to forces under conditions that simulated those existing at various depths within the crust.

Although each rock type deforms somewhat differently, the general characteristics of rock deformation were determined from these experiments. Geologists discovered that when stress is gradually applied, rocks first respond by deforming elastically. Changes that result from *elastic deformation* are recoverable; that is, like a rubber band, the rock will return to nearly its original size and shape when the force is removed. (As you saw in Chapter 7, the energy for most earthquakes comes from stored elastic energy that is released as rock snaps back to its original shape.)

Once the elastic limit (strength) of a rock is surpassed, it either flows (*ductile deformation*) or fractures (*brittle deformation*). The factors that influence the strength of a rock and thus how it will deform include temperature, confining pressure, rock type, and time.

Temperature and Confining Pressure

Rocks near the surface, where temperatures and confining pressures are low, tend to behave like a brittle solid and fracture once their strength is exceeded. This

▼ **FIGURE 10.1** Mount Sneffels in the Colorado Rockies. (Photo by Gavrel Jecan/Art Wolfe, Inc.)

◄ **FIGURE 10.2** Folded sedimentary layers exposed on the face of Mount Kidd, Alberta, Canada. (Photo by Peter French/DRK Photo)

type of deformation is called **brittle** (*bryttian* = to shatter) **failure** or **brittle deformation**. From our everyday experience, we know that glass objects, wooden pencils, china plates, and even our bones exhibit brittle failure once their strength is surpassed. By contrast at depth, where temperatures and confining pressures are high, rocks exhibit *ductile* behavior. **Ductile deformation** is a type of solid-state flow that produces a change in the size and shape of an object without fracturing. Ordinary objects that display ductile behavior include modeling clay, bee's wax, caramel candy, and most metals. For example, a copper penny placed on a railroad track will be flattened and deformed (without breaking) by the force applied by a passing train.

Ductile deformation of a rock—strongly aided by high temperature and high confining pressure—is somewhat similar to the deformation of a penny flattened by a train. Rocks that display evidence of ductile flow usually were deformed at great depth and may exhibit contorted folds that give the impression that the strength of the rock was akin to soft putty (Figure 10.2).

Rock Type

In addition to the physical environment, the mineral composition and texture of a rock greatly influence how it will deform. For example, crystalline rocks, such as granite, basalt, and quartzite, that are composed of minerals that have strong internal molecular bonds tend to fail by brittle fracture. By contrast, sedimentary rocks that are weakly cemented, or metamorphic rocks that contain zones of weakness, such as foliation, are more susceptible to ductile flow. Rocks that are weak and thus most likely to behave in a ductile manner when subjected to differential forces include rock salt, gypsum, and shale; limestone,

schist, and marble are of intermediate strength. In fact, rock salt is so weak that it deforms under small amounts of differential stress and rises like stone pillars through beds of sediment that lie in and around the Gulf of Mexico.

Time

One key factor that researchers are unable to duplicate in the laboratory is how rocks respond to small amounts of force applied over long spans of *geologic time*. However, insights into the effects of time on deformation are provided in everyday settings. For example, marble benches have been known to sag under their own weight over a span of a hundred years or so, and wooden bookshelves may bend after being loaded with books for a relatively short period. In nature small stresses applied over long time spans surely play an important role in the deformation of rock. Forces that are unable to deform rock when initially applied may cause rock to flow if the force is maintained over an extended period of time.

Folds

During mountain building, flat-lying sedimentary and volcanic rocks are often bent into a series of wavelike undulations called **folds**. Folds in sedimentary strata are much like those that would form if you were to hold the ends of a sheet of paper and then push them together. In nature, folds come in a wide variety of sizes and configurations. Some folds are broad flexures in which rock units hundreds of meters thick have been slightly warped. Others are very tight microscopic structures found in metamorphic rocks. Size differences notwithstanding, most folds are the result of *compressional forces* that result in the shortening and thickening of the crust.

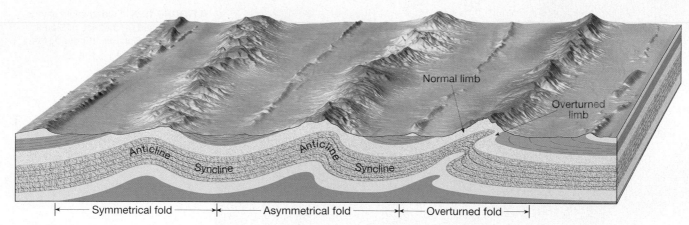

▲ **FIGURE 10.3** Block diagram of principal types of folded strata. The upfolded or arched structures are anticlines. The downfolds or troughs are synclines. Notice that the limb of an anticline is also the limb of the adjacent syncline.

Types of Folds

The two most common types of folds are anticlines and synclines (Figure 10.3). An **anticline** is most commonly formed by the upfolding, or arching, of rock layers*. Anticlines are sometimes spectacularly displayed where highways have been cut through deformed strata (Figure 10.4). Often found in association with anticlines are downfolds, or troughs, called **synclines**. Notice in Figure 10.3 that the limb of an anticline is also a limb of the adjacent syncline.

---

*By strict definition, an anticline is a structure in which the oldest strata are found in the center. This most typically occurs when strata are upfolded. Further, a syncline is strictly defined as a structure in which the youngest strata are found in the center. This occurs most commonly when strata are downfolded.

▼ **FIGURE 10.4** Anticline in folded sedimentary rocks, Devon, England. (Photo by Tom Bean/DRK Photo)

Depending on their orientation, these basic folds are described as *symmetrical* when the limbs are mirror images of each other and *asymmetrical* when they are not. An asymmetrical fold is said to be *overturned* if one limb is tilted beyond the vertical (Figure 10.3). An overturned fold can also "lie on its side" so that a plane extending through the axis of the fold would be horizontal. These *recumbent* folds are common in mountainous regions such as the Alps.

Folds do not continue forever; rather, their ends die out much like the wrinkles in cloth. Some folds *plunge* because the axis of the fold penetrates into the ground (Figure 10.5). As the figure shows, both anti-

▼ **FIGURE 10.5** Plunging folds. **A.** Idealized view of plunging folds in which a horizontal surface has been added. **B.** View of plunging folds as they might appear after extensive erosion. Notice that in a plunging anticline the outcrop pattern "points" in the direction of the plunge, while the opposite is true of plunging synclines.

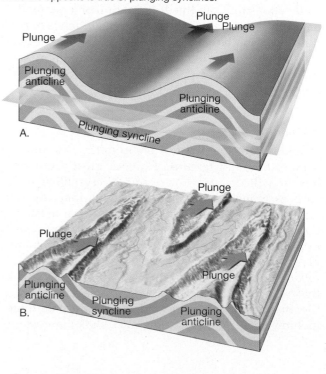

▲ **FIGURE 10.6** Sheep Mountain, a doubly plunging anticline. Note that erosion has cut the flanking sedimentary beds into low ridges that make a "V" pointing in the direction of plunge. (Photo by John S. Shelton)

clines and synclines can plunge. Figure 10.6 shows an example of a plunging anticline and the pattern produced when erosion removes the upper layers of the structure and exposes its interior. Note that the outcrop pattern of an anticline points in the direction it is plunging, whereas the opposite is true for a syncline. A good example of the kind of topography that results when erosional forces attack folded sedimentary strata is found in the Valley and Ridge Province of the Appalachians.

Although we have separated our discussion of folds and faults, in the real world folds are generally intimately coupled with faults. Examples of this close association are broad, regional features called *monoclines*. Particularly prominent features of the Colorado Plateau,

monoclines are large, steplike folds in otherwise horizontal sedimentary strata (Figure 10.7). These folds appear to be the result of the reactivating of steeply dipping fault zones located in basement rocks beneath the plateau. As large blocks of basement rock were displaced upward along ancient faults, the comparatively ductile sedimentary strata above responded by folding. On the Colorado Plateau, monoclines display a narrow zone of steeply inclined beds that flatten out to form the uppermost layers of large elevated areas, including the Zuni Uplift, Echo Cliffs Uplift, and San Rafael Swell. Displacement along these reactivated faults often exceeds 1 kilometer (0.6 mile).

Domes and Basins

Broad upwarps in basement rock may deform the overlying cover of sedimentary strata and generate large folds. When this upwarping produces a circular or elongated structure, the feature is called a **dome**. Downwarped structures having a similar shape are termed **basins**.

The Black Hills of western South Dakota is a large domed structure thought to be generated by upwarping. Here erosion has stripped away the highest portions of the upwarped sedimentary beds, exposing older igneous and metamorphic rocks in the center (Figure 10.8). Remnants of these once continuous sedimentary layers are visible, flanking the crystalline core of these mountains.

Several large basins exist in the United States (Figure 10.9). The basins of Michigan and Illinois have very gently sloping beds similar to saucers. These basins are thought to be the result of large accumulations of sediment, whose weight caused the crust to subside.

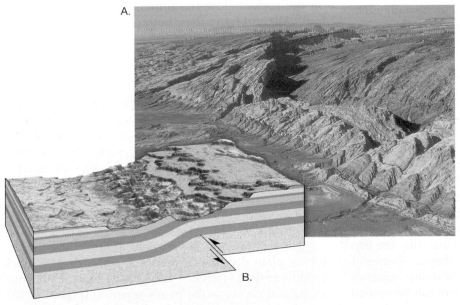

◄ **FIGURE 10.7** Monocline. **A.** Monocline located near Mexican Hat, Utah. (Photo by Stephen Trimble) **B.** Monocline consisting of bent sedimentary beds that were deformed by faulting in the bedrock below.

A.

B.

287

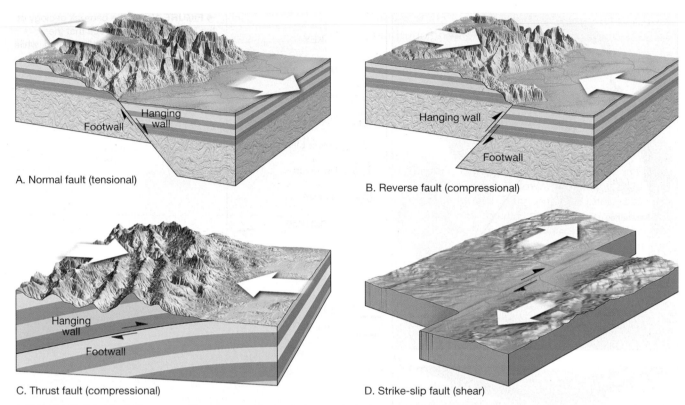

A. Normal fault (tensional)

B. Reverse fault (compressional)

C. Thrust fault (compressional)

D. Strike-slip fault (shear)

▲ **FIGURE 10.12** Block diagrams of four types of faults. **A.** Normal fault. **B.** Reverse fault. **C.** Thrust fault. **D.** Strike-slip fault.

▶ **FIGURE 10.13** Normal faulting in the Basin and Range Province. Here, tensional stresses have elongated and fractured the crust into numerous blocks. Movement along these fractures has tilted the blocks producing parallel mountain ranges called fault-block mountains. (Photo by Michael Collier)

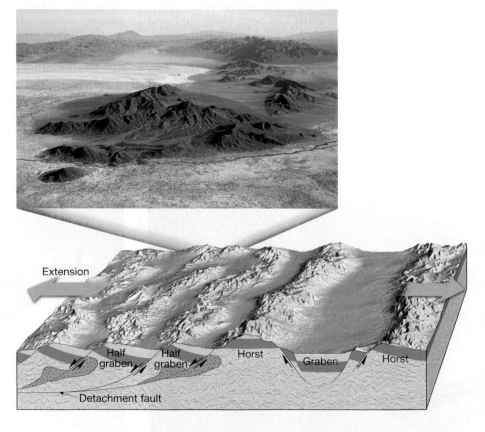

surfaces to stretch and break or by opposing horizontal forces.

Reverse and Thrust Faults.   **Reverse faults** and **thrust faults** are dip-slip faults in which the hanging wall block moves up relative to the footwall block (Figure 10.12B and C). Reverse faults have dips greater than 45°, and thrust faults have dips less than 45°. Because the hanging wall block moves up and over the footwall block, reverse and thrust faults accommodate shortening of the crust.

Most high-angle reverse faults are small and accommodate local displacements in regions dominated by other types of faulting. Thrust faults, on the other hand, exist at all scales. In mountainous regions such as the Alps, Northern Rockies, Himalayas, and Appalachians, thrust faults have displaced strata as far as 50 kilometers (about 30 miles) over adjacent rock units. The result of this large-scale movement is that older strata end up overlying younger rocks.

Whereas normal faults occur in tensional environments, reverse and thrust faults result from strong compressional stresses. In these settings, crustal blocks are displaced *toward* one another, with the hanging wall being displaced upward relative to the footwall. Thrust faulting is most pronounced in subduction zones and other convergent boundaries where plates are colliding. Compressional forces generally produce folds as well as faults and result in a thickening and shortening of the material involved.

Strike-Slip Faults

Faults in which the dominant displacement is horizontal and parallel to the trend, or strike, of the fault surface are called **strike-slip faults** (Figure 10.12D). Because of their large size and linear nature, many strike-slip faults produce a trace that is visible over a great distance (Figure 10.14). Rather than a single fracture along which movement takes place, large strike-slip faults consist of a zone of roughly parallel fractures. The zone may be up to several kilometers wide. The most recent movement, however, is often along a strand only a few meters wide, which may offset features such as stream channels (see Figure 10.A in Box 10.1). Furthermore, crushed and broken rocks produced during faulting are more easily eroded, often producing linear valleys or troughs that mark the locations of strike-slip faults.

▲ **FIGURE 10.14**   Aerial view of strike-slip (right-lateral) fault in Southern Nevada. (Photo by Martin G. Miller)

The earliest scientific records of strike-slip faulting were made following surface ruptures that produced large earthquakes. One of the most noteworthy of these was the great San Francisco earthquake of 1906. During this strong earthquake, structures such as fences that were built across the San Andreas Fault were displaced as much as 4.7 meters (15 feet). Because the movement along the San Andreas causes the crustal block on the opposite side of the fault to move to the right as you face the fault, it is called a *right-lateral* strike-slip fault. The Great Glen fault in Scotland is a well-known example of a *left-lateral* strike-slip fault, which exhibits the opposite sense of displacement.

Many major strike-slip faults cut through the lithosphere and accommodate motion between two large crustal plates. This special kind of strike-slip fault is called a **transform** (*trans* = across, *forma* = form) **fault.** Numerous transform faults cut the oceanic lithosphere and link offset segments of oceanic ridges. Others accommodate displacement between continental plates that move horizontally with respect to each other. One of the best-known transform faults is California's San Andreas Fault (see Box 10.1). This plate-bounding fault can be traced for about 950 kilometers (600 miles) from the Gulf of California to a point along the Pacific Coast north of San Francisco, where it heads out to sea. Ever since its formation, about 29 million years ago, displacement along the San Andreas Fault has exceeded 560 kilometers (340 miles). This movement has accommodated the northward displacement of southwestern California and the Baja Peninsula of Mexico in relation to the remainder of North America.

Joints

Among the most common rock structures are fractures called **joints.** Unlike faults, joints are fractures along which *no appreciable-displacement* has occurred. Although some joints have a random orientation, most occur in roughly parallel groups (Figure 10.15).

We have already considered two types of joints. Earlier we learned that *columnar joints* form when igneous rocks cool and develop shrinkage fractures that produce elongated, pillarlike columns (see Figure 9.24, p. 270). Also recall that sheeting produces a pattern of gently curved joints that develop more or less parallel to the surface of large exposed igneous bodies such as batholiths. Here the jointing results from the gradual expansion that occurs when erosion removes the overlying load (see Figure 4.4, p. 87).

In contrast to the situations just described, most joints are produced when rocks in the outermost crust are deformed. Here forces associated with crustal movements cause the rock to fail by brittle fracture. For example, when folding occurs, rocks situated at the axes of the folds are elongated and pulled apart to produce tensional joints. Extensive joint patterns can also develop in response to relatively subtle and often barely perceptible regional upwarping and downwarping of the crust. In many cases, the cause for jointing at a particular locale is not readily apparent.

Many rocks are broken by two or even three sets of intersecting joints that slice the rock into numerous regularly shaped blocks. These joint sets often exert a strong influence on other geologic processes. For example, chemical weathering tends to be concentrated along joints, and in many areas groundwater movement and the resulting dissolution in soluble rocks is controlled by the joint pattern (Figure 10.15). Moreover, a system of joints can influence the direction that stream courses follow. The rectangular drainage pattern described in Chapter 5 is such a case.

Mountain Building

Like other people, geologists have been inspired more by Earth's mountains than by any other landforms (Figure 10.16). Through extensive scientific exploration over the last 150 years, much has been learned about the internal processes that generate these often spectacular terrains. The name for the processes that collectively produce a mountain belt is **orogenesis,** (*oros* = mountain, *genesis* = to come into being). The rocks comprising mountains provide striking visual evidence of the enormous compressional forces that have deformed large sections of Earth's crust and subsequently elevated them to their present positions. Although folding is often the most conspicuous sign of these forces, thrust faulting, metamorphism, and igneous activity are always present in varying degrees.

Mountain building has occurred during the recent geologic past in several locations around the world. These young mountainous belts include the American Cordillera, which runs along the western margin of the Americas from Cape Horn to Alaska and includes the Andes and Rocky mountains; the Alpine-Himalaya chain, which extends from the Mediterranean through Iran to northern India and into Indochina; and the mountainous terrains of the western Pacific, which include volcanic island arcs such as Japan, the Philippines, and Sumatra. Most of these young mountain belts have come into existence within the last 100 million years. Some, including the Himalayas, began their growth as recently as 45 million years ago.

In addition to these relatively young mountain belts, several chains of older mountains exist on Earth as well. Although these older structures are deeply eroded and topographically less prominent, they clearly possess the same structural features found in younger mountains. Typical of this older group are the Appalachians in the eastern United States and the Urals in Russia.

Over the years, several hypotheses have been put forward regarding the formation of Earth's major mountain belts. One early proposal suggested that mountains are simply wrinkles in Earth's crust, produced as the planet cooled from its original semi-molten state. As Earth lost heat, it contracted and shrank. In response to this process, the crust was deformed similar to when the peel of an orange wrinkles as the fruit dries out. However, neither this nor any other early hypothesis was able to withstand careful scrutiny and had to be discarded.

▼ **FIGURE 10.15**   Chemical weathering is enhanced along joints in sandstone, near Moab, Utah. (Photo by Michael Collier)

▲ **FIGURE 10.16** This peak is part of the Karakoram Range in Pakistan. The Karakoram are part of the Himalayan system. (Photo by Art Wolfe)

Mountain Building at Subduction Zones

With the development of the theory of plate tectonics, a model for orogenesis with excellent explanatory power has emerged. According to this model, most mountain building occurs at convergent plate boundaries. Here, the subduction of oceanic lithosphere triggers partial melting of mantle rock, providing a source of magma that intrudes the crustal rocks that form the margin of the overlying plate. In addition, colliding plates provide the tectonic forces that fold, fault, and metamorphose the thick accumulations of sediments that have been deposited along the flanks of landmasses. Together, these processes thicken and shorten the continental crust, thereby elevating rocks that may have formed near the ocean floor, to lofty heights.

To unravel the events that produce mountains, researchers examine ancient mountain structures as well as sites where orogenesis is currently active. Of particular interest are active subduction zones, where lithospheric plates are converging. Here the subduction of oceanic lithosphere generates Earth's strongest earthquakes and most explosive volcanic eruptions, as well as playing a pivotal role in generating many of Earth's mountain belts.

The subduction of oceanic lithosphere gives rise to two different types of mountain belts. Where *oceanic lithosphere* subducts beneath an *oceanic plate*, an *island arc* and related tectonic features develop. Subduction beneath a *continental block*, on the other hand, results in the formation of a *continental volcanic arc* along the margin of the adjacent landmass. Plate boundaries that generate continental volcanic arcs are often referred to as *Andean-type plate margins*.

Island Arcs

Island arcs form where two oceanic plates converge and one is subducted beneath the other (Figure 10.17). This activity results in partial melting of the mantle wedge located above the subducting plate and eventually leads to the growth of a volcanic island arc on the ocean floor. Because they are associated with subducting oceanic lithosphere, island arcs are typically found on the margins of an ocean basin, such as the Pacific— where the majority of volcanic island arcs are found. Examples of active island arcs include the Mariana, New Hebrides, Tonga, and Aleutian arcs.

Island arcs represent what are perhaps the simplest mountain belts. These structures result from the steady subduction of oceanic lithosphere, which may last for 100 million years or more. Somewhat sporadic volcanic activity, the emplacement of plutonic bodies at depth, and the accumulation of sediment that is scraped from the subducting plate gradually increase the volume of crustal material capping the upper

▶ **FIGURE 10.17** The development of a volcanic island arc by the convergence of two oceanic plates. Continuous subduction along these Aleutian-type convergent zones results in the development of thick units of continental-type crust.

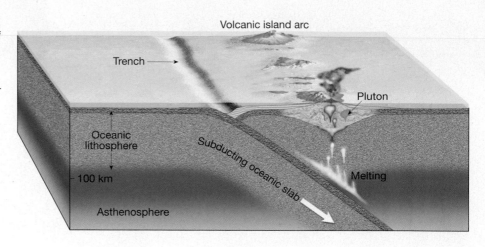

plate. Some mature volcanic island arcs, such as Japan, appear to have been built upon a preexisting fragment of crustal material.

The continued development of a mature volcanic island arc can result in the formation of mountainous topography consisting of belts of igneous and metamorphic rocks. This activity, however, is viewed as just one phase in the development of a major mountain belt. As you will see later, some volcanic arcs are carried by a subducting plate to the margin of a large continental block, where they become involved in a major mountain-building episode.

Mountain Building Along Andean-Type Margins

Mountain building along continental margins involves the convergence of an oceanic plate and a plate whose leading edge contains continental crust. Exemplified by the Andes Mountains, an *Andean-type convergent zone* results in the formation of a continental volcanic arc and related tectonic features inland of the continental margin.

The first stage in the development of an idealized Andean-type mountain belt occurs prior to the formation of the subduction zone. During this period the continental margin is a **passive continental margin**; that is, it is not a plate boundary but a part of the same plate as the adjoining oceanic crust. The East Coast of North America provides a present-day example of a passive continental margin. Here, as at other passive continental margins surrounding the Atlantic, deposition of sediment on the continental shelf is producing a thick wedge of shallow-water sandstones, limestones, and shales (Figure 10.18A). Beyond the continental shelf, turbidity currents are depositing sediments on the continental slope and rise (see Chapter 13).

At some point the continental margin becomes active. A subduction zone forms and the deformation process begins (Figure 10.18B). A good place to examine

an **active continental margin** is the west coast of South America. Here the Nazca plate is being subducted beneath the South American plate along the Peru–Chile trench. This subduction zone probably formed prior to the breakup of the supercontinent of Pangaea.

In an idealized Andean-type subduction, convergence of the continental block and the subducting oceanic plate leads to deformation and metamorphism of the continental margin. Once the oceanic plate descends to about 100 kilometers (60 miles), partial melting of mantle rock above the subducting slab generates magma that migrates upward (Figure 10.18B).

Thick continental crust greatly impedes the ascent of magma. Consequently, a high percentage of the magma that intrudes the crust never reaches the surface—instead, it crystallizes at depth to form plutons. Eventually, uplifting and erosion exhume these igneous bodies and associated metamorphic rocks. Once they are exposed at the surface, these massive structures are called *batholiths* (Figure 10.18C). Composed of numerous plutons, batholiths form the core of the Sierra Nevada in California and are prevalent in the Peruvian Andes.

During the development of this continental volcanic arc, sediment derived from the land and scraped from the subducting plate is plastered against the landward side of the trench like piles of dirt in front of a bulldozer. This chaotic accumulation of sedimentary and metamorphic rocks with occasional scraps of ocean crust is called an **accretionary wedge** (Figure 10.18B). Prolonged subduction can build an accretionary wedge that is large enough to stand above sea level (Figure 10.18C).

Andean-type mountain belts are composed of two roughly parallel zones. The volcanic arc develops on the continental block. It consists of volcanoes and large intrusive bodies intermixed with high-temperature metamorphic rocks. The seaward segment is the accretionary wedge. It consists of folded, faulted sedimentary and metamorphic rocks (Figure 10.18C).

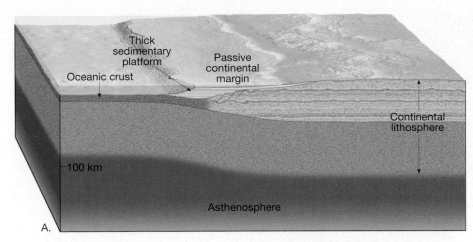

◀ **FIGURE 10.18** Mountain building
along an Andean-type subduction zone.
A. Passive continental margin with
extensive wedge of sediments. **B.** Plate
convergence generates a subduction
zone, and partial melting produces a
developing continental volcanic arc.
Continued convergence and igneous
activity further deform and thicken the
crust, elevating the mountain belt, while
the accretionary wedge grows. **C.**
Subduction ends and is followed by a
period of uplift and erosion.

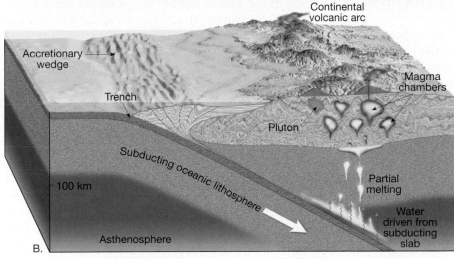

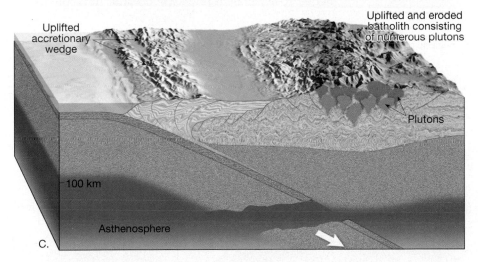

Sierra Nevada and Coast Ranges. One of the best
examples of an inactive Andean-type orogenic belt
is found in the western United States. It includes the
Sierra Nevada and the Coast Ranges in California.
These parallel mountain belts were produced by the
subduction of a portion of the Pacific Basin under
the western edge of the North American plate. The
Sierra Nevada batholith is a remnant of a portion of
the continental volcanic arc that was produced by
several surges of magma over tens of millions of
years. Subsequent uplifting and erosion have removed most of the evidence of past volcanic activity
and exposed a core of crystalline, igneous, and associated metamorphic rocks.

In the trench region, sediments scraped from the subducting plate, plus those provided by the eroding continental volcanic arc, were intensely folded and faulted into an accretionary wedge. This chaotic mixture of rocks presently constitutes the Franciscan Formation of California's Coast Ranges. Uplifting of the Coast Ranges took place only recently, as evidenced by the young unconsolidated sediments that still mantle portions of these highlands.

Collisional Mountain Ranges

As you have seen, when a slab of oceanic lithosphere subducts beneath a continental margin, an Andean-type mountain belt develops. If the subducting plate also contains a slab of continental lithosphere, continued subduction eventually carries the continental block to the trench. Oceanic lithosphere is relatively dense and readily subducts, but continental crust is composed of low-density material that is too buoyant to undergo subduction. Consequently, the arrival of the continental block at the trench results in a collision with the overriding continent. The result is crustal shortening and thickening to produce a mountain belt.

Mountain belts can develop as a result of the collision and merger of an island arc, or some other small crustal fragment with a continental block, as well as from the collision and joining of two or more continents.

Terranes and Mountain Building

The process of collision and accretion (joining together) of comparatively small crustal fragments to a continental margin has generated many of the mountainous regions rimming the Pacific. Geologists refer to these accreted crustal blocks as *terranes*. Simply, the term **terrane** refers to any crustal fragment that has a geologic history distinct from that of adjoining terranes. Terranes come in various shapes and sizes.

What is the nature of these crustal fragments, and from where do they originate? Research suggests that prior to their accretion to a continental block, some of the fragments may have been *microcontinents* similar to the present-day island of Madagascar, located east of Africa in the Indian Ocean. Many others were island arcs similar to Japan, the Philippines, and the Aleutian Islands. Still others are submerged crustal fragments, such as *oceanic plateaus*, which were created by massive outpourings of basaltic lavas associated with hot-spot activity (Figure 10.19).

Accretion and Orogenesis. The widely accepted view is that as oceanic plates move, they carry embedded oceanic plateaus, volcanic island arcs, and microcontinents to an Andean-type subduction zone. When an oceanic plate contains a chain of small seamounts, these structures are generally subducted along with the descending oceanic slab. However, thick units of oceanic crust, such as the Ontong Java Plateau, or a mature island arc composed of abundant "light" igneous rocks may render the oceanic lithosphere too

▼ **FIGURE 10.19** Distribution of present-day oceanic plateaus and other submerged crustal fragments. (Data from Ben-Avraham and others)

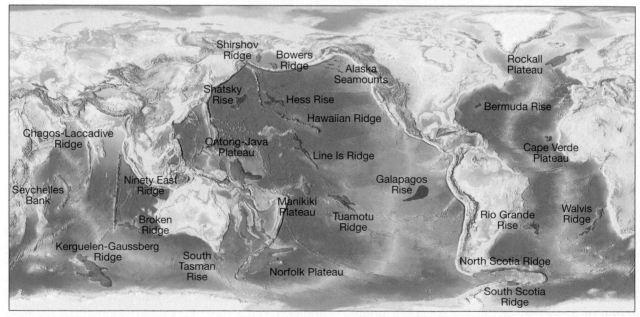

buoyant to subduct. In these situations, a collision between the crustal fragment and the continent occurs.

The sequence of events that occurs when a mature
island arc reaches an Andean-type margin is shown in
Figure 10.20. Because of its buoyancy, a mature island
arc will not subduct beneath the continental plate. Instead, the upper portions of these thickened zones are
peeled from the descending plate and thrust in relatively thin sheets upon the adjacent continental block.
In some settings continued subduction may carry another crustal fragment to the continental margin.
When this fragment collides with the continental margin, it displaces the accreted island arc further inland,

▼ **FIGURE 10.20** This sequence illustrates the collision of an
inactive volcanic island arc with an Andean-type plate margin.

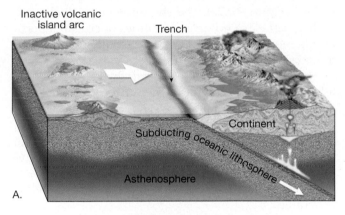

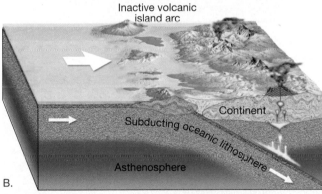

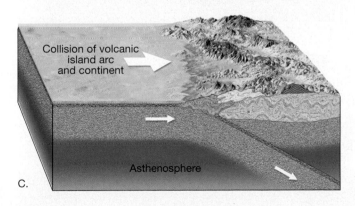

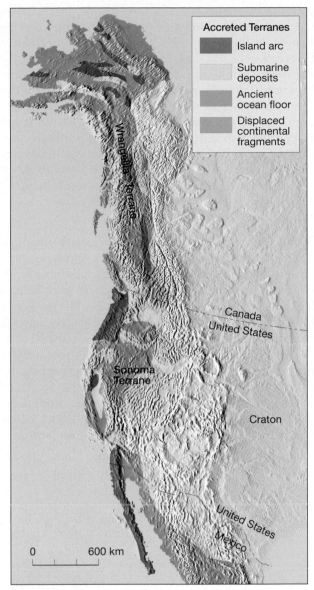

▲ **FIGURE 10.21** Map showing terranes thought to have been
added to western North America during the past 200 million years.
(Redrawn after D. R. Hutchinson and others)

adding to the zone of deformation and to the thickness
and lateral extent of the continental margin.

The North American Cordillera. The idea that mountain building occurs in association with the accretion
of crustal fragments to a continental mass arose principally from studies conducted in the North American
Cordillera (Figure 10.21). Here it was determined that
some mountainous areas, principally those in the orogenic belts of Alaska and British Columbia, contain
fossil and paleomagnetic evidence indicating that
these strata once lay nearer the equator.

It is now assumed that many of the other terranes
found in the North American Cordillera were once

scattered throughout the eastern Pacific, much as we find island arcs and oceanic plateaus distributed in the western Pacific today (see Figure 10.19). Since before the breakup of Pangaea, the eastern portion of the Pacific basin (Farallon plate) has been subducting under the western margin of North America. Apparently, this activity resulted in the piecemeal addition of crustal fragments to the entire Pacific margin of the continent—from Mexico's Baja Peninsula to northern Alaska (see Figure 10.21). In a like manner, many modern microcontinents will eventually be accreted to active continental margins, producing new orogenic belts.

Continental Collisions

Continental collisions result in the development of mountains that are characterized by shortened and thickened crust. Thicknesses of 50 kilometers (30 miles) are common, and some regions have crustal thicknesses in excess of 70 kilometers (40 miles). In these settings, crustal thickening is achieved through folding and faulting.

We will take a closer look at two examples of collision mountains—the Himalayas and the Appalachians. The Himalayas are the youngest collision mountains on Earth and are still rising. The Appalachians are a much older mountain belt, in which active mountain building ceased about 250 million years ago.

The Himalayas. The mountain-building episode that created the Himalayas began roughly 45 million years ago when India began to collide with Asia. Prior to the breakup of Pangaea, India was part of a Southern Hemisphere landmass that also included Australia. Upon splitting from that continent, India moved rapidly, geologically speaking, a few thousand kilometers in a northward direction (see Figure 8.15, p. 232).

The subduction zone that facilitated India's northward migration was located near the southern margin of Asia (Figure 10.22). Ongoing subduction along Asia's margin created an Andean-type plate margin that contained a well-developed volcanic arc and accretionary wedge. Eventually, the intervening ocean basin was consumed at the subduction zone and India collided with the Eurasian plate. The tectonic forces involved in the collision were immense and caused the more deformable materials located on the seaward edges of these landmasses to be highly folded and faulted (Figure 10.22). The shortening and thickening of the crust elevated great quantities of crustal material, thereby generating the spectacular Himalayan mountains (see Figure 10.16).

In addition to uplift, crustal thickening caused lower layers to become deeply buried and to experience elevated temperatures and pressures. Partial melting within the deepest and most deformed region of the developing mountain belt produced magma bodies that intruded and further deformed the overlying rocks. It is in such environments where the metamorphic and igneous core of a major mountain belt is generated.

▶ **FIGURE 10.22** Illustration showing the collision of India with the Eurasian plate, producing the spectacular Himalayas.

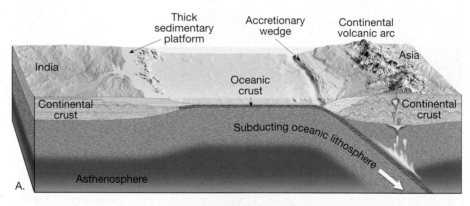

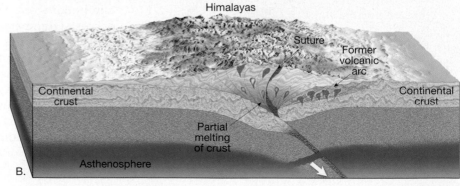

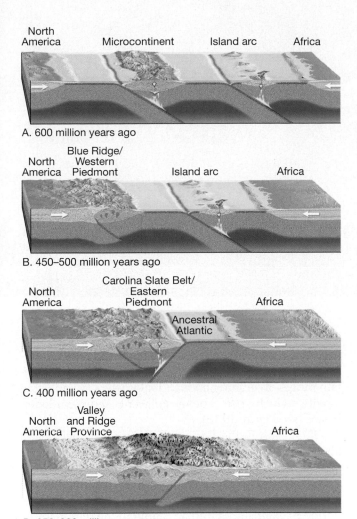

A. 600 million years ago

B. 450–500 million years ago

C. 400 million years ago

D. 250–300 million years ago

▲ **FIGURE 10.23** These simplified diagrams depict the development of the southern Appalachians as the ancient North Atlantic was closed during the formation of Pangaea. Three separate stages of mountain-building activity spanned more than 300 million years. (After Zve Ben-Avraham, Jack Oliver, Larry Brown, and Frederick Cook)

The Appalachians. The Appalachian Mountains provide great scenic beauty near the eastern margin of North America from Alabama to Newfoundland. The orogeny that generated this extensive mountain system lasted a few hundred million years and was one of the stages in the assembling of Pangaea. Our simplified scenario begins roughly 600 million years ago when an ocean body, which predated the North Atlantic (referred to as the ancestral North Atlantic), began to close. Two subduction zones probably formed. One was located seaward of the coast of Africa and gave rise to a volcanic arc similar to those that presently rim the western Pacific. The other developed adjacent to a continental fragment that lay off the coast of North America, as shown in Figure 10.23A.

Between 450 and 500 million years ago, the marginal sea located between this crustal fragment and

North America began to close. The ensuing collision deformed the continental shelf and sutured the crustal fragment to the North American plate. The metamorphosed remnants of the continental fragment are recognized today as the crystalline rocks of the Blue Ridge and western Piedmont regions of the Appalachians (Figure 10.23B). In addition to the pervasive metamorphism, igneous activity placed numerous plutonic bodies along the continental margin, particularly in New England.

A second episode of mountain building occurred about 400 million years ago. The continued closing of the ancestral North Atlantic resulted in the collision of the developing volcanic arc with North America (Figure 10.23C). Evidence for this event is visible in the Carolina Slate Belt of the eastern Piedmont, which contains metamorphosed sedimentary and volcanic rocks characteristic of an island arc.

The final orogeny occurred somewhere between 250 and 300 million years ago, when Africa collided with North America. This event displaced and further deformed the shelf sediments and sedimentary rocks that had once flanked the eastern margin of North America (Figure 10.23D). Today these folded and thrust-faulted sandstones, limestones, and shales make up the largely unmetamorphosed rocks of the Valley and Ridge Province.

Geologically speaking, shortly after the formation of the Appalachian Mountains, the newly formed supercontinent of Pangaea began to break into smaller fragments. Because the zone of rifting occurred east of the location where Africa collided with North America, a remnant of Africa remains "welded" to the North American plate.

Other mountain ranges that exhibit evidence of continental collisions include the Alps and the Urals. The Alps are thought to have formed as a result of a collision between Africa and Europe. The Urals, on the other hand, formed during the assembly of Pangaea when Northern Europe and northern Asia collided.

Fault-Block Mountains

Most mountain belts, including the Alps, Himalayas, and Appalachians, form in compressional environments, as evidenced by the predominance of large thrust faults and folded strata. However, other tectonic processes, such as continental rifting, can also produce uplift and the formation of topographic mountains. The mountains that form in these settings, termed **fault-block mountains**, are bounded by high-angle normal faults that gradually flatten with depth. Most fault-block mountains form in response to broad uplifting, which causes elongation and faulting. Such a situation is exemplified by the fault blocks that rise high above the rift valleys of East Africa.

► **FIGURE 10.24** The Grand Tetons of Wyoming are an example of fault-block mountains. (Photo by Art Wolfe, Inc.)

Mountains in the United States in which faulting and gradual uplift have contributed to their lofty stature include the Sierra Nevada of California and the Grand Tetons of Wyoming. Both are faulted along their eastern flanks, which were uplifted as the blocks tilted downward to the west. Looking west from Owens Valley, California, and Jackson Hole, Wyoming, the eastern fronts of these ranges (the Sierra Nevada and the Tetons, respectively) rise over 2 kilometers (1.2 miles) making them two of the most imposing mountain fronts in the United States (Figure 10.24).

One of Earth's largest regions of fault-block mountains is the Basin and Range Province. This region extends in a roughly north to south direction for nearly 3000 kilometers (2000 miles) and encompasses all of Nevada and portions of the surrounding states, as well as parts of southern Canada and western Mexico. Here, the brittle upper crust has literally been broken into hundreds of fault blocks. Tilting of these faulted structures (half-grabens) gave rise to nearly parallel mountain ranges, averaging about 80 kilometers (50 miles) in length, which rise above adjacent sediment-laden basins (see Figure 10.13).

Extension in the Basin and Range Province began about 20 million years ago and appears to have "stretched" the crust as much as twice its original width. Figure 10.25 shows a rough outline of the boundaries of the western states before and after this period of extension. High heat flow in the region, three times average, and several episodes of volcanism provide strong evidence that mantle upwelling caused doming of the crust, which in turn contributed to extension in the region.

Vertical Movements of the Crust

In addition to the large crustal displacements driven mainly by plate tectonics, gradual up-and-down motions of the continental crust are observed at many locations around the globe. Although much of this vertical movement occurs along plate margins and is associated with active mountain building, some of it is not.

Evidence for crustal uplift occurs along the West Coast of the United States. When the elevation of a coastal area remains unchanged for an extended period, a wavecut platform develops (see Figure 15.15, p. 419). In parts of California, ancient wave-cut platforms can now be found as terraces hundreds of meters above sea level (Figure 10.26). Such evidence of crustal uplift is easy to find; unfortunately, the reason for uplift is not always as easy to determine.

▼ **FIGURE 10.25** Extension in the Basin and Range Province has "stretched" the crust in some locations by as much as twice its original width. Shown here is a rough outline of the western states before (left) and after (right) extension.

▲ **FIGURE 10.26** Former wave-cut platforms now exist as a series of elevated terraces on the west side of San Clemente Island off the southern California coast. Once at sea level, the highest terraces are now about 400 meters (1320 feet) above it. (Photo by John S. Shelton)

Isostasy

Early workers discovered that Earth's less-dense crust floats on top of the denser and deformable rocks of the mantle. The concept of a floating crust in gravitational balance is called **isostasy** (*iso* = equal, *stasis* = standing). One way to grasp the concept of isostasy is to envision a series of wooden blocks of different heights floating in water, as shown in Figure 10.27. Note that the thicker wooden blocks float higher than the thinner blocks.

Similarly, many mountain belts stand high above the surrounding terrain because of crustal thickening. These compressional mountains have buoyant crustal "roots" that extend deep into the supporting material below, just like the thicker wooden blocks shown in Figure 10.27.

Visualize what would happen if another small block of wood were placed atop one of the blocks in Figure 10.27. The combined block would sink until a new isostatic (gravitational) balance was reached. However, the top of the combined block would actually be higher than before, and the bottom would be lower. This process of establishing a new level of gravitational equilibrium is called **isostatic adjustment**.

▼ **FIGURE 10.27** This drawing illustrates how wooden blocks of different thicknesses float in water. In a similar manner, thick sections of crustal material float higher than thinner crustal slabs.

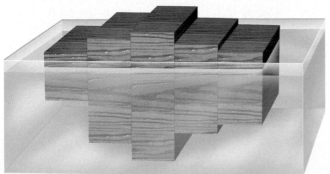

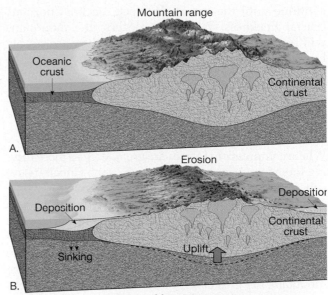
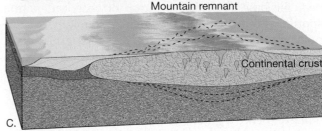

▲ **FIGURE 10.28** This sequence illustrates how the combined effect of erosion and isostatic adjustment results in a thinning of the crust in mountainous regions. **A.** When mountains are young, the continental crust is thickest. **B.** As erosion lowers the mountains, the crust rises in response to the reduced load. **C.** Erosion and uplift continue until the mountains reach "normal" crustal thickness.

Applying the concept of isostatic adjustment, we should expect that when weight is added to the crust, it will respond by subsiding, and when weight is removed, the crust will rebound. (Visualize what happens to a ship as cargo is being loaded and unloaded.) Evidence for crustal subsidence followed by crustal rebound is provided by Ice Age glaciers. When continental ice sheets occupied portions of North America during the Pleistocene epoch, the added weight of 3-kilometer-thick (nearly 2-mile-thick) masses of ice caused downwarping of Earth's crust by hundreds of meters. In the 8000 years since the last ice sheet melted, uplifting of as much as 330 meters (1000 feet) has occurred in Canada's Hudson Bay region, where the thickest ice had accumulated.

One of the consequences of isostatic adjustment is that as erosion lowers the summits of mountains, the crust will rise in response to the reduced load (Figure 10.28). However, each episode of isostatic uplift is somewhat less than the elevation loss due to erosion. The processes of uplifting and erosion will

303

continue until the mountain block reaches "normal" crustal thickness. When this occurs, the mountains will be eroded to near sea level, and the once deeply buried interior of the mountain will be exposed at the surface. In addition, as mountains are worn down, the eroded sediment is deposited on adjacent landscapes, causing these areas to subside (Figure 10.28).

How High Is Too High?

Where compressional forces are great, such as those driving India into Asia, mountains such as the Himalayas result. But is there a limit on how high a mountain can rise? As mountaintops are elevated, gravity-driven processes such as erosion and mass wasting accelerate, carving the deformed strata into rugged landscapes. Just as important, however, is the fact that gravity also acts on the rocks within these mountainous masses. The higher the mountain, the greater the downward force on the rocks near the base. (Visualize a group of cheerleaders at a sporting event building a human pyramid.) At some point the rocks deep within the developing mountain, which are comparatively warm and weak, will begin to flow laterally, as shown in Figure 10.29. This is analogous to what happens when a ladle of very thick pancake batter is poured on a hot griddle. As a result, the mountain will experience a **gravitational collapse**, which involves normal faulting and subsidence in the upper, brittle portion of the crust and ductile spreading at depth.

You then might ask, What keeps the Himalayas standing? Simply, the horizontal compressional forces that are driving India into Asia are greater than the vertical force of gravity. However, once India's northward trek ends, the downward pull of gravity will become the dominant force acting on this mountainous region.

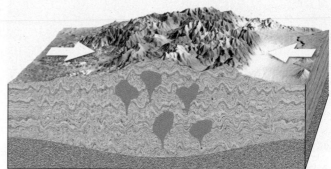

A. Horizontal compressional forces dominate causing shortening and thickening of the crust

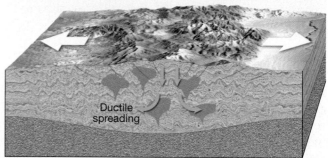

B. Gravitational forces dominate resulting in stretching and thinning of the crust

▲ **FIGURE 10.29** Block diagram of a mountain belt that is collapsing under its own "weight." Gravitational collapse involves normal faulting in the upper, brittle portion of the crust and ductile spreading at depth.

STUDENTS SOMETIMES ASK...

You mentioned that most mountains are the result of crustal deformation. Are there areas that exhibit mountainous topography but have been produced without crustal deformation?

Yes. Plateaus—areas of high-standing rocks that are essentially horizontal—are one example of a feature that can be deeply dissected by erosional forces into rugged, mountain-like landscapes. Although these highlands resemble mountains topographically, they lack the structures associated with orogenesis. The opposite situation also exists. For instance, the Piedmont section of the eastern Appalachians exhibits topography that is nearly as subdued as that seen in the Great Plains. Yet, because this region is composed of deformed metamorphic rocks, it is clearly part of the Appalachian Mountains.

STUDENTS SOMETIMES ASK...

What's the difference between a terrane and a terrain?

The term *terrane* is used to designate a distinct and recognizable series of rock formations that has been transported by plate tectonic processes. Since geologists who mapped these rocks were unsure where they came from, these rocks were sometimes called "exotic," "suspect," "accreted," or "foreign" terranes. Don't confuse this with the term terrain, which describes the shape of the surface topography or "lay of the land."

Chapter Summary

• *Deformation* refers to changes in the shape and/or volume of a rock body. Rocks deform differently depending on the environment (temperature and confining pressure), the composition of the rock, and the length of time stress is maintained. Rocks first respond by deforming *elastically* and will return to their original shape when the stress is removed. Once their elastic limit (strength) is surpassed, rocks either deform by ductile flow or they fracture. *Ductile deformation* is a solid-state flow that results in a change in size and shape of rocks without fracturing. Ductile deformation occurs in a high-temperature/high pressure environment. In a near-surface environment, most rocks deform by *brittle failure.*

• Among the most basic geologic structures associated with rock deformation are *folds* (flat-lying sedimentary and volcanic rocks bent into a series of wavelike undulations). The two most common types of folds are *anticlines,* formed by the upfolding, or arching, of rock layers, and *synclines,* which are downfolds. Most folds are the result of horizontal *compressional stresses.* *Domes* (upwarped structures) and *basins* (downwarped structures) are circular or somewhat elongated folds formed by vertical displacements of strata.

• Faults are fractures in the crust along which appreciable displacement has occurred. Faults in which the movement is primarily vertical are called *dip-slip faults.* Dip-slip faults include both *normal* and *reverse faults.* Low-angle reverse faults are called *thrust faults.* Normal faults indicate *tensional stresses* that pull the crust apart. Along spreading centers, divergence can cause a central block called a *graben,* bounded by normal faults, to drop as the plates separate.

• Reverse and thrust faulting indicate that *compressional forces* are at work. Large *thrust faults* are found along subduction zones and other convergent boundaries where plates are colliding.

• *Strike-slip faults* exhibit mainly horizontal displacement parallel to the fault surface. Large strike-slip faults, called *transform faults,* accommodate displacement between plate boundaries. Most transform faults cut the oceanic lithosphere and link spreading centers. The San Andreas Fault cuts the continental lithosphere and accommodates the northward displacement of southwestern California.

• *Joints* are fractures along which no appreciable displacement has occurred. Joints generally occur in groups with roughly parallel orientations and are the result of brittle failure of rock units located in the outermost crust.

• The name for the processes that collectively produce a mountain system is *orogenesis.* Most mountains consist of roughly parallel ridges of folded and faulted sedimentary and volcanic rocks, portions of which have been strongly metamorphosed and intruded by younger igneous bodies.

• Subduction of oceanic lithosphere under a continental block gives rise to an *Andean-type plate margin* that is characterized by a continental volcanic arc and associated igneous plutons. In addition, sediment derived from the land, as well as material scraped from the subducting plate, becomes plastered against the landward side of the trench, forming an *accretionary wedge.* An excellent example of an inactive Andean-type mountain belt is found in the western United States and includes the Sierra Nevada and the Coast Range in California.

• Mountain belts can develop as a result of the collision and merger of an island arc, oceanic plateau, or some other small crustal fragment to a continental block. Many of the mountain belts of the North American Cordillera, were generated in this manner.

• Continued subduction of oceanic lithosphere beneath an Andean-type continental margin will eventually close an ocean basin. The result will be a *continental collision* and the development of compressional mountains that are characterized by shortened and thickened crust as exhibited by the Himalayas. The development of a major mountain belt is often complex involving two or more distinct episodes of mountain building. Continental collisions have generated many mountain belts, including the Alps, Urals, and Appalachians.

• Although most mountains form along convergent plate boundaries, other tectonic processes, such as continental rifting can produce uplift and the formation of topographic mountains. The mountains that form in these settings, termed *fault-block mountains,* are bounded by high-angle normal faults that gradually flatten with depth. The Basin and Range Province in the western United States consists of hundreds of faulted blocks that give rise to nearly parallel mountain ranges that stand above sediment-laden basins.

• Earth's less dense crust floats on top of the denser and deformable rocks of the mantle, much like wooden blocks floating in water. The concept of a floating crust in gravitational balance is called *isostasy.* Most mountainous topography is located where the crust has been shortened and thickened. Therefore, mountains have deep crustal roots that isostatically support them. As erosion lowers the peaks, *isostatic adjustment* gradually raises the mountains in response. The processes of uplifting and erosion will continue until the mountain block reaches "normal" crustal thickness. Gravity also causes elevated mountainous structures to collapse under their own "weight."

Key Terms

accretionary wedge (p. 296)
active continental margin (p. 296)
anticline (p. 286)
basin (p. 287)

brittle failure (brittle deformation) (p. 286)
deformation (p. 284)
dip-slip fault (p. 288)
dome (p. 287)

ductile deformation (p. 286)
fault (p. 288)
fault-block mountains (p. 301)
fault scarp (p. 288)

fold (p.285)
graben (p. 288)
gravitational collapse (p. 304)
horst (p. 289)
isostasy (p. 303)
isostatic adjustment (p. 303)

joint (p. 294)
monocline (p. 287)
normal fault (p. 288)
orogenesis (p. 294)
passive continental margin (p. 296)
reverse fault (p. 291)

strike-slip fault (p. 291)
syncline (p. 286)
terrane (p. 298)
thrust fault (p. 291)
transform fault (p. 291)

Review Questions

1. What is rock *deformation*?

2. How is *brittle deformation* different from *ductile deformation*?

3. List three factors that determine how rocks will behave when exposed to stresses that exceed their strength. Briefly explain the role of each.

4. Distinguish between *anticlines* and *synclines, domes* and *basins, anticlines* and *domes*.

5. How is a *monocline* different from an *anticline*?

6. The Black Hills of South Dakota are a good example of what type of structural feature?

7. Contrast the movements that occur along normal and reverse faults. What type of force is indicated by each fault?

8. Is the fault shown in Figure 10.10 a normal or reverse fault?

9. Describe a *horst* and a *graben*. Explain how a graben valley forms, and name one.

10. What type of faults are associated with fault-block mountains?

11. How are reverse faults different from thrust faults? In what way are they the same?

12. The San Andreas Fault is an excellent example of a _____ fault.

13. How are joints different from faults?

14. In the plate tectonics model, which type of plate boundary is most directly associated with mountain building?

15. Briefly describe the development of a volcanic island arc.

16. The formation of mountainous topography at a volcanic island arc, such as Japan, is considered just one phase in the development of a major mountain belt. Explain.

17. What is an *accretionary wedge*? Briefly describe its formation.

18. What is a *passive margin*? Give an example. Then give an example of an *active continental margin*.

19. In what way are the Sierra Nevada and the Andes similar?

20. How can the Appalachian Mountains be considered a collision-type mountain range when the nearest continent is 5000 kilometers (3000 miles) away?

21. How does the plate tectonics theory help explain the existence of fossil marine life in rocks atop mountains formed by continental collisions?

22. Define the term *terrane*. How is it different from the term *terrain*?

23. In addition to microcontinents, what other structures are thought to be carried by the oceanic lithosphere and eventually accreted to a continent?

24. Briefly describe the major differences between the evolution of the Appalachian Mountains and the North American Cordillera.

25. Compare the processes that generate fault-block mountains to those associated with most other major mountain belts.

26. Give one example of evidence that supports the concept of crustal uplift.

27. What happens to a floating object when weight is added? Subtracted? How does this principal apply to changes in the elevation of mountains? What term is applied to the adjustment that causes crustal uplift of this type?

28. How does the formation and melting of Pleistocene ice sheets support the idea that the lithosphere tries to remain in isostatic balance?

Examining the Earth System

1. A good example of the interactions between Earth's spheres is the influence of mountains on climate. Examine the temperature graphs and annual precipitation amounts for the cities of Seattle and Spokane, Washington (Figure 10.30). Notice on the inset map that mountains (the Cascades) separate the two cities and that the general wind direction in the region is from west to east (the prevailing westerlies of the mid-latitudes). Which city receives the greatest annual precipitation? Contrast the summer and winter temperatures that occur at each city. What influence have the mountains apparently had on Spokane's rainfall and temperature?

2. The Cascades have had a profound effect on the amount and type of plant and animal life (biosphere) that inhabit the region around Spokane and Seattle. Provide several specific examples to verify this statement.

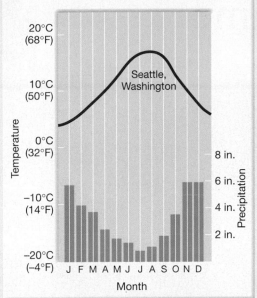
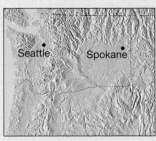
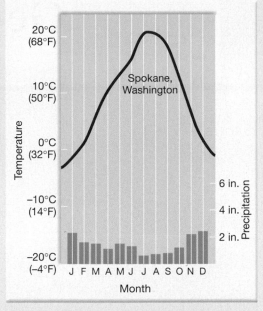

▲ **FIGURE 10.30** Comparison of the average monthly temperatures and precipitation for Seattle and Spokane. Note that Seattle receives an average of 38.4 inches of precipitation per year while Spokane receives only 17 inches of precipitation on average.

Online Study Guide

The *Earth Science* Website uses the resources and flexibility of the Internet to aid in your study of the topics in this chapter. Written and developed by Earth science instructors, this site will help improve your understanding of Earth science. Visit **http://www.prenhall.com/tarbuck** and click on the cover of *Earth Science 11e* to find:

- **On-line review quizzes.**
- **Critical thinking exercises.**
- **Links to chapter-specific Web resources.**
- **Internet-wide key term searches.**

http://www.prenhall.com/tarbuck

CHAPTER 11

Geologic Time

The strata exposed in Arizona's Grand Canyon contain
clues to hundreds of millions of years of Earth history.
(Photo © by David Muench Photography).

n the eighteenth century, James Hutton recognized the immensity of Earth history and the importance of time as a component in all geological processes. In the nineteenth century, others effectively demonstrated that Earth had experienced many episodes of mountain building and erosion, which must have required great spans of geologic time. Although these pioneering scientists understood that Earth was very old, they had no way of knowing its true age. Was it tens of millions, hundreds of millions, or even billions of years old? Rather, a geologic time scale was developed that showed the sequence of events based on relative dating principles. What are these principles? What part do fossils play? With the discovery of radioactivity and the development of radiometric dating techniques, geologists now can assign fairly accurate dates to many of the events in Earth history. What is radioactivity? Why is it a good "clock" for dating the geologic past? In this chapter we shall answer these questions.

Geology Needs a Time Scale

In 1869 John Wesley Powell, who was later to head the U.S. Geological Survey, led a pioneering expedition down the Colorado River and through the Grand Canyon (Figure 11.1). Writing about the rock layers that were exposed by the downcutting of the river, Powell said that "the canyons of this region would be a Book of Revelations in the rock-leaved Bible of geology." He was undoubtedly impressed with the millions of years of Earth history exposed along the walls of the Grand Canyon (see chapter-opening photo).

Powell realized that the evidence for an ancient Earth is concealed in its rocks. Like the pages in a long and complicated history book, rocks record the geological events and changing life forms of the past. The book, however, is not complete. Many pages, especially in the early chapters, are missing. Others are tattered, torn, or smudged. Yet enough of the book remains to allow much of the story to be deciphered.

Interpreting Earth history is a prime goal of the science of geology. Like a modern-day sleuth, the geologist must interpret clues found preserved in the rocks. By studying rocks, especially sedimentary rocks, and the features they contain, geologists can unravel the complexities of the past.

Geological events by themselves, however, have little meaning until they are put into a time perspective. Studying history, whether it be the Civil War or the Age of Dinosaurs, requires a calendar. Among geology's major contributions to human knowledge is the *geologic time scale* and the discovery that Earth history is exceedingly long.

The geologists who developed the geologic time scale revolutionized the way people think about time and how they perceive our planet. They learned that

▼ **FIGURE 11.1** **A.** Start of the expedition from Green River station. A drawing from Powell's 1875 book. **B.** Major John Wesley Powell, pioneering geologist and the second director of the U.S. Geological Survey. (Courtesy of the U.S. Geological Survey, Denver)

A.

B.

Earth is much older than anyone had previously imagined and that its surface and interior have been changed over and over again by the same geological processes that operate today.

A Brief History of Geology

In the mid-1600s, James Ussher, Anglican Archbishop of Armagh, Primate of all Ireland, published a work that had immediate and profound influence on people's view of Earth's age. A respected scholar of the Bible, Ussher constructed a chronology of human and Earth history in which he determined that Earth was only a few thousand years old, having been created in 4004 B.C. Ussher's treatise earned widespread acceptance among Europe's scientific and religious leaders, and his chronology was soon printed in the margins of the Bible itself.

During the seventeenth and eighteenth centuries the doctrine of **catastrophism** strongly influenced people's thinking about Earth. Briefly stated, catastrophists believed that Earth's landscapes had been developed primarily by great catastrophes. Features such as mountains and canyons, which today we know take great periods of time to form, were explained as having been produced by sudden and often worldwide disasters triggered by unknowable causes that no longer operate. This philosophy was an attempt to fit the rate of Earth processes to the prevailing ideas on the age of Earth.

Birth of Modern Geology

Modern geology began in the late 1700s when James Hutton, a Scottish physician and gentleman farmer, published his *Theory of the Earth*. In this work, Hutton put forth a fundamental principle that is a pillar of geology today: **uniformitarianism.** It simply states that *the physical, chemical, and biological laws that operate today have also operated in the geologic past.* This means that the forces and processes that we observe presently shaping our planet have been at work for a very long time. Thus, to understand ancient rocks, we must first understand present-day processes and their results. This idea is commonly expressed by saying "The present is the key to the past" (see Box 11.1).

Box 11.1 Understanding Earth: Deciphering the Past by Understanding the Present

Louis Agassiz, a Swiss scientist born in 1807, was instrumental in formulating modern ideas about the Ice Age (Figure 11.A). The development of this knowledge provides an excellent example of the application of the principle of uniformitarianism.

In 1821 Agassiz heard another scientist present a paper in which he indicated that glacial features occurred in places that were a significant distance from existing glaciers in the Alps. This, of course, implied that the glaciers had once occupied areas considerably beyond their present limits. Agassiz was skeptical about this hypothesis and set out to invalidate it. Ironically, his fieldwork in the Alps convinced him of the merits of his colleague's hypothesis. Agassiz found the same unique deposits and features that can be seen forming in association with active glaciers in places far beyond the limits of the ice. Subsequent work led Agassiz to hypothesize that a great ice age had occurred in response to a period of worldwide climate change and had affected large parts of the globe. Agassiz's ideas eventually developed into our present-day glacial theory.

The proof of the glacial theory proposed by Agassiz and others constitutes a classic example of applying the principle of uniformitarianism. Realizing that certain landforms and other features are produced by no other known process but glacial activity, they were able to reconstruct the extent of now vanished ice sheets. Clearly, understanding the present was the key to deciphering the past.

Figure 11.A Louis Agassiz (1807–1873) played a major role in the development of glacial theory. (Courtesy of Harvard University Archives)

Prior to Hutton's *Theory of the Earth,* no one had effectively demonstrated that geological processes occur over extremely long periods of time. However, Hutton persuasively argued that weak, slow-acting processes could, over long spans of time, produce effects just as great as those resulting from sudden catastrophic events. Unlike his predecessors, Hutton carefully cited verifiable observations to support his ideas.

For example, when he argued that mountains are sculpted and ultimately destroyed by weathering and the work of running water, and that their wastes are carried to the oceans by processes that can be observed, Hutton said, "We have a chain of facts which clearly demonstrates that the materials of the wasted mountains have traveled through the rivers"; and further, "There is not one step in all this progress that is not to be actually perceived." He then went on to summarize this thought by asking a question and immediately providing the answer: "What more can we require? Nothing but time."

Geology Today

Today the basic tenets of uniformitarianism are just as viable as in Hutton's day. Indeed, we realize more strongly than ever that the present gives us insight into the past and that the physical, chemical, and biological laws that govern geological processes remain unchanging through time. However, we also understand that the doctrine should not be taken too literally. To say that geological processes in the past were the same as those occurring today is not to suggest that they always had the same relative importance or that they operated at precisely the same rate. Moreover, some important geologic processes are not currently observable, but evidence that they occur is well established. For example, we know that Earth has experienced impacts from large meteorites even though we have no human witnesses. Such events altered Earth's crust, modified its climate, and strongly influenced life on the planet.

The acceptance of uniformitarianism meant the acceptance of a very long history for Earth. Although Earth's processes vary in intensity, they still take a very long time to create or destroy major landscape features.

For example, geologists have established that mountains once existed in portions of present-day Minnesota, Wisconsin, and Michigan. Today the region consists of low hills and plains. Erosion gradually destroyed these peaks. Estimates indicate that the North American continent is being lowered at a rate of about 3 centimeters per 1000 years. At this rate, it would take 100 million years for water, wind, and ice to lower mountains that were 300 meters (10,000 feet) high.

But even this time span is relatively short on the time scale of Earth history, for the rock record contains evidence that shows Earth has experienced many cycles of mountain building and erosion. Concerning the ever-changing nature of Earth through great expanses of geologic time, Hutton made a statement that was to become his most famous. In concluding his classic 1788 paper published in the *Transactions of the Royal Society of Edinburgh,* he stated, "The results, therefore, of our present enquiry is, that we find no vestige of a beginning—no prospect of an end."

It is important to remember that although many features of our physical landscape may seem to be unchanging over our lifetimes, they are nevertheless changing, but on time scales of hundreds, thousands, or even many millions of years.

Relative Dating—Key Principles

 GEODe
Deciphering Earth's History
▼ Relative Dating

The geologists who developed the geologic time scale revolutionized the way people think about time and perceive our planet. They learned that Earth is much older than anyone had previously imagined and that its surface and interior have been changed over and over again by the same geological processes that operate today.

During the late 1800s and early 1900s, various attempts were made to determine the age of Earth. Although some of the methods appeared promising at the time, none proved reliable. What these scientists were seeking was a **numerical date.** Such dates specify the actual number of years that have passed since an event occurred—for example, the extinction of the dinosaurs about 65 million years ago. Today our understanding of radioactivity allows us to accurately determine numerical dates for rocks that represent important events in Earth's distant past. We will study radioactivity later in this chapter. Prior to the discovery of radioactivity, geologists had no accurate and dependable method of numerical dating and had to rely solely on relative dating.

Relative dating means placing rocks in their proper *sequence of formation*—which ones formed first, second, third, and so on. Relative dating cannot tell us how long ago something took place, only that it followed one event and preceded another. The relative dating techniques that were developed are valuable and still widely used. Numerical dating methods did not replace these techniques; they simply supplemented them. To establish a relative time scale, a few basic principles or rules had to be discovered and applied. Although they may seem obvious to us today, they were major breakthroughs in thinking at the time, and their discovery and acceptance was an important scientific achievement.

Law of Superposition

Nicolaus Steno, a Danish anatomist, geologist, and priest (1636–1686), is credited with being the first to recognize a sequence of historical events in an outcrop of sedimentary rock layers. Working in the mountains of western Italy, Steno applied a very simple rule that has come to be the most basic principle of relative dating—the **law of superposition.** The law simply states that in an undeformed sequence of sedimentary rocks, each bed is older than the one above it and younger than the one below. Although it may seem obvious that a rock layer could not be deposited unless it had something older beneath it for support, it was not until 1669 that Steno clearly stated the principle.

This rule also applies to other surface-deposited materials, such as lava flows and beds of ash from volcanic eruptions. Applying the law of superposition to the beds exposed in the upper portion of the Grand Canyon (Figure 11.2), you can easily place the layers in their proper order. Among those that are shown, the sedimentary rocks in the Supai Group must be the oldest, followed in order by the Hermit Shale, Coconino Sandstone, Toroweap Formation, and Kaibab Limestone.

Principle of Original Horizontality

Steno is also credited with recognizing the importance of another basic principle, called the **principle of original horizontality.** Simply stated, it means that layers of sediment are generally deposited in a horizontal position. Thus, if we observe rock layers that are flat, it means they have not been disturbed and thus still have their *original* horizontality. The

A.

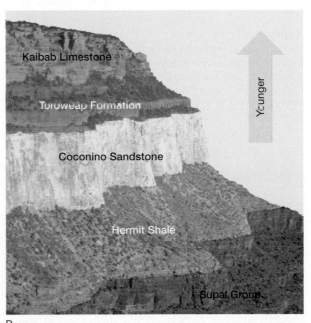

B.

▲ **FIGURE 11.2** Applying the law of superposition to these layers exposed in the upper portion of the Grand Canyon, the Supai Group is oldest and the Kaibab Limestone is youngest. (Photo by E. J. Tarbuck)

▲ FIGURE 11.3 Most layers of sediment are deposited in a nearly horizontal position. Thus, when we see rock layers that are folded or tilted, we can assume that they must have been moved into that position by crustal disturbances *after* their deposition. These folded layers are exposed at Stair Hole, near Lulworth, Dorset, England. (Photo by Tom and Susan Bean, Inc.)

layers in the Grand Canyon illustrate this in the chapter-opening photo and in Figure 11.2. But if they are folded or inclined at a steep angle, they must have been moved into that position by crustal disturbances sometime *after* their deposition (Figure 11.3).

Principle of Cross-Cutting Relationships

When a fault cuts through other rocks, or when magma intrudes and crystallizes, we can assume that the fault or intrusion is younger than the rocks affected. For example, in Figure 11.4, the faults and dikes clearly must have occurred *after* the sedimentary layers were deposited.

This is the **principle of cross-cutting relationships.** By applying the cross-cutting principle, you can see that fault *A* occurred *after* the sandstone layer was deposited, because it "broke" the layer. However, fault *A* occurred *before* the conglomerate was laid down, because that layer is unbroken.

We can also state that dike *B* and its associated sill are older than dike *A*, because dike *A* cuts the sill. In the same manner, we know that the batholith was emplaced after movement occurred along fault *B*, but before dike *B* was formed. This is true because the batholith cuts across fault *B*, and dike *B* cuts across the batholith.

Inclusions

Sometimes inclusions can aid the relative dating process. **Inclusions** are pieces of one rock unit that are contained within another. The basic principle is logical and straightforward. The rock mass adjacent to the one containing the inclusions must have been there first in order to provide the rock fragments. Therefore, the rock mass containing inclusions is the younger of the two. Figure 11.5 provides an example. Here the inclusions of intrusive igneous rock in the

► FIGURE 11.4 Cross-cutting relationships are an important principle used in relative dating. An intrusive rock body is younger than the rocks it intrudes. A fault is younger than the rock layers it cuts.

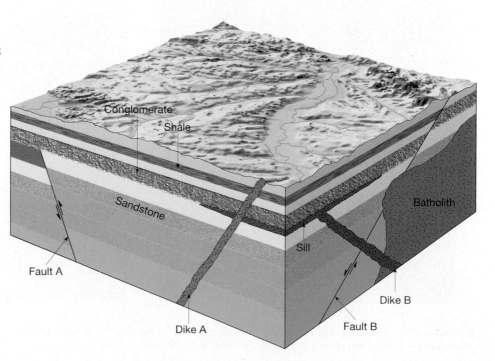

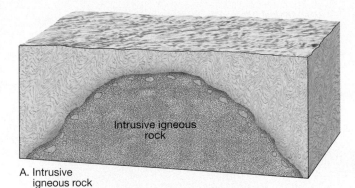

A. Intrusive
igneous rock

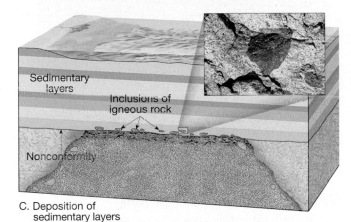

B. Exposure and
weathering of intrusive igneous rock

C. Deposition of
sedimentary layers

▲ **FIGURE 11.5** These diagrams illustrate two ways that inclusions can form, as well as a type of unconformity termed a nonconformity. In part **A,** the inclusions in the igneous mass represent unmelted remnants of the surrounding host rock that were broken off and incorporated at the time the magma was intruded. In part **C,** the igneous rock must be older than the overlying sedimentary beds because the sedimentary beds contain inclusions of the igneous rock. When older intrusive igneous rocks are overlain by younger sedimentary layers, a nonconformity is said to exist. The photo shows an inclusion of dark igneous rock in a lighter-colored and younger host rock. (Photo by Tom Bean)

adjacent sedimentary layer indicate that the sedimentary layer was deposited on top of a weathered igneous mass rather than being intruded from below by magma that later crystallized.

Unconformities

When we observe layers of rock that have been deposited essentially without interruption, we call them **conformable.** Particular sites exhibit conformable beds representing certain spans of geologic time. However, no place on Earth has a complete set of conformable strata.

Throughout Earth history, the deposition of sediment has been interrupted again and again. All such breaks in the rock record are termed unconformities. An **unconformity** represents a long period during which deposition ceased, erosion removed previously formed rocks, and then deposition resumed. In each case uplift and erosion are followed by subsidence and renewed sedimentation. Unconformities are important features because they represent significant geologic events in Earth history. Moreover, their recognition helps us identify what intervals of time are not represented by strata and thus are missing from the geologic record.

The rocks exposed in the Grand Canyon of the Colorado River represent a tremendous span of geologic history. It is a wonderful place to take a trip through time. The canyon's colorful strata record a long history of sedimentation in a variety of environments—advancing seas, rivers and deltas, tidal flats, and sand dunes. But the record is not continuous. Unconformities represent vast amounts of time that have not been recorded in the canyon's layers. Figure 11.6 is a geologic cross section of the Grand Canyon. Refer to it as you read about the three basic types of unconformities: angular unconformities, disconformities, and nonconformities.

Angular Unconformity. Perhaps the most easily recognized unconformity is an **angular unconformity.** It consists of tilted or folded sedimentary rocks that are overlain by younger, more flat-lying strata. An angular unconformity indicates that during the pause in deposition, a period of deformation (folding or tilting) and erosion occurred (Figure 11.7).

When James Hutton studied an angular unconformity in Scotland more than 200 years ago, it was clear to him that it represented a major episode of geologic activity (Figure 11.7E). He also appreciated the immense time span implied by such relationships. When a companion later wrote of their visit to the site, he stated that "the mind seemed to grow giddy by looking so far into the abyss of time."

Disconformity. When contrasted with angular unconformities, **disconformities** are more common, but usually far less conspicuous because the strata on either side are essentially parallel. For example, look at the disconformities in the cross section of the Grand Canyon in Figure 11.6. Many disconformities

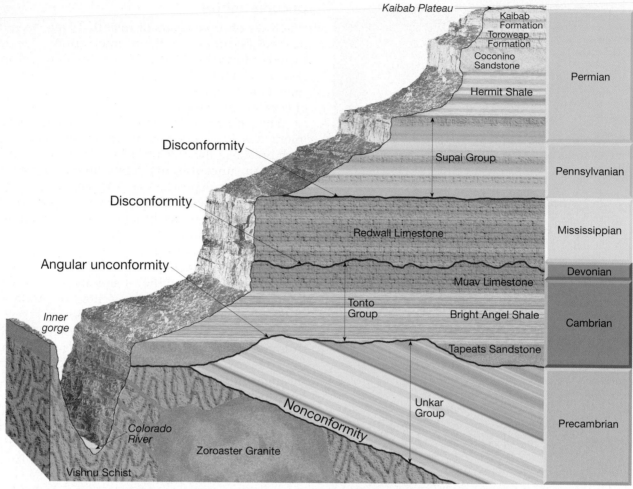

▲ **FIGURE 11.6** This cross section through the Grand Canyon illustrates the three basic types of unconformities. An angular unconformity can be seen between the tilted Precambrian Unkar Group and the horizontal Cambrian-age layers above. Two disconformities are marked, above and below the Redwall Limestone. A nonconformity occurs between the igneous and metamorphic rocks of the Inner Gorge and the sedimentary strata of the Unkar Group.

are difficult to identify because the rocks above and below are similar and there is little evidence of erosion. Such a break often resembles an ordinary bedding plane. Other disconformities are easier to identify because the ancient erosion surface is cut deeply into the older rocks below.

Nonconformity. The third basic type of unconformity is a **nonconformity**. Here the break separates older metamorphic or intrusive igneous rocks from younger sedimentary strata (Figures 11.5 and 11.6). Just as angular unconformities and disconformities imply crustal movements, so too do nonconformities. Intrusive igneous masses and metamorphic rocks originate far below the surface. Thus, for a nonconformity to develop, there must be a period of uplift and the erosion of overlying rocks. Once exposed at the surface, the igneous or metamorphic rocks are subjected to weathering and erosion prior to subsidence and the renewal of sedimentation.

Using Relative Dating Principles

If you apply the principles of relative dating to the hypothetical geologic cross section in Figure 11.8, you can place in proper sequence the rocks and the events they represent. The statements within the figure summarize the logic used to interpret the cross section.

In this example, we establish a relative time scale for the rocks and events in the area of the cross section. Remember that this method gives us no indication as to how many years of Earth history are represented, for we have no numerical dates. Nor do we know how this area compares to any other.

Correlation of Rock Layers

To develop a geologic time scale that is applicable to the entire Earth, rocks of similar age in different regions must be matched up. Such a task is referred to as **correlation.**

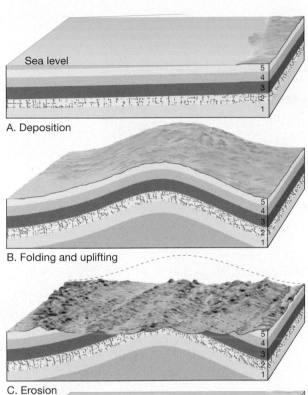

A. Deposition

B. Folding and uplifting

C. Erosion

6 (Angular unconformity)

Sea level

D. Subsidence and renewed deposition

E.

▲ **FIGURE 11.7** Formation of an angular unconformity. An angular unconformity represents an extended period during which deformation and erosion occurred. Part E shows an angular unconformity at Siccar Point, Scotland, that was first described by James Hutton more than 200 years ago. (Photo by Edward Hay)

Within a limited area, correlating the rocks of one locality with those of another may be done simply by walking along the outcropping edges. However, this might not be possible when the rocks are mostly con-

cealed by soil and vegetation. Correlation over short distances is often achieved by noting the position of a distinctive rock layer in a sequence of strata. Or, a layer may be identified in another location if it is composed of very distinctive or uncommon minerals.

By correlating the rocks from one place to another, a more comprehensive view of the geologic history of a region is possible. Figure 11.9, for example, shows the correlation of strata at three sites on the Colorado Plateau in southern Utah and northern Arizona. No single locale exhibits the entire sequence, but correlation reveals a more complete picture of the sedimentary rock record.

Many geologic studies involve relatively small areas. Such studies are important in their own right, but their full value is realized only when the rocks are correlated with those of other regions. Although the methods just described are sufficient to trace a rock formation over relatively short distances, they are not adequate for matching rocks that are separated by great distances. When correlation between widely separated areas or between continents is the objective, geologists must rely on fossils.

Fossils: Evidence of Past Life

Fossils, the remains or traces of prehistoric life, are important inclusions in sediment and sedimentary rocks. They are important basic tools for interpreting the geologic past. The scientific study of fossils is called **paleontology.** It is an interdisciplinary science that blends geology and biology in an attempt to understand all aspects of the succession of life over the vast expanse of geologic time. Knowing the nature of the life forms that existed at a particular time helps researchers understand past environmental conditions. Further, fossils are important time indicators and play a key role in correlating rocks of similar ages that are from different places.

Types of Fossils

Fossils are of many types. The remains of relatively recent organisms may not have been altered at all. Such objects as teeth, bones, and shells are common examples. Far less common are entire animals, flesh included, that have been preserved because of rather unusual circumstances. Remains of prehistoric elephants called mammoths that were frozen in the Arctic tundra of Siberia and Alaska are examples, as are the mummified remains of sloths preserved in a dry cave in Nevada.

Given enough time, the remains of an organism are likely to be modified. Often fossils become *petrified* (literally, "turned into stone"), meaning that the small internal cavities and pores of the original structure are filled with precipitated mineral matter

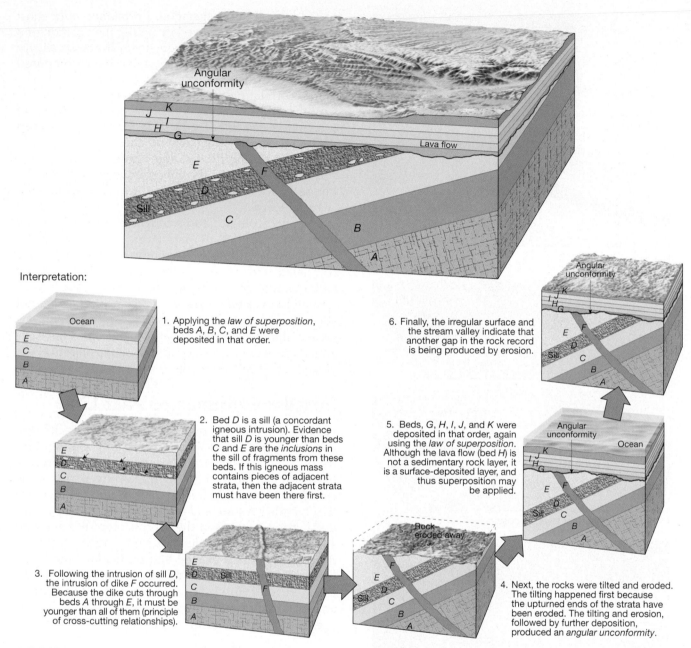

Interpretation:

1. Applying the *law of superposition*, beds *A*, *B*, *C*, and *E* were deposited in that order.

2. Bed *D* is a sill (a concordant igneous intrusion). Evidence that sill *D* is younger than beds *C* and *E* is the *inclusions* in the sill of fragments from these beds. If this igneous mass contains pieces of adjacent strata, then the adjacent strata must have been there first.

3. Following the intrusion of sill *D*, the intrusion of dike *F* occurred. Because the dike cuts through beds *A* through *E*, it must be younger than all of them (principle of cross-cutting relationships).

4. Next, the rocks were tilted and eroded. The tilting happened first because the upturned ends of the strata have been eroded. The tilting and erosion, followed by further deposition, produced an *angular unconformity*.

5. Beds, *G*, *H*, *I*, *J*, and *K* were deposited in that order, again using the *law of superposition*. Although the lava flow (bed *H*) is not a sedimentary rock layer, it is a surface-deposited layer, and thus superposition may be applied.

6. Finally, the irregular surface and the stream valley indicate that another gap in the rock record is being produced by erosion.

▲ **FIGURE 11.8** Geologic cross section of a hypothetical region.

(Figure 11.10A). In other instances *replacement* may occur. Here the cell walls and other solid material are removed and replaced with mineral matter. Sometimes the microscopic details of the replaced structure are faithfully retained.

Molds and casts constitute another common class of fossils. When a shell or other structure is buried in sediment and then dissolved by underground water, a *mold* is created. The mold faithfully reflects only the shape and surface marking of the organism; it does not reveal any information concerning its internal structure. If these hollow spaces are subsequently filled with mineral matter, *casts* are created (Figure 11.10B).

A type of fossilization called *carbonization* is particularly effective in preserving leaves and delicate animal

forms. It occurs when fine sediment encases the remains of an organism. As time passes, pressure squeezes out the liquid and gaseous components and leaves behind a thin residue of carbon (Figure 11.10C). Black shales deposited as organic-rich mud in oxygen-poor environments often contain abundant carbonized remains. If the film of carbon is lost from a fossil preserved in fine-grained sediment, a replica of the surface, called an *impression*, may still show considerable detail (Figure 11.10D).

Delicate organisms, such as insects, are difficult to preserve, and consequently they are relatively rare in the fossil record. Not only must they be protected from decay but they must not be subjected to any pressure that would crush them. One way in which some insects have been preserved is in *amber*, the hardened resin of ancient

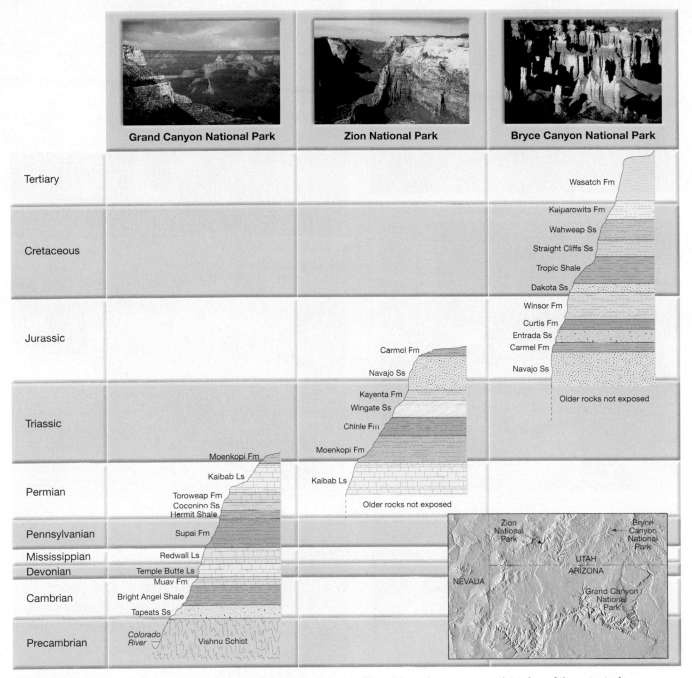

▲ FIGURE 11.9   Correlation of strata at three locations on the Colorado Plateau reveals a more complete view of the extent of sedimentary rocks in the region. (After U.S. Geological Survey; photos by E. J. Tarbuck)

trees. The fly in Figure 11.10E was preserved after being trapped in a drop of sticky resin. Resin sealed off the insect from the atmosphere and protected the remains from damage by water and air. As the resin hardened, a protective pressure-resistant case was formed.

In addition to the fossils already mentioned, there are numerous other types, many of them only traces of prehistoric life. Examples of such indirect evidence include:

1. Tracks—animal footprints made in soft sediment that was later lithified (Figure 11.10 F).

2. Burrows—tubes in sediment, wood, or rock made by an animal. These holes may later become filled with mineral matter and preserved. Some of the oldest-known fossils are believed to be worm burrows.

3. Coprolites—fossil dung and stomach contents that can provide useful information pertaining to food habits of organisms.

4. Gastroliths—highly polished stomach stones that were used in the grinding of food by some extinct reptiles.

▲ **FIGURE 11.10** There are many types of fossilization. Six examples are shown here. **A.** Petrified wood in Petrified Forest National Park, Arizona. **B.** Natural casts of shelled invertebrates. **C.** A fossil bee preserved as a thin carbon film. **D.** Impressions are common fossils and often show considerable detail. **E.** Insect in amber. **F.** Dinosaur footprint in fine-grained limestone near Tuba City, Arizona. (Photo A by David Muench; Photos B, D, and F by E. J. Tarbuck; Photo C courtesy of the National Park Service; Photo E by Breck P. Kent)

Conditions Favoring Preservation

Only a tiny fraction of the organisms that have lived during the geologic past have been preserved as fossils. Normally the remains of an animal or plant are destroyed. Under what circumstances are they preserved? Two special conditions appear to be necessary: rapid burial and the possession of hard parts.

When an organism perishes, its soft parts usually are quickly eaten by scavengers or decomposed by bacteria. Occasionally, however, the remains are buried by sediment. When this occurs, the remains are protected from the environment, where destructive processes operate. Rapid burial therefore is an important condition favoring preservation.

In addition, animals and plants have a much better chance of being preserved as part of the fossil record if they have hard parts. Although traces and imprints of soft-bodied animals such as jellyfish, worms, and insects exist, they are not common. Flesh usually decays so rapidly that preservation is exceedingly unlikely. Hard parts such as shells, bones, and teeth predominate in the record of past life.

Because preservation is contingent on special conditions, the record of life in the geologic past is biased. The fossil record of those organisms with hard parts that lived in areas of sedimentation is quite abundant. However, we get only an occasional glimpse of the vast array of other life forms that did not meet the special conditions favoring preservation.

STUDENTS SOMETIMES ASK...
*How is paleontology different
from archaeology?*

People frequently confuse these two areas of study because a common perception of both paleontologists and archaeologists is of scientists carefully extracting important clues about the past from layers of rock or sediment. While it is true that scientists in both disciplines "dig" a lot, the focus of each is different. Paleontologists study fossils and are concerned with *all* life forms in the geologic past. By contrast, archaeologists focus on the material remains of past human life. These remains include both the objects used by people long ago, called *artifacts,* and the buildings and other structures associated with where people lived, called *sites.* Archaeologists help us learn about how our human ancestors met the challenges of life in the past.

Fossils and Correlation

The existence of fossils had been known for centuries, yet it was not until the late 1700s and early 1800s that their significance as geologic tools was made evident. During this period an English engineer and canal builder, William Smith, discovered that each rock formation in the canals he worked on contained fossils unlike those in the beds either above or below. Further, he noted that sedimentary strata in widely separated areas could be identified and correlated by their distinctive fossil content.

Based on Smith's classic observations and the findings of many geologists who followed, one of the most important and basic principles in historical geology was formulated: *Fossil organisms succeed one another in a definite and determinable order, and therefore any time period can be recognized by its fossil content.* This has come to be known as the **principle of fossil succession.** In other words, when fossils are arranged according to their age by applying the law of superposition to the rocks in which they are found, they do not present a random or haphazard picture. To the contrary, fossils show changes that document the evolution of life through time.

For example, an Age of Trilobites is recognized quite early in the fossil record. Then, in succession, paleontologists recognize an Age of Fishes, an Age of Coal Swamps, an Age of Reptiles, and an Age of Mammals. These "ages" pertain to groups that were especially plentiful and characteristic during particular time periods. Within each of the ages, there are many subdivisions based, for example, on certain species of trilobites and certain types of fish, reptiles, and so on. This same succession of dominant organisms, never out of order, is found on every continent.

Once fossils were recognized as time indicators, they became the most useful means of correlating rocks of similar age in different regions. Geologists pay particular attention to certain fossils called **index fossils.** These fossils are widespread geographically and are limited to a short span of geologic time, so their presence provides an important method of matching rocks of the same age. Rock formations, however, do not always contain a specific index fossil. In such situations, groups of fossils are used to establish the age of the bed. Figure 11.11 illustrates how an assemblage of fossils can be used to date rocks more precisely than could be accomplished by the use of only one of the fossils.

In addition to being important and often essential tools for correlation, fossils are important environmental indicators. Although much can be deduced about past environments by studying the nature and characteristics of sedimentary rocks, a close examination of any fossils present can usually provide a great deal more information.

For example, when the remains of certain clam shells are found in limestone, the geologist can assume that the region was once covered by a shallow sea, because that is where clams live today. Also, by using what we know of living organisms, we can conclude that fossil animals with thick shells capable of withstanding pounding and surging waves must have inhabited shorelines. Conversely, animals with thin, delicate shells probably indicate deep, calm offshore

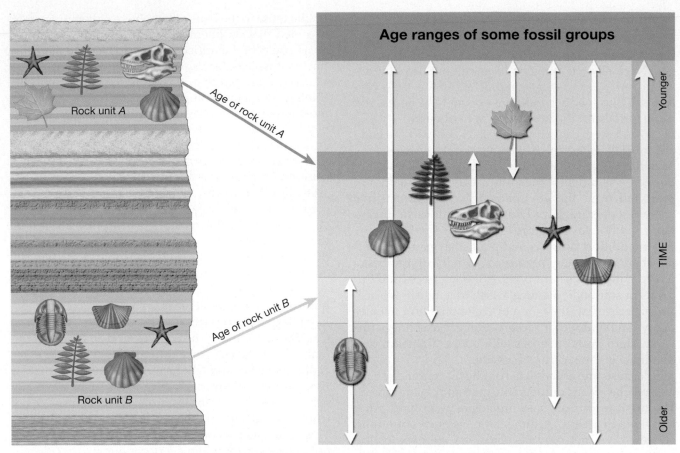

▲ **FIGURE 11.11** Overlapping ranges of fossils help date rocks more exactly than using a single fossil.

waters. Hence, by looking closely at the types of fossils, the approximate position of an ancient shoreline may be identified.

Further, fossils can indicate the former temperature of the water. Certain present-day corals require warm and shallow tropical seas like those around Florida and the Bahamas. When similar corals are found in ancient limestones, they indicate that a Florida-like marine environment must have existed when the corals were alive. These examples illustrate how fossils can help unravel the complex story of Earth history.

Dating with Radioactivity

GEODe
EARTH SCIENCE

Deciphering Earth's History
▼ Radiometric Dating

In addition to establishing relative dates by using the principles described in the preceding sections, it is also possible to obtain reliable numerical dates for events in the geologic past. For example, we know that Earth is about 4.5 billion years old and that the dinosaurs became extinct about 65 million years ago. Dates that are expressed in millions and billions of years truly stretch our imagination because our personal calendars involve time

measured in hours, weeks, and years. Nevertheless, the vast expanse of geologic time is a reality, and it is radiometric dating that allows us to measure it. In this section you will learn about radioactivity and its application in radiometric dating.

Reviewing Basic Atomic Structure

Recall from Chapter 1 that each atom has a *nucleus* containing protons and neutrons and that the nucleus is orbited by electrons. *Electrons* have a negative electrical charge, and *protons* have a positive charge. A *neutron* is actually a proton and an electron combined, so it has no charge (it is neutral).

The *atomic number* (each element's identifying number) is the number of protons in the nucleus. Every element has a different number of protons and thus a different atomic number (hydrogen = 1, carbon = 6, oxygen = 8, uranium = 92, etc.). Atoms of the same element always have the same number of protons, so the atomic number stays constant.

Practically all of an atom's mass (99.9%) is in the nucleus, indicating that electrons have virtually no mass at all. So, by adding the protons and neutrons in an atom's nucleus, we derive the atom's *mass number*. The number of neutrons can vary, and these variants, or *isotopes,* have different mass numbers.

To summarize with an example, uranium's nucleus always has 92 protons, so its atomic number always is 92. But its neutron poplation varies, so uranium has three isotopes: uranium-234, (protons + neutrons = 234), uranium-235, and uranium-238. All three isotopes are mixed in nature. They look the same and behave the same in chemical reactions.

Radioactivity

The forces that bind protons and neutrons together in the nucleus usually are strong. However, in some isotopes, the nuclei are unstable because the forces binding protons and neutrons together are not strong enough. As a result, the nuclei spontaneously break apart (decay), a process called **radioactivity.**

What happens when unstable nuclei break apart? Three common types of radioactive decay are illustrated in Figure 11.12. and are summarized as follows:

1. Alpha particles (α particles) may be emitted from the nucleus. An alpha particle consists of 2 protons and 2 neutrons. Consequently, the emission of an alpha particle means that the mass number of the isotope is reduced by 4 and the atomic number is decreased by 2.

2. When a beta particle (β particle), or electron, is given off from a nucleus, the mass number remains unchanged, because electrons have practically no mass. However, because the electron has come from a neutron (remember, a neutron is a combination of a proton and an electron), the nucleus contains one more proton than before. Therefore, the atomic number increases by 1.

3. Sometimes an electron is captured by the nucleus. The electron combines with a proton and forms an additional neutron. As in the last example, the mass number remains unchanged. However, as the nucleus now contains one less proton, the atomic number decreases by 1.

An unstable (radioactive) isotope of an element is called the *parent*. The isotopes resulting from the decay

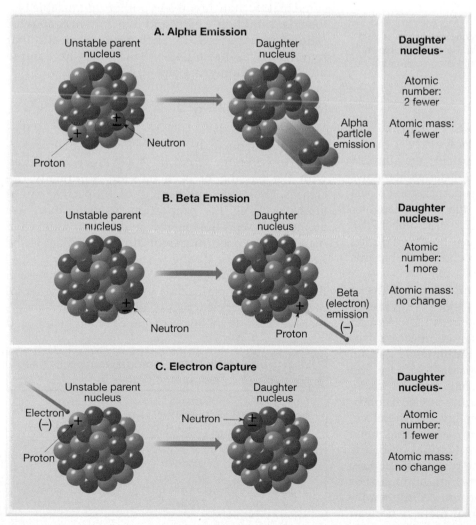

◄ **FIGURE 11.12** Common types of radioactive decay. Notice that in each case the number of protons (atomic number) in the nucleus changes, thus producing a different element.

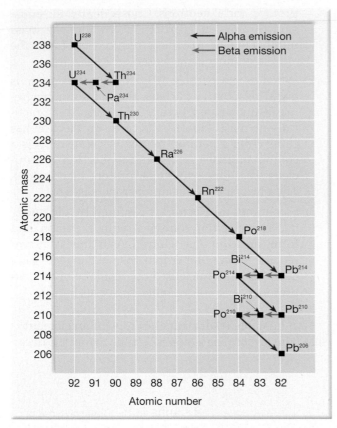

▲ **FIGURE 11.13**   The most common isotope of uranium (U-238) is an example of a radioactive decay series. Before the stable end product (Pb-206) is reached, many different isotopes are produced as intermediate steps.

of the parent are the *daughter products*. Figure 11.13 provides an example of radioactive decay. Here it can be seen that when the radioactive parent, uranium-238 (atomic number 92, mass number 238), decays, it follows a number of steps, emitting 8 alpha particles and 6 beta particles before finally becoming the stable daughter product lead-206 (atomic number 82, mass number 206). One of the unstable daughter products produced during this decay series is radon. Box 11.2 examines the hazards associated with this radioactive gas.

Certainly among the most important results of the discovery of radioactivity is that it provided a reliable means of calculating the ages of rocks and minerals that contain particular radioactive isotopes. The procedure is called **radiometric dating.** Why is radiometric dating reliable? Because the rates of decay for many isotopes have been precisely measured and do not vary under the physical conditions that exist in Earth's outer layers. Therefore, each radioactive isotope used for dating has been decaying at a fixed rate since the formation of the rocks in which it occurs, and the products of decay have been accumulating at a corresponding rate. For example, when uranium is incorporated into a mineral that crystallizes from magma, there is no lead

(the stable daughter product) from previous decay. The radiometric "clock" starts at this point. As the uranium in this newly formed mineral disintegrates, atoms of the daughter product are trapped, and measurable amounts of lead eventually accumulate.

Half-Life

The time required for one half of the nuclei in a sample to decay is called the **half-life** of the isotope. Half-life is a common way of expressing the rate of radioactive disintegration. Figure 11.14 illustrates what occurs when a radioactive parent decays directly into its stable daughter product. When the quantities of parent and daughter are equal (ratio 1:1), we know that one half-life has transpired. When one-quarter of the original parent atoms remain and three-quarters have decayed to the daughter product, the parent/daughter ratio is 1:3 and we know that two half-lives have passed. After three half-lives, the ratio of parent atoms to daughter atoms is 1:7 (one parent for every seven daughter atoms).

If the half-life of a radioactive isotope is known and the parent/daughter ratio can be measured, the age of the sample can be calculated. For example, assume that the half-life of a hypothetical unstable isotope is 1 million years and the parent/daughter ratio in a sample is 1:15. Such a ratio indicates that four half-lives have passed and that the sample must be 4 million years old.

Radiometric Dating

Notice that the *percentage* of radioactive atoms that decay during one half-life is always the same: 50 percent. However, the *actual number* of atoms that decay with the passing of each half-life continually decreases. Thus, as the percentage of radioactive parent atoms declines, the proportion of stable daughter atoms rises, with the increase in daughter atoms just matching the drop in parent atoms. This fact is the key to radiometric dating.

Of the many radioactive isotopes that exist in nature, five have proved particularly useful in providing radiometric ages for ancient rocks (Table 11.1). Rubidium-87, thorium-232, and the two isotopes of uranium are used

TABLE 11.1	Radioactive isotopes frequently used in radiometric dating.	
Radioactive Parent	**Stable Daughter Product**	**Currently Accepted Half-Life Values**
Uranium-238	Lead-206	4.5 billion years
Uranium-235	Lead-207	713 million years
Thorium-232	Lead-208	14.1 billion years
Rubidium-87	Strontium-87	47.0 billion years
Potassium-40	Argon-40	1.3 billion years

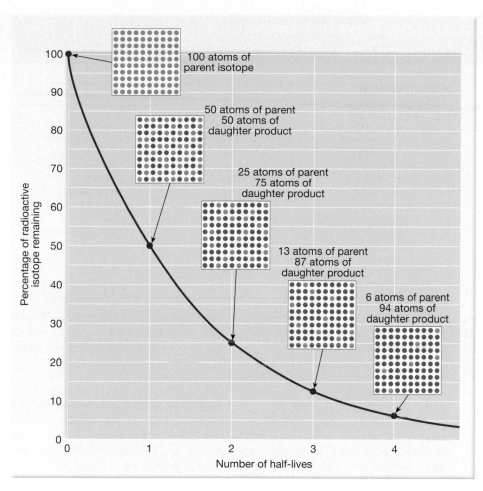

◄ **FIGURE 11.14** The radioactive decay curve shows change that is exponential. Half of the radioactive parent remains after one half-life. After a second half-life one-quarter of the parent remains, and so forth.

only for dating rocks that are millions of years old, but potassium-40 is more versatile. Although the half-life of potassium-40 is 1.3 billion years, analytical techniques make possible the detection of tiny amounts of its stable daughter product, argon-40, in some rocks that are younger than 100,000 years.

It is important to realize that an accurate radiometric date can be obtained only if the mineral remained a closed system during the entire period since its formation. A correct date is not possible unless there was neither the addition nor loss of parent or daughter isotopes. This is not always the case. In fact, an important limitation of the potassium-argon method arises from the fact that argon is a gas, and it may leak from minerals, throwing off measurements. Cross-checking of samples, using two different radiometric methods, is done where possible to ensure accurate age determinations.

Theoretically, no. During each half-life, half of the parent material is converted to daughter product. Then half again is converted after another half-life, and so on.

(Figure 11.14 shows how this logarithmic relationship works—notice that the red line becomes nearly parallel to the horizontal axis after several half-lives.) By converting only half of the remaining parent material to daughter product, there is never a time when all the parent material would be converted. Think about it this way. If you kept cutting a cake in half and eating only half, would you ever eat all of it? (The answer is no, assuming you had a sharp enough knife to slice the cake at an atomic scale!) However, after many half-lives, the parent material can exist in such small amounts that it is essentially undetectable.

Dating with Carbon-14

To date very recent events, carbon-14 is used. Carbon-14 is the radioactive isotope of carbon. The process is often called **radiocarbon dating.** Because the half-life of carbon-14 is only 5730 years, it can be used for dating events from the historic past as well as those from very recent geologic history. In some cases carbon-14 can be used to date events as far back as 75,000 years.

Carbon-14 is continuously produced in the upper atmosphere as a consequence of cosmic-ray bombardment. Cosmic rays, which are high-energy particles, shatter the nuclei of gas atoms, releasing neutrons.

Box 11.2 People and the Environment:
Radon—A Threat to Human Health

Richard L.
Hoffman*

Radioactivity is defined as the spontaneous emission of atomic particles and/or electromagnetic waves from unstable atomic nuclei. For example, in a sample of uranium-238, unstable nuclei decay and produce a variety of radioactive progeny or "daughter" products as well as energetic forms of radiation (Table 11.A). One of its radioactive decay products is radon—a colorless, odorless, invisible gas.

Radon gained public attention in 1984 when a worker in a Pennsylvania nuclear power plant set off radiation alarms not when he left work, but when he first arrived. His clothing and hair were contaminated with radon decay products. Investigation revealed that his basement at home had a radon level 2800 times the average level in indoor air. The home was located along a geological formation known as the Reading Prong—a mass of uranium-bearing black rock that runs from near Reading, Pennsylvania, to near Trenton, New Jersey.

Originating in the radio decay of traces of uranium and thorium found in almost all soils, radon isotopes (Rn-222 and Rn-220) are continually renewed in an ongoing, natural process. Geologists estimate that the top six feet of soil from an average acre of land contains about 50 pounds of uranium (about 2 to 3 parts per million); some types of rocks contain more. Radon is continually generated by the gradual decay of this uranium. Because uranium has a half-life of about 4.5 billion years, radon will be with us forever.

Radon itself decays, having a half-life of only about four days. Its decay products (except lead-206) are all radioactive solids that adhere to dust particles, many of which we inhale. During prolonged exposure to a radon-contaminated environment, some decay will occur while the gas is in the lungs, thereby placing the radioactive radon progeny in direct contact with delicate lung tissue. Steadily accumulating evidence indicates radon to be a significant cause of lung cancer second only to smoking.

A house with a radon level of 4.0 picocuries per liter of air has about eight to nine atoms of radon decaying every minute in every liter of air. The EPA suggests indoor radon levels be kept below this level. EPA risk estimates are conservative; they are based on an assumption that one would spend 75 percent of a 70-year time span (about 52 years) in the contaminated space, which most people would not.

Once radon is produced in the soil, it diffuses throughout the tiny spaces between soil particles. Some radon ultimately reaches the soil surface, where it dissipates into the air. Radon enters buildings and homes through holes and cracks in basement floors and walls. Radon's density is greater than air, so it tends to remain in basements during its short decay cycle.

The source of radon is as enduring as its generation mechanism within Earth; radon will never go away. However, cost-effective mitigation strategies are available to reduce radon to acceptable levels, generally without great expense.

*Dr. Hoffman is Professor of Chemistry, Emeritus, Illinois Central College.

TABLE 11.A	Decay Products of Uranium-238.	
Some Decay Products of Uranium-238	**Decay Particle Produced**	**Half-Life**
Uranium-238	alpha	4.5 billion years
Radium-226	alpha	1600 years
Radon-222	**alpha**	**3.82 days**
Polonium-218	alpha	3.1 minutes
Lead-214	beta	26.8 minutes
Bismuth-214	beta	19.7 minutes
Polonium-214	alpha	1.6×10^{-4} second
Lead-210	beta	20.4 years
Bismuth-210	beta	5.0 days
Polonium-210	alpha	138 days
Lead-206	none	stable

Some of the neutrons are absorbed by nitrogen atoms (atomic number 7), causing their nuclei to emit a proton. As a result, the atomic number decreases by 1 (to 6), and a different element, carbon-14, is created (Figure 11.15A). This isotope of carbon quickly becomes incorporated into carbon dioxide, which circulates in the atmosphere and is absorbed by living matter. As a result, all organisms contain a small amount of carbon-14, including yourself.

While an organism is alive, the decaying radiocarbon is continually replaced, and the proportions of carbon-14 and carbon-12 remain constant. Carbon-12 is the stable and most common isotope of carbon. However, when any plant or animal dies, the amount of carbon-14 gradually decreases as it decays to nitrogen-14 by beta emission (Figure 11.15B). By comparing the proportions of carbon-14 and carbon-12 in a sample, radiocarbon dates can be determined.

Although carbon-14 is useful in dating only the last small fraction of geologic time, it has become a very valuable tool for anthropologists, archaeologists, and historians, as well as for geologists who study very recent Earth history. (Box 11.3 explores another method of studying and dating recent events.) In fact, the development of radiocarbon dat-

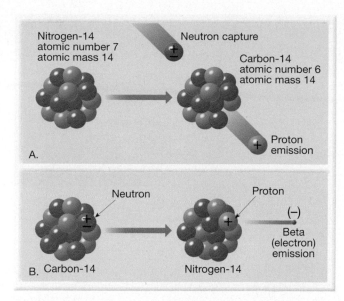

▲ **FIGURE 11.15** **A.** Production and **B.** decay of carbon-14. These sketches represent the nuclei of the respective atoms.

ing was considered so important that the chemist who discovered this application, Willard F. Libby, received a Nobel prize.

Importance of Radiometric Dating

Bear in mind that although the basic principle of radiometric dating is simple, the actual procedure is quite complex. The analysis that determines the quantities of parent and daughter must be painstakingly precise. In addition, some radioactive materials do not decay directly into the stable daughter product. As you saw in Figure 11.13, uranium-238 produces 13 intermediate unstable daughter products before the fourteenth and final daughter product, the stable isotope lead-206, is produced.

Radiometric dating methods have produced literally thousands of dates for events in Earth history. Rocks from several localities have been dated at more than 3 billion years, and geologists realize that still older rocks exist. For example, a granite from South Africa has been dated at 3.2 billion years, and it contains inclusions of quartzite. (Remember that inclusions are older than the rock containing them.) Quartzite itself is a metamorphic rock that originally was the sedimentary rock sandstone. Sandstone, in turn, is the product of the lithification of sediments produced by the weathering of existing rocks. Thus, we have a positive indication that much older rocks existed.

Radiometric dating has vindicated the ideas of James Hutton, Charles Darwin, and others who inferred that geologic time must be immense. Indeed, modern dating methods have proved that there has been enough time for the processes we observe to have accomplished tremendous tasks.

The Geologic Time Scale

GEODe
Deciphering Earth's History
▼ Geologic Time Scale

Geologists have divided the whole of geologic history into units of varying magnitude. Together they comprise the **geologic time scale** of Earth history (Figure 11.16). The major units of the time scale were delineated during the nineteenth century, principally by scientists working in Western Europe and Great Britain. Because radiometric dating was unavailable at that time, the entire time scale was created using methods of relative dating. It was only in the twentieth century that radiometric dating permitted numerical dates to be added.

Structure of the Time Scale

The geologic time scale subdivides the 4.5-billion-year history of Earth into many different units and provides a meaningful time frame within which the events of the geologic past are arranged. As shown in Figure 11.16, **eons** represent the greatest expanses of time. The eon that began about 540 million years ago is the **Phanerozoic**, a term derived from Greek words meaning *visible life*. It is an appropriate description because the rocks and deposits of the Phanerozoic eon contain abundant fossils that document major evolutionary trends.

Another glance at the time scale reveals that the Phanerozoic eon is divided into **eras**. The three eras within the Phanerozoic are the **Paleozoic**, (*paleo* = ancient, *zoe* = life), the **Mesozoic** (*meso* = middle, *zoe* = life), and the **Cenozoic** (*ceno* = recent, *zoe* = life). As the names imply, the eras are bounded by profound worldwide changes in life forms. Each era is subdivided into **periods**. The Paleozoic has seven, the Mesozoic three, and the Cenozoic two. Each of these 12 periods is characterized by a somewhat less profound change in life forms as compared with the eras.

Finally, periods are divided into still smaller units called **epochs**. As you can see in Figure 11.16, seven epochs have been named for the periods of the Cenozoic. The epochs of other periods, however, are not usually referred to by specific names. Instead, the terms *early, middle,* and *late* are generally applied to the epochs of these earlier periods.

Precambrian Time

Notice that the detail of the geologic time scale does not begin until about 540 million years ago, the date for the beginning of the Cambrian period. The more than 4 billion years prior to the Cambrian is divided into three eons, the *Hadean,* the *Archean,* and the *Proterozoic*. It is also common for this vast expanse of time

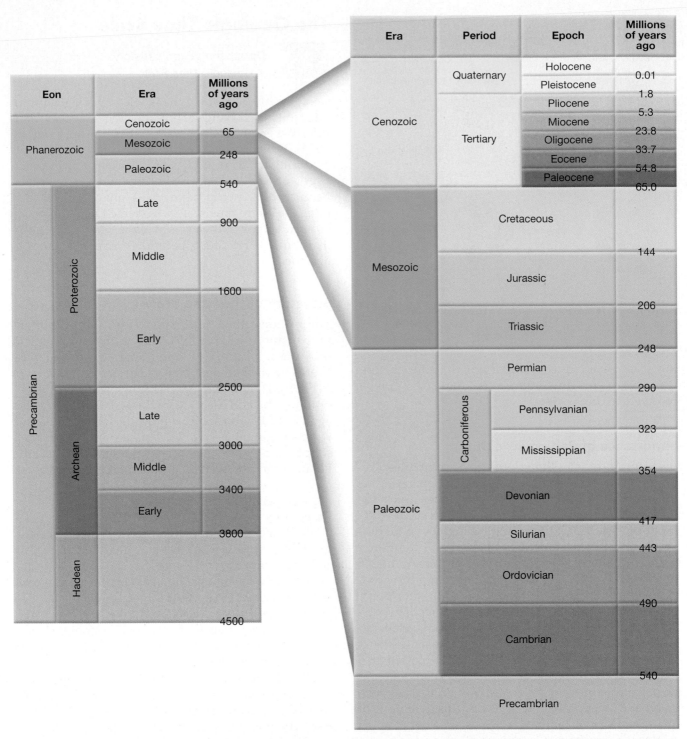

▲ FIGURE 11.16 The geologic time scale. The numerical dates were added long after the time scale had been established using relative dating techniques. (Data from Geological Society of America)

to simply be referred to as the **Precambrian.** Although it represents about 88 percent of Earth history, the Precambrian is not divided into nearly as many smaller time units as is the Phanerozoic eon.

The quantity of information geologists have deciphered about Earth's past is somewhat analogous to the detail of human history. The further back we go, the less we know. Certainly more data and informa-

tion exist about the past 10 years than for the first decade of the twentieth century; the events of the nineteenth century have been documented much better than the events of the first century A.D., and so on. Thus it is with Earth history. The more recent past has the freshest, least disturbed, and most observable record. The further back in time the geologist goes, the more fragmented the record and clues become.

Difficulties in Dating the Geologic Time Scale

Although reasonably accurate numerical dates have been worked out for the periods of the geologic time scale (Figure 11.16), the task is not without difficulty. The primary problem in assigning numerical dates to units of time is the fact that not all rocks can be dated by radiometric methods. Recall that for a radiometric date to be useful, all minerals in the rock must have formed at about the same time. For this reason, radioactive isotopes can be used to determine when minerals in an igneous rock crystallized and when pressure and heat created new minerals in a metamorphic rock.

However, samples of sedimentary rock can only rarely be dated directly by radiometric means. A sedimentary rock may include particles that contain radioactive isotopes, but the rock's age cannot be accurately determined because the grains making up the rock are not the same age as the rock in which they occur. Rather, the sediments have been weathered from rocks of diverse ages.

Radiometric dates obtained from metamorphic rocks may also be difficult to interpret, because the age of a particular mineral in a metamorphic rock does not necessarily represent the time when the rock initially formed. Instead, the date may indicate any one of a number of subsequent metamorphic phases.

If samples of sedimentary rocks rarely yield reliable radiometric ages, how can numerical dates be assigned to sedimentary layers? Usually the geologist must relate them to datable igneous masses, as in Figure 11.17. In this example, radiometric dating has determined the ages of the volcanic ash bed within the Morrison Formation and the dike cutting the Mancos Shale and Mesaverde Formation. The sedimentary beds below the ash are obviously older than the ash, and all the layers above the ash are younger (principle of superposition). The dike is younger than the Mancos Shale and the Mesaverde Formation but older than the Wasatch Formation because the dike does not intrude the Tertiary rocks (cross-cutting relationships).

From this kind of evidence, geologists estimate that a part of the Morrison Formation was deposited about 160 million years ago, as indicated by the ash bed. Further, they conclude that the Tertiary period began after the intrusion of the dike, 66 million years ago. This is one example of literally thousands that illustrates how datable materials are used to *bracket* the various episodes in Earth history within specific time periods. It shows the necessity of combining laboratory dating methods with field observations of rocks.

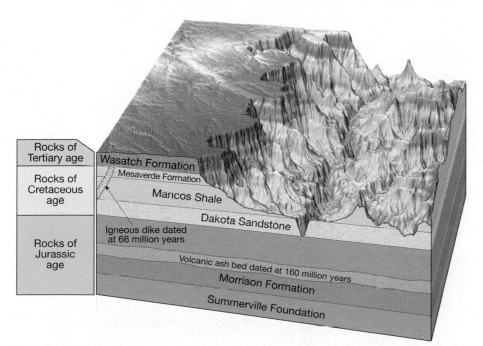

◀ **FIGURE 11.17** Numerical dates for sedimentary layers are often determined by examining their relationship to igneous rocks. (After U.S. Geological Survey)

If you look at the top of a tree stump or at the end of a log, you will see that it is composed of a series of concentric rings. Each of these *tree rings* becomes larger in diameter outward from the center (Figure 11.B). Every year in temperate regions trees add a layer of new wood under the bark. Characteristics of each tree ring, such as size and density, reflect the environmental conditions (especially climate) that prevailed during the year when the ring formed. Favorable growth conditions produce a wide ring; unfavorable ones produce a narrow ring. Trees growing at the same time in the same region show similar tree-ring patterns.

Because a single growth ring is usually added each year, the age of the tree when it was cut can be determined by counting the rings. If the year of cutting is known, the age of the tree and the year in which each ring formed can be determined by counting back from the outside ring.* This procedure can be used to determine the dates of recent geologic events. For example, the minimum number of years since a new land surface was created by a landslide or a flood. The dating and study of annual rings in trees is called *dendrochronology*.

To make the most effective use of tree rings, extended patterns known as

*Scientists are not limited to working with trees that have been cut down. Small, nondestructive core samples can be taken from living trees.

Figure 11.B Each year a growing tree produces a layer of new cells beneath the bark. If the tree is felled and the trunk examined (or if a core is taken, to avoid cutting the tree), each year's growth can be seen as a ring. Because the amount of growth (thickness of a ring) depends upon precipitation and temperature, tree rings are useful records of past climates.(Photo by Stephen J. Krasemann/DRK Photo)

ring chronologies are established. They are produced by comparing the patterns of rings among trees in an area. If the same pattern can be identified in two samples, one of which has been dated, the second sample can be dated from the first by matching the ring pattern common to both. This technique, called *cross dating*, is illustrated in Figure 11.C. Tree-ring chronologies extending back for thousands of years have been established for some regions. To date a timber sample of unknown age, its ring pattern is matched against the reference chronology.

Tree-ring chronologies are unique archives of environmental history and have important applications in such disciplines as climate, geology, ecology, and archaeology. For example, tree rings are used to reconstruct climate variations within a region for spans of thousands of years prior to human historical records. Knowledge of such long-term variations is of great value in making judgments regarding the recent record of climate change.

In summary, dendrochronology provides useful numerical dates for events in the historic and recent prehistoric past. Moreover, because tree rings are a storehouse of data, they are a valuable tool in the reconstruction of past environments.

Figure 11.C Cross dating is a basic principle in dendrochronology. Here it was used to date an archaeological site by correlating tree-ring patterns for wood from trees of three different ages. First, a tree-ring chronology for the area is established using cores extracted from living trees. This chronology is extended further back in time by matching overlapping patterns from older, dead trees. Finally, cores taken from beams inside the ruin are dated using the chronology established from the other two sites.

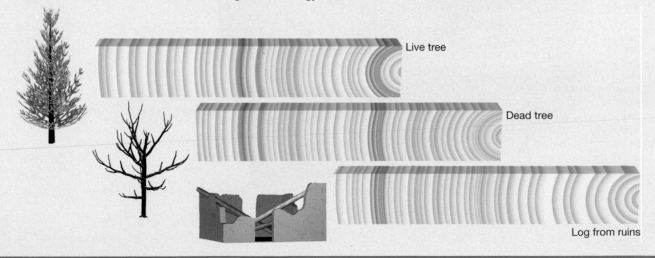

Live tree

Dead tree

Log from ruins

Chapter Summary

• During the seventeenth and eighteenth centuries, *catastrophism* influenced the formulation of explanations about Earth. Catastrophism states that Earth's landscapes have been developed primarily by great catastrophes. By contrast, *uniformitarianism*, one of the fundamental principles of modern geology advanced by *James Hutton* in the late 1700s, states that the physical, chemical, and biological laws that operate today have also operated in the geologic past. The idea is often summarized as "The present is the key to the past." Hutton argued that processes that appear to be slow-acting could, over long spans of time, produce effects that were just as great as those resulting from sudden catastrophic events.

• The two types of dates used by geologists to interpret Earth history are (1) *relative dates*, which put events in their *proper sequence of formation*, and (2) *numerical dates*, which pinpoint the *time in years* when an event took place.

• Relative dates can be established using the *law of superposition, principle of original horizontality, principle of cross-cutting relationships, inclusions,* and *unconformities.*

• *Correlation*, the matching up of two or more geologic phenomena in different areas, is used to develop a geologic time scale that applies to the entire Earth.

• *Fossils* are the remains or traces of prehistoric life. The special conditions that favor preservation are *rapid burial* and the possession of *hard parts* such as shells, bones, or teeth.

• Fossils are used to *correlate* sedimentary rocks from different regions by using the rocks' distinctive fossil content and applying the *principle of fossil succession*. It states that fossil organisms succeed one another in a definite and determinable order, and therefore any time period can be recognized by its fossil content.

• Each atom has a nucleus containing *protons* (positively charged particles) and *neutrons* (neutral particles). Orbiting the nucleus are negatively charged *electrons*. The *atomic number* of an atom is the number of protons in the nucleus. The *mass number* is the number of protons plus the number of neutrons in an atom's nucleus. *Isotopes* are variants of the same atom, but with a different number of neutrons and hence a different mass number.

• *Radioactivity* is the spontaneous breaking apart (decay) of certain unstable atomic nuclei. Three common types of radioactive decay are (1) emission of alpha particles from the nucleus, (2) emission of beta particles (electrons) from the nucleus, and (3) capture of electrons by the nucleus.

• An unstable *radioactive isotope*, called the *parent*, will decay and form stable *daughter products*. The length of time for half of the nuclei of a radioactive isotope to decay is called the *half-life* of the isotope. If the half-life of the isotope is known and the parent/daugher ratio can be measured, the age of a sample can be calculated.

• The *geologic time scale* divides Earth's history into units of varying magnitude. It is commonly presented in chart form, with the oldest time and event at the bottom and the youngest at the top. The principal subdivisions of the geologic time scale, called *eons*, include the *Hadean, Archean, Proterozoic* (together, these three eons are commonly referred to as the *Precambrian*), and, beginning about 540 million years ago, the *Phanerozoic*. The Phanerozoic (meaning "visible life") eon is divided into the following *eras: Paleozoic* ("ancient life"), *Mesozoic* ("middle life"), and *Cenozoic* ("recent life").

• A significant problem in assigning numerical dates to units of time is that *not all rocks can be dated radiometrically*. A sedimentary rock may contain particles of many ages that have been weathered from different rocks that formed at various times. One way geologists assign numerical dates to sedimentary rocks is to relate them to datable igneous masses, such as dikes and volcanic ash beds.

Key Terms

angular unconformity (p. 315)
catastrophism (p. 311)
Cenozoic era (p. 327)
conformable (p. 315)
correlation (p. 316)
cross-cutting relationships, principle of (p. 314)
disconformity (p. 315)
eon (p. 327)
epoch (p. 327)
era (p. 327)
fossil (p. 317)

fossil succession, principle of (p. 321)
geologic time scale (p. 327)
half-life (p. 324)
inclusions (p. 314)
index fossil (p. 321)
Mesozoic era (p. 327)
nonconformity (p. 312)
numerical date (p. 313)
original horizontality, principle of (p. 313)
paleontology (p317)
Paleozoic era (p. 327)

period (p. 327)
Phanerozoic eon (p. 327)
Precambrian (p. 328)
radioactivity (p. 323)
radiocarbon dating (p. 325)
radiometric dating (p. 324)
relative dating (p. 312)
superposition, law of (p. 313)
unconformity (p. 315)
uniformitarianism (p. 311)

Review Questions

1. Contrast catastrophism and uniformitarianism. How did the proponents of each perceive the age of Earth?

2. Distinguish between numerical and relative dating.

3. What is the law of superposition? How are cross-cutting relationships used in relative dating?

4. When you observe an outcrop of steeply inclined sedimentary layers, what principle allows you to assume that the beds became tilted *after* they were deposited?

5. Refer to Figure 11.4 to answer the following questions:
   (a) Is fault *A* older or younger than the sandstone layer?
   (b) Is dike *A* older or younger than the sandstone layer?
   (c) Was the conglomerate deposited before or after fault *A*?
   (d) Was the conglomerate deposited before or after fault *B*?
   (e) Which fault is older, *A* or *B*?
   (f) Is dike *A* older or younger than the batholith?

6. A mass of granite is in contact with a layer of sandstone. Using a principle described in this chapter, explain how you might determine whether the sandstone was deposited on top of the granite or the granite was intruded from below after the sandstone was deposited.

7. Distinguish among angular unconformity, disconformity, and nonconformity.

8. What is meant by the term *correlation?*

9. List and briefly describe at least five different types of fossils.

10. List two conditions that improve an organism's chances of being preserved as a fossil.

11. Why are fossils such useful tools in correlation?

12. In addition to being important aids in dating and correlating rocks, how else are fossils helpful in geologic investigations?

13. If a radioactive isotope of thorium (atomic number 90, mass number 232) emits 6 alpha particles and 4 beta particles during the course of radioactive decay, what are the atomic number and mass number of the stable daughter product?

14. Why is radiometric dating the most reliable method of dating the geologic past?

15. Assume that a hypothetical radioactive isotope has a half-life of 10,000 years. If the ratio of radioactive parent to stable daughter product is $1:3$, how old is the rock containing the radioactive material? What if the ratio were $1:15$?

16. To make calculations easier, let us round the age of Earth to 5 billion years.
   (a) What fraction of geologic time is represented by recorded history (assume 5000 years for the length of recorded history)?
   (b) The first abundant fossil evidence does not appear until the beginning of the Cambrian period (approximately 550 million years ago). What percentage of geologic time is represented by abundant fossil evidence?

17. What subdivisions make up the geologic time scale? What is the primary basis for differentiating the eras?

18. Briefly describe the difficulties in assigning numerical dates to layers of sedimentary rock.

Examining the Earth System

1. Figure 11.11A is a large petrified log in Arizona's Petrified Forest National Park. Describe the transition of this tree from being part of the biosphere to being a component of the solid Earth. How might the hydrosphere and/or atmosphere have played a role in the transition?

2. The famous angular unconformity at Scotland's Siccar Point was originally studied by James Hutton in the late 1700s (see Figure 11.7E, p. 317). Can you describe in a general way what occurred to produce this feature? Could all of the spheres of the Earth system have been involved? The Earth system is powered by energy from two sources. How are both sources represented here?

Online Study Guide

The *Earth Science* Website uses the resources and flexibility of the Internet to aid in your study of the topics in this chapter. Written and developed by Earth science instructors, this site will help improve your understanding of Earth science. Visit **http://www.prenhall.com/tarbuck** and click on the cover of *Earth Science* 11e to find:

- **Online review quizzes.**
- **Critical thinking exercises.**
- **Links to chapter-specific Web resources.**
- **Internet-wide key term searches.**

http://www.prenhall.com/tarbuck

CHAPTER 12

Earth's History: A Brief Summary

Fossil of an ancient amphibian from the Permian period.
(Photo by P. Degginger/RoberStock/Retrofile)

335

Earth has a long and complex history. The splitting and colliding of continents has resulted in the formation of new ocean basins and the creation of Earth's great mountain ranges. Furthermore, the nature of life on our planet has experienced dramatic changes through time.

Many of the changes on planet Earth occur at a "snail's pace," generally too slow for people to perceive. Thus, human awareness of evolutionary change is fairly recent. Evolution is not confined to life forms, for all of Earth's "spheres" have evolved together: the atmosphere, hydrosphere, solid Earth, and biosphere. Examples are evolutionary changes in the air we breathe, evolution of the world oceans, the rise of mountains, ponderous movements of crustal plates, the comings and goings of vast ice sheets, and the evolution of a vast array of life forms. As each facet of Earth has evolved, it has powerfully influenced the others.

As we saw in Chapter 11, geologists have many tools at their disposal for interpreting the clues about Earth's past. Using these tools, and clues that are contained in the rock record, geologists have been able to unravel many of the complex events of the geological past. The goal of this chapter is to provide a brief overview of the history of our planet and its life forms (Figure 12.1). We will describe how our physical world assumed its present form and how Earth's inhabitants changed through time.*

Precambrian Time: Vast and Enigmatic

The Precambrian encompasses immense geological time, from Earth's distant beginnings 4.5 billion years ago until the start of the Cambrian period, over 4 billion years later. Thus, the Precambrian spans about 88 percent of Earth's history. To get a visual sense of this proportion, look at the right side of Figure 12.2, which shows the relative time span of eras. Our knowledge of this ancient time is sketchy, for much of the early rock record has been obscured by the very Earth processes you have been studying, especially plate tectonics, erosion, and deposition.

Untangling the long, complex Precambrian rock record is a formidable task, and it is far from complete. Most Precambrian rocks are devoid of fossils, which hinders correlation of rocks. Rocks of this great age are metamorphosed and deformed, extensively eroded, and obscured by overlying strata. Consequently, this least-understood span of Earth's history has not been

---

*You may want to refer to the section on the "Early Evolution of Earth" on p. 10. It sets the stage by providing a brief but useful summary of the formation of Earth and the other members of our solar system.

▼ **FIGURE 12.1** Here scientists are excavating the remains of *Albertasaurus*, a large carnivore similiar to *Tyrannosaurus* rex, that lived during the late Cretaceous period. The excavation site is near Red Deer River, Alberta, Canada. (Photo by Richard T. Nowitz/Science Source/Photo Researchers, Inc.)

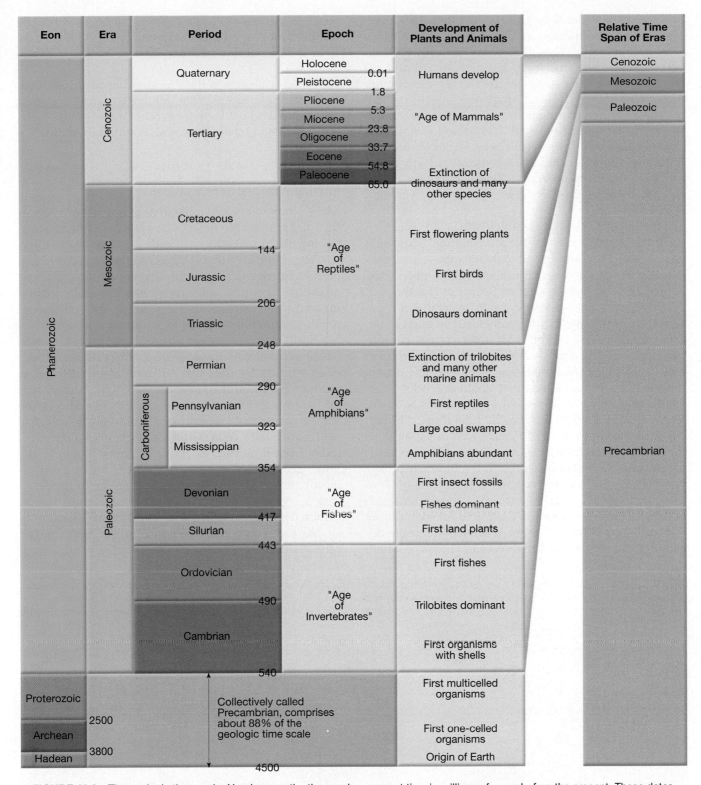

▲ FIGURE 12.2 The geologic time scale. Numbers on the time scale represent time in millions of years before the present. These dates were added long after the time scale had been established using relative dating techniques. The Precambrian accounts for more than 88 percent of geologic time. (Data from Geological Society of America)

successfully divided into briefer time units, as have later intervals. Indeed, Precambrian history is written in scattered, speculative episodes, like a long book with many missing chapters.

Precambrian Rocks

Looking at Earth from the Space Shuttle, astronauts see plenty of ocean (71 percent) and much less land area (29 percent). Over large expanses of the continents, the orbiting space scientists gaze upon many Paleozoic, Mesozoic, and Cenozoic rock surfaces, but fewer Precambrian surfaces. This demonstrates the law of superposition: Precambrian rocks in these regions are buried from view beneath varying thicknesses of more recent rocks. Here, Precambrian rocks peek through the surface where younger strata are extensively eroded, as in the Grand Canyon and in some mountain ranges. However, on each continent, large core areas of Precambrian rocks dominate the surface, mostly as deformed metamorphic rocks. These areas are called **shields** because they roughly resemble a warrior's shield in shape.

Figure 12.3 shows these shield areas of Precambrian rocks worldwide. In North America (including Greenland), the Canadian Shield encompasses 7.2 million square kilometers (2.8 million square miles), the equivalent of about 10 states of Texas put together.

Much of what we know about Precambrian rocks comes from mining the ores contained in such rocks. The mining of iron, nickel, gold, silver, copper, chromium, uranium, and diamonds has provided Precambrian rock samples for study, and surveys to locate valuable ore deposits have revealed much about the rocks.

Noteworthy are extensive iron-ore deposits. Rocks from the middle Precambrian (1.2 billion to 2.5 billion years ago) contain most of Earth's iron ore, mainly as the mineral hematite (Fe_2O_3). These iron-rich sedimentary rocks probably represent the time when oxygen became sufficiently abundant to react with iron dissolved in shallow lakes and seas. Later, after much of the iron was oxidized and deposited on lake and sea bottoms, formation of these iron-rich deposits declined and oxygen levels in the ocean and atmosphere began to increase. Because most of Earth's free oxygen results from plant photosynthesis, the formation of extensive Precambrian iron-ore deposits is linked to life in the sea.

Notably absent in the Precambrian are fossil fuels (coal, oil, natural gas). The reason is clear—a virtual absence of land plants to form coal swamps and of certain animals to form petroleum. Fossil fuels are from a later time.

Earth's Atmosphere Evolves

Earth's atmosphere is unlike that of any other body in the solar system. No other planet has the same life-sustaining mixture of gases as Earth.

Today, the air you breathe is a stable mixture of 78 percent nitrogen, 21 percent oxygen, about 1 percent argon (an inert gas), and trace gases like carbon dioxide and water vapor. But our planet's original atmosphere, several billion years ago, was far different.

Earth's very earliest atmosphere probably was swept into space by the *solar wind*, a vast stream of particles emitted by the Sun. As Earth slowly cooled, a more enduring atmosphere formed. The molten surface solidified into a crust, and gases that had been dissolved in the molten rock were gradually released, a process called **outgassing.** Outgassing continues today from hundreds of active volcanoes worldwide. Thus, geologists hypothesize that Earth's original atmosphere was made up of gases similar to those released in volcanic emissions today: water vapor, carbon dioxide, nitrogen, and several trace gases.

As the planet continued to cool, the water vapor condensed to form clouds, and great rains commenced.

▼ **FIGURE 12.3** Today's continents look very different from the Precambrian. Remnants of Precambrian rocks are the continental shields, composed largely of metamorphosed igneous and sedimentary rocks.

Key

- Shields
- Stable platforms (shields covered by sedimentary rock)
- Young mountain belts (less than 100 million years old)
- Old mountain belts

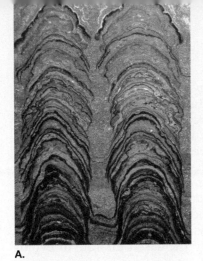

A.

B.

▲ **FIGURE 12.4** Stromatolites are among the most common Precambrian fossils. **A.** These Precambrian fossil stromatolites from Argentina are more than 2 billion years old. (Photo by Sinclair Stammers/Photo Researchers, Inc.) **B.** Modern stromatolites growing in saline area, Western Australia. (Photo by Bill Bachman/Photo Researchers, Inc.)

At first, the water evaporated in the hot air before reaching the ground, or quickly boiled away upon contacting the surface, just like water sprayed on a hot grill. This accelerated the cooling of Earth's crust. When the surface had cooled below water's boiling point (100°C or 212°F), torrential rains slowly filled low areas, forming the oceans. This reduced not only the water vapor in the air but also the amount of carbon dioxide, for it became dissolved in the water. What remained was a nitrogen-rich atmosphere.

If Earth's primitive atmosphere resulted from volcanic outgassing, we have a problem, because volcanoes do not emit free oxygen. Where did the very significant percentage of oxygen in our present atmosphere (nearly 21 percent) come from?

The major source of oxygen is green plants. Put another way, *life itself* has strongly influenced the composition of our present atmosphere. Plants did not just adapt to their environment; they actually influenced it, dramatically altering the composition of the entire planet's atmosphere by using carbon dioxide and releasing oxygen. This is a good example of how Earth operates as a giant system in which living things interact with their environment.

How did plants come to alter the atmosphere? The key is the way in which plants create their own food. They employ *photosynthesis*, in which they use light energy to synthesize food sugars from carbon dioxide and water. The process releases a waste gas: oxygen. Those of us in the animal kingdom rely on oxygen to metabolize our food, and we in turn exhale carbon dioxide as a waste gas. The plants use this carbon dioxide for more photosynthesis, and so on, in a continuing system.

The first life forms on Earth, probably bacteria, did not need oxygen. Their life processes were geared to the earlier, oxygenless atmosphere. Even today, many *anaerobic* bacteria thrive in environments that lack free oxygen. Later, primitive plants evolved that used photosynthesis and released oxygen. Slowly, the oxygen content of Earth's atmosphere increased. The Precambrian rock record suggests that much of the first free oxygen did not remain free because it combined with (oxidized) other substances dissolved in water, especially iron. Iron has tremendous affinity for oxygen, and the two elements combine to form iron oxides (rust) at any opportunity.

Then, once the available iron satisfied its need for oxygen, substantial quantities of oxygen accumulated in the atmosphere. By the beginning of the Paleozoic era, about 4 billion years into Earth's existence (after seven-eighths of Earth's history had transpired), the fossil record reveals abundant ocean-dwelling organisms that require oxygen to live. Hence, the composition of Earth's atmosphere has evolved together with its life forms, from an oxygenless envelope to today's oxygen-rich environment.

Precambrian Fossils

A century ago, the earliest fossils known dated from the Cambrian period, about 540 million years ago. None were known from the Precambrian. This created a major problem for science at that time: How could complex organisms like trilobites abruptly appear in the geologic record? The answer, of course, was that there were fossils in Precambrian rocks, but they were rare, small, obscure, and simply had not been discovered yet. Today, our knowledge of Precambrian life, although far from complete, is quite extensive.

Precambrian fossils are disappointing if you are expecting to see fascinating plants and large animals, for these had not yet evolved. Instead, the most common Precambrian fossils are **stromatolites.** These are distinctively layered mounds or columns of calcium carbonate (Figure 12.4). Stromatolites are not the remains of actual organisms but are material deposited by algae. They are indirect evidence of algae because they closely resemble similar deposits made by modern algae.

Stromatolites did not become common until the middle Precambrian, around 2 billion years ago. Stromatolites are large, but most actual organisms preserved in Precambrian rocks are microscopic. Well-preserved remains of many tiny organisms have been discovered, extending the record of life back beyond 3.5 billion years.

Many of these most ancient fossils are preserved in *chert*, a hard, dense chemical sedimentary rock. Chert must be very thinly sliced and studied under powerful microscopes to observe bacteria and algae fossils within it.

Microfossils have been found at several locations worldwide. Two notable areas are in southern Africa, where the rocks date to more than 3.1 billion years, and in the Gunflint Chert (named for its use in flintlock rifles) of Lake Superior, which dates to 1.7 billion years. In both places, bacteria and blue-green algae have been discovered. The fossils are of the most primitive organisms, *prokaryotes*. Their cells lack organized nuclei, and they reproduce asexually.

More advanced organisms, *eukaryotes*, have cells that contain nuclei. Eukaryotes are among billion-year-old fossils discovered at Bitter Springs in Australia, such as green algae. Unlike prokaryotes, eukaryotes reproduce sexually, which means that genetic material is exchanged between organisms. This reproductive mode permits greatly increased genetic variation. Thus, development of eukaryotes may have dramatically increased the rate of evolutionary change.

Plant fossils date from the middle Precambrian, but animal fossils came a bit later, in the late Precambrian. Many of these fossils are *trace fossils*, meaning that they are not of the animals themselves but of their activities, such as trails and worm holes. Areas in Australia and Newfoundland have yielded hun-

Box 12.1 Understanding Earth:
The Burgess Shale

The possession of hard parts greatly enhances the likelihood of organisms being preserved in the fossil record. Nevertheless, there have been rare occasions in geologic history when large numbers of soft-bodied organisms have been preserved. The Burgess Shale is one well-known example. Located in the Canadian Rockies near the town of Field in southeastern British Columbia, the site was discovered in 1909 by Charles D. Walcott of the Smithsonian Institution.

The Burgess Shale is a site of exceptional fossil preservation and records a diversity of animals found nowhere else (Figure 12.A). The animals of the Burgess Shale lived shortly after the *Cambrian explosion*, a time when there had been a huge expansion of marine biodiversity. Its beautifully preserved fossils represent our most complete and authoritative snapshot of Cambrian life, far better than deposits containing only fossils of organisms with hard parts. To date, more than 100,000 unique fossils have been found.

The animals preserved in the Burgess Shale inhabited a warm shallow sea adjacent to a large reef that was part of the continental margin of North America. During the Cambrian, the North American continent was in the tropics astride the equator. Life was restricted to the ocean, and the land was barren and uninhabited.

FIGURE 12.A Two examples of Burgess Shale Fossils. *Thaumaptilon walcotti* (at left) was a relatively large (up to 20 centimeters, or 8 inches, long) leaflike animal. (Photo by Simon Conway Morris, University of Cambridge) *Aysheaia pedunculata* (on right) was an ancient relative of modern velvet worms and may have clung to soft sponges with tiny hooks on its feet. (With permission of the Royal Ontario Museum (C)Rom. Photo Credit: D.H. Collins)

What were the circumstances that led to the preservation of the many life forms found in the Burgess Shale? The animals lived in and on underwater mud banks that formed as sediment accumulated on the outer margins of a reef adjacent to a steep escarpment (cliff). Periodically the accumulation of muds became unstable and the slumping and sliding sediments moved down the escarpment as turbidity currents. These flows transported the animals in a turbulent cloud of sediment to the base of the reef where they were buried. Here, in an environment lacking oxygen, the buried carcasses were pro-

tected from scavengers and decomposing bacteria. This process occurred again and again, building a thick sequence of fossil-rich sedimentary layers. Beginning about 175 million years ago, mountain-building forces elevated these strata from the seafloor and moved them many kilometers eastward along huge faults to their present location in the Canadian Rockies.

The Burgess Shale is one of the most important fossil discoveries of the twentieth century. Its layers preserve for us an intriguing glimpse of early animal life that is more than a half billion years old.

A.

B.

▲ **FIGURE 12.5** Fossils of common Paleozoic life forms. **A.** Natural cast of a trilobite. Trilobites dominated the Paleozoic ocean, scavenging food from the bottom. **B.** Extinct coiled cephalopods. Like their modern descendants, these were highly developed marine organisms. (Photos by E. J. Tarbuck)

dreds of fossil impressions of soft-bodied creatures. Most, if not all, of the Precambrian fauna lacked shells, which would develop as protective armor during the Paleozoic.

As the Precambrian came to a close, the fossil record disclosed diverse and complete multicelled organisms. This set the stage for more complex plants and animals to evolve at the dawn of the Paleozoic era (see Box 12.1).

Paleozoic Era: Life Explodes

Following the long Precambrian, the most recent 540 million years of Earth history are divided into three eras: Paleozoic, Mesozoic, and Cenozoic. The Paleozoic era encompasses about 292 million years and is by far the longest of the three. Seven periods make up the Paleozoic era (see Figure 12.2).

Before the Paleozoic, life forms possessed no hard parts—shells, scales, bones, or teeth. The beginning of the Paleozoic is marked by the appearance of the first life forms with hard parts (Figure 12.5). Hard parts greatly enhanced their chance of being preserved as part of the fossil record. Therefore, our knowledge of life's diversification improves greatly from the Paleozoic onward. This diversity is demonstrated in Figure 12.6.

Abundant Paleozoic fossils have allowed geologists to construct a far more detailed time scale for the last one-eighth of geologic time than for the preceding seven-eighths, the Precambrian. Moreover, because every organism is associated with a particular environment, the greatly improved fossil record provided invaluable information for deciphering ancient environments. To facilitate our brief tour of the Paleozoic, we divide it into Early Paleozoic (Cambrian, Ordovician, Silurian periods) and Late Paleozoic (Devonian, Mississippian, Pennsylvanian, Permian periods).

Early Paleozoic History

The early Paleozoic consists of a 123-million-year span that embraces the Cambrian, Ordovician, and Silurian periods. Anyone approaching Earth from space at this time would have seen the familiar blue planet with plentiful white clouds, but the arrangement of continents would have looked very different from today (Figure 12.7). At this time, the vast southern continent of Gondwana encompassed five continents (South America, Africa, Australia, Antarctica, India, and perhaps China). Evidence of an extensive continental glaciation places western Africa near the South Pole!

Landmasses that were not part of Gondwana existed as five separate units and some scattered fragments. Although the exact position of these northern continents is uncertain, ancestral North America and Europe are thought to have been near the equator and separated by a narrow sea, as shown on the map in Figure 12.7.

As the Paleozoic opened, North America was a land with no living things, plant or animal. There were no Appalachian or Rocky Mountains; the continent was largely a barren lowland. Several times during the

341

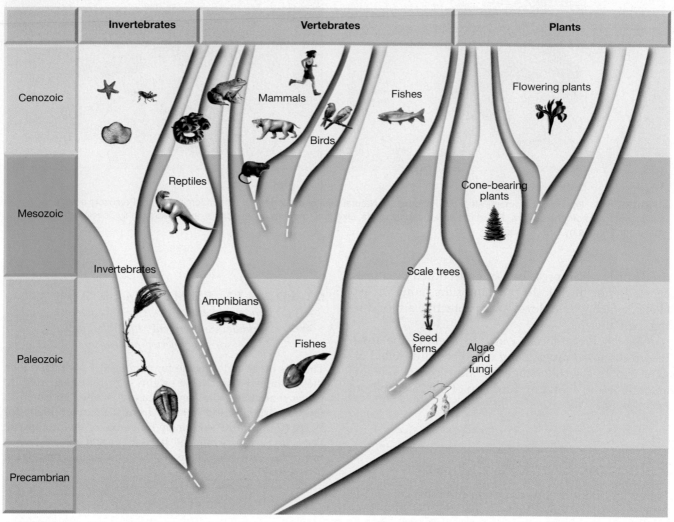

▲ FIGURE 12.6 This chart indicates the times of appearance and relative abundance of major groups of organisms. The wider the band, the more dominant the group.

Cambrian and Ordovician periods, shallow seas moved inland and then receded from the interior of the continent. Deposits of clean sandstones, used today to make glass, mark the edge of these shallow seas in the mid-continent.

Early in the Paleozoic, a mountain-building event affected eastern North America from the present-day central Appalachians to Newfoundland. The mountains produced during this event, known as the Taconic Orogeny, have since eroded away, leaving behind deformed strata and a large volume of detrital sedimentary rocks that were derived from the weathering of these mountains.

During the Silurian period, much of North America was once again inundated by shallow seas. This time large barrier reefs restricted circulation between shallow marine basins and the open ocean. Water in these basins evaporated, causing deposition of large quantities of rock salt and gypsum. Today these thick *evaporite beds* are important resources for the chemical, rubber, plasterboard, and photographic industries in Ohio, Michigan, and western New York State.

Early Paleozoic Life

Life in early Paleozoic time was restricted to the seas. Vertebrates had not yet evolved, so life consisted of several invertebrate groups. The Cambrian period was the golden age of *trilobites*. More than 600 genera of these mud-burrowing scavengers flourished worldwide. By Ordovician times, *brachiopods* outnumbered the trilobites. Brachiopods are among the most widespread Paleozoic fossils and, except for one modern group, are now extinct. Although the adults lived attached to the seafloor, the young larvae were free-swimming. This mobility accounts for the group's wide geographic distribution.

The Ordovician also marked the appearance of abundant *cephalopods*—mobile, highly developed mollusks that became the major predators of their time. The descendents of cephalopods include the modern squid, octopus, and nautilus. Cephalopods were the

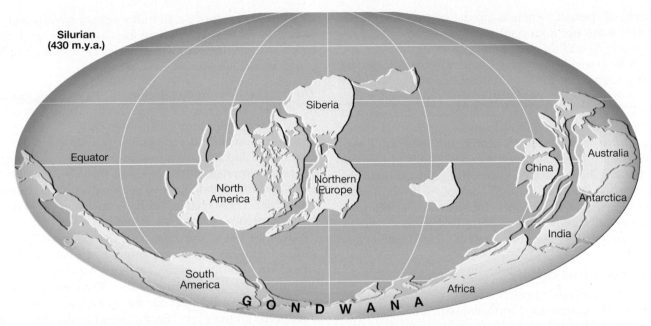

▲ FIGURE 12.7 Reconstruction of Earth as it may have appeared in early Paleozoic time. The southern continents were joined into a single landmass called Gondwana. Four of the five landmasses that would later join to form the northern continent of Laurasia lay scattered roughly along the equator. (After C. Scotese, R. K. Bambach, C. Barton, R. VanderVoo, and A. Ziegler)

first truly large organisms on Earth (Figure 12.8). Whereas the largest trilobites seldom exceeded 30 centimeters (12 inches) in length and the biggest brachiopods were no more than about 20 centimeters (8 inches) across, one species of cephalopod reached a length of nearly 10 meters (30 feet).

The beginning of the Cambrian period marks an important event in animal evolution. For the first time,

organisms appeared that secreted material that formed *hard parts*, such as shells. Why several diverse life forms began to develop hard parts about the same time remains unanswered. One proposal suggests that because an external skeleton provides protection from predators, hard parts evolved for survival. Yet, the fossil record does not seem to support this hypothesis. Organisms with hard parts were plentiful in the

▼ FIGURE 12.8 During the Ordovician period (490–443 million years ago), the shallow waters of an inland sea over central North America contained an abundance of marine invertebrates. Shown in this reconstruction are straight-shelled cephalopods, trilobites, brachiopods, snails, and corals. (© The Field Museum, Neg. #GEO80820c, Chicago)

Cambrian period, whereas predators such as cephalopods were not abundant until the Ordovician period, some 70 million years later.

Whatever the answer, hard parts clearly served many useful purposes and aided adaptations to new ways of life. Sponges, for example, developed a network of fine interwoven silica spicules that allowed them to grow larger and more erect, capable of extending above the surface in search for food. Mollusks (clams and snails) secreted external shells of calcium carbonate that protected them and allowed body organs to function in a more controlled environment. The successful trilobites developed an exoskeleton of a protein called *chitin*, which permitted them to burrow through soft sediment in search of food (Figure 12.9).

Late Paleozoic History

The late Paleozoic consists of four periods—the Devonian, Mississippian, Pennsylvanian, and Permian—that span about 160 million years. Tectonic forces reorganized Earth's landmasses during this time, culminating with the formation of the supercontinent *Pangaea* (Figure 12.10).

Forming Pangaea, the Supercontinent. As ancestral North America collided with Africa, the narrow sea that separated these landmasses began to close slowly (compare Figure 12.10B and 12.10C). Strong compressional forces from this collision deformed the rocks to produce the original northern Appalachian Mountains of eastern North America.

During the fusion of North America and Africa, the other northern continents began to converge (Figure 12.10). By the Permian period, this newly formed landmass had collided with western Asia and the Siberian landmass along the line of the Ural Mountains. Through this union, the northern continent of *Laurasia* was born, encompassing present-day North America, Europe, western Asia, and Siberia.

As Laurasia was forming, Gondwana migrated northward. By the Pennsylvanian period, Gondwana collided with Laurasia, forming a mountainous belt through central Europe. Simultaneously, a collision between the African fragment of Gondwana and the southeastern edge of North America created the southern Appalachian Mountains.

The Great Paleozoic Extinction The Paleozoic ended at a time when Earth's major landmasses joined to form the supercontinent Pangaea (Figure 12.10). This redistribution of land and water and changes in the elevations of landmasses brought pronounced changes in world climates. Broad areas of the northern continents became elevated above sea level, and the climate grew drier. These changes apparently triggered extinctions of many species on land and sea.

By the close of the Paleozoic, 75 percent of the amphibian families had disappeared, and plants had declined in number and variety. Although many amphibian groups became extinct, their descendants, the reptiles, would become the most successful and advanced animals on Earth. Marine life was not spared. At least 80 percent, and perhaps as much as 95 percent, of marine life disappeared. Many marine invertebrates that had been dominant during the Paleozoic, including all the remaining trilobites as well as some types of corals and brachiopods, failed to adapt to the wide-spread environmental changes.

The late Paleozoic extinction was the greatest of at least five mass extinctions to occur over the past 500 million years. Each extinction wreaked havoc with the existing biosphere, wiping out large numbers of species. In each case, however, the survivors formed new biological communities that were more diverse than their predecessors. Thus, mass extinctions actually invigorated life on Earth, as the few hardy survivors eventually filled more niches than the ones left by the victims.

The cause of the great Paleozoic extinction is uncertain. The climate changes from the formation of Pangaea and the associated drop in sea level undoubtedly stressed many species. In addition, at least 2 million cubic kilometers of lava flowed across Siberia to produce what is called the Siberian Traps. Perhaps debris from these eruptions blocked incoming sunlight, or perhaps enough sulfuric acid was emitted to make the seas virtually uninhabitable. A recently discovered impact crater off the northwest coast of Australia has been cited as a potential contributing cause of the late Paleozoic extinction. Whatever caused this extinction, it is clear that without it a very different population of organisms would today inhabit this planet.

Late Paleozoic Life

During most of the late Paleozoic, organisms diversified dramatically. Some 400 million years ago, plants that had adapted to survive at the water's edge began to move inland, becoming *land plants*. These earliest

▼ **FIGURE 12.9** Fossils of trilobites, a common Paleozoic life form that would burrow through soft sediment in search of food. (Photo by Phil Degginger/Getty Images Inc—Stone Allstock)

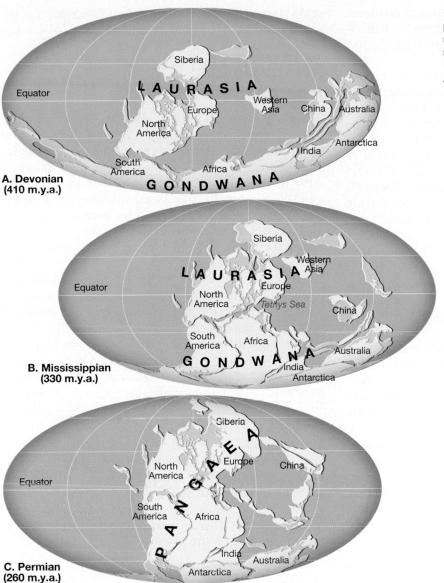

**A. Devonian
(410 m.y.a.)**

**B. Mississippian
(330 m.y.a.)**

**C. Permian
(260 m.y.a.)**

◄ **FIGURE 12.10** During the late Paleozoic, plate movements were joining together the major landmasses to produce the supercontinent of Pangaea. (After C. Scotese, R. K. Bambach, C. Barton, R. VanderVoo, and A. Ziegler)

land plants were leafless vertical spikes about the size of your index finger. However, by the end of the Devonian, 40 million years later, the fossil record indicates the existence of forests with trees tens of meters high.

In the oceans, armor-plated fishes that had evolved during the Ordovician continued to adapt. Their armor plates thinned to lightweight scales that increased their speed and mobility (Figure 12.11). Other fishes evolved during the Devonian, including

◄ **FIGURE 12.11** These placoderms or "plate-skinned" fish were abundant during the Devonian (417–354 million years ago). (Drawing after A. S. Romer)

primitive sharks that had a skeleton made of cartilage, and bony fishes, the groups to which virtually all modern fishes belong. Because of this, the Devonian period is often called the "age of fishes."

By late Devonian time, two groups of bony fishes—the lung fish and the lobe-finned fish—became adapted to land environments. Not unlike their modern relatives, these fishes had primitive lungs that supplemented their breathing through gills. Lobe-finned fish likely occupied tidal flats or small ponds. In times of drought, they may have used their bony fins to "walk" from dried-up pools in search of other ponds. Through time, the lobe-finned fish began to rely more on their lungs and less on their gills. By late Devonian time, they had evolved into true air-breathing amphibians with fishlike heads and tails. It should be noted that insects had already invaded the land.

Modern amphibians, like frogs, toads, and salamanders, are small and occupy limited biological niches. But conditions during the remainder of the Paleozoic were ideal for these newcomers to the land. Plants and insects, which were their main diet, already were very abundant and large. Having only minimal competition from other land dwellers, the amphibians rapidly diversified. Some groups took on roles and forms that were more similar to modern reptiles, such as crocodiles, than to modern amphibians.

By the Pennsylvanian period, large tropical swamps extended across North America, Europe, and Siberia (Figure 12.12). Trees approached 30 meters (100 feet), with trunks over a meter across. The coal deposits that fueled the Industrial Revolution, and which provide a substantial portion of our electric power today, originated in these vast swamps. Further, it was in the lush coal swamp environment of the late Paleozoic that the amphibians evolved quickly into a variety of species.

Mesozoic Era: Age of the Dinosaurs

Spanning about 183 million years, the Mesozoic era is divided into three periods: the Triassic, Jurassic, and Cretaceous. The Mesozoic era witnessed the beginning of the breakup of the supercontinent Pangaea. Also in this era, organisms that had survived the great Permian extinction began to diversify in spectacular ways. On land, dinosaurs became dominant and remained unchallenged for over 100 million years. Because of this, the Mesozoic era is often called the "age of dinosaurs."

Early geologists recognized a profound difference between the fossils in Permian strata and those in younger Triassic rocks, as if someone had drawn a bold line to separate the two time periods. Clearly one half of the fossil groups that occurred in late Paleozoic rocks were missing in Mesozoic rocks. On this basis, it was decided to separate the Paleozoic and Mesozoic at the Permian–Triassic boundary.

Mesozoic History

The Mesozoic era began with much of the world's land above sea level. In fact, in North America no period exhibits a more meager marine sedimentary record than does the Triassic period. Of the exposed Triassic strata, most are red sandstones and mudstones that lack fossils and contain features indicating a terrestrial environment.

▼ **FIGURE 12.12**   Restoration of a Pennsylvanian-age coal swamp (323–290 million years ago). Shown are scale trees (left), seed ferns (lower left), and scouring rushes (right). Also note the large dragonfly. (© The Field Museum, Neg. #GEO85637, Chicago. Photographer: John Weinstein.)

As the second period opened, the Jurassic, the sea invaded western North America. Adjacent to this shallow sea, extensive continental sediments were deposited on what is now the Colorado Plateau. The most prominent is the Navajo Sandstone, a windblown, white quartz sandstone that in places approaches a thickness of 300 meters (1000 feet). These massive dunes indicate that a major desert occupied much of the American Southwest during early Jurassic times.

A well-known Jurassic deposit is the Morrison Formation, within which is preserved the world's richest storehouse of dinosaur fossils. Included are fossilized bones of huge dinosaurs such as *Apatosaurus* (formerly *Brontosaurus*), *Brachiosaurus*, and *Stegosaurus*.

As the Jurassic period gave way to the Cretaceous, shallow seas once again invaded much of western North America, the Atlantic, and Gulf coastal regions. This created great swamps like those of the Paleozoic era, forming Cretaceous coal deposits that are very important economically in the western United States and Canada. For example, on the Crow Indian reservation in Montana, there exists nearly 20 billion tons of high-quality coal of Cretaceous age.

A major event of the Mesozoic era was the breakup of Pangaea (Figure 12.13). A rift developed between what is now the eastern United States and western Africa, marking the birth of the Atlantic Ocean. It also represents the beginning of the breakup of Pangaea, a process that continued for 200 million years, through the Mesozoic and into the Cenozoic.

As Pangaea fragmented, the westward-moving North American plate began to override the Pacific plate. This tectonic event began a continuous wave of deformation that moved inland along the entire western margin of the continent. By Jurassic times, subduction of the Pacific plate had begun to produce the

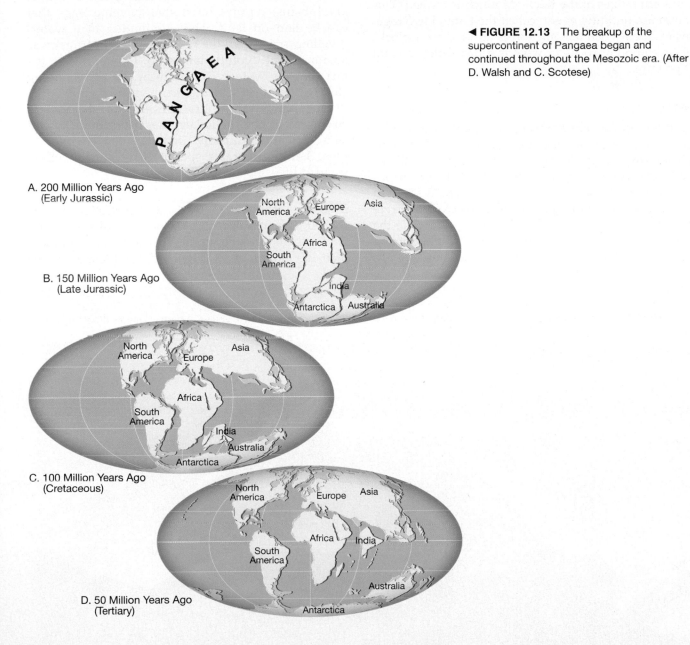

◄ **FIGURE 12.13** The breakup of the supercontinent of Pangaea began and continued throughout the Mesozoic era. (After D. Walsh and C. Scotese)

A. 200 Million Years Ago (Early Jurassic)

B. 150 Million Years Ago (Late Jurassic)

C. 100 Million Years Ago (Cretaceous)

D. 50 Million Years Ago (Tertiary)

chaotic mixture of rocks that exists today in the Coast Ranges of California. Further inland, igneous activity was widespread, and for nearly 60 million years huge masses of magma rose to within a few miles of the surface, where they cooled and solidified. The remnants of this intrusive activity include the granitic rocks of the Sierra Nevada, the Idaho batholith, and the Coast Range batholith of British Columbia.

Tectonic activity that began in the Jurassic continued throughout the Cretaceous, ultimately forming the vast mountains of western North America (Figure 12.14). Compressional forces moved huge rock units in a shinglelike fashion toward the east. Throughout much of the western margin of North America, older rocks were thrust eastward over younger strata, for a distance exceeding 150 kilometers (90 miles).

Toward the end of the Mesozoic, the middle and southern ranges of the Rocky Mountains formed. This mountain-building event, called the Laramide Orogeny, resulted when tectonic forces uplifted large blocks of Precambrian rocks in what is today Colorado and Wyoming.

Mesozoic Life

As the Mesozoic era dawned, its life forms were survivors of the great Paleozoic extinction. These survivors diversified in many new ways to fill the biological voids created at the close of the Paleozoic. On land, conditions favored those that could adapt to drier climates. Among plants, the gymnosperms were one such group. Unlike the first plants to invade the land, the seed-bearing gymnosperms did not depend on free-standing water for fertilization. Consequently, these plants were not restricted to a life near the water's edge.

The gymnosperms quickly became the dominant trees of the Mesozoic. They included the cycads, the conifers, and the ginkgoes. The cycads resembled a large pineapple plant. The ginkgoes had fan-shaped leaves, much like their modern relatives. Largest were the conifers, whose modern descendants include the pines, firs, and junipers. The best-known fossil occurrence of these ancient trees is in northern Arizona's Petrified Forest National Park. Here, huge petrified logs lie exposed at the surface, having been weathered from rocks of the Triassic Chinle Formation (see Figure 11.10A, p. 320).

The Shelled Egg. Among the animals, reptiles readily adapted to the drier Mesozoic environment. They were the first true terrestrial animals. Unlike amphibians, reptiles have shell-covered eggs that can be laid on land. The elimination of a water-dwelling stage (like the tadpole stage in frogs) was an important evolutionary step. Of interest is the fact that the watery fluid within the reptilian egg closely resembles seawater in chemical composition. Because the reptile embryo develops in this watery environment, the shelled egg has been characterized as a "private aquarium" in which the embryos of these land vertebrates spend their water-dwelling stage of life.

Dinosaurs Dominate. With the perfection of the shelled egg, reptiles quickly became the dominant land animals. They continued this dominance for more than 160 million years (Figure 12.15). Most awesome

▼ **FIGURE 12.14** Uplifted and deformed sedimentary strata in the Canadian Rockies near Lake Louise, Alberta, Canada. (Photo by Carr Clifton/Minden Pictures)

▲ FIGURE 12.15 Fossil skull of *Sarcosuchus imperator*, or "flesh crocodile emperor." Remains of this large crocodile have been uncovered in the desert of Niger. These animals were river dwellers, indicating that the climate of this region during the Cretaceous period was very different from today. For more about this amazing animal, see the *Students Sometimes Ask* to the right of this photo. (Photo by Project Exploration P.A.S.T.)

of the Mesozoic reptiles were the dinosaurs. Some of the huge dinosaurs were carnivorous (*Tyrannosaurus*), whereas others were herbivorous (like the ponderous *Apatosaurus*, formerly *Brontosaurus*). The extremely long neck of *Apatosaurus* may have been an adaptation for feeding on tall conifer trees. However, not all dinosaurs were large. In fact, certain small forms closely resembled modern fleet-footed lizards. Further, evidence indicates that some dinosaurs, unlike their present-day reptile relatives, were warm blooded.

The reptiles made one of the most spectacular adaptive radiations in all of Earth history (Figure 12.16). One group, the pterosaurs, took to the air. These "dragons of the sky" possessed huge membranous wings that allowed them rudimentary flight (Figure 12.17). Another group of reptiles, exemplified by the fossil *Archaeopteryx*, led to more successful flyers: the birds. Whereas some reptiles took to the skies, others returned to the sea, including the fish-eating plesiosaurs and ichthyosaurs. These reptiles became proficient swimmers but retained their reptilian teeth and breathed by means of lungs.

At the close of the Mesozoic, many reptile groups became extinct. Only a few types survived to recent times, including the turtles, snakes, crocodiles, and lizards. The huge land-dwelling dinosaurs, the marine plesiosaurs, and the flying pterosaurs all are known only through the fossil record. What caused this great extinction (Box 12.2.)?

STUDENTS SOMETIMES ASK...

Many dinosaurs were very large. Were they the only large reptiles?

No, indeed. One well-publicized example is a crocodile known as *Sarcosuchus imperator* (see Figure 12.15). This huge river dweller lived in Africa about 110 million years ago (Cretaceous period). At maturity (50–60 years) the animal weighed 8 metric tons (more than 17,600 pounds!) and was about 12 meters (40 feet) long—as long as *Tyrannosaurus rex* and much heavier. Its jaws were roughly as long as an adult human. This animal has appropriately been dubbed "supercroc." Paleontologists indicate that the teeth and jaw suggest a diet of large vertebrates, including fish and dinosaurs.

Cenozoic Era: Age of Mammals

The Cenozoic era, or "era of recent life," encompasses the past 65 million years of Earth history. It is the "postdinosaur" era, the time of mammals, including humans. It is during this span that the physical landscapes and life forms of our modern world came into being. The Cenozoic era represents a much smaller fraction of geologic time than either the Paleozoic or the Mesozoic. Although shorter, it nevertheless possesses a rich history, because the completeness of the geologic record improves as time approaches the present. The rock formations of this time span are more widespread and less disturbed than those of any preceding time.

The Cenozoic era is divided into two periods of very unequal duration, the Tertiary period and the Quaternary period. The Tertiary period includes five epochs and embraces about 63 million years, practically all of the Cenozoic era. The Quaternary period consists of two epochs that represent only the last 2 million years of geologic time.

Cenozoic North America

Most of North America was above sea level throughout the Cenozoic era. However, the eastern and western margins of the continent experienced markedly contrasting events, because of their different relationships with plate boundaries. The Atlantic and Gulf coastal regions, far removed from an active plate boundary, were tectonically stable. In contrast, western North America was the leading edge of the North American plate. As a result, plate interactions during the Cenozoic gave rise to many events of mountain building, volcanism, and earthquakes in the West.

Eastern North America. The stable continental margin of eastern North America was the site of abundant marine sedimentation. The most extensive deposition surrounded the Gulf of Mexico, from the Yucatan

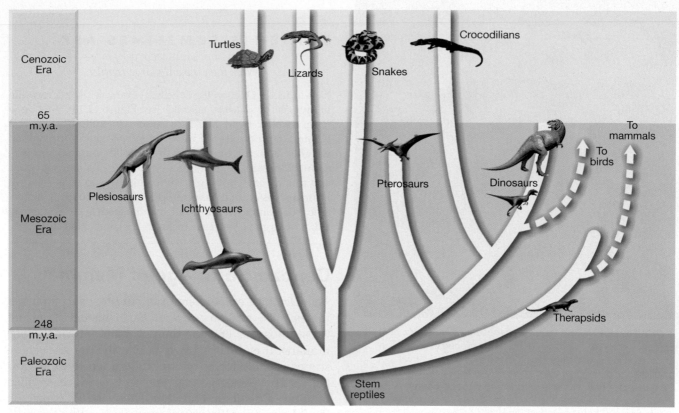

▲ **FIGURE 12.16** Origin and development of the major reptile groups and their descendants. Dinosaurs, pterosaurs (flying reptiles), and large marine reptiles all became extinct by the end of the Mesozoic.

Peninsula in present-day Mexico to Florida. Here, the great buildup of sediment caused the crust to downwarp and produced numerous faults. In many instances, the faults created traps in which oil and natural gas accumulated. Today, these and other petroleum traps are the most economically important re-

▼ **FIGURE 12.17** Fossils of the great flying reptile Pteranodon have been recovered from Cretaceaous chalk deposits located in Kansas. Pteranodon had a wingspan of 7 meters (22 feet), but flying reptiles with twice this wingspan have been discovered in strata of west Texas.

source in Cenozoic strata of the Gulf Coast, as evidenced by the Gulf's well-known off-shore drilling platforms.

By early Cenozoic time, most of the original Appalachians had been eroded to a low plain. Then, by the mid-Cenozoic, isostatic adjustments raised the region once again, changing its orientation to base level and rejuvenating its rivers.* Streams eroded with renewed vigor, gradually sculpturing the surface into its present-day topography (Figure 12.18). Sediments from all of this erosion were deposited along the eastern margin of the continent, where they attained a thickness of many kilometers. Today, portions of the strata deposited during the Cenozoic are exposed as the gently sloping Atlantic and Gulf coastal plains.

Western North America. In the West, the formation of the Rocky Mountains was coming to an end. As erosion lowered the mountains, the basins between uplifted ranges filled with sediments. Eastward, a great wedge of sediment from the eroding Rockies was building the Great Plains.

Beginning in the Miocene epoch, a broad region from northern Nevada into Mexico experienced crustal movements that formed more than 150 fault-block mountain ranges. They rise abruptly above the

*To review the concept of base level, see Chapter 5.

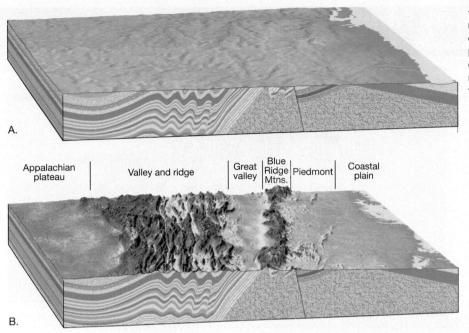

◄ **FIGURE 12.18** The formation of the modern Appalachian Mountains. **A.** The original Appalachians eroded to a low plain. **B.** The recent upwarping and erosion of the Appalachians began nearly 30 million years ago to produce the present topography.

adjacent basins, creating the Basin and Range Province (see Chapter 10).

As the Basin and Range Province was forming, the entire western interior of the continent was gradually uplifted. This uplift reelevated the Rockies and caused many of the West's major rivers to vigorously downcut. As the rivers became entrenched, many spectacular gorges were formed, including the Grand Canyon of the Colorado River, the Grand Canyon of the Snake River, and the Black Canyon of the Gunnison. The present topography of the Rocky Mountains is in large measure the result of this late Tertiary uplift and the subsequent excavation of the early Tertiary basin deposits by reinvigorated streams.

Volcanic activity was common in the West during much of the Cenozoic. Beginning in the Miocene epoch, great volumes of fluid basaltic lava flowed from fissures in portions of present-day Washington, Oregon, and Idaho. These eruptions built the extensive (1.3 million square kilometers, or 507,000 square miles) Columbia Plateau (see Figure 9.19, p. 266). Immediately west of the Columbia Plateau, volcanic activity was quite different. Here, thicker magmas with a higher silica content erupted explosively, creating the Cascade Range, a chain of stratovolcanoes from northern California to the Canadian border, some of which remain active—like Washington's Mount St. Helens (Figure 12.19).

A final episode of folding occurred in the West in late Tertiary time, creating the Coast Ranges, which stretch along the Pacific Coast. Meanwhile, the Sierra Nevada became fault-block mountains as they were uplifted along their eastern flank, creating the imposing mountain front we know today.

As the Tertiary period drew to a close, the effects of mountain building, volcanic activity, isostatic adjustments, and extensive erosion and sedimentation

▼ **FIGURE 12.19** Mount Hood, Oregon. This volcano is one of several large composite cones that comprise the Cascade Range. (Photo by John M. Roberts/Corbis/Stock Market)

BOX 12.2 EARTH AS A SYSTEM
Demise of the Dinosaurs

The boundaries between divisions on the geologic time scale represent times of significant geological and/or biological charge. Of special interest is the boundary between the Mesozoic era ("middle life") and Cenozoic era ("recent life"), about 65 million years ago. Around this time, about three-quarters of all plant and animal species died out in a *mass extinction*. This boundary marks the end of the era in which dinosaurs and other reptiles dominated the landscape and the beginning of the era when mammals become very important (Figure 12.B). Because the last period of the Mesozoic is the Cretaceous (abbreviated K to avoid confusion with other "C" periods), and the first period of the Cenozoic is the Tertiary (abbreviated T), the time of this mass extinction is called the *Cretaceous-Tertiary* or *KT boundary*.

The extinction of the dinosaurs is generally attributed to this group's inability to adapt to some radical change in environmental conditions. What event could have triggered the rapid extinction of the dinosaurs—one of the most successful groups of land animals ever to have lived?

FIGURE 12.B Dinosaurs dominated the Mesozoic landscape until their extinction at the close of the Cretaceous period. This dinosaur fossil is in New York's Museum of Natural History. (Photo © by Gail Mooney/CORBIS)

The most strongly supported hypothesis proposes that about 65 million years ago our planet was struck by a large carbonaceous meteorite, a relic from the formation of the solar system. The errant mass of rock was approximately 10 kilometers in diameter and was traveling at about 90,000 kilometers per hour at impact. It collided with the southern portion of North America in what is now Mexico's Yucatán Peninsula but at the time was a shallow tropical sea (Figure 12.C). The energy released by the impact is estimated to have been equivalent to 100 million megatons (*mega* = million) of high explosives.

For a year or two after the impact, suspended dust greatly reduced the sunlight reaching Earth's surface. This caused global cooling ("impact winter") and inhibited photosynthesis, greatly disrupting food production. Long after the dust settled, carbon dioxide, water vapor, and

---

had created a physical landscape very similar to the configuration of today. All that remained of Cenozoic time was the final 2-million-year episode called the Quaternary period. During this most recent (and current) phase of Earth history, in which humans evolved, the action of glacial ice and other erosional agents added the finishing touches.

Cenozoic Life

Mammals replaced reptiles as the dominant land animals in the Cenozoic. Angiosperms (flowering plants with covered seeds) replaced gymnosperms as the dominant land plants. Marine invertebrates took on a modern look. Microscopic animals called *foraminifera* became especially important. Today foraminifera are among the most intensely studied of all fossils because their widespread occurrence makes them invaluable in correlating Tertiary sediments. Tertiary strata are very important to the modern world, for they yield more oil than do rocks of any other age.

The Cenozoic is often called the "age of mammals," because these animals came to dominate land life. It could also be called the "age of flowering plants," for the angiosperms enjoy a similar status in the plant world. As a result of advances in seed fertilization and dispersal, angiosperms experienced rapid development and expansion as the Mesozoic drew to a close. Thus, as the Cenozoic era began, angiosperms were already the dominant land plants.

Development of the flowering plants, in turn, strongly influenced the evolution of both birds and mammals. Birds that feed on seeds and fruits, for example, evolved rapidly during the Cenozoic in close association with the flowering plants. During the middle Tertiary, grasses developed rapidly and spread over the plains. This fostered the emergence of herbivorous (plant-eating) mammals that were mainly grazers. In turn, the development and spread of grazing animals established the setting for the evolution of the carnivorous mammals that preyed upon them.

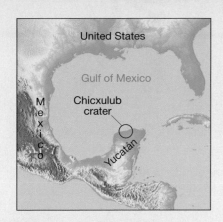

FIGURE 12.C Employing gravitational measurements, scientists have located a giant impact crater that formed about 65 million years ago and has since been filled with sediment. About 180 kilometers (112 miles) in diameter, Chicxulub crater is regarded by some researchers as the impact site that resulted in the demise of the dinosaurs.

sulfur oxides that had been added to the atmosphere by the blast remained. If significant quantities of sulfate aerosols formed, their high reflectivity would have helped to perpetuate the cooler surface temperatures for a few more years.*

*These aerosols are tiny droplets created by the combining of sulfur oxides and water that can remain suspended for long spans. For more on aerosols, see the section on "How Aerosols Influence Climate" in Chapter 20.

Eventually sulfate aerosols leave the atmosphere as acid precipitation. By contrast, carbon dioxide has a much longer residence time in the atmosphere. Carbon dioxide is a *greenhouse gas*, a gas that traps a portion of the radiation emitted by Earth's surface.** With the aerosols gone, the enhanced greenhouse effect caused by the carbon dioxide would have led to a long-term rise in average global temperatures. The likely result was that some of the plant and animal life that had survived the initial environmental assault finally fell victim to stresses associated with global cooling, followed by acid precipitation and global warming.

The extinction of the dinosaurs opened up habitats for the small mammals that survived. These new habitats, along with evolutionary forces, led to the development of the large mammals that occupy our modern world.

What evidence points to such a catastrophic collision 65 million years ago? First, a thin layer of sediment nearly 1 centimeter thick has been discovered at the KT boundary, worldwide. This sedi-

**For more about carbon dioxide and the greenhouse effect, see the section on "Heating the Atmosphere: The Greenhouse Effect" in Chapter 16.

ment contains a high level of the element *iridium*, rare in Earth's crust but found in high proportions in stony meteorites. Could this layer be the scattered remains of the meteorite that was responsible for the environmental changes that led to the demise of many reptile groups?

Despite its growing support, some scientists disagree with the impact hypothesis. They suggest instead that huge volcanic eruptions may have led to a breakdown in the food chain. To support this hypothesis, they site enormous outpourings of lavas in the Deccan Plateau of northern India about 65 million years ago.

Whatever caused the KT extinction, we now have a greater appreciation of the role of catastrophic events in shaping the history of our planet and the life that occupies it. Could a catastrophic event having similar results occur today? This possibility may explain why an event that occurred 65 million years ago has captured the interest of so many.

Mammals Replace Reptiles. Back in the Mesozoic, an important evolutionary event was the appearance of primitive mammals in the late Triassic, about the same time the dinosaurs emerged. Yet throughout the period of dinosaur dominance, mammals remained in the background as small and inconspicuous animals. By the close of the Mesozoic era, dinosaurs and other reptiles no longer dominated the land. It was only after these large reptiles became extinct that mammals came into their own as the dominant land animals. The transition is a major example in the fossil record of the replacement of one large group by another.

Mammals are distinct from reptiles in important respects. Mammalian young are born live, and mammals maintain a steady body temperature—that is, they are "warm-blooded." This latter adaptation allowed mammals to lead more active and diversified lives than reptiles because they could survive in cold regions and search for food during any season or time of day. Other mammalian adaptations included the de-

velopment of insulating body hair and more efficient heart and lungs.

It is worth noting that perfect boundaries cannot be drawn between mammalian and reptilian traits. One minor group of animals, the monotremes, still lay eggs. The two species in this group, the duck-billed platypus and the spiny anteater, are found only in Australia. Moreover, although modern reptiles are "cold-blooded," some paleontologists believe that dinosaurs may have been "warm-blooded."

With the demise of most Mesozoic reptiles, Cenozoic mammals diversified rapidly. The many forms that exist today evolved from small primitive mammals that were characterized by short legs, flat five-toed feet, and small brains. Their development and specialization took four principal directions: (1) increase in size, (2) increase in brain capacity, (3) specialization of teeth to better accommodate a particular diet, and (4) specialization of limbs to better equip the animal for life in a particular environment.

A.

B.

▲ **FIGURE 12.20** La Brea tar pits in Los Angeles, California. **A.** Fossils being excavated from the thick tar in 1914. **B.** Skeleton of an extinct saber-toothed cat. These bones came from the La Brea tar pits. (Photos courtesy of The George C. Page Museum)

Marsupials, Placentals, and Diversity. Following the reptilian extinctions at the close of the Mesozoic, two groups of mammals, the marsupials and the placentals, evolved and expanded to dominate the Cenozoic. The groups differ principally in their modes of reproduction. Young marsupials are born live but at a very early stage of development. After birth, the tiny and immature young crawl into the mother's external stomach pouch to complete their development. Examples are kangaroos, opossums, and koala bears.

Placental mammals, in contrast, develop within the mother's body for a much longer period, so that birth occurs after the young are relatively mature and independent. Most mammals are placental, including humans.

Today marsupials are found primarily in Australia, where they went through a separate evolutionary expansion during the Cenozoic, largely isolated from placental mammals.

In South America, both primitive marsupials and placentals coexisted before that landmass became completely isolated during the breakup of Pangaea. Evolution and specialization of both groups continued undisturbed for approximately 40 million years until the close of the Pliocene epoch, when the Central American land bridge emerged, connecting the two American continents. Then an invasion of advanced carnivores from North America brought the extinction of many hoofed mammals that had persisted in South America for millions of years. The marsupials, except for opossums, also could not compete and became extinct. Both Australia and South America provide excellent examples of how isolation caused by the separation of continents increased the diversity of animals in the world.

Large Mammals and Extinction. As we have seen, mammals diversified quite rapidly during the Cenozoic era. One tendency was for some groups to became very large. For example, by the Oligocene epoch a hornless

rhinoceros that stood nearly 5 meters (16 feet) high had evolved. It is the largest land mammal known to have existed. As time approached the present, many other types evolved to a large size as well—more, in fact, than now exist. Many of these large forms were common as recently as 11,000 years ago. However, a wave of late Pleistocene extinctions rapidly eliminated these animals from the landscape.

In North America, the mastodon and mammoth, both huge relatives of the elephant, became extinct. In addition, saber-toothed cats, giant beavers, large ground sloths, horses, camels, giant bison, and others died out (Figure 12.20). In Europe, late Pleistocene extinctions included woolly rhinos, large cave bears, and the Irish elk. The reason for this recent wave of large animal extinctions puzzles scientists. These animals had survived several major glacial advances and interglacial periods, so it is difficult to ascribe these extinctions to climatic change. Some scientists believe that early humans hastened the decline of these mammals by selectively hunting large forms. Although this hypothesis is preferred by many, it is not yet accepted by all.

? STUDENTS SOMETIMES ASK...
What are the La Brea tar pits?

The La Brea tar pits, located in downtown Los Angeles, are famous because they contain a rich and very well preserved assemblage of Pleistocene vertebrate fossils (see Figure 12.20). These organisms roamed southern California from 40,000 to 8000 years ago. The collection includes 59 species of mammals and more than 130 species of birds. Hundreds of invertebrate and plant fossils are also preserved. Tar pits form when crude oil seeps to the surface and the light portion evaporates, leaving behind sticky pools of heavy tar.

Chapter Summary

• The *Precambrian* spans about 88 percent of Earth history, beginning with the formation of Earth about 4.5 billion years ago and ending 540 million years ago with the diversification of life that marks the start of the Paleozoic era. It is the least understood span of Earth's history because most Precambrian rocks are buried from view. However, on each continent there is a "core area" of Precambrian rocks called the *shield*. The iron-ore deposits of Precambrian age represent the time when oxygen became abundant in the atmosphere and combined with iron to form iron oxide.

• Earth's primitive atmosphere consisted of such gases as water vapor, carbon dioxide, nitrogen, and several trace gases that were released in volcanic emissions, a process called *outgassing*. The first life forms on Earth, probably *anaerobic bacteria*, did not require oxygen. As life evolved, plants, through the process of *photosynthesis*, used carbon dioxide and water and released oxygen into the atmosphere. Once the available iron on Earth was oxidized (combined with oxygen), substantial quantities of oxygen accumulated in the atmosphere. About 4 billion years into Earth's existence, the fossil record reveals abundant ocean-dwelling organisms that require oxygen to live.

• The most common middle Precambrian fossils are *stromatolites*. Microfossils of bacteria and blue-green algae, both primitive *prokaryotes* whose cells lack organized nuclei, have been found in chert, a hard, dense, chemical sedimentary rock in southern Africa (3.1 billion years of age) and near Lake Superior (1.7 billion years of age). *Eukaryotes*, with cells containing organized nuclei, are among billion-year-old fossils discovered in Australia. Plant fossils date from the middle Precambrian, but animal fossils came a bit later, in the late Precambrian. Many of these fossils are *trace fossils*, and not of the animals themselves.

• The *Paleozoic era* extends from 540 million years ago to about 248 million years ago. The beginning of the Paleozoic is marked by the *appearance of the first life forms with hard parts*, such as shells. Therefore, abundant Paleozoic fossils occur, and a far more detailed record of Paleozoic events can be constructed. During the early Paleozoic (the Cambrian, Ordovician, and Silurian periods) the vast southern continent of *Gondwana* existed. Seas inundated and receded from North America several times, leaving thick evaporite beds of rock salt and gypsum. Life in the early Paleozoic was restricted to the seas and consisted of several invertebrate groups. During the late Paleozoic (the Devonian, Mississippian, Pennsylvanian, and Permian periods), ancestral North America collided with Africa to produce the original northern Appalachian Mountains, and the northern continent of *Laurasia* formed. By the close of the Paleozoic, all the continents had fused into the supercontinent of *Pangaea*. During most of the Paleozoic, organisms diversified dramatically. Insects and plants moved onto the land, and amphibians evolved and diversi-

fied quickly. By the Pennsylvanian period, large tropical swamps, which became the major coal deposits of today, extended across North America, Europe, and Siberia. At the close of the Paleozoic, altered climatic conditions caused one of the most dramatic biological declines in all of Earth history.

• The *Mesozoic era*, often called the *"age of dinosaurs,"* began about 248 million years ago and ended approximately 65 million years ago. Early in the Mesozoic much of the land was above sea level. However, by the middle Mesozoic, seas invaded western North America. As Pangaea began to break up, the westward-moving North American plate began to override the Pacific plate, causing crustal deformation along the entire western margin of the continent. Organisms that had survived extinction at the end of the Paleozoic began to diversify in spectacular ways. *Gymnosperms* (cycads, conifers, and ginkgoes) quickly became the dominant trees of the Mesozoic because they could adapt to the drier climates. Reptiles quickly became the dominant land animals, with one group eventually becoming the birds. The most awesome of the Mesozoic reptiles were the *dinosaurs*. At the close of the Mesozoic, many reptile groups, including the dinosaurs, became extinct.

• The *Cenozoic era*, or *"era of recent life,"* began approximately 65 million years ago and continues today. It is *the time of mammals*, including humans. The widespread, less disturbed rock formations of the Cenozoic provide a rich geologic record. Most of North America was above sea level throughout the Cenozoic. Because of their different relations with tectonic plate boundaries, the eastern and western margins of the North American continent experienced contrasting events. The stable eastern margin was the site of abundant sedimentation as isostatic adjustment raised the eroded Appalachians, causing the streams to downcut with renewed vigor and to deposit their sediment along the continental margin. In the west, building of the Rocky Mountains was coming to an end, the Basin and Range Province was forming, and volcanic activity was extensive. The Cenozoic is often called *"the age of mammals"* because these animals replaced the reptiles as the dominant land life. Two groups of mammals, the *marsupials* and the *placentals*, evolved and expanded to dominate the era. One tendency was for some mammal groups to become very large. However, a wave of late *Pleistocene* extinctions rapidly eliminated these animals from the landscape. Some scientists believe that humans hastened the decline of these animals by selectively hunting the larger species. The Cenozoic could also be called the *"age of flowering plants."* As a source of food, flowering plants strongly influenced the evolution of both birds and herbivorous (plant-eating) mammals throughout the Cenozoic era.

Key Terms

outgassing (p. 338)   shields (p. 338)   stromatolites (p. 339)

Review Questions

1. Explain why Precambrian history is more difficult to decipher than more recent geological history.

2. What is the major source of free oxygen in Earth's atmosphere?

3. Match the following words and phrases to the most appropriate time span. Select among the following: Precambrian, early Paleozoic, late Paleozoic, Mesozoic, or Cenozoic.

   (a) Pangaea came into existence.
   (b) Bacteria and blue-green algae preserved in chert.
   (c) The era that encompasses the least amount of time.
   (d) Shields.
   (e) Age of dinosaurs.
   (f) Formation of the original northern Appalachian mountains.
   (g) Mastodons and mammoths.
   (h) Extensive deposits of rock salt.
   (i) Triassic, Jurassic, and Cretaceous.
   (j) Coal swamps extended across North America, Europe, and Siberia.
   (k) Gulf Coast oil deposits formed.
   (l) Formation of most of the world's major iron-ore deposits.
   (m) Massive sand dunes covered large portions of the Colorado Plateau region.
   (n) The "age of fishes" occurred during this span.
   (o) Cambrian, Ordovician, and Silurian.
   (p) Pangaea began to break apart.
   (q) "Age of mammals."
   (r) Animals with hard parts first appeared in abundance.
   (s) Gymnosperms were the dominant trees.
   (t) Columbia Plateau formed.
   (u) Stromatolites are among its more common fossils.
   (v) "Golden age of trilobites" occurred during this span.
   (w) Fault-block mountains form in the Basin and Range Province.

4. Briefly discuss two proposals that attempt to explain why several groups developed hard parts at the beginning of the Cambrian period. Do these proposals appear to provide a satisfactory explanation? Why or why not?

5. List some differences between amphibians and reptiles. List differences between reptiles and mammals.

6. Describe a hypothesis that attempts to explain the extinction of the dinosaurs.

7. Contrast the eastern and western margins of North America during the Cenozoic era in terms of their relationships to plate boundaries.

Examining the Earth System

1. The Earth system has been responsible for both the conditions that favored the evolution of life on this planet and for the mass extinctions that have occurred throughout geological time. Describe the role of the biosphere, hydrosphere, and solid Earth in forming the current level of atmospheric oxygen. How did Earth's near-space environment interact with the atmosphere and biosphere to contribute to the great mass extinction that marked the end of the dinosaurs?

2. Most of the vast North American coal resources located from Pennsylvania to Illinois began forming during the Pennsylvanian and Mississippian periods of Earth history. (This time period is also referred to as the Carboniferous period.) Using Figure 12.12, a restoration of a Pennsylvanian period coal swamp, describe the climatic and biological conditions associated with this unique environment. Next, examine Figure 12.10B, C. The maps show the formation of the supercontinent of Pangaea and illustrate the geographic position of North America during the period of coal formation. Where, relative to the equator, was North America located during the time of coal formation? What role did plate tectonics play in determining the conditions that eventually produced North America's eastern coal reserves? Why is it unlikely that the coal-forming environment will repeat itself in North America in the near future? (You may find it helpful to visit the Illinois State Museum Mazon Creek Fossils exhibit at **http://www.museum.state.il.us/exhibits/mazon_creek**, and/or the University of California Time Machine Exhibit at **http://www.ucmp.berkeley.edu/carboniferous/carboniferous.html**.

Online Study Guide

The *Earth Science* Website uses the resources and flexibility of the Internet to aid in your study of the topics in this chapter. Written and developed by Earth science instructors, this site will help improve your understanding of Earth science. Visit **http://www.prenhall.com/tarbuck** and click on the cover of *Earth Science 11e* to find:

- **On-line review quizzes.**
- **Critical thinking exercises.**
- **Links to chapter-specific Web resources.**
- **Internet-wide key term searches.**

http://www.prenhall.com/tarbuck

CHAPTER 13

The Ocean Floor*

*Professor Alan P. Trujillo, Palomar College, aided in the revision of this chapter.

The **Thomas G. Thompson** *is the University of Washington's oceanographic research vessel. (Photo by Paul Souders/CORBIS)*

How deep is the ocean? How much of Earth is covered by the global sea? What does the seafloor look like? Humans have long been interested in finding answers to these questions, but it was not until rather recently that these seemingly simple questions could be answered. Suppose, for example, that all of the water were drained from the ocean. What would we see? Plains? Mountains? Canyons? Plateaus? Indeed, the ocean conceals all of these features, and more. And what about the carpet of sediment that covers much of the seafloor? Where did it come from, and what can be learned by examining it? This chapter provides answers to these questions.

Calling Earth the "water planet" is certainly appropriate, because nearly 71 percent of its surface is covered by the global ocean. Although the ocean comprises a much greater percentage of Earth's surface than the continents, it has only been in the relatively recent past that the ocean became an important focus of study. Recently there has been a virtual explosion of data about the oceans, and with it oceanography has grown dramatically. **Oceanography** is an interdisciplinary science that draws on the methods and knowledge of geology, chemistry, physics, and biology to study all aspects of the world ocean.

The Vast World Ocean

A glance at a globe or a view of Earth from space reveals a planet dominated by an interconnected single world ocean (Figure 13.1). Indeed, it is for this reason that Earth is often referred to as the *blue planet*.

Geography of the Oceans

The area of Earth is about 510 million square kilometers (197 million square miles). Of this total, approximately 360 million square kilometers (140 million square miles), or 71 percent, is represented by oceans and marginal seas (meaning seas around the ocean's margin, like the Mediterranean Sea and Caribbean Sea). Continents and islands comprise the remaining 29 percent, or 150 million square kilometers (58 million square miles).

Even though oceans dominate Earth's surface, is the distribution of land and water similar in the Northern and Southern hemispheres? By studying a world map or globe (Figure 13.1), it is readily apparent that the continents and oceans are not evenly divided between the two hemispheres. In the Northern Hemisphere, for instance, nearly 61 percent of the surface is water, whereas about 39 percent is land. In the Southern Hemisphere, on the other hand, almost 81 percent of the surface is water, and only 19 percent is land. It is

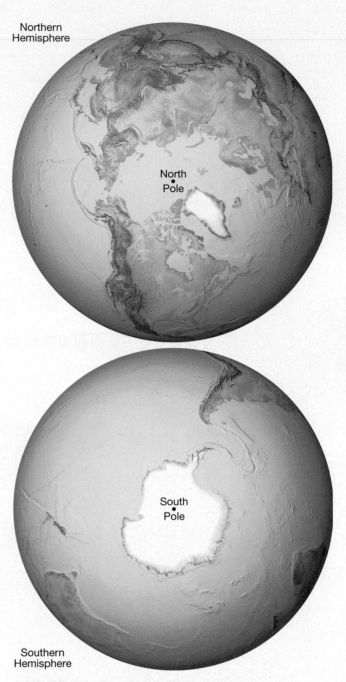

▲ **FIGURE 13.1** These views of Earth show the uneven distribution of land and water between the Northern and Southern Hemispheres. Almost 81 percent of the Southern Hemisphere is covered by the oceans—20 percent more than the Northern Hemisphere.

no wonder then that the Northern Hemisphere is called the *land hemisphere* and the Southern Hemisphere the *water hemisphere*.

Figure 13.2A shows the distribution of land and water in the Northern and Southern hemispheres. Between latitudes 45 degrees north and 70 degrees north, there is actually more land than water, whereas between 40 degrees south and 65 degrees south there is

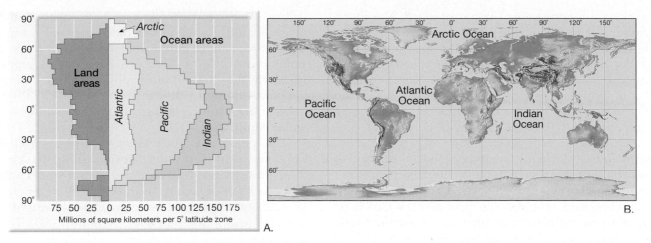

▲ **FIGURE 13.2** Distribution of land and water. **A.** The graph shows the amount of land and water in each 5-degree latitude belt. **B.** The world map provides a more familiar view.

almost no land to interrupt the oceanic and atmospheric circulation.

The world ocean can be divided into four main ocean basins (Figure 13.2B):

1. The *Pacific Ocean*, which is the largest ocean (and the largest single geographic feature on the planet), covering over half of the ocean surface area on Earth. In fact, the Pacific Ocean is so large that all of the continents could fit into the space occupied by it—with room left over! It is also the world's deepest ocean, with an average depth of 3940 meters (12,927 feet).

2. The *Atlantic Ocean*, which is about half the size of the Pacific Ocean but is not quite as deep. It is a relatively narrow ocean as compared to the Pacific and is bounded by almost parallel continental margins.

3. The *Indian Ocean*, which is slightly smaller than the Atlantic Ocean but has about the same average depth. Unlike the Pacific and Atlantic oceans, it is largely a Southern Hemisphere water body.

4. The *Arctic Ocean*, which is about 7 percent the size of the Pacific Ocean and is only a little more than one-quarter as deep as the rest of the oceans.

Comparing the Oceans to the Continents

A major difference between continents and the ocean basins is their relative levels. The average elevation of the continents above sea level is about 840 meters (2756 feet), whereas the average depth of the oceans is nearly four and a half times this amount—3729 meters (12,234 feet). The volume of ocean water is so large

that if Earth's solid mass were perfectly smooth (level) and spherical, the oceans would cover Earth's entire surface to a uniform depth of more than 2000 meters (1.2 miles)!

An Emerging Picture of the Ocean Floor

GEODe The Global Ocean
▼ Floor of the Ocean

If all water were drained from the ocean basins, a great variety of features would be seen, including broad volcanic peaks, deep trenches, extensive plains, linear mountain chains, and large plateaus. In fact, the scenery would be nearly as diverse as that on the continents.

An understanding of seafloor features came with the development of techniques to measure the depth of the oceans. **Bathymetry** (*bathos* = depth, *metry* = measurement) is the measurement of ocean depths and the charting of the shape or topography of the ocean floor.

Mapping the Seafloor

The first understanding of the ocean floor's varied topography did not unfold until the historic three-and-a-half-year voyage of the HMS *Challenger* (Figure 13.3A). From December 1872 to May 1876, the *Challenger* expedition made the first—and perhaps still most comprehensive—study of the global ocean ever attempted by one agency. The 127,500-kilometer (79,200-mile) trip took the ship and its crew of scientists to every ocean except the Arctic. Throughout the voyage, they sampled various ocean properties, including water depth, which was accomplished by laboriously lowering a long weighted line overboard. Not many years later the knowledge gained by the *Challenger* of the ocean's great depth and varied topography was further expanded with the laying of transatlantic communication cables, especially in the North Atlantic Ocean. However, as long as a weighted line was the only way to measure ocean depths, knowledge of seafloor features remained limited (Figure 13.3B).

Bathymetric Techniques Today sound energy is used to measure water depths. The basic approach employs some type of **sonar** an acronym for sound navigation and ranging. The first devices that used sound to measure water depth, called **echo sounders**, were developed early in the twentieth century. Echo sounders work by transmitting a sound wave (called a *ping*) into the water in order to produce an echo when it bounces off any object, such as a marine organism or the ocean floor (Figure 13.4A). A sensitive receiver intercepts the echo reflected from the bottom, and a clock precisely measures the travel time to fractions of a second. By knowing the speed of sound waves in water—about 1500 meters (4900 feet) per second—and the time required for the energy pulse to reach the ocean floor and return, depth can be calculated. The depths determined from continuous monitoring of these echoes are plotted so a profile of the ocean floor is obtained. By laboriously combining profiles from several adjacent traverses, a chart of the seafloor can be produced.

Following World War II, the U.S. Navy developed *sidescan sonar* to look for mines and other explosive devices. These torpedo-shaped instruments can be towed behind a ship where they send out a fan of sound extending to either side of the ship's track. By combining swaths of sidescan sonar data, researchers produced the first photograph-like images of the seafloor. Although sidescan sonar provides valuable views of the seafloor, it does not provide bathymetric (water depth) data.

This problem is not present in the *high-resolution multibeam* instruments developed during the 1990s. These systems use hull-mounted sound sources that send out a fan of sound, then record reflections from the seafloor through a set of narrowly focused receivers aimed at different angles (Figure 13.4B). Thus, rather than obtaining the depth of a single point every few seconds, this technique makes it possible for a survey ship to map the features of the ocean floor along a strip tens of kilometers wide (Figure 13.5). When a ship uses multibeam sonar to make a map of a section of seafloor, it travels through the area in a regularly spaced back-and-forth pattern known as, appropriately enough, "mowing the lawn." Furthermore, these systems can collect bathymetric data of such high resolution that they can distinguish depths that differ by less than a meter.

Despite their greater efficiency and enhanced detail, research vessels equipped with multibeam sonar travel at a mere 10 to 20 kilometers (6 to 12 miles) per hour. It would take at least 100 vessels outfitted with this equipment hundreds of years to map the entire seafloor. This explains why only about 5 percent of the seafloor has been mapped in detail—and why large

▼ **FIGURE 13.3** A comparison between one of the first oceanographic research vessels and an ultramodern research vessel. **A.** The HMS. *Challenger*. (From C.W. Thompson and Sir John Murray, Report on the Scientific Results of the Voyage of the HMS *Challenger*, Vol. 1. Great Britain: Challenger Office, 1895, Plate 1, Library of Congress) **B.** The research vessel *Kilo Moana* entering the port of Kodiak, Alaska. This ultramodern ship, operated by the University of Hawaii with the oversight of NOAA, is capable of simultaneously conducting a variety of oceanographic and meteorological studies. Equipped with multibeam sonar, it is capable of charting the ocean floor to depths as great as 11,000 meters (36,000 feet). (AP/Wide World Photos)

A.

B.

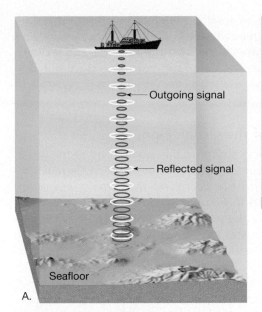 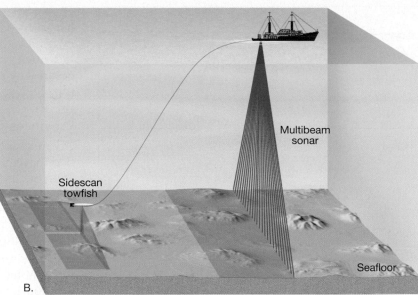

▲ **FIGURE 13.4** Various types of sonar. **A.** An echo sounder determines the water depth by measuring the time interval required for an acoustic wave to travel from a ship to the seafloor and back. The speed of sound in water is 1500 m/sec. Therefore, depth = $\frac{1}{2}$(1500 m/sec × echo travel time). **B.** Modern multibeam sonar and sidescan sonar obtain an "image" of a narrow swath of seafloor every few seconds.

areas of the seafloor have not yet been mapped with sonar at all.

Seismic Reflection Profiles   Marine geologists are also interested in viewing the rock structure beneath the sediments that blanket much of the seafloor. This can be accomplished by making a **seismic reflection profile**. To construct such a profile, strong low-frequency sounds are produced by explosions (depth charges) or air guns. These sound waves penetrate beneath the seafloor and reflect off the contacts between rock layers and fault zones, just like sonar reflects off the bottom of the sea. Figure 13.6 shows a seismic profile of a portion of the Madeira abyssal plain in the eastern Atlantic. Although

the seafloor is flat, notice the irregular ocean crust buried by a thick accumulation of sediments.

Viewing the Ocean Floor from Space

Another technological breakthrough that has led to an enhanced understanding of the seafloor involves measuring the shape of the ocean surface from space. After compensating for waves, tides, currents, and atmospheric effects, it was discovered that the ocean surface is not perfectly flat. This is because gravity attracts water toward regions where massive seafloor features occur. Therefore, mountains and ridges produce elevated areas on the ocean surface, and, conversely,

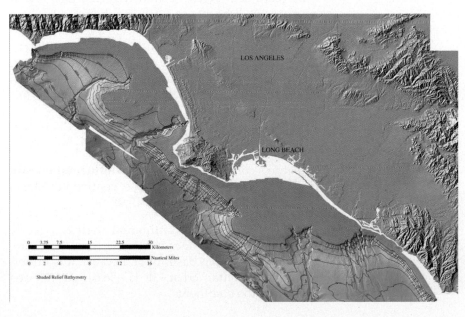

◄ **FIGURE 13.5**   Color-enhanced perspective map of the seafloor and coastal landforms in the Los Angeles area of California. The ocean floor portion of this map was constructed from data collected using a high-resolution mapping system. (U.S. Geological Survey)

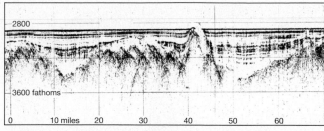

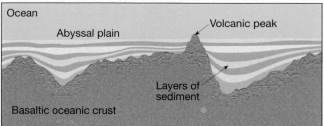

▲ **FIGURE 13.6** Seismic cross section and matching sketch across a portion of the Madeira abyssal plain in the eastern Atlantic Ocean, showing the irregular oceanic crust buried by sediments. (Image courtesy of Charles Hollister, Woods Hole Oceanographic Institution)

canyons and trenches cause slight depressions. Satellites equipped with *radar altimeters* are able to measure these subtle differences by bouncing microwaves off the sea surface (Figure 13.7). These devices can meas-

▼ **FIGURE 13.7** A satellite altimeter measures the variation in sea surface elevation, which is caused by gravitational attraction and mimics the shape of the seafloor. The sea surface anomaly is the difference between the measured and theoretical ocean surface.

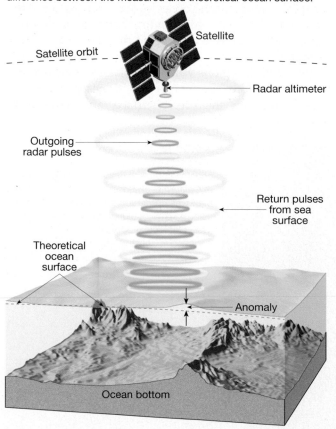

ure variations as small as 3 to 6 centimeters. Such data have added greatly to the knowledge of ocean-floor topography. Cross-checked with traditional sonar depth measurements, the data are used to produce detailed ocean-floor maps, such as the one shown in Figure 1.15 (pp. 20–21).

Provinces of the Ocean Floor

Oceanographers studying the topography of the ocean floor have delineated three major units: *continental margins*, the *ocean basin floor*, and the *oceanic (mid-ocean) ridge*. The map in Figure 13.8 outlines these provinces for the North Atlantic Ocean, and the profile at the bottom of the illustration shows the varied topography. Such profiles usually have their vertical dimension exaggerated many times—40 times in this case—to make topographic features more conspicuous. Vertical exaggeration, however, makes slopes shown in seafloor profiles appear to be *much* steeper than they actually are.

Continental Margins

The Global Ocean
▼ **Floor of the Ocean**

Two main types of **continental margins** have been identified—*passive* and *active*. Passive margins are found along most of the coastal areas that surround the Atlantic Ocean, including the east coasts of North and South America, as well as the coastal areas of Western Europe and Africa. Passive margins are *not* associated with plate boundaries and therefore experience very little volcanism and few earthquakes. Here, weathered materials eroded from the adjacent landmass accumulate to form a thick, broad wedge of relatively undisturbed sediments.

By contrast, active continental margins occur where oceanic lithosphere is being subducted beneath the edge of a continent. The result is a relatively narrow margin, consisting of highly deformed sediments that were scraped from the descending lithospheric slab. Active continental margins are common around the Pacific Rim, where they parallel deep oceanic trenches.

Passive Continental Margins

The features comprising a **passive continental margin** include the continental shelf, the continental slope, and the continental rise (Figure 13.9).

Continental Shelf. The **continental shelf** is a gently sloping submerged surface extending from the shoreline toward the ocean basin floor. Because it is underlain by continental crust, it is clearly a flooded extension of the continents.

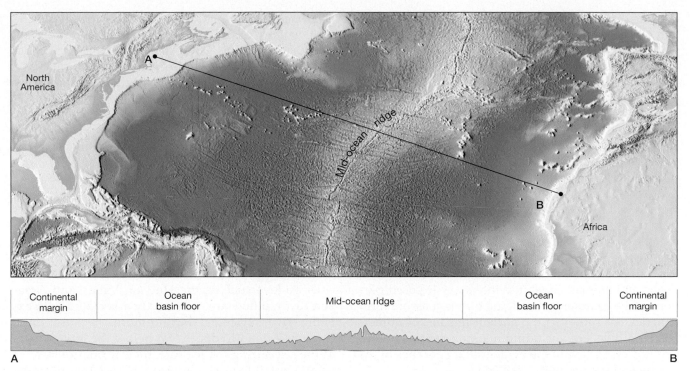

▲ **FIGURE 13.8** Map view (*above*) and corresponding profile view (*below*) showing the major topographic divisions of the North Atlantic Ocean. On the profile, the vertical scale has been expanded (exaggerated) by 40 times to make topographic features more conspicuous.

The continental shelf varies greatly in width. Although almost nonexistent along some continents, the shelf may extend seaward as far as 1500 kilometers (930 miles) along others. On average, the continental shelf is about 80 kilometers (50 miles) wide and 130 meters (425 feet) deep at its seaward edge. The average inclination of the continental shelf is only about one-tenth of 1 degree, a drop of only about 2 meters per kilometer (10 feet per mile). The slope is so slight that it would appear to an observer to be a horizontal surface.

Although continental shelves represent only 7.5 percent of the total ocean area, they have economic and political significance because they contain important mineral deposits, including large reservoirs of oil and natural gas, as well as huge sand and gravel

▼ **FIGURE 13.9** Features of a passive continental margin, including the continental shelf, continental slope, and continental rise. The Atlantic coast of North America is a good example of a passive margin. Note that the steepness of the slopes shown for the continental shelf and continental slope are greatly exaggerated. The continental shelf has an average slope of one-tenth of 1 degree, whereas the continental slope has an average slope of about 5 degrees.

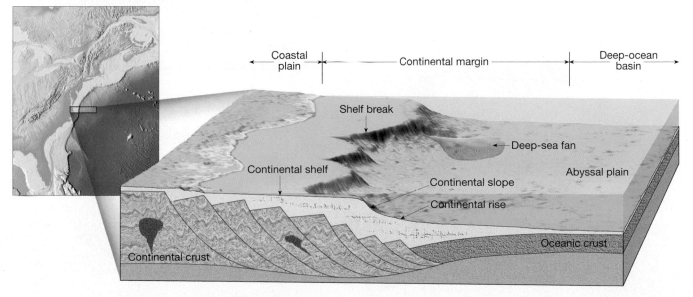

deposits. The waters of the continental shelf also contain many important fishing grounds, which are significant sources of food.

Even though the continental shelf is relatively featureless, some areas are mantled by extensive glacial deposits and are thus quite rugged. In addition, some continental shelves are dissected by large valleys running from the coastline into deeper waters. Many of these *shelf valleys* are the seaward extensions of river valleys on the adjacent landmass. Such valleys appear to have been excavated during the Pleistocene epoch (Ice Age). During this time great quantities of water were stored in vast ice sheets on the continents. This caused sea level to drop by 100 meters (330 feet) or more, exposing large areas of the continental shelves (see Figure 6.2, p. 155). Because of this drop in sea level, rivers extended their courses, and land-dwelling plants and animals inhabited the newly exposed portions of the continents. Dredging off the coast of North America has retrieved the ancient remains of numerous land dwellers, including mammoths, mastodons, and horses, adding to the evidence that portions of the continental shelves were once above sea level.

Most passive continental shelves, such as those along the East Coast of the United States, consist of shallow-water deposits that can reach several kilometers in thickness. Such deposits have led researchers to conclude that these thick accumulations of sediment are produced along a gradually subsiding continental margin.

Continental Slope. Marking the seaward edge of the continental shelf is the **continental slope**, a relatively steep structure (as compared with the shelf) that marks the boundary between continental crust and oceanic crust (Figure 13.9). Although the inclination of the continental slope varies greatly from place to place, it averages about 5 degrees and in places may exceed 25 degrees. Further, the continental slope is a relatively narrow feature, averaging only about 20 kilometers (12 miles) in width.

Submarine Canyons and Turbidity Currents. Deep, steep-sided valleys known as **submarine canyons** are cut into the continental slope and may extend across the entire continental rise to the ocean basin floor (Figure 13.10). Although some of these canyons appear to be the seaward extensions of river valleys such as the Hudson Valley and Amazon River, others are not directly associated with existing river systems. Furthermore, because these canyons extend to depths far below the lowest level of the sea during the Ice Age, their formation cannot be attributed to stream erosion. These features must be created by some process that operates far below the ocean surface.

Most available information suggests that submarine canyons have been eroded, at least in part, by turbidity currents (Figure 13.10). **Turbidity currents** are periodic downslope movements of dense, sediment-

▼ **FIGURE 13.10** Turbidity currents are high-density flows of sediment and water that move downslope, eroding the continental margin and enlarging submarine canyons. Turbidity currents eventually lose momentum and deposit their loads of sediment as deep-sea fans. Beds deposited by these currents are called *turbidites*. Each event produces a single bed characterized by a decrease in sediment size from bottom to top, a feature known as a *graded bed*.

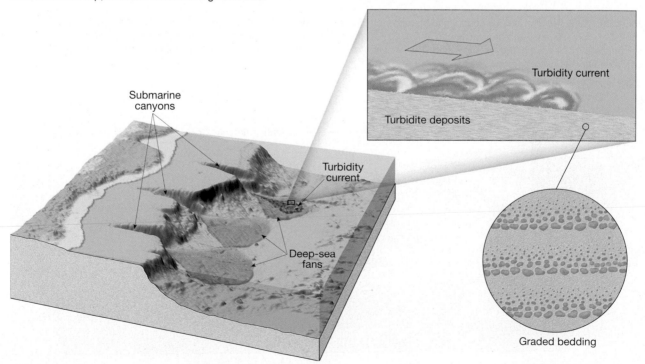

laden water. They are created when sand and mud on the continental shelf and slope are dislodged, perhaps by an earthquake, and are thrown into suspension. Because such mud-choked water is denser than normal seawater, it flows downslope, eroding and accumulating more sediment. The erosional work accomplished by these muddy torrents is thought to be the major force in the excavation of most submarine canyons (Box 13.1).

Narrow continental margins, such as the one located along the California coast, are dissected by numerous submarine canyons. Here erosion has extended many of these canyons landward into shallow water. Sediments carried to the coasts by rivers are transported along the shore by wave activity until they reach a submarine canyon. This steady supply of sediment collects until it becomes unstable and moves as a massive underwater landslide (turbidity current) to the deep-ocean floor.

Turbidity currents eventually lose momentum and come to rest along the floor of the ocean basin (Figure 13.10). As these currents slow, the suspended sediments begin to settle out. First, the coarser, heavier sand is dropped, followed by successively finer deposits of silt and then clay. These deposits, called **turbidites**, are characterized by a decrease in sediment grain size from bottom to top, a phenomenon known as **graded bedding** (Figure 13.10, *inset*).

Although there is still more to be learned about the complex workings of turbidity currents, it has been well established that they are an important mechanism of sediment transport in the ocean. By the action of turbidity currents, submarine canyons are eroded and sediments are deposited on the deep-ocean floor.

Continental Rise. In regions where trenches do not exist, the steep continental slope merges into a more gradual incline known as the **continental rise**. Here the slope drops to about one-third degree, or about 6 meters per kilometer (32 feet per mile). Whereas the width of the continental slope averages about 20 kilometers (12 miles), the continental rise may extend for hundreds of kilometers into the deep-ocean basin.

The continental rise consists of a thick accumulation of sediment that moved downslope from the continental shelf to the deep-ocean floor. The sediments are delivered to the base of the continental slope by turbidity currents that periodically flow down submarine canyons. When these muddy currents emerge from the mouth of a canyon onto the relatively flat ocean floor, they deposit sediment that forms a **deep-sea fan** (Figure 13.10). Deep-sea fans have the same basic shape as the *alluvial fans* that form at the foot of steep mountain slopes on land. As fans from adjacent submarine canyons grow, they merge laterally with one another to produce a continuous covering of sediment at the base of the continental slope called the continental rise.

Active Continental Margins

Along some coasts the continental slope descends abruptly into a deep-ocean trench located between the continent and ocean basin. Thus, the landward wall of the trench and the continental slope are essentially the same feature. In such locations, the continental shelf is very narrow, if it exists at all.

Active continental margins are located primarily around the Pacific Ocean in areas where oceanic lithosphere is being subducted beneath the leading edge of a continent (Figure 13.11). Here sediments from the ocean floor and pieces of oceanic crust are scraped from the descending oceanic plate and plastered against the edge of the overriding continent. This chaotic accumulation of deformed sediment

▲ **FIGURE 13.11** Active continental margin. Here sediments from the ocean floor are scraped from the descending plate and added to the continental crust as an accretionary wedge. The Pacific margin of South America is an excellent example.

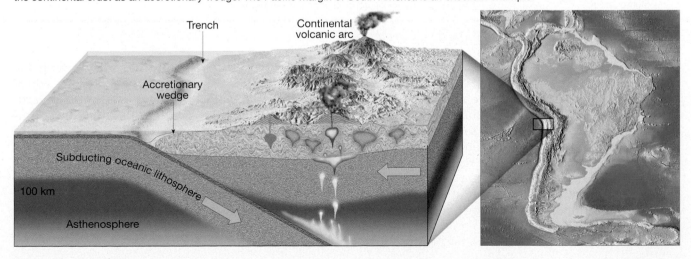

BOX 13.1 UNDERSTANDING EARTH:
A Grand Break—Evidence for Turbidity Currents

How do earthquakes and telephone cables help explain how turbidity currents move across the ocean floor and carve submarine canyons? In 1929, the Grand Banks earthquake in the North Atlantic Ocean severed some of the trans-Atlantic telephone and telegraph cables that lay across the seafloor south of Newfoundland near the earthquake epicenter (Figure 13.A). At first it was assumed that seafloor movement caused all of these breaks. However, analysis of the data revealed that the cables closest to the epicenter broke simultaneously during the earthquake, but cables that crossed the slope and deeper ocean floor at greater distances from the epicenter were broken progressively later in time. It seemed unusual that certain cables were affected by the failure of the slope due to ground shaking, but others were broken several hours later.

Following a reanalysis of the data several years later, a hypothesis was formulated to explain the distribution of cable breaks. The hypothesis suggested that a turbidity current moving down the continental slope and rise could account for the pattern of cable breaks. Based on the sequence of breaks, the turbidity current must have reached speeds approaching 80 kilometers (50 miles) per hour on the steep portions of the continental slope, and about 24 kilometers (15 miles) per hour on the more gently sloping continental rise. Thus, turbidity currents reach high speeds and are strong enough to break underwater cables, suggesting that they must be powerful enough to erode submarine canyons.

Further evidence of turbidity currents comes from several studies that have documented turbidity currents using sonar. For instance, a study of Rupert Inlet in British Columbia, Canada, monitored turbidity currents moving through an underwater channel. Such studies indicate that submarine canyons are gradually eroded by turbidity currents over long periods of time, just as canyons on land are carved by running water.

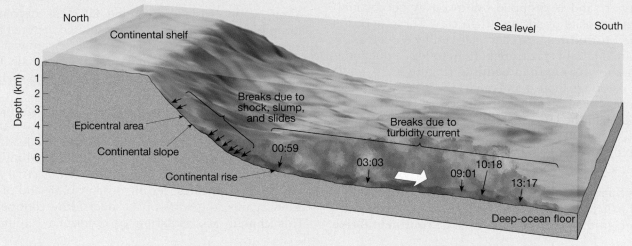

Figure 13.A Profile of the seafloor showing the events of the November 18, 1929, earthquake off the coast of Newfoundland. The arrows point to cable breaks; the numbers show times of breaks in hours and minutes after the earthquake. Vertical scale is greatly exaggerated. (After B. C. Heezen and M. Ewing, "Turbidity Currents and Submarine Slump and the 1929 Grand Banks Earthquake," *American Journal of Science* 250:867)

and scraps of oceanic crust is called an *accretionary* (*ad* = toward, *crescere* = to grow) *wedge*. Prolonged plate subduction, along with the accretion of sediments on the landward side of the trench, can produce a large accumulation of sediments along a continental margin. A large accretionary wedge, for example, is found along the northern coast of Japan's Honshu Island.*

Some subduction zones have little or no accumulation of sediments, indicating that ocean sediments are being carried into the mantle with the subducting plate. These tend to be regions where old oceanic lithosphere is subducting nearly vertically into the mantle. In these locations the continental margin is very narrow, as the trench may lie a mere 50 kilometers (31 miles) offshore.

Some subduction zones have little or no accumulation of sediments, indicating that ocean sediments are being carried into the mantle with the subducting plate. Here the continental margin is very narrow, as the trench may lie a mere 50 kilometers (31 miles) offshore.

*For more on accretionary wedges and subduction, see Chapter 10.

The Ocean Basin Floor

GEODe
The Global Ocean
▼ Floor of the Ocean

Between the continental margin and the oceanic ridge lies the **ocean basin floor** (see Figure 13.8). The size of this region—almost 30 percent of Earth's surface—is roughly comparable to the percentage of land above sea level. This region includes remarkably flat areas known as *abyssal plains*; tall volcanic peaks called *seamounts* and *guyots*; *deep-ocean trenches*, which are extremely deep linear depressions in the ocean floor; and extensive areas of lava flows piled one atop the other called *oceanic plateaus*.

Deep-Ocean Trenches

Deep-ocean trenches are long, relatively narrow creases in the seafloor that form the deepest parts of the ocean. Most trenches are located along the margins of the Pacific Ocean (Figure 13.12) where many exceed 10,000 meters (33,000 feet) in depth. A portion of one trench—the Challenger Deep in the Mariana Trench—has been measured at a record 11,022 meters (36,163 feet) below sea level, making it the deepest known part of the world ocean. Only two trenches are located in the Atlantic—the Puerto Rico Trench and the South Sandwich Trench (see Figure 1.15, pp. 20-21)

STUDENTS SOMETIMES ASK...

*Have humans ever explored
the deepest ocean trenches?
Could anything live there?*

Humans have indeed visited the deepest part of the oceans—where there is crushing high pressure, complete darkness, and near-freezing water temperatures—over 40 years ago! In January 1960, U.S. Navy Lt. Don Walsh and explorer Jacques Piccard descended to the bottom of the Challenger Deep region of the Mariana Trench in the deep diving bathyscaphe *Trieste*. At 9906 meters (32,500 feet), the men heard a loud cracking sound that shook the cabin. They were unable to see that a 7.6-centimeter (3-inch) Plexiglas viewing port had cracked (miraculously, it held for the rest of the dive). More than five hours after leaving the surface, they reached the bottom at 10,912 meters (35,800 feet)—a record depth of human descent that has not been broken since. They did see some life forms that are adapted to life in the deep: a small flatfish, a shrimp, and some jellyfish.

Although deep-ocean trenches represent only a very small portion of the area of the ocean floor, they are nevertheless significant geologic features. Trenches are sites of plate convergence where a moving crustal plate subducts and plunges back

into the mantle*. In addition to earthquakes being created as one plate "scrapes" beneath another, volcanic activity is also associated with these regions. The release of volatiles—especially water—from a descending plate triggers melting in the wedge of asthenosphere above it. This buoyant material slowly migrates upward and gives rise to volcanic activity at the surface. Thus, trenches are often paralleled by an arc-shaped row of active volcanoes called a **volcanic island arc**. Furthermore, **continental volcanic arcs**, such as those making up portions of the Andes and Cascades, are located parallel to trenches that lie adjacent to continental margins (see Figure 13.11). The large number of trenches and associated volcanic activity along the margins of the Pacific Ocean is why the region is known as the *Ring of Fire*.

Abyssal Plains

Abyssal (*a* = without, *byssus* = bottom) **plains** are deep, incredibly flat features; in fact, these regions are likely the most level places on Earth. The abyssal plain found off the coast of Argentina, for example, has less than 3 meters (10 feet) of relief over a distance exceeding 1300 kilometers (800 miles). The monotonous topography of abyssal plains is occasionally interrupted by the protruding summit of a partially buried volcanic peak.

Using *seismic profilers* (instruments that generate signals designed to penetrate far below the ocean floor), researchers have determined that abyssal plains owe their relatively featureless topography to thick accumulations of sediment that have buried an otherwise rugged ocean floor (see Figure 13.6). The nature of the sediment indicates that these plains consist primarily of fine sediments transported far out to sea by turbidity currents or deposited through suspension settling.

Abyssal plains are found in all oceans of the world. However, the Atlantic Ocean has the most extensive abyssal plains because it has few trenches to act as traps for sediment carried down the continental slope.

Seamounts, Guyots, and Oceanic Plateaus

Dotting the ocean floor are shield volcanoes called **seamounts**, which may rise hundreds of meters above the surrounding topography. Although these steep-sided conical peaks are found on the floors of all the oceans, the greatest number have been identified in the Pacific.

Some, like the Hawaiian Island–Emperor Seamount chain in the Pacific, which stretches from the Hawaiian Islands to the Aleutian trench, form in association with volcanic hot spots (see Figure 1.15, pp 20-21). Others are born near oceanic ridges, divergent plate boundaries

---

*For more on convergent plate boundaries, see Chapter 8.

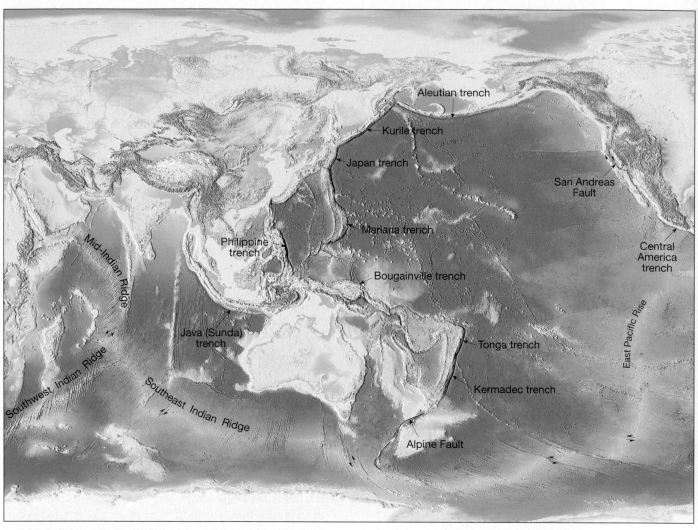

▲ **FIGURE 13.12** Distribution of the world's deep-ocean trenches and oceanic ridges.

where plates move apart.* If the volcano grows large enough before being carried from the magma source by plate movement, the structure emerges as an island. Examples of volcanic islands in the Atlantic include the Azores, Ascension, Tristan da Cunha, and St. Helena.

During the time they exist as islands, some of these volcanoes are lowered to near sea level by weathering, mass wasting, running water, and wave action. Over a span of millions of years, the islands gradually sink and disappear below the water surface as the moving plate slowly carries them away from the elevated oceanic ridge or hot spot where they originated (see Box 13.2). These submerged, flat-topped seamounts are called **guyots** or **tablemounts**.**

Mantle plumes have also generated several large **oceanic plateaus**, which resemble the flood basalt provinces found on the continents. Examples of these extensive volcanic structures include the Ontong Java and

Rockall plateaus (see Figure 10.19, p. 298) which formed from vast outpourings of fluid basaltic lavas onto the ocean floor. Hence, oceanic plateaus are composed mostly of pillow basalt and other mafic rocks that in some cases exceed 30 kilometers in thickness.

The Oceanic Ridge

The Global Ocean
▼ Floor of the Ocean

Along well-developed divergent plate boundaries, the seafloor is elevated, forming a broad linear swell called the **oceanic ridge**, or **mid-ocean ridge**. Our knowledge of the oceanic ridge system comes from soundings taken of the ocean floor, core samples obtained from deep-sea drilling, visual inspection using deep-diving submersibles, and even firsthand inspection of slices of ocean floor that have been displaced onto dry land along convergent plate boundaries (Figure 13.13). An elevated position, extensive faulting and associated earthquakes,

---

*For more on hot spots and divergent plate boundaries, see Chapter 8.
**The term *guyot* is named after Princeton University's first geology professor Arnold Guyot. It is pronounced "GEE-oh" with a hard *g* as in "give."

▲ FIGURE 13.12 *(Con't)*

The term *ridge* may be misleading, because these features are not narrow and steep as the term implies, but have widths of from 1000 to 4000 kilometers and the appearance of a broad elongated swell that often exhibits rugged topography. Furthermore, careful examination of Figure 13.12 shows that the ridge system is broken into segments that range from a few tens to hundreds of kilometers in length. Although each segment is offset from the adjacent segment, they are generally connected, one to the next, by a transform fault.

Oceanic ridges are as high as some mountains found on the continents, and thus they are often described as mountainous in nature. However, the similarity ends there. Whereas most continental mountains form when compressional forces fold and metamorphose thick sequences of sedimentary rocks along convergent plate boundaries, oceanic ridges form where tensional forces fracture and pull the ocean crust apart. The oceanic ridge consists of layers and piles of newly formed basaltic rocks that have been faulted into elongated blocks that are buoyantly uplifted.

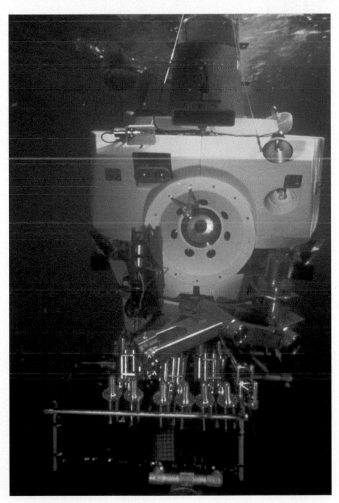

▲ FIGURE 13.13 The deep-diving submersible *Alvin* is 7.6 meters long, weighs 16 tons, has a cruising speed of 1 knot, and can reach depths as great as 4000 meters (2.5 miles). A pilot and two scientific observers are along during a normal 6- to 10-hour dive. (Courtesy of Rod Catanach/Woods Hole Oceanographic Institution)

high heat flow, and numerous volcanic structures characterize the oceanic ridge.

The interconnected oceanic ridge system is the longest topographic feature on Earth's surface, exceeding 70,000 kilometers (43,000 miles) in length. Representing 20 percent of Earth's surface, the oceanic ridge winds through all major oceans in a manner similar to the seam on a baseball (Figure 13.14). The crest of this linear structure typically stands 2 to 3 kilometers above the adjacent ocean basin floor and marks the plate margins where new oceanic crust is created.

Notice in Figure 13.14 that large sections of the oceanic ridge system have been named based on their locations within the various ocean basins. Ideally, ridges run along the middle of ocean basins, where they are called *mid-ocean* ridges. This holds true for the Mid-Atlantic Ridge, which is positioned in the middle of the Atlantic, roughly paralleling the margins of the continents on either side (Figure 13.14A). This is also true for the Mid-Indian Ridge, but note that the East Pacific Rise is displaced to the eastern side of the Pacific Ocean (Figure 13.14B, C).

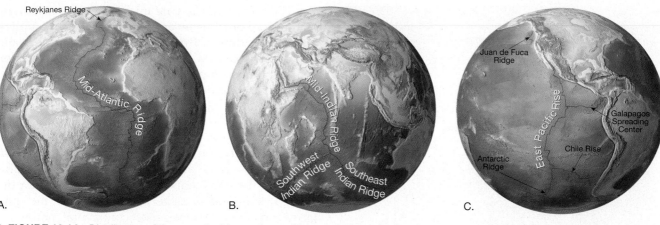

▲ FIGURE 13.14 Distribution of the oceanic ridge system, which winds through all major ocean basins, like the seam on a baseball.

Box 13.2 Understanding Earth:

Explaining Coral Atolls—Darwin's Hypothesis

Coral *atolls* are ring-shaped structures that often extend several thousand meters below sea level (Figure 13.B). What causes atolls to form, and how do they attain such a great thickness?

Corals are colonial animals about the size of an ant that feed with stinging tentacles and are related to jellyfish. Most corals protect themselves by growing a hard external skeleton made of calcium carbonate. Where corals reproduce and grow over many centuries, their skeletons fuse into large structures called *coral reefs*. Other corals—as well as sponges

and algae—begin to attach to the reef, enlarging it further. Eventually fishes, sea slugs, octopus, and other organisms are attracted to these diverse and productive habitats.

Corals require specific environmental conditions to grow. For example, reef-building corals grow best in waters with an average annual temperature of about 24°C (75°F). They cannot survive prolonged exposure to temperature below 18°C (64°F) or above 30°C (86°F). In addition, reef-builders require an attachment site (usually other corals) and clear sunlit water. Consequently, the limiting depth

of most active reef growth is only about 45 meters (150 feet).

The restricted environmental conditions required for coral growth create an interesting paradox: How can corals—which require warm, shallow, sunlit water no deeper than a few dozen meters to live—create thick structures such as coral atolls that extend into deep water?

The naturalist Charles Darwin was one of the first to formulate a hypothesis on the origin of atolls. From 1831 to 1836 he sailed aboard the British ship HMS *Beagle* during its famous circumnavigation of the globe. In various places that Darwin visited, he noticed a progression of stages in coral reef development from (1) a *fringing reef* along the margins of a volcano to (2) a *barrier reef* with a volcano in the middle to (3) an *atoll*, which consists of a continuous or broken ring of coral reef surrounded by a central lagoon (Figure 13.C). The essence of Darwin's hypothesis was that as a volcanic island slowly sinks, the corals continue to build the reef complex upward.

Darwin's hypothesis explained how coral reefs, which are restricted to shallow water, can build structures that now exist in much deeper water. During Darwin's time, however, there was no plausible mechanism to account for how an island might sink.

Today, plate tectonics helps explain how a volcanic island can become extinct and sink to great depths over long periods of time. Volcanic islands often form

Figure 13.B An aerial view of Tetiaroa Atoll in the Pacific. The light blue waters of the relatively shallow lagoon contrast with the dark blue color of the deep ocean surrounding the atoll. (Photo by Douglas Peebles Photography)

over a relatively stationary mantle plume, which causes the lithosphere to be buoyantly uplifted. Over a span of millions of years, these volcanic islands become inactive and gradually sink as the moving plate carries them away from the region of hot-spot volcanism (Figure 13.C).

Drilling through atolls has revealed that volcanic rock does indeed underlie the oldest (and deepest) coral reef structures, confirming Darwin's hypothesis. Thus, atolls owe their existence to the gradual sinking of volcanic islands containing coral reefs that build upward through time.

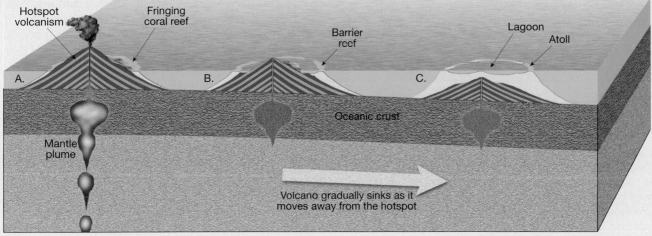

Figure 13.C Formation of a coral atoll due to the gradual sinking of oceanic crust and upward growth of the coral reef. A. A fringing coral reef forms around an active volcanic island. B. As the volcanic island moves away from the region of hotspot activity it sinks, and the fringing reef gradually becomes a barrier reef farther from shore. C. Eventually, the volcano is completely submerged and an atoll remains.

Along the axis of some segments of the oceanic ridge system are deep down-faulted structures called **rift valleys** (Figure 13.15). These features may exceed 50 kilometers in width and 2000 meters in depth. Because they contain faulted and tilted blocks of oceanic crust, as well as volcanic cones that have grown upon the newly formed seafloor, rift valleys usually exhibit rugged topography. The name *rift valley* has been applied to these features because they are so strikingly similar to continental rift valleys as exemplified by the East African Rift.

Topographically, the outermost flanks of most ridges are relatively subdued (except for isolated volcanic peaks) and rise very gradually (slope less than 1 degree) toward the ridge axis. Approaching the ridge crest, the topography becomes more rugged as volcanic structures, and faulted valleys that tend to parallel the ridge axis become more prominent. The most rugged topography is found on those ridges that exhibit large rift valleys.

Because of its accessibility to both American and European scientists, parts of the Mid-Atlantic Ridge have been studied in considerable detail. It is a broad, submerged structure standing 2500 to 3000 meters (8200 to 9800 feet) above the adjacent ocean basin floor. In a few places, such as Iceland, the ridge has actually grown

▼ FIGURE 13.15 Rift valleys are deep, down-faulted structures found along portions of the oceanic ridge system. They usually exhibit rugged topography.

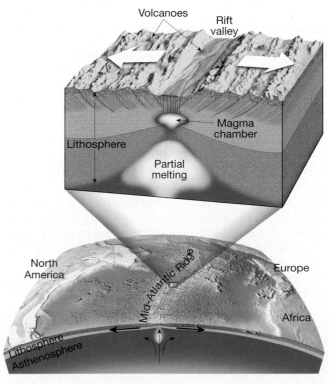

above sea level. Throughout most of its length, however, this divergent plate boundary lies far below sea level. Another prominent feature of the Mid-Atlantic Ridge is its deep linear rift valley extending along the ridge axis (Figure 13.15). Using both surface ships and submersibles, as well as sophisticated side-scanning sonar equipment, "images" of this rift valley have been obtained for the benefit of current and future investigations. In places, this rift valley is more than 30 kilometers wide and bounded by walls that are about 1500 meters high. This makes it comparable to the deepest and widest part of Arizona's Grand Canyon.

STUDENTS SOMETIMES ASK...
What causes oceanic ridges to be elevated?

The primary reason for the elevated position of a ridge system is the fact that newly created oceanic lithosphere is hot, occupies more volume, and is therefore less dense than cooler rocks of the ocean basin floor. As the newly formed basaltic crust travels away from the ridge crest, it is cooled from above as seawater circulates through the pore spaces and fractures in the rock. It also cools because it is moving away from the zone of upwelling, which is the main source of heat. As a result, the lithosphere gradually cools, contracts, and becomes more dense. This thermal contraction accounts for the greater ocean depths that exist away from the ridge. It takes almost 80 million years before cooling and contraction cease completely. By this time, rock that was once part of an elevated ocean-ridge system is part of the ocean basin floor, where it is covered by thick accumulations of sediment.

Seafloor Sediments

Except for steep areas of the continental slope and areas near the crest of the mid-ocean ridge, most of the ocean floor is covered with sediment. Part of this material has been deposited by turbidity currents, and the rest has slowly settled onto the seafloor from above. The thickness of this carpet of debris varies greatly. In some trenches, which act as traps for sediment originating on the continental margin, accumulations may approach 10 kilometers (6 miles) in thickness. In general, however, sediment accumulations are considerably less. In the Pacific Ocean, for example, sediment thickness is about 600 meters (2000 feet) or less, whereas on the floor of the Atlantic, the thickness varies from about 500 to 1000 meters (1500 to 3000 feet).

Although deposits of sand-size particles are found on the deep-ocean floor, *mud* (composed of fine clay-size particles) is the most common sediment covering this region. Muds also predominate on the continental shelf and slope, but the sediments in these areas are coarser overall because of greater quantities of sand.

Types of Seafloor Sediments

Seafloor sediments can be classified according to their origin into three broad categories: (1) **terrigenous** (*terra* = land, *generare* = to produce); (2) **biogenous** (*bio* = life, *generare* = to produce); and (3) **hydrogenous** (*hydro* = water, *generare* = to produce). Although each category is discussed separately, seafloor sediments usually come from a variety of sources and are thus mixtures of the various sediment types.

Terrigenous Sediment. Terrigenous sediment consists primarily of mineral grains that were weathered from continental rocks and transported to the ocean. Larger particles (gravel and sand) usually settle rapidly near shore, whereas finer particles (microscopic clay-size particles) can take years to settle to the ocean floor and may be carried thousands of kilometers by ocean currents. As a consequence, virtually every part of the ocean receives some terrigenous sediment. The rate at which this sediment accumulates on the deep-ocean floor, though, is very slow. To form a 1-centimeter (0.4-inch) *abyssal clay* layer, for example, requires as much as 50,000 years. Conversely, on the continental margins near the mouths of large rivers, terrigenous sediment accumulates rapidly and forms thick deposits. In the Gulf of Mexico, for instance, the sediment has reached a depth of many kilometers.

Because fine particles usually remain suspended in the water for a very long time, ample opportunity exists for chemical reactions to occur. Because of this, the colors of the deep-sea sediments are often red or brown. This results when iron in the particle or in the water reacts with dissolved oxygen in the water and produces a coating of iron oxide (rust).

Biogenous Sediment.   Biogenous sediment consists of shells and skeletons of marine animals and algae (Figure 13.16). This debris is produced mostly by microscopic organisms living in sunlit waters near the ocean surface. Once these organisms die, their hard *tests* (*testa* = shell) constantly "rain" down and accumulate on the seafloor.

The most common biogenous sediment is *calcareous* ($CaCO_3$) *ooze*, which, as its name implies, has the consistency of thick mud. This sediment is produced from the tests of organisms such as *coccolithophores* (single-celled algae) and *foraminifers* (small organisms) that inhabit warm surface waters. When calcareous tests slowly sink into deeper parts of the ocean, they begin to dissolve. This results because the deeper, cold seawater is rich in carbon dioxide and is thus more acidic than warm water. In seawater deeper than about 4500 meters (15,000 feet), calcareous tests will completely dissolve before they reach bottom. Consequently, calcareous ooze does not accumulate in the deep-ocean basins.

Other biogenous sediments include *siliceous* (SiO_2) *ooze* and phosphate-rich material. The former is

▲ FIGURE 13.16 Microscopic hard parts of radiolarians and foraminifers are examples of biogenous sediments. This photomicrograph has been enlarged hundreds of times. (Photo courtesy of Deep Sea Drilling Project, Scripps Institution of Oceanography, University of California, San Diego)

composed primarily of tests of *diatoms* (single-celled algae) and *radiolarians* (single-celled animals) that prefer cooler surface waters, whereas the latter is derived from the bones, teeth, and scales of fish and other marine organisms.

STUDENTS SOMETIMES ASK...

Are diatoms an ingredient in diatomaceous earth, which is used in swimming pool filters?

Not only are diatoms used as swimming pool filters, they are also used in a variety of everyday products, including toothpaste (yes, you're brushing your teeth with the remains of dead microscopic organisms!). Diatoms secrete walls of silica in a great variety of forms that accumulate as sediments in enormous quantities. Because it is lightweight, chemically stable, has high surface area, and is highly absorbent, diatomaceous earth has many practical uses. The main uses of diatoms include filters (for refining sugar, straining yeast from beer, and filtering swimming pool water); mild abrasives (in household cleaning and polishing compounds and facial scrubs); and absorbents (for chemical spills).

Hydrogenous Sediment. Hydrogenous sediment consists of minerals that crystallize directly from seawater through various chemical reactions. Hydrogenous sediments represent a relatively small portion of the overall sediment in the ocean. They do, however, have many different compositions and are distributed in diverse environments of deposition.

Some of the most common types of hydrogenous sediment include:

• *Manganese nodules*, which are rounded, hard lumps of manganese, iron, and other metals that precipitate in concentric layers around a central object (such as a volcanic pebble or a grain of sand). The nodules can be up to 20 centimeters (8 inches) in diameter and are often littered across large areas of the deep seafloor.

• *Calcium carbonates*, which form by precipitation directly from seawater in warm climates. If this material is buried and hardened, it forms limestone. Most limestone, however, is composed of biogenous sediment.

• *Metal sulfides*, which are usually precipitated as coatings on rocks near black smokers associated with the crest of the mid-ocean ridge. These deposits contain iron, nickel, copper, zinc, silver, and other metals in varying proportions.

• *Evaporites*, which form where evaporation rates are high and there is restricted open-ocean circulation. As water evaporates from such areas, the remaining seawater becomes saturated with dissolved minerals, which then begin to precipitate. Heavier than seawater, they sink to the bottom or form a characteristic white crust of evaporite minerals around the edges of these areas. Collectively termed "salts," some evaporite minerals taste salty, such as *halite* (common table salt, NaCl), and some do not, such as the calcium sulfate minerals *anhydrite* ($CaSO_2$) and *gypsum* ($CaSO_4 \cdot 2\,H_2O$).

Distribution of Seafloor Sediments

Figure 13.17 shows the distribution of seafloor sediments. Coarse-grained terrigenous deposits dominate continental margin areas, whereas fine-grained terrigenous material (abyssal clay) is common in deeper areas of the ocean basins. However, deep-ocean deposits are dominated by calcareous oozes, which are found on the shallower portions of deep-ocean areas along the mid-ocean ridge. Siliceous oozes are found beneath areas of unusually high biologic productivity such as the Antarctic and the equatorial Pacific Ocean. Hydrogenous sediment comprises only a small proportion of deposits in the ocean.

Various types of sediment accumulate on nearly all areas of the ocean floor in the same way dust accumulates in all parts of your home (which is why seafloor sediment is often referred to as "marine dust"). Even the deep-ocean floor far from land receives small amounts of windblown material, microscopic biogenous particles, and even space dust.

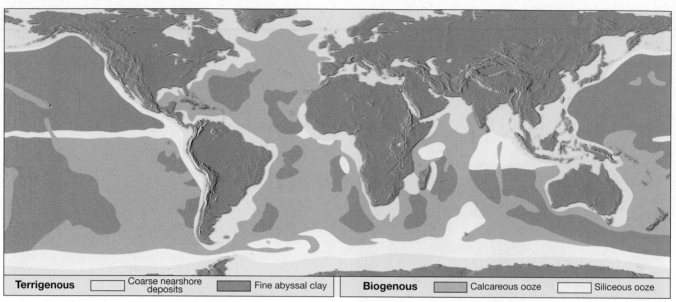

| Terrigenous | | Coarse nearshore deposits | | Fine abyssal clay | Biogenous | | Calcareous ooze | | Siliceous ooze |

▲ **FIGURE 13.17** Distribution of marine sediment.

There are a few places in the ocean, however, where very little sediment accumulates. One such place is along the continental slope, where there is active erosion by turbidity and other deep-ocean currents. Another place where very little sediment can be found is along the oceanic ridge. Here, the seafloor along the crest of the ridge is so young (because of seafloor spreading), and the rates of sediment accumulation far from land are so slow that there has not been enough time for sediments to accumulate.

Seafloor Sediments and Climate Change

Reliable climate records go back only a couple of hundred years, at best. How do scientists learn about climates and climate changes prior to that time? The answer is that they must reconstruct past climates from *indirect* evidence; that is, they must examine and analyze clues from phenomena that responded to and reflect changing atmospheric conditions. An important technique for analyzing Earth's climate history is the study of seafloor sediments.

Most seafloor sediments contain the remains of microscopic organisms that once lived near the sea surface (the ocean–atmosphere interface). When such near-surface organisms die, their tests slowly settle onto the ocean floor, where they can become buried and preserved over time. Thus, the deep-ocean floor has become a repository for sediment representing millions of years of Earth history.

Seafloor sediments are useful recorders of worldwide climate change because the numbers and types of organisms living near the sea surface change with the climate:

> . . . [W]e would expect that in any area of the ocean/atmosphere interface the average annual temperature of the surface water of the ocean would approximate that of the contiguous atmosphere. The temperature equilibrium established between surface seawater and the air above it should mean that . . . changes in climate should be reflected in changes in organisms living near the surface of the deep sea. . . . When we recall that the seafloor sediments in vast areas of the ocean consist mainly of shells of [oceanic] foraminifers, and that these animals are sensitive to variations in water temperature, the connection between such sediments and climatic change becomes obvious.*

In seeking to understand climate and other environmental changes, scientists are tapping the huge reservoir of data in seafloor sediments. Sediment cores gathered by drilling ships and other research vessels have provided invaluable data that have significantly expanded scientific knowledge and understanding of past climates (Box 13.3). Analysis of seafloor sediments, for instance, has revealed periods of Ice Ages and global warming, ocean circulation changes, the timing of major extinction events, and the movement of Earth's plates.

---

*Richard F. Flint, *Glacial and Quaternary Geology* (New York: Wiley, 1971), p. 718.

BOX 13.3 UNDERSTANDING EARTH:
Collecting Geologic History from the Deep-Ocean Floor

A fundamental aspect of scientific inquiry is the gathering of basic facts through observation and measurement. Formulating and testing hypotheses requires reliable data. The acquisition of such information is no easy task when it comes to sampling the vast storehouse of data contained in seafloor sediments and oceanic crust. Acquiring samples is technically challenging and very expensive.

The ship in Figure 13.D, the *JOIDES Resolution*, is a research vessel that is capable of drilling into the seafloor and collecting long cylinders (cores) of sediment and rock. The "*JOIDES*" in the ship's name stands for *Joint Oceanographic Institutions for Deep Earth Sampling* and reflects the international commitment from the program's 22 participating countries. The "*Resolution*" honors the ship HMS *Resolution*, commanded more than 200 years ago by the prolific English explorer Captain James Cook.

The *JOIDES Resolution* has a tall metal derrick that is used to conduct *rotary drilling*, while the ship's thrusters hold it in a fixed position at sea. Individual sections of drill pipe are fitted together to make a single string of pipe up to 8200 meters (27,000 feet) long. The drill bit, located at the end of the pipe string, rotates as it is pressed against the ocean bottom and can drill up to 2100 meters (6900 feet) below the seafloor. Like twirling a soda straw into a layer cake, the drilling operation crushes the rock around the

outside and retains a cylinder of rock (a core sample) on the inside of the hollow pipe, which can then be raised onboard the ship and analyzed in state-of-the-art laboratory facilities.

Since 1985, the ship has drilled more than 1700 holes worldwide. The result has been the recovery of more than 210,000 meters (about 133 miles!) of core samples that represent millions of years of Earth history and are used by scientists to study many aspects of Earth science, including changes in global climate. Although the number of holes drilled into the ocean floor is impressive, it represents just one hole per area about the size of Colorado.

In September 2003, the *JOIDES Resolution* completed its 110th and final expedition as part of the highly successful Ocean Drilling Program (ODP). In October 2003, it became part of a new program, the Integrated Ocean Drilling Program (IODP). This new international effort will not rely on just one drilling ship, but will use multiple vessels for exploration. One of the new additions is the massive 210-meter- (nearly 700-foot-) long *Chikyu*, which begins operations in 2006. The primary objective of the new program is to collect cores that will allow scientists to better understand Earth history and Earth system processes, including the properties of the deep crust, climate change patterns, earthquake mechanisms, and the microbiology of the deep-ocean floor.

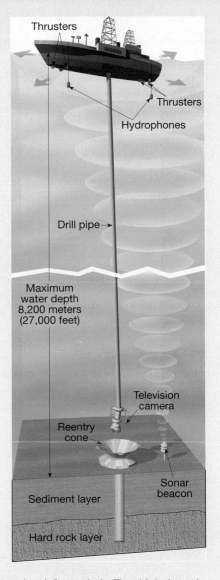

Thrusters

Thrusters

Hydrophones

Drill pipe →

Maximum water depth 8,200 meters (27,000 feet)

Television camera

Reentry cone

Sediment layer

Sonar beacon

Hard rock layer

Figure 13.D The *JOIDES Resolution* drills into the ocean floor and collects cores of sediment and rock for analysis. The ship's dynamic positioning system consists of powerful thrusters (small propellers) that allow it to remain stationary above a drill site. Previous drill sites can be reused years later and are located by bouncing sound waves between the ship's hydrophones and sonar beacons. A remote television camera aids in positioning the drill pipe into the reeentry cone.

Resources from the Seafloor

The seafloor is rich in mineral and organic resources. Most, however, are not easily accessible, and recovery involves technological challenges and high cost. Nevertheless, certain resources have high value and thus make appealing exploration targets.

Energy Resources

Among the nonliving resources extracted from the oceans, more than 95 percent of the economic value

comes from energy products. The main energy products are oil and natural gas, which are currently being extracted, and gas hydrates, which are not yet utilized but have vast potential.

Oil and Natural Gas. The ancient remains of microscopic organisms, buried within marine sediments before they could decompose, are the source of today's deposits of oil and natural gas. The percentage of world oil produced from offshore regions has increased from trace amounts in the 1930s to over 30 percent today. Most of this increase results from continuing technological advancements employed by offshore drilling platforms (Figure 13.18).

Major offshore reserves exist in the Persian Gulf, in the Gulf of Mexico, off the coast of southern California, in the North Sea, and in the East Indies. Additional reserves are probably located off the north coast of Alaska and in the Canadian Arctic, Asian seas, Africa, and Brazil. Because the likelihood of finding major new reserves on land is small, future offshore exploration will continue to be important, especially in deeper waters of the continental margins. A major environmental concern about offshore petroleum exploration is the possibility of oil spills caused by inadvertent leaks or blowouts during the drilling process.

Gas Hydrates. **Gas hydrates** are unusually compact chemical structures made of water and natural gas. The most common type of natural gas is methane, which produces *methane hydrate*. Gas hydrates occur beneath permafrost areas on land and under the ocean floor at depths below 525 meters (1720 feet).

Most oceanic gas hydrates are created when bacteria break down organic matter trapped in seafloor sediments, producing methane gas with minor amounts of ethane and propane. These gases combine with water in deep-ocean sediments (where pressures are high and temperatures are low) in such a way that the gas is trapped inside a latticelike cage of water molecules.

Vessels that have drilled into gas hydrates have retrieved cores of mud mixed with chunks or layers of gas hydrates (Figure 13.19A) that fizzle and evaporate quickly when they are exposed to the relatively warm, low-pressure conditions at the ocean surface. Gas hydrates resemble chunks of ice but ignite when lit by a flame because methane and other flammable gases are released as gas hydrates vaporize (Figure 13.19B).

Some estimates indicate that as much as 20 quadrillion cubic meters (700 quadrillion cubic feet) of methane are locked up in sediments containing gas hydrates. This is equivalent to about *twice* as much carbon as Earth's coal, oil, and conventional gas reserves combined, so gas hydrates would seem to have great potential. However, research suggests that the potential is modest. A recent article states that

> . . . all but a few percent of the great vastness of gas hydrates will likely remain beyond reach indefinitely. Most deposits are simply spread too thinly for economical recovery.*

It goes on to say that commercial production of gas from hydrates may begin within the next 10 to 15 years but will probably not make a significant contribution for at least 30 years.

Other Resources

Other major resources from the seafloor include sand and gravel, evaporative salts, and manganese nodules.

Sand and Gravel. The offshore sand-and-gravel industry is second in economic value only to the petroleum industry. Sand and gravel, which includes rock fragments that are washed out to sea and shells of marine organisms, are mined by offshore barges using suction dredges. Sand and gravel are primarily used as an aggregate in concrete, as a fill material in grading projects, and on recreational beaches.

In some cases, materials of high economic value are associated with offshore sand and gravel deposits. Gem-quality diamonds, for example, are recovered from gravels on the continental shelf offshore of South Africa and Australia. Sediments rich in tin have been mined from some offshore areas of Southeast Asia. Platinum and gold have been found in deposits in gold-mining areas throughout the world, and some Florida beach sands are rich in titanium.

Evaporative Salts. When seawater evaporates, the salts increase in concentration until they can no longer remain dissolved, so they precipitate out of solution and form salt deposits, which can then be harvested.

▼ **FIGURE 13.18** Offshore drilling rigs are used to tap the oil and natural gas reserves of the continental shelf. These platforms are near Santa Barbara, California. (Photo by Gregory Ochocki/Photo Researchers, Inc.)

*Kerr, Richard A. "Gas Hydrate Resource: Smaller But Sooner," *Science*, Vol. 303, 13 February 2004, pp. 946–47.

▲ **FIGURE 13.19** Gas hydrates. **A.** A sample retrieved from the ocean floor shows layers of white icelike gas hydrate mixed with mud. **B.** Gas hydrates evaporate when exposed to surface conditions and release natural gas, which can be ignited. (Photos courtesy of GEOMAR Research Center, Kiel, Germany)

▲ **FIGURE 13.20** Harvesting salt produced by the evaporation of seawater. (Photo by William E. Townsend, Jr./Photo Researchers, Inc.)

▲ **FIGURE 13.21** Manganese nodules photographed at a depth of 2909 fathoms (5323 meters or 3.3 miles) beneath the *Robert Conrad*, south of Tahiti. (Photo courtesy of Lawrence Sullivan, Lamont-Doherty Earth Observatory/Columbia University))

The most economically important salt is *halite* (common table salt). Halite is widely used for seasoning, curing, and preserving foods. It is also used in water conditioners, in agriculture, in the clothing industry for dying fabric, and to de-ice roads.

Ever since ancient times, the ocean has been an important source of salt for human consumption, and the sea remains a significant supplier. Today, about 30 percent of the world's salt is produced by evaporating seawater (Figure 13.20).

Manganese Nodules. Manganese nodules contain significant concentrations of manganese, iron, and smaller concentrations of copper, nickel, and cobalt, all of which have a variety of economic uses. Cobalt, for example, is deemed "strategic" (essential to U.S. national security) because it is required to produce dense, strong alloys with other metals and is used in high-speed cutting tools, powerful permanent magnets, and jet engine parts. Technologically, mining the deep-ocean floor for manganese nodules is possible but economically not profitable (Figure 13.21).

Nodules are widely distributed, but not all regions have the same potential for mining. Good locations have abundant nodules that contain the economically optimum mix of copper, nickel, and cobalt. Sites meeting these criteria, however, are relatively limited. Additionally, there are political problems of establishing mining rights far from land and environmental concerns about disturbing large portions of the deep-ocean floor.

Chapter Summary

• *Oceanography is an interdisciplinary science* that draws on the methods and knowledge of geology, chemistry, physics, and biology to study all aspects of the world ocean.

• *Earth is a planet dominated by oceans.* Seventy-one percent of Earth's area consists of oceans and marginal seas. In the Southern Hemisphere, often called the *water hemisphere*, about 81 percent of the surface is water. The world ocean can be divided into four main ocean basins: the *Pacific Ocean* (largest and deepest ocean), the *Atlantic Ocean* (about half the size of the Pacific), the *Indian Ocean* (slightly smaller than the Atlantic and mostly in the Southern Hemisphere), and the *Arctic Ocean* (smallest and shallowest ocean). The average depth of the oceans is 3729 meters (12,234 feet).

• *Ocean bathymetry is determined using echo sounders and multibeam sonars*, which bounce sonic signals off the ocean floor. Ship-based receivers record the reflected echoes and accurately measure the time interval of the signals. With this information, ocean depths are calculated and plotted to produce maps of ocean-floor topography. Recently, *satellite measurements* of the shape of the ocean surface have added data for mapping ocean-floor features.

• The zones that collectively make up a *passive continental margin* include the *continental shelf* (a gently sloping, submerged surface extending from the shoreline toward the deep-ocean basin); the *continental slope* (the true edge of the continent, which has a steep slope that leads from the continental shelf into deep water); and in regions where trenches do not exist, the steep continental slope merges into a more gradual incline known as the *continental rise* (which consists of sediments that have moved downslope from the continental shelf to the deep-ocean floor).

• *Submarine canyons* are deep, steep-sided valleys that originate on the continental slope and may extend to the deep-ocean basin. Many submarine canyons have been excavated by *turbidity currents*, which are downslope movements of dense, sediment-laden water.

• *Active continental margins* are located primarily around the Pacific Rim in areas where the leading edge of a continent is overrunning oceanic lithosphere. Here sediment scraped from the descending oceanic plate is plastered against the continent to form a collection of sediments called an *accretionary wedge.* An active continental margin generally has a narrow continental shelf, which grades into a steep continental slope and deep-ocean trench.

• The *ocean basin floor* lies between the continental margin and the oceanic ridge system. The features of the ocean basin floor include *deep-ocean trenches* (the deepest parts of the ocean, where moving crustal plates descend into the mantle), *abyssal plains* (the most level places on Earth, consisting of thick accumulations of sediments that were deposited atop the low, rough portions of the ocean floor), *seamounts* and *guyots* (isolated volcanic peaks on the ocean floor that originate near the mid-ocean ridge or in association with volcanic hot spots) and *oceanic plateaus* (vast accumulations of basaltic lava flows).

• *Atolls* form from corals that grow on the flanks of sinking volcanic islands, where the corals continue to build the reef complex upward as the island sinks.

• The *oceanic (mid-ocean) ridge* winds through the middle of most ocean basins. Seafloor spreading occurs along this broad feature, which is characterized by an elevated position, extensive faulting, and volcanic structures that have developed on newly formed oceanic crust. Most of the geologic activity associated with ridges occurs along a narrow region on the ridge crest, called the *rift valley*, where magma moves upward to create new slivers of oceanic crust.

• *There are three broad categories of seafloor sediments. Terrigenous sediment* consists primarily of mineral grains that were weathered from continental rocks and transported to the ocean; *biogenous sediment* consists of shells and skeletons of marine animals and plants; and *hydrogenous sediment* includes minerals that crystallize directly from seawater through various chemical reactions. The global distribution of marine sediments is affected by proximity to source areas and water temperatures that favor the growth of certain marine organisms.

• *Seafloor sediments are helpful when studying worldwide climate changes* because they often contain the remains of organisms that once lived near the sea surface. The numbers and types of these organisms change as the climate changes, and their remains in seafloor sediments record these changes.

• *Energy resources* from the seafloor include *oil and natural gas* and large untapped deposits of *gas hydrates.* Other seafloor resources include *sand and gravel, evaporative salts,* and metals within *manganese nodules.*

Key Terms

abyssal plain (p. 369)
active continental margin (p. 367)
bathymetry (p. 361)
biogenous sediment (p. 374)
continental margin (p. 364)
continental rise (p. 367)
continental shelf (p. 364)
continental slope (p. 366)
continental volcanic arc (p. 369)
deep-ocean trench (p. 369)
deep-sea fan (p. 367)

echo sounder (p. 362)
gas hydrate (p. 378)
graded bedding (p. 367)
guyot (p. 370)
hydrogenous sediment (p. 374)
mid-ocean ridge (p. 370)
ocean basin floor (p. 369)
oceanic plateau (p. 370)
oceanic ridge (p. 370)
oceanography (p. 360)
passive continental margin (p. 364)

rift valley (p. 373)
seamount (p. 369)
seismic reflection profile (p. 363)
sonar (p. 362)
submarine canyon (p. 366)
tablemount (p. 370)
terrigenous sediment (p. 374)
turbidite (p. 367)
turbidity current (p. 366)
volcanic island arc (p. 369)

Review Questions

1. How does the area of Earth's surface covered by the oceans compare with that of the continents? Describe the distribution of land and water on Earth.

2. Name the four main oceans basins. Of the four:
   (a) Which one is the largest in area? Which one is smallest?
   (b) Which one is the deepest? Which one is shallowest?
   (c) Which one is almost entirely within the Southern Hemisphere?
   (d) Which one is exclusively in the Northern Hemisphere?

3. How does the average depth of the ocean compare to the average elevation of the continents?

4. Assuming that the average speed of sound waves in water is 1500 meters per second, determine the water depth in meters if a signal sent out by an echo sounder requires 6 seconds to strike bottom and return to the recorder (see Figure 13.4).

5. Describe how satellites orbiting Earth can determine features on the seafloor without being able to directly observe them beneath several kilometers of seawater.

6. List the three major subdivisions of a passive continental margin. Which subdivision is considered a flooded extension of the continent? Which has the steepest slope?

7. Describe the differences between active and passive continental margins. Be sure to include how various features relate to plate tectonics and give a geographic example of each type of margin.

8. Defend or rebut the statement "Most submarine canyons found on the continental slope and rise were formed during the Ice Age when rivers extended their valleys seaward."

9. What are turbidity currents? How do they differ from turbidites? What is meant by the term *graded bedding?*

10. What are differences between a submarine canyon and a deep-ocean trench?

11. Describe the process by which abyssal plains are created. Why are abyssal plains more extensive on the floor of the Atlantic than on the floor of the Pacific?

12. Discuss how seamounts and guyots are created and why they have a different shape.

13. Describe the environmental conditions required for the development of coral reefs.

14. What paradox was there concerning atolls? Describe Darwin's proposal on the origin of atolls. How was his proposal confirmed or disproved?

15. Describe features associated with the oceanic ridge.

16. How are oceanic ridges and deep-ocean trenches related to plate tectonics?

17. Distinguish among the three basic types of seafloor sediment, and give an example of each.

18. Describe how seafloor sediments are useful in studying past climates. What kind of information can sediments reveal?

19. Discuss the present importance and future prospects for the production of offshore petroleum, gas hydrates, sand and gravel, evaporative salts, and manganese nodules.

Examining the Earth System

1. Describe some of the material and energy exchanges that take place at the interface between the (a) ocean surface and atmosphere, (b) ocean water and ocean floor, and (c) ocean biosphere and ocean water.

2. Sediment on the seafloor often leaves clues about various conditions that existed during deposition. What do the following layers in a seafloor core tell about the environment in which each layer was deposited?

   • Layer #5 (top): Red layer composed of fine clays

   • Layer #4: Siliceous ooze

   • Layer #3: Calcareous ooze

   • Layer #2: Fragments of coral reef

   • Layer #1 (bottom): Rocks of basaltic composition with some metal sulfide coatings

Explain how one area of the seafloor could experience such varied conditions of deposition.

Online Study Guide

The *Earth Science* Website uses the resources and flexibility of the Internet to aid in your study of the topics in this chapter. Written and developed by Earth science instructors, this site will help improve your understanding of Earth science. Visit **http://www.prenhall.com/tarbuck** and click on the cover of *Earth Science 11e* to find:

• **On-line review quizzes.**
• **Critical thinking exercises.**
• **Links to chapter-specific Web resources.**
• **Internet-wide key term searches.**
http://www.prenhall.com/tarbuck

CHAPTER 14

Ocean Water
and Ocean Life*

*This chapter was prepared by Professor Alan P. Trujillo, Palomar College.

Underwater seascape in the South Pacific near New Guinea.
(Photo by Jeff Hunter/Photographer's Choice/Getty)

383

What is the difference between pure water and seawater? One of the most obvious differences is that seawater contains dissolved substances that give it a distinctly salty taste. These dissolved substances are not simply sodium chloride (table salt)—they include various other salts, metals, and even dissolved gases. In fact, every known naturally occurring element is found dissolved in at least trace amounts in seawater. Unfortunately, the salt content of seawater makes it unsuitable for drinking or for irrigating most crops and causes it to be highly corrosive to many materials. Yet, many parts of the ocean are teeming with life that is superbly adapted to the marine environment.

There is an amazing variety of life in the ocean, from microscopic bacteria and algae to the largest organism alive today (the blue whale). Water is the major component of nearly every life form on Earth, and our own body fluid chemistry is remarkably similar to the chemistry of seawater.

Composition of Seawater

Seawater consists of about 3.5 percent (by weight) dissolved mineral substances that are collectively termed "salts." Although the percentage of dissolved components may seem small, the actual quantity is huge because the ocean is so vast.

Salinity

Salinity (*salinus* = salt) is the total amount of solid material dissolved in water. More specifically, it is the ratio of the mass of dissolved substances to the mass of the water sample. Many common quantities are expressed in percent (%), which is really *parts per hundred*. Because the proportion of dissolved substances in seawater is such a small number, oceanographers typically express salinity in *parts per thousand* ‰. Thus, the average salinity of seawater is 3.5% or 35‰.

Figure 14.1 shows the principal elements that contribute to the ocean's salinity. If one wanted to make artificial seawater, it could be approximated by following the recipe shown in Table 14.1. From this table it is evident that most of the salt in seawater is sodium chloride—common table salt. Sodium chloride together with the next four most abundant salts comprise over 99 percent of all dissolved substances in the sea. Although only eight elements make up these five most abundant salts, seawater contains all of Earth's other naturally occurring elements. Despite their presence in minute quantities, many of these elements are very important in maintaining the necessary chemical environment for life in the sea.

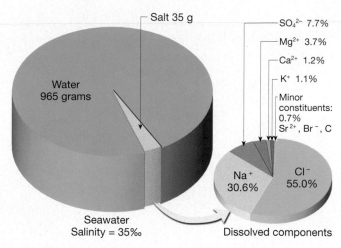

▲ **FIGURE 14.1** Relative proportions of water and dissolved components in seawater. Components shown by chemical symbol are chlorine (Cl^-), sodium (Na^+), sulfate (SO_4^{2-}), magnesium (Mg^{2+}), calcium (Ca^{2+}), potassium (K^+), strontium (Sr^{2+}), bromine (Br^-), and carbon (C).

Sources of Sea Salts

What are the primary sources for the vast quantities of dissolved substances in the ocean? Chemical weathering of rocks on the continents is one source. These dissolved materials are delivered to the oceans by streams at an estimated rate of more than 2.3 billion metric tons (2.5 billion short tons) annually.* The second major source of elements found in seawater is from Earth's interior. Through volcanic eruptions, large quantities of water vapor and other gases have been emitted during much of geologic time. This process, called *outgassing*, is the principal source of water in the oceans. Certain elements—notably chlorine, bromine, sulfur, and boron—were outgassed along with water and exist in the ocean in much greater abundance than could be explained by weathering of rocks alone.

*A metric ton equals 1000 kilograms or 2205 pounds. Thus, it is larger than the "ton" that most Americans are familiar with (called a *short ton*). There are 2000 pounds in a short ton.

TABLE 14.1	Recipe for artificial seawater.
To make seawater, combine:	**Amount (grams)**
Sodium chloride (NaCl)	23.48
Magnesium chloride ($MgCl_2$)	4.98
Sodium sulfate (Na_2SO_4)	3.92
Calcium chloride ($CaCl_2$)	1.10
Potassium chloride (KCl)	0.66
Sodium bicarbonate ($NaHCO_3$)	0.192
Potassium bromide (KBr)	0.096
Hydrogen borate (H_3BO_3)	0.026
Strontium chloride ($SrCl_2$)	0.024
Sodium fluoride (NaF)	0.003
Then add:	
Pure water (H_2O) to form 1000 grams of solution.	

Although rivers and volcanic activity continually contribute dissolved substances to the oceans, the salinity of seawater is not increasing. In fact, evidence suggests that the composition of seawater has been relatively stable for millions of years. Why doesn't the sea get saltier? The answer is because material is being removed just as rapidly as it is added. For example, some dissolved components are withdrawn from seawater by organisms as they build hard parts. Other components are removed when they chemically precipitate out of the water as sediment. Still others are exchanged at the oceanic ridge by *hydrothermal* (*hydro* = water, *thermos* = hot) *activity*. The net effect is that the overall makeup of seawater has remained relatively constant through time.

Processes Affecting Seawater Salinity

Because the ocean is well mixed, the relative abundances of the major components in seawater are essentially constant, no matter where the ocean is sampled. Variations in salinity, therefore, are primarily a consequence of changes in the water content of the solution.

Various surface processes alter the amount of water in seawater, thereby affecting salinity (Figure 14.2). Processes that add large amounts of fresh water to seawater—and thereby decrease salinity—include precipitation, runoff from land, icebergs melting, and sea ice melting. Processes that remove large amounts of fresh water from seawater—and thereby increase seawater salinity—include evaporation and the formation of sea ice. High salinities, for example, are found where evaporation rates are high, as is the case in the dry subtropical regions (roughly between 25 and 35 degrees north or south latitude). Conversely, where large amounts of precipitation dilute ocean waters, as in the midlatitudes (between 35 and 60 degrees north or south latitude) and near the equator, lower salinities prevail (Figure 14.3).

Surface salinity in polar regions varies seasonally due to the formation and melting of sea ice. When seawater freezes in winter, sea salts do not become part of the ice. Therefore, the salinity of the remaining seawater increases. In summer when sea ice melts, the addition of the relatively fresh water dilutes the solution and salinity decreases.

Surface salinity variation in the open ocean normally ranges from 33‰ to 38‰. Some marginal seas, however, demonstrate extraordinary extremes. For example, in the restricted waters of the Middle East's Persian Gulf and Red Sea—where evaporation far exceeds precipitation—salinity may exceed 42‰. Conversely, very low salinities occur where large quantities of fresh water are supplied by rivers and precipitation. Such is the case for northern Europe's Baltic Sea, where salinity is often below 10‰.

▼ **FIGURE 14.2** Processes affecting seawater salinity. Processes that *decrease* seawater salinity include precipitation, runoff, icebergs melting, and sea ice melting. Processes that *increase* seawater salinity include formation of sea ice and evaporation. (Upper left, Tom & Susan Bean; upper right, Wolfgang Kaehler Photography; lower left, NASA; lower right, Paul Steele/CORBIS/Stock Market)

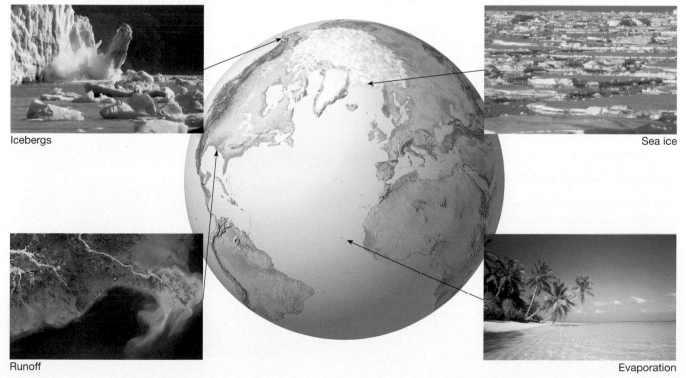

Icebergs

Sea ice

Runoff

Evaporation

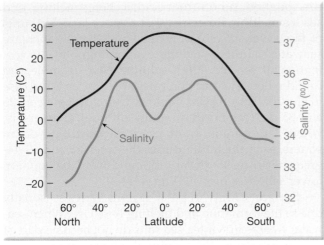

▲ **FIGURE 14.3** This graph shows variations in the ocean's surface temperature (top curve) and surface salinity (lower curve) with latitude. As you might expect, average temperatures are highest near the equator and get colder moving poleward. An important factor influencing differences in surface salinity is variations in rainfall and evaporation rates. For example, in the dry subtropics in the vicinity of the Tropics of Cancer and Capricorn, high evaporation rates remove more water than is replaced by the meager rainfall, resulting in high surface salinities. In the wet equatorial region, abundant rainfall reduces surface salinities.

STUDENTS SOMETIMES ASK...
*What would happen to a person
if he or she drank seawater?*

It depends on the quantity. The salinity of seawater is about four times greater than that of your body fluids. In your body, seawater causes your internal membranes to lose water through *osmosis*, which transports water molecules from higher concentrations (the normal body chemistry of your internal fluids) to areas of lower concentrations (your digestive tract containing seawater). Thus, your natural body fluids would move into your digestive tract and eventually be expelled, causing dehydration if seawater is consumed in large amounts. If you've inadvertently swallowed a little bit of seawater, don't worry too much.

Ocean Temperature Variation

The ocean's surface water temperature varies with the amount of solar radiation received, which is primarily a function of latitude (Figure 14.3). The intensity of solar radiation in high latitudes is significantly less than that received in tropical latitudes.* Therefore, much lower sea surface temperatures are found in high-latitude regions, and much higher sea surface temperatures are found in low-latitude regions.

---

*For more on this subject, see the section on "Earth-Sun Relationships" in Chapter 16.

STUDENTS SOMETIMES ASK...
*Where is the saltiest water
in the world?*

Some of the most highly saline water in the world is found in arid regions that have inland lakes, which are often called "seas" because they are so salty. The Great Salt Lake in Utah, for example, has a salinity of 280‰, and the Dead Sea on the border of Israel and Jordan has a salinity of 330‰. The water in the Dead Sea, therefore, contains 33 percent dissolved solids and is almost *10 times saltier than seawater.* As a result, these waters have such high density and are so buoyant that while lying down in the water, you can easily float—with arms and legs sticking up above water level!

Temperature Variation with Depth

If a thermometer were lowered from the surface of the ocean into deeper water, what temperature pattern would be found? Surface waters are warmed by the Sun, so they generally have higher temperatures than deeper waters. However, the observed temperature pattern depends on the latitude.

Figure 14.4 shows two graphs of temperature versus depth: one for high-latitude regions and one for low-latitude regions. The low-latitude curve begins at the surface with high temperature, but the temperature decreases rapidly with depth because of the inability of the Sun's rays to penetrate very far into the ocean. At a depth of about 1000 meters (3300 feet), the temperature remains just a few degrees above freezing and is relatively constant from this level down to the ocean floor. The layer of ocean water between about 300 meters (980 feet) and 1000 meters (3300 feet), where there is a rapid change of temperature with depth, is called the **thermocline** (*thermo* = heat, *cline* = slope). The thermocline is a very important structure in the ocean be-

▼ **FIGURE 14.4** Variations in ocean water temperature with depth for low- and high-latitude regions. The layer of rapidly changing temperature, called the *thermocline*, is present only in the low latitudes.

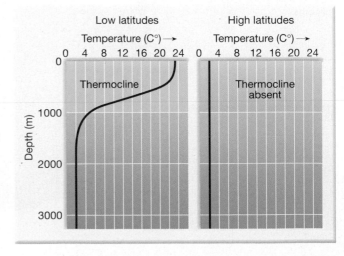

cause it creates a vertical barrier to many types of marine life.

The high-latitude curve in Figure 14.4 displays a pattern quite different from the low-latitude curve. Surface water temperatures in high latitudes are much cooler than in low latitudes, so the curve begins at the surface with low temperature. Deeper in the ocean, the temperature of the water is similar to that at the surface (just a few degrees above freezing), so the curve remains vertical and there is no rapid change of temperature with depth. A thermocline is not present in high latitudes; instead, the water column is *isothermal* (*iso* = same, *thermo* = heat).

Some high-latitude waters can experience minor warming during the summer months. Thus, certain high-latitude regions experience an extremely weak seasonal thermocline. Midlatitude waters, on the other hand, experience a more dramatic seasonal thermocline and exhibit characteristics intermediate between high-and low-latitude regions.

Ocean Temperature Change over Time

Seawater has many unique thermal properties that make it resistant to changes in temperature. Marine species are well adapted to life in the ocean, and many cannot withstand rapid changes in temperature.* In

*This is discussed more fully in the section on "Controls of Temperature" in Chapter 16.

Box 14.1 People and the Environment:
Desalination of Seawater—Fresh Water from the Sea

Earth's growing population uses fresh water in greater volumes each year. As fresh water becomes increasingly scarce, several countries have begun using the ocean as a source of water. The removal of salts and other chemicals to extract low-salinity ("fresh") water from seawater is termed *desalination*.

Worldwide, there are more than 12,500 desalination plants (Figure 14.A), with the majority located in arid regions of the Middle East, Caribbean, and Mediterranean. The United States produces only about 10 percent of the world's desalted water, with the majority produced in Florida and California. Because desalinated water is expensive to produce, most desalination plants are small-scale operations. In fact, desalination plants provide only about 1 percent of the world's drinking water. More than half of the world's desalination plants use *distillation* to purify water, while most of the remaining plants use *membrane processes*.

In distillation, saltwater is evaporated, and the resulting water vapor is collected and condensed to produce fresh water. This simple procedure is very efficient at purifying seawater. For instance, distillation of 35‰ seawater produces fresh water with a salinity of only 0.03‰, which is about 10 times fresher than bottled water.

Membrane processes such as *electrolysis* or *reverse osmosis* use specialized semipermeable membranes to separate dissolved components from water molecules, thereby purifying water. Worldwide, at least 30 countries are operating reverse osmosis units. Santa Barbara, California, for example, operates a reverse osmosis plant that produces up to 34 million liters (9 million gallons) of fresh water daily, which supplies up to 60 percent of its municipal water needs. When technical difficulties are overcome, a facility in Florida's Tampa Bay area could produce up to 94.5 million liters (25 million gallons) of fresh water per day, making it the largest plant in the nation. This method is also used in many household water purification units and aquariums.

Other methods of desalination include freeze separation, crystallization of dissolved components directly from seawater, solvent demineralization using chemical catalysts, and even making use of salt-eating bacteria!

Although fresh water produced by various desalination methods is becoming more important as a source of water for human and even industrial use, it is unlikely to be an important supply for agricultural purposes because of the enormous quantities of water necessary to support agriculture. Consequently, making the deserts "bloom" by irrigating them with fresh water produced by desalination is only a dream and, for economic reasons, is likely to remain so for the foreseeable future.

Figure 14.A Desalination plants such as this one in Saudi Arabia make fresh water from seawater.(Photo by Anthony Howarth/Woodfin Camp & Associates)

fact, researchers have concluded that the ocean's stability as a habitat has been instrumental in the development of life on Earth.

Scientists who study the effects of global warming on the ocean indicate that atmospheric warming would eventually be transferred to the ocean, potentially impacting marine organisms. How can we determine if the average temperature of the global ocean is actually changing? Accurately measuring such changes with traditional thermometers would be extremely difficult. One group of researchers, led by Walter Munk of Scripps Institution of Oceanography, has initiated an experiment to determine the amount of ocean warming by using sound as a thermometer.

The scientists have used a worldwide sound channel—called the *SOFAR channel* (an acronym for *SOund Fixing And Ranging*)—to transmit low frequency sound across an ocean basin to a distant receiver. The experiment—which is called *ATOC (Acoustic Thermometry of Ocean Climate)*—is designed to accurately measure the travel time of similar sound signals through the SOFAR channel now and in the future. The speed of sound in seawater increases as temperature increases, so sound should take less time to travel the same distance in the future if, in fact, the oceans are warming.

The group successfully tested the experiment by transmitting signals across the Pacific Ocean in 1991 and 1995. The transmissions were halted, however, because of concern about the sound's effect on whales, some of which may also use the SOFAR channel for communication. Nonetheless, the transmissions that have been conducted have established an important baseline for comparison with future measurements.

Ocean Density Variation

Density is defined as mass per unit volume but can be thought of as a measure of *how heavy something is for its size*. For instance, an object that has low density is lightweight for its size, such as a dry sponge, foam packing, or a surfboard. Conversely, an object that has high density is heavy for its size, such as cement, most metals, or a large container full of water.

Density is an important property of ocean water because it determines the water's vertical position in the ocean. Furthermore, density differences cause large areas of ocean water to sink or float. For example, when high-density seawater is added to low-density fresh water, the denser seawater sinks below the fresh water.

Factors Affecting Seawater Density

Seawater density is influenced by two main factors: *salinity* and *temperature*. An increase in salinity adds dissolved substances and results in an increase in seawater density. An increase in temperature, on the other hand, causes thermal expansion and results in a de-

crease in seawater density. Such a relationship where one variable decreases as a result of another variable's increase is known as an *inverse relationship*, where one variable is *inversely proportional* with the other.

Temperature has the greatest influence on surface seawater density because variations in surface seawater temperature are greater than salinity variations. In fact, only in the extreme polar areas of the ocean, where temperatures are low and remain relatively constant, does salinity significantly affect density. Cold water that also has high salinity is some of the highest-density water in the world.

Density Variation with Depth

By extensively sampling ocean waters, oceanographers have learned that temperature and salinity—and the water's resulting density—vary with depth. Figure 14.5 shows two graphs of density versus depth: one for high-latitude regions and one for low-latitude regions. Not surprisingly, the curves in Figure 14.5 are a mirror image of the temperature curves in Figure 14.4. This similarity demonstrates that temperature is the most important factor affecting seawater density and that temperature is inversely proportional with density.

The low-latitude curve in Figure 14.5 begins at the surface with low density (related to high surface water temperatures). However, density increases rapidly with depth because the water temperature is getting colder. At a depth of about 1000 meters (3300 feet), seawater density reaches a maximum value related to the water's low temperature. From this depth to the ocean floor, density remains constant and high. The layer of ocean water between about 300 meters (980 feet) and 1000 meters (3300 feet), where there is a rapid change of density with depth, is called the **pycnocline**

▼ **FIGURE 14.5** Variations in ocean water density with depth for low- and high-latitude regions. The layer of rapidly changing density, called the *pycnocline*, is present in the low latitudes but absent in the high latitudes.

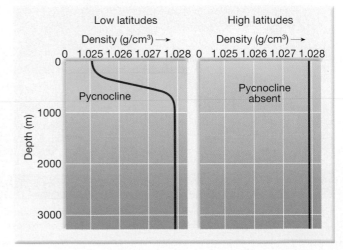

(*pycno* = density, *cline* = slope). A pycnocline has a high gravitational stability and presents a significant barrier to mixing between low-density water above and high-density water below.

The high-latitude curve in Figure 14.5 is also related to the temperature curve for high latitudes shown in Figure 14.4. Figure 14.5 shows that in high latitudes, there is high-density (cold) water at the surface and high-density (cold) water below. Thus, the high-latitude density curve remains vertical, and there is no rapid change of density with depth. A pycnocline is not present in high latitudes; instead, the water column is *isopycnal* (*iso* = same, *pycno* = density).

Ocean Layering

The ocean, like Earth's interior, is layered according to density. Low-density water exists near the surface, and higher-density water occurs below. Except for some shallow inland seas with a high rate of evaporation, the highest-density water is found at the greatest ocean depths. Oceanographers generally recognize a three-layered structure in most parts of the open ocean: a shallow surface mixed zone, a transition zone, and a deep zone (Figure 14.6).

Because solar energy is received at the ocean surface, it is here that water temperatures are warmest. The mixing of these waters by waves as well as the turbulence from currents and tides creates a rapid vertical heat transfer. Hence, this *surface mixed zone* has nearly uniform temperatures. The thickness and temperature of this layer vary, depending on latitude and season. The zone usually extends to about 300 meters (980 feet) but may attain a thickness of 450 meters (1500 feet). The surface mixed zone accounts for only about 2 percent of ocean water.

Below the Sun-warmed zone of mixing, the temperature falls abruptly with depth (see Figure 14.4). Here, a distinct layer called the *transition zone* exists between the warm surface layer above and the deep zone of cold water below. The transition zone includes a prominent thermocline and associated pycnocline and accounts for about 18 percent of ocean water.

Below the transition zone is the *deep zone*, where sunlight never reaches and water temperatures are just a few degrees above freezing. As a result, water density remains constant and high. Remarkably, the deep zone includes about 80 percent of ocean water, indicating the immense depth of the ocean (the average depth of the ocean is 3729 meters, or 12,234 feet).

In high latitudes, the three-layer structure of ocean layering does not exist (Figure 14.6). This is because the water column is isothermal (and isopycnal), which indicates that there is no rapid change in temperature (or density) with depth. Consequently, good vertical mixing between surface and deep waters can occur in high-latitude regions. Here, cold high-density water forms at the surface, sinks, and initiates deep-ocean currents, which are discussed in Chapter 15.

The Diversity of Ocean Life

A wide variety of organisms inhabit the marine environment. These organisms range in size from microscopic bacteria and algae to blue whales, which are as long as three buses lined up end to end. Marine biologists have identified over 250,000 marine species, a number that is constantly increasing as new organisms are discovered.

Most marine organisms live within the sunlit surface waters of the ocean. Strong sunlight supports **photosynthesis** (*photo* = light, *syn* = with, *thesis* = an *arranging*) by marine algae, which either directly or indirectly provide food for the vast majority of marine organisms. All marine algae live near the surface because they need sunlight; most marine animals also live near the surface because this is where food can be obtained. In shallow water areas close to land, sunlight reaches all the way to the bottom, resulting in an abundance of marine life on the ocean floor.

There are advantages and disadvantages to living in the marine environment. One advantage is that there is an abundance of water available, which is necessary for supporting all types of life. One disadvantage is that maneuvering in water, which has high

▼ **FIGURE 14.6** Oceanographers recognize three main "layers" in the ocean based on water density, which varies with temperature and salinity. The warm surface mixed layer accounts for only 2 percent of ocean water; the transition zone includes the thermocline and pycnocline and accounts for 18 percent of ocean water; the deep zone contains cold, high-density water that accounts for 80 percent of ocean water.

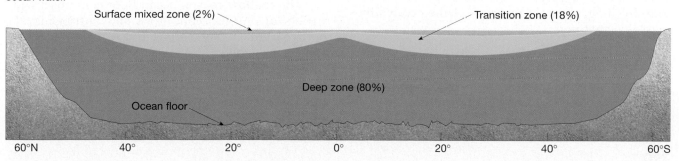

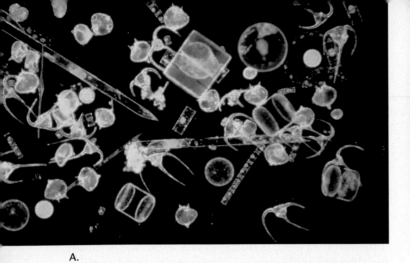

A.

B.

▲ **FIGURE 14.7** **A.** A variety of live *phytoplankton* from the Atlantic Ocean. They include various diatoms and dinoflagellates. (Copyright N. T. Nicoll) **B.** This image of *zooplankton* includes the most common forms known as copepods and their larvae. The picture also includes larval forms of other common marine organisms. (Copyright N. T. Nicoll)

density and impedes movement, can be difficult. The individual success of species depends on their ability to avoid predators, find food, and cope with the physical challenges of their environment.

Classification of Marine Organisms

Marine organisms can be classified according to where they live (their habitat) and how they move (their mobility). Organisms that inhabit the water column can be classified as either *plankton* (floaters) or *nekton* (swimmers). All other organisms are *benthos* (bottom dwellers).

Plankton (Floaters) **Plankton** (*planktos* = wandering) include all organisms—algae, animals, and bacteria—that drift with ocean currents. Just because plankton drift does not mean they are unable to swim. Many plankton can swim but either move very weakly or move only vertically within the water column.

Among plankton, the algae (photosynthetic cells, most of which are microscopic) are called **phytoplankton** and the animals are called **zooplankton** (*zoo* = animal, *planktos* = wandering). Representative members of each group are shown in Figure 14.7.

Plankton are extremely abundant and very important within the marine environment. In fact, most of Earth's **biomass**—the mass of all living organisms—consists of plankton adrift in the oceans. Even though 98 percent of marine species are bottom-dwelling, the vast majority of the ocean's biomass is planktonic.

Nekton (Swimmers) **Nekton** (*nektos* = swimming) include all animals capable of moving independently of the ocean currents, by swimming or other means of propulsion. They are capable not only of determining

their position within the ocean but also, in many cases, of long migrations. Nekton include most adult fish and squid, marine mammals, and marine reptiles (Figure 14.8).

Although nekton move freely, they are unable to move throughout the breadth of the ocean. Gradual changes in temperature, salinity, density, and availability of nutrients effectively limit their lateral range. The deaths of large numbers of fish, for example, can be caused by temporary shifts of water masses in the ocean. High water pressure at depth normally limits the vertical range of nekton.

Fish may appear to exist everywhere in the oceans, but they are more abundant near continents and islands and in colder waters. Some fish, such as salmon, ascend fresh water rivers to spawn. Many eels do just the reverse, growing to maturity in fresh water and then descending the streams to breed in the great depths of the ocean.

Benthos (Bottom Dwellers) The term **benthos** (*benthos* = bottom) describes organisms living on or in the ocean bottom. *Epifauna* (*epi* = upon, *fauna* = animal) live on the surface of the seafloor, either attached to rocks or moving along the bottom. *Infauna* (*in* = inside, *fauna* = animal) live buried in the sand or mud. Some benthos, called *nektobenthos*, live on the bottom but also swim or crawl through the water above the ocean floor. Examples of benthos are shown in Figure 14.9.

The shallow coastal ocean floor contains a wide variety of physical conditions and nutrient levels, both of which have allowed a great number of species to evolve. Moving across the bottom from the shore into deeper water, the number of benthos species may remain relatively constant, but the biomass of benthos organisms decreases. In addition, shallow coastal areas are the only locations where

▲ **FIGURE 14.8** Nekton includes all animals capable of moving independently of ocean currents. **A.** Gray reef shark, Bikini Atoll. (Photo by Doug Perrine/DRK Photo); **B.** California market squid. (Photo by Tom McHugh/Photo Researchers, Inc.); **C.** School of grunts, Florida Keys. (Photo by Larry Lipsky/DRK Photo); **D.** Yellow-head moray eel. (© by David B. Fleetham/Seapics.com)

large marine algae (often called "seaweeds") are found attached to the bottom. This is the case because these are the only areas of the seafloor that receive sufficient sunlight.

Throughout most of the deeper parts of the seafloor, animals live in perpetual darkness, where photosynthesis cannot occur. They must feed on each other, or on whatever nutrients fall from the productive surface waters.

The deep-sea bottom is an environment of coldness, stillness, and darkness. Under these conditions, life progresses slowly, and organisms that live in the deep sea usually are widely distributed because physical conditions vary little on the deep-ocean floor, even over great distances.

Marine Life Zones

The distribution of marine organisms is affected by the chemistry, physics, and geology of the oceans. Marine organisms are influenced by a variety of physical oceanographic factors. Some of these factors—such as availability of sunlight, distance from shore, and water

depth—are used to divide the ocean into distinct marine life zones (Table 14.2 and Figure 14.10).

Availability of Sunlight The upper part of the ocean into which sunlight penetrates is called the **photic** (*photos* = light) **zone**. The clarity of seawater is affected by many factors, such as the amount of plankton, suspended sediment, and decaying organic particles in the water. In addition, the amount of sunlight varies with atmospheric conditions, time of day, season of the year, and latitude.

The **euphotic** (*eu* = good, *photos* = light) **zone** is the portion of the photic zone near the surface where light is strong enough for photosynthesis to occur. In the open ocean, this zone can reach a depth of 100 meters (330 feet), but the zone will be much shallower close to shore where water clarity is typically reduced. In the euphotic zone, phytoplankton use sunlight to produce food molecules and become the basis of most oceanic food webs.

Different wavelengths of sunlight are absorbed as they pass through seawater. The longer wavelength red

▲ **FIGURE 14.9** Benthos describes organisms living on or in the ocean bottom. **A.** Sea star. (Photo by David Hall/Photo Researchers, Inc.); **B.** Yellow tube sponge. (Photo by Andrew Martinez/Photo Researchers, Inc.); **C.** Green sea urchin. (Photo by Andrew Martinez/Photo Researchers, Inc.); **D.** Coral crab. (Photo by Fred Bavendam/Peter Arnold, Inc.)

and orange colors are absorbed first, the greens and yellows next, and the shorter wavelength blues and violets penetrate the farthest. In fact, faint traces of blue and violet light can still be measured in extremely clear water at depths of 1000 meters (3300 feet).

Although photosynthesis cannot occur much below 100 meters (330 feet), there is enough light in the lower photic zone for marine animals to avoid predators, find food, recognize their species, and locate mates. Even deeper is the **aphotic** (*a* = without, *photos* = light) **zone**, where there is no sunlight.

Distance from Shore Marine life zones can also be subdivided based on distance from shore. The area

Basis	Marine Life Zone	Subdivision	Characteristics
Available sunlight	Photic		Sunlit surface waters.
		Euphotic	Has enough sunlight to support photosynthesis.
	Aphotic		No sunlight; many organisms have bioluminescent capabilities.
Distance from shore	Intertidal		Narrow strip of land between high and low tides; dynamic area.
	Neritic		Above continental shelf; high biomass and diversity of species.
	Oceanic		Open ocean beyond the shelf break; low nutrient concentrations.
Depth	Pelagic		All water above the ocean floor; organisms swim or float.
	Benthic		Bottom of ocean; organisms attach to, burrow into, or crawl on seafloor.
		Abyssal	Deep-sea bottom; dark, cold, high pressure; sparse life.

TABLE 14.2 Marine life zones.

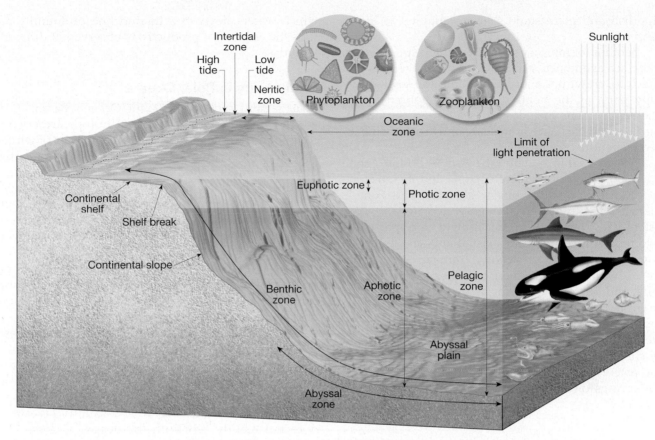

▲ **FIGURE 14.10**   The ocean is divided into marine life zones, based on availability of light, distance from shore, and water depth.

where the land and ocean meet and overlap is called the **intertidal zone**. This narrow strip of land between high and low tides is alternately covered and uncovered by seawater with each tidal change. Even though it appears to be a harsh place to live with crashing waves, periodic drying out, and rapid changes in temperature, salinity, and oxygen concentrations, many species live here that are superbly adapted to the dramatic environmental changes.

STUDENTS SOMETIMES ASK...

Do any deep-sea organisms produce light themselves?

Yes. Well over half of deep-sea organisms—including fish, jellies, crustaceans, and deep-sea squid—can *bioluminesce*, which means they can produce light organically. These organisms create light through a chemical reaction in specially designed structures or cells called *photophores*, some of which contain luminescent bacteria that live symbiotically within the organism. In a world of darkness, the ability to produce light can be used to attract prey, define territory, communicate with others, or avoid predators.

Seaward from the low-tide line is the **neritic** (*neritos* = of the coast) **zone**. This covers the gently sloping continental shelf out to the *shelf break* (see

Figure 13.9, p. 365). This zone can be very narrow or may extend hundreds of kilometers from shore. The neritic zone is often shallow enough for sunlight to reach all the way to the ocean floor, putting it entirely within the photic zone.

Although the neritic zone covers only about 5 percent of the world ocean, it is rich in both biomass and number of species. Many organisms find the conditions here ideal because photosynthesis occurs readily, nutrients wash in from the land, and the bottom provides shelter and habitat. This zone is so rich, in fact, that it supports 90 percent of the world's commercial fisheries.

Beyond the continental shelf is the **oceanic zone**. The open ocean reaches great depths, and as a result, surface waters typically have lower nutrient concentrations because nutrients tend to sink out of the photic zone to the deep-ocean floor. This low nutrient concentration usually results in much smaller populations than the more productive neritic zone.

Water Depth   A third method of classifying marine habitats is based on water depth. Open ocean of *any* depth is called the **pelagic** (*pelagios* = of the sea) **zone**. Animals in this zone swim or float freely. The photic part of the pelagic zone is home to phytoplankton, zooplankton, and nekton, such as tuna, sea turtles, and dolphins. The aphotic part has strange species like viperfish and giant squid that are adapted to life in deep water.

Benthos organisms such as giant kelp, sponges, crabs, sea anemones, sea stars, and marine worms that attach to, crawl upon, or burrow into the seafloor occupy parts of the **benthic** (*benthos* = bottom) **zone**. The benthic zone includes any sea-bottom surface regardless of its distance from shore and is mostly inhabited by benthos organisms.

The **abyssal** (*a* = without, *byssus* = bottom) **zone** is a subdivision of the benthic zone and includes the deep-ocean floor, such as *abyssal plains*. This zone is characterized by extremely high water pressure, consistently low temperature, no sunlight, and sparse life. Two food sources typically exist at abyssal depths: (1) tiny decaying particles steadily "raining" down from above, which provide food for filter-feeders, brittle stars, and burrowing worms; and (2) large fragments or entire dead bodies falling at scattered sites, which supply meals for actively searching fish, such as the grenadier, tripodfish, and hagfish, which locate food by chemical sensing. However, a third source of food has recently been discovered that is associated with hot springs on the seafloor, called *hydrothermal vents* (Box 14.2).

Oceanic Productivity

Why are some regions of the ocean teeming with life, while other areas seem barren? The answer is related to the amount of primary productivity in various parts of the ocean. **Primary productivity** is the amount of carbon fixed by organisms through the synthesis of organic matter using energy derived from solar radiation (*photosynthesis*) or chemical reactions (*chemosynthesis*). Although chemosynthesis supports hydrothermal vent biocommunities along the oceanic ridge (Box 14.2), it is much less significant than photosynthesis in worldwide oceanic productivity.

Two factors influence a region's photosynthetic productivity: *availability of nutrients* (such as nitrates, phosphorus, iron, and silica) and the *amount of solar radiation* (sunlight). Thus, the most abundant marine life exists where there are ample nutrients and good sunlight. Oceanic productivity, however, varies dramatically because of the uneven distribution of nutrients throughout the photosynthetic zone and seasonal changes in the availability of solar energy.

A permanent *thermocline* (and resulting *pycnocline*) develops nearly everywhere in the oceans (see Figure 14.6). This layer forms a barrier to vertical mixing and prevents the resupply of nutrients to sunlit surface waters. In the midlatitudes, a thermocline develops only during the summer season, and in polar regions a thermocline does not usually develop at all. The degree to which waters develop a thermocline profoundly affects the amount of productivity observed at different latitudes.

Productivity in Polar Oceans

Polar regions such as the Arctic Ocean's Barents Sea, which is off the northern coast of Europe, experience continuous darkness for about three months of winter and continuous illumination for about three months during summer. Productivity of phytoplankton—mostly single-celled algae called *diatoms*—peaks there during May (Figure 14.11), when the Sun rises high enough in the sky so that there is deep penetration of sunlight into the water. As soon as the diatoms develop, zooplankton—mostly small crustaceans called *copepods* (Figure 14.11) and larger *krill*—begin feeding on them. The zooplankton biomass peaks in June and continues at a relatively high level until winter darkness begins in October.

Recall that density and temperature change very little with depth in polar regions (see Figures 14.4 and 14.5), so these waters are *isothermal*, and there is no barrier to mixing between surface waters and deeper, nutrient-rich waters. In the summer, however, melting ice creates a thin, low-salinity layer that does not readily mix with the deeper waters. This stratification is crucial to summer production, because it helps prevent phytoplankton from being carried into deeper, darker waters. Instead, they are concentrated in the sunlit surface waters where they reproduce continuously.

Because of the constant supply of nutrients rising from deeper waters below, high-latitude surface waters typically have high nutrient concentrations. The availability of solar energy, however, is what limits photosynthetic productivity in these areas.

Productivity in Tropical Oceans

You may be surprised to learn that productivity is low in tropical regions of the open ocean. Because the Sun is more directly overhead, light penetrates much

▼ **FIGURE 14.11** One example of productivity in polar oceans is illustrated by the Barents Sea. A springtime increase in diatom mass is followed closely by an increase in zooplankton abundance.

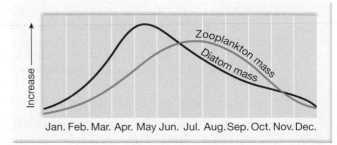

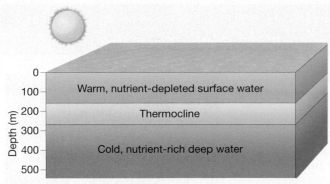

▲ **FIGURE 14.12** Productivity in tropical oceans. Although tropical regions receive adequate sunlight year-round, a permanent thermocline prevents the mixing of surface and deep water. As phytoplankton consume nutrients in the surface layer, productivity is limited because the thermocline prevents replenishment of nutrients from deeper water. Thus, productivity remains at a steady, low level.

deeper into tropical oceans than in temperate and polar waters, and solar energy is available year-round. However, productivity is low in tropical regions of the open ocean because a permanent thermocline produces a stratification of water masses that prevents mixing between surface waters and nutrient-rich deeper waters (Figure 14.12). In essence, the thermocline is a barrier that eliminates the supply of nutrients from deeper waters below. So, productivity in tropical regions is limited by the lack of nutrients (unlike polar regions, where productivity is limited by the lack of sunlight). In fact, these areas have so few organisms that they are considered biological deserts.

Productivity in Temperate Oceans

Productivity is limited by available sunlight in polar regions and by nutrient supply in the tropics. In temperate (midlatitude) regions, a combination of these two limiting factors controls productivity as shown in Figure 14.13 (which shows the pattern for the Northern Hemisphere; in the Southern Hemisphere, the seasons are reversed).

Winter. Productivity in temperate oceans is very low during winter, even though nutrient concentration is highest at this time. The reason is that solar energy is limited because days are short and the Sun angle is low. As a result, the depth at which photosynthesis can occur is so shallow that phytoplankton do not grow much.

Spring. The Sun rises higher in the sky during spring, creating a greater depth at which photosynthesis can occur. A *spring bloom* of phytoplankton occurs because solar energy and nutrients are available, and a seasonal thermocline develops (due to increased solar heating) that traps algae in the euphotic zone (Figure 14.13). This creates a tremendous demand for nutrients in the euphotic zone, so the supply is quickly depleted, causing productivity to decrease sharply. Even though the days are lengthening and sunlight is increasing, productivity during the spring bloom is limited by the lack of nutrients.

Summer. The Sun rises even higher in the summer, so surface waters in temperate parts of the ocean continue to warm. A strong seasonal thermocline is created that in turn prevents vertical mixing, so nutrients depleted from surface waters cannot be replaced by those from deeper waters. Throughout summer, the phytoplankton population remains relatively low (Figure 14.13).

Fall. Solar radiation diminishes in the fall as the Sun moves lower in the sky, so surface temperatures drop and the summer thermocline breaks down. Nutrients return to the surface layer as increased wind strength mixes surface waters with deeper waters. These conditions create a *fall bloom* of phytoplankton, which is much less dramatic than the spring bloom (Figure 14.13). The fall bloom is very short-lived because sunlight (not nutrient supply, as in the spring bloom) becomes the limiting factor as winter approaches to repeat the seasonal cycle.

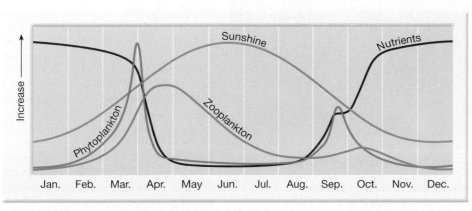

◄ **FIGURE 14.13** Productivity in temperate oceans (Northern Hemisphere). The graph shows the relationship among phytoplankton, zooplankton, amount of sunshine, and nutrient levels for surface waters.

BOX 14.2 EARTH AS A SYSTEM:
Deep-Sea Hydrothermal Vent Biocommunities—Earth's First Life?

Deep-sea hydrothermal vents form along many active rift zones of the oceanic ridge. Here seawater percolates down into the hot and newly formed oceanic crust. During its journey, the water can become saturated with minerals before it spews back into the ocean as a *black smoker* (Figure 14.B). Black smokers usually emit from tall chimneys composed of metal sulfide minerals that have precipitated as the hot ventwater is exposed to cold seawater.

Water temperatures at some vents reach as high as 350°C (660°F), which is far too hot for anything to live. At some vents, however, water temperatures of 100°C (212°F) or lower nourish dense, exotic *hydrothermal vent biocommunities* of organisms found nowhere else in the world. In fact, hundreds of new species—and even new genera and families—have been discovered surrounding these deep-sea habitats since their discovery by scientists along the Galápagos Rift in 1977. Additional hydrothermal vent biocommunities exist in discrete patches along the oceanic ridge and have been visited by scientists in deep-sea submersibles along the East Pacific Rise, Mid-Atlantic Ridge, Mid-Indian Ocean Ridge, and the Juan de Fuca Ridge.

How do these organisms survive in this dark, hot, sulfur-rich environment where photosynthesis cannot occur? Study of hydrothermal vent organisms reveals that microscopic bacteria-like organisms called *archaea* (*archaeo* = ancient) living in and near the vents perform *chemosynthesis* (*chemo* = chemistry, *syn* = with, *thesis* = an arranging) and constitute the base of the food web. Hydrothermal vents provide heat energy for archaea to oxidize hydrogen sulfide (H_2S) which is formed by the reaction of hot water with dissolved sulfate (SO_4^{-2}). Through chemosynthesis, archaea produce sugars and other foods that enable them and many other organisms to live in this very unusual and extreme environment.

Some archaea live symbiotically inside giant gutless tube-dwelling worms (Figure 14.C). These archaea provide food for the tube worms to grow as rapidly as 1 meter (3.3 feet) each year and up to 3 meters (10 feet) in length. Other archaea are consumed by specialized yel-

Figure 14.C Tube worms up to 3 meters (10 feet) in length are among the organisms found in the extreme environment of hydrothermal vents along the crest of the oceanic ridge where sunlight is nonexistent. These organisms obtain their food from internal microscopic bacteria-like archaea, which acquire their nourishment and energy through the processes of chemosynthesis. (Photo by Al Giddings Images, Inc.)

low mussels, giant white clams, and pink sea urchins. These in turn are eaten by unique crabs and fishes. Thus, archaea are the foundation of a living ecosystem that does not require sunlight.

It is very likely that environments similar to those of hydrothermal vents were present during the early history of the planet. Some scientists have suggested that the uniformity of conditions and abundant energy of the vents would have provided an ideal habitat for the origin of life. In fact, hydrothermal vents may represent one of the oldest life-sustaining environments, because hydrothermal activity occurs wherever there are both volcanoes and water. An additional line of evidence in support of hydrothermal vents harboring some of Earth's first life exists in the fact that archaea contain ancient genetic makeup.

Figure 14.B View from the submersible *Alvin* of a black smoker spewing hot, mineral-rich water along the East Pacific Rise. When heated solutions meet cold seawater, metal sulfides precipitate and form mounds of minerals around these hydrothermal vents. (Photo by Dudley Foster, © Woods Hole Oceanographic Institution)

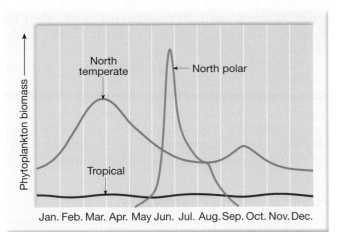

▲ FIGURE 14.14 Comparison of productivity in tropical, temperate, and polar oceans (Northern Hemisphere), showing seasonal variation in phytoplankton biomass. The total area under each curve represents annual photosynthetic productivity.

Figure 14.14 compares the seasonal variation in phytoplankton biomass of tropical, north polar, and north temperate regions, where the total area under each curve represents photosynthetic productivity. The figure shows the dramatic peak in productivity in polar oceans during the summer; the steady, low rate of productivity year-round in the tropical oceans; and the seasonal productivity that occurs in temperate oceans. It also shows that the highest overall productivity occurs in temperate regions.

Oceanic Feeding Relationships

Marine algae, plants, bacteria, and bacteria-like archaea are the main oceanic producers. As these producers make food (organic matter) available to the consuming animals of the ocean, it passes from one feeding population to the next. Only a small percentage of the energy taken in at any level is passed on to the next because energy is consumed and lost at each level. As a result, the producers' biomass in the ocean is many times greater than the mass of the top consumers, such as sharks or whales.

STUDENTS SOMETIMES ASK...
Are all sharks dangerous?

Of the more than 350 shark species, about 80 percent are unable to hurt people or rarely encounter people. The largest shark—also the largest fish in the world—is the whale shark (*Rhincodon typus*), which reaches lengths of up to 15 meters (50 feet) but eats only plankton and is therefore not considered dangerous.

BOX 14.3 PEOPLE AND THE ENVIRONMENT:
The Iron Hypothesis—Fertilizing the Ocean to Reduce Global Warming

The documented increase in Earth's average surface temperature over the last 130 years—likely as a result of human activities that have increased the amount of carbon dioxide in the atmosphere—has been a topic of much debate and discussion.* Nevertheless, scientists are seeking ways to prevent further alteration of Earth's climate by reducing the amount of atmospheric carbon dioxide. Oceanographer John Martin was one of the first to propose increasing oceanic productivity as a way to remove carbon dioxide from the atmosphere, thereby reducing global warming.

How does an increase in ocean productivity remove carbon dioxide from the atmosphere? Through photosynthesis, microscopic marine algae convert carbon dioxide dissolved in the ocean to carbohydrates and oxygen gas. As the number of photosynthetic marine organisms increases, additional carbon dioxide is removed from the ocean, which in turn causes the ocean to absorb more carbon dioxide from the atmosphere.

Areas of the ocean that have relatively low productivity, such as the tropical oceans, are a good place to stimulate productivity and thus increase the amount of carbon dioxide removed from the atmosphere. In 1987, Martin determined that the absence of iron limited productivity in tropical oceans, so he proposed fertilizing the ocean with iron—the only element that seemed to be lacking—in order to increase its productivity (the *"iron hypothesis"*). In 1993, Martin's associates added finely ground iron to a test area of the ocean near the Galápagos Islands in the Pacific Ocean. Their results and the results of other "Iron Ex" open-ocean experiments in 1995 and 1999 showed that adding iron to the ocean increased productivity up to 30 times. Although these results are promising, there were problems grinding the iron fine enough, dispersing the iron, and keeping it in suspension for long periods of time.

An even larger problem may be the long-term global environmental effects of adding additional iron and large amounts of carbon dioxide to the ocean. Many scientists fear that increased amounts of carbon dioxide in the ocean would upset the ocean's natural chemical balance and alter the entire global marine ecosystem. Proponents of iron fertilization, however, claim that stimulating productivity mimics what is done naturally in the ocean and that iron fertilization may be one of the most promising solutions for reducing atmospheric carbon dioxide.

*This topic is explored more fully in "Human Impact on Global Climate" in Chapter 20.

Trophic Levels

Chemical energy stored in the mass of the ocean's algae (the "grass of the sea") is transferred to the animal community mostly through feeding. Zooplankton are *herbivores* (*herba* = grass, *vora* = eat), so they eat diatoms and other microscopic marine algae. Larger herbivores feed on the larger algae and marine plants that grow attached to the ocean bottom near shore.

The herbivores (grazers) are then eaten by larger animals, the *carnivores* (*carni* = meat, *vora* = eat). They in turn are eaten by another population of larger carnivores, and so on. Each of these feeding stages is called a **trophic** (*tropho* = nourishment) **level**.

Generally, individual members of a feeding population are larger—but not too much larger—than the organisms they eat. There are conspicuous exceptions, however, such as the blue whale. Up to 30 meters (100 feet) long, it is possibly the largest animal that has ever existed on Earth, yet it feeds mostly on krill, which have a maximum length of only 6 centimeters (2.4 inches).

Transfer Efficiency

The transfer of energy between trophic levels is very inefficient. The efficiencies of different algal species vary, but the average is only about *2 percent*, which means that 2 percent of the light energy absorbed by algae is ultimately synthesized into food and made available to herbivores.

Figure 14.15 shows the passage of energy between trophic levels through an entire ecosystem, from the solar energy assimilated by phytoplankton through all trophic levels to the ultimate carnivore—humans. Because energy is lost at each trophic level, it takes thousands of smaller marine organisms to produce a single fish that is so easily consumed during a meal!

Food Chains and Food Webs

A **food chain** is a sequence of organisms through which energy is transferred, starting with an organism that is the primary producer, then an herbivore, then one or more carnivores, finally culminating with the "top carnivore," which is not usually preyed upon by any other organism.

Because energy transfer between trophic levels is inefficient, it is advantageous for fishers to choose a population that feeds as close to the primary producing population as possible. This increases the biomass available for food and the number of individuals available to be taken by the fishery. Newfoundland herring, for example, are an important fishery that usually rep-

▼ **FIGURE 14.15** Ecosystem energy flow and efficiency. For every 500,000 units of radiant energy input available to the producers (phytoplankton), only one unit of mass is added to the fifth trophic level (humans). Average phytoplankton transfer efficiency is 2 percent (98 percent loss) and all other trophic levels average 10 percent efficiency (90 percent loss). The ultimate effect of energy transfer between trophic levels is that the number of individuals and the total biomass decrease at successive trophic levels because the amount of available energy decreases.

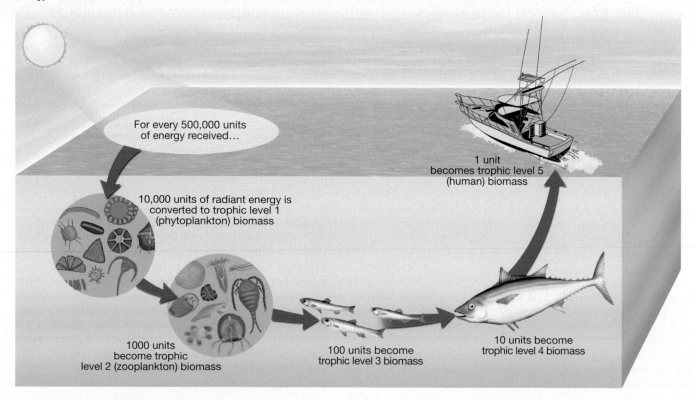

For every 500,000 units of energy received…

10,000 units of radiant energy is converted to trophic level 1 (phytoplankton) biomass

1000 units become trophic level 2 (zooplankton) biomass

100 units become trophic level 3 biomass

10 units become trophic level 4 biomass

1 unit becomes trophic level 5 (human) biomass

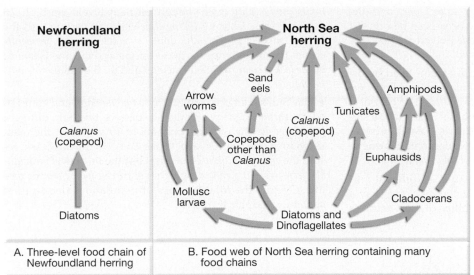

◄ FIGURE 14.16 Comparison between a food chain and a food web. **A.** A food chain is the passage of energy along a single path, such as from diatoms to copepods to Newfoundland herring. Feeding relationships are rarely this simple. **B.** A food web, showing multiple paths for food sources of the North Sea herring.

A. Three-level food chain of Newfoundland herring

B. Food web of North Sea herring containing many food chains

resents the third trophic level in a food chain. They feed primarily on small copepods that feed, in turn, upon diatoms (Figure 14.16A).

Feeding relationships are rarely as simple as that of the Newfoundland herring. More often, top carnivores in a food chain feed on a number of different animals, each of which has its own simple or complex feeding relationships. This constitutes a **food web**, as shown in Figure 14.16B for North Sea herring.

Animals that feed through a food web rather than a food chain are more likely to survive because they have alternative foods to eat should one of their food sources diminish in quantity or even disappear. Newfoundland herring, on the other hand, eat only copepods, so the disappearance of copepods would catastrophically affect their population.

Chapter Summary

• *Salinity* is the amount of dissolved substances in water, usually expressed in parts per thousand ‰. Seawater salinity in the open ocean *averages 35‰*. The principal elements that contribute to the ocean's salinity are *chlorine* (55 percent) and *sodium* (31 percent), which combine to produce table salt. The primary *sources of the elements in sea salt* in the ocean are *chemical weathering* of rocks on the continents and *volcanic outgassing*.

• *Variations in seawater salinity* are primarily caused by changing the *water content*. Natural processes that add large amounts of fresh water to seawater and *decrease salinity* include *precipitation, runoff from land, icebergs melting*, and *sea ice melting*. Processes that remove large amounts of fresh water from seawater and *increase salinity* include the *formation of sea ice* and *evaporation*. Seawater salinity in the open ocean *ranges from 33‰ to 38‰*, with some marginal seas experiencing considerably more variation.

• The ocean's *surface temperature* is related to the amount of solar energy received and *varies as a function of latitude*. Low-latitude regions have distinctly colder water at depth, creating a *thermocline*, which is a layer of rapidly changing temperature. No thermocline exists in high-latitude regions, because the water column is *isothermal*.

• Water's unique thermal properties have caused the *ocean's temperature to remain stable* for long periods of time, facilitat-

ing the development of life on Earth. Experiments have been conducted that *send sound through the ocean* to determine if the ocean's temperature is increasing as a result of global warming.

• Seawater *density is mostly affected by water temperature* but also by salinity. Low-latitude regions have distinctly denser (colder) water at depth, creating a *pycnocline*, which is a layer of rapidly changing density. No pycnocline exists in high-latitude regions because the water column is *isopycnal*.

• Most open-ocean regions exhibit a *three-layered structure based on water density*. The shallow *surface mixed zone* has warm and nearly uniform temperatures. The *transition zone* includes a prominent thermocline and associated pycnocline. The *deep zone* is continually dark and cold and accounts for 80 percent of the water in the ocean. In high latitudes, the three-layered structure does not exist.

• *Marine life is superbly adapted to the oceans*. Marine organisms can be *classified into* one of three groups based on *habitat* and *mobility*. *Plankton* are free-floating forms with little power of locomotion, *nekton* are swimmers, and *benthos* are bottom dwellers. *Most of the ocean's biomass is planktonic*.

• Three criteria are frequently used to establish *marine life zones*. Based on *availability of sunlight*, the ocean can be divided into the *photic zone* (which includes the *euphotic zone*) and the *aphotic zone*. Based on *distance from shore*, the ocean can be divided into the *intertidal zone*, the *neritic zone*, and the *oceanic zone*. Based on *water depth*, the ocean can be divided into the *pelagic zone* and the *benthic zone* (which includes the *abyssal zone*).

• *Primary productivity* is the amount of carbon fixed by organisms through the synthesis of organic matter using energy derived from solar radiation (*photosynthesis*) or chemical reactions (*chemosynthesis*). Chemosynthesis is much less significant than photosynthesis in worldwide oceanic productivity. Photosynthetic productivity in the ocean varies due to the *availability of nutrients* and *amount of solar radiation*.

• *Oceanic photosynthetic productivity* varies at different latitudes because of *seasonal changes* and the *development of a thermocline*. In *polar oceans*, the availability of *solar radiation*

limits productivity even though nutrient levels are high. In *tropical oceans*, a strong *thermocline* exists year-round, so the *lack of nutrients generally limits productivity*. In *temperate oceans*, productivity peaks in the spring and fall and is *limited by the lack of solar radiation in winter* and by the *lack of nutrients in summer*.

• The Sun's energy is utilized by *phytoplankton* and converted to *chemical energy*, which is passed through different *trophic levels*. On average, only about *10 percent* of the mass taken in at one trophic level is passed on to the next. As a result, the *size of individuals increases* but the *number of individuals decreases* with each trophic level of the *food chain* or *food web*. Overall, the total biomass of populations decreases at successive trophic levels.

Key Terms

abyssal zone (p. 394)
aphotic zone (p. 392)
benthic zone (p. 394)
benthos (p. 390)
biomass (p. 390)
density (p. 388)
euphotic zone (p. 391)
food chain (p. 393)

food web (p. 399)
intertidal zone (p. 393)
nekton (p. 390)
neritic zone (p. 393)
oceanic zone (p. 393)
pelagic zone (p. 393)
photic zone (p. 391)
photosynthesis (p. 389)

phytoplankton (p. 390)
plankton (p. 390)
primary productivity (p. 394)
pycnocline (p. 388)
salinity (p. 384)
thermocline (p. 386)
trophic level (p. 398)
zooplankton (p. 390)

Review Questions

1. Define *salinity*. What is the average salinity of the ocean? Why do oceanographers typically express salinity in parts per thousand?

2. What are the six most abundant components (elements) dissolved in seawater? What is produced when the two most abundant elements are combined?

3. What are the two primary sources for the materials that comprise the dissolved components in seawater?

4. Describe the processes that affect seawater salinity. For each process, indicate whether water is added or removed and if it decreases or increases salinity. What physical conditions create high salinity water in the Red Sea and low salinity water in the Baltic Sea?

5. Why is desalination not likely to be a significant source of water for agriculture in the foreseeable future?

6. What one factor is primarily responsible for influencing seawater temperature?

7. Describe temperature variation with depth in both high and low latitudes. Why do high-latitude waters generally lack a thermocline?

8. What is the goal of the ATOC experiment? How do scientists propose achieving their goal? What problems could the experiment create?

9. Which two factors influence seawater density? Of the two, which one has the greatest influence on surface seawater density?

10. Describe density variation with depth in both high and low latitudes. Why do high-latitude waters generally lack a pycnocline?

11. Describe the ocean's layered structure. Why does the three-layer structure not exist in high latitudes?

12. Describe the lifestyles of plankton, nekton, and benthos, and give examples of each.

13. Of plankton, nekton, and benthos, which group comprises the largest biomass? Explain why.

14. List three physical factors that are used to divide the ocean into marine life zones. Describe how each factor influences the abundance and distribution of marine life.

15. Why are there greater numbers and types of organisms in the neritic zone than in the oceanic zone?

16. What is the base of the food web for hydrothermal vent biocommunities? What is unusual about this food source?

17. List the two methods in which primary productivity is accomplished in the ocean. Of these two methods, which one is the most significant? What two factors influence it?

18. Compare the biological productivity of polar, temperate, and tropical regions of the oceans. Consider seasonal changes, the development of a thermocline, the availability of nutrients, and variations in solar radiation.

19. What is the average efficiency of energy transfer between trophic levels? Use this efficiency to determine how much phytoplankton mass is required to add *1 gram* of new mass to a killer whale, which is a third-level carnivore. Include a diagram that shows the different trophic levels and the relative size and abundance of organisms at different levels. How would your answer change if the efficiency were half the average rate, or twice the average rate?

20. Describe the advantage that a top carnivore gains by eating from a food web as compared to a single food chain.

Examining the Earth System

1. Discuss the ocean's importance in the hydrologic cycle (see Figures 5.1 and 5.2). If Earth did not have oceans, how might this affect (a) the global biosphere, and (b) the rock cycle?

2. Describe the iron hypothesis and discuss the relative merits and dangers of undertaking a project that could cause dramatic changes in the global environment. For details about the iron hypothesis and its inventor, oceanographer John Martin, visit NASA's Earth Observatory site at: **http://earthobservatory.nasa.gov/Library/ Giants/Martin/**

Online Study Guide

The *Earth Science* Website uses the resources and flexibility of the Internet to aid in your study of the topics in this chapter. Written and developed by Earth science instructors, this site will help improve your understanding of Earth science. Visit **http://www.prenhall.com/tarbuck** and click on the cover of *Earth Science* 11e to find:

- **Online review quizzes.**
- **Critical thinking exercises.**
- **Links to chapter-specific Web resources.**
- **Internet-wide key term searches.**

http://www.prenhall.com/tarbuck

CHAPTER 15

The Dynamic Ocean*

*This chapter was revised with the assistance of Professor Alan P. Trujillo,
Palomar College.

Coastal erosion by Pacific storm waves destroyed these homes at Laguna Niguel, California in 1998. (Photo by A. Ramsey/Photo Edit)

he restless waters of the ocean are constantly in motion, powered by many different forces. Winds, for example, generate surface currents, which influence coastal climate and provide nutrients that affect the abundance of algae and other marine life in surface waters. Winds also produce waves that carry energy from storms to distant shores, where their impact erodes the land. In some areas, density differences create deep-ocean circulation, which is important for ocean mixing and recycling nutrients. In addition, the Moon and the Sun produce tides, which periodically raise and lower average sea level. This chapter examines these movements of ocean waters and their effect upon coastal regions (Figure 15.1).

Surface Circulation

Ocean currents are masses of ocean water that flow from one place to another. The amount of water can be large or small, currents can be at the surface or deep below, and the phenomena that create them can be simple or complex. In all cases, however, the currents that are generated involve water masses in motion.

Surface currents develop from friction between the ocean and the wind that blows across its surface. Some of these currents are short-lived and affect only small areas. Such water movements are responses to local or seasonal influences. Other surface currents are relatively permanent phenomena that extend over large portions of the oceans. These major horizontal movements of surface waters are closely related to the general circulation pattern of the atmosphere. This is clearly illustrated by comparing the pattern of global winds shown in Figure 18.16 (p. 513) and the position of Earth's principal surface ocean currents shown in Figure 15.2.

Ocean Circulation Patterns

Huge circular-moving current systems dominate the surfaces of the oceans. These large whirls of water within an ocean basin are called **gyres** (*gyros* = a circle). Figure 15.2 shows the world's five main gyres: the *North Pacific Gyre*, the *South Pacific Gyre*, the *North Atlantic Gyre*, the *South Atlantic Gyre*, and the *Indian Ocean Gyre* (which exists mostly within the Southern Hemisphere). The center of each gyre coincides with the subtropics at about 30 degrees north or south latitude, so they are often called *subtropical gyres*. Comparing Figure 18.16 (p. 513) to Figure 15.2 shows that there is a striking correspondence between the direction of surface-current flow and the major wind belts of the world.

As shown in Figure 15.2, subtropical gyres rotate clockwise in the Northern Hemisphere and counterclockwise in the Southern Hemisphere. Why do the gyres flow in different directions in the two hemi-

▼ **FIGURE 15.1** Wind is responsible for creating the ocean waves that modify shorelines such as California's Big Sur Coast.. (Photo by Richard Price/Getty)

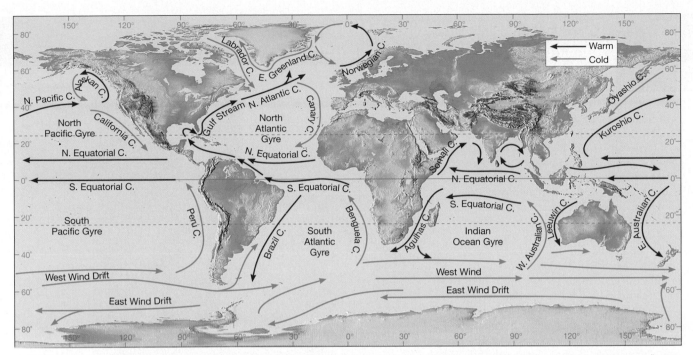

▲ **FIGURE 15.2** Average ocean surface currents February–March. The ocean's circulation is organized into five major current gyres (large circular-moving loops of water), which exist in the North Pacific, South Pacific, North Atlantic, South Atlantic, and Indian oceans.

spheres? Although wind is the force that generates surface currents, other factors also influence the movement of ocean waters. The most significant of these is the **Coriolis effect**. Because of Earth's rotation, currents are deflected to the *right* in the Northern Hemisphere and to the *left* in the Southern Hemisphere. (The Coriolis effect is more fully explained in Chapter 18.) As a consequence, gyres flow in opposite directions in the two different hemispheres.

Four main currents generally exist within each gyre (Figure 15.2). The North Pacific Gyre, for example, consists of the North Equatorial Current, the Kuroshio* Current, the North Pacific Current, and the California Current. The tracking of floating objects that are released into the ocean intentionally or accidentally reveal that it takes about six years for the objects to go all the way around the loop (see Box 15.1).

In the North Atlantic, the North Equatorial Current is deflected northward through the Caribbean, where it becomes the Gulf Stream (Figure 15.2). As the Gulf Stream moves along the East Coast of the United States, it is strengthened by the prevailing westerly winds and is deflected to the east (to the right) between the Carolinas and New England. As it continues northeastward, it gradually widens and slows until it becomes a vast, slowly moving current known as the North Atlantic Current, which, because of its sluggish nature, is also known as the North Atlantic Drift.

As the North Atlantic Current approaches Western Europe, it splits, part of it moving northward past Great Britain, Norway, and Iceland, carrying heat to these otherwise chilly areas. The other part is deflected southward as the cool Canary Current. As the Canary Current moves southward, it eventually merges into the North Equatorial Current, completing the gyre. Because the North Atlantic Ocean basin is about half the size of the North Pacific, it takes floating objects about three years to go completely around this gyre.

The circular motion of gyres leaves a large central area that has no well-defined currents. In the North Atlantic, this zone of calmer waters is known as the Sargasso Sea, named for the large quantities of *Sargassum*, a type of floating seaweed encountered there.

The ocean basins in the Southern Hemisphere exhibit a similar pattern of flow as the Northern Hemisphere basins, with surface currents that are influenced by wind belts, the position of continents, and the Coriolis effect. In the South Atlantic and South Pacific, for example, surface ocean circulation is very much the same as in their Northern Hemisphere counterparts except that the direction of flow is counterclockwise (Figure 15.2).

The Indian Ocean exists mostly in the Southern Hemisphere, so it follows a surface circulation pattern similar to other Southern Hemisphere ocean basins (Figure 15.2). The small portion of the Indian Ocean in the Northern Hemisphere, however, is influenced by the seasonal wind shifts known as the summer and winter *monsoons* (*mausim* = season). When the winds change direction, the surface currents also reverse direction.

---

*Kuroshio is pronounced "kuhr-ROH-shee-oh" and is sometimes known as the Japan Current. The term *Kuroshio* is Japanese for "black tide" in reference to its clear, lifeless waters.

BOX 15.1 UNDERSTANDING EARTH:
Running Shoes as Drift Meters—Just Do It

Any floating object can serve as a makeshift drift meter, as long as it is known where the object entered the ocean and where it was retrieved. The path of the object can then be inferred, providing information about the movement of surface currents. If the times of release and retrieval are known, the speed of currents can also be determined. Oceanographers have long used *drift bottles* (a floating "message in a bottle" or a radio-transmitting device set adrift in the ocean) to track the movement of currents and, more recently, to refine computer models of ocean circulation.

Many objects have inadvertently become drift meters when ships have lost some (or all) of their cargo at sea. In this way, Nike athletic shoes and colorful floating bathtub toys (Figure 15.A, inset) have helped oceanographers advance the understanding of surface circulation in the North Pacific Ocean.

In May 1990 the container vessel *Hansa Carrier* was en route from Korea to Seattle, Washington, when it encountered a severe North Pacific storm. During the storm the ship lost 21 deck containers overboard, including five that held Nike athletic shoes. The shoes that were released from their containers floated and were carried east by the North Pacific Current. Within six months thousands of the shoes began to wash up along the beaches of Alaska, Canada, Washington, and Oregon (Figure 15.A), over 2400 kilometers (1500 miles) from the site of the spill. A few shoes were found on beaches in northern California, and over two years later shoes from the spill were even recovered from the north end of the Big Island of Hawaii!

With help from the beachcombing public and remotely based lighthouse operators, information on the location and number of shoes collected was compiled during the months following the spill. Serial numbers inside the shoes were traced to individual containers, indicating that only four of the five containers had released their shoes; evidently, one entire

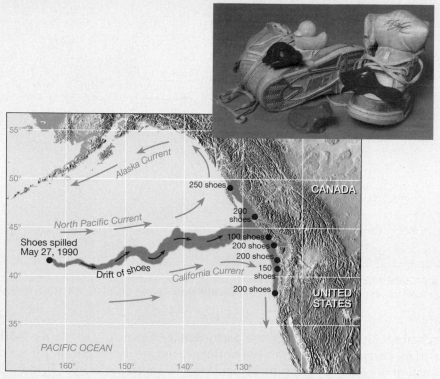

Figure 15.A Large ships crossing the ocean have lost entire containers overboard. If the containers release floating items, inadvertent float meters are launched that help oceanographers track ocean surface currents. Map shows path of drifting shoes and recovery locations from a spill in 1990; inset shows recovered shoes from the 1990 spill and plastic bathtub toys from a spill in 1992. (Copyright by the American Geophysical Union)

container sank without opening. Thus, a maximum of 30,910 pairs of shoes (61,820 individual shoes) were released. Before the shoe spill, the largest number of drift bottles purposefully released at one time by oceanographers was about 30,000. Although only 2.6 percent of the shoes were recovered, this compares favorably with the 2.4 percent recovery rate of drift bottles released by oceanographers conducting research.

In January 1992 another cargo ship lost 12 containers overboard during a storm to the north of where the shoes had previously spilled. One of these containers held 29,000 packages of small, floatable, colorful plastic bathtub toys in the shapes of blue turtles, yellow ducks, red beavers, and green frogs (Figure 15.A, inset). Even though the toys were housed

in plastic packaging glued to a cardboard backing, studies showed that after 24 hours in seawater, the glue deteriorated, thereby releasing over 100,000 individual floating toys.

The floating bathtub toys began to come ashore in southeast Alaska 10 months later, verifying computer models of North Pacific circulation. The models indicate that many of the bathtub toys will continue to be carried by the Alaska Current, eventually dispersing throughout the North Pacific Ocean.

Since 1992 oceanographers have continued to study ocean currents by tracking other floating items spilled from cargo ships, including 34,000 hockey gloves, 5 million plastic Lego pieces, and an unidentified number of small plastic doll parts.

The West Wind Drift is the only current that completely encircles Earth (Figure 15.2). It flows around the ice-covered continent of Antarctica, where no large landmasses are in the way, so its cold surface waters circulate in a continuous loop. It moves in response to the Southern Hemisphere prevailing westerly winds, and portions of it split off into the adjoining southern ocean basins.

The Gulf Stream

Anyone who navigates the oceans needs to be aware of currents. By understanding the direction and strength of ocean currents, sailors soon realize that their voyage time can be reduced if they travel with a current, or increased if they travel against a current. Such is the case for the strong Gulf Stream current, which is the best known and most studied of all ocean currents.

The Gulf Stream flows northward along the East Coast of the United States. It was given the name Gulf Stream because it carries warm water from the Gulf of Mexico and because it was narrow and well defined—similar to a stream, but in the ocean. As deputy postmaster general for the new colonies, Benjamin Franklin was the first to produce a map of the Gulf Stream (Figure 15.3A). His interest in the North Atlantic began when it was brought to his attention that ships carrying the mail took two weeks longer going from England to America than in the other direction. P. L. Richardson* describes how Franklin discovered that an ocean current was causing the delay:

> While he was in London as Deputy Postmaster General for the American colonies, Franklin was consulted on the question of why the mail packets took a fortnight longer to sail to America than the merchant ships. In October 1768 Franklin discussed this problem with his cousin Timothy Folger, a Nantucket ship captain then visiting London. Folger told him the packet captains were ignorant of the Gulf Stream and frequently sailed in this current. . . . Folger sketched the Gulf Stream on a chart and added written notes on how to avoid the Gulf Stream and Franklin had the chart printed in 1769 or 1770.

Even though the Franklin-Folger chart of the Gulf Stream was widely distributed, it was initially ignored by the captains of the mail ships because they thought that simple American fishers could not possess superior knowledge of the sea.

Nearly 100 years later Matthew Fontaine Maury, the founder of physical oceanography, described the Gulf Stream in his 1855 book, *The Physical Geography of the Sea*:

> There is a river in the ocean. In the severest droughts it never fails, and in the mightiest floods . . . it never overflows. Its banks and its bottoms are of cold water, while its current is of warm. The Gulf of Mexico is its fountain, and its mouth is in the Arctic Sea. It is the Gulf Stream. There is in the world no other such majestic flow of waters.

Today the warm waters of the Gulf Stream can be mapped by Earth-orbiting satellites (Figure 15.3B). Such satellite maps reveal that the Gulf Stream has many complexities, including prominent bends called

---
*P. L. Richardson, 1980, "Benjamin Franklin and Timothy Folger's First Printed Chart of the Gulf Stream," *Science* 207 (4431): 643.

A.

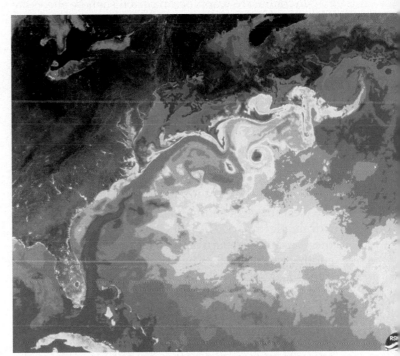

B.

▲ **FIGURE 15.3** **A.** To aid ships crossing the Atlantic, Benjamin Franklin, with the assistance of his cousin, Timothy Folger, produced the first detailed chart of the Gulf Stream about 1769 or 1770. (Courtesy NOAA) **B.** Today satellite images such as this false-color image of sea-surface temperature provide views of the Gulf Stream's complexities. The warm waters of the Gulf Stream are shown in red and orange; colder waters are shown in green, blue, and purple. As the Gulf Stream meanders northward, some of its meanders pinch off to form large circular eddies. (Courtesy of O. Brown, R. Evans, and M. Carle/University of Miami Rosenstiel School of Marine and Atmospheric Science, Miami, Florida)

meanders, which produce large circular eddies that spin off from the main current and can last for up to two years before dissipating. Although the Franklin-Folger chart was created long before the advent of modern technology, which allows navigators to map ocean currents from space, it still remains a good summary of the average path of the Gulf Stream.

Ocean Currents and Climate

Ocean currents have an important effect on climates. When currents from low-latitude regions move into higher latitudes, they transfer heat from warmer to cooler areas on Earth. In fact, the North Atlantic Current—an extension of the warm Gulf Stream—keeps Great Britain and much of northwestern Europe warmer during the winter than one would expect for their latitudes, which are similar to the latitudes of Alaska and Labrador. The prevailing westerly winds carry the moderating effects far inland. For example, Berlin, Germany (52 degrees north latitude), has an average January temperature similar to that experienced at New York City, which lies 12 degrees latitude farther south.

In contrast to warm ocean currents whose effects are felt mostly in the middle latitudes in winter, the influence of cold currents is most pronounced in the tropics or during summer months in the middle latitudes. Cold currents originate in cold high-latitude regions. As these currents travel equatorward, they tend to moderate the warm temperatures of adjacent land areas. Such is the case for the Benguela Current along western Africa, the Peru Current along the west coast of South America, and the California Current (see Figure 15.2).

In addition to influencing temperatures of adjacent land areas, cold currents have other climatic influences. For example, where tropical deserts exist along the west coasts of continents, cold ocean currents have a dramatic impact. The principal west coast deserts are the Atacama in Peru and Chile, and the Namib in southwestern Africa. The aridity along these coasts is intensified because the lower atmosphere is chilled by cold offshore waters. When this occurs, the air becomes very stable and resists the upward movement necessary to create precipitation-producing clouds. In addition, the presence of cold currents causes temperatures to approach and often reach the dew point, the temperature at which water vapor condenses. As a result, these areas are characterized by high relative humidities and much fog. Thus, not all tropical deserts are hot with low humidities and clear skies. Rather, the presence of cold currents transforms some tropical deserts into relatively cool, damp places that are often shrouded in fog.

Ocean currents also play a major role in maintaining Earth's heat balance. They accomplish this task by transferring heat from the tropics, where there is an excess of heat, to the polar regions, where a heat deficit exists. Ocean water movement accounts for about a quarter of this heat transport, and winds transport the remaining three-quarters.

Upwelling

In addition to producing surface currents, winds can also cause *vertical* water movements. **Upwelling**, the rising of cold water from deeper layers to replace

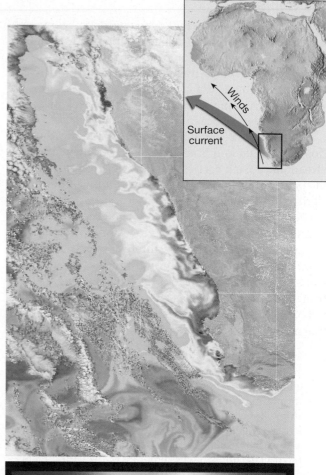

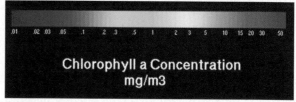

**Chlorophyll a Concentration
mg/m3**

▲ **FIGURE 15.4** Coastal upwelling occurs along the west coasts of continents where winds blow toward the equator and parallel to the coast. Owing to the Coriolis effect (deflection to the left in the Southern Hemisphere), surface water moves away from the shore, which brings cold, nutrient-rich water to the surface. This image from the SeaStar satellite shows chlorophyll concentration along the southwest coast of Africa (February 21, 2001). An instrument aboard the satellite detects changes in seawater color caused by changing concentrations of chlorophyll. High chlorophyll concentrations indicate high amounts of photosynthesis, which is linked to the upwelling nutrients. Red indicates high concentrations and blue, low concentrations. (Provided by the SeaWiFS Project, NASA/Goddard Space Flight Center and ORBIMAGE)

warmer surface water, is a common wind-induced vertical movement. One type of upwelling, called *coastal upwelling*, is most characteristic along the west coasts of continents, most notably along California, western South America, and West Africa.

Coastal upwelling occurs in these areas when winds blow toward the equator and parallel to the coast (Figure 15.4). Coastal winds combined with the Coriolis effect cause surface water to move away from shore. As the surface layer moves away from the coast, it is replaced by water that "upwells" from below the surface. This slow upward movement of water from depths of 50 to 300 meters (165 to 1000 feet) brings water that is cooler than the original surface water and results in lower surface water temperatures near the shore.

For swimmers who are accustomed to the warm waters along the mid-Atlantic shore of the United States, a swim in the Pacific off the coast of central California can be a chilling surprise. In August, when temperatures in the Atlantic are 21°C (70°F) or higher, central California's surf is only about 15°C (60°F).

Upwelling brings greater concentrations of dissolved nutrients, such as nitrates and phosphates, to the ocean surface. These nutrient-enriched waters from below promote the growth of microscopic plankton, which in turn support extensive populations of fish and other marine organisms. Figure 15.4 is a satellite image that shows high productivity due to coastal upwelling off the southwest coast of Africa.

Deep-Ocean Circulation

In contrast to the largely horizontal movements of surface currents, deep-ocean circulation has a significant vertical component and accounts for the thorough mixing of deep-water masses. This component of ocean circulation is a response to density differences among water masses that cause denser water to sink and slowly spread out beneath the surface. Because the density variations that cause deep-ocean circulation are caused by differences in temperature and salinity, deep-ocean circula-

tion is also referred to as **thermohaline** (*thermo* = heat, *haline* = salt), **circulation**.

Recall from Chapter 14 that an increase in seawater density can be caused by a decrease in temperature or an increase in salinity. Density changes due to salinity variations are important in very high latitudes, where water temperature remains low and relatively constant.

Most water involved in deep-ocean currents (thermohaline circulation) begins in high latitudes at the surface. In these regions, surface water becomes cold and its salinity increases as sea ice forms (Figure 15.5). When this surface water becomes dense enough, it sinks, initiating deep-ocean currents. Once this water sinks, it is removed from the physical processes that increased its density in the first place, and so its temperature and salinity remain largely unchanged for the duration it spends in the deep ocean.

Near Antarctica, surface conditions create the highest density water in the world. This cold saline brine slowly sinks to the seafloor, where it moves throughout the ocean basins in sluggish currents. After sinking from the surface of the ocean, deep waters will not reappear at the surface for an average of 500 to 2000 years.

A simplified model of ocean circulation is similar to a conveyor belt that travels from the Atlantic Ocean through the Indian and Pacific Oceans and back again (Figure 15.6). In this model, warm water in the ocean's upper layers flows poleward, converts to dense water, and returns equatorward as cold deep water that eventually upwells to complete the circuit. As this "conveyor belt" moves around the globe, it influences global climate by converting warm water to cold and liberating heat to the atmosphere.

◀ **FIGURE 15.5** Sea ice in the Arctic Ocean. When seawater freezes, sea salts do not become part of the ice. Consequently, the salt content of the remaining seawater becomes more concentrated, which makes it denser and prone to sink. (Photo by Wayne Lynch/DRK Photo)

▶ **FIGURE 15.6** Idealized "conveyor belt" model of ocean circulation, which is initiated in the North Atlantic Ocean when warm water transfers its heat to the atmosphere, cools, and sinks below the surface. This water moves southward as a subsurface flow and joins water that encircles Antarctica. From here, this deep water spreads into the Indian and Pacific Oceans, where it slowly rises and completes the conveyer as it travels along the surface into the North Atlantic Ocean.

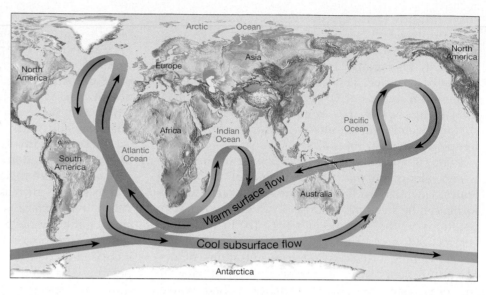

The Shoreline: A Dynamic Interface

Nowhere is the restless nature of the ocean's water more noticeable than along the shore—the dynamic interface among air, land, and sea. An *interface* is a common boundary where different parts of a system interact. This is certainly an appropriate designation for the coastal zone. Here we can see the rhythmic rise and fall of tides and observe waves rolling in and breaking. Sometimes the waves are low and gentle. At other times they pound the shore with awesome fury.

Shorelines are dynamic environments. Their topography, geologic makeup, and climate vary greatly from place to place. Continental and oceanic processes converge along coasts to create landscapes that frequently undergo rapid change. When it comes to the deposition of sediment, they are transition zones between marine and continental environments.

Although it may not be obvious, the shoreline is constantly being modified by waves. Crashing surf can erode the adjacent land (see chapter opening photo). Wave activity also moves sediment toward and away from the shore as well as along it. Such activity sometimes produces narrow sandbars that frequently change size and shape as storm waves come and go.

The nature of present-day shorelines is not just the result of the relentless attack of the land by the sea. Indeed, the shore has a complex character that results from multiple geologic processes. For example, practically all coastal areas were affected by the worldwide rise in sea level that accompanied the melting of glaciers at the close of the Pleistocene epoch. As the sea encroached landward, the shoreline retreated, becoming superimposed upon existing landscapes that had resulted from such diverse processes as stream erosion, glaciation, volcanic activity, and the forces of mountain building.

Today the coastal zone is experiencing intensive human activity. Unfortunately, people often treat the shoreline as if it were a stable platform on which structures can safely be built. This attitude inevitably leads to conflicts between people and nature. As you will see, many coastal landforms, especially beaches and barrier islands, are relatively fragile, short-lived features that are inappropriate sites for development.

The Coastal Zone

In general conversation a number of terms are used when referring to the boundary between land and sea. In the preceding section, the terms *shore, shoreline, coastal zone*, and *coast* were all used. Moreover, when many think of the land-sea interface, the word *beach* comes to mind. Let's take a moment to clarify these terms and introduce some other terminology used by those who study the land-sea boundary zone. You will find it helpful to refer to Figure 15.7, which is an idealized profile of the coastal zone.

The **shoreline** is the line that marks the contact between land and sea. Each day, as tides rise and fall, the position of the shoreline migrates. Over longer time spans, the average position of the shoreline gradually shifts as sea level rises or falls.

The **shore** is the area that extends between the lowest tide level and the highest elevation on land that is affected by storm waves. By contrast, the coast extends inland from the shore as far as ocean-related features can be found. The **coastline** marks the coast's seaward edge, whereas the inland boundary is not always obvious or easy to determine.

As Figure 15.7 illustrates, the shore is divided into the *foreshore* and the *backshore*. The **foreshore** is the area exposed when the tide is out (low tide) and sub-

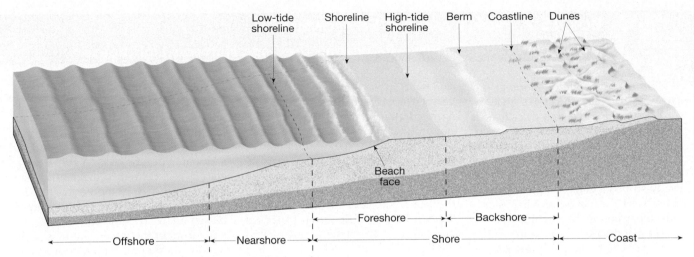

▲ FIGURE 15.7 The coastal zone consists of several parts. The beach is an accumulation of sediment on the landward margin of the ocean or a lake. It can be thought of as material in transit along the shore.

merged when the tide is in (high tide). The **backshore** is landward of the high-tide shoreline. It is usually dry, being affected by waves only during storms. Two other zones are commonly identified. The **nearshore zone** lies between the low-tide shoreline and the line where waves break at low tide. Seaward of the nearshore zone is the **offshore zone**.

For many a beach is the sandy area where people lie in the sun and walk along the water's edge. Technically, a **beach** is an accumulation of sediment found along the landward margin of the ocean or a lake. Along straight coasts, beaches may extend for tens or hundreds of kilometers. Where coasts are irregular, beach formation may be confined to the relatively quiet waters of bays.

Beaches consist of one or more **berms**, which are relatively flat platforms often composed of sand that are adjacent to coastal dunes or cliffs and marked by a change in slope at the seaward edge. Another part of the beach is the **beach face**, which is the wet sloping surface that extends from the berm to the shoreline. Where beaches are sandy, sunbathers usually prefer the berm, whereas joggers prefer the wet, hard-packed sand of the beach face.

Beaches are composed of whatever material is locally abundant. The sediment for some beaches is derived from the erosion of adjacent cliffs or nearby coastal mountains. Other beaches are built from sediment delivered to the coast by rivers.

Although the mineral makeup of many beaches is dominated by durable quartz grains, other minerals may be dominant. For example, in areas such as southern Florida, where there are no mountains or other sources of rock-forming minerals nearby, most beaches are composed of shell fragments and the remains of organisms that live in coastal waters. Some beaches on volcanic islands in the open ocean are composed of weathered grains of the basaltic lava

that comprise the islands, or of coarse debris eroded from coral reefs that develop around islands in low latitudes.

Regardless of the composition, the material that comprises the beach does not stay in one place. Instead, crashing waves are constantly moving it. Thus, beaches can be thought of as material in transit along the shore.

Waves

 The Global Ocean
▼ Coastal Processes

Ocean waves are energy traveling along the interface between ocean and atmosphere, often transferring energy from a storm far out at sea over distances of several thousand kilometers. That's why even on calm days the ocean still has waves that travel across its surface. When observing waves, always remember that you are watching *energy* travel through a medium (water). If you make waves by tossing a pebble into a pond, or by splashing in a pool, or by blowing across the surface of a cup of coffee, you are imparting *energy* to the water, and the waves you see are just the visible evidence of the energy passing through.

Wind-generated waves provide most of the energy that shapes and modifies shorelines. Where the land and sea meet, waves that may have traveled unimpeded for hundreds or thousands of kilometers suddenly encounter a barrier that will not allow them to advance farther and must absorb their energy. Stated another way, the shore is the location where a practically irresistible force confronts an almost immovable object. The conflict that results is never-ending and sometimes dramatic (see Box 15.2).

BOX 15.2 PEOPLE AND THE ENVIRONMENT:
The Move of the Century—Relocating the Cape Hatteras Lighthouse

In spite of efforts to protect structures that are too close to the shore, they can still be in danger of being destroyed by receding shorelines and the destructive power of waves. Such was the case for one of the nation's most prominent landmarks, the candy-striped lighthouse at Cape Hatteras, North Carolina, which is 21 stories tall—the nation's tallest lighthouse.

The lighthouse was built in 1870 on the Cape Hatteras barrier island 457 meters (1500 feet) from the shoreline to guide mariners through the dangerous offshore shoals known as the "Graveyard of the Atlantic." As the barrier island began migrating toward land, its beach narrowed. When the waves began to lap just 37 meters (120 feet) from its brick and granite base, there was concern that even a moderate-strength hurricane could trigger beach erosion sufficient to topple the lighthouse.

In 1970 the U.S. Navy built three groins in front of the lighthouse in an effort to protect the lighthouse from further erosion. The groins initially slowed erosion but disrupted sand flow in the surf zone, which caused the flattening of nearby dunes and the formation of a bay south of the lighthouse. Attempts to increase the width of the beach in front of the lighthouse included beach nourishment and artificial offshore beds of seaweed, both of which failed to widen the beach substantially. In the 1980s the Army Corps of Engineers proposed building a massive stone seawall around the lighthouse but decided the eroding coast would eventually move out from under the structure, leaving it stranded at sea on its own island. In 1988 the National Academy of Sciences determined that the shoreline in front of the lighthouse would retreat so far as to destroy the lighthouse and recommended relocation of the tower as had been done with smaller lighthouses. In 1999 the National Park Service, which owns the lighthouse, finally authorized moving the structure to a safer location.

Moving the lighthouse, which weighs 4395 metric tons (4830 short tons), was accomplished by severing it from its foundation and carefully hoisting it onto a platform of steel beams fitted with roller dollies. Once on the platform, it was slowly rolled along a specially designed steel track using a series of hydraulic jacks. A strip of vegetation was cleared to make a runway along which the lighthouse traveled 1.5 meters (5 feet) at a time, with the track picked up from behind and reconstructed in front of the tower as it moved. In less than a month, the lighthouse was gingerly transported 884 meters (2900 feet) from its original location, making it one of the largest structures ever successfully moved.

After its $12 million move, the lighthouse now resides in a scrub oak and pine woodland (Figure 15.B). Although it now stands farther inland, the light's slightly higher elevation makes it visible just as far out to sea, where it continues to warn mariners of the hazardous shoals. At the current rate of shoreline retreat, the lighthouse should be safe from the threat of waves for at least another century.

Figure 15.B When North Carolina's Cape Hatteras Lighthouse was threatened by shoreline erosion in 1999, it was relocated 488 meters (1600 feet) from the shore. (Photo courtesy of Drew Wilson, *Virginian–Pilot*, copyright 1999)

Wave Characteristics

Most ocean waves derive their energy and motion from the wind. When a breeze is less than 3 kilometers (2 miles) per hour, only wavelets appear. At greater wind speeds, more stable waves gradually form and advance with the wind.

Characteristics of ocean waves are illustrated in Figure 15.8, which shows a simple, nonbreaking

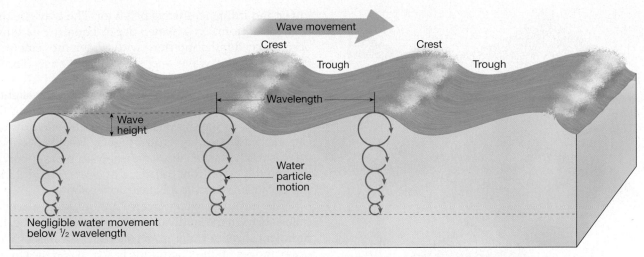

▲ FIGURE 15.8 Diagrammatic view of an idealized non-breaking ocean wave showing the basic parts of a wave as well as the movement of water particles at depth. Negligible water movement occurs below a depth equal to one half the wavelength (*lower dashed line*).

waveform. The tops of the waves are the *crests*, which are separated by *troughs*. Halfway between the crests and troughs is the *still water level*, which is the level that the water would occupy if there were no waves. The vertical distance between trough and crest is called the **wave height**, and the horizontal distance between successive crests (or troughs) is the **wavelength**. The time it takes one full wave—one wavelength—to pass a fixed position is the **wave period**.

The height, length, and period that are eventually achieved by a wave depend on three factors: (1) wind speed; (2) length of time the wind has blown; and (3) **fetch**, the distance that the wind has traveled across open water. As the quantity of energy transferred from the wind to the water increases, both the height and steepness of the waves also increase. Eventually, a critical point is reached where waves grow so tall that they topple over, forming ocean breakers called *whitecaps*.

For a particular wind speed, there is a maximum fetch and duration of wind beyond which waves will no longer increase in size. When the maximum fetch and duration are reached for a given wind velocity, the waves are said to be "fully developed." The reason that waves can grow no further is that they are losing as much energy through the breaking of whitecaps as they are receiving energy from the wind.

When the wind stops or changes direction, or the waves leave the storm area where they were created, they continue on without relation to local winds. The waves also undergo a gradual change to *swells*, which are lower in height and longer in length and may carry a storm's energy to distant shores. Because many independent wave systems exist at the same time, the sea surface acquires a complex and irregular pattern, sometimes producing very large waves. The sea waves that are seen

from shore are usually a mixture of swells from faraway storms and waves created by local winds.

Circular Orbital Motion

Waves can travel great distances across ocean basins. In one study, waves generated near Antarctica were tracked as they traveled through the Pacific Ocean basin. After more than 10,000 kilometers (over 6000 miles), the waves finally expended their energy a week later along the shoreline of the Aleutian Islands of Alaska. The water itself doesn't travel the entire distance, but the wave form does. As the wave travels, the water passes the energy along by moving in a circle. This movement is called *circular orbital motion*.

Observation of an object floating in waves reveals that it moves not only up and down but also slightly forward and backward with each successive wave. Figure 15.9 shows that a floating object moves up and backward as the crest approaches, up and forward as the crest passes, down and forward after the crest, down and backward as the trough approaches, and rises and moves backward again as the next crest advances. When the movement of the toy boat shown in Figure 15.9 is traced as a wave passes, it can be seen that the boat moves in a circle and it returns to essentially the same place. Circular orbital motion allows a waveform (the wave's shape) to move forward *through the water* while the individual water particles that transmit the wave move around in a circle. Wind moving across a field of wheat causes a similar phenomenon: The wheat itself doesn't travel across the field, but the waves do.

The energy contributed by the wind to the water is transmitted not only along the surface of the sea but also downward. However, beneath the surface the circular motion rapidly diminishes until, at a depth equal

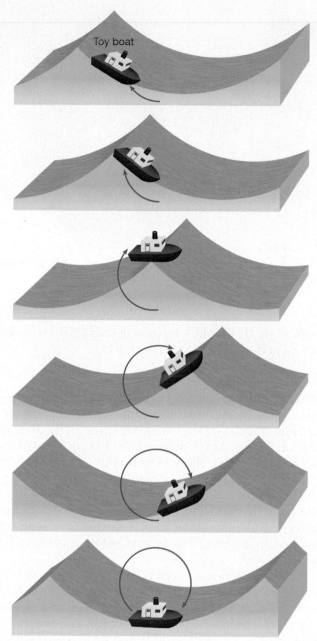

▲ **FIGURE 15.9** The movements of the toy boat show that the wave form advances, but the water does not advance appreciably from its original position. In this sequence, the wave moves from left to right as the boat (and the water in which it is floating) rotates in an imaginary circle.

to one half the wavelength measured from still water level, the movement of water particles becomes negligible. This depth is known as the *wave base*. The dramatic decrease of wave energy with depth is shown by the rapidly diminishing diameters of water-particle orbits in Figure 15.8.

Waves in the Surf Zone

As long as a wave is in deep water, it is unaffected by water depth (Figure 15.10, left). However, when a wave approaches the shore, the water becomes shal-

lower and influences wave behavior. The wave begins to "feel bottom" at a water depth equal to its wave base. Such depths interfere with water movement at the base of the wave and slow its advance (Figure 15.10, center).

As a wave advances toward the shore, the slightly faster waves farther out to sea catch up, decreasing the wavelength. As the speed and length of the wave diminish, the wave steadily grows higher. Finally, a critical point is reached when the wave is too steep to support itself and the wave front collapses, or *breaks* (Figure 15.10, right), causing water to advance up the shore.

The turbulent water created by breaking waves is called **surf**. On the landward margin of the surf zone, the turbulent sheet of water from collapsing breakers, called *swash*, moves up the slope of the beach. When the energy of the swash has been expended, the water flows back down the beach toward the surf zone as *backwash*.

STUDENTS SOMETIMES ASK...
What are tidal waves?

Tidal waves, more accurately known as *tsunami* (*tsu* = harbor, *nami* = wave), have *nothing* to do with the tides. They are long-wavelength, fast-moving, often large, and sometimes destructive waves that originate from sudden changes in the topography of the seafloor. Tsunami are initiated by underwater faulting, landslides, or volcanic eruptions. Since the mechanisms that trigger these dangerous waves are frequently seismic events, tsunami are appropriately termed *seismic sea waves*. For more information about tsunami and their destructive effects, see "Tsunami" in Chapter 7.

Wave Erosion

GEODe   The Global Ocean
▼ Coastal Processes

During calm weather, wave action is minimal. During storms, however, waves are capable of causing much erosion. The impact of large, high-energy waves against the shore can be awesome in its violence. Each breaking wave may hurl thousands of tons of water against the land, sometimes causing the ground literally to tremble. The pressures exerted by Atlantic waves in wintertime, for example, average nearly 10,000 kilograms per square meter (more than 2000 pounds per square foot). The force during storms is even greater.

It is no wonder that cracks and crevices are quickly opened in cliffs, coastal structures, and anything else that is subjected to these enormous shocks (Figure 15.11). Water is forced into every opening, causing air in the cracks to become highly compressed by the

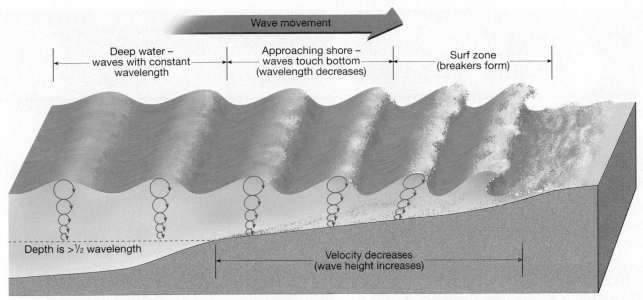

Wave movement

Deep water –
waves with constant
wavelength

Approaching shore –
waves touch bottom
(wavelength decreases)

Surf zone
(breakers form)

Depth is >½ wavelength

Velocity decreases
(wave height increases)

▲ **FIGURE 15.10**   Changes that occur when a wave moves onto shore. The waves touch bottom as they encounter water depths less than half a wavelength. The wave speed decreases, and the waves stack up against the shore, causing the wavelength to decrease. This results in an increase in wave height to the point where the waves pitch forward and break in the surf zone.

thrust of crashing waves. When the wave subsides, the air expands rapidly, dislodging rock fragments and enlarging and extending fractures.

In addition to the erosion caused by wave impact and pressure, **abrasion**—the sawing and grinding action of the water armed with rock fragments—is also important. In fact, abrasion is probably more intense in the surf zone than in any other environment. Smooth, rounded stones and pebbles along the shore are obvious reminders of the relentless grinding action of rock against rock in the surf zone (Figure 15.12A). Such fragments are also used as "tools" by the waves as they cut horizontally into the land (Figure 15.12B).

Moreover, waves are very effective at breaking down rock material and supplying sand to beaches.

Sand Movement on the Beach

Beaches are sometimes called "rivers of sand." The reason is that the energy from breaking waves often causes large quantities of sand to move along the beach face and in the surf zone roughly parallel to the shoreline. Wave energy also causes sand to move perpendicular to (toward and away from) the shoreline.

▼ **FIGURE 15.11**   The force of breaking waves can be powerful. Here waves batter a seawall at Sea Bright, New Jersey. A seawall 5 to 6 meters (16 to 20 feet) high and 8 kilometers (5 miles) long was built to protect the town and the railroad that brought tourists to the beach. As you can see, after the wall was built, the beach narrowed dramatically. (Photo by Rafael Macia/Photo Researchers, Inc.)

A.

B.

▲ **FIGURE 15.12 A.** Abrasion can be intense in the surf zone. Smooth rounded stones along the shore are an obvious reminder of this fact. Garrapata State Park, California. (Photo © by Carr Clifton) **B.** Sandstone cliff undercut by wave erosion at Gabriola Island, British Columbia, Canada. (Photo by Fletcher and Baylis/Photo Researchers, Inc.)

Movement Perpendicular to the Shoreline

If you stand ankle-deep in water at the beach, you will see that swash and backwash move sand toward and away from the shoreline. Whether there is a net loss or addition of sand depends on the level of wave activity. When wave activity is relatively light (less energetic waves), much of the swash soaks into the beach, which reduces the backwash. Consquently, the swash dominates and causes a net movement of sand up the beach face toward the berm.

When high-energy waves prevail, the beach is saturated from previous waves, so much less of the swash soaks in. As a result, the berm erodes because backwash is strong and causes a net movement of sand down the beach face.

Along many beaches, light wave activity is the rule during the summer. Therefore, a wide sand berm gradually develops. During winter, when storms are frequent and more powerful, strong wave activity erodes and narrows the berm. A wide berm that may have taken months to build can be dramatically narrowed in just a few hours by the high-energy waves created by a strong winter storm.

STUDENTS SOMETIMES ASK...

During heavy wave activity, where does the sand from the berm go?

The orbital motion of waves is too shallow to move sand very far offshore. Consequently, the sand accumulates just beyond where the surf zone ends and forms one or more offshore sandbars called longshore bars.

Wave Refraction

The bending of waves, called **wave refraction** (*refringere* = to break up) plays an important part in shoreline processes. It affects the distribution of energy along the shore and thus strongly influences where and to what degree erosion, sediment transport, and deposition will take place.

Waves seldom approach the shore straight on. Rather, most waves move toward the shore at a slight angle. However, when they reach the shallow water of a smoothly sloping bottom, the wave crests are refracted (bent) and tend to line up nearly parallel to the shore. Such bending occurs because the part of the wave nearest the shore touches bottom and slows first, whereas the part of the wave that is still in deep water continues forward at its full speed. This causes wave crests to become nearly parallel to the shore regardless of their original orientation.

Because of refraction, wave energy is concentrated against the sides and ends of headlands that project into the water, whereas wave attack is weakened in bays. This differential wave attack along irregular coastlines is illustrated in Figure 15.13. Because the waves reach the shallow water in front of the headland sooner than they do in adjacent bays, they are bent more nearly parallel to the protruding land and strike it from all three sides. By contrast, refraction in the bays causes waves to diverge and expend less energy. In these zones of weakened wave activity, sediments can accumulate and form sandy beaches. Over a long period, erosion of the headlands and deposition in the bays will straighten an irregular shoreline.

Longshore Transport

Although waves are refracted, most still reach the shore at a slight angle. Consequently, the uprush of water from each breaking wave (the swash) is at an oblique angle to the shoreline. However, the backwash is straight down the slope of the beach. The effect of this pattern of water movement is to transport sediment in a zigzag pattern along the beach face (Figure 15.14). This movement is called **beach drift**, and it can transport sand and pebbles hundreds or even thousands of meters daily. However, a more typical rate is 5 to 10 meters per day.

Oblique waves also produce currents within the surf zone that flow parallel to the shore and move substantially more sediment than beach drift (Figure 15.14). Because the water here is turbulent, these **longshore currents** easily move the fine suspended sand and roll larger sand and gravel along the bottom. When the sediment transported by longshore currents is added to the quantity moved by beach drift, the total amount can be very large. At Sandy Hook, New Jersey, for example, the quantity of sand transported along the shore over a 48-year period averaged almost 680,000 metric tons (750,000 short tons) annually. For a 10-year period at Oxnard, California, more than 1.4 million metric tons (1.5 million short tons) of sediment moved along the shore each year.

Both rivers and coastal zones move water and sediment from one area (*upstream*) to another (*downstream*). As a result, the beach has often been characterized as a "river of sand." Beach drift and longshore currents, however, move in a zigzag pattern, whereas rivers flow mostly in a turbulent, swirling fashion. Additionally, the direction of flow of longshore currents along a shoreline can change, whereas rivers flow in the same direction (downhill). Longshore currents change direction because the direction that waves approach the beach changes seasonally. Nevertheless, longshore currents generally flow

southward along both the Atlantic and Pacific shores of the United States

Shoreline Features

GEODe The Global Ocean
▼ Coastal Processes

A fascinating assortment of shoreline features can be observed along the world's coastal regions. These shoreline features vary depending on the type of rocks exposed along the shore, the intensity of waves, the nature of coastal currents, and whether the coast is stable, sinking, or rising. Features that owe their origin primarily to the work of erosion are called *erosional features*, while deposits of sediment produce *depositional features*.

Erosional Features

Many coastal landforms owe their origin to erosional processes. Such erosional features are common along the rugged and irregular New England coast and along the steep shorelines of the West Coast of the United States.

Wave-Cut Cliffs, Wave-Cut Platforms, and Marine Terraces. **Wave-cut cliffs**, as the name implies, originate by the cutting action of the surf against the base of coastal land. As erosion progresses, rocks overhanging the notch at the base of the cliff crumble into the surf and the cliff retreats. A relatively flat, benchlike surface, called a **wave-cut platform**, is left behind by the receding cliff (Figure 15.15, left). The platform broadens as wave attack continues. Some debris produced by the breaking waves remains along the water's edge as sediment on the beach, while the remainder is transported farther seaward. If a wave-cut platform is uplifted above sea level by tectonic forces, it becomes a

▶ **FIGURE 15.13** Wave refraction along an irregular coastline. As waves first touch bottom in the shallows off the headlands, they are slowed, causing the waves to refract (bend) and align nearly parallel to the shoreline. This causes wave energy to be concentrated at headlands (resulting in erosion) and dispersed in bays (resulting in deposition).

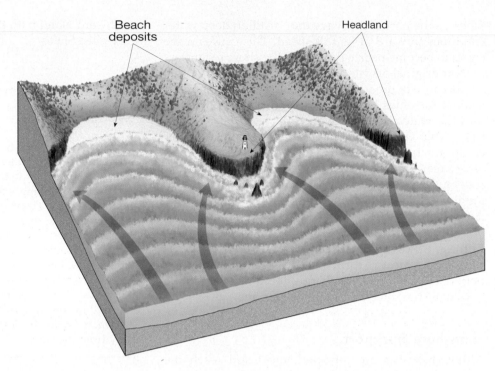

marine terrace (Figure 15.15, right). Marine terraces are easily recognized by their gentle seaward-sloping shape and are often desirable sites for coastal roads, buildings, or agriculture.

Sea Arches and Sea Stacks. Headlands that extend into the sea are vigorously attacked by waves because of refraction. The surf erodes the rock selectively, wearing away the softer or more highly fractured rock at the fastest rate. At first, sea caves may form. When two caves on opposite sides of a headland unite, a **sea arch** results (Figure 15.16). Eventually, the arch falls in, leaving an isolated remnant, or **sea stack**, on the wave-cut platform (Figure 15.16). In time, it too will be consumed by the action of the waves.

Depositional Features

Sediment eroded from the beach is transported along the shore and deposited in areas where wave energy is low. Such processes produce a variety of depositional features.

Spits, Bars, and Tombolos. Where beach drift and longshore currents are active, several features related to the movement of sediment along the shore may develop. A **spit** (*spit* = spine) is an elongated ridge of

▶ **FIGURE 15.14** Beach drift and longshore currents are created by obliquely breaking waves. Beach drift occurs as incoming waves carry sand obliquely up the beach, while the water from spent waves carries it directly down the slope of the beach. Similar movements occur offshore in the surf zone to create the longshore current. These processes transport large quantities of material along the beach and in the surf zone.

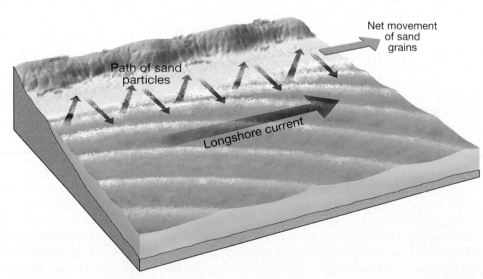

▲ **FIGURE 15.15** Wave-cut platform and marine terrace. A wave-cut platform is exposed at low tide along the California coast at Bolinas Point near San Francisco. On the right is an elevated wave-cut platform, called a marine terrace. (Photo by John S. Shelton)

sand that projects from the land into the mouth of an adjacent bay. Often the end in the water hooks landward in response to the dominant direction of the longshore current (Figure 15.17). The term **baymouth bar** is applied to a sandbar that completely crosses a bay, sealing it off from the open ocean (Figure 15.17). Such a feature tends to form across bays where currents are weak, allowing a spit to extend to the other side. A **tombolo** (*tombolo* = mound), a ridge of sand that connects an island to the mainland or to another island, forms in much the same manner as a spit.

Barrier Islands. The Atlantic and Gulf Coastal Plains are relatively flat and slope gently seaward. The shore zone is characterized by **barrier islands**. These low

ridges of sand parallel the coast at distances from 3 to 30 kilometers (1.9 to 19 miles) offshore. From Cape Cod, Massachusetts, to Padre Island, Texas, nearly 300 barrier islands rim the coast (Figure 15.18).

Most barrier islands are from 1 to 5 kilometers (0.6 to 3 miles) wide and between 15 and 30 kilometers (9 to 18 miles) long. The tallest features are sand dunes, which usually reach heights of 5 to 10 meters (16 to 33 feet). The lagoons that separate these narrow islands from the shore are zones of relatively quiet water that allow small craft traveling between New York and northern Florida to avoid the rough waters of the North Atlantic.

Barrier islands probably formed in several ways. Some originated as spits that were subsequently

◀ **FIGURE 15.16** Sea arch and sea stack at the tip of Mexico's Baja Peninsula. (Photo by Mark A. Johnson/Corbis/Stock Market)

419

▲ **FIGURE 15.17** High-altitude image of a well-developed spit and baymouth bar along the coast of Martha's Vineyard, Massachusetts. Also notice the tidal delta in the lagoon adjacent to the inlet through the baymouth bar. (Photo courtesy of USDA-ASCS)

severed from the mainland by wave erosion or by the general rise in sea level following the last episode of glaciation. Others were created when turbulent waters in the line of breakers heaped up sand that had been scoured from the bottom. Finally, some barrier islands

may be former sand-dune ridges that originated along the shore during the last glacial period, when sea level was lower. As the ice sheets melted, sea level rose and flooded the area behind the beach-dune complex.

The Evolving Shore

A shoreline continually undergoes modification regardless of its initial configuration. At first most coastlines are irregular, although the degree of and reason for the irregularity may differ considerably from place to place. Along a coastline that is characterized by varied geology, the pounding surf may initially increase its irregularity because the waves will erode the weaker rocks more easily than the stronger ones. However, if a shoreline remains stable, marine erosion and deposition will eventually produce a straighter, more regular coast.

Figure 15.19 illustrates the evolution of an initially irregular coast that remains relatively stable and shows many of the coastal features discussed in the previous section. As headlands are eroded and erosional features such as wave-cut cliffs and wave-cut platforms are created, sediment is produced that is carried along the shore by beach drift and longshore currents. Some material is deposited in the bays, while other debris is formed into depositional features such as spits and baymouth bars. At the same time, rivers fill the bays with sediment. Ultimately, a smooth coast results.

▼ **FIGURE 15.18** Nearly 300 barrier islands rim the Gulf and Atlantic Coasts. The islands along the south Texas coast and along the coast of North Carolina are excellent examples.

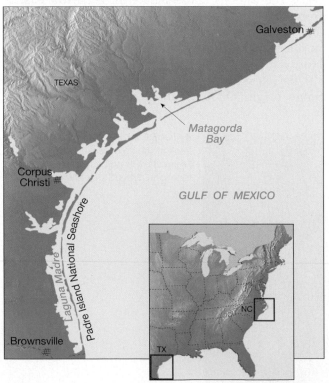

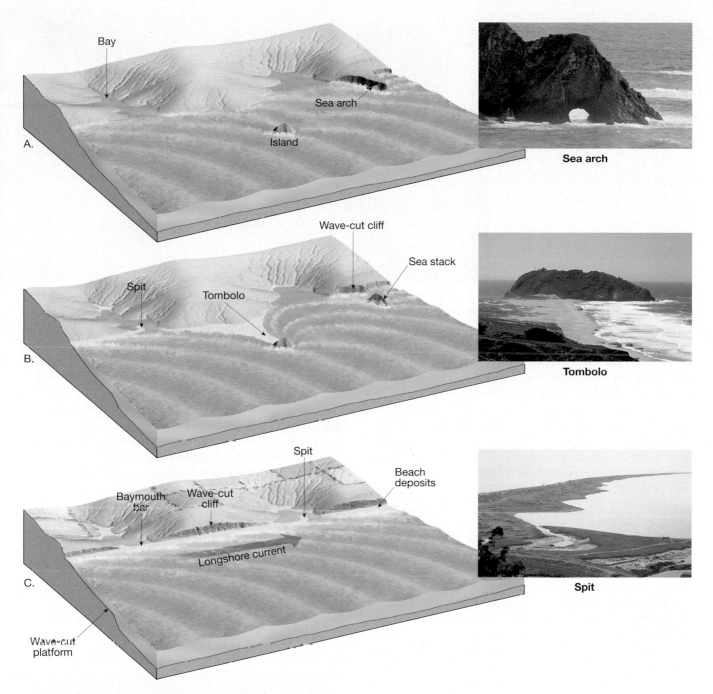

▲ **FIGURE 15.19**   These diagrams illustrate the changes that can take place through time along an initially irregular coastline that remains relatively stable. The coastline shown in part **A** gradually evolves to **B**, and then **C**. The diagrams also serve to illustrate many of the features described in the section on shoreline features. (Photos by E. J. Tarbuck)

Stabilizing the Shore

The Global Ocean
▼ Coastal Processes

Today the coastal zone teems with human activity. Unfortunately, people often treat the shoreline as if it were a stable platform on which structures can be built safely. This approach jeopardizes both people and the shoreline because many coastal landforms are relative-ly fragile, short-lived features that are easily damaged by development. As anyone who has endured a strong coastal storm knows, the shoreline is not always a safe place to live (Figure 15.20).

Compared with other natural hazards, such as earthquakes, volcanic eruptions, and landslides, shoreline erosion appears to be a more continuous and predictable process that causes relatively modest damage to limited areas. In reality, the shoreline is one of Earth's most dynamic places that changes rapidly in

► **FIGURE 15.20** Erosion caused by strong storms forced abondonment of this highly developed shoreline area in Long Island, New York. Coastal areas are dynamic places that can change rapidly in response to natural forces. (Photo by Mark Wexler/Woodfin Camp & Associates)

response to natural forces. Storms, for example, are capable of eroding beaches and cliffs at rates that far exceed the long-term average. Such bursts of accelerated erosion not only have a significant impact on the natural evolution of a coast but can also have a profound impact on people who reside in the coastal zone. Erosion along the coast causes significant property damage. Huge sums are spent annually not only to repair damage but also to prevent or control erosion. Already a problem at many sites, shoreline erosion is certain to become increasingly serious as extensive coastal development continues.

Although the same processes cause change along every coast, not all coasts respond in the same way. Interactions among different processes and the relative importance of each process depend on local factors. The factors include: (1) the proximity of a coast to sediment-laden rivers, (2) the degree of tectonic activity, (3) the topography and composition of the land, (4) prevailing winds and weather patterns, and (5) the configuration of the coastline and nearshore areas.

During the past 100 years, growing affluence and increasing demands for recreation have brought unprecedented development to many coastal areas. As both the number and the value of buildings have increased, so too have efforts to protect property from storm waves by stabilizing the shore.

Hard Stabilization

Structures built to protect a coast from erosion or to prevent the movement of sand along a beach are known as **hard stabilization**. Hard stabilization can take many forms and often results in predictable yet unwanted outcomes. Hard stabilization includes groins, breakwaters, and seawalls.

Groins. To maintain or widen beaches that are losing sand, groins are sometimes constructed. A **groin** (*groin* = ground) is a barrier built at a right angle to the beach to trap sand that is moving parallel to the shore. Groins are usually constructed of large rocks but may also be composed of wood. These structures often do their job so effectively that the longshore current beyond the groin becomes sand-starved. As a result, the current erodes sand from the beach on the downstream side of the groin.

To offset this effect, property owners downstream from the structure may erect a groin on their property. In this manner, the number of groins multiplies, resulting in a *groin field* (Figure 15.21). An example of such proliferation is the shoreline of New Jersey, where hundreds of these structures have been built. Because it has been shown that groins often do not provide a satisfactory solution, they are no longer the preferred method of keeping beach erosion in check.

▼ **FIGURE 15.21** A series of groins along the shoreline near Chichester, Sussex, England. Because groins trap sand on the upcurrent side, the movement of sand along this coast caused by the longshore current must be from lower right toward upper left. (Photo by Sandy Stockwell/London Aerial Photo Library/CORBIS)

STUDENTS SOMETIMES ASK...
Is it true that sea level is rising due to global warming?

Yes. One probable impact of a human-induced global warming is a rise in sea level. Some of the rise comes from water added by shrinking glaciers. A more significant factor is that a warmer atmosphere causes an increase in ocean volume due to thermal expansion. Higher air temperatures warm the upper layers of the ocean, which causes the water to expand and sea level to rise. For more on this, see the discussion of "Human Impact on Global Climate" in Chapter 20.

Breakwaters and Seawalls. Hard stabilization can also be built parallel to the shoreline. One such structure is a **breakwater**, the purpose of which is to protect boats from the force of large breaking waves by creating a quiet water zone near the shore. However, when this is done, the reduced wave activity along the shore behind the structure may allow sand to accumulate. If this happens, the boat anchorage will eventually fill with sand, while the downstream beach erodes and retreats. At Santa Monica, California, where the building of a breakwater created such a problem, the city uses a dredge to remove sand from the protected quiet water zone and deposit it farther downstream where longshore currents continue to move the sand down the coast (Figure 15.22).

Another type of hard stabilization built parallel to the shore is a **seawall**, which is designed to armor the coast and defend property from the force of breaking waves (see Figure 15.11, p. 415). Waves ex-

▼ **FIGURE 15.22** Aerial view of a breakwater at Santa Monica, California. The breakwater appears as a faint line in the water behind which many boats are anchored. The construction of the breakwater disrupted longshore transport and caused the seaward growth of the beach. (Photo by John S. Shelton)

pend much of their energy as they move across an open beach. Seawalls cut this process short by reflecting the force of unspent waves seaward. As a consequence, the beach to the seaward side of the seawall experiences significant erosion and may, in some instances, be eliminated entirely. Once the width of the beach is reduced, the seawall is subjected to even greater pounding by the waves. Eventually this battering takes its toll on the seawall such that the seawall will fail and a larger, more expensive structure must be built to take its place.

The wisdom of building temporary protective structures along shorelines is increasingly questioned. The opinion of many coastal scientists and engineers is expressed in the following excerpt from a position paper that was developed as a result of a conference on America's Eroding Shoreline:

> It is now clear that halting the receding shoreline with protective structures benefits only a few and seriously degrades or destroys the natural beach and the value it holds for the majority. Protective structures divert the ocean's energy temporarily from private properties, but usually refocus the energy on the adjacent natural beaches. Many interrupt the natural sand flow in coastal currents, robbing many beaches of vital sand replacement.*

Alternatives to Hard Stabilization

Armoring the coast with hard stabilization has several potential drawbacks, including the cost of the structure and the loss of sand on the beach. Alternatives to hard stabilization include beach nourishment and relocation.

Beach Nourishment. **Beach nourishment** represents an approach to stabilizing shoreline sands without hard stabilization. As the term implies, this practice simply involves the addition of large quantities of sand to the beach system (Figure 15.23). By building the beaches seaward, beach quality and storm protection are both improved. Beach nourishment, however, is not a permanent solution to the problem of shrinking beaches because the same processes that removed the sand in the first place will eventually remove the replacement sand as well. In addition, it can be very expensive because huge volumes of sand must be transported to the beach from offshore areas, nearby rivers, or other source areas.

In some instances, beach nourishment can lead to unwanted environmental effects. For example, beach replenishment at Waikiki Beach, Hawaii, involved replacing the natural coarse calcareous beach sand with softer, muddier sand. Destruction of the softer sand by breaking waves increased the water's turbidity and killed offshore coral reefs. Similar damage to local

*"Strategy for Beach Preservation Proposed," *Geotimes* 30 (No. 12, December 1985): 15.

A.   B.

▲ FIGURE 15.23   Miami Beach. **A.** Before beach nourishment and **B.** After beach nourishment. (Courtesy of the U.S. Army Corps of Engineers, Vicksburg District)

coral communities occurred after beach replenishment at Miami Beach.

Beach nourishment appears to be an economically viable long-range solution to the beach-preservation problem only in areas where there exists dense development, large supplies of sand, relatively low wave energy, and reconcilable environmental issues. Unfortunately, few areas possess all these attributes.

Relocation. Instead of building structures such as groins and seawalls to hold the shoreline in place or adding sand to replenish eroding beaches, another option is also available. Many coastal scientists and planners are calling for a policy shift from defending and rebuilding beaches and coastal property in high hazard areas to *relocating* storm-damaged or at-risk buildings (Box 15.2) and letting nature reclaim the beach. This approach is similar to that adopted by the federal government for river floodplains following the devastating 1993 Mississippi River floods in which vulnerable structures are abandoned and relocated on higher, safer ground.

Such proposals, of course, are controversial. People with significant shoreline investments shudder at the thought of not rebuilding and defending coastal developments from the erosional wrath of the sea. Others, however, argue that with sea level rising, the impact of coastal storms will only get worse in the decades to come. This group advocates that oft-damaged structures be relocated or abandoned to improve personal safety and to reduce costs. Such ideas will no doubt be the focus of much study and debate as states and communities evaluate and revise coastal land-use policies.

Erosion Problems Along U.S. Coasts

The shoreline along the Pacific Coast of the United States is strikingly different from that characterizing the Atlantic and Gulf Coast regions. Some of the differences are related to plate tectonics. The West Coast represents the leading edge of the North American plate, and because of this, it experiences active uplift and deformation. By contrast, the East Coast is a tectonically quiet region that is far from any active plate margin. Because of this basic geological difference, the nature of shoreline erosion problems along America's opposite coasts is different.

Atlantic and Gulf Coasts.   Much of the coastal development along the Atlantic and Gulf coasts has occurred on barrier islands. Typically, barrier islands consist of a wide beach that is backed by dunes and separated from the mainland by marshy lagoons. The broad expanses of sand and exposure to the ocean have made barrier islands exceedingly attractive sites for development. Unfortunately, development has taken place more rapidly than has our understanding of barrier island dynamics.

Because barrier islands face the open ocean, they receive the full force of major storms that strike the coast. When a storm occurs, the barriers absorb the energy of the waves primarily through the movement of sand. This process and the dilemma that results have been described as follows:

> Waves may move sand from the beach to offshore areas or, conversely, into the dunes; they may erode the dunes, depositing sand onto the beach or carrying it out to sea; or they may carry sand from the beach and the dunes into the marshes behind the barrier, a process known as overwash. The common factor is movement. Just as a flexible reed may survive a wind that destroys an oak tree, so the barriers survive hurricanes and nor'easters not through unyielding strength but by giving before the storm.

424

This picture changes when a barrier is developed for homes or a resort. Storm waves that previously rushed harmlessly through gaps between the dunes now encounter buildings and roadways. Moreover, since the dynamic nature of the barriers is readily perceived only during storms, homeowners tend to attribute damage to a particular storm, rather than to the basic mobility of coastal barriers. With their homes or investments at stake, local residents are more likely to seek to hold the sand in place and the waves at bay than to admit that development was improperly placed to begin with.*

Pacific Coast. In contrast to the broad, gently sloping coastal plains of the Atlantic and Gulf coasts, much of the Pacific Coast is characterized by relatively narrow beaches that are backed by steep cliffs and mountain ranges. Recall that America's western margin is a more rugged and tectonically active region than the eastern margin. Because uplift continues, the rise in sea level in the West is not so readily apparent. Nevertheless, like the shoreline erosion problems facing the East's barrier islands, West Coast difficulties also stem largely from the alteration of natural systems by people.

A major problem facing the Pacific shoreline—particularly along southern California—is a significant narrowing of many beaches. The bulk of the sand on many of these beaches is supplied by rivers that transport it from the mountainous regions to the coast. Over the years this natural flow of material to the coast has been interrupted by dams built for irrigation and flood control. The reservoirs effectively trap the sand that would otherwise nourish the beach environment. When the beaches were wider, they served to protect the cliffs behind them from the force of storm waves. Now, however, the waves move across the narrowed beaches without losing much energy and cause more rapid erosion of the sea cliffs.

Although the retreat of the cliffs provides material to replace some of the sand impounded behind dams, it also endangers homes and roads built on the bluffs. In addition, development atop the cliffs aggravates the problem. Urbanization increases runoff, which, if not carefully controlled, can result in serious bluff erosion. Watering lawns and gardens adds significant quantities of water to the slope. This water percolates downward toward the base of the cliff, where it may emerge in small seeps. This action reduces the slope's stability and facilitates mass wasting.

Shoreline erosion along the Pacific Coast varies considerably from one year to the next, largely because of the sporadic occurrence of storms. As a consequence, when the infrequent but serious episodes of erosion occur, the damage is often blamed on the unusual storms and not

on coastal development or the sediment-trapping dams that may be great distances away. If, as predicted, sea level rises at an increasing rate in the years to come, increased shoreline erosion and sea-cliff retreat should be expected along many parts of the Pacific Coast.

Coastal Classification

The great variety of shorelines demonstrates their complexity. Indeed, to understand any particular coastal area, many factors must be considered, including rock types, size and direction of waves, frequency of storms, tidal range, and offshore topography. In addition, practically all coastal areas were affected by the worldwide rise in sea level that accompanied the melting of Ice Age glaciers at the close of the Pleistocene epoch. Finally, tectonic events that uplift or downdrop the land or change the volume of ocean basins must be taken into account. The large number of factors that influence coastal areas make shoreline classification difficult.

Many geologists classify coasts based on changes that have occurred with respect to sea level. This commonly used classification system divides coasts into two very general categories: emergent and submergent. **Emergent coasts** develop either because an area experiences uplift or as a result of a drop in sea level. Conversely, **submergent coasts** are created when sea level rises or the land adjacent to the sea subsides.

Emergent Coasts

In some areas, the coast is clearly emergent because rising land or falling water levels expose wave-cut cliffs and marine terraces above sea level. Excellent examples include portions of coastal California where uplift has occurred in the recent geological past (see Figure 10.26, p. 303). The elevated wave-cut platform in Figure 15.15 also illustrates this situation. In the case of the Palos Verdes Hills, south of Los Angeles, California, seven different terrace levels exist, indicating at least seven episodes of uplift. The ever persistent sea is now cutting a new platform at the base of the cliff. If uplift follows, it too will become an elevated marine terrace.

Other examples of emergent shores include regions that were once buried beneath great ice sheets. When glaciers were present, their weight depressed the crust, and when the ice melted, the crust began gradually to spring back. Consequently, prehistoric shoreline features may now be found high above sea level. The Hudson Bay region of Canada is such an area, portions of which are still rising at a rate of more than a centimeter (0.4 inch) annually.

Submergent Coasts

In contrast to the preceding examples, other coastal areas show definite signs of submergence. Shorelines that have been submerged in the relatively recent past

---

*Frank Lowenstein, "Beaches or Bedrooms—The Choice as Sea Level Rises," *Oceanus* 28 (No. 3, Fall 1985): 22.

are often highly irregular because the sea typically floods the lower reaches of river valleys flowing into the ocean. The ridges separating the valleys, however, remain above sea level and project into the sea as headlands. These drowned river mouths, which are called **estuaries** (*aestus* = tide), characterize many coasts today. Along the Atlantic coastline, Chesapeake and Delaware bays are examples of estuaries created by submergence (Figure 15.24). The picturesque coast of Maine, particularly in the vicinity of Acadia National Park, is another excellent example of an area that was flooded by the postglacial rise in sea level and transformed into a highly irregular submerged coastline.

Keep in mind, however, that most coasts have a complicated geologic history. With respect to sea level, many coasts have at various times emerged and then submerged again. Each time they retain some of the features created during the previous situation.

Tides

Tides are daily changes in the elevation of the ocean surface. Their rhythmic rise and fall along coastlines have been known since antiquity. Other than waves, they are the easiest ocean movements to observe (Figure 15.25).

Although known for centuries, tides were not explained satisfactorily until Sir Isaac Newton applied the law of gravitation to them. Newton showed that there is a mutual attractive force between two bodies, as between Earth and the Moon. Because both the atmosphere and the ocean are fluids and are free to move, both are deformed by this force. Hence, ocean tides result from the gravitational attraction exerted upon Earth by the Moon and, to a lesser extent, by the Sun.

Causes of Tides

To illustrate how tides are produced, consider the Earth as a rotating sphere covered to a uniform depth with water (Figure 15.26). Further, ignore the effect of the Sun for now. It is easy to see how the Moon's gravitational force can cause the water to bulge on the side of Earth nearest the Moon. In addition, however, an equally large tidal bulge is produced on the side of Earth directly *opposite* the Moon.

Both tidal bulges are caused, as Newton discovered, by the pull of gravity. Gravity is inversely proportional to the square of the distance between two objects, meaning simply that it quickly weakens with distance. In this case, the two objects are the Moon and Earth. Because the force of gravity decreases with distance, the Moon's gravitational pull on Earth is slightly greater on the near side of Earth than on the far side. The result of this differential pulling is to stretch (elongate) the "solid" Earth very slightly. In contrast, the world ocean, which is mobile, is deformed quite dramatically by this effect to produce the two opposing tidal bulges.

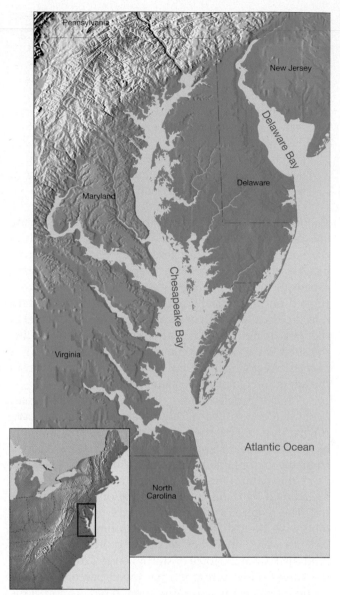

▲ FIGURE 15.24 Major estuaries along the East Coast of the United States. The lower portions of many river valleys were flooded by the rise in sea level that followed the end of the last Ice Age, creating large estuaries such as Chesapeake and Delaware bays.

STUDENTS SOMETIMES ASK...
Where are the world's largest tides?

The world's largest *tidal range* (the difference between successive high and low tides) is found in the northern end of Nova Scotia's 258-kilometer- (160-mile-) long Bay of Fundy. During maximum spring tide conditions, the tidal range at the mouth of the bay (where it opens to the ocean) is only about 2 meters (6.6 feet). However, the tidal range progressively increases from the mouth of the bay northward because the natural geometry of the bay concentrates tidal energy. In the northern end of Minas Basin, the maximum spring tidal range is about 17 meters (56 feet). This extreme tidal range leaves boats high and dry during low tide (see Figure 15.25).

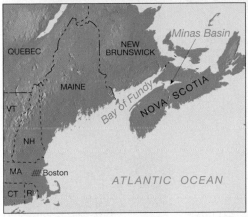

▲ **FIGURE 15.25** High tide and low tide on Nova Scotia's Minas Basin in the Bay of Fundy. The areas exposed during low tide and flooded during high tide are called *tidal flats.* Tidal flats here are extensive. (Courtesy of Nova Scotia Department of Tourism and Culture)

Because the position of the Moon changes only moderately in a single day, the tidal bulges remain in place while Earth rotates "through" them. For this reason, if you stand on the seashore for 24 hours, Earth will rotate you through alternating areas of higher and lower water. As you are carried into each tidal bulge, the tide rises, and as you are carried into the intervening troughs between the tidal bulges, the tide falls. Therefore, most places on Earth experience two high tides and two low tides each day.

In addition, the tidal bulges migrate as the Moon revolves around Earth about every 29 days. As a result, the tides—like the time of moonrise—shift about 50 minutes later each day. In essence, the tidal bulges exist in fixed positions relative to the Moon, which is slowly moving eastward.

In many locations, there may be an inequality between the high tides during a given day. Depending on the Moon's position, the tidal bulges may be inclined to the equator as in Figure 15.26. This figure illustrates that one high tide experienced by an observer in the Northern Hemisphere is considerably higher than the high tide half a day later. In contrast, a Southern Hemisphere observer would experience the opposite effect.

Monthly Tidal Cycle

The primary body that influences the tides is the Moon, which makes one complete revolution around Earth every 29 and a half days. The Sun,

▼ **FIGURE 15.26** Idealized tidal bulges on Earth caused by the Moon. If Earth were covered to a uniform depth with water, there would be two tidal bulges: one on the side of Earth facing the Moon (right) and the other on the opposite side of Earth (left). Depending on the Moon's position, tidal bulges may be inclined relative to Earth's equator. In this situation, Earth's rotation causes an observer to experience two unequal high tides during a day.

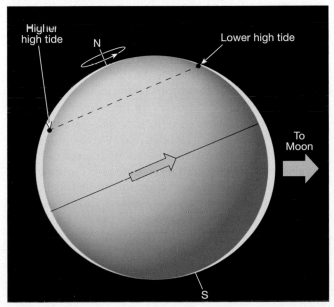

however, also influences the tides. It is far larger than the Moon, but because it is much farther away, its effect is considerably less. In fact, the Sun's tide-generating effect is only about 46 percent that of the Moon's.

Near the times of new and full moons, the Sun and Moon are aligned and their forces are added together (Figure 15.27A). Accordingly, the combined gravity of these two tide-producing bodies causes larger tidal bulges (higher high tides) and larger tidal troughs (lower low tides), producing a large tidal range. These are called the **spring** (*springen* = to rise up) **tides**, which have no connection with the spring season but occur twice a month during the time when the Earth–Moon–Sun system is aligned. Conversely, at about the time of the first and third quarters of the Moon, the gravitational forces of the Moon and Sun act on Earth at right angles, and each partially offsets the influence of the other (Figure 15.27B). As a result, the daily tidal range is less. These are called **neap** (*nep* = scarcely or barely touching) **tides**, which also occur twice each month. Each month, then, there are two spring tides and two neap tides, each about one week apart.

Tidal Patterns

So far, the basic causes and types of tides have been explained. Keep in mind, however, that these theoretical considerations cannot be used to predict either the height or the time of actual tides at a particular place. This is because many factors—including the shape of the coastline, the configuration of ocean basins, and water depth—greatly influence the tides. Consequently, tides at various locations respond differently to the tide-producing forces. This being the case, the nature of the tide at any coastal location can be determined most accurately by actual observation. The predictions in tidal tables and tidal data on nautical charts are based on such observations.

Three main tidal patterns exist worldwide. A **diurnal** (*diurnal* = daily) **tidal pattern** is characterized by a single high tide and a single low tide each tidal day (Figure 15.28). Tides of this type occur along the northern shore of the Gulf of Mexico, among other locations. A **semidiurnal** (*semi* = twice, *diurnal* = daily) **tidal pattern** exhibits two high tides and two low tides each tidal day, with the two highs about the same height and the two lows about the same height (Figure 15.28). This type of tidal pattern is common along the Atlantic Coast of the United States. A **mixed tidal pattern** is similar to a semidiurnal pattern except that it is characterized by a large inequality in high water heights, low water heights, or both (Figure 15.28). In this case, there are usually two high and two low tides each day, with high tides of different heights and low tides of

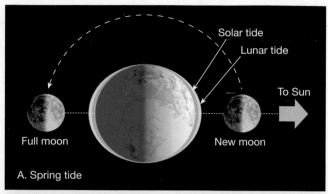

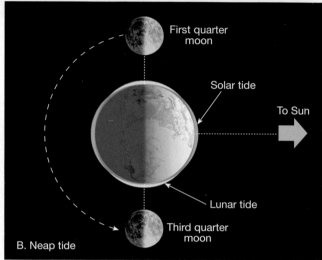

▲ **FIGURE 15.27** Earth–Moon–Sun positions and the tides. **A.** When the Moon is in the full or new position, the tidal bulges created by the Sun and Moon are aligned, there is a large tidal range on Earth, and spring tides are experienced. **B.** When the Moon is in the first- or third-quarter position, the tidal bulges produced by the Moon are at right angles to the bulges created by the Sun. Tidal ranges are smaller, and neap tides are experienced.

different heights. Such tides are prevalent along the Pacific Coast of the United States and in many other parts of the world.

Tidal Currents

Tidal current is the term used to describe the *horizontal* flow of water accompanying the rise and fall of the tides. These water movements induced by tidal forces can be important in some coastal areas. Tidal currents that advance into the coastal zone as the tide rises are called *flood currents*. As the tide falls, seaward-moving water generates *ebb currents*. Periods of little or no current, called *slack water*, separate flood and ebb. The areas affected by these alternating tidal currents are called **tidal flats** (see Figure 15.25). Depending on the nature of the coastal zone, tidal flats vary from narrow strips seaward of the beach to zones that may extend for several kilometers.

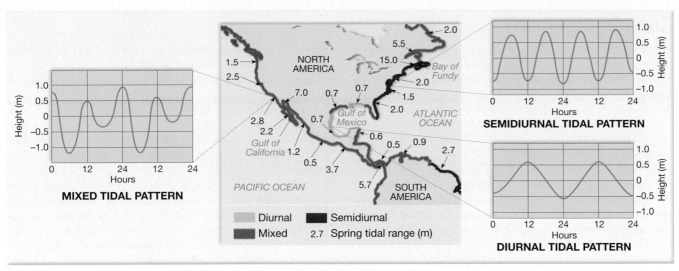

▲ **FIGURE 15.28** Tidal patterns and their occurrence along North and Central American coasts. A diurnal tidal pattern (lower right) shows one high and low tide each tidal day. A semidiurnal pattern (upper right) shows two highs and lows of approximately equal heights during each tidal day. A mixed tidal pattern (left) shows two highs and lows of unequal heights during each tidal day.

Although tidal currents are not important in the open sea, they can be rapid in bays, river estuaries, straits, and other narrow places. Off the coast of Brittany in France, for example, tidal currents that accompany a high tide of 12 meters (40 feet) may attain a speed of 20 kilometers (12 miles) per hour. Tidal currents are not generally considered to be major agents of erosion and sediment transport, but notable exceptions occur where tides move through narrow inlets. Here they scour the narrow entrances to many harbors that would otherwise be blocked.

Sometimes deposits called **tidal deltas** are created by tidal currents (Figure 15.29). They may develop either as *flood deltas* landward of an inlet or as *ebb deltas* on the seaward side of an inlet. Because wave activity and longshore currents are reduced on the sheltered landward side, flood deltas are more common and actually more prominent (see Figure 15.17, p. 420). They form after the tidal current moves rapidly through an inlet. As the current emerges into more open waters from the narrow passage, it slows and deposits its load of sediment.

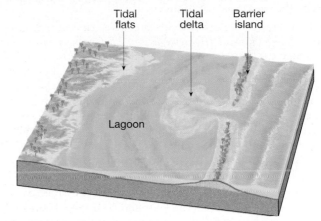

▲ **FIGURE 15.29** As a rapidly moving tidal current moves through a barrier island's inlet into the quiet waters of the lagoon, the current slows and deposits sediment, creating a tidal delta. Because this tidal delta has developed on the landward side of the inlet, it is called a *flood delta*.

Chapter Summary

• *The ocean's surface currents follow the general pattern of the world's major wind belts.* Surface currents are parts of huge, slowly moving loops of water called *gyres* that are centered in the subtropics of each ocean basin. The *positions of the continents* and the *Coriolis effect* also influence the movement of ocean water within gyres. Because of the Coriolis effect, subtropical gyres move *clockwise in the Northern Hemisphere* and *counterclockwise in the Southern Hemisphere.* Generally, *four main currents* comprise each subtropical gyre.

• *Ocean currents are important in navigation and for the effect they have on climates.* Poleward-moving *warm ocean currents moderate winter temperatures* in the middle latitudes. Cold currents exert their greatest influence during summer in middle latitudes and year-round in the tropics. In addition to cooler temperatures, *cold currents are associated with greater fog frequency and drought.*

• *Upwelling,* the rising of colder water from deeper layers, is a wind-induced movement that brings cold, nutrient-rich water to the surface. *Coastal upwelling* is most characteristic along the west coasts of continents.

• In contrast to surface currents, *deep-ocean circulation* is governed by *gravity* and driven by *density differences.* The two factors that are most significant in creating a dense mass of water are *temperature* and *salinity,* so the movement of deep-ocean water is often termed *thermohaline circulation.* Most water involved in thermohaline circulation begins in high latitudes at the surface when the salinity of the cold water increases due to sea ice formation. This dense water sinks, initiating deep-ocean currents.

• The *shore* is the area extending between the lowest tide level and the highest elevation on land that is affected by storm waves. The *coast* extends inland from the shore as far as ocean-related features can be found. The shore is divided into the *foreshore* and *backshore,* Seaward of the foreshore are the *nearshore* and *offshore* zones.

• A *beach* is an accumulation of sediment found along the landward margin of the ocean or a lake. Among its parts are one or more *berms* and the *beach face.* Beaches are composed of whatever material is locally abundant and should be thought of as material in transit along the shore.

• *Waves are moving energy* and *most ocean waves are initiated by the wind.* The three factors that influence the *height, wavelength,* and *period* of a wave are (1) *wind speed,* (2) *length of time the wind has blown,* and (3) *fetch,* the distance that the wind has traveled across open water. Once waves leave a storm area, they are termed *swells,* which are symmetrical, longer-wavelength waves.

• As waves travel, *water particles transmit energy by circular orbital motion,* which extends to a depth equal to one half the wavelength. When a wave travels into shallow water, it experiences physical changes that can cause the wave to collapse, or *break,* and form *surf.*

• Wave erosion is caused by *wave impact pressure* and *abrasion* (the sawing and grinding action of water armed with rock fragments). The bending of waves is called *wave re-fraction.* Owing to refraction, wave impact is concentrated against the sides and ends of headlands.

• Most waves reach the shore at an angle. The uprush (swash) and backwash of water from each breaking wave moves the sediment in a zigzag pattern along the beach. This movement is called *beach drift.* Oblique waves also produce *longshore currents* within the surf zone that flow parallel to the shore and transport more sediment than beach drift.

• *Erosional features* include *wave-cut cliffs* (which originate from the cutting action of the surf against the base of coastal land), *wave-cut platforms* (relatively flat, benchlike surfaces left behind by receding cliffs), and *marine terraces* (uplifted wave-cut platforms). Erosional features also include *sea arches* (formed when a headland is eroded and two sea caves from opposite sides unite), and *sea stacks* (formed when the roof of a sea arch collapses).

• Some of the *depositional features* that form when sediment is moved by beach drift and longshore currents are *spits* (elongated ridges of sand that project from the land into the mouth of an adjacent bay), *baymouth bars* (sandbars that completely cross a bay), and *tombolos* (ridges of sand that connect an island to the mainland or to another island). Along the Atlantic and Gulf Coastal Plains, the coastal region is characterized by offshore *barrier islands,* which are low ridges of sand that parallel the coast.

• Local *factors that influence shoreline erosion* are (1) the proximity of a coast to sediment-laden rivers, (2) the degree of tectonic activity, (3) the topography and composition of the land, (4) prevailing winds and weather patterns, and (5) the configuration of the coastline and nearshore areas.

• *Hard stabilization* involves building hard, massive structures in an attempt to protect a coast from erosion or prevent the movement of sand along the beach. Hard stabilization includes *groins* (short walls constructed at a right angle to the shore to trap moving sand), *breakwaters* (structures built parallel to the shore to protect it from the force of large breaking waves), and *seawalls* (armoring the coast to prevent waves from reaching the area behind the wall). *Alternatives to hard stabilization* include *beach nourishment,* which involves the addition of sand to replenish eroding beaches, and *relocation* of damaged or threatened buildings.

• Because of basic geological differences, the *nature of shoreline erosion problems along America's Pacific and Atlantic/Gulf Coasts is very different.* Much of the development along the Atlantic and Gulf Coasts has occurred on barrier islands, which receive the full force of major storms. Much of the Pacific Coast is characterized by narrow beaches backed by steep cliffs and mountain ranges. A major problem facing the Pacific shoreline is the narrowing of beaches caused by irrigation and flood control dams that interrupt the natural flow of sand to the coast.

• One commonly used classification of coasts is based upon changes that have occurred with respect to sea level.

Emergent coasts often exhibit wave-cut cliffs and marine terraces and develop either because an area experiences uplift or as a result of a drop in sea level. Conversely, *submergent coasts* commonly display drowned river mouths called *estuaries* and are created when sea level rises or the land adjacent to the sea subsides.

• *Tides*, the daily rise and fall in the elevation of the ocean surface at a specific location, are caused by the *gravitational attraction* of the Moon and, to a lesser extent, the Sun. The Moon and the Sun each produce a pair of *tidal bulges* on Earth. These tidal bulges remain in fixed positions relative to the generating bodies as Earth rotates through them, resulting in alternating high and low tides. *Spring tides* occur near the times of new and full moons when the Sun and Moon are aligned and their bulges are added together to produce especially high and low tides (a *large daily tidal range*). Conversely, *neap tides* occur at about the times of the first and third quarters of the Moon when the bulges of the Moon and Sun are at right angles, producing a *smaller daily tidal range*.

• *Three main tidal patterns exist worldwide.* A *diurnal tidal pattern* exhibits one high and low tide daily; a *semidiurnal tidal pattern* exhibits two high and low tides daily of about the same height; and a *mixed tidal pattern* usually has two high and low tides daily of different heights.

• *Tidal currents* are horizontal movements of water that accompany the rise and fall of the tides. *Tidal flats* are the areas that are affected by the advancing and retreating tidal currents. When tidal currents slow after emerging from narrow inlets, they deposit sediment that may eventually create *tidal deltas*.

Key Terms

abrasion (p. 415)
backshore (p. 411)
barrier island (p. 419)
baymouth bar (p. 419)
beach (p. 411)
beach drift (p. 417)
beach face (p. 411)
beach nourishment (p. 423)
berm (p. 411)
breakwater (p. 423)
coastline (p. 410)
Coriolis effect (p. 405)
diurnal tidal pattern (p. 428)
emergent coast (p. 425)
estuary (p. 426)
fetch (p. 413)
foreshore (p. 410)

groin (p. 422)
gyre (p. 404)
hard stabilization (p. 422)
longshore current (p. 416)
marine terrace (p. 418)
mixed tidal pattern (p. 428)
neap tide (p. 428)
nearshore zone (p. 411)
offshore zone (p. 411)
sea arch (p. 418)
sea stack (p. 418)
seawall (p. 423)
semidiurnal tidal pattern (p. 428)
shore (p. 410)
shoreline (p. 410)
spit (p. 418)
spring tide (p. 428)

submergent coast (p. 425)
surf (p. 414)
thermohaline circulation (p. 409)
tidal current (p. 428)
tidal delta (p. 428)
tidal flat (p. 428)
tide (p. 426)
tombolo (p. 419)
upwelling (p. 408)
wave-cut cliff (p. 417)
wave-cut platform (p. 417)
wave height (p. 413)
wavelength (p. 413)
wave period (p.413)
wave refraction (p. 417)

Review Questions

1. What is the primary driving force of surface ocean currents? How do the distribution of continents on Earth and the Coriolis effect influence these currents?

2. What is a *gyre?* How many currents exist within each gyre? Name the five subtropical gyres, and identify the primary surface currents that comprise each gyre.

3. Why did Benjamin Franklin want to know about the surface current pattern in the North Atlantic Ocean?

4. How do ocean currents influence climate? Give at least one example.

5. Describe the process of coastal upwelling. Why is an abundance of marine life associated with these areas?

6. What is the driving force of deep-ocean circulation? Why is the movement of deep currents often termed "thermohaline circulation"?

7. Distinguish among *shore, shoreline, coast,* and *coastline.*

8. What is a beach? Briefly distinguish between *beach face* and *berm*. What are the sources of beach sediment?

9. List three factors that determine the height, length, and period of a wave.

10. Describe the motion of a floating object as a wave passes (see Figure 15.9).

11. Describe the physical changes that occur to a wave's speed, wavelength, and height as a wave moves into shallow water and breaks on the shore.

12. How do waves cause erosion?

13. What is *wave refraction*? What is the effect of this process along irregular coastlines?

14. What is *beach drift*, and how is it related to a longshore current? Why are beaches often called "rivers of sand"?

15. Describe the formation of the following shoreline features: *wave-cut cliff, wave-cut platform, marine terrace, sea stack, spit, baymouth bar, tombolo*.

16. List three ways that barrier islands may form.

17. List the types of hard stabilization and describe what each is intended to do. What effect does each one have on the distribution of sand on the beach?

18. List two alternatives to hard stabilization, indicating potential problems with each one.

19. Relate the damming of rivers to the shrinking of beaches at some locations along the West Coast of the United States.

20. What observable features would lead you to classify a coastal area as emergent?

21. Are estuaries associated with submergent or emergent coasts? Explain.

22. Discuss the origin of ocean tides. Explain why the Sun's influence on Earth's tides is only about half that of the Moon's, even though the Sun is so much more massive than the Moon.

23. Explain how an observer can experience two unequal high tides during a day (see Figure 15.26).

24. How do diurnal, semidiurnal, and mixed tidal patterns differ?

25. Distinguish between a flood current and an ebb current. Of flood current, ebb current, high slack water, and low slack water, when is the best time to navigate a boat in a shallow, rocky harbor?

Examining the Earth System

1. Describe the exchanges of energy between the ocean and the atmosphere that are responsible for (1) surface ocean currents and (2) the ocean's deep circulation.

2. If the warm North Atlantic Drift were to cease, how might the climate of Western Europe change? Speculate about how such a climate change might influence the biosphere.

3. In this chapter, the shoreline was described as a "dynamic interface." What is an interface? List and briefly describe some other interfaces in the Earth system. You need not confine yourself to examples from this chapter.

4. It is believed that global temperatures will be increasing in the decades to come. How can a warmer atmosphere lead to a rise in sea level? How could geologic events associated with the solid Earth contribute to a rise in sea level? Describe some possible changes in coastal areas if sea level were to rise. You may find some interesting and useful information at this Environmental Protection Agency Website: **http://yosemite.epa.gov/oar/globalwarming.nsf/content/ImpactsCoastalZones.html**.

Online Study Guide

The *Earth Science* Website uses the resources and flexibility of the Internet to aid in your study of the topics in this chapter. Written and developed by Earth science instructors, this site will help improve your understanding of Earth science. Visit **http://www.prenhall.com/tarbuck** and click on the cover of *Earth Science 11e* to find:

- **Online review quizzes.**
- **Critical thinking exercises.**
- **Links to chapter-specific Web resources.**
- **Internet-wide key term searches.**

http://www.prenhall.com/tarbuck

CHAPTER 16

The Atmosphere: Composition, Structure, and Temperature

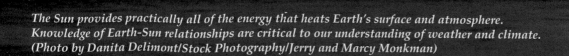

The Sun provides practically all of the energy that heats Earth's surface and atmosphere. Knowledge of Earth-Sun relationships are critical to our understanding of weather and climate. (Photo by Danita Delimont/Stock Photography/Jerry and Marcy Monkman)

rth's atmosphere is unique. No other planet in our solar system has an atmosphere with the exact mixture of gases or the heat and moisture conditions necessary to sustain life as we know it. The gases that make up Earth's atmosphere and the controls to which they are subject are vital to our existence. In this chapter we begin our examination of the ocean of air in which we all must live. We will try to answer a number of basic questions. What is the composition of the atmosphere? At what point do we leave the atmosphere and enter outer space? What causes the seasons? How is air heated? What factors control temperature variations over the globe?

Weather influences our everyday activities, our jobs, and our health and comfort (Figure 16.1). Many of us pay little attention to the weather unless we are inconvenienced by it or when it adds to our enjoyment outdoors. Nevertheless, there are few other aspects of our physical environment that affect our lives more than the phenomena we collectively call the weather.

The United States has the greatest variety of weather of any country in the world. Severe weather events such as tornadoes, flash floods, and intense thunderstorms, as well as hurricanes and blizzards, are collectively more frequent and more damaging in the United States than in any other nation. Beyond its direct impact on the lives of individuals, the weather has a strong effect on the world economy by influencing agriculture, energy use, water resources, transportation, and industry.

Weather clearly influences our lives a great deal. Yet it is also important to realize that people influence the atmosphere and its behavior as well. There are, and will continue to be, significant political and scientific decisions that must be made involving these impacts. Answers to questions regarding air pollution and its control and the effects of various emissions on global climate and the atmosphere's ozone layer are important examples. So there is a need for increased awareness and understanding of our atmosphere and its behavior.

Weather and Climate

Acted on by the combined effects of Earth's motions and energy from the Sun, our planet's formless and invisible envelope of air reacts by producing an infinite variety of weather, which in turn creates the basic pattern of global climates. Although not identical, weather and climate have much in common.

Weather is constantly changing, sometimes from hour to hour and at other times from day to day. It is a term that refers to the state of the atmosphere at a given time and place. Whereas changes in the weather are continuous and sometimes seemingly erratic, it is

▼ **FIGURE 16.1**   North America has a great variety of weather. Severe weather events are frequent and costly occurrences. One of the heaviest snowstorms to hit New York City since the "Blizzard of '93" occurred on February 17, 2003. The blizzard shut down much of the Northeast with blinding windblown snow that piled up to depths as great as four feet and left nearly a quarter million homes and businesses without power.   (Photo by Benjamin Lowy/CORBIS)

nevertheless possible to arrive at a generalization of these variations. Such a description of aggregate weather conditions is termed **climate**. It is based on observations that have been accumulated over many years. Climate is often defined simply as "average weather," but this is an inadequate definition. In order to more accurately portray the character of an area, variations and extremes must also be included, as well as the probabilities that such departures will take place. For example, it is not only necessary for farmers to know the average rainfall during the growing season, but it is also important to know the frequency of extremely wet and extremely dry years. Thus, climate is the sum of all statistical weather information that helps describe a place or region.

Suppose you were planning a vacation trip to an unfamiliar place. You would probably want to know what kind of weather to expect. Such information would help as you selected clothes to pack and could influence decisions regarding activities you might engage in during your stay. Unfortunately, weather forecasts that go beyond a few days are not very dependable. Thus, it would not be possible to get a reliable weather report about the conditions you are likely to encounter during your vacation.

Instead, you might ask someone who is familiar with the area about what kind of weather to expect. "Are thunderstorms common?" "Does it get cold at night?" "Are the afternoons sunny?" What you are seeking is information about the climate, the conditions that are typical for that place. Another useful source of such information is the great variety of climate tables, maps, and graphs that are available. For example, the graph in Figure 16.2 shows average daily high and low temperatures for each month, as well as extremes for New York City.

Such information could no doubt help as you planned your trip. But it is important to realize that *climate data cannot predict the weather.* Although the place may usually (climatically) be warm, sunny, and dry during the time of your planned vacation you may actually experience cool, overcast, and rainy weather. There is a well-known saying that summarizes this idea: "Climate is what you expect, but weather is what you get."

The nature of weather and climate is expressed in terms of the same basic **elements**, those quantities or properties that are measured regularly. The most important are (1) air temperature, (2) humidity, (3) type and amount of cloudiness, (4) type and amount of precipitation, (5) air pressure, and (6) the speed and direction of the wind. These elements are the major variables from which weather patterns and climate types are deciphered. Although you will study these elements separately at first, keep in mind that they are very much interrelated. A change in any one of the elements will often bring about changes in the others.

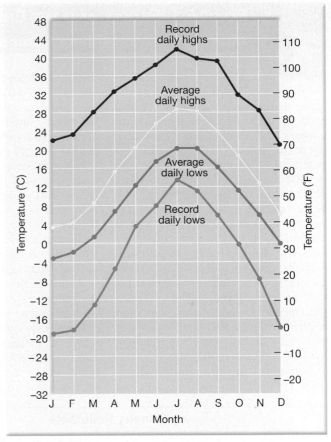

▲ **FIGURE 16.2** Graph showing daily temperature data for New York City. In addition to the average daily maximum and minimum temperatures for each month, extremes are also shown. As this graph shows, there can be significant departures from average.

? STUDENTS SOMETIMES ASK...
Who provides all of the data needed to prepare a weather forecast?

Data from every part of the globe are needed to produce accurate weather forecasts. The World Meteorological Organization (WMO) was established by the United Nations to coordinate scientific activity related to weather and climate. It consists of 185 countries and territories representing all parts of the globe. Its World Weather Watch provides up-to-the-minute standardized observations through member-operated observation systems. This global system involves 10 satellites, 10,000 land-observation and 7000 ship stations, as well as hundreds of automated data buoys and thousands of aircraft.

Composition of the Atmosphere

Air is *not* a unique element or compound. Rather, **air** is a *mixture* of many discrete gases, each with its own physical properties, in which varying quantities of tiny solid and liquid particles are suspended.

BOX 16.1 PEOPLE AND THE ENVIRONMENT:
Altering the Atmosphere's Composition—Sources and Types of Air Pollution

Air pollutants are airborne particles and gases that occur in concentrations that endanger the health and well-being of organisms or disrupt the orderly functioning of the environment. One category of pollutants, the *primary pollutants*, are emitted directly from identifiable sources. They pollute the air immediately upon being emitted. Figure 16.A shows the major primary pollutants and the sources that produce them. When the sources are examined, the significance of the transportation category is obvious. It accounts for nearly half of our air pollution (by weight). In addition to highway vehicles, this category includes trains, ships, and airplanes. Still, the tens of millions of cars and trucks on U.S. roads are, without a doubt, the greatest contributors.

Sometimes the direct impact of primary pollutants on human health and the environment is less severe than the effects of the secondary pollutants they form. *Secondary pollutants* are not emitted directly into the air, but form in the atmosphere when reactions take place among primary pollutants. The chemicals that make up smog are important examples, as is the sulfuric acid that falls as acid precipitation. After the primary pollutant, sulfur dioxide, is emitted into the atmosphere, it combines with oxygen to produce sulfur trioxide, which then combines with water to create this irritating and corrosive acid.

Many reactions that produce secondary pollutants are triggered by strong sunlight and so are called *photochemical reactions*. One common example occurs when nitrogen oxides absorb solar radiation, initiating a chain of complex reactions. When certain volatile organic compounds are present, the result is the

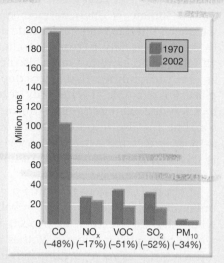

Figure 16.B Comparison of 1970 and 2002 emissions. The 2002 total is about 48 percent lower than 1970.

formation of a number of undesirable secondary products that are very reactive, irritating, and toxic. Collectively, this noxious mixture of gases and particles is called *photochemical smog*.

Between 1970 and 2002, emissions of the six principal pollutants were cut by 48 percent. During that same span, U.S. population increased 38 percent, energy consumption increased 42 percent, and vehicle miles traveled increased 155 percent (Figure 16.B). Despite this progress, about 160 million tons of pollutants are emitted each year in the United States and approximately 146 million people live in counties where monitored air in 2002 was unhealthy at times because of high levels of at least one of the principal air pollutants.

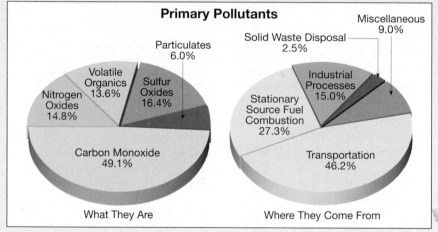

Figure 16.A Major primary pollutants and their sources. Percentages are calculated on the basis of weight. (Data from the U.S. Environmental Protection Agency)

Major Components

The composition of air is not constant; it varies from time to time and from place to place (Box 16.1). If the water vapor, dust, and other variable components were removed from the atmosphere, we would find that its makeup is very stable worldwide up to an altitude of about 80 kilometers (50 miles).

As you can see in Figure 16.3, two gases—nitrogen and oxygen—make up 99 percent of the volume of clean, dry air. Although these gases are the most plentiful components of air and are of great significance to life on Earth, they are of minor importance in affecting weather phenomena. The remaining 1 percent of dry air is mostly

the inert gas argon (0.93 percent) plus tiny quantities of a number of other gases. Carbon dioxide, although present in only minute amounts (0.037 percent), is nevertheless an important constituent of air. Carbon dioxide is of great interest to meteorologists because it is an efficient absorber of energy emitted by Earth and thus influences the heating of the atmosphere.

Variable Components

Air includes many gases and particles that vary significantly from time to time and place to place. Important examples include water vapor, dust particles, and ozone. Although usually present in small per-

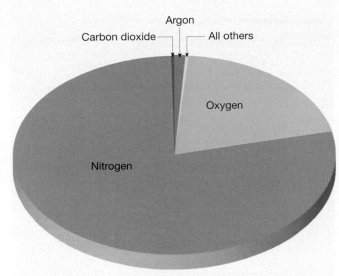

▲ **FIGURE 16.3** Proportional volume of gases composing dry air. Nitrogen and oxygen clearly dominate.

centages, they can have significant effects on weather and climate.

Water Vapor. The amount of water vapor in the air varies considerably, from practically none at all up to about 4 percent by volume. Why is such a small fraction of the atmosphere so significant? Certainly the fact that water vapor is the source of all clouds and precipitation would be enough to explain its importance. However, water vapor has other roles. Like carbon dioxide, it has the ability to absorb heat given off by Earth as well as some solar energy. It is therefore important when we examine the heating of the atmosphere.

When water changes from one state to another (see Figure 17.2, p. 467), it absorbs or releases heat. This energy is termed *latent heat*, which means "hidden heat." As we shall see in later chapters, water vapor in the atmosphere transports this latent heat from one region to another, and it is the energy source that helps drive many storms.

Aerosols. The movements of the atmosphere are sufficient to keep a large quantity of solid and liquid particles suspended within it. Although visible dust sometimes clouds the sky, these relatively large particles are too heavy to stay in the air for very long. Still, many particles are microscopic and remain suspended for considerable periods of time. They may originate from many sources, both natural and human made, and include sea salts from breaking waves, fine soil blown into the air, smoke and soot from fires, pollen and microorganisms lifted by the wind, ash and dust from volcanic eruptions, and more (Figure 16.4A). Collectively, these tiny solid and liquid particles are called **aerosols.**

From a meteorological standpoint, these tiny, often invisible particles can be significant. First, many act as

surfaces on which water vapor can condense, an important function in the formation of clouds and fog. Second, aerosols can absorb, reflect, and scatter incoming solar radiation. Thus, when an air-pollution episode is occurring or when ash fills the sky following a volcanic eruption, the amount of sunlight reaching Earth's surface can be measurably reduced. Finally, aerosols contribute to an optical phenomenon we have all observed—the varied hues of red and orange at sunrise and sunset (Figure 16.4B).

Ozone. Another important component of the atmosphere is **ozone**. It is a form of oxygen that combines three oxygen atoms into each molecule (O_3). Ozone is not the same as oxygen we breathe, which has two atoms per molecule (O_2). There is very little of this gas in the atmosphere, and its distribution is not uniform. In the lowest portion of the atmosphere, ozone represents less than one part in 100 million. It is concentrated well above the surface in a layer called the *stratosphere*, between 10 and 50 kilometers (6 and 31 miles).

In this altitude range, oxygen molecules (O_2) are split into single atoms of oxygen (O) when they absorb ultraviolet radiation emitted by the Sun. Ozone is then created when a single atom of oxygen (O) and a molecule of oxygen (O_2) collide. This must happen in the presence of a third, neutral molecule that acts as a *catalyst* by allowing the reaction to take place without itself being consumed in the process. Ozone is concentrated in the 10- to 50-kilometer height range because a crucial balance exists there: The ultraviolet radiation from the Sun is sufficient to produce single atoms of oxygen, and there are enough gas molecules to bring about the required collisions.

The presence of the ozone layer in our atmosphere is crucial to those of us who dwell on Earth. The reason is that ozone absorbs the potentially harmful ultraviolet (UV) radiation from the Sun. If ozone did not filter a great deal of the ultraviolet radiation, and if the Sun's UV rays reached the surface of Earth undiminished, our planet would be uninhabitable for most life as we know it. Thus, anything that reduces the amount of ozone in the atmosphere could affect the well-being of life on Earth. Just such a problem exists and is described in Box 16.2.

STUDENTS SOMETIMES ASK...
Isn't ozone some sort of pollutant?

Yes, you're right. Although the naturally occurring ozone in the stratosphere is critical to life on Earth, it is regarded as a pollutant when produced at ground level because it can damage vegetation and be harmful to human health. Ozone is a major component in a noxious mixture of gases and particles called photochemical smog. It forms as a result of reactions triggered by sunlight that occur among pollutants emitted by motor vehicles and industries.

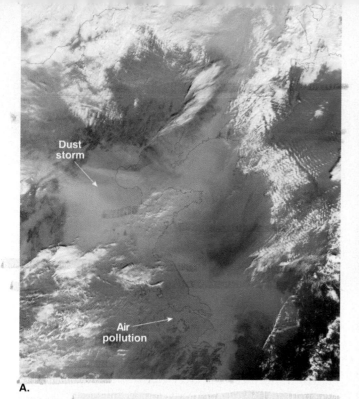

A.

B.

▲ **FIGURE 16.4** **A.** This satellite image from November 11, 2002, shows two examples of aerosols. First, a large dust storm is blowing across northeastern China toward the Korean Peninsula. Second, a dense haze toward the south (bottom center) is human-generated air pollution. (NASA Image) **B.** Dust in the air can cause sunsets to be especially colorful. (Photo by Steve Elmore/CORBIS/The Stock Market)

Height and Structure of the Atmosphere

Earth's Dynamic Atmosphere
▼ Heating the Atmosphere

To say that the atmosphere begins at Earth's surface and extends upward is obvious. However, where does the atmosphere end and outer space begin? There is no sharp boundary; the atmosphere rapidly thins as you travel away from Earth, until there are too few gas molecules to detect.

Pressure Changes

To understand the vertical extent of the atmosphere, let us examine the changes in the atmospheric pressure with height. Atmospheric pressure is simply the weight of the air above. At sea level, the average pressure is slightly more than 1000 millibars. This corresponds to a weight of slightly more than 1 kilogram per square centimeter (14.7 pounds per square inch). Obviously the pressure at higher altitudes is less (Figure 16.5).

One half of the atmosphere lies below an altitude of 5.6 kilometers (3.5 miles). At about 16 kilometers (10 miles), 90 percent of the atmosphere has been traversed, and above 100 kilometers (62 miles), only 0.00003 percent of all the gases making up the atmosphere remains. Even so, traces of our atmosphere extend far beyond this altitude, gradually merging with the emptiness of space.

▼ **FIGURE 16.5** Atmospheric pressure variation with altitude. The rate of pressure decrease with an increase in altitude is not constant. Rather, pressure decreases rapidly near Earth's surface and more gradually at greater heights.

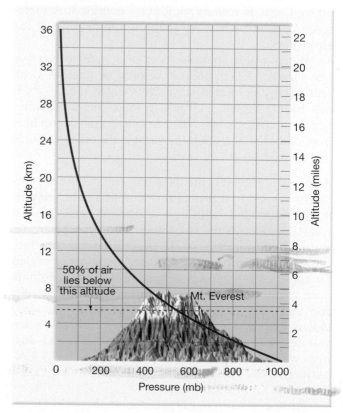

BOX 16.2 PEOPLE AND THE ENVIRONMENT:
Ozone Depletion—A Global Issue

The loss of ozone high in the atmosphere as a consequence of human activities is a serious global-scale environmental problem (Figure 16.C). For nearly a billion years Earth's ozone layer has protected life on the planet. However, over the past half century, people have unintentionally placed the ozone layer in jeopardy by polluting the atmosphere. The offending chemicals are known as chlorofluorocarbons (CFCs for short). They are versatile compounds that are chemically stable, odorless, nontoxic, noncorrosive, and inexpensive to produce. Over the decades many uses were developed for CFCs, including as coolants for airconditioning and refrigeration equipment, cleaning solvents for electronic components, propellants for aerosol sprays, and the production of certain plastic foams.

No one worried about how CFCs might affect the atmosphere until three scientists, Paul Crutzen, F. Sherwood Rowland, and Mario Molina, studied the relationship. In 1974 they alerted the world when they reported that CFCs were probably reducing the average concentration of ozone in the stratosphere. In 1995 these scientists were awarded the Nobel Prize in chemistry for their pioneering work.

They discovered that because CFCs are practically inert (that is, not chemically active) in the lower atmosphere, a portion of these gases gradually makes its way to the ozone layer, where sunlight separates the chemicals into their constituent atoms. The chlorine atoms released this way, through a complicated series of reactions, have the net effect of removing some of the ozone.

Because ozone filters out most of the ultraviolet (UV) radiation from the Sun, a decrease in its concentration permits more of these harmful wavelengths to reach Earth's surface. The most serious threat to human health is an increased risk of skin cancer. An increase in damaging UV radiation also can impair the human immune system as well as promote cataracts, a clouding of the eye lens that reduces vision and may cause blindness if not treated.

In response to this problem, an international agreement known as the *Montreal Protocol* was developed under the sponsorship of the United Nations to eliminate the production and use of CFCs.

Although relatively strong action was taken, CFC levels in the atmosphere will not drop rapidly. Once in the atmosphere, CFC molecules can take many years to reach the ozone layer and once there, they can remain active for decades. This does not promise a near-term reprieve for the ozone layer. Even after CFC levels begin to decline, the ozone layer probably will not cleanse itself of most of the pollutants until sometime in the second half of the twenty-first century.

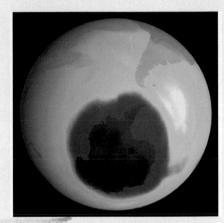

Figure 16.C This satellite image shows ozone distribution in the Southern Hemisphere on September 17, 2001. The area of greatest depletion is called the "ozone hole." Dark blue colors correspond to the region with the sparsest ozone. Light blue, green, and yellow indicate progressively more ozone. The ozone hole forms over Antarctica during the Southern Hemisphere spring. In 2001 it extended over about 26 million square kilometers (10 million square miles). For comparison, the area of the continental United States is about 9.4 million square kilometers (3.7 million square miles). (Data from NOAA)

Temperature Changes

By the early twentieth century, much had been learned about the lower atmosphere. The upper atmosphere was partly known from indirect methods. Data from balloons and kites had revealed that the air temperature dropped with increasing height above Earth's surface. This phenomenon is felt by anyone who has climbed a high mountain and is obvious in pictures of snowcapped mountaintops rising above snow-free lowlands (Figure 16.6). We divide the atmosphere vertically into four layers on the basis of temperature (Figure 16.7).

Troposphere. The bottom layer in which we live, where temperature decreases with an increase in altitude, is the **troposphere**. The term literally means the region where air "turns over," a reference to the appreciable vertical mixing of air in this lowermost zone. The troposphere is the chief focus of meteorologists, because it is in this layer that essentially all important weather phenomena occur.

The temperature decrease in the troposphere is called the **environmental lapse rate**. Its average value is 6.5°C per kilometer (3.5°F per 1000 feet), a figure known as the *normal lapse rate*. It should be emphasized, however, that the environmental lapse rate is not a constant, but rather can be highly variable, and must be regularly measured. To determine the actual environmental lapse rate as well as gather information about vertical changes in pressure, wind, and humidity, radiosondes are used. The *radiosonde* is an instrument package that is attached to a balloon and transmits data by radio as it ascends through the atmosphere (Figure 16.8).

▲ **FIGURE 16.6** Temperatures drop with an increase in altitude in the troposphere. Therefore, it is possible to have snow on a mountaintop and warmer, snow-free lowlands below. Mount Kerkeslin, Jasper National Park, Alberta, Canada. (Photo by Carr Clifton/Minden Pictures)

The thickness of the troposphere is not the same everywhere; it varies with latitude and the season. On the average, the temperature drop continues to a

height of about 12 kilometers (7.4 miles). The outer boundary of the troposphere is the *tropopause.*

Stratosphere. Beyond the tropopause is the **stratosphere**. In the stratosphere, the temperature remains constant to a height of about 20 kilometers (12 miles) and then begins a gradual increase that continues until the *stratopause,* at a height of nearly 50 kilometers (30 miles) above Earth's surface. Below the tropopause, atmospheric properties like temperature and humidity are readily transferred by large-scale turbulence and mixing. Above the tropopause, in the stratosphere, they are not. Temperatures increase in the stratosphere because it is in this layer that the atmosphere's ozone is concentrated. Recall that ozone absorbs ultraviolet radiation from the Sun. As a consequence, the stratosphere is heated.

Mesosphere. In the third layer, the **mesosphere**, temperatures again decrease with height until, at the *mesopause,* more than 80 kilometers (50 miles) above the surface, the temperature approaches −90°C (−130°F). The coldest temperatures anywhere in the atmosphere occur at the mesopause. Because accessibility is difficult, the mesosphere is one of the least explored regions of the atmosphere. The reason is that it cannot be

► **FIGURE 16.7** Thermal structure of the atmosphere.

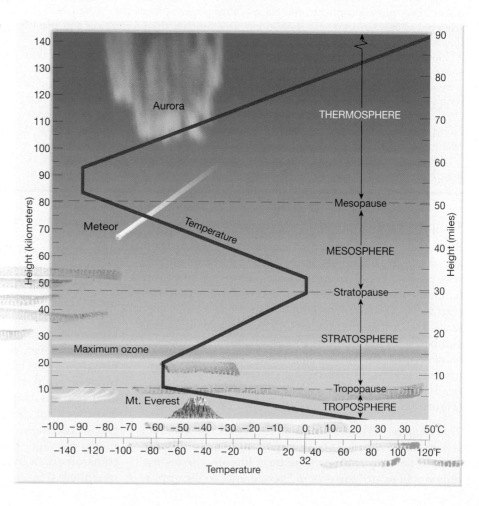

reached by the highest research balloons nor is it accessible to the lowest orbiting satellites. Recent technical developments are just beginning to fill this knowledge gap.

Thermosphere.   The fourth layer extends outward from the mesopause and has no well-defined upper limit. It is the **thermosphere**, a layer that contains only a tiny fraction of the atmosphere's mass. In the extremely rarefied air of this outermost layer, temperatures again increase, owing to the absorption of very short-wave, high-energy solar radiation by atoms of oxygen and nitrogen.

Temperatures rise to extremely high values of more than 1000°C in the thermosphere. But such temperatures are not comparable to those experienced near Earth's surface. Temperature is defined in terms of the average speed at which molecules move. Because the gases of the thermosphere are moving at very high speeds, the temperature is very high. But the gases are so sparse that, collectively, they possess only an insignificant quantity of heat. For this reason, the temperature of a satellite orbiting Earth in the thermosphere is determined chiefly by the amount of solar radiation it absorbs and not by the high temperature of the almost nonexistent surrounding air. If an astronaut inside were to expose his or her hand, it would not feel hot.

▼ **FIGURE 16.8**   A lightweight package of instruments, *the radiosonde*, is carried aloft by a small weather balloon. Radiosondes supply data on vertical changes in temperature, pressure, and humidity. (Photo by AP/Wide World Photos)

Earth–Sun Relationships

Always remember that nearly all of the energy that drives Earth's variable weather and climate comes from the Sun. Earth intercepts only a minute percentage of the energy given off by the Sun—less than one twobillionth. This may seem to be an insignificant amount until we realize that it is several hundred thousand times the electrical-generating capacity of the United States.

Solar energy is not distributed evenly over Earth's land–sea surface. The amount of energy received varies with latitude, time of day, and season of the year. Contrasting images of polar bears on ice rafts and palm trees along a remote tropical beach serve to illustrate the extremes. It is the unequal heating of Earth that creates winds and drives the ocean's currents. These movements, in turn, transport heat from the tropics toward the poles in an unending attempt to balance energy inequalities. The consequences of these processes are the phenomena we call weather. If the Sun were "turned off," global winds and ocean currents would quickly cease. Yet as long as the Sun shines, the winds *will* blow and weather *will* persist. So to understand how the atmosphere's dynamic weather machine works, we must first know why different latitudes receive varying quantities of solar energy and why the amount of solar energy changes to produce the seasons. As you will see, the variations in solar heating are caused by the motions of Earth relative to the Sun and by variations in Earth's land–sea surface.

Earth's Motions

Earth has two principal motions—rotation and revolution. **Rotation** is the spinning of Earth about its axis. The axis is an imaginary line running through the poles. Our planet rotates once every 24 hours, producing the daily cycle of daylight and darkness. At any moment, half of Earth is experiencing daylight, and the other half darkness. The line separating the dark half of Earth from the lighted half is called the **circle of illumination**.

Revolution refers to the movement of Earth in its orbit around the Sun. Hundreds of years ago, most people believed that Earth was stationary in space and that the Sun and stars revolved around our planet. Today, we know that Earth is traveling at nearly 113,000 kilometers (70,000 miles) per hour in an elliptical orbit about the Sun.

Seasons

We know that it is colder in winter than in summer. But why? Length of daylight certainly accounts for some of the difference. Long summer days expose us to more solar radiation, whereas short winter days expose us to less.

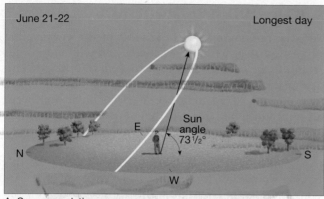

A. Summer solstice

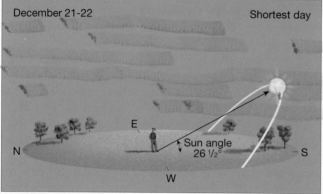

B. Winter solstice

▲ **FIGURE 16.9** Daily paths of the Sun for a place located at 40°N latitude for: **A.** summer solstice and **B.** winter solstice. As we move from summer to winter, the angle of the noon Sun decreases from 73½ to 26½ degrees—a difference of 47 degrees. Notice also how the location of sunrise (east) and sunset (west) change during a year.

Furthermore, a gradual change in the angle of the noon Sun above the horizon is quite noticeable (Figure 16.9). At midsummer, the noon Sun is seen high above the horizon. But as summer gives way to autumn, the noon Sun appears lower in the sky and

sunset occurs earlier each evening. What we observe here is the annual shifting of the solar angle or *altitude* of the Sun.

The seasonal variation in the altitude of the Sun affects the amount of energy received at Earth's surface in two ways. First, when the Sun is high in the sky, the solar rays are most concentrated (you can see this in Figure 16.10A). The lower the angle, the more spread out and less intense is the solar radiation reaching the surface (Figure 16.10B, C). To illustrate this principle, hold a flashlight at a right angle to a surface and then change the angle.

Second, and of lesser importance, the angle of the Sun determines the amount of atmosphere the rays must penetrate (Figure 16.11). When the Sun is directly overhead, the rays pass through a thickness of only 1 atmosphere, whereas rays entering at a 30-degree angle travel through twice this amount, and 5-degree rays travel through a thickness roughly equal to 11 atmospheres. The longer the path, the greater the chances for absorption, reflection, and scattering by the atmosphere, all of which reduce the intensity at the surface. The same effects account for the fact that we cannot look directly at the midday Sun, but we can enjoy gazing at a sunset.

It is also important to remember that Earth has a spherical shape. Hence, on any given day only places located at a particular latitude receive vertical (90-degree) rays from the Sun. As we move either north or south of this location, the Sun's rays strike at an ever decreasing angle. Thus, the nearer a place is to the latitude receiving vertical rays of the Sun, the higher will be its noon Sun and the more intense will be the radiation it receives.

Earth's Orientation

What causes the fluctuations in the Sun angle and length of daylight that occur during the course of a year? They occur because *Earth's orientation to the Sun*

▼ **FIGURE 16.10** Changes in the Sun's angle cause variations in the amount of solar energy reaching Earth's surface. The higher the angle, the more intense the solar radiation.

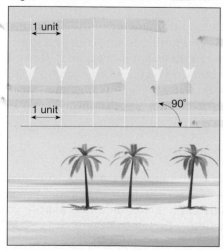

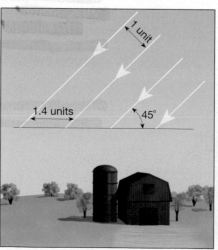

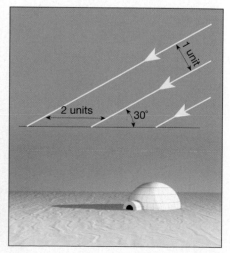

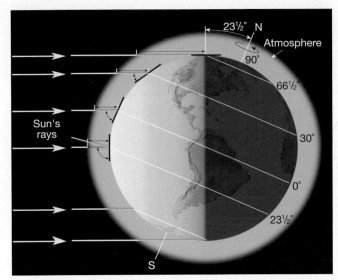

▲ **FIGURE 16.11** Notice that rays striking Earth at a low angle (toward the poles) must travel through more of the atmosphere than rays striking at a high angle (around the equator) and thus are subject to greater depletion by reflection and absorption.

continually changes as it travels along its orbit. Earth's axis is not perpendicular to the plane of its orbit around the Sun. Instead it is tilted $23\frac{1}{2}$ degrees from the perpendicular, as shown in Figure 16.12. This is called the **inclination of the axis.** As you will see, if the axis were not inclined, we would have no seasonal changes. In addition, because the axis remains pointed

in the same direction (toward the North Star) as Earth journeys around the Sun, the orientation of Earth's axis to the Sun's rays is constantly changing (Figure 16.12).

For example, on one day in June each year the axis is such that the Northern Hemisphere is "leaning" $23\frac{1}{2}$ degrees *toward* the Sun. Six months later, in December, when Earth has moved to the opposite side of its orbit, the Northern Hemisphere "leans" $23\frac{1}{2}$ degrees *away* from the Sun. On days between these extremes, Earth's axis is leaning at amounts less than $23\frac{1}{2}$ degrees to the rays of the Sun. This change in orientation causes the spot where the Sun's rays are vertical to make an annual migration from $23\frac{1}{2}$ degrees north of the equator to $23\frac{1}{2}$ degrees south of the equator.

In turn, this migration causes the angle of the noon Sun to vary by up to 47 degrees $(23\frac{1}{2} + 23\frac{1}{2})$ for many locations during the year. For example, a midlatitude city like New York (about 40 degrees north latitude) has a maximum noon Sun angle of $73\frac{1}{2}$ degrees when the Sun's vertical rays reach their farthest northward location in June and a minimum noon Sun angle of $26\frac{1}{2}$ degrees six months later.

Solstices and Equinoxes

Historically, four days a year have been given special significance based on the annual migration of the direct rays of the Sun and its importance to the yearly

▼ **FIGURE 16.12** Earth–Sun relationships.

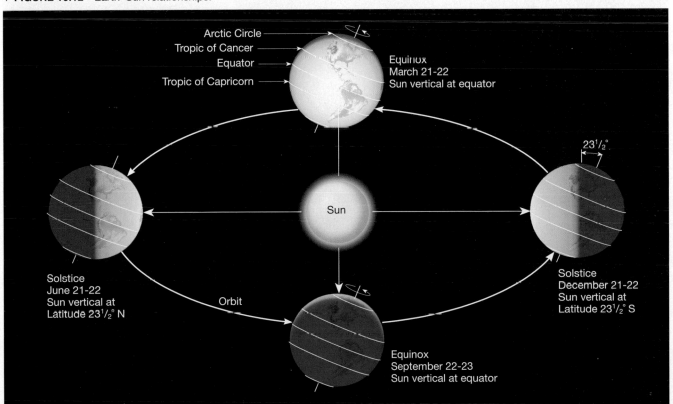

weather cycle. On June 21 or 22, Earth is in a position where the axis in the Northern Hemisphere is tilted $23\frac{1}{2}$ degrees toward the Sun (Figure 16.13A). At this time the vertical rays of the Sun strike $23\frac{1}{2}$ degrees north latitude ($23\frac{1}{2}$ degrees north of the equator), a latitude known as the **Tropic of Cancer**. For people in the Northern Hemisphere, June 21 or 22 is known as the **summer solstice**, the first "official" day of summer.

Six months later, on about December 21 or 22, Earth is in an opposite position, with the Sun's vertical rays striking at $23\frac{1}{2}$ degrees south latitude (Figure 16.13B). This parallel is known as the **Tropic of Capricorn**. For those in the Northern Hemisphere, December 21 and 22 is the **winter solstice**, the first day of winter. However, at the same time in the Southern Hemisphere, people are experiencing just the opposite: the summer solstice.

The equinoxes occur midway between the solstices. September 22 or 23 is the date of the **autumnal equinox** in the Northern Hemisphere, and March 21 or 22 is the date of the **spring equinox**. On these dates, the vertical rays of the Sun strike the equator (0 degrees latitude) because Earth is in such a position in its orbit that the axis is tilted neither toward nor away from the Sun (Figure 16.13C).

The length of daylight versus darkness is also determined by Earth's position in orbit. The length of daylight on June 21, the summer solstice in the Northern Hemisphere, is greater than the length of night. This fact can be established from Figure 16.13A by comparing the fraction of a given latitude that is on the "day" side of the circle of illumination with the fraction on the "night" side. The opposite is true for the winter solstice, when the nights are longer than the days. Again for comparison let us consider New York City, which has 15 hours of daylight on June 21 and only nine hours on December 21 (you can see this in Figure 16.13 and Table 16.1). Also note from Table 16.1 that the farther you are north of the equator on June 21, the longer the period of daylight. When you reach the Arctic Circle ($66\frac{1}{2}°$N), the length of daylight is 24 hours. This is the land of the "midnight Sun," which does not set for about six months at the North Pole (Figure 16.14).

▼ **FIGURE 16.13** Characteristics of the solstices and equinoxes.

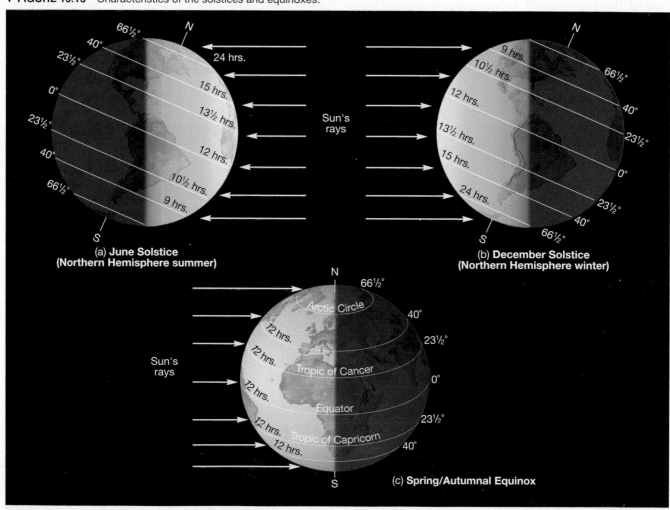

(a) **June Solstice** (Northern Hemisphere summer)

(b) **December Solstice** (Northern Hemisphere winter)

(c) **Spring/Autumnal Equinox**

TABLE 16.1	Length of daylight.		
Latitude (degrees)	Summer Solstice	Winter Solstice	Equinoxes
0	12 h	12 h	12 h
10	12 h 35 min	11 h 25 min	12
20	13 h 12 min	10 h 48 min	12
30	13 h 56 min	10 h 04 min	12
40	14 h 52 min	9 h 08 min	12
50	16 h 18 min	7 h 42 min	12
60	18 h 27 min	5 h 33 min	12
70	24 h (for 2 mo)	0 h 00 min	12
80	24 h (for 4 mo)	0 h 00 min	12
90	24 h (for 6 mo)	0 h 00 min	12

During an equinox (meaning "equal night"), the length of daylight is 12 hours everywhere on Earth, because the circle of illumination passes directly through the poles, dividing the latitudes in half.

As a review of the characteristics of the summer solstice for the Northern Hemisphere, examine Figure 16.13A and Table 16.1 and consider the following facts:

1. The solstice occurs on June 21 or 22.

2. The vertical rays of the Sun are striking the Tropic of Cancer ($23\frac{1}{2}$ degrees north latitude).

3. Locations in the Northern Hemisphere are experiencing their greatest length of daylight (opposite for the Southern Hemisphere).

4. Locations north of the Tropic of Cancer are experiencing their highest noon Sun angles (opposite for places south of the Tropic of Capricorn).

5. The farther you are north of the equator, the longer the period of daylight, until the Arctic Circle is reached, where daylight lasts for 24 hours (opposite for the Southern Hemisphere).

The facts about the winter solstice are just the opposite. It should now be apparent why a midlatitude location is warmest in the summer. It is then that the days are longest and the Sun's altitude is highest.

In summary, seasonal variations in the amount of solar energy reaching places on Earth's surface are caused by the migrating vertical rays of the Sun and the resulting variations in Sun angle and length of daylight.

These changes in turn cause the month-to-month variations in temperature observed at most locations outside the tropics. Figure 16.15 shows mean monthly temperatures for selected cities at different latitudes. Notice that the cities located at more poleward latitudes experience larger temperature differences from summer to winter than do cities located nearer the equator. Also notice that temperature minimums for Southern Hemisphere locations occur in July, whereas they occur in January for most places in the Northern Hemisphere.

All places at the same latitude have identical Sun angles and lengths of daylight. If the Earth–Sun relationships just described were the only controls of

▼ FIGURE 16.14 Multiple exposures of the midnight Sun in late June or July in high northern latitudes—Alaska, Scandinavia, northern Canada, etc. (Photo by Brian Stablyk/Getty Images, Inc.—Stone Allstock)

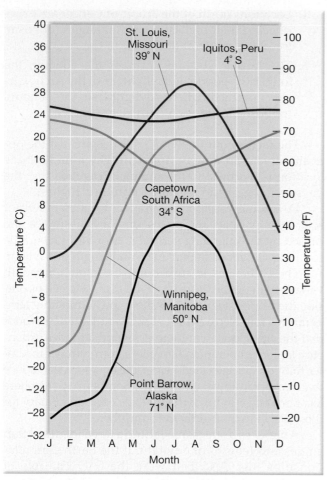

▲ **FIGURE 16.15** Mean monthly temperatures for six cities located at different latitudes. Note that Capetown, South Africa, experiences winter in June, July, and August.

temperature, we would expect these places to have identical temperatures as well. Obviously this is not the case. Although the altitude of the Sun is an extremely important control of temperature, it is not the only control, as you will see.

STUDENTS SOMETIMES ASK...

The "official" first day of winter isn't until December 21 or 22. Realistically, doesn't winter start sooner?

The most publicized dates for the seasons are based on the astronomical definition, which uses the dates of the solstices and equinoxes as the "first day" of each of the seasons. Because the weather phenomena we normally associate with each season do not coincide well with this definition, meteorologists prefer to divide the year into four three-month periods based primarily on temperature. Thus, winter is defined as December, January, and February, the three coldest months of the year in the Northern Hemisphere. Summer is defined as the three warmest months, June, July, and August. Spring and autumn are the transitional periods between these two seasons. This definition is more useful for meteorological purposes.

Energy, Heat, and Temperature

The universe is made up of a combination of matter and energy. The concept of matter is easy to grasp because it is the "stuff" we can see, smell, and touch. Energy, on the other hand, is abstract and therefore more difficult to describe. For our purposes we will define energy simply as *the capacity to do work*. We can think of work as being accomplished whenever matter is moved. You are likely familiar with some of the common forms of energy, such as thermal, chemical, nuclear, radiant (light), and gravitational energy. One type of energy is described as *kinetic energy*, which is energy of motion. Recall that matter is composed of atoms or molecules that are constantly in motion and therefore possesses kinetic energy.

Heat is a term that is commonly used synonymously with *thermal energy*. In this usage, heat is energy possessed by a material arising from the internal motions of its atoms or molecules. Whenever a substance is heated, its atoms move faster and faster which leads to an increase in its heat content. **Temperature**, on the other hand, is related to the average kinetic energy of a material's atoms or molecules. Stated another way, the term *heat* generally refers to the quantity of energy present, whereas the word *temperature* refers to the intensity, that is the degree of "hotness."

Heat and temperature are closely related concepts. Heat is the energy that flows because of temperature differences. In all situations, *heat is transferred from warmer to cooler objects*. Thus, if two objects of different temperature are in contact, the warmer object will become cooler and the cooler object will become warmer until they both reach the same temperature.

Mechanisms of Heat Transfer

 GEODe Earth's Dynamic Atmosphere
▼ Heating the Atmosphere

Three mechanisms of heat transfer are recognized: conduction, convection, and radiation. Although we present them separately, all three processes go on simultaneously in the atmosphere. In addition, these mechanisms operate to transfer heat between Earth's surface (both land and water) and the atmosphere.

Conduction

Conduction is familiar to all of us. Anyone who has touched a metal spoon that was left in a hot pan has discovered that heat was conducted through the spoon. **Conduction** is *the transfer of heat through matter by molecular activity*. The energy of molecules is transferred through collisions from one molecule to another, with the heat flowing from the higher temperature to the lower temperature.

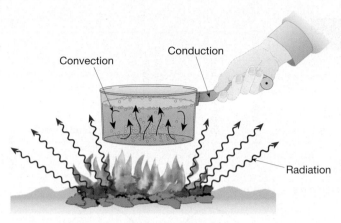

▲ **FIGURE 16.16** The three mechanisms of heat transfer: conduction, convection, and radiation.

The ability of substances to conduct heat varies considerably. Metals are good conductors, as those of us who have touched hot metal have quickly learned (Figure 16.16). Air, conversely, is a very poor conductor of heat. Consequently, conduction is important only between Earth's surface and the air directly in contact with the surface. As a means of heat transfer for the atmosphere as a whole, conduction is the least significant.

STUDENTS SOMETIMES ASK...

In the morning when I get out of bed why does the tile flooring in the bathroom feel much colder than the carpeted area, even though both materials are the same temperature?

The difference you feel is due manly to the fact that floor tile is a much better conductor of heat than carpet. Hence, heat is more rapidly conducted from your bare feet when you are standing on the tile floor than when you are on the carpeted floor. Even at room temperature (20°C or 68°F) objects that are good conductors can feel chilly to the touch. (Remember, body temperature is about 98.6°F).

Convection

Much of the heat transport that occurs in the atmosphere is carried on by convection. **Convection** is *the transfer of heat by mass movement or circulation within a substance*. It takes place in fluids (for example, liquids like the ocean and gases like air) where the atoms and molecules are free to move about.

The pan of water in Figure 16.16 illustrates the nature of simple convective circulation. Radiation from the fire warms the bottom of the pan, which conducts heat to the water near the bottom of the container. As the water is heated, it expands and becomes less dense than the water above. Because of this new buoyancy, the warmer water rises. At the same time, cooler, denser water near the top of the pan sinks to the bottom, where it becomes heated. As long as the water is heated unequally—that is, from

the bottom up—the water will continue to "turn over," producing a *convective circulation*. In a similar manner, most of the heat acquired in the lowest layer of the atmosphere by way of radiation and conduction is transferred by convective flow.

On a global scale, convection in the atmosphere creates a huge, worldwide air circulation. This is responsible for the redistribution of heat between hot equatorial regions and the frigid poles. This important process will be discussed in detail in Chapter 18.

Radiation

The third mechanism of heat transfer is radiation. As shown in Figure 16.16, radiation travels out in all directions from its source. Unlike conduction and convection, which need a medium to travel through, radiant energy readily travels through the vacuum of space. Thus, radiation is the heat-transfer mechanism by which solar energy reaches our planet.

Because the Sun is the ultimate source of energy that creates our weather, we will consider the nature of solar radiation in more detail. From our everyday experience we know that the Sun emits light and heat as well as the ultraviolet rays that cause suntan. Although these forms of energy comprise a major portion of the total energy that radiates from the Sun, they are only part of a large array of energy called **radiation** or **electromagnetic radiation**. This array or spectrum of electromagnetic energy is shown in Figure 16.17. All radiation, whether X rays, radio waves, or heat waves, travel through the vacuum of space at 300,000 kilometers (186,000 miles) per second and only slightly slower through our atmosphere.

It helps to understand radiant energy by picturing ripples made in a pond by tossing in a pebble or by observing ocean waves. Like these waves, *electromagnetic waves*, as waves of radiant energy are called, come in various sizes. For our purpose, the most important difference among electromagnetic waves is their *wavelength*, or the distance from one crest to the next. Radio waves have the longest wavelengths, ranging to tens of kilometers. Gamma waves are the shortest, being less than a billionth of a centimeter long.

Visible light, as the name implies, is the only portion of the spectrum we can see. We often refer to visible light as "white" light because it appears "white" in color. However, it is easy to show that white light is really a mixture of colors, each corresponding to a specific wavelength (Figure 16.18). By using a prism, white light can be divided into the colors of the rainbow, from violet with the shortest wavelength—0.4 micrometer (1 micrometer is 0.0001 centimeter)—to red with the longest wavelength—0.7 micrometer.

Located adjacent to red, and having a longer wavelength, is **infrared** radiation, which we cannot see but which we can detect as heat. The closest invisible waves

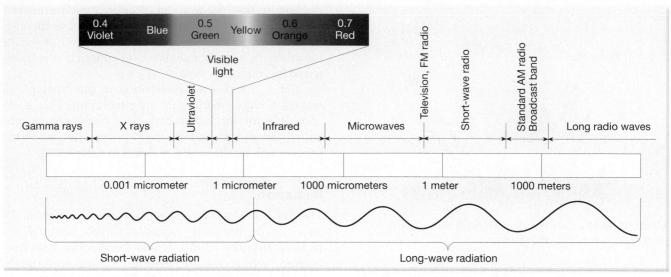

▲ **FIGURE 16.17** The electromagnetic spectrum, illustrating the wavelengths and names of various types of radiation.

to violet are called **ultraviolet** rays. They are responsible for the sunburn that can occur after an intense exposure to the Sun. Although we divide radiant energy into groups based on our ability to perceive them, all forms of radiation are basically the same. When any form of radiant energy is absorbed by an object, the result is an increase in molecular motion, which causes a corresponding increase in temperature.

To understand better how the atmosphere is heated, it is useful to have a general understanding of the basic laws governing radiation.

1. *All objects, at whatever temperature, emit radiant energy.* Thus, not only hot objects like the Sun but

▼ **FIGURE 16.18** Visible light consists of an array of colors we commonly call the "colors of the rainbow." Rainbows are relatively common optical phenomena produced by the bending and reflection of light by drops of water. Denali National Park, Alaska. (Photo © by Carr Clifton)

also Earth, including its polar ice caps, continually emit energy.

2. *Hotter objects radiate more total energy per unit area than do colder objects.*

3. *The hotter the radiating body, the shorter the wavelength of maximum radiation.* The Sun, with a surface temperature of nearly 6000°C, radiates maximum energy at 0.5 micrometer, which is in the visible range. The maximum radiation for Earth occurs at a wavelength of 10 micrometers, well within the infrared (heat) range. Because the maximum Earth radiation is roughly 20 times longer than the maximum solar radiation, it is often called long-wave radiation, and solar radiation is called short-wave radiation.

4. *Objects that are good absorbers of radiation are good emitters as well.* Earth's surface and the Sun are nearly perfect radiators because they absorb and radiate with nearly 100 percent efficiency for their respective temperatures. On the other hand, *gases are selective absorbers and radiators.* Thus, the atmosphere, which is nearly transparent to (does not absorb) certain wavelengths of radiation, is nearly opaque (a good absorber) to others. Our experience tells us that the atmosphere is transparent to visible light; hence, it readily reaches Earth's surface. This is not the case for the longer wavelength radiation emitted by Earth.

The Fate of Incoming Solar Radiation

  Earth's Dynamic Atmosphere
▼ Heating the Atmosphere

When radiation strikes an object, there are usually three different results. First, some of the energy is *absorbed* by the object. Recall that when radiant energy is absorbed, it is converted to heat, which causes an increase in temperature. Second, substances such as water and air are transparent to certain wavelengths of radiation. Such materials simply *transmit* this energy. Radiation that is transmitted does not contribute energy to the object. Third, some radiation may "bounce off" the object without being absorbed or transmitted. *Reflection* and *scattering* are responsible for redirecting incoming solar radiation. In summary, *radiation may be absorbed, transmitted, or redirected (reflected or scattered).*

Figure 16.19 shows the fate of incoming solar radiation averaged for the entire globe. Notice that the atmosphere is quite transparent to incoming solar radiation. On average, about 50 percent of the solar energy reaching the top of the atmosphere is absorbed at Earth's surface. Another 30 percent is reflected back to space by the atmosphere, clouds, and reflective surfaces such as snow and water. The remaining 20 percent is absorbed by clouds and the atmosphere's gases.

What determines whether solar radiation will be transmitted to the surface, scattered, reflected outward, or absorbed by the atmosphere? As you will see, it depends greatly on the wavelength of the energy being transmitted, as well as on the nature of the intervening material.

Reflection and Scattering

Reflection is the process whereby light bounces back from an object at the same angle at which it encounters a surface and with the same intensity (Figure 16.20A).

By contrast, **scattering** produces a larger number of weaker rays that travel in different directions. Although scattering disperses light both forward and backward (*backscattering*), more energy is dispersed in the forward direction (Figure 16.20B).

Reflection and Earth's Albedo. Energy is returned to space from Earth in two ways: reflection and emission of radiant energy. The portion of solar energy that is reflected back to space leaves in the same short wavelengths in which it came to Earth. About 30 percent of the solar energy reaching the outer atmosphere is reflected back to space. Included in this figure is the amount sent skyward by backscattering. This energy is lost to Earth and does not play a role in heating the atmosphere.

The fraction of the total radiation that is reflected by a surface is called its **albedo**. Thus, the albedo for Earth as a whole (the *planetary albedo*) is 30 percent. However, the albedo from place to place as well as from time to time in the same locale varies considerably, depending on the amount of cloud cover and particulate matter in the air, as well as on the angle of the Sun's rays and the nature of the surface. A lower Sun angle means that more atmosphere must be penetrated, thus making the "obstacle course" longer and therefore the loss of solar radiation greater (see Figure 16.11, p. 445). Table 16.2 gives the albedo for various surfaces. Note that the angle at which the Sun's rays strike a water surface greatly affects its albedo.

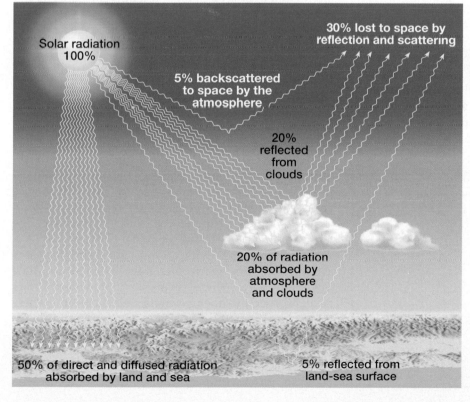

◀ **FIGURE 16.19** Average distribution of incoming solar radiation by percentage. More solar energy is absorbed by Earth's surface than by the atmosphere. Consequently, the air is not heated directly by the Sun, but is heated indirectly from Earth's surface.

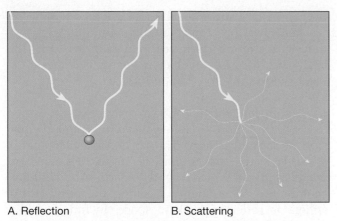

A. Reflection   B. Scattering

▲ **FIGURE 16.20**   *Reflection and scattering.* **A.** Reflected light bounces back from a surface at the same angle at which it strikes that surface and with the same intensity. **B.** When a beam of light is scattered, it results in a larger number of weaker rays, traveling in all different directions. Usually more energy is scattered in the forward direction than is backscattered.

Scattering.   Although incoming solar radiation travels in a straight line, small dust particles and gas molecules in the atmosphere scatter some of this energy in all directions. The result, called **diffused light**, explains how light reaches into the area beneath a shade tree, and how a room is lit in the absence of direct sunlight. Further, scattering accounts for the brightness and even the blue color of the daytime sky (see Box 16.3). In contrast, bodies like the Moon and Mercury, which are without atmospheres, have dark skies and "pitch-black" shadows, even during daylight hours. Overall, about half of the solar radiation that is absorbed at Earth's surface arrives as diffused (scattered) light.

Absorption

As stated earlier, gases are selective absorbers, meaning that they absorb strongly in some wavelengths, moderately in others, and only slightly in still others. When a gas molecule absorbs light waves, this energy is transformed into internal molecular motion, which is detectable as a rise in temperature.

TABLE 16.2	Albedo (reflectivity) of various surfaces.
Surface	**Percent Reflected**
Fresh snow	80–90
Old snow	50–60
Sand (beach, desert)	20–40
Grass	5–25
Dry soil (plowed field)	15–25
Wet earth (plowed field)	10
Forest	5–10
Water (Sun near horizon)	50–80
Water (Sun near zenith)	5–10
Thick cloud	70–85
Thin cloud	25–30
Earth and atmosphere (overall total)	30

Nitrogen, the most abundant constituent in the atmosphere, is a poor absorber of all types of incoming radiation. Oxygen and ozone are efficient absorbers of ultraviolet radiation. Oxygen removes most of the shorter ultraviolet radiation high in the atmosphere, and ozone absorbs most of the remaining ultraviolet rays in the stratosphere. The absorption of UV radiation in the stratosphere accounts for the high temperatures experienced there. The only other significant absorber of incoming solar radiation is water vapor, which, along with oxygen and ozone, accounts for most of the solar radiation absorbed within the atmosphere.

For the atmosphere as a whole, none of the gases are effective absorbers of visible radiation. This explains why most visible radiation reaches Earth's surface and why we say that the atmosphere is *transparent* to incoming solar radiation. Thus, the atmosphere does not acquire the bulk of its energy directly from the Sun. Rather, it is heated chiefly by energy that is first absorbed by Earth's surface and then reradiated to the sky.

Heating the Atmosphere: The Greenhouse Effect

GEODe   Earth's Dynamic Atmosphere
▼ Heating the Atmosphere

Approximately 50 percent of the solar energy that strikes the top of the atmosphere reaches Earth's surface and is absorbed. Most of this energy is then reradiated skyward. Because Earth has a much lower surface temperature than the Sun, the radiation that it emits has longer wavelengths than solar radiation.

The atmosphere as a whole is an efficient absorber of the longer wavelengths emitted by Earth (*terrestrial radiation*). Water vapor and carbon dioxide are the principal absorbing gases. Water vapor absorbs roughly five times more terrestrial radiation than do all other gases combined and accounts for the warm temperatures found in the lower troposphere, where it is most highly concentrated. Because the atmosphere is quite transparent to shorter-wavelength solar radiation and more readily absorbs longer-wavelength terrestrial radiation, the atmosphere is heated from the ground up rather than vice versa. This explains the general drop in temperature with increasing altitude experienced in the troposphere. The farther from the "radiator," the colder it becomes.

When the gases in the atmosphere absorb terrestrial radiation, they warm, but they eventually radiate this energy away. Some travels skyward, where it may be reabsorbed by other gas molecules, a possibility less likely with increasing height because the concentration of water vapor decreases with altitude. The remainder travels Earthward and is again absorbed by Earth. For this reason, Earth's surface is continually being supplied with heat from the atmosphere as well

BOX 16.3 UNDERSTANDING EARTH
Blue Skies and Red Sunsets

Gas molecules more effectively scatter the shorter wavelengths (blue and violet) of visible light than the longer wavelengths (red and orange). This fact explains the blue color of the sky and the orange and red colors seen at sunrise and sunset (see Figure 16.4B). Remember, sunlight is composed of all colors. When the Sun is overhead you can look in any direction away from the direct Sun and see predominantly blue light, which is the wavelength more readily scattered by the atmosphere.

Conversely, the Sun appears to have an orangish-to-reddish tint when viewed near the horizon (Figure 16.D). This is because solar radiation must travel through a greater thickness of atmosphere before it reaches your eyes (see Figure 16.11). As a consequence, most of the blue and violet wavelengths will be scattered out, leaving light that consists mostly of reds and oranges. The reddish appearance of clouds during sunrise and sunset also results because the clouds are illuminated by light from which the blue color has been subtracted by scattering.

The most spectacular sunsets occur when large quantities of fine dust or smoke particles penetrate into the stratosphere. For three years after the great eruption of the Indonesian volcano Krakatau in 1883, brilliant sunsets occurred worldwide. The European summer that followed this colossal explosion was cooler than normal, a fact that has been attributed to the greater loss of radiation caused by backscattering.

Large particles associated with haze, fog, or smog scatter light more equally in all wavelengths. Because no color is predominant over any other, the sky appears white or gray on days when large particles are abundant.

In summary, the color of the sky gives an indication of the number of large or small particles present. Lots of small particles produce red sunsets, whereas large particles produce a white sky. Furthermore, the bluer the sky, the cleaner the air.

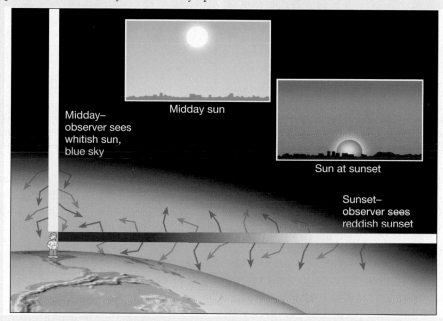

Midday–observer sees whitish sun, blue sky

Midday sun

Sun at sunset

Sunset–observer sees reddish sunset

Figure 16.D Short wavelengths (blue and violet) of visible light are scattered more effectively than are longer wavelengths (red, orange). Therefore, when the Sun is overhead an observer can look in any direction and see predominantly blue light that was selectively scattered by the gases in the atmosphere. By contrast, at sunset, the path that light must take through the atmosphere is much longer. Consequently, most of the blue light is scattered before it reaches an observer. Thus, the Sun appears reddish in color.

---

as from the Sun. Without these absorptive gases in our atmosphere, Earth would not be a suitable habitat for humans and numerous other life forms.

This very important phenomenon has been termed the **greenhouse effect** because it was once thought that greenhouses were heated in a similar manner (Figure 16.21). The gases of our atmosphere, especially water vapor and carbon dioxide, act very much like the glass in the greenhouse. They allow shorter-wavelength solar radiation to enter, where it is absorbed by the objects inside. These objects in turn reradiate the heat, but at longer wavelengths, to which glass is nearly opaque. The heat therefore is trapped in the greenhouse. However, a more important factor in keeping a greenhouse warm is the fact that the greenhouse itself prevents mixing of air inside with cooler air outside. Nevertheless, the term *greenhouse effect* is still used.

? STUDENTS SOMETIMES ASK...
Isn't the greenhouse effect responsible for global warming?

The popular press frequently points to the greenhouse effect as the "villain" of the global-warming problem. It is important to note that the greenhouse effect and global warming *are not* the same thing. Without the greenhouse effect, Earth would be uninhabitable. We do have mounting evidence that human activity (particularly the release of carbon dioxide into the atmosphere) is responsible for a rise in global temperatures (see Chapter 19). Thus, human activities seem to be enhancing an otherwise natural process (the greenhouse effect) to increase Earth's temperature. Nevertheless, to equate the greenhouse effect, which makes life possible, with undesirable changes to our atmosphere caused by human activity is incorrect.

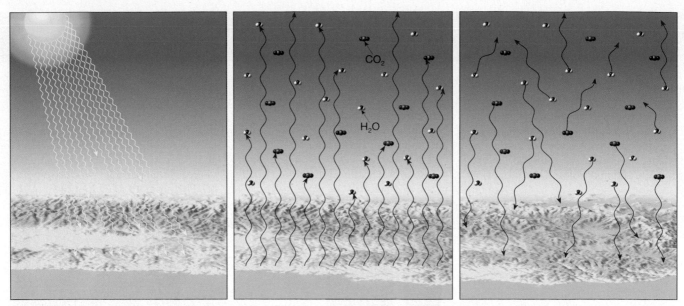

▲ **FIGURE 16.21** The heating of the atmosphere. Most of the short-wavelength radiation from the Sun passes through the atmosphere and is absorbed by Earth's land-sea surface. This energy is then emitted from the surface as longer-wavelength radiation, much of which is absorbed by certain gases in the atmosphere. Some of the energy absorbed by the atmosphere will be reradiated Earthward. This so-called greenhouse effect is responsible for keeping Earth's surface much warmer than it would be otherwise.

For the Record: Air Temperature Data

Temperature is one of the basic elements of weather and climate. When someone asks what it is like outside, air temperature is often the first element we mention. At a weather station, the temperature is read on a regular basis from instruments mounted in an instrument shelter (Figure 16.22). The shelter protects the instruments from direct sunlight and allows a free flow of air. In addition to a standard mercury thermometer, the shelter is likely to contain a thermograph to continuously record temperature and a set of maximum–minimum thermometers. As their name implies, these thermometers record the highest and lowest temperatures during a measurement period, usually 24 hours.

The daily maximum and minimum temperatures are the bases for many of the temperature data compiled by meteorologists:

1. By adding the maximum and minimum temperatures and then dividing by two, the *daily mean temperature* is calculated.

2. The *daily range* of temperature is computed by finding the difference between the maximum and minimum temperatures for a given day.

3. The *monthly mean* is calculated by adding together the daily means for each day of the month and dividing by the number of days in the month.

4. The *annual mean* is an average of the 12 monthly means.

5. The *annual temperature range* is computed by finding the difference between the highest and lowest monthly means.

Mean temperatures are particularly useful for making comparisons, whether on a daily, monthly, or annual basis. It is common to hear a weather reporter state, "Last month was the hottest July on record," or "Today Chicago was 10 degrees warmer than Miami." Temperature ranges are also useful statistics, because they give an indication of extremes.

To examine the distribution of air temperatures over large areas, isotherms are commonly used. An **isotherm** is a line that connects points on a map that have the same temperature (*iso* = equal, *therm* = temperature). Therefore, all points through which an isotherm passes have identical temperatures for the time period indicated. Generally, isotherms representing 5° or 10° temperature differences are used, but any interval may be chosen. Figure 16.23 illustrates how isotherms are drawn on a map. Notice that most isotherms do not pass directly through the observing stations, because the station readings may not coincide with the values chosen for the isotherms. Only an occasional station temperature will be exactly the same as the value of the isotherm, so it is usually necessary to draw the lines by estimating the proper position between stations.

Isothermal maps are valuable tools because they clearly make temperature distribution visible at a glance. Areas of low and high temperatures are easy to pick out. In addition, the amount of temperature change per unit of distance, called the *temperature gradient*, is easy to visualize. Closely spaced isotherms indicate a

A.   **B.**

▲ **FIGURE 16.22**   **A**. Standard instrument shelter. A shelter protects instruments from direct sunlight and allows for the free flow of air. (Courtesy of Qualimetrics, Inc.) **B**. This modern shelter contains an electrical thermometer called a *thermistor*. (Photo by Bobbé Christopherson)

rapid rate of temperature change, whereas more widely spaced lines indicate a more gradual rate of change. You can see this in Figure 16.23. The isotherms are closer in Colorado and Utah (steeper temperature gradient), whereas the isotherms are spread farther in Texas (gentler temperature gradient). Without isotherms, a map would be covered with numbers representing temperatures at dozens or hundreds of places, which would make patterns difficult to see.

Why Temperatures Vary: The Controls of Temperature

A *temperature control* is any factor that causes temperature to vary from place to place and from time to time. Earlier in this chapter we examined the most important cause for temperature variations—differences in the receipt of solar radiation. Because variations in Sun angle and length of daylight depend on latitude, they are responsible for warm temperatures in the tropics and colder temperatures at more poleward locations. Of course, seasonal temperature changes at a given lat-

itude occur as the Sun's vertical rays migrate toward and away from a place during the year.

However, latitude is not the only control of temperature. If it were, we would expect all places along the same parallel of latitude to have identical temperatures. This is clearly not the case. For example, Eureka, California, and New York City are both coastal cities at about the same latitude and both have an annual mean temperature of 11°C (52°F). However, New York City is 9°C (16°F) warmer than Eureka in July and 10°C (18°F) cooler in January. In another example, two cities in Ecuador—Quito and Guayaquil—are relatively close to each other, yet the annual mean temperatures of these two cities differ by 12°C (21°F). To explain these situations and countless others, we must realize that factors other than latitude also exert a strong influence on temperature. In the next sections, we will examine these other controls, which include differential heating of land and water, altitude, geographic position, cloud cover and albedo, and ocean currents.*

---

*For a discussion of the effects of ocean currents on temperature, see Chapter 15.

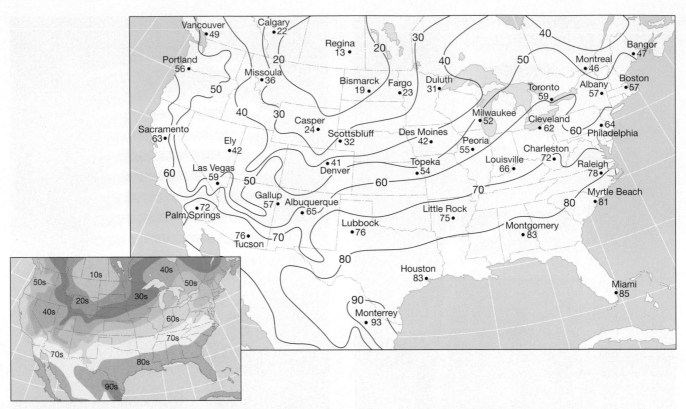

▲ **FIGURE 16.23** Temperature distribution using isotherms. Isotherms are lines that connect points of equal temperature. Showing temperature distribution in this way makes patterns easier to see. On television, and in many newspapers, temperature maps are in color. Rather than labeling isotherms, the area *between* isotherms is labeled. For example, the zone between the 60° and 70° isotherms is labeled "60s."

STUDENTS SOMETIMES ASK...

What are the highest and lowest temperatures ever recorded at Earth's surface?

The world's record-high temperature is nearly 59°C (136°F). It was recorded on September 13, 1922, at Azizia, Libya, in North Africa's Sahara Desert. The lowest recorded temperature is −89°C (−129°F). It should come as no surprise that this incredibly frigid temperature was recorded in Antarctica, at the Russian Vostok Station, on August 24, 1960.

Land and Water

The heating of Earth's surface controls the heating of the air above it. Therefore, to understand variations in air temperature, we must examine the nature of the surface. Different land surfaces absorb varying amounts of incoming solar energy, which in turn cause variations in the temperature of the air above. The largest contrast, however, is not between different land surfaces, but between land and water. *Land heats more rapidly and to higher temperatures than water and cools more rapidly and to lower temperatures than water.* Temperature variations, therefore, are considerably greater over land than over water.

Among the reasons for the differential heating of land and water are the following:

1. The *specific heat* (amount of energy needed to raise 1 gram of a substance 1°C) is far greater for water than for land. Thus, water requires a great deal more heat to raise its temperature the same amount as an equal quantity of land.

2. Land surfaces are opaque, so heat is absorbed only at the surface. Water, being more transparent, allows some solar radiation to penetrate to a depth of many meters.

3. The water that is heated often mixes with water below, thus distributing the heat through an even larger mass.

4. Evaporation (a cooling process) from water bodies is greater than that from land surfaces.

All these factors collectively cause water to warm more slowly, store greater quantities of heat, and cool more slowly than land.

Monthly temperature data for two cities will demonstrate the moderating influence of a large water body and the extremes associated with land (Figure 16.24). Vancouver, British Columbia, is located along

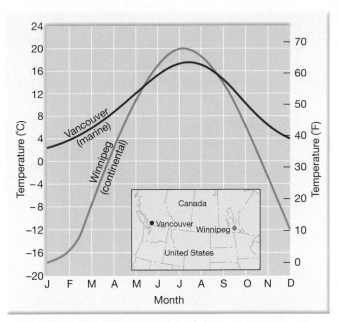

▲ FIGURE 16.24 Mean monthly temperatures for Vancouver, British Columbia, and Winnipeg, Manitoba. Vancouver has a much smaller annual temperature range owing to the strong marine influence of the Pacific Ocean. Winnipeg illustrates the greater extremes associated with an interior location.

the windward Pacific coast, whereas Winnipeg, Manitoba, is in a continental position far from the influence of water. Both cities are at about the same latitude and thus experience similar Sun angles and lengths of daylight. Winnipeg, however, has a mean January temperature that is 20°C lower than Vancouver's. Conversely, Winnipeg's July mean is 2.6°C higher than Vancouver's. Although their latitudes are nearly the same, Winnipeg, which has no water influence, experiences much greater temperature extremes than does Vancouver, which does. The key to Vancouver's moderate year-round climate is the Pacific Ocean.

On a different scale, the moderating influence of water may also be demonstrated when temperature variations in the Northern and Southern hemispheres are compared. In the Northern Hemisphere, 61 percent is covered by water, and land accounts for the remaining 39 percent. However, in the Southern Hemisphere, 81 percent is covered by water and 19 percent by land. The Southern Hemisphere is correctly called the *water hemisphere* (see Figure 13.1, p. 360). Table 16.3 portrays the considerably smaller annual temperature variations in the water-dominated Southern Hemisphere as compared with the Northern Hemisphere.

Altitude

The two cities in Ecuador mentioned earlier—Quito and Guayaquil—demonstrate the influence of altitude on mean temperature (Figure 16.25). Both cities are near the equator and relatively close to one another,

TABLE 16.3	Variation in annual mean temperature range (°C) with latitude.	
Latitude	Northern Hemisphere	Southern Hemisphere
0	0	0
15	3	4
30	13	7
45	23	6
60	30	11
75	32	26
90	40	31

but the annual mean temperature at Guayaquil is 25°C (77°F) compared to Quito's mean of 13°C (55°F). The difference may be understood when the cities' elevations are noted. Guayaquil is only 12 meters (40 feet) above sea level, whereas Quito is high in the Andes Mountains at 2800 meters (9200 feet).

Recall that temperatures drop an average of 6.5°C per kilometer in the troposphere; thus, cooler temperatures are to be expected at greater heights (see Figure 16.6, p. 442). Yet, the magnitude of the difference is not explained completely by the normal lapse rate. If the normal lapse rate is used, we would expect Quito to be about 18°C cooler than Guayaquil, but the difference is only 12°C. The fact that high-altitude places such as Quito are warmer than the value calculated using the normal lapse rate results from the absorption and reradiation of solar energy by the ground surface.

▼ FIGURE 16.25 Because Quito is high in the Andes Mountains, it experiences much cooler temperatures than Guayaquil, which is near sea level. Because both cities are near the equator, they have a negligible annual temperature range.

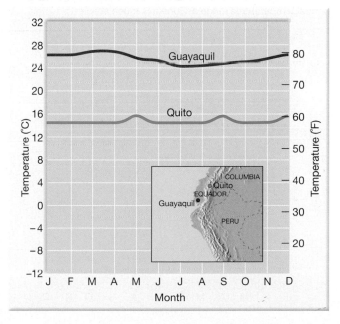

Geographic Position

The geographic setting can greatly influence the temperatures experienced at a specific location. A coastal location where prevailing winds blow from the ocean onto the shore (a *windward* coast) experiences considerably different temperatures than does a coastal location where the prevailing winds blow from the land toward the ocean (a *leeward* coast). In the first situation, the windward coast will experience the full moderating influence of the ocean—cool summers and mild winters, compared to an inland station at the same latitude.

A leeward coast, on the other hand, will have a more continental temperature pattern because the winds do not carry the ocean's influence onshore. Eureka, California, and New York City, the two cities mentioned earlier, illustrate this aspect of geographic position. The annual temperature range at New York City is 19°C (34°F) greater than Eureka's (Figure 16.26).

Seattle and Spokane, both in the state of Washington, illustrate a second aspect of geographic position—mountains that act as barriers. Although Spokane is only about 360 kilometers (220 miles) east of Seattle, the towering Cascade Range separates the cities. Consequently, Seattle's temperatures show a marked marine influence, but Spokane's are more typically continental (Figure 16.27). Spokane is 7°C (13°F) cooler than Seattle in January and 4°C (7°F) warmer than

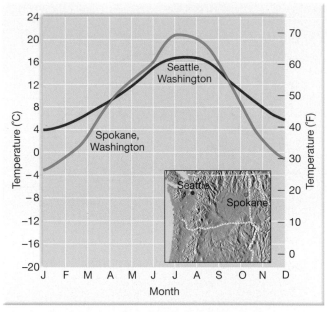

▲ **FIGURE 16.27**   Monthly mean temperatures for Seattle and Spokane, Washington. Because the Cascade Mountains cut off Spokane from the moderating influence of the Pacific Ocean, its annual temperature range is greater than Seattle's.

Seattle in July. The annual range at Spokane is 11°C (20°F) greater than at Seattle. The Cascade Range effectively cuts Spokane off from the moderating influence of the Pacific Ocean.

Cloud Cover and Albedo

The extent of cloud cover is a factor that influences temperatures in the lower atmosphere. Cloud cover is important because many clouds have a high albedo and therefore reflect a significant portion of the sunlight that strikes them back to space. By reducing the amount of incoming solar radiation, daytime temperatures will be lower than if the clouds were absent and the sky were clear (Figure 16.28A).

At night, clouds have the opposite effect as during daylight. They act as a blanket by absorbing outgoing terrestrial radiation and reradiating a portion of it back to the surface (Figure 16.28B). Consequently, some of the heat that otherwise would have been lost remains near the ground. Thus, nighttime air temperatures do not drop as low as they would on a clear night. The effect of cloud cover is to reduce the daily temperature range by lowering the daytime maximum and raising the nighttime minimum.

Clouds are not the only phenomenon that increase albedo and thereby reduce air temperatures. We also recognize that snow- and ice-covered surfaces have high albedos. This is one reason why mountain glaciers do not melt away in the summer

▼ **FIGURE 16.26**   Monthly mean temperatures for Eureka, California, and New York City. Both cities are coastal and located at about the same latitude. Because Eureka is strongly influenced by prevailing winds from the ocean and New York City is not, the annual temperatures range at Eureka is much smaller.

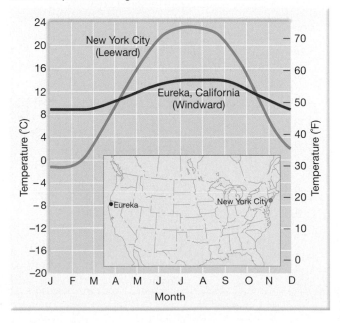

▲ **FIGURE 16.28** Clouds reduce the daily temperature range.
A. During daylight hours, clouds reflect solar radiation back to space. Therefore, the maximum temperature is lower than if the sky were clear. **B.** At night, the minimum temperature will not fall as low because clouds retard the loss of heat.

and why snow may still be present on a mild spring day. In addition, during the winter, when snow covers the ground, daytime maximums on a sunny day are less than they otherwise would be because energy that the land would have absorbed and used to heat the air is reflected and lost.

World Distribution of Temperature

Take a moment to study the two world isothermal maps (Figures 16.29 and 16.30). From hot colors near the equator to cool colors toward the poles, these maps portray sea-level temperatures in the seasonally extreme months of January and July. On these maps you can study global temperature patterns and the effects of the controlling factors of temperature, especially latitude, the distribution of land and water, and ocean currents. Like most isothermal maps of large regions,

all temperatures on these world maps have been reduced to sea level to eliminate the complications caused by differences in altitude.

On both maps, the isotherms generally trend east and west and show a decrease in temperatures poleward from the tropics. They illustrate one of the most fundamental aspects of world temperature distribution: that the effectiveness of incoming solar radiation in heating Earth's surface and the atmosphere above it is largely a function of latitude.

Moreover, there is a latitudinal shifting of temperatures caused by the seasonal migration of the Sun's vertical rays. To see this, compare the color bands by latitude on the two maps. For example, on the January map, the hot spots of 30°C are *south* of the equator, but in July they have shifted *north* of the equator.

If latitude were the only control of temperature distribution, our analysis could end here, but this is not the case. The added effect of the differential heating of land and water is clearly reflected on the January and July temperature maps. The warmest and coldest temperatures are found over land—note the coldest area, a purple oval in Siberia, and the hottest areas, the deep orange ovals—all over land. Consequently, because temperatures do not fluctuate as much over water as over land, the north-south migration of isotherms is greater over the continents than over the oceans.

In addition, it is clear that the isotherms in the Southern Hemisphere, where there is little land and where the oceans predominate, are much more regular than in the Northern Hemisphere, where they bend sharply northward in July and southward in January over the continents.

Isotherms also reveal the presence of ocean currents. Warm currents cause isotherms to be deflected toward the poles, whereas cold currents cause an equatorward bending. The horizontal transport of water poleward warms the overlying air and results in air temperatures that are higher than would otherwise be expected for the latitude. Conversely, currents moving toward the equator produce cooler than expected air temperatures.

Figures 16.29 and 16.30 show the seasonal extremes of temperature, so comparing them enables us to see the annual range of temperature from place to place. Comparing the two maps shows that a station near the equator has a very small annual range because it experiences little variation in the length of daylight and it always has a relatively high Sun angle. A station in the middle latitudes, however, experiences wide variations in Sun angle and length of daylight and hence large variations in temperature. Therefore, we can state that the annual temperature range increases with an increase in latitude.

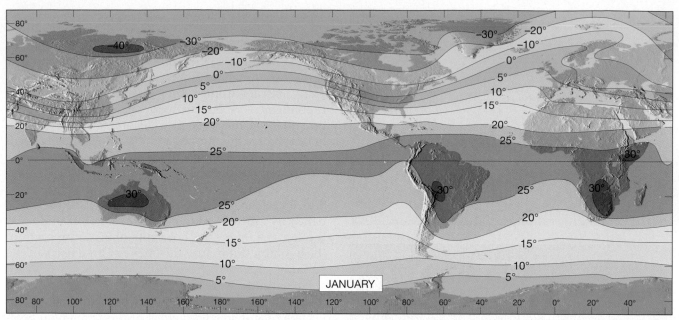

▲ **FIGURE 16.29**   World mean sea-level temperatures in January in degrees Celsius.

Moreover, land and water also affect seasonal temperature variations, especially outside the tropics. A continental location must endure hotter summers and colder winters than a coastal location. Consequently, outside the tropics the annual temperature range will increase with an increase in continentality.

A classic example of the effect of latitude and continentality on annual temperature range is Yakutsk, Russia. This city in the heart of Siberia is just a few degrees south of the Arctic Circle and far from the influence of water. As a result, Yakutsk has an average annual temperature range of 62.2°C (112°F), among the greatest ranges in the world.

▼ **FIGURE 16.30**   World mean sea-level temperatures in July in degrees Celsius.

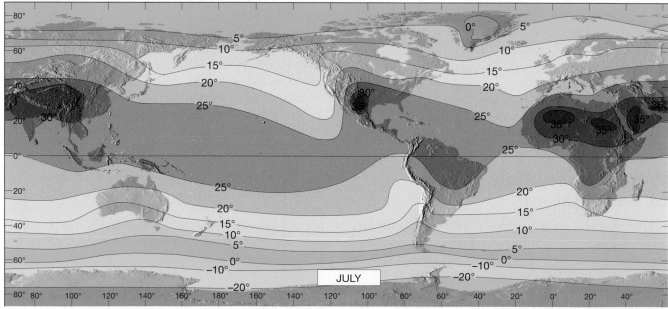

Chapter Summary

- *Weather* is the state of the atmosphere at a particular place for a short period of time. *Climate*, on the other hand, is a generalization of the weather conditions of a place over a long period of time.

- The most important *elements*, those quantities or properties that are measured regularly, of weather and climate are (1) *air temperature*, (2) *humidity*, (3) type and amount of *cloudiness*, (4) type and amount of *precipitation*, (5) *air pressure*, and (6) the speed and direction of the *wind*.

- If water vapor, dust, and other variable components of the atmosphere were removed, clean, dry air would be composed almost entirely of *nitrogen* (N), about 78% of the atmosphere by volume, and *oxygen* (O_2) about 21%. *Carbon dioxide* (CO_2) although present only in minute amounts (0.036%), is important because it has the ability to absorb heat radiated by Earth and thus helps keep the atmosphere warm. Among the variable components of air, *water vapor* is very important because it is the source of all clouds and precipitation and, like carbon dioxide, it is also a heat absorber.

- *Ozone* (O_3) the triatomic form of oxygen, is concentrated in the 10- to 50-kilometer altitude range of the atmosphere, and is important to life because of its ability to absorb potentially harmful ultraviolet radiation from the Sun.

- Because the atmosphere gradually thins with increasing altitude, it has no sharp upper boundary but simply blends into outer space. Based on temperature, the atmosphere is divided vertically into four layers. The *troposphere* is the lowermost layer. In the troposphere, temperature usually decreases with increasing altitude. This *environmental lapse rate* is variable, but averages about 6.5°C per kilometer (3.5°F per 1000 feet). Essentially all important weather phenomena occur in the troposphere. Above the troposphere is the *stratosphere*, which exhibits warming because of absorption of ultraviolet radiation by ozone. In the *mesosphere*, temperatures again decrease. Upward from the mesosphere is the *thermosphere*, a layer with only a tiny fraction of the atmosphere's mass and no well-defined upper limit.

- The two principal motions of Earth are (1) *rotation*, the spinning of Earth about its axis, which produces the daily cycle of daylight and darkness; and (2) *revolution*, the movement of Earth in its orbit around the Sun.

- *Several factors act together to cause the seasons*. Earth's axis is inclined $23\frac{1}{2}°$ from the perpendicular to the plane of its orbit around the Sun and remains pointed in the same direction (toward the North Star) as Earth's journeys around the Sun. As a consequence, Earth's orientation to the Sun continually changes. The yearly fluctuations in the angle of the Sun and length of daylight brought about by Earth's changing orientation to the Sun cause the seasons.

- The three mechanisms of heat transfer are (1) *conduction*, the transfer of heat through matter by molecular activity; (2) *convection*, the transfer of heat by the movement of a mass or substance from one place to another; and (3) *radiation*, the transfer of heat by electromagnetic waves.

- *Electromagnetic radiation* is energy emitted in the form of rays, called electromagnetic waves. All radiation is capable of transmitting energy through the vacuum of space. An important difference among electromagnetic waves is their *wavelengths*, which range from very long *radio waves* to very short *gamma rays*. *Visible light* is the only portion of the electromagnetic spectrum we can see. Some of the basic laws that govern radiation as it heats the atmosphere are (1) all objects emit radiant energy; (2) hotter objects radiate more total energy than do colder objects; (3) the hotter the radiating body, the shorter the wavelengths of maximum radiation; and (4) objects that are good absorbers of radiation are good emitters as well.

- The general drop in temperature with increasing altitude in the troposphere supports the fact that the *atmosphere is heated from the ground up*. Approximately 50 percent of the solar energy that strikes the top of the atmosphere is ultimately absorbed at Earth's surface. Earth emits the absorbed radiation in the form of long-wave radiation. The atmospheric absorption of this long-wave *terrestrial radiation*, primarily by water vapor and carbon dioxide, is responsible for heating the atmosphere.

- The factors that cause temperature to vary from place to place, also called the *controls of temperature*, are (1) *differences in the receipt of solar radiation*—the greatest single cause; (2) the unequal heating and cooling of *land and water*, in which land heats more rapidly and to higher temperatures than water and cools more rapidly and to lower temperatures than water; (3) *altitude*; (4) *geographic position*; (5) *cloud cover and albedo*; and (6) *ocean currents*.

- Temperature distribution is shown on a map by using *isotherms*, which are lines that connect equal temperatures. Differences between January and July temperatures around the world can be explained in terms of the basic controls of temperature.

Key Terms

aerosols (p. 439)
air (p. 437)
albedo (p. 451)
autumnal equinox (p. 446)
circle of illumination (p. 443)
climate (p. 437)
conduction (p. 448)
convection (p. 448)
diffused light (p. 452)
electromagnetic radiation (p. 449)
element (of weather and climate) (p. 437)
environmental lapse rate (p. 441)

greenhouse effect (p. 453)
heat (p. 448)
inclination of the axis (p. 445)
infrared (p. 449)
isotherm (p.454)
mesosphere (p. 442)
ozone (p. 439)
radiation (p. 449)
reflection (p. 451)
revolution (p. 443)
rotation (p. 443)
scattering (p. 451)
spring equinox (p. 446)

stratosphere (p. 442)
summer solstice (p. 446)
temperature (p. 448)
thermosphere (p. 443)
Tropic of Cancer (p. 446)
Tropic of Capricorn (p. 446)
troposphere (p. 441)
ultraviolet (p. 450)
visible light (p. 449)
weather (p. 436)
winter solstice (p. 446)

Review Questions

1. Distinguish between weather and climate.

2. List the basic elements of weather and climate.

3. What are the two major components of clean, dry air? Are they important meteorologically?

4. Why are water vapor and aerosols important constituents of our atmosphere?

5. What source is responsible for the most pollution? (See Figure 16.A.)

6. What is the difference between primary and secondary pollutants? (See Box 16.1.)

7. (a) Why is ozone important to life on Earth? (b) What is the most serious threat to human health of a decrease in the stratosphere's ozone? (See Box 16.2.)

8. The atmosphere is divided vertically into four layers on the basis of temperature. List the names of these layers in order (from lowest to highest) and describe how temperature changes in each layer.

9. If the temperature at sea level were 23°C, what would the air temperature be at a height of 2 kilometers *under average conditions*?

10. Why do temperatures increase in the stratosphere?

11. Briefly explain the primary cause of the seasons.

12. After examining Table 16.1, write a general statement that relates the season, the latitude, and the length of daylight.

13. Describe the relationship between the temperature of a radiating body and the wavelengths it emits.

14. Distinguish between heat and temperature.

15. Describe the three basic mechanisms of heat transfer.

16. Figure 16.19 illustrates what happens to incoming solar radiation. The percentages shown, however, are only global averages. In particular, the amount of solar radiation reflected (albedo) may vary considerably. What factors might cause variations in albedo?

17. Why does the daytime sky usually appear blue? Why may the sky appear red or orange near sunrise or sunset? (See Box 16.3.)

18. Describe the process by which Earth's atmosphere is heated.

19. How are the following temperature data computed: daily mean, daily range, monthly mean, annual mean, annual range?

20. Quito, Ecuador, is located on the equator and is not a coastal city. It has an average annual temperature of only 13°C (55°F). What is the likely cause for this low average temperature?

21. In what ways can geographic position be considered a control of temperature?

22. How does cloud cover influence the maximum temperature on an overcast day? How is the nighttime minimum influenced by clouds?

23. Yakutsk is located in Siberia at about 60°N latitude. This Russian city has one of the highest average annual temperature ranges in the world: 62.2°C (112°F). Explain the reasons for the very high annual temperature range.

Examining the Earth System

1. Give an example or two of how the Earth system might be affected if Earth's axis were perpendicular to the plane of its orbit instead of being inclined $23\frac{1}{2}°$.

2. Speculate on the changes in global temperatures that might occur if Earth had substantially more land area and less ocean area than at present. How might such changes influence the biosphere?

Online Study Guide

The *Earth Science* Web site uses the resources and flexibility of the Internet to aid in your study of the topics in this chapter. Written and developed by Earth science instructors, this site will help improve your understanding of Earth science. Visit **http://www.prenhall.com/tarbuck** and click on the cover of *Earth Science* 11e to find:

- **Online review quizzes.**
- **Critical thinking exercises.**
- **Links to chapter-specific Web resources.**
- **Internet-wide key term searches.**

http://www.prenhall.com/tarbuck

Lightning display associated with a thunderstorm (cumulonimbus clouds) near Colorado Springs, Colorado. (Photo by Sean Cayton/The Image Works)

CHAPTER 17

Moisture, Clouds, and Precipitation

Water vapor is an odorless, colorless gas that mixes freely with the other gases of the atmosphere. Unlike oxygen and nitrogen—the two most abundant components of the atmosphere—water can change from one state of matter to another (solid, liquid, or gas) at the temperatures and pressures experienced on Earth. (By contrast, nitrogen will not condense to a liquid unless its temperature is lowered to −196°C (−371°F). Because of this unique property, water freely leaves the oceans as a gas and returns again as a liquid (Figure 17.1)

As you observe day-to-day weather changes, you might ask: Why is it generally more humid in the summer than in the winter? Why do clouds form on some occasions but not on others? Why do some clouds look thin and harmless whereas others form gray and ominous towers? Answers to these questions involve the role of water vapor in the atmosphere, the central theme of this chapter.

Water's Changes of State

Earth's Dynamic Atmosphere
▼ Moisture and Cloud Formation

Water is the only substance that exists in the atmosphere as a solid (ice), liquid, and gas (water vapor). It is made of hydrogen and oxygen atoms that are bonded together to form water molecules (H_2O). In all three states of matter (even ice) these molecules are in constant motion—the higher the temperature, the more vigorous the movement. The chief difference among liquid water, ice, and water vapor is the arrangement of the water molecules.

Ice, Liquid Water, and Water Vapor

Ice is composed of water molecules that are held together by mutual molecular attractions. Here the molecules form a tight, orderly network as shown in Figure 17.2. As a consequence, the water molecules in ice are not free to move relative to each other but rather vibrate about fixed sites. When ice is heated, the molecules oscillate more rapidly. When the rate of molecular movement increases sufficiently, the bonds between some of the water molecules are broken, resulting in melting.

In the liquid state, water molecules are still tightly packed but are moving fast enough that they are able to slide past one another. As a result, liquid water is fluid and will take the shape of its container.

As liquid water gains heat from its environment, some of the molecules will acquire enough energy to break the remaining molecular attractions and escape from the surface, becoming water vapor. Water-vapor molecules are widely spaced compared to liquid water and exhibit very energetic random motion. What distinguishes a gas from a liquid is its compressibility (and expandability). For example, you can easily put more and more air into a tire and increase its volume only slightly. However, don't try to put 10 gallons of gasoline into a five-gallon can.

▼ **FIGURE 17.1** People caught in a downpour. (Photo by Mary Fulton/Getty Images, Inc.—Liaison)

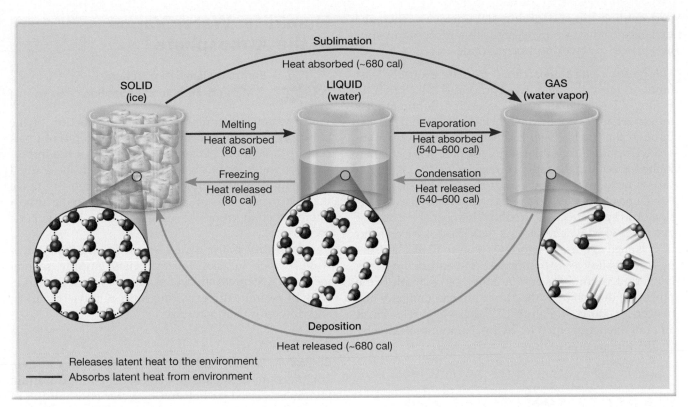

▲ **FIGURE 17.2** Changes of state.

To summarize, when water changes state, it does not turn into a different substance; only the distances and interactions among the water molecules change.

Latent Heat

Whenever water changes state, heat is exchanged between water and its surroundings. When water evaporates, heat is absorbed (Figure 17.2). Meteorologists often measure heat energy in calories. One **calorie** is the amount of heat required to raise the temperature of 1 gram of water 1°C (1.8°F). Thus, when 10 calories of heat are absorbed by 1 gram of water, the molecules vibrate faster and a 10°C (18°F) temperature rise occurs.

Under certain conditions, heat may be added to a substance without an accompanying rise in temperature. For example, when a glass of ice water is warmed, the temperature of the ice-water mixture remains a constant 0°C (32°F) until all the ice has melted. If adding heat does not raise the temperature, where does this energy go? In this case, the added energy went to break the molecular attractions between the water molecules in the ice cubes.

Because the heat used to melt ice does not produce a temperature change, it is referred to as **latent heat**. (Latent means *hidden*, like the latent fingerprints hidden at a crime scene.) This energy can be thought of as being stored in liquid water, and it is not released to its surroundings as heat until the liquid returns to the solid state.

It requires 80 calories to melt one gram of ice, an amount referred to as *latent heat of melting*. *Freezing*, the reverse process, releases these 80 calories per gram to the environment as *latent heat of fusion*.

Evaporation and Condensation.   We saw that heat is absorbed when ice is converted to liquid water. Heat is also absorbed during **evaporation**, the process of converting a liquid to a gas (vapor). The energy absorbed by water molecules during evaporation is used to give them the motion needed to escape the surface of the liquid and become a gas. This energy is referred to as the *latent heat of vaporization*. During the process of evaporation, it is the higher-temperature (faster-moving) molecules that escape the surface. As a result, the average molecular motion (temperature) of the remaining water is reduced—hence, the common expression "evaporation is a cooling process." You have undoubtedly experienced this cooling effect on stepping dripping wet from a swimming pool or bathtub. In this situation the energy used to evaporate water comes from your skin—hence, you feel cool.

Condensation, the reverse process, occurs when water vapor changes to the liquid state. During condensation, water-vapor molecules release energy (*latent heat of condensation*) in an amount equivalent to what was absorbed during evaporation. When condensation occurs in the atmosphere, it results in the formation of such phenomena as fog and clouds.

As you will see, latent heat plays an important role in many atmospheric processes. In particular, when water vapor condenses to form cloud droplets, latent heat of condensation is released, warming the surrounding air and giving it buoyancy. When the moisture content of air is high, this process can spur the growth of towering storm clouds.

Sublimation and Deposition. You are probably least familiar with the last two processes illustrated in Figure 17.2—sublimation and deposition. **Sublimation** is the conversion of a solid directly to a gas without passing through the liquid state. Examples you may have observed include the gradual shrinking of unused ice cubes in the freezer and the rapid conversion of dry ice (frozen carbon dioxide) to wispy clouds that quickly disappear.

Deposition refers to the reverse process, the conversion of a vapor directly to a solid. This change occurs, for example, when water vapor is deposited as ice on solid objects such as grass or windows (Figure 17.3). These deposits are called *white frost* or *hoar frost* and are frequently referred to simply as *frost*. A household example of the process of deposition is the "frost" that accumulates in a freezer. As shown in Figure 17.2, deposition releases an amount of energy equal to the total amount released by condensation and freezing.

? STUDENTS SOMETIMES ASK...
What is freezer burn?

"Freezer burn" is a common expression used to describe foods that have been left in a frost-free refrigerator for extended periods of time. Frost-free refrigerators circulate comparatively dry air through the freezer compartments. This causes ice on the freezer walls to sublimate (change from a solid to a gas) so it can be removed by the circulating air. Unfortunately, this process also removes moisture from those frozen foods that are not in airtight containers. Consequently, over a period of a few months these foods begin to dry out rather than actually being burned.

Humidity: Water Vapor in the Atmosphere

 Earth's Dynamic Atmosphere
▼ Moisture and Cloud Formation

Water vapor constitutes only a small fraction of the atmosphere, varying from as little as one-tenth of 1 percent up to about 4 percent by volume. But the importance of water in the air is far greater than these small percentages would indicate. Indeed, scientists agree that *water vapor* is the most important gas in the atmosphere when it comes to understanding atmospheric processes.

Humidity is the general term for the amount of water vapor in air. Meteorologists employ several methods to express the water-vapor content of the air; we will examine three: mixing ratio, relative humidity, and dew-point temperature.

Saturation

Before we consider these humidity measures further, you need to understand the concept of **saturation.** Imagine a closed jar half full of water and half full of dry air, both at the same temperature. As the water begins to evaporate from the water surface, a small increase in pressure can be detected in the air above. This increase is the result of the motion of the water-vapor molecules that were added to the air through evaporation. In the open atmosphere, this pressure is termed **vapor pressure** and is defined as that part of the total atmospheric pressure that can be attributed to the water-vapor content.

In the closed container, as more and more molecules escape from the water surface, the steadily increasing vapor pressure in the air above forces more and more of these molecules to return to the liquid.

▶ **FIGURE 17.3** Frost on a window pane is an example of deposition. (Photo by D. Cavagnaro/DRK Photo)

Eventually the number of vapor molecules returning to the surface will balance the number leaving. At that point, the air is said to be *saturated*. However, if we add heat to the container, increasing the temperature of the water and air, more water will evaporate before a balance is reached. Consequently, at higher temperatures, more moisture is required for saturation. The amount of water vapor required for saturation at various temperatures is shown in Table 17.1

Mixing Ratio

Not all air is saturated, of course. Thus, we need ways to express how humid a parcel of air is. One method specifies the amount of water vapor contained in a unit of air. The **mixing ratio** is the mass of water vapor in a unit of air compared to the remaining mass of dry air.

$$\text{mixing ratio} = \frac{\text{mass of water vapor (grams)}}{\text{mass of dry air (kilograms)}}$$

Table 17.1 shows the mixing ratios of saturated air at various temperatures. For example, at 25°C (77°F) a saturated parcel of air (one kilogram) would contain 20 grams of water vapor.

Because the mixing ratio is expressed in units of mass (usually in grams per kilogram), it is not affected by changes in pressure or temperature. However, the mixing ratio is time consuming to measure by direct sampling. Thus, other methods are employed to express the moisture content of the air. These include relative humidity and dew-point temperature.

Relative Humidity

The most familiar and, unfortunately, the most misunderstood term used to describe the moisture content of air is relative humidity. **Relative humidity** *is a ratio of the air's actual water-vapor content compared with the amount of water vapor required for saturation at that temperature (and pressure).* Thus, relative humidity indicates how near the air is to saturation, rather than the actual quantity of water vapor in the air.

To illustrate, we see from Table 17.1 that at 25°C, air is saturated when it contains 20 grams of water vapor per kilogram of air. Thus, if the air contains 10 grams per kilogram on a 25°C day, the relative humidity is expressed as 10/20, or 50 percent. Further, if air with a temperature of 25°C had a water-vapor content of 20 grams per kilogram, the relative humidity would be expressed as 20/20 or 100 percent. On those occasions when the relative humidity reaches 100 percent, the air is said to be saturated.

Because relative humidity is based on the air's water-vapor content, as well as the amount of moisture required for saturation, it can be changed in either of two ways. First, relative humidity can be changed by the addition or removal of water vapor. Second, because the amount of moisture required for saturation is a function of air temperature, relative humidity varies with temperature. (Recall that the amount of water vapor required for saturation is temperature dependent, such that at higher temperatures it takes more water vapor to saturate air than at lower temperatures.)

Adding or Subtracting Moisture. Notice in Figure 17.4 that when water vapor is added to a parcel of air, its relative humidity increases until saturation occurs (100 percent relative humidity). What if even more moisture is added to this parcel of saturated air? Does the relative humidity exceed 100 percent? Normally, this situation does not occur. Instead, the excess water vapor condenses to form liquid water.

In nature, moisture is added to the air mainly via evaporation from the oceans. However, plants, soil, and smaller bodies of water do make substantial contributions.

Changes with Temperature. The second condition that affects relative humidity is air temperature. Examine Figure 17.5 carefully. Note that in Figure 17.5A when air at 20°C contains 7 grams of water vapor per kilogram, it has a relative humidity of 50 percent. This can be verified by referring to Table 17.1. Here we can see that at 20°C, air is saturated when it contains 14 grams of water vapor per kilogram of air. Because the air in Figure 17.5A contains 7 grams of water vapor, its relative humidity is 7/14, or 50 percent.

How does cooling affect relative humidity? When the flask in Figure 17.5A is cooled from 20° to 10°C, as shown in Figure 17.5B, the relative humidity increases from 50 to 100 percent. We can conclude from this that when the water-vapor content remains constant, a decrease in temperature results in an increase in relative humidity.

But there is no reason to assume that cooling would cease the moment the air reached saturation.

TABLE 17.1	Amount of water vapor needed to saturate a kilogram of air at various temperatures.

Temperature°C (°F)	Water-Vapor Content at Saturation (grams)
−40 (−40)	0.1
−30 (−22)	0.3
−20 (−4)	0.75
−10 (14)	2
0 (32)	3.5
5 (41)	5
10 (50)	7
15 (59)	10
20 (68)	14
25 (77)	20
30 (86)	26.5
35 (95)	35
40 (104)	47

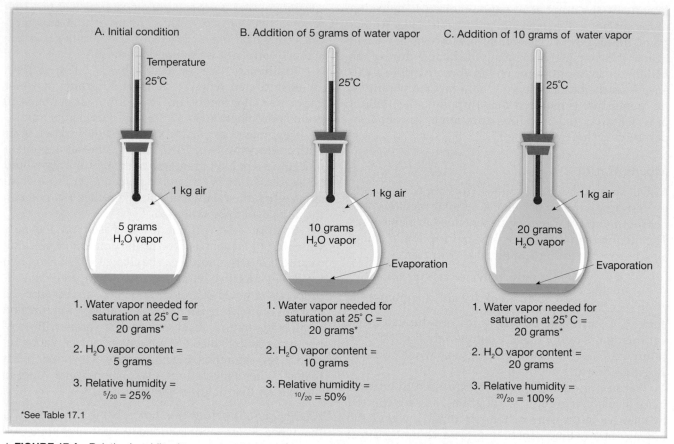

A. Initial condition

Temperature
25°C

1 kg air

5 grams
H₂O vapor

1. Water vapor needed for
saturation at 25° C =
20 grams*

2. H₂O vapor content =
5 grams

3. Relative humidity =
⁵/₂₀ = 25%

*See Table 17.1

B. Addition of 5 grams of water vapor

25°C

1 kg air

10 grams
H₂O vapor

Evaporation

1. Water vapor needed for
saturation at 25° C =
20 grams*

2. H₂O vapor content =
10 grams

3. Relative humidity =
¹⁰/₂₀ = 50%

C. Addition of 10 grams of water vapor

25°C

1 kg air

20 grams
H₂O vapor

Evaporation

1. Water vapor needed for
saturation at 25° C =
20 grams*

2. H₂O vapor content =
20 grams

3. Relative humidity =
²⁰/₂₀ = 100%

▲ **FIGURE 17.4** Relative humidity. At a constant temperature, the relative humidity will increase as water vapor is added to the air. Here, the capacity remains constant at 20 grams per kilogram and the relative humidity rises from 25 to 100 percent as the water-vapor content increases.

What happens when the air is cooled below the temperature at which saturation occurs? Figure 17.5C illustrates this situation. Notice from Table 17.1 that when the flask is cooled to 0°C, the air is saturated at 3.5 grams of water vapor per kilogram of air. Because this flask originally contained 7 grams of water vapor, 3.5 grams of water vapor will condense to form liquid droplets that collect on the walls of the container. In the meantime, the relative humidity of the air inside remains at 100 percent. This raises an important concept. When air aloft is cooled below its saturation level, some of the water vapor condenses to form clouds. As clouds are made of liquid droplets, this moisture is no longer part of the *water-vapor* content of the air.

We can summarize the effects of temperature on relative humidity as follows. When the water-vapor content of air remains at a constant level, a decrease in air temperature results in an increase in relative humidity and an increase in temperature causes a decrease in relative humidity. In Figure 17.6 the variations in temperature and relative humidity during a typical day demonstrate the relationship just described.

Dew-Point Temperature

Another important measure of humidity is the dew-point temperature. The **dew-point temperature** or simply the **dew point** is the temperature to which a parcel of air would need to be cooled to reach saturation. Note that in Figure 17.5, unsaturated air at 20°C must be cooled to 10°C before saturation occurs. Therefore, 10°C would be the dew-point temperature for this air. If the same air were cooled further, the air's saturation mixing ratio would be exceeded and the excess water vapor would condense, typically as dew, fog, or clouds. The term *dew point* stems from the fact that during nighttime hours, objects near the ground often cool below the dew-point temperature and become coated with dew* (Figure 17.7).

Unlike relative humidity, which is a measure of how near the air is to being saturated, dew-point temperature is a measure of its *actual moisture* content. Because the dew-point temperature is directly related to

*Normally, we associate dew with grass. Because of transpiration by the blades of grass, the relative humidity on a calm night is much higher near the grass than a few inches above the surface. Consequently, dew forms on grass before it does on most other objects.

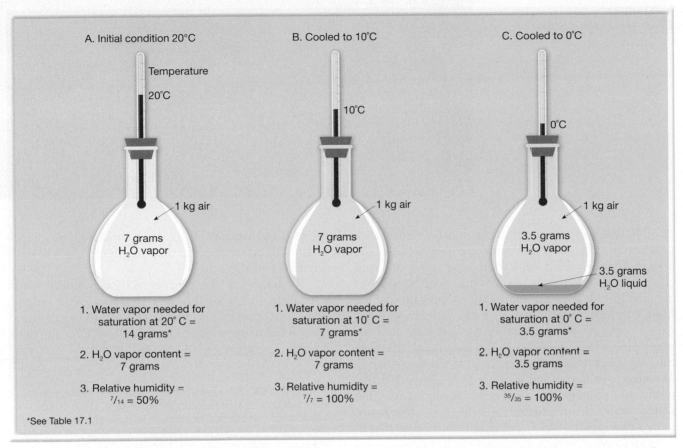

A. Initial condition 20°C

Temperature

20°C

1 kg air

7 grams
H₂O vapor

1. Water vapor needed for
   saturation at 20° C =
   14 grams*

2. H₂O vapor content =
   7 grams

3. Relative humidity =
   ⁷/₁₄ = 50%

*See Table 17.1

B. Cooled to 10°C

10°C

1 kg air

7 grams
H₂O vapor

1. Water vapor needed for
   saturation at 10° C =
   7 grams*

2. H₂O vapor content =
   7 grams

3. Relative humidity =
   ⁷/₇ = 100%

C. Cooled to 0°C

0°C

1 kg air

3.5 grams
H₂O vapor

3.5 grams
H₂O liquid

1. Water vapor needed for
   saturation at 0° C =
   3.5 grams*

2. H₂O vapor content =
   3.5 grams

3. Relative humidity =
   ³⁵/₃₅ = 100%

▲ **FIGURE 17.5**   Relative humidity varies when temperatures change. When the water-vapor content (mixing ratio) remains constant, the relative humidity can be changed by increasing or decreasing the air temperature. In this example, when the temperature of the air in the flask was lowered from 20° to 10°C, the relative humidity increased from 50 to 100 percent. Further cooling (from 10° to 0°C) causes one-half of the water vapor to condense. In nature, cooling of air below its saturated mixing ratio generally causes condensation in the form of clouds, dew, or fog.

the amount of water vapor in the air, and because it is easy to determine, it is one of the most widely used measures of humidity.

Recall that the amount of water vapor needed for saturation is temperature dependent and that for every 10°C (18°F) increase in temperature, the amount of water vapor needed for saturation doubles (see Table 17.1). Therefore, relatively cold *saturated air* (0°C or 32°F) contains about half the water vapor of *saturated air* having a temperature of 10°C (50°F) and

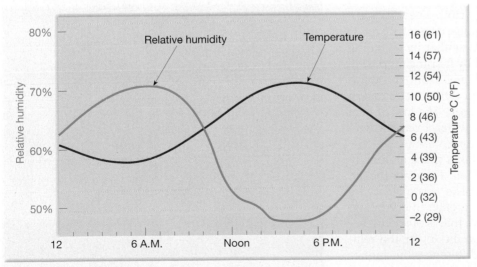

◀ **FIGURE 17.6**   Typical daily variations in temperature and relative humidity during a spring day at Washington, D.C. When temperature increases, relative humidity drops (see midafternoon) and vice versa.

► **FIGURE 17.7** Dew on a spider web. (Photo by Wolfgang Kaehler)

roughly one-fourth that of hot *saturated air* with a temperature of 20°C (68°F). Because the dew point is the temperature at which saturation occurs, we can conclude that high dew-point temperatures equate to moist air, and low dew-point temperatures indicate dry air. More precisely, based on what we have learned about vapor pressure and saturation, we can state that for every 10°C (18°F) increase in the dew-point temperature, the air contains about twice as much water vapor. Therefore, we know that air over Fort Myers, Florida, with a dew point of 25°C (77°F) contains about twice the water vapor of air situated over St. Louis, Missouri, with a dew point of 15°C (59°F) and four times that of Tucson, Arizona, with a dew point of 5°C (41°F).

Measuring Humidity

Relative humidity is commonly measured using a **hygrometer** (*hygro* = moisture, *metron* = measing instrument). One type of hygrometer, called a **psychrometer**, consists of two identical thermometers mounted side by side (Figure 17.8). One thermometer, the *dry-bulb*, gives the present air temperature. The other, called the *wet-bulb* thermometer, has a thin muslin wick tied around the end (see end of thermometers in the photo).

To use the psychrometer, the cloth sleeve is saturated with water and a continuous current of air is passed over the wick. This is done either by swinging the instrument freely in the air or by fanning air past it. As a consequence, water evaporates from the wick, and the heat absorbed by the evaporating water makes the temperature of the wet bulb drop. The loss of heat that was required to evaporate water from the wet bulb lowers the thermometer reading.

The amount of cooling that takes place is directly proportional to the dryness of the air. The drier the air, the more moisture evaporates. The more heat the evaporating water absorbs, the greater the cooling. Therefore, the larger the difference that is observed between the thermometer readings, the lower the relative humidity; the smaller the difference, the higher

the relative humidity. If the air is saturated, no evaporation will occur, and the two thermometers will have identical readings.

To determine the precise relative humidity from the thermometer readings, a standard table is used (refer to Appendix C, Table C.1). With the same information but using a different table (Table C.2), the dew-point temperature can also be calculated.

▼ **FIGURE 17.8** Sling psychrometer. This instrument is used to determine both relative humidity and dew point. The dry-bulb thermometer gives the current air temperature. The web-bulb thermometer is covered with a cloth wick that is dipped in water. The thermometers are spun until the temperature of the wet-bulb thermometer stops declining. Then the thermometers are read and the data used in conjunction with the tables in Appendix C. (Photo by E. J. Tarbuck)

Another instrument used for measuring relative humidity, the *hair hygrometer*, can be read directly, without the use of tables. The hair hygrometer operates on the principle that hair or certain synthetic fibers change their length in proportion to changes in the relative humidity, lengthening as relative humidity increases, and shrinking as the relative humidity drops. The tension of a bundle of hairs is linked mechanically to an indicator that is calibrated between 0 and 100 percent. Thus, we need only glance at the dial to determine the relative humidity. Unfortunately, the hair hygrometer is less accurate than the psychrometer. Further, it requires frequent calibration and is slow in responding to changes in humidity, especially at low temperatures.

A different type of hygrometer is used in remote-sensing instrument packages such as radiosondes that transmit upper-air observations back to ground stations. The *electric hygrometer* contains an electrical conductor coated with a moisture-absorbing chemical. It works on the principle that the passage of current varies as the relative humidity varies.

STUDENTS SOMETIMES ASK...

Why is the air in buildings so dry in the winter?

The answer lies in the relationship between temperature and relative humidity. Recall that if the water-vapor content of air remains at a constant level, an increase in temperature lowers the relative humidity, and a drop in temperature raises the relative humidity. During the winter months, outside air is comparatively cool and has a low mixing ratio. When this air is drawn into the home, it is heated to room temperature. This process causes the relative humidity to plunge, often to uncomfortably low levels of 10 percent or lower. Living with dry air can mean static electrical shocks, dry skin, sinus headaches, or even nosebleeds. Consequently, the homeowner may install a humidifier, which adds water to the air and increases the relative humidity to a more comfortable level.

The Basis of Cloud Formation: Adiabatic Cooling

 GEODe Earth's Dynamic Atmosphere
▼ Moisture and Cloud Formation

Up to this point, we have considered basic properties of water vapor and how its variability is measured. We are now ready to examine some of the important roles that water vapor plays in weather, especially the formation of clouds.

Fog and Dew vs. Cloud Formation

Recall that condensation occurs when water vapor changes to a liquid. Condensation may form dew, fog, or clouds. Although these three forms are different, all require saturated air to develop. As indicated earlier, saturation occurs either when sufficient water vapor is added to the air or, more commonly, when the air is cooled to its dew point.

Near Earth's surface, heat is readily exchanged between the ground and the air above. During evening hours, the surface radiates heat away, causing the surface and adjacent air to cool rapidly. This radiation cooling accounts for the formation of dew and some types of fog. Thus, surface cooling that occurs after sunset accounts for some condensation. However, cloud formation often takes place during the warmest part of the day. Clearly some other mechanism must operate aloft that cools air sufficiently to generate clouds.

Adiabatic Temperature Changes

The process that is responsible for most cloud formation is easily visualized if you have ever pumped up a bicycle tire and noticed that the pump barrel became quite warm. The heat you felt was the consequence of the work you did on the air to compress it. When energy is used to compress air, the motion of the gas molecules increases and therefore the temperature of the air rises. Conversely, air that is allowed to escape from a bicycle tire *expands and cools*. This results because the expanding air pushes (does work on) the surrounding air and must cool by an amount equivalent to the energy expended.

You have probably felt the cooling effect of the propellant gas expanding as you applied hair spray or spray deodorant. As the compressed gas in the aerosol can is released, it quickly expands and cools. This drop in temperature occurs *even though heat is neither added nor subtracted*. Such variations are known as **adiabatic temperature changes** and result when air is compressed or allowed to expand. In summary, *when air is allowed to expand, it cools, and when it is compressed, it warms.*

Adiabatic Cooling and Condensation

To simplify the following discussion, it helps if we imagine a volume of air enclosed in a thin elastic cover. Meteorologists call this imaginary volume of air a **parcel**. Typically, we consider a parcel to be a few hundred cubic meters in volume, and we assume that it acts independently of the surrounding air. It is also assumed that no heat is transferred into, or out of, the parcel. Although highly idealized, over short time spans a parcel of air behaves in a manner much like an actual volume of air moving vertically in the atmosphere.

Dry Adiabatic Rate. As you travel from Earth's surface upward through the atmosphere, the atmospheric pressure rapidly diminishes, because there are fewer and fewer gas molecules. Thus, any time a parcel of air moves upward, it passes through regions of successively lower pressure. As a result, the ascending air expands. As it expands, it cools adiabatically. Unsaturated air cools at the constant rate of 10°C for every 1000 meters of ascent (5.5°F per 1000 feet).

Conversely, descending air comes under increasingly higher pressures, compresses, and is heated 10°C for every 1000 meters of descent. This rate of cooling or heating applies only to *unsaturated air* and is known as the **dry adiabatic rate**.

Wet Adiabatic Rate. If a parcel of air rises high enough, it will eventually cool to its dew point. Here the process of condensation begins. From this point on along its ascent, *latent heat of condensation* stored in the water vapor will be liberated. Although the air will continue to cool after condensation begins, the released latent heat works against the adiabatic process, thereby reducing the rate at which the air cools. This slower rate of cooling caused by the addition of latent heat is called the **wet adiabatic rate** of cooling. Because the amount of latent heat released depends on the quantity of moisture present in the air, the wet adiabatic rate varies from 5°C per 1000 meters for air with a high moisture content to 9°C per 1000 meters for dry air.

Figure 17.9 illustrates the role of adiabatic cooling in the formation of clouds. Note that from the surface up to the condensation level the air cools at the dry adiabatic rate. The wet adiabatic rate commences at the condensation level.

Processes that Lift Air

GEODe Earth's Dynamic Atmosphere
▼ Moisture and Cloud Formation

To review, when air rises, it expands and cools adiabatically. If air is lifted sufficiently, it will eventually cool to its dew-point temperature, saturation will occur, and clouds will develop. But why does air rise on some occasions and not on others?

It turns out that, in general, the tendency is for air to resist vertical movement. Therefore, air located near the surface tends to stay near the surface, and air aloft tends to remain aloft. Exceptions to this rule, as we shall see, include conditions in the atmosphere that give air sufficient buoyancy to rise without the aid of outside forces. In many situations, however, when you see clouds forming there is some mechanical phenomenon at work that forces the air to rise (at least initially).

We will be looking at four mechanisms that cause air to rise. These are:

1. *Orographic lifting*—air is forced to rise over a mountainous barrier.

2. *Frontal wedging*—warmer, less dense air, is forced over cooler, denser air.

3. *Convergence*—a pileup of horizontal air flow results in upward movement.

4. *Localized convective lifting*—unequal surface heating causes localized pockets of air to rise because of their buoyancy.

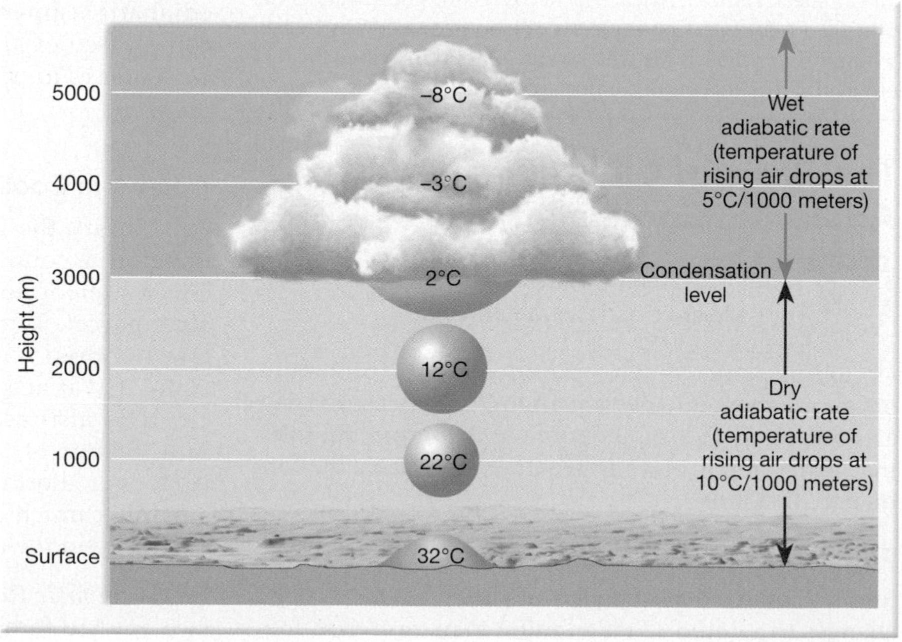

▶ **FIGURE 17.9** Rising air cools at the dry adiabatic rate of 10° per 1000 meters, until the air reaches the dew point and condensation (cloud formation) begins. As air continues to rise, the latent heat released by condensation reduces the rate of cooling. The wet adiabatic rate is therefore always less than the dry adiabatic rate.

▲ **FIGURE 17.10** Orographic lifting occurs where air is forced over a topographic barrier.

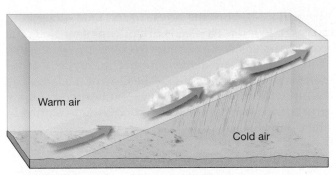

▲ **FIGURE 17.12** Frontal wedging. Colder, denser air acts as a barrier over which warmer, less dense air rises.

Orographic Lifting

Orographic lifting occurs when elevated terrains, such as mountains, act as barriers to the flow of air (Figure 17.10). As air ascends a mountain slope, adiabatic cooling often generates clouds and copious precipitation. In fact, many of the rainiest places in the world are located on windward mountain slopes.

By the time air reaches the leeward side of a mountain, much of its moisture has been lost. If the air descends, it warms adiabatically, making condensation and precipitation even less likely. As shown in Figure 17.10, the result can be a **rain shadow desert**. The Great Basin Desert of the western United States lies only a few hundred kilometers from the Pacific Ocean, but it is effectively cut off from the ocean's moisture by the imposing Sierra Nevada (Figure 17.11). The Gobi Desert of Mongolia, the Takla Makan of China, and the Patagonia Desert of Argentina are other examples of deserts that exist because they are on the leeward sides of mountains (for a map showing deserts, see Figure 20.6, p. 567).

▼ **FIGURE 17.11** Rain shadow desert. The arid conditions in California's Death Valley can be partially attributed to the adjacent mountains, which orographically remove the moisture from air originating over the Pacific. (Photo by James E. Patterson)

Frontal Wedging

If orographic lifting were the only mechanism that forced air aloft, the relatively flat central portion of North America would be an expansive desert instead of the nation's breadbasket. Fortunately, this is not the case.

In central North America, masses of warm and cold air collide producing a **front**. Here the cooler, denser air acts as a barrier over which the warmer, less dense air rises. This process, called **frontal wedging**, is illustrated in Figure 17.12.

It should be noted that weather-producing fronts are associated with storm systems called *middle-latitude cyclones*. Because these storms are responsible for producing a high percentage of the precipitation in the middle latitudes, we will examine them closely in Chapter 19.

Convergence

We saw that the collision of contrasting air masses forces air to rise. In a more general sense, whenever air in the lower atmosphere flows together, lifting results. This phenomenon is called **convergence**. When air flows in from more than one direction, it must go somewhere. As it cannot go down, it goes up (Figure 17.13). This, of course, leads to adiabatic cooling and possibly cloud formation.

The Florida peninsula provides an excellent example of the role that convergence can play in initiating cloud development and precipitation. On warm days,

▼ **FIGURE 17.13** Convergence. When surface air converges, it increases in height to allow for the decreased area it occupies.

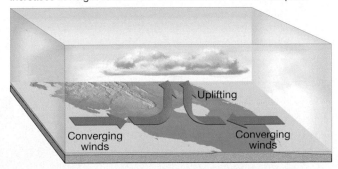

the airflow is from the ocean to the land along both coasts of Florida. This leads to a pileup of air along the coasts and general convergence over the peninsula. This pattern of air movement and the uplift that results is aided by intense solar heating of the land. The result is that the peninsula of Florida experiences the greatest frequency of midafternoon thunderstorms in the United States.

Localized Convective Lifting

On warm summer days, unequal heating of Earth's surface may cause pockets of air to be warmed more than the surrounding air. For instance, air above a paved parking lot will be warmed more than the air above an adjacent wooded park. Consequently, the parcel of air above the parking lot, which is warmer (less dense) than the surrounding air, will be buoyed upward (Figure 17.14). These rising parcels of warmer air are called *thermals*. Birds such as hawks and eagles use these thermals to carry them to great heights where they can gaze down on unsuspecting prey. People have learned to employ these rising parcels using hang gliders as a way to "fly."

The phenomenon that produces rising thermals is called **localized convective lifting**. When these warm parcels of air rise above the lifting condensation level, clouds form, which on occasion produce midafternoon rain showers. The accompanying rains, although occasionally heavy, are of short duration and widely scattered.

The Weathermaker: Atmospheric Stability

GEODe Earth's Dynamic Atmosphere
▼ Moisture and Cloud Formation

When air rises, it cools and eventually produces clouds. Why do clouds vary so much in size, and why does the resulting precipitation vary so much? The answers are closely related to the *stability* of the air.

Recall that a parcel of air can be thought of as having a thin flexible cover that allows it to expand but prevents it from mixing with the surrounding air (picture a hot-air balloon). If this parcel were forced to rise, its temperature would decrease because of expansion. By comparing the parcel's temperature to that of the surrounding air, we can determine its stability. If the parcel were *cooler* than the surrounding environment, it would be more dense; and if allowed to do so, it would sink to its original position. Air of this type, called **stable air**, resists vertical movement.

If, however, our imaginary rising parcel were *warmer* and hence less dense than the surrounding air, it would continue to rise until it reached an altitude where its temperature equalled that of its surroundings. This is exactly how a hot-air balloon works, rising as long as it is warmer and less dense than the surrounding air (Figure 17.15). This type of air is classified as **unstable air**. In summary, stability is a property of air that describes its tendency to remain in its original position (stable), or to rise (unstable).

STUDENTS SOMETIMES ASK...
What are the wettest places on Earth?

Many of the rainiest places in the world are located on windward mountain slopes. A station at Mount Waialeale, Hawaii, records the highest average annual rainfall, some 1234 centimeters (486 inches). The greatest recorded rainfall for a single 12-month period occurred at Cherrapunji, India, where an astounding 2647 centimeters (1042 inches, over 86 feet) fell. Much of this rainfall occurred in the month of July—a record 930 centimeters (366 inches, over 30 feet). This is 10 times more rain than Chicago receives in an average year.

Types of Stability

The stability of air is determined by measuring the temperature of the atmosphere at various heights. Recall from Chapter 16 that this measure is termed the

▼ **FIGURE 17.14** Localized convective lifting. Unequal heating of Earth's surface causes pockets of air to be warmed more than the surrounding air. These buoyant parcels of hot air rise, producing thermals, and if they reach the condensation level, clouds form.

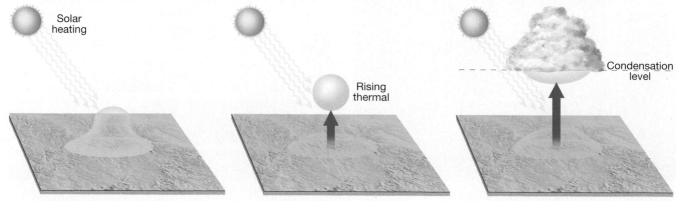

▲ **FIGURE 17.15** As long as air is warmer than its surroundings, it will rise. Hot-air balloons rise up through the atmosphere for this reason. (Photo by Barbara Cushman Rowell/Mountain Light Photography, Inc.)

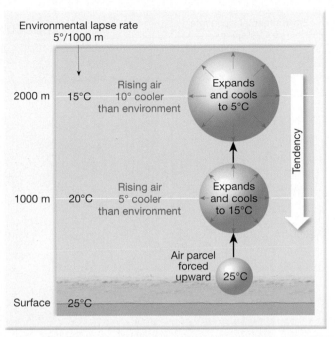

▲ **FIGURE 17.16** In a stable atmosphere, as an unsaturated parcel of air is lifted, it expands and cools at the dry adiabatic rate of 10°C per 1000 meters. Because the temperature of the rising parcel of air is lower than that of the surrounding environment, it will be heavier and, if allowed to do so, will sink to its original position.

environmental lapse rate. The environmental lapse rate is the temperature of the atmosphere as determined from observations made by radiosondes and aircraft. It is important not to confuse this with *adiabatic temperature changes,* which are changes in temperature caused by expansion or compression as a parcel of air rises or descends.

For illustration, we will examine a situation in which the environmental lapse rate is 5°C per 1000 meters (Figure 17.16). Under this condition, when air at the surface has a temperature of 25°C, the air at 1000 meters will be 5 degrees cooler, or 20°C, the air at 2000 meters will have a temperature of 15°C, and so forth. At first glance it appears that the air at the surface is less dense than the air at 1000 meters, because it is 5 degrees warmer. However, if the air near the surface is unsaturated and were to rise to 1000 meters, it would expand and cool at the dry adiabatic rate of 10°C per 1000 meters. Therefore, upon reaching 1000 meters, its temperature would have dropped 10°C. Being 5 degrees cooler than its environment, it would be denser and tend to sink to its original position. Hence, we say that the air near the surface is potentially cooler than the air aloft and therefore will not rise on its own. The air just described is *stable* and resists vertical movement.

With this as background, we will now look at three fundamental conditions of the atmosphere: absolute stability, absolute instability, and conditional instability.

Absolute Stability. Stated quantitatively, **absolute stability** prevails when the environmental lapse rate is

less than the wet adiabatic rate. Figure 17.17 depicts this situation using an environmental lapse rate of 5°C per 1000 meters and a wet adiabatic rate of 6°C per 1000 meters. Note that at 1000 meters the temperature of the surrounding air is 15°C, while the rising parcel of air has cooled to 10°C and is therefore the denser air. Even if this stable air were to be forced above the condensation level, it would remain cooler and denser than its environment, and thus it would tend to return to the surface.

The most stable conditions occur when the temperature in a layer of air actually increases with altitude. When such a reversal occurs, a *temperature inversion* is said to exist. Temperature inversions frequently occur on clear nights as a result of radiation cooling of Earth's surface. Under these conditions, an inversion is created because the ground and the air immediately above will cool more rapidly than the air aloft. When warm air overlies cooler air, it acts as a lid and prevents appreciable vertical mixing. Because of this, temperature inversions are responsible for trapping pollutants in a narrow zone near Earth's surface. This idea is explored more fully in Box 17.1.

Absolute Instability. At the other extreme, air is said to exhibit **absolute instability** when the environmental lapse rate is greater than the dry adiabatic rate. As shown in Figure 17.18, the ascending parcel of air is always warmer than its environment and will continue to rise because of its own buoyancy. However,

477

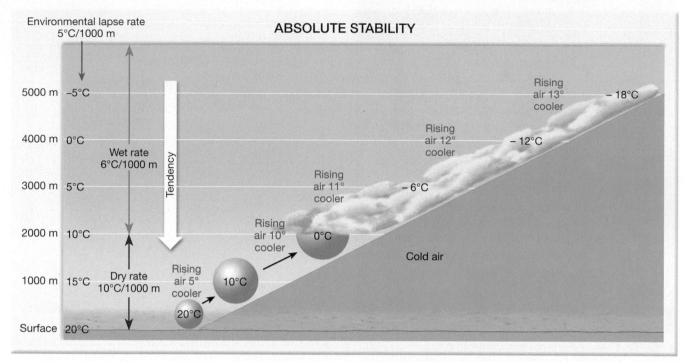

▲ **FIGURE 17.17** Absolute stability prevails when the environmental lapse rate is less than the wet adiabatic rate. The rising parcel of air is therefore always cooler and denser than the surrounding air. When stable air is forced to rise, it spreads out, producing flat, layered clouds.

absolute instability is generally limited to near Earth's surface. On hot, sunny days the air above some surfaces, like paved shopping malls, is heated more than the air over adjacent surfaces. These invisible pockets of more intensely heated air, being less dense than the air aloft, will rise like a hot-air balloon. This phenomenon produces the small fluffy clouds we associate with

fair weather. Occasionally, when the surface air is considerably warmer than the air aloft, clouds with considerable vertical development can form.

Conditional Instability. A more common type of atmospheric instability is called **conditional instability**. This situation prevails when moist air has an environ-

▼ **FIGURE 17.18** Absolute instability illustrated using an environmental lapse rate of 12°C per 1000 meters. The rising air is always warmer and therefore lighter than the surrounding air.

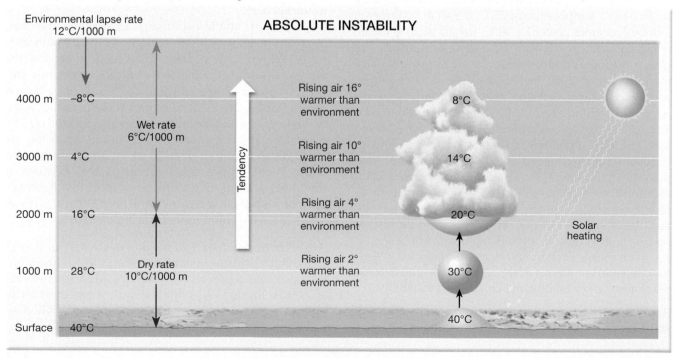

mental lapse rate between the dry and wet adiabatic rates (between 5 and 10°C per 1000 meters). Simply, the atmosphere is said to be conditionally unstable when it is *stable* for an *unsaturated* parcel of air, but *unstable* for a *saturated* parcel of air. Notice in Figure 17.19 that the rising parcel of air is cooler than the surrounding air for nearly 3000 meters. With the addition of latent heat above the lifting condensation level, the parcel becomes warmer than the surrounding air. From this point along its ascent the parcel will continue to rise because of its own buoyancy, without an outside force. Thus, conditional instability depends on whether or not the rising air is saturated. The word *conditional* is used because the air must be forced upward, such as over mountainous terrain, before it becomes unstable and rises because of its own buoyancy.

Summary. In summary, the stability of air is determined by measuring the temperature of the atmosphere at various heights. In simple terms, a column of air is deemed unstable when the air near the bottom of this layer is significantly warmer (less dense) than the air aloft, indicating a steep environmental lapse rate. Under these conditions the air actually turns over, as the warm air below rises and displaces the colder air aloft. Conversely, the air is considered to be stable when the temperature drops gradually with increasing altitude. The most stable conditions occur during a temperature inversion when the temperature actually increases with height. Under these conditions, there is very little vertical air movement.

Stability and Daily Weather

From the previous discussion, we can conclude that stable air resists vertical movement, whereas unstable air ascends freely because of its own buoyancy. But how do these facts manifest themselves in our daily weather?

Because stable air resists upward movement, we might conclude that clouds will not form when stable conditions prevail in the atmosphere. Although this seems reasonable, recall that processes exist that *force* air aloft. These include orographic lifting, frontal wedging, and convergence. On those occasions when stable air is forced aloft, the clouds that form are widespread and have little vertical thickness when compared to their horizontal dimension, and precipitation, if any, is light to moderate.

▼ **FIGURE 17.19** Illustration of conditional instability where warm air is forced to rise along a frontal boundary. Note that the environmental lapse rate of 9°C per 1000 meters lies between the dry and wet adiabatic rates. The parcel of air is cooler than the surrounding air up to nearly 3000 meters, where its tendency is to sink toward the surface (stable). Above this level, however, the parcel is warmer than its environment and will rise because of its own buoyancy (unstable). Thus when conditionally unstable air is forced to rise, the result can be towering cumulus clouds.

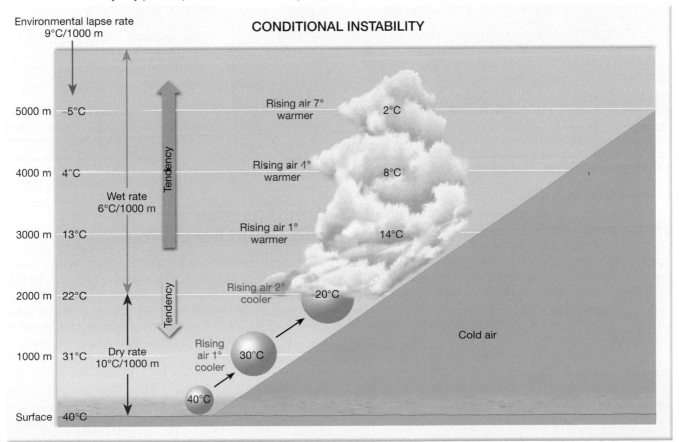

Box 17.1 People and the Environment:
Atmospheric Stability and Air Pollution

Air quality is not just a function of the quantity and types of pollutants emitted into the air, but it is also closely linked to the atmosphere's ability to disperse these noxious substances. Perhaps you have heard this following well-known phrase: "The solution to pollution is dilution." To a significant degree this is true. If the air into which the pollution is released is not dispersed, the air will become more toxic. Two of the most important atmospheric conditions affecting the dispersion of pollutants are the strength of the wind and the stability of the air. These factors are critical because they determine how rapidly pollutants are diluted by mixing with the surrounding air after leaving the source.

The way in which wind speed influences the concentration of pollutants is straightforward. When winds are weak or calm, the concentration of pollutants is higher than when winds are strong. High wind speeds mix polluted air into a greater volume of surrounding air and therefore cause the pollution to be more diluted. When winds are light, there is less turbulence and mixing, so the concentration of pollutants is higher.

Whereas wind speed governs the amount of air into which pollutants are initially mixed, atmospheric stability determines the extent to which vertical motions will mix the pollution with cleaner air above. The distance between Earth's surface and the height to which vertical air movements extend is termed the *mixing depth*. Generally, the greater the mixing depth, the better the air quality. When the mixing depth is several kilometers, pollutants are mixed through a large volume of cleaner air and dilute rapidly. When the mixing depth is shallow, pollutants are confined to a much smaller volume of air and concentrations can reach unhealthy levels. When air is stable, vertical motions are suppressed and mixing depths are small. Conversely, an unstable atmosphere promotes vertical air movements and greater mixing depths. Because heating of Earth's surface by the Sun enhances convectional movements, mixing depths are usually greater during the afternoon hours. For the same reason, mixing depths during the summer months are typically greater than during the winter months.

Temperature inversion represents a situation in which the atmosphere is very stable and the mixing depth is significantly restricted. Warm air overlying cooler air acts as a lid and prevents upward movement, leaving the pollutants trapped in a relatively narrow zone near the ground (Figure 17.A).

Inversions are generally classified into one of two categories—those that form near the ground and those that

Figure 17.A Air pollution in downtown Los Angeles. Temperature inversions act as lids to trap pollutants below. (Photo by Ted Spiegel/Black Star)

By contrast, clouds associated with the lifting of unstable air are towering and often generate thunderstorms and occasionally even a tornado. For this reason, we can conclude that on a dreary, overcast day with light drizzle, stable air has been forced aloft. On the other hand, during a day when cauliflower-shaped clouds appear to be growing as if bubbles of hot air are surging upward, we can be fairly certain that the ascending air is unstable.

In summary, the role of stability in determining our daily weather is very important. To a large degree, stability determines the type of clouds that develop and whether precipitation will come as a gentle shower or a heavy downpour.

Condensation and Cloud Formation

To review briefly, condensation occurs when water vapor in the air changes to a liquid. The result of this process may be dew, fog, or clouds. For any of these forms of condensation to occur, the air must be saturated. Saturation occurs most commonly when air is cooled to its dew point, or less often when water vapor is added to the air.

Generally, there must be a surface on which the water vapor can condense. When dew occurs, objects at or near the ground serve this purpose, like grass and car windows. But when condensation occurs in the air

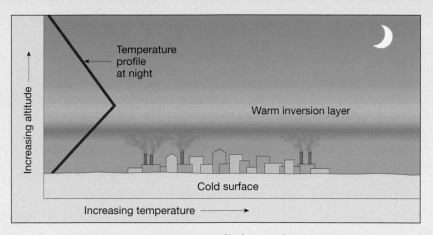

Figure 17.B Generalized temperature profile for a surface inversion.

these thicker surface inversions will not dissipate as quickly after sunrise.

Many extensive and long-lived air-pollution episodes are linked to temperature inversions that develop in association with the sinking air that characterizes slow-moving centers of high pressure (Figure 17.C). As the air sinks to lower altitudes, it is compressed and so its temperature rises. Because turbulence is almost always present near the ground, this lowermost portion of the atmosphere is generally prevented from participating in the general subsidence. Thus, an inversion develops aloft between the lower turbulent zone and the subsiding warmed layers above.

form aloft. A *surface inversion* develops close to the ground on clear and relatively calm nights. It forms because the ground is a more effective radiator than the air above. This being the case, radiation from the ground to the clear night sky causes more rapid cooling at the surface than higher in the atmosphere. The result is that the air close to the ground is cooled more than the air above, yielding a temperature profile similar to the one shown in Figure 17.B. After sunrise the ground is heated and the inversion disappears.

Although surface inversions are usually shallow, they may be quite thick in regions where the land surface is uneven. Because cold air is denser (heavier) than warm air, the chilled air near the surface gradually drains from the uplands and slopes into adjacent lowlands and valleys. As might be expected,

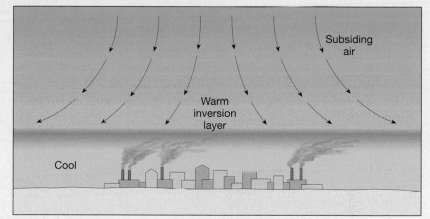

Figure 17.C Inversions aloft frequently develop in association with slow moving centers of high pressure where the air aloft subsides and warms by compression. The turbulent surface zone does not subside as much. Thus, an inversion often forms between the lower turbulent zone and the subsiding layers above.

above the ground, tiny bits of particulate matter, known as **condensation nuclei**, serve as surfaces for water-vapor condensation. These nuclei are very important, for in their absence a relative humidity well in excess of 100 percent is needed to produce clouds.

Condensation nuclei such as microscopic dust, smoke, and salt particles (from the ocean) are profuse in the lower atmosphere. Because of this abundance of particles, relative humidity rarely exceeds 101 percent. Some particles, such as ocean salt, are particularly good nuclei because they absorb water. These particles are termed **hygroscopic** (*hygro* = moisture, *scopic* = to seek) **nuclei**. When condensation takes place, the initial growth rate of

cloud droplets is rapid. It diminishes quickly because the excess water vapor is quickly absorbed by the numerous competing particles. This results in the formation of a cloud consisting of millions upon millions of tiny water droplets, all so fine that they remain suspended in air. When cloud formation occurs at below-freezing temperatures, tiny ice crystals form. Thus, a cloud might consist of water droplets, ice crystals, or both.

The slow growth of cloud droplets by additional condensation and the immense size difference between cloud droplets and raindrops suggest that condensation alone is not responsible for the formation of drops large enough to fall as rain. We will first

▶ **FIGURE 17.20** Classification of clouds according to height and form.

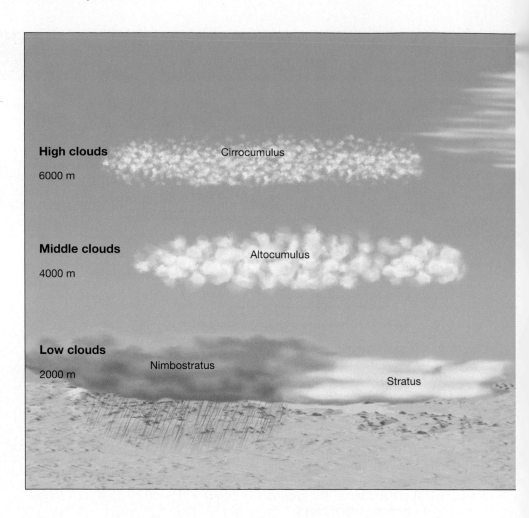

examine clouds and then return to the questions of how precipitation forms.

Types of Clouds

Clouds are among the most conspicuous and observable aspects of the atmosphere and its weather. **Clouds** are a form of condensation best described as *visible aggregates of minute droplets of water or tiny crystals of ice.* In addition to being prominent and sometimes spectacular features in the sky, clouds are of continual interest to meteorologists, because they provide a visible indication of what is going on in the atmosphere. Anyone who observes clouds with the hope of recognizing different types often finds that there is a bewildering variety of these familiar white and gray masses streaming across the sky. Still, once one comes to know the basic classification scheme for clouds, most of the confusion vanishes.

Clouds are classified on the basis of their *form* and *height* (Figure 17.20). Three basic forms are recognized: cirrus, cumulus, and stratus.

- **Cirrus** (*cirrus* = a curl of hair) clouds are high, white, and thin. They can occur as patches or as delicate veil-like sheets or extended wispy fibers that often have a feathery appearance.
- **Cumulus** (*cummulus* = a pile) clouds consist of globular individual cloud masses. Normally they exhibit a flat base and have the appearance of rising domes or towers. Such clouds are frequently described as having a cauliflower structure.
- **Stratus** (*stratum* = a layer) clouds are best described as sheets or layers that cover much or all of the sky. While there may be minor breaks, there are no distinct individual cloud units.

All other clouds reflect one of these three basic forms or are combinations or modifications of them.

Three levels of cloud heights are recognized: high, middle, and low (Figure 17.20). **High clouds** normally have bases above 6000 meters (20,000 feet), **middle clouds** generally occupy heights from 2000 to 6000 meters (6500 to 20,000 feet), and **low clouds** form below 2000 meters (6,500 feet). The altitudes listed for each height category are not hard and fast. There is some seasonal as well as latitudinal variation. For example, at high latitudes or during cold winter months in the midlatitudes, high clouds are often found at lower altitudes.

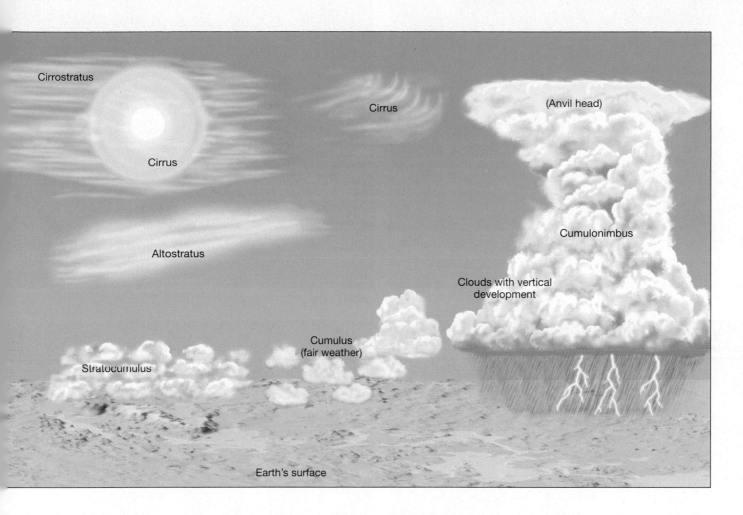

Cirrostratus

Cirrus

Cirrus

(Anvil head)

Cumulonimbus

Clouds with vertical development

Altostratus

Cumulus (fair weather)

Stratocumulus

Earth's surface

High Clouds. Three cloud types make up the family of high clouds (above 6000 meters): cirrus, cirrostratus, and cirrocumulus. *Cirrus* clouds are thin and delicate and sometimes appear as hooked filaments called "mares tails" (Figure 17.21A). As the names suggest, *cirrocumulus* clouds consist of fluffy masses (Figure 17.21B), whereas *cirrostratus* clouds are flat layers (Figure 17.21C). Because of the low temperatures and small quantities of water vapor present at high altitudes, all high clouds are thin and white and are made up of ice crystals. Further, these clouds are not considered precipitation makers. However, when cirrus clouds are followed by cirrocumulus clouds and increased sky coverage, they may warn of impending stormy weather.

Middle Clouds. Clouds that appear in the middle range (2000 to 6000 meters) have the prefix *alto* as part of their name. *Altocumulus* clouds are composed of globular masses that differ from cirrocumulus clouds in that they are larger and denser (Figure 17.21D). *Altostratus* clouds create a uniform white to grayish sheet covering the sky with the Sun or Moon visible as a bright spot (Figure 17.21E). Infrequent light snow or drizzle may accompany these clouds.

Low Clouds. There are three members in the family of low clouds: stratus, stratocumulus, and nimbostratus. *Stratus* are a uniform foglike layer of clouds that frequently covers much of the sky. On occasions these clouds may produce light precipitation. When stratus clouds develop a scalloped bottom that appears as long parallel rolls or broken globular patches, they are called *stratocumulus* clouds.

Nimbostratus clouds derive their name from the Latin *nimbus,* which means rainy cloud, and *stratus,* which means to cover with a layer (Figure 17.21F). As the name suggests, nimbostratus clouds are one of the chief precipitation producers. Nimbostratus clouds form in association with stable conditions. We might not expect clouds to grow or persist in stable air, yet cloud growth of this type is common when air is forced to rise, as occurs along a mountain range, a front, or near the center of a cyclone where converging winds cause air to ascend. Such forced ascent of stable air leads to the formation of a stratified cloud layer that is large horizontally compared to its depth.

Clouds of Vertical Development. Some clouds do not fit into any one of the three height categories mentioned. Such clouds have their bases in the low

A. Cirrus

B. Cirrocumulus

C. Cirrostratus

D. Altocumulus

▲ **FIGURE 17.21** These photos depict common forms of several different cloud types. (Photos A, B, D, E, F, and G by E. J. Tarbuck. Photo C by A. and J. Verkaik/Corbis/The Stock Market. Photo H by Doug Millar/Science Source/Photo Researchers, Inc.)

E. Altostratus

F. Nimbostratus

G. Cumulus

H. Cumulonimbus

▲ **FIGURE 17.22** Advection fog rolling into San Francisco Bay. (Photo by Ed Pritchard/Getty Images, Inc.—Stone Allstock)

▼ **FIGURE 17.23** Satellite image of dense fog in California's San Joaquin Valley on November 20, 2002. This early morning radiation fog was responsible for several car accidents in the region, including a 14-car pileup. The white areas to the east of the fog are the snowcapped Sierra Nevadas. (NASA image)

Snow in Sierra Nevada

Fog

Pacific Ocean

of westward-moving air is less than 13°C, an extensive fog can result in the western plains.

Evaporation Fogs

When the saturation of air occurs primarily because of the addition of water vapor, the resulting fogs are called *evaporation fogs*. Two types of evaporation fogs are recognized: steam fog and frontal, or precipitation, fog.

Steam Fog. When cool air moves over warm water, enough moisture may evaporate from the water surface to produce saturation. As the rising water vapor meets the cold air, it immediately recondenses and rises with the air that is being warmed from below. Because the water has a steaming appearance, the phenomenon is called **steam fog**. Steam fog is fairly common over lakes and rivers in the fall and early winter, when the water may still be relatively warm and the air is rather crisp. Steam fog is often shallow because as the steam rises, it reevaporates in the unsaturated air above.

Frontal Fog When frontal wedging occurs, warm air is lifted over colder air. If the resulting clouds yield rain, and the cold air below is near the dew point, enough rain will evaporate to produce fog. A fog formed in this manner is called **frontal fog**, or **precipitation fog**. The result is a more or less continuous zone of condensed water droplets reaching from the ground up through the clouds.

In summary, both steam fog and frontal fog result from the addition of moisture to a layer of air. As you saw, the air is usually cool or cold and already near saturation. Because air's capacity for water vapor at low temperatures is small, only a relatively modest amount of evaporation is necessary to produce saturated conditions and fog.

How Precipitation Forms

Although all clouds contain water, why do some produce precipitation and others drift placidly overhead? This seemingly simple question perplexed meteorologists for many years. Before examining the processes that generate precipitation, we need to examine a couple of facts.

First, cloud droplets are very tiny, averaging under 20 micrometers (0.02 millimeter) in diameter (Figure 17.24). For comparison, a human hair is about 75 micrometers in diameter. The small size of cloud droplets results mainly because condensation nuclei are usually very abundant and the available water is distributed among numerous droplets rather than concentrated into fewer large droplets.

Second, because of their small size, the rate at which cloud droplets fall is incredibly slow. An average cloud droplet falling from a cloud base at 1000 meters would require several hours to reach the ground. However, it would never complete its journey. This cloud droplet would evaporate before it fell a few meters from the cloud base into the unsaturated air below.

▼ **FIGURE 17.24** Comparative diameters of particles involved in condensation and precipitation processes.

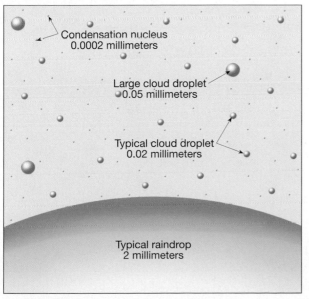

Condensation nucleus
0.0002 millimeters

Large cloud droplet
0.05 millimeters

Typical cloud droplet
0.02 millimeters

Typical raindrop
2 millimeters

How large must a droplet grow in order to fall as precipitation? A typical raindrop has a diameter of about 2000 micrometers (2 millimeters) or 100 times that of the average cloud droplet having a diameter of 20 micrometers (0.02 millimeter). However, the *volume* of a typical raindrop is a million times that of a cloud droplet. Thus, for precipitation to form, cloud droplets must grow in volume by roughly one million times. Two mechanisms that give rise to these "massive" drops are: the Bergeron process and the collision-coalescence process.

Precipitation from Cold Clouds: The Bergeron Process

You have probably watched a TV documentary in which mountain climbers brave intense cold and a ferocious snowstorm to scale an ice-covered peak. Although it is hard to imagine, very similar conditions exist in the upper portions of towering cumulonimbus clouds, even on sweltering summer days. It turns out that the frigid conditions high in the troposphere provide an ideal environment to initiate precipitation. In fact, in the middle latitudes much of the rain that falls begins with the birth of snowflakes high in the cloud tops where temperatures are considerably below freezing. Obviously, in the winter, even low clouds are cold enough to trigger precipitation.

The process that generates much of the precipitation in the middle latitudes is named the **Bergeron process** for its discoverer, the highly respected Swedish meteorologist, Tor Bergeron (see Box 17.2). The process relies on two interesting phenomena: supercooling and supersaturation.

Supercooling. Cloud droplets do not freeze at 0°C (32°F) as expected. In fact, pure water suspended in air does not freeze until it reaches a temperature of nearly −40°C (−40°F). Water in the liquid state below 0°C is referred to as **supercooled**. Supercooled water will readily freeze if it impacts an object. This explains why airplanes collect ice when they pass through a cloud composed of supercooled droplets.

In addition, supercooled droplets will freeze upon contact with solid particles that have a crystal form closely resembling that of ice. Such materials are termed **freezing nuclei**. The need for freezing nuclei to initiate the freezing process is similar to the requirement for condensation nuclei in the process of condensation.

In contrast to condensation nuclei, freezing nuclei are very sparse in the atmosphere and do not generally become active until the temperature reaches −10°C (14°F) or less. Only at temperatures well below freezing will ice crystals begin to form in clouds, and even at that, they will be few and far between. Once ice

TABLE 17.3	Relative humidity with respect to ice when relative humidity with respect to water is 100 percent.	
Temperature (°C)	**Relative Humidity with Respect to: Water (%)**	**Ice (%)**
0	100	100
−5	100	105
−10	100	110
−15	100	116
−20	100	121

crystals form, they are in direct competition with the supercooled droplets for the available water vapor.

Supersaturation. When air is saturated (100 percent relative humidity) with respect to water, it is **supersaturated** (relative humidity is greater than 100 percent) with respect to ice. Table 17.3 shows that at −10°C (14°F), when the relative humidity is 100 percent with respect to water, the relative humidity with respect to ice is nearly 110 percent. Thus, ice crystals cannot coexist with water droplets, because the air always "appears" supersaturated to the ice crystals. Hence, the ice crystals begin to consume the "excess" water vapor, which lowers the relative humidity near

the surrounding droplets. In turn, the water droplets evaporate to replenish the diminishing water vapor, thereby providing a continual source of vapor for the growth of the ice crystals (Figure 17.25).

Because the level of supersaturation with respect to ice can be quite great, the growth of ice crystals is generally rapid enough to generate crystals large enough to fall. During their descent, these ice crystals enlarge as they intercept cloud drops, which freeze upon them. Air movement will sometimes break up these delicate crystals and the fragments will serve as freezing nuclei. A chain reaction develops, producing many ice crystals, which by accretion form into large crystals called snowflakes.

In summary, the Bergeron process can produce precipitation throughout the year in the middle latitudes, provided at least the upper portions of clouds are cold enough to generate ice crystals. The type of precipitation (snow, sleet, rain or freezing rain) that reaches the ground depends on the temperature profile in the lower few kilometers of the atmosphere. When the surface temperature is above 4°C (39°F), snowflakes usually melt before they reach the ground and continue their descent as rain. Even on a hot summer day, a heavy downpour may have begun as a snowstorm high in the clouds overhead.

▶ **FIGURE 17.25** The Bergeron process. Ice crystals grow at the expense of cloud droplets until they are large enough to fall. The size of these particles has been greatly exaggerated.

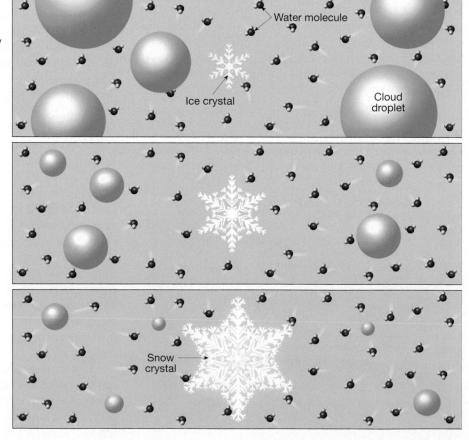

Precipitation from Warm Clouds: The Collision-Coalescence Process

A few decades ago, meteorologists believed that the Bergeron process was responsible for the formation of most precipitation. However, it was discovered that copious rainfall can be associated with clouds located well below the freezing level (*warm clouds*), particularly in the tropics. This led to the proposal of a second mechanism thought to produce precipitation—the **collision-coalescence process**.

Research has shown that clouds composed entirely of liquid droplets must contain some droplets larger than 20 micrometers (0.02 millimeters) if precipitation is to form. These large droplets form when "giant" condensation nuclei are present, or when hygroscopic particles such as sea salt exist. Hygroscopic particles begin to remove water vapor from the air at relative humidities under 100 percent and can grow quite large. Because the rate at which drops fall is size-dependent, these "giant" droplets fall most rapidly. As they plummet, they collide with smaller, slower droplets and coalesce. Growing larger in the process, they fall even more rapidly (or in an updraft, rise more slowly), increasing their chances of collision and rate of growth (Figure 17.26). After a great many such collisions they are large enough to fall to the surface without completely evaporating.

Because of the number of collisions required for growth to raindrop size, droplets in clouds that have great vertical thickness and abundant moisture have a better chance of reaching the required size. Updrafts also aid this process, for they allow the droplets to traverse the cloud repeatedly.

Raindrops can grow to a maximum size of 5 millimeters, at which point they fall at the rate of 33 kilometers (20 miles) per hour. At this size and speed, the water's surface tension, which holds the drop together, is overcome by the drag imposed by the air, which in turn pulls the drops apart. The resulting breakup of a large raindrop produces numerous smaller drops that begin anew the task of sweeping up cloud droplets. Drops that are less than 0.5 millimeter upon reaching the ground are termed *drizzle* and require about 10 minutes to fall from a cloud 1000 meters (3300 feet) overhead.

The collision-coalescence process is not quite as simple as described. First, as the large droplets descend, they produce an air stream around them similar to that produced by a fast-moving automobile. If an automobile is driven at night and we use the bugs that are often out to be analogous to the cloud droplets, it is easy to visualize how most cloud droplets are swept aside. The larger the cloud droplet (or bug), the better its chance of colliding with the giant droplet (or car). Second, collision

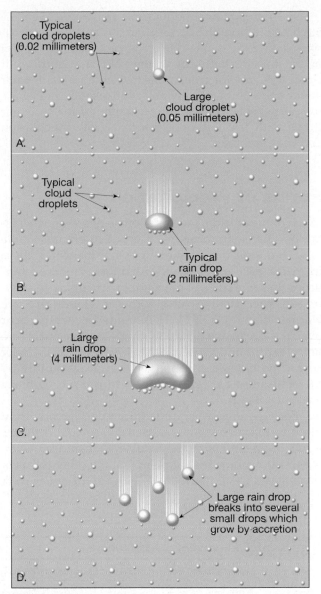

▲ **FIGURE 17.26** The collision-coalescence process. **A.** Most cloud droplets are so small that the motion of the air keeps them suspended. Because large cloud droplets fall more rapidly than smaller droplets, they are able to sweep up the smaller ones in their path and grow. **B.** As these drops increase in size, their fall velocity increases, resulting in increased air resistance, which causes the raindrop to flatten. **C.** As the raindrop approaches 4 millimeters in size it develops a depression in the bottom. **D.** Finally, when the diameter exceeds about 5 millimeters, the depression grows upward almost explosively, forming a donutlike ring of water that immediately breaks into smaller drops. (Note that the drops are not drawn to scale—a typical raindrop has a volume equal to roughly 1 million cloud droplets.)

does not guarantee coalescence. Experiments indicate that atmospheric electricity may hold these droplets together once they collide. If a droplet with a negative charge should collide with a positively charged droplet, electrical attraction of opposite charges may bind them together.

In summary, two mechanisms are known to generate precipitation: the *Bergeron process* and the *collision-coalescence process*. The Bergeron process is dominant in the middle latitudes where cold clouds (or cold cloud tops) are the rule. In the tropics, abundant water vapor and comparatively few condensation nuclei are the norm. This leads to the formation of fewer, larger drops with fast fall velocities that grow by collision and coalescence.

Forms of Precipitation

Much of the world's precipitation begins as snow crystals or other solid forms, such as hail or graupel (Table 17.4). Entering the warmer air below the cloud, these ice particles often melt and reach the ground as raindrops. In some parts of the world, particularly the subtropics, precipitation often forms in clouds that are warmer than 0°C (32°F). These rains frequently occur over the ocean where cloud condensation nuclei are not plentiful and those that exist vary in size. Under such conditions, cloud droplets can grow rapidly by the collision-coalescence process to produce copious amounts of rain.

Because atmospheric conditions vary greatly from place to place as well as seasonally, several different forms of precipitation are possible. Rain and snow are the most common and familiar, but the other forms of precipitation listed in Table 17.4 are important as well. The occurrence of sleet, glaze, or hail is often associated with important weather events. Although limited in occurrence and sporadic in both time and space, these forms, especially glaze and hail, can cause considerable damage.

Rain

We are familiar with rain. In meteorology, the term **rain** is restricted to drops of water that fall from a cloud and have a diameter of at least 0.5 millimeter (0.002 inch). Most rain originates either in nimbostratus clouds or in towering cumulonimbus clouds that are capable of producing unusually heavy rainfalls known as *cloudbursts*. Raindrops rarely exceed about 5 millimeters (0.2 inch) in diameter. Larger drops do not survive, because surface tension, which holds the drops together, is exceeded by the frictional drag of the air. Consequently, large raindrops regularly break apart into smaller ones.

Fine, uniform drops of water having a diameter less than 0.5 millimeter (0.002 inch) are called *drizzle*. Drizzle can be so fine that the tiny drops appear to float and their impact is almost imperceptible. Drizzle and small raindrops generally are produced in stratus or nimbostratus clouds where precipitation can be continuous for several hours, or on rare occasions for days.

Snow

Snow is precipitation in the form of ice crystals (snowflakes) or, more often, aggregates of crystals. The size, shape, and concentration of snowflakes depend to a great extent on the temperature at which they form.

TABLE 17.4 Forms of precipitation.			
Type	**Appropriate Size**	**State of Matter**	**Description**
Mist	0.005 to 0.05 mm	Liquid	Droplets large enough to be felt on the face when air is moving 1 meter/second. Associated with stratus clouds.
Drizzle	Less than 0.5 mm	Liquid	Small uniform drops that fall from stratus clouds, generally for several hours.
Rain	0.5 to 5 mm	Liquid	Generally produced by nimbostratus or cumulonimbus clouds. When heavy, it can show high variability from one place to another.
Sleet	0.5 to 5 mm	Solid	Small, spherical to lumpy ice particles that form when raindrops freeze while falling through a layer of sub-freezing air. Because the ice particles are small, damage, if any, is generally minor. Sleet can make travel hazardous.
Glaze	Layers 1 mm to 2 cm thick	Solid	Produced when supercooled raindrops freeze on contact with solid objects. Glaze can form a thick coating of ice having sufficient weight to seriously damage trees and power lines
Rime	Variable	Solid	Deposits usually consisting of ice feathers that point into the wind. These delicate, frostlike accumulations form as supercooled cloud or fog droplets encounter objects and freeze on contact.
Snow	1 mm to 2 cm	Solid	The crystalline nature of snow allows it to assume many shapes including six-sided crystals, plates, and needles. Produced in supercooled clouds where water vapor is deposited as ice crystals that remain frozen during their descent.
Hail	5 mm to 10 cm	Solid	Precipitation in the form of hard, rounded pellets or irregular lumps of ice. Produced in large convective, cumulonimbus clouds, where frozen ice particles and supercooled water coexist.
Graupel	2 to 5 mm	Solid	Sometimes called soft hail, graupel forms when rime collects on snow crystals to produce irregular masses of "soft" ice. Because these particles are softer than hailstones, they normally flatten out upon impact.

Recall that at very low temperatures, the moisture content of air is small. The result is the formation of very light, fluffy snow made up of individual six-sided ice crystals. This is the "powder" that downhill skiers talk so much about. By contrast, at temperatures warmer than about −5°C (23°F), the ice crystals join together into larger clumps consisting of tangled aggregates of crystals. Snowfalls composed of these composite snowflakes are generally heavy and have a high moisture content, which makes them ideal for making snowballs.

STUDENTS SOMETIMES ASK...
What is the record snowfall in the U.S.?

Although the lake effect snows of upstate New York are legendary, the greatest snowfalls usually occur in sparsely inhabited mountainous regions. The record for the largest snowfall from a single snowstorm in the United States goes to the Mt. Shasta Ski Bowl in California. This storm, which raged from February 13–19, 1959, dropped nearly 5 meters (16 feet) of snow!

Sleet and Glaze

Sleet is a wintertime phenomenon and refers to the fall of small particles of ice that are clear to translucent. For sleet to be produced, a layer of air with temperatures above freezing must overlie a subfreezing layer near the ground. When raindrops, which are often melted snow, leave the warmer air and encounter the colder air below, they freeze and reach the ground as small pellets of ice the size of the raindrops from which they formed.

On some occasions, when the vertical distribution of temperatures is similar to that associated with the formation of sleet, freezing rain or **glaze** results instead. In such situations, the subfreezing air near the ground is not thick enough to allow the raindrops to freeze. The raindrops, however, do become supercooled as they fall through the cold air and turn to ice upon colliding with solid objects. The result can be a thick coating of ice having sufficient weight to break tree limbs, down power lines, and make walking or driving extremely hazardous (Figure 17.27).

Hail

Hail is precipitation in the form of hard, rounded pellets or irregular lumps of ice. Moreover, large hailstones often consist of a series of nearly concentric shells of differing densities and degrees of opaqueness (Figure 17.28). Most hailstones have diameters between 1 centimeter (pea size) and 5 centimeters (golf ball–size), although some can be as big as an orange or larger. Occasionally, hailstones weighing a pound or more have been reported.

▲ **FIGURE 17.27** Glaze forms when supercooled raindrops freeze on contact with objects. In January 1998 an ice storm of historic proportions caused enormous damage in New England and southeastern Canada. Nearly five days of freezing rain (glaze) left millions without electricity—some for as long as a month. (Photo by Syracuse Newspapers/The Image Works)

▲ **FIGURE 17.28** A cross-section of the Coffeyville hailstone. This largest recorded hailstone fell over Kansas in 1970 and weighed 766 grams (1.67 pounds). (University Corporation for Atmospheric Research/National Science Foundation/Visual Communications NCAR)

► **FIGURE 17.29** Hail damage to a used car lot in Fort Worth, Texas, on May 6, 1995. This storm, which packed high winds and hail the size of baseballs, killed at least nine people and injured more than 100 as it swept through the northern part of the state. (Photo by Ron Heflin/AP Wide World Photos)

Many of these were probably composites of several stones frozen together.

The heaviest authenticated hailstone in the United States fell on Coffeyville, Kansas, September 3, 1970 (Figure 17.28). With a 14-centimeter (5.5-inch) diameter, this "giant" weighed 766 grams (1.67 pounds). It is estimated that this stone hit the ground at a speed in excess of 160 kilometers (100 miles) per hour.

The destructive effects of large hailstones are well known, especially to farmers whose crops have been devastated in a few minutes and to people whose windows and roofs have been damaged (Figure 17.29). In the United States, hail damage each year is in the hundreds of millions of dollars.

Hail is produced only in large cumulonimbus clouds where updrafts can sometimes reach speeds approaching 160 kilometers (100 miles) per hour, and where there is an abundant supply of supercooled water. Hailstones begin as small embryonic ice pellets that grow by collecting supercooled water droplets as they fall through the cloud. If they encounter a strong updraft, they may be carried upward again and begin the downward journey anew. Each trip through the supercooled portion of the cloud may be represented by an additional layer of ice. Hailstones can also form from a single descent through an updraft. Either way, the process continues until the hailstone encounters a downdraft or grows too heavy to remain suspended by the thunderstorm's updraft.

Rime

Rime is a deposit of ice crystals formed by the freezing of supercooled fog or cloud droplets on objects whose surface temperature is below freezing. When rime forms on trees, it adorns them with its characteristic ice feathers, which can be spectacular to behold (Figure 17.30). In these situations, objects such as pine needles act as freezing nuclei, causing the supercooled droplets to freeze on contact (see Box 17.2). On occasions when the wind is blowing, only the windward surfaces of objects will accumulate the layer of rime.

Measuring Precipitation

The most common form of precipitation—rain—is the easiest to measure. Any open container having a consistent cross section throughout can be a rain gauge. In general practice, however, more sophisticated devices are used so that small amounts of rainfall can be measured more accurately, and to reduce losses from evaporation. The *standard rain gauge* (Figure 17.31) has a diameter of about 20 centimeters (8 inches) at the top. Once the water is caught, a funnel conducts the rain into a cylindrical measuring tube that has a cross-sectional area only one-tenth as large as the receiver. Consequently, rainfall depth is magnified 10 times, which allows for accurate measurements to the nearest 0.025 centimeter (0.01 inch). The narrow opening also minimizes evaporation. When the amount of rain is less than 0.025 centimeter, it is reported as a *trace* of precipitation.

Measurement Errors

In addition to the standard rain gauge, several types of recording gauges are routinely used. These instruments record not only the amount of rain but also its time of occurrence and intensity (amount per unit of time).

No matter which rain gauge is used, proper exposure is critical. Errors arise when the gauge is shielded from obliquely falling rain by buildings, trees, or other high objects. Hence, the instrument should be at least as far away from such obstructions as the objects are

▲ FIGURE 17.30 Rime consists of delicate ice crystals that form when supercooled fog or cloud droplets freeze on contact with objects. (Photo by John Cancalosi/Stock Boston)

high. Another cause of error is the wind. It has been shown that with increasing wind and turbulence, it becomes more difficult to collect a representative quantity of rain.

▼ FIGURE 17.31 Precipitation measurement. The standard rain gauge allows for accurate rainfall measurement to the nearest 0.025 centimeter (0.01 inch). Because the cross-sectional area of the measuring tube is only one-tenth as large as the collector, rainfall is magnified 10 times.

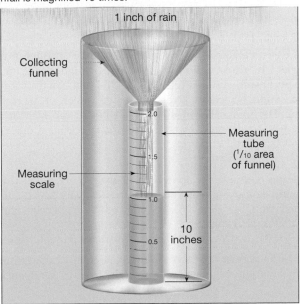

Measuring Snowfall

When snow records are kept, two measurements are normally taken: depth and water equivalent. Usually the depth of snow is measured with a calibrated stick. The actual measurement is simple, but choosing a *representative* spot often poses a dilemma. Even when winds are light or moderate, snow drifts freely. As a rule, it is best to take several measurements in an open place away from trees and obstructions and then average them. To obtain the water equivalent, samples can be melted and then weighed or measured as rain.

The quantity of water in a given volume of snow is not constant. A general ratio of 10 units of snow to 1 unit of water is often used when exact information is not available, but the actual water content of snow may deviate widely from this figure. It may take as much as 30 centimeters of light, fluffy dry snow or as little as 4 centimeters of wet snow to produce 1 centimeter of water.

Precipitation Measurement by Weather Radar

Today's TV weathercasts show helpful maps like the one in Figure 17.32 to depict precipitation patterns. The instrument that produces these images is the *weather radar*.

▼ FIGURE 17.32 A color weather radar display, commonly seen on TV weathercasts. Colors indicate different intensities of precipitation. This National Weather Service Image shows the St. Louis, Missouri, region on the morning of December 18, 2002. The unusual springlike weather included heavy rain, severe thunderstorms, and a tornado watch. (National Weather Service)

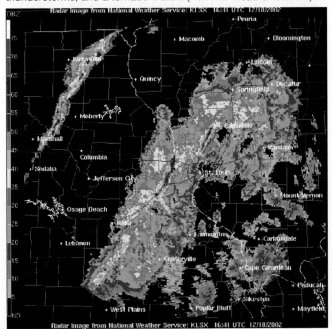

The development of radar has given meteorologists an important tool to probe storm systems that may be up to a few hundred kilometers away. All radar units have a transmitter that sends out short pulses of radio waves. The specific wavelengths that are used depend on the objects the user wants to detect. When radar is used to monitor precipitation, wavelengths between 3 and 10 centimeters are employed.

These wavelengths can penetrate small cloud droplets, but are reflected by larger raindrops, ice crystals, or hailstones. The reflected signal, called an *echo*, is received and displayed on a TV monitor. Because the echo is "brighter" when the precipitation is more intense, modern radar is able to depict not only the regional extent of the precipitation but also the rate of rainfall. Figure 17.32 is a typical radar display in which colors show precipitation intensity. Weather radar is also an important tool for determining the rate and direction of storm movement.

Chapter Summary

• *Water vapor*, an odorless, colorless gas, changes from one state of matter (solid, liquid, or gas) to another at the temperatures and pressures experienced near Earth's surface. The processes involved are *evaporation, condensation, melting, freezing, sublimation,* and *deposition*. During each change, *latent* (hidden) *heat* is either absorbed or released.

• *Humidity* is the general term to describe the amount of water vapor in the air. The methods used to express humidity quantitatively include (1) *mixing ratio*, the mass of water vapor in a unit of air compared to the remaining mass of dry air; (2) *vapor pressure*, that part of the total atmospheric pressure attributable to its water-vapor content; (3) *relative humidity*, the ratio of the air's actual water-vapor content compared with the amount of water vapor required for saturation at that temperature; and (4) *dew point*, that temperature to which a parcel of air would need to be cooled to reach saturation. When air is saturated, the pressure exerted by the water vapor, called the *saturation vapor pressure*, produces a balance between the number of water molecules leaving the surface of the water and the number returning. Because the saturation vapor pressure is temperature-dependent, at higher temperatures more water vapor is required for saturation to occur.

• *Relative humidity can be changed in two ways*. One is by *adding or subtracting water vapor*. The second is by *changing the air's temperature*. When air is cooled, its relative humidity increases.

• The cooling of air as it rises and expands owing to successively lower air pressure is the basic cloud-forming process. Temperature changes in air brought about by compressing or expanding the air are called *adiabatic temperature changes*. Unsaturated air warms by compression and cools by expansion at the rather constant rate of 10°C per 1000 meters of altitude change, a figure called the *dry adiabatic rate*. If air rises high enough, it will cool sufficiently to cause condensation and form a cloud. From this point on, air that continues to rise will cool at the *wet adiabatic rate*, which varies from 5°C to 9°C per 1000 meters of ascent. The difference in the wet and dry adiabatic rates is caused by the condensing water vapor releasing *latent heat*, thereby reducing the rate at which the air cools.

• Four mechanisms that can initiate the vertical movement of air are (1) *orographic lifting*, which occurs when elevated terrains, such as mountains, act as barriers to the flow of air; (2) *frontal wedging*, when cool air acts as a barrier over which warmer, less dense air rises; (3) *convergence*, which happens when air flows together and a general upward movement of air occurs; and (4) *localized convective lifting*, which occurs when unequal surface heating causes pockets of air to rise because of their buoyancy.

• The *stability of air* is determined by examining the temperature of the atmosphere at various altitudes. Air is said to be *unstable* when the *environmental lapse rate* (the rate of temperature decrease with increasing altitude in the troposphere) is greater than the *dry adiabatic rate*. Stated differently, a column of air is unstable when the air near the bottom is significantly warmer (less dense) than the air aloft.

• *For condensation to occur, air must be saturated*. Saturation takes place either when air is cooled to its dew point, which most commonly happens, or when water vapor is added to the air. There must also be a surface on which the water vapor can condense. In cloud and fog formation, tiny particles called *condensation nuclei* serve this purpose.

• *Clouds* are classified on the basis of their *appearance* and *height*. The three basic forms are *cirrus* (high, white, thin, wispy fibers), *cumulus* (globular, individual cloud masses), and *stratus* (sheets or layers that cover much or all of the sky). The four categories based on height are *high clouds* (bases normally above 6000 meters), *middle clouds* (from 2000 to 6000 meters), *low clouds* (below 2000 meters), and *clouds of vertical development*.

• *Fog* is defined as a cloud with its base at or very near the ground. Fogs form when air is cooled below its dew point or when enough water vapor is added to the air to bring about saturation. Various types of fog include *advection fog, radiation fog, upslope fog, steam fog,* and *frontal* (or *precipitation*) *fog*.

• For *precipitation* to form, millions of cloud droplets must somehow join together into large drops. Two mechanisms for the formation of precipitation have been proposed. (1) In clouds where the temperatures are below freezing, ice crystals form and fall as snowflakes. At lower altitudes the snowflakes melt and become raindrops before they reach the ground. (2) *Large* droplets form in warm clouds that contain large *hygroscopic* ("water-seeking") *nuclei*, such as salt particles. As these big droplets descend, they

collide and join with smaller water droplets. After many collisions, the droplets are large enough to fall to the ground as rain.

• The forms of precipitation include *rain, snow, sleet, freezing rain (glaze), hail*, and *rime*.

Key Terms

absolute instability (p. 477)
absolute stability (p. 477)
adiabatic temperature change (p. 473)
advection fog (p. 487)
Bergeron process (p. 489)
calorie (p. 467)
cirrus (p. 482)
cloud (p. 482)
cloud of vertical development (p. 480)
collision-coalescence process (p. 491)
condensation (p. 467)
condensation nuclei (p. 481)
conditional instability (p. 478)
convergence (p. 475)
cumulus (p. 482)
deposition (p. 468)
dew-point temperature (p. 470)
dry adiabatic rate (p. 474)
evaporation (p.467)

fog (p. 487)
freezing nuclei (p. 489)
frontal fog (p. 488)
front (p. 475)
frontal wedging (p. 475)
glaze (p. 493)
hail (p. 492)
high cloud (p. 482)
humidity (p. 468)
hygrometer (p. 472)
hygroscopic nuclei (p. 481)
latent heat (p. 467)
localized convective lifting (p. 476)
low cloud (p. 482)
middle cloud (p. 482)
mixing ratio (p. 469)
orographic lifting (p. 475)
parcel (p. 473)
precipitation fog (p. 488)

psychrometer (p. 472)
radiation fog (p. 487)
rain (p.492)
rain shadow desert (p. 475)
relative humidity (p. 469)
rime (p. 494)
saturation (p. 468)
sleet (p. 492)
snow (p. 492)
stable air (p. 476)
steam fog (p. 488)
stratus (p. 482)
sublimation (p. 468)
supercooled (p. 498)
supersaturation (p. 460)
unstable air (p. 476)
upslope fog (p. 487)
vapor pressure (p. 468)
wet adiabatic rate (p. 474)

Review Questions

1. Summarize the processes by which water changes from one state of matter to another. Indicate whether heat energy is absorbed or liberated.

2. After studying Table 17.1, write a generalization relating temperature and the amount of water vapor needed to saturate the air.

3. How do relative humidity and mixing ratio differ?

4. Referring to Figure 17.6, answer the following questions and then write a generalization relating changes in air temperature to changes in relative humidity.
   (a) During a typical day, when is the relative humidity highest? Lowest?
   (b) At what time of day would dew most likely form?

5. If the temperature remains unchanged and the mixing ratio decreases, how will relative humidity change?

6. On a cold winter day when the temperature is −10°C and the relative humidity is 50 percent, what is the mixing ratio (refer to Table 17.1)? What is the mixing ratio

for a day when the temperature is 20°C and the relative humidity is 50 percent?

7. Explain the principle of the sling psychrometer and the hair hygrometer.

8. Using the standard tables in Appendix D, determine the relative humidity and dew-point temperature if the dry-bulb thermometer reads 16°C and the wet-bulb thermometer reads 12°C. How would the relative humidity and dew point change if the wet-bulb thermometer read 8°C?

9. On a warm summer day when the relative humidity is high, it may seem even warmer than the thermometer indicates. Why do we feel so uncomfortable on a muggy day?

10. Why does air cool when it rises through the atmosphere?

11. Explain the difference between environmental lapse rate and adiabatic cooling.

12. If unsaturated air at 23°C were to rise, what would its temperature be at 500 meters? If the dew-point temperature at

the condensation level were 13°C, at what altitude would clouds begin to form?

13. Why does the adiabatic rate of cooling change when condensation begins? Why is the wet adiabatic rate not a constant figure?

14. The contents of an aerosol can are under very high pressure. When you push the nozzle on such a can, the spray feels cold. Explain.

15. How do orographic lifting and frontal wedging act to force air to rise?

16. Explain why the Great Basin area of the western United States is so dry. What term is applied to such a situation?

17. How does stable air differ from unstable air? Describe the general nature of the clouds and precipitation expected with each.

18. How do temperature inversions influence air pollution? (See Box 17.1.)

19. What is the function of condensation nuclei in cloud formation? The function of the dew point?

20. As you drink an ice-cold beverage on a warm day, the outside of the glass or bottle becomes wet. Explain.

21. What is the basis for the classification of clouds?

22. Why are high clouds always thin?

23. Which cloud types are associated with the following characteristics: thunder, halos, precipitation, hail, mackerel sky, lightning, mares' tails?

24. List five types of fog and discuss the details of their formation.

25. What is the difference between precipitation and condensation?

26. List the forms of precipitation and the circumstances of their formation.

27. Sometimes, when rainfall is light, the amount is reported as a trace. When this occurs, how much (or little) rain has fallen?

Examining the Earth System

1. The interrelationships among Earth's spheres have produced the Great Basin area of the western United States, which includes some of the driest areas in the world. Examine the map of the region in Appendix D. Although the area is only a few hundred miles from the Pacific Ocean, it is a desert. Why? Did any geologic factor(s) contribute to the formation of this desert environment? Do any major rivers have their source in the Great Basin? Explain. (For information about deserts in the United States, visit the United States Geological Survey (USGS) Deserts: Geology and Resources Website at **http://pubs.usgs.gov/gip/deserts/**)

2. Phoenix and Flagstaff, Arizona, are located nearby each other in the southwestern United States. Using the climate diagrams for the two cities found in Figure 20.13, describe the impact that elevation has on the precipitation and temperature of each city. Compare the natural vegetation of these nearby locations. (To examine the current weather conditions and forecasts for Phoenix and Flagstaff, Arizona, contact *USA Today* at **http://www.usatoday.com/weather/** and/or *The Weather Channel* at **http://www.weather.com**)

3. The amount of precipitation that falls at any particular place and time is controlled by the quantity of moisture in the air and many other factors. How might each of the following alter the precipitation at a particular locale? (a) An increase in the elevation of the land. (b) A decrease in the area covered by forests and other types of vegetation. (c) Lowering of average ocean-surface temperatures. (d) An increase in the percentage of time that the winds blow from an adjacent body of water. (e) A major episode of global volcanism lasting for decades.

Online Study Guide

The *Earth Science* Website uses the resources and flexibility of the Internet to aid in your study of the topics in this chapter. Written and developed by Earth science instructors, this site will help improve your understanding of Earth science. Visit **http://www.prenhall.com/tarbuck** and click on the cover of *Earth Science 11e* to find:

- **On-line review quizzes.**
- **Critical thinking exercises.**
- **Links to chapter-specific Web resources.**
- **Internet-wide key term searches.**

http://www.prenhall.com/tarbuck

Wind turbines in California. U.S. installed wind-power capacity is growing rapidly.
(Photo by Lester Lefkowitz/CORBIS)

CHAPTER 18

Air Pressure and Wind

Of the various elements of weather and climate, changes in air pressure are the least noticeable. In listening to a weather report, generally we are interested in moisture conditions (humidity and precipitation), temperature, and perhaps wind. It is the rare person, however, who wonders about air pressure. Although the hour-to-hour and day-to-day variations in air pressure are not perceptible to human beings, they are very important in producing changes in our weather. For example, it is variations in air pressure from place to place that generate winds that in turn can bring changes in temperature and humidity (Figure 18.1). Air pressure is one of the basic weather elements and is a significant factor in weather forecasting. As you will see, air pressure is closely tied to the other elements of weather in a cause-and-effect relationship.

Understanding Air Pressure

In Chapter 16 we noted that **air pressure** is simply the pressure exerted by the weight of air above. Average air pressure at sea level is about 1 kilogram per square centimeter, or 14.7 pounds per square inch. This is roughly the same pressure that is produced by a column of water 10 meters (33 feet) in height. With some simple arithmetic you can calculate that the air pressure exerted on the top of a small (50 centimeter by 100 centimeter) school desk exceeds 5000 kilograms (11,000 pounds), or about the weight of a 50-passenger school bus. Why doesn't the desk collapse under the weight of the ocean of air above? Simply, air pressure is exerted in all directions—down, up, and sideways. Thus, the air pressure pushing down on the desk exactly balances the air pressure pushing up on the desk.

You might be able to visualize this phenomenon better if you imagine a tall aquarium that has the same dimensions as the desktop. When this aquarium is filled to a height of 10 meters (33 feet), the water pressure at the bottom equals 1 atmosphere (14.7 pounds per square inch). Now, imagine what will happen if this aquarium is placed on top of our student desk so that all the force is directed downward. Compare this to what results when the desk is placed inside the aquarium and allowed to sink to the bottom. In the latter situation the desk survives because the water pressure is exerted in all directions, not just downward as in our earlier example. The desk, like your body, is "built" to withstand the pressure of 1 atmosphere. It is important to note that although we do not generally notice the pressure exerted by the ocean of air around us, except when ascending or descending in an elevator or airplane, it is nonetheless substantial. The pressurized suits used by astronauts on space walks are designed to duplicate the atmospheric pressure experienced at Earth's surface. Without these protective suits to keep body fluids from boiling away, astronauts would perish in minutes.

The concept of air pressure can be better understood if we examine the behavior of gases. Gas molecules, un-

▼ FIGURE 18.1 Palm trees buffeted by hurricane-force winds, Corpus Christi, Texas. (National Geographic Image Collection/Getty Images, Inc.—Hulton Archive Photos)

like those of the liquid and solid phases, are not "bound" to one another but are freely moving about, filling all space available to them. When two gas molecules collide, which happens frequently under normal atmospheric conditions, they bounce off each other like very elastic balls. If a gas is confined to a container, this motion is restricted by its sides, much like the walls of a handball court redirect the motion of the handball. The continuous bombardment of gas molecules against the sides of the container exerts an outward push that we call air pressure. Although the atmosphere is without walls, it is confined from below by Earth's surface and effectively from above because the force of gravity prevents its escape. Here we define *air pressure* as the force exerted against a surface by the continuous collision of gas molecules.

Measuring Air Pressure

 GEODe   Earth's Dynamic Atmosphere
▼ Air Pressure and Wind

When meteorologists measure atmospheric pressure, they employ a unit called the *millibar*. Standard sea-level pressure is 1013.2 millibars. Although the millibar has been the unit of measure on all U.S. weather maps since January 1940, the media use "inches of mercury" to describe atmospheric pressure. In the United States, the National Weather Service converts millibar values to inches of mercury for public and aviation use.

Inches of mercury are easy to understand. The use of mercury for measuring air pressure dates from 1643, when Torricelli, a student of the famous Italian scientist Galileo, invented the **mercury barometer** (*bar* = pressure, *metron* = measuring instrument). Torricelli correctly described the atmosphere as a vast ocean of air that exerts pressure on us and all objects about us. To measure this force, he filled a glass tube, which was closed at one end, with mercury. He then inverted the tube into a dish of mercury (Figure 18.2). Torricelli found that the mercury flowed out of the tube until the weight of the column was balanced by the pressure that the atmosphere exerted on the surface of the mercury in the dish. In other words, the weight of mercury in the column equaled the weight of the same diameter column of air that extended from the ground to the top of the atmosphere.

When air pressure increases, the mercury in the tube rises. Conversely, when air pressure decreases, so does the height of the mercury column. With some refinements the mercurial barometer invented by Torricelli is still the standard pressure-measuring instrument used today. Standard atmospheric pressure at sea level equals 29.92 inches of mercury.

The need for a smaller and more portable instrument for measuring air pressure led to the development of the **aneroid**, (*an* = without, *ner* = fluid) **barometer** (Figure 18.3). Instead of having a mercury

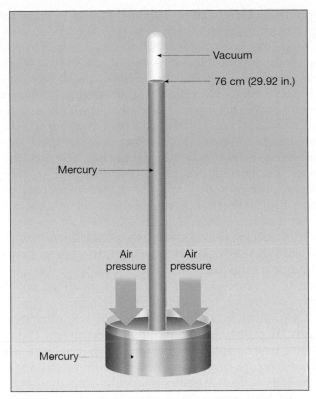

▲ **FIGURE 18.2** Simple mercury barometer. The weight of the column of mercury is balanced by the pressure exerted on the dish of mercury by the air above. If the pressure decreases, the column of mercury falls; if the pressure increases, the column rises.

column held up by air pressure, the aneroid barometer uses a partially evacuated metal chamber. The chamber, being very sensitive to variations in air pressure, changes shape, compressing as the pressure increases and expanding as the pressure decreases. A series of levers transmits the movements of the chamber to a

▼ **FIGURE 18.3** Aneroid barometer. This instrument has a partially evacuated chamber that changes shape, compressing as atmospheric pressure increases, and expanding as pressure decreases.

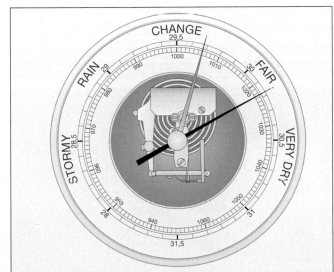

pointer on a dial that is calibrated to read in inches of mercury and/or millibars.

As shown in Figure 18.3, the face of an aneroid barometer intended for home use is inscribed with words like *fair, change, rain*, and *stormy*. Notice that "fair weather" corresponds with high-pressure readings, whereas "rain" is associated with low pressures. Although barometric readings may indicate the present weather, this is not always the case. The dial may point to "fair" on a rainy day, or you may be experiencing "fair" weather when the dial indicates "rainy." If you want to "predict" the weather in a local area, the change in air pressure over the past few hours is more important than the current pressure reading. Falling pressure is often associated with increasing cloudiness and the possibility of precipitation, whereas rising air pressure generally indicates clearing conditions. It is useful to remember, however, that particular barometer readings or trends do not always correspond to specific types of weather.

One advantage of the aneroid barometer is that it can easily be connected to a recording mechanism. The resulting instrument is a **barograph**, which provides a continuous record of pressure changes with the passage of time (Figure 18.4). Another important adaptation of the aneroid barometer is its use to indicate altitude for aircraft, mountain climbers, and mapmakers.

STUDENTS SOMETIMES ASK...

What is the lowest barometric pressure ever recorded?

All of the lowest-recorded barometric pressures have been associated with strong hurricanes. The record for the United States is 888 millibars (26.20 inches) measured during Hurricane Gilbert in September 1988. The world's record, 870 millibars (25.70 inches), occurred during Typhoon Tip (a Pacific hurricane), in October 1979. Although tornadoes undoubtedly have produced even lower pressures, none have been accurately measured.

▼ **FIGURE 18.4** An aneroid barograph makes a continuous record of pressure changes. (Photo courtesy of Qualimetrics, Inc., Sacramento, California)

Factors Affecting Wind

GEODe
Earth's Dynamic Atmosphere
▼ Air Pressure and Wind

In Chapter 17 we examined the upward movement of air and its role in cloud formation. As important as vertical motion is, far more air moves horizontally, the phenomenon we call **wind.** What causes wind?

Simply stated, wind is the result of horizontal differences in air pressure. *Air flows from areas of higher pressure to areas of lower pressure.* You may have experienced this when opening a vacuum-packed can of coffee. The noise you hear is caused by air rushing from the higher pressure outside the can to the lower pressure inside. Wind is nature's attempt to balance such inequalities in air pressure. Because unequal heating of Earth's surface generates these pressure differences, *solar radiation is the ultimate energy source for most wind.*

If Earth did not rotate, and if there were no friction between moving air and Earth's surface, air would flow in a straight line from areas of higher pressure to areas of lower pressure. But because both factors exist, wind is controlled by a combination of forces, including: (1) the pressure-gradient force, (2) Coriolis effect, and (3) friction. We will now examine each of these factors.

Pressure-Gradient Force

Pressure differences create wind, and the greater these differences, the greater the wind speed. Over Earth's surface, variations in air pressure are determined from barometric readings taken at hundreds of weather stations. These pressure data are shown on a weather map using **isobars,** lines that connect places of equal air pressure (Figure 18.5). The spacing of isobars indicates the amount of pressure change occurring over a given distance and is expressed as the **pressure gradient** (*gradus* = slope).

You might find it easier to visualize a pressure gradient if you think of it as being similar to the slope of a hill. A steep pressure gradient, like a steep hill, causes greater acceleration of an air parcel than does a weak pressure gradient (a gentle hill). Thus, the relationship between wind speed and the pressure gradient is straightforward: *Closely spaced isobars indicate a steep pressure gradient and high winds, whereas widely spaced isobars indicate a weak pressure gradient and light winds.* Figure 18.5 illustrates the relationship between the spacing of isobars and wind speed. Notice that wind speeds are greater in Ohio, Kentucky, Michigan, and Illinois, where isobars are more closely spaced, than in the western states where isobars are more widely spaced.

The pressure gradient is the driving force of wind, and it has both magnitude and direction. Its magni-

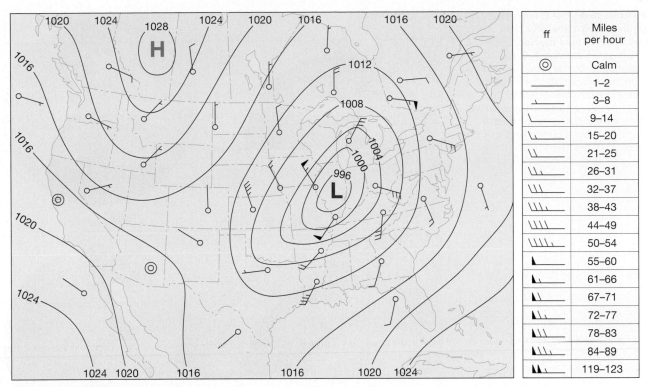

ff	Miles per hour
◎	Calm
—	1–2
⌐	3–8
⌐	9–14
⌐⌐	15–20
⌐⌐	21–25
⌐⌐⌐	26–31
⌐⌐⌐	32–37
⌐⌐⌐⌐	38–43
⌐⌐⌐⌐	44–49
⌐⌐⌐⌐⌐	50–54
◣	55–60
◣⌐	61–66
◣⌐	67–71
◣⌐⌐	72–77
◣⌐⌐	78–83
◣⌐⌐⌐	84–89
◣◣⌐	119–123

▲ **FIGURE 18.5** Isobars are lines connecting places of equal sea-level pressure. They are used to show the distribution of pressure on daily weather maps. Isobars are seldom straight, but usually form broad curves. Concentric rings of isobars indicate cells of high and low pressure. The "wind flags" indicate the expected airflow surrounding pressure cells and are plotted as "flying" with the wind (that is, the wind blows toward the station circle). Notice on this map that the isobars are more closely spaced and the wind speed is faster around the low-pressure center than around the high.

tude is determined from the spacing of isobars. The direction of force is always from areas of higher pressure to areas of lower pressure and at right angles to the isobars. Once the air starts to move, the Coriolis effect and friction come into play, but then only to modify the movement, not to produce it.

Coriolis Effect

The weather map in Figure 18.5 shows the typical air movements associated with high- and low-pressure systems. As expected, the air moves out of the regions of higher pressure and into the regions of lower pressure. However, the wind does not cross the isobars at right angles as the pressure-gradient force directs it. This deviation is the result of Earth's rotation and has been named the **Coriolis effect** after the French scientist who first thoroughly described it.

All free-moving objects or fluids, including the wind, are deflected to the *right* of their path of motion in the Northern Hemisphere and to the *left* in the Southern Hemisphere. The reason for this deflection can be illustrated by imagining the path of a rocket launched from the North Pole toward a target located on the equator (Figure 18.6). If the rocket took an hour to reach its target, Earth would have rotated 15 degrees to the east during its flight. To someone standing on Earth it would look as if the rocket veered off its

path and hit Earth 15 degrees west of its target. The true path of the rocket is straight and would appear so to someone out in space looking down at Earth. It was Earth turning under the rocket that gave it its *apparent* deflection.

Note that the rocket was deflected to the right of its path of motion because of the counterclockwise rotation of the Northern Hemisphere. In the Southern Hemisphere, the effect is reversed. Clockwise rotation produces a similar deflection, but to the left of the path of motion. The same deflection is experienced by wind regardless of the direction it is moving.

We attribute the apparent shift in wind direction to the Coriolis effect. This deflection (1) is always directed at right angles to the direction of airflow; (2) affects only wind direction, not wind speed; (3) is affected by wind speed (the stronger the wind, the greater the deflection); and (4) is strongest at the poles and weakens equatorward, becoming nonexistent at the equator.

It is of interest to point out that any "free-moving" object will experience a deflection caused by the Coriolis effect. This fact was dramatically discovered by the United States Navy in World War II. During target practice long-range guns on battleships continually missed their targets by as much as several hundred yards until ballistic corrections were made for the changing position of a seemingly stationary target.

► **FIGURE 18.6** The Coriolis effect illustrated using a one-hour flight of a rocket traveling from the North Pole to a location on the equator. **A.** On a nonrotating Earth, the rocket would travel straight to its target. **B.** However, Earth rotates 15° each hour. Thus, although the rocket travels in a straight line, when we plot the path of the rocket on Earth's surface, it follows a curved path that veers to the right of the target.

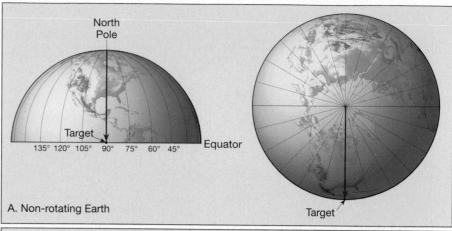

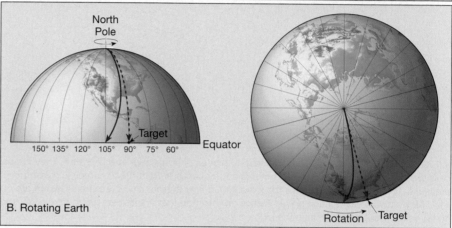

Over a short distance, however, the Coriolis effect is relatively small.

Friction with Earth's Surface

The effect of friction on wind is important only within a few kilometers of Earth's surface. Friction acts to slow air movement and, as a consequence, alters wind direction. To illustrate friction's effect on wind direction, let us look at a situation in which it has no role. Above the friction layer, the pressure-gradient force and Coriolis effect work together to direct the flow of air. Under these conditions, the pressure-gradient force causes air to start moving across the isobars. As soon as the air starts to move, the Coriolis effect acts at right angles to this motion. The faster the wind speed, the greater the deflection.

Eventually, the Coriolis effect will balance the pressure-gradient force, and the wind will blow parallel to the isobars (Figure 18.7). Upper-air winds generally take this path and are called **geostrophic winds**. Because of the lack of friction with Earth's surface, geostrophic winds travel at higher speeds than do surface winds. This can be observed in Figure 18.8 by noting the wind flags, many of which indicate winds of 50 to 100 miles per hour.

The most prominent features of upper-level flow are the **jet streams**. First encountered by high-flying bombers during World War II, these fast-moving rivers of air travel between 120 and 240 kilometers (75 and 150 miles) per hour in a west-to-east direction. One such stream is situated over the polar front, which is the zone separating cool polar air from warm subtropical air.

Below 600 meters (2000 feet), friction complicates the airflow just described. Recall that the Coriolis effect is proportional to wind speed. Friction lowers the wind speed, so it reduces the Coriolis effect. Because the pressure-gradient force is not affected by wind speed, it wins the tug of war shown in Figure 18.9. The result is a movement of air at an angle across the isobars toward the area of lower pressure.

STUDENTS SOMETIMES ASK...

Why doesn't the Coriolis effect cause a baseball to be deflected when you are playing catch?

Over very short distances the Coriolis deflection is too small to be noticed. Nevertheless, in the middle latitudes the Coriolis effect is great enough to potentially affect the outcome of a baseball game. A ball hit a horizontal distance of 100 meters (330 feet) in 4 seconds down the right field line will be deflected 1.5 centimeters (more than 1/2 inch) to the right by the Coriolis effect. This could be just enough to turn a potential home run into a foul ball!

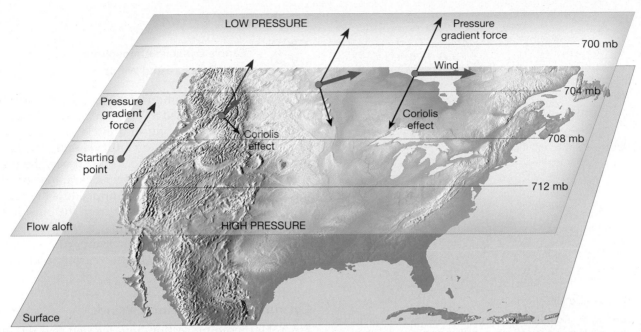

▲ **FIGURE 18.7** The geostrophic wind. The only force acting on a stationary parcel of air is the pressure-gradient force. Once the air begins to accelerate, the Coriolis effect deflects it to the right in the Northern Hemisphere. Greater wind speeds result in a stronger Coriolis effect (deflection) until the flow is parallel to the isobars. At this point the pressure-gradient force and Coriolis effect are in balance and the flow is called a *geostrophic wind*. It is important to note that in the "real" atmosphere, airflow is continually adjusting for variations in the pressure field. As a result, the adjustment to geostrophic equilibrium is much more irregular than shown.

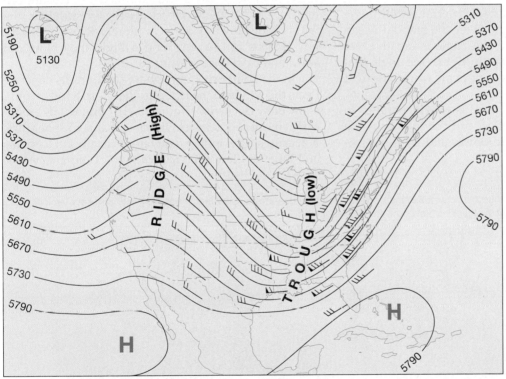

A. Upper-level weather chart

ff	Miles per hour
◎	Calm
	1–2
	3–8
	9–14
	15–20
	21–25
	26–31
	32–37
	38–43
	44–49
	50–54
	55–60
	61–66
	67–71
	72–77
	78–83
	84–89
	119–123

B. Representation of upper-level chart

▲ **FIGURE 18.8** Upper-air winds. This map shows the direction and speed for the upper-air wind for a particular day. Note that the airflow is nearly parallel to the contours. These isolines are height contours for the 500-millibar level.

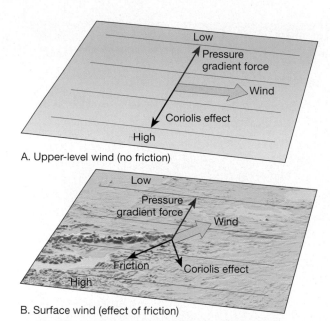

A. Upper-level wind (no friction)

B. Surface wind (effect of friction)

▲ **FIGURE 18.9** Comparison between upper-level winds and surface winds showing the effects of friction on airflow. Friction slows surface wind speed, which weakens the Coriolis effect, causing the winds to cross the isobars and move toward the lower pressure.

The roughness of the terrain determines the angle of airflow across the isobars. Over the smooth ocean surface, friction is low and the angle is small. Over rugged terrain, where friction is higher, the angle that air makes as it flows across the isobars can be as great as 45 degrees.

In summary, upper airflow is nearly parallel to the isobars, whereas the effect of friction causes the surface winds to move more slowly and cross the isobars at an angle.

Highs and Lows

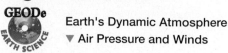

Earth's Dynamic Atmosphere
▼ Air Pressure and Winds

Among the most common features on any weather map are areas designated as pressure centers. **Lows**, or **cyclones** (*kyklon* = moving in a circle) are centers of low pressure, and **highs**, or **anticyclones**, are high-pressure centers. As Figure 18.10 illustrates, the pressure decreases from the outer isobars toward the center in a low. In a high, just the opposite is the case—the values of the isobars increase from the outside toward the center. By knowing just a few basic facts about centers of high and low pressure, you can greatly increase your understanding of current and forthcoming weather.

Cyclonic and Anticyclonic Winds

From the preceding section, you learned that the two most significant factors that affect wind are the pressure-gradient force and the Coriolis effect. Winds move from higher pressure to lower pressure and are deflected to the right or left by Earth's rotation. When these controls of airflow are applied to pressure centers in the Northern Hemisphere, the result is that winds blow inward and counterclockwise around a low (Figure 18.11A). Around a high, they blow outward and clockwise (Figure 18.10).

In the Southern Hemisphere the Coriolis effect deflects the winds to the left, and therefore winds around a low blow clockwise (Figure 18.11B), and winds around a high move counterclockwise. In either hemisphere, friction causes a net inflow (**convergence**) around a cyclone and a net outflow (**divergence**) around an anticyclone.

▼ **FIGURE 18.10** Cyclonic and anticyclonic winds in the Northern Hemisphere. Arrows show that winds blow into and counterclockwise around a low. By contrast, around a high, winds blow outward and clockwise.

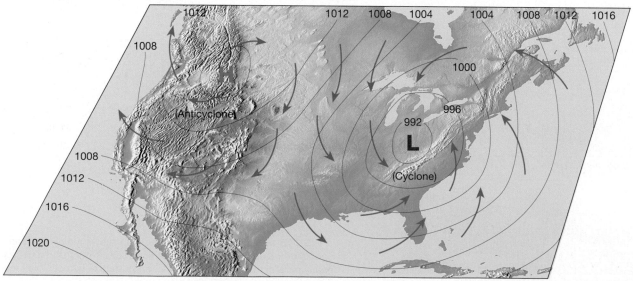

A.

B.

▲ **FIGURE 18.11** Cyclonic circulation in the Northern and Southern hemispheres. The cloud patterns in these images allow us to "see" the circulation pattern in the lower atmosphere. **A.** This satellite image shows a large low-pressure center in the Gulf of Alaska on August 17, 2004. The cloud pattern clearly shows an inward and *counterclockwise* spiral. **B.** This image from March 26, 2004, shows a strong cyclonic storm in the South Atlantic near the coast of Brazil. The cloud pattern reveals an inward and *clockwise* circulation. (NASA images)

Weather Generalizations About Highs and Lows

Rising air is associated with cloud formation and precipitation, whereas subsidence produces clear skies. In this section you will learn how the movement of air can itself create pressure change and hence generate winds. In addition, you will examine the relationship between horizontal and vertical flow and its effect on the weather.

Let us first consider the situation around a surface low-pressure system where the air is spiraling inward. Here the net inward transport of air causes a shrinking of the area occupied by the air mass, a process that is termed *horizontal convergence*. Whenever air converges horizontally, it must pile up, that is, increase in height to allow for the decreased area it now occupies. This generates a taller and therefore heavier air column. Yet a surface low can exist only as long as the column of air above exerts less pressure than that occurring in surrounding regions. We seem to have encountered a paradox—a low-pressure center causes a net accumulation of air, which increases its pressure. Consequently, a surface cyclone should quickly eradicate itself in a manner not unlike what happens when a vacuum-packed can is opened.

You can see that for a surface low to exist for very long, compensation must occur aloft. For example, surface convergence could be maintained if divergence (spreading out) aloft occurred at a rate equal to the in-

flow below. Figure 18.12 shows the relationship between surface convergence (inflow) and divergence (outflow) aloft that is needed to maintain a low-pressure center.

Divergence aloft may even exceed surface convergence, thereby resulting in intensified surface inflow and accelerated vertical motion. Thus, divergence aloft can intensify storm centers as well as maintain them. On the other hand, inadequate divergence aloft permits surface flow to "fill" and weaken the accompanying cyclone.

Note that surface convergence about a cyclone causes a net upward movement. The rate of this vertical movement is slow, generally less than 1 kilometer per day. Nevertheless, because rising air often results in cloud formation and precipitation, a low-pressure center is generally related to unstable conditions and stormy weather (Figure 18.13A).

As often as not, it is divergence aloft that creates a surface low. Spreading out aloft initiates upflow in the atmosphere directly below, eventually working its way to the surface, where inflow is encouraged.

Like their cyclonic counterparts, anticyclones must be maintained from above. Outflow near the surface is accompanied by convergence aloft and general subsidence of the air column (Figure 18.12). Because descending air is compressed and warmed, cloud formation and precipitation are unlikely in an anticyclone. Thus, "fair" weather can usually be expected with the approach of a high pressure center (Figure 18.13B).

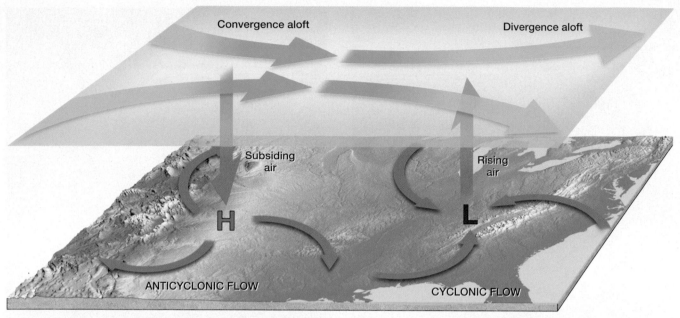

▲ **FIGURE 18.12** Airflow associated with surface cyclones and anticyclones. A low, or cyclone, has converging surface winds and rising air causing cloudy conditions. A high, or anticyclone, has diverging surface winds and descending air, which lead to clear skies and fair weather.

For reasons that should now be obvious, it has been common practice to print on household barometers the words "stormy" at the low-pressure end and "fair" on the high-pressure end. By noting whether the pressure is rising, falling, or steady, we have a good indication of what the forthcoming weather will be. Such a determination, called the **pressure**, or **barometric tendency**, is a very useful aid in short-range weather prediction.

You should now be better able to understand why television weather reporters emphasize the locations and projected paths of cyclones and anticyclones. The "villain" on these weather programs is always the low-pressure center, which produces "bad" weather in any season. Lows move in roughly a west-to-east direction across the United States and require a few days to more than a week for the journey. Because their paths can be somewhat erratic, accurate prediction of their migration is difficult, although essential, for short-range forecasting.

Meteorologists must also determine if the flow aloft will intensify an embryo storm or act to suppress its development. Because of the close tie between conditions at the surface and those aloft, a great deal of emphasis has been placed on the importance and understanding of the total atmospheric circulation, particularly in the midlatitudes. We will now examine the workings of Earth's general atmospheric circulation, and then again consider the structure of the cyclone in light of this knowledge.

General Circulation of the Atmosphere

As noted, the underlying cause of wind is unequal heating of Earth's surface (see Box 18.1). In tropical regions, more solar radiation is received than is radiated back to space. In polar regions the opposite is true: Less solar energy is received than is lost. Attempting to balance these differences, the atmosphere acts as a giant heat-transfer system, moving warm air poleward and cool air equatorward. On a smaller scale, but for the same reason, ocean currents also contribute to this global heat transfer. The general circulation is very complex, and there is a great deal that has yet to be explained. We can, however, develop a general understanding by first considering the circulation that would occur on a nonrotating Earth having a uniform surface. We will then modify this system to fit observed patterns.

Circulation on a Nonrotating Earth

On a hypothetical nonrotating planet with a smooth surface of either all land or all water, two large thermally produced cells would form (Figure 18.14). The heated equatorial air would rise until it reached the tropopause, which, acting like a lid, would deflect the air poleward. Eventually, this upper-level airflow would reach the poles, sink, spread out in all

A.

B.

▲ **FIGURE 18.13** These two photographs illustrate the basic weather generalizations associated with pressure centers. **A.** A sea of umbrellas on a rainy day in Shanghai, China. Centers of low pressure are frequently associated with cloudy conditions and precipitation. (Stone/Getty Images Inc.—Stone Allstock) **B.** By contrast, clear skies and "fair" weather may be expected when an area is under the influence of high pressure. Sunbathers on a beach at Cape Henlopen, Delaware. (Photo by Mark Gibson/DRK Photo)

directions at the surface, and move back toward the equator. Once there, it would be reheated and start its journey over again. This hypothetical circulation system has upper-level air flowing poleward and surface air flowing equatorward.

If we add the effect of rotation, this simple convection system will break down into smaller cells. Figure 18.15 illustrates the three pairs of cells proposed to carry on the task of heat redistribution on a rotating planet. The polar and tropical cells retain the characteristics of the thermally generated convection described earlier. The nature of the midlatitude circu-

▼ **FIGURE 18.14** Global circulation on a nonrotating Earth. A simple convection system is produced by unequal heating of the atmosphere.

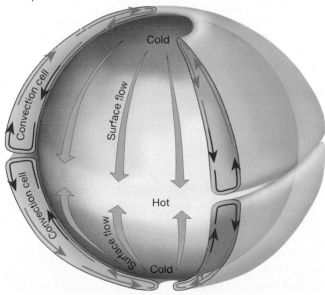

lation is more complex and will be discussed in more detail in a later section.

Idealized Global Circulation

Near the equator, the rising air is associated with the pressure zone known as the **equatorial low**—a region marked by abundant precipitation. As the upper-level flow from the equatorial low reaches 20 to 30 degrees latitude, north or south, it sinks back toward the surface. This subsidence and associated adiabatic heating produce hot, arid conditions. The center of this zone of subsiding dry air is the **subtropical high**, which encircles the globe near 30 degrees latitude, north and south (Figure 18.15). The great deserts of Australia, Arabia, and North Africa exist because of the stable dry conditions associated with the subtropical highs.

At the surface, airflow is outward from the center of the subtropical high. Some of the air travels equatorward and is deflected by the Coriolis effect, producing the reliable **trade winds**. The remainder travels poleward and is also deflected, generating the prevailing **westerlies** of the midlatitudes. As the westerlies move poleward, they encounter the cool **polar easterlies** in the region of the **subpolar low**. The interaction of these warm and cool winds produces the stormy belt known as the **polar front**. The source region for the variable polar easterlies is the **polar high**. Here, cold polar air is subsiding and spreading equatorward.

In summary, this simplified global circulation is dominated by four pressure zones. The subtropical and polar highs are areas of dry subsiding air that flows outward at the surface, producing the prevailing winds. The low-pressure zones of the equatorial and subpolar regions are associated with inward and upward airflow accompanied by clouds and precipitation.

511

▶ **FIGURE 18.15** Idealized global circulation proposed for the three-cell circulation model of a rotating Earth.

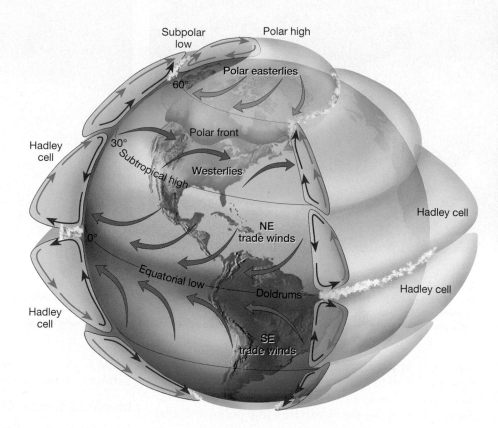

Influence of Continents

Up to this point, we have described the surface pressure and associated winds as continuous belts around Earth. However, the only truly continuous pressure belt is the subpolar low in the Southern Hemisphere. Here the ocean is uninterrupted by landmasses. At other latitudes, particularly in the Northern Hemisphere where landmasses break up the ocean surface, large seasonal temperature differences disrupt the pattern. Figure 18.16 shows the resulting pressure and wind patterns for January and July. The circulation over the oceans is dominated by semipermanent cells of high pressure in the subtropics and cells of low pressure over the subpolar regions. The subtropical highs are responsible for the trade winds and westerlies, as mentioned earlier.

The large landmasses, on the other hand, particularly Asia, become cold in the winter and develop a seasonal high-pressure system from which surface flow is directed off the land (Figure 18.16A). In the summer, the opposite occurs; the landmasses are heated and develop low-pressure cells, which permit air to flow onto the land (Figure 18.16B). These seasonal changes in wind direction are known as the **monsoons**. During warm months, areas such as India experience a flow of warm, water-laden air from the Indian Ocean, which produces the rainy summer monsoon. The winter monsoon is dominated by dry continental air. A similar situation exists, but to a lesser extent, over North America.

In summary, the general circulation is produced by semipermanent cells of high and low pressure over the oceans and is complicated by seasonal pressure changes over land.

STUDENTS SOMETIMES ASK...
Does monsoon mean "rainy season"?

No. Regions that experience monsoons typically have both a wet and a dry season. Monsoon refers to a wind system that exhibits a pronounced season reversal in direction. In general, winter is associated with winds that blow predominantly off the continents and produce a dry winter monsoon. By contrast, in summer, warm moisture-laden air blows from the sea toward the land. Thus, the summer monsoon, which is usually associated with abundant precipitation, is the source of the misconception.

The Westerlies

The circulation in the midlatitudes, the zone of the westerlies, is complex and does not fit the convection system proposed for the tropics. Between about 30 and 60 degrees latitude, the general west-to-east flow is interrupted by migrating cyclones and anticyclones. In the Northern Hemisphere these cells move from west to east around the globe, creating an anticyclonic (clockwise) flow or a cyclonic (counter-

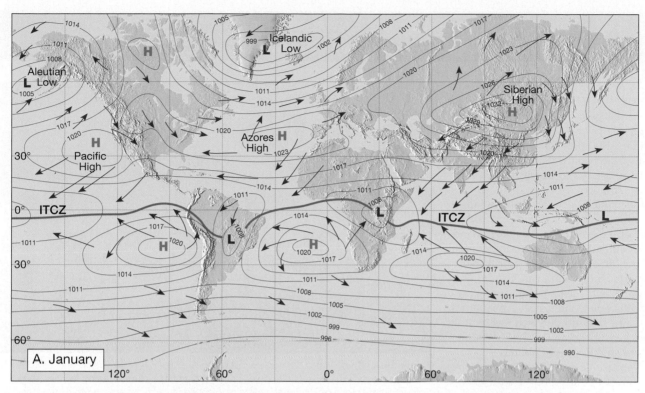

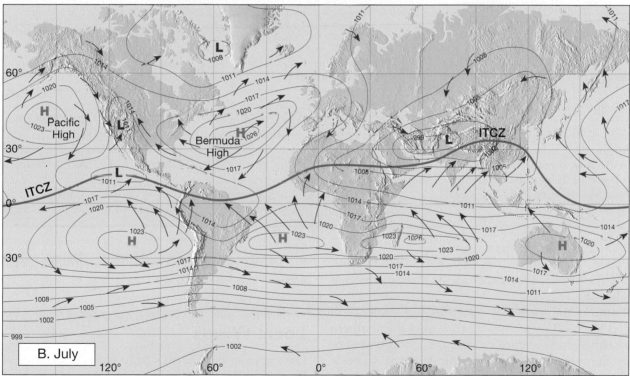

▲ **FIGURE 18.16** Average surface pressure in millibars for **A.** January and **B.** July, with associated winds.

clockwise) flow in their area of influence. A close correlation exists between the paths taken by these surface pressure systems and the position of the upper-level airflow, indicating that the upper air strongly influences the movement of cyclonic and anticyclonic systems.

Among the most obvious features of the flow aloft are the seasonal changes. The steep temperature gradient across the middle latitudes in the winter months corresponds to a stronger flow aloft. In addition, the polar jet stream fluctuates seasonally such that its average position migrates southward with the approach

Box 18.1 People and the Environment
Wind Energy—An Alternative with Potential

Air has mass, and when it moves (that is, when the wind blows), it contains the energy of that motion—kinetic energy. A portion of that energy can be converted into other forms—mechanical force or electricity—that we can use to perform work (Figure 18.A).

Mechanical energy from wind is commonly used for pumping water in rural or remote places. The "farm windmill," still a familiar site in many rural areas, is an example. Mechanical energy converted from wind can also be used for other purposes, such as sawing logs, grinding grain, and propelling sailboats. By contrast, wind-powered electric turbines generate electricity for homes, businesses, and for sale to utilities.

Approximately 0.25 percent (one-quarter of 1 percent) of the solar energy that reaches the lower atmosphere is transformed into wind. Although it is just a minuscule percentage, the absolute amount of energy is enormous. According to one estimate, North Dakota alone is theoretically capable of producing enough wind-generated power to meet more than one-third of U.S. electricity demand. Wind speed is a crucial element in determining whether a place is a suitable site for installing a wind-energy facility. Generally a minimum, annual average wind speed of 21 kilometers (13 miles) per hour is necessary for a utility-scale wind-power plant.

Figure 18.A These wind turbines are operating near Palm Springs, California. California is the state in which the most wind-power development has occurred. As of January 2004, California had a total of 2043 megawatts of installed capacity (nearly one-third of total U.S. capacity). That's enough electricity to supply between 500,000 and 600,000 average American households. (Photo by John Mead/Science Photo Library/Photo Researchers, Inc.)

The power available in the wind is proportional to the cube of its speed. Thus, a turbine operating at a site with an average wind speed of 12 mph could in theory generate about 33 percent more electricity than one at an 11-mph site, because the cube of 12 (1,768) is 33 percent larger than the cube of 11 (1,331). (In the real world, the turbine will not produce quite that much more electricity, but it will still generate much more than the 9 percent difference in wind speed.) The important thing to understand is that what seems like a small difference in wind speed can mean a large difference in available energy and in electricity produced, and therefore a large

---

of winter and northward as summer nears. By midwinter, the jet core may penetrate as far south as central Florida.

Because the paths of low-pressure centers are guided by the flow aloft, we can expect the southern tier of states to experience more of their stormy weather in the winter season. During the hot summer months, the storm track is across the northern states, and some cyclones never leave Canada. The northerly storm track associated with summer applies also to Pacific storms, which move toward Alaska during the warm months, thus producing an extended dry season for much of the West Coast. The number of cyclones generated is seasonal as well, with the largest number occurring in the cooler months when the temperature gradients are greatest. This fact is in agreement with

the role of cyclonic storms in the distribution of heat across the midlatitudes.

Local Winds

Having examined Earth's large-scale circulation, let us turn briefly to winds that influence much smaller areas. Remember that all winds are produced for the same reason: pressure differences that arise because of temperature differences that are caused by unequal heating of Earth's surface. *Local winds* are simply small-scale winds produced by a locally generated pressure gradient. Those described here are caused either by topographic effects or variations in surface composition in the immediate area.

difference in the cost of the electricity generated. Also, there is little energy to be harvested at very low wind speeds (6-mph winds contain less than one-eighth the energy of 12-mph winds).*

As technology has improved, efficiency has increased, and the costs of wind-generated electricity have become more competitive. Between 1983 and 2004, technological advances cut the cost of wind power by more than 85

*American Wind Energy Association. "Wind Energy Basics" **http://www.awea.org/faq/tutorial/wwt_basics.html**

percent. As a result, the growth of installed capacity has grown dramatically. Worldwide, the total amount of installed wind power grew more than 500 percent from 7636 megawatts in 1997 to 39,294 megawatts in 2004 (Table 18.A). 39,294 megawatts is enough to supply 9 million average American households, or as much as could be generated by a dozen large nuclear power plants. By January 2004, U.S. capacity reached nearly 6400 megawatts (Figure 18.B). By 2008 or 2009 the United States is expected to add at least 3000 megawatts of new utility wind-power projects.

The U.S. Department of Energy has announced a goal of obtaining 5 percent

of U.S. electricity from wind by the year 2020—a goal that seems consistent with the current growth rate of wind energy nationwide. Thus, wind-generated electricity seems to be shifting from being an "alternative" to being a "mainstream" energy source.

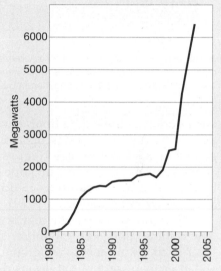

Figure 18.B U.S.-installed wind-power capacity (in megawatts). Growth in recent years has been dramatic. By 2004, U.S. capacity reached 6374 megawatts. Utility wind-power projects now under construction or in the planning stages will add an additional 3000 megawatts of wind capacity in the United States by the year 2009.(Data from U.S. Department of Energy and American Wind Energy Association)

TABLE 18.A	World Leaders in Wind Capacity (January 2004).
Country	**Capacity (megawatts*)**
Germany	14,609
United States	6,374
Spain	6,202
Denmark	3,110
India	2,110
Netherlands	912
Italy	904
Japan	686
United Kingdom	649
China	568

*1 megawatt is enough electricity to supply 250–300 average American households. The "top 10" nations listed in this table account for over 95 percent of the total wind energy produced.

Land and Sea Breezes

In coastal areas during the warm summer months, the land surface is heated more intensely during the daylight hours than is the adjacent body of water (see the section on Land and Water in Chapter 16). As a result, the air above the land surface heats, expands, and rises, creating an area of lower pressure. A **sea breeze** then develops, because cooler air over the water (higher pressure) moves toward the warmer land (lower pressure) (Figure 18.17A). The sea breeze begins to develop shortly before noon and generally reaches its greatest intensity during the mid- to late afternoon. These relatively cool winds can be a significant moderating influence on afternoon temperatures in coastal areas.

At night, the reverse may take place. The land cools more rapidly than the sea, and the **land breeze** develops (Figure 18.17B). Small-scale sea breezes can also develop along the shores of large lakes. People who live in a city near the Great Lakes, such as Chicago, recognize this lake effect, especially in the summer. They are reminded daily by weather reports of the cool temperatures near the lake as compared to warmer outlying areas.

Mountain Valley and Breezes

A daily wind similar to land and sea breezes occurs in many mountainous regions. During daylight hours, the air along the slopes of the mountains is heated more intensely than the air at the same

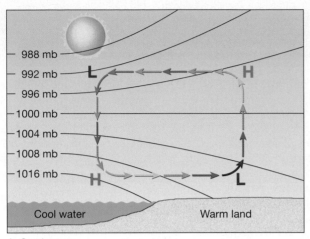

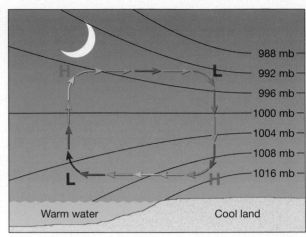

A. Sea breeze

B. Land breeze

▲ **FIGURE 18.17**   Illustration of a sea breeze and a land breeze. **A.** During the daylight hours the air above the land heats and expands, creating an area of lower pressure. Cooler and denser air over the water moves onto the land, generating a sea breeze. **B.** At night the land cools more rapidly than the sea, generating an offshore flow called a *land breeze.*

elevation over the valley floor. Because this warmer air is less dense, it glides up along the slope and generates a **valley breeze** (Figure 18.18A). The occurrence of these daytime upslope breezes can often be identified by the cumulus clouds that develop on adjacent mountain peaks.

After sunset, the pattern may reverse. Rapid radiation cooling along the mountain slopes produces a layer of cooler air next to the ground. Because cool air is denser than warm air, it drains downslope into the valley. Such a movement of air is called a **mountain breeze** (Figure 18.18B). The same type of cool air drainage can occur in places that have very modest slopes. The result is that the coldest pockets of air are usually found in the lowest spots. Like many other winds, mountain and valley breezes have seasonal preferences. Although valley breezes are most common during the warm season when solar heating is

most intense, mountain breezes tend to be more dominant in the cold season.

Chinook and Santa Ana Winds

Warm, dry winds are common on the eastern slopes of the Rockies, where they are called **chinooks**. Such winds are created when air descends the leeward (sheltered) side of a mountain and warms by compression. Because condensation may have occurred as the air ascended the windward side, releasing latent heat, the air descending the leeward slope will be warmer and drier than it was at a similar elevation on the windward side. Although the temperature of these winds is generally less than 10°C (50°F) which is not particularly warm, they occur mostly in the winter and spring when the affected areas may be experiencing below-freezing temperatures. Thus, by comparison, these dry, warm winds often bring a drastic

▼ **FIGURE 18.18**   Valley and mountain breezes. **A.** Heating during the daylight hours warms the air along the mountain slopes. This warm air rises, generating a valley breeze. **B.** After sunset, cooling of the air near the mountain can result in cool air drainage into the valley, producing the mountain breeze.

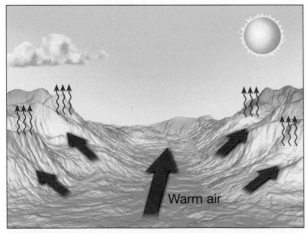

A. Valley breeze

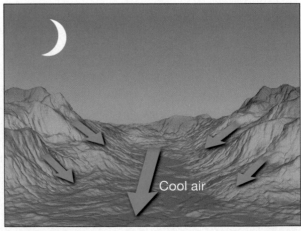

B. Mountain breeze

▲ FIGURE 18.19 Wind vane (right) and cup anemometer (left). The wind vane shows wind direction and the anemometer measures wind speed. (Photo by Belfort Instrument Company)

How Wind Is Measured

Two basic wind measurements, direction and speed, are particularly significant to the weather observer. *Winds are always labeled by the direction from which they blow.* A north wind blows *from* the north *toward* the south, an east wind *from* the east *toward* the west. The instrument most commonly used to determine wind direction is the **wind vane** (Figure 18.19, upper right). This instrument, a common sight on many buildings, always points *into* the wind. Often the wind direction is shown on a dial that is connected to the wind vane. The dial indicates wind direction, either by points of the compass (N, NE, E, SE, etc.) or by a scale of 0° to 360°. On the latter scale, 0 degrees or 360 degrees are both north, 90 degrees is east, 180 degrees is south, and 270 degrees is west.

When the wind consistently blows more often from one direction than from any other, it is called a **prevailing wind**. You may be familiar with the prevailing westerlies that dominate the circulation in the midlatitudes. In the United States, for example, these winds consistently move the "weather" from west to east across the continent. Embedded within this general eastward flow are cells of high and low pressure with the characteristic clockwise and counterclockwise flow. As a result, the winds associated with the westerlies, as measured at the surface, often vary considerably from day to day and from place to place. By contrast, the direction of airflow associated with the belt of trade winds is much more consistent, as can be seen in Figure 18.20.

Wind speed is commonly measured using a **cup anemometer** (*anemo* = wind, *metron* = measuring instrument) (Figure 18.19, upper left). The wind speed is read from a dial much like the speedometer of an automobile. Places where winds are steady and speeds are relatively high are potential sites for tapping wind energy (see Box 18.1).

By knowing the locations of cyclones and anticyclones in relation to where you are, you can predict

change. When the ground has a snow cover, these winds are known to melt it in short order.

A chinooklike wind that occurs in southern California is the **Santa Ana**. These hot, desiccating winds greatly increase the threat of fire in this already dry area.

STUDENTS SOMETIMES ASK...

A friend who lives in Colorado talks about "snow eaters." What are they?

"Snow eaters" is a local term for chinooks, the warm, dry winds that descend the eastern slopes of the Rockies. These winds have been known to melt more than a foot of snow in a single day. A chinook that moved through Granville, North Dakota, on February 21, 1918, caused the temperature to rise from −33°F to 50°F, an increase of 83°F!

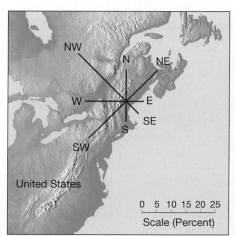

A. Westerlies (winter)

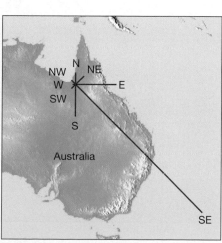

B. Southeast Trades (winter)

◄ FIGURE 18.20 Wind roses showing the percentage of time airflow is from various directions. **A.** Wind frequency for the winter in the eastern United States. **B.** Wind frequency for the winter in northern Australia. Note the reliability of the southeast trades in Australia as compared to the westerlies in the eastern United States. (Data from G. T. Trewartha)

the changes in wind direction that will be experienced as a pressure center moves past. Because changes in wind direction often bring changes in temperature and moisture conditions, the ability to predict the winds can be very useful. In the Midwest, for example, a north wind may bring cool, dry air from Canada, whereas a south wind may bring warm, humid air from the Gulf of Mexico. Sir Francis Bacon summed it up nicely when he wrote, "Every wind has its weather."

El Niño and La Niña

As can be seen in Figure 18.21A, the cold Peruvian current flows equatorward along the coast of Ecuador and Peru. This flow encourages upwelling of cold nutrient-filled waters that serve as the primary food source for millions of fish, particularly anchovies. Near the end of each year, however, a warm current that flows southward along the coasts of Ecuador and Peru replaces the cold Peruvian current. During the

▼ **FIGURE 18.21** The relationship between the Southern Oscillation and El Niño is illustrated on these simplified maps. **A.** Normally, the trade winds and strong equatorial currents flow toward the west. At the same time, the strong Peruvian current causes upwelling of cold water along the west coast of South America. **B.** When the Southern Oscillation occurs, the pressure over the eastern and western Pacific flip-flops. This causes the trade winds to diminish, leading to an eastward movement of warm water along the equator. As a result, the surface waters of the central and eastern Pacific warm, with far-reaching consequences to weather patterns.

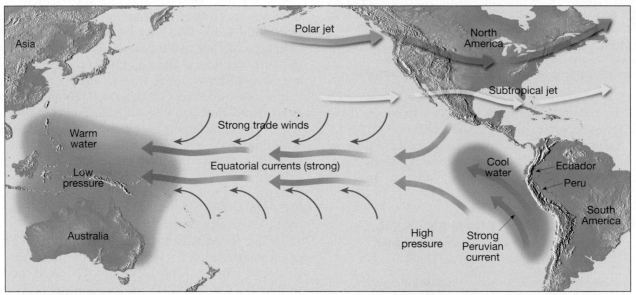

A. Normal conditions

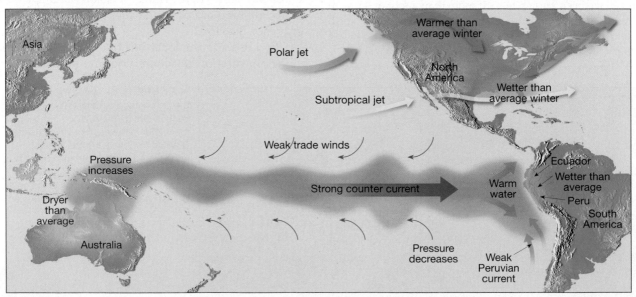

B. El Niño

nineteenth century the local residents named this warm countercurrent El Niño ("the child") after the Christ child because it usually appeared during the Christmas season. Normally, these warm countercurrents last for at most a few weeks when they again give way to the cold Peruvian flow. However, at irregular intervals of three to seven years, these countercurrents become unusually strong and replace normally cold offshore waters with warm equatorial waters (Figure 18.21B). Today, scientists use the term **El Niño** for these episodes of ocean warming that affect the eastern tropical Pacific.

The onset of El Niño is marked by abnormal weather patterns that drastically affect the economies of Ecuador and Peru. As shown in Figure 18.21B, these unusually strong undercurrents amass large quantities of warm water that block the upwelling of colder, nutrient-filled water. As a result, the anchovies starve, devastating the fishing industry. At the same time, some inland areas that are normally arid receive an abnormal amount of rain. Here, pastures and cotton fields have yields far above the average. These climatic fluctuations have been known for years, but they were originally considered local phenomena. Today, we know that El Niño is part of the global circulation and affects the weather at great distances from Peru and Ecuador.

Two of the strongest El Niño events on record occurred between 1982–83 and 1997–98 and were re-sponsible for weather extremes of a variety of types in many parts of the world. The 1997–98 El Niño brought ferocious storms that struck the California coast, causing unprecedented beach erosion, landslides, and floods. In the southern United States, heavy rains also brought floods to Texas and the Gulf states. The same energized jet stream that produced storms in the South, upon reaching the Atlantic, sheared off the northern portions of hurricanes, destroying the storms. It was one of the quietest Atlantic hurricane seasons in years.

Major El Niño events, such as the one in 1997 and 1998, are intimately related to the large-scale atmospheric circulation. Each time an El Niño occurs, the barometric pressure drops over large portions of the southeastern Pacific, whereas in the western Pacific, near Indonesia and northern Australia, the pressure rises (Figure 18.22). Then, as a major El Niño event comes to an end, the pressure difference between these two regions swings back in the opposite direction. This seesaw pattern of atmospheric pressure between the eastern and western Pacific is called the **Southern Oscillation.** It is an inseparable part of the El Niño warmings that occur in the central and eastern Pacific every three to seven years. Therefore, this phenomenon is often termed El Niño/Southern Oscillation, or ENSO for short.

Winds in the lower atmosphere are the link between the pressure change associated with the Southern

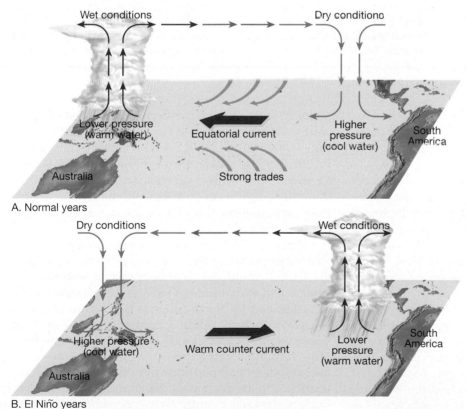

A. Normal years

B. El Niño years

◀ **FIGURE 18.22** Simplified illustration of the see-saw pattern of atmospheric pressure between the eastern and western Pacific, called the *Southern Oscillation.*
A. During average years, high pressure over the eastern Pacific causes surface winds and warm equatorial waters to flow westward. The result is a pileup of warm water in the western Pacific, which promotes the lowering of pressure.
B. An El Niño event begins as surface pressure increases in the western Pacific and decreases in the eastern Pacific. This air pressure reversal weakens, or may even reverse the trade winds, and results in an eastward movement of the warm waters that had accumulated in the western Pacific.

BOX 18.2 UNDERSTANDING EARTH
Monitoring Ocean Winds from Space

The global ocean makes up 71 percent of Earth's surface. Winds moving over the sea surface effectively move heat and moisture from place to place, while driving ocean currents and ultimately weather and climate. Surface-based methods of monitoring winds using ships and data buoys provide only a glimpse of the wind patterns over the ocean. Today instruments aboard NASA's *QuickSCAT* spacecraft can continuously and accurately map wind speed and direction under most atmospheric conditions over 90 percent of Earth's ice-free oceans.

QuickSCAT carries the *SeaWinds* scatterometer, a specialized type of radar that operates by transmitting microwave pulses to the ocean surface and measuring the amount of energy that is bounced back (or echoed) to the satellite. Smooth ocean surfaces return weaker signals because less energy is reflected, while rough water returns a stronger signal. From such data, scientists can compute wind speed and direction (Figure 18.C). By measuring global sea-surface winds, *SeaWinds* helps researchers more accurately predict marine phenomena that have the potential to affect human endeavors.

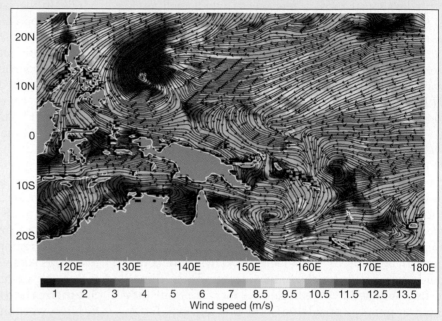

Figure 18.C Map produced from *SeaWinds* data showing wind patterns over a portion of the western Pacific on March 4, 2002. The strongest winds in this image (red area) are about 13.6 meters per second (49 kilometers, or 30 miles per hour).

For example, the satellite's instruments provide meteorologists with a method to identify areas of gale-force winds over the entire ocean. These real-time data can be used to give advance warning of high waves to vessels at sea and to coastal communities.

In addition, these snapshots of ocean winds help researchers better understand and predict the development of large storm systems, such as midlatitide cyclones and hurricanes, as well as aid in our understanding of global weather events such as El Niño (Figure 18.D).

Oscillation and the extensive ocean warming associated with El Niño (see Box 18.2). During a typical year, the trade winds converge near the equator and flow westward toward Indonesia (Figure 18.22A). This steady westward flow creates a warm surface current that moves from east to west along the equator. The result is a "piling up" of a thick layer of warm surface water that produces higher sea levels (by 30 centimeters) in the western Pacific. Meanwhile, the eastern Pacific is characterized by a strong Peruvian current, upwelling of cold water, and lower sea levels.

Then when the Southern Oscillation occurs, the normal situation just described changes dramatically. Barometric pressure rises in the Indonesian region, causing the pressure gradient along the equator to weaken or even to reverse. As a consequence, the once-steady trade winds diminish and may even change direction. This reversal creates a major change in the equatorial current system, with warm water flowing eastward (Figure 18.22B). With time, water temperatures in the central and eastern Pacific increase and sea level in the region rises. This eastward shift of the warmest surface water marks the onset of El Niño and sets up changes in atmospheric circulation that affect areas far outside the tropical Pacific.

When an El Niño began in the summer of 1997, forecasters predicted that the pool of warm water over the Pacific would displace the paths of both the subtropical and midlatitude jet streams, which steer weather systems across North America (see Figure 18.21). As predicted, the subtropical jet brought rain to the Gulf Coast, where Tampa, Florida, received more than three times its normal winter precipitation. Furthermore, the midlatitude jet pumped warm air far north into the continent. As a result, winter temperatures west of the Rockies were significantly above normal.

Figure 18.D This image shows Hurricane Ivan as it roared through the Caribbean as a deadly Category 5 storm early on September 9, 2004. Data from the *SeaWinds* scatterometer augments traditional satellite images of clouds (as in Figure 18.11, p. 509) by providing direct measurements of surface winds to compare with observed cloud patterns in an effort to better determine a storm's structure and strength.

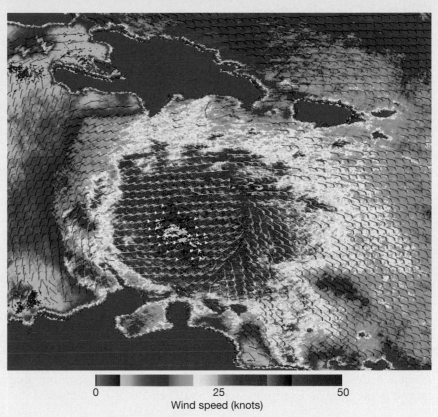

Wind speed (knots)

The effects of El Niño are somewhat variable depending in part on the temperatures and size of the warm pools. Nevertheless, some locales appear to be affected more consistently. In particular, during most El Niños, warmer-than-normal winters occur in the northern United States and Canada. In addition, normally arid portions of Peru and Ecuador, as well as the eastern United States, experience wet conditions. By contrast, drought conditions are generally observed in Indonesia, Australia, and the Philippines. One major benefit of the circulation associated with El Niño is a suppression of the number of Atlantic hurricanes.

The opposite of El Niño is an atmospheric phenomenon known as **La Niña.** Once thought to be the normal conditions that occur between two El Niño events, meteorologists now consider La Niña an important atmospheric phenomenon in its own right. Researchers have come to recognize that when sur-

face temperatures in the eastern Pacific are *colder than average*, a La Niña event is triggered that has a distinctive set of weather patterns. A typical La Niña winter blows colder than normal air over the Pacific Northwest and the northern Great Plains while warming much of the rest of the United States. Further, greater precipitation is expected in the Northwest. During the La Niña winter of 1998–99, a world-record snowfall for one season occurred in Washington State. Another La Niña impact is greater hurricane activity. A recent study concluded that the cost of hurricane damages in the United States is 20 times greater in La Niña years as compared to El Niño years.

In summary, the effects of El Niño and La Niña on world climate are widespread and variable. There is no place on Earth where the weather is indifferent to air and ocean conditions in the tropical Pacific. Events associated with El Niño and La Niña are now understood to have a significant influence on the state of weather and climate almost everywhere.

Global Distribution of Precipitation

A casual glance at Figure 18.23 shows a relatively complex pattern for the distribution of precipitation. Although the map appears to be complicated, the general features of the map can be explained by applying our knowledge of global winds and pressure systems.

In general, regions influenced by high pressure, with its associated subsidence and diverging winds, experience relatively dry conditions. On the other hand, regions under the influence of low pressure and its converging winds and ascending air receive ample precipitation. This pattern is illustrated by noting that the tropical region dominated by the equatorial low is the rainiest region on Earth. It includes the rain forests of the Amazon basin in South America and the Congo basin in Africa. Here the warm, humid trade winds converge to yield abundant rainfall throughout the year. By way of contrast, areas dominated by the subtropical high-pressure cells clearly receive much smaller amounts of precipitation. These are regions of extensive deserts. In the Northern Hemisphere the largest is the Sahara. Examples in the Southern Hemisphere include the Kalahari in southern Africa and the dry lands of Australia.

If Earth's pressure and wind belts were the only factors controlling precipitation distribution, the pattern shown in Figure 18.23 would be simpler. The inherent nature of the air is also an important factor in determining precipitation potential. Because cold air has a low capacity for moisture compared with warm air, we would expect a latitudinal variation in precipitation, with low latitudes receiving the greatest amounts of precipitation and high latitudes receiving the smallest amounts. Figure 18.23 indeed reveals heavy rainfall in equatorial regions and meager precipitation in high-latitude areas. Recall that the dry region in the warm subtropics is explained by the presence of the subtropical high.

In addition to latitudinal variations in precipitation, the distribution of land and water complicates the precipitation pattern. Large landmasses in the middle latitudes commonly experience decreased precipitation toward their interiors. For example, central North America and central Eurasia receive considerably less precipitation than do coastal regions at the same latitude. Furthermore, the effects of mountain barriers alter the idealized precipitation patterns we would expect solely from global wind and pressure systems. Windward mountain slopes receive abundant rainfall resulting from orographic lifting, whereas leeward slopes and adjacent lowlands are usually deficient in moisture.

▼ **FIGURE 18.23** Average annual precipitation in millimeters. (**Note:** 400 mm is equal to 15.6 inches.)

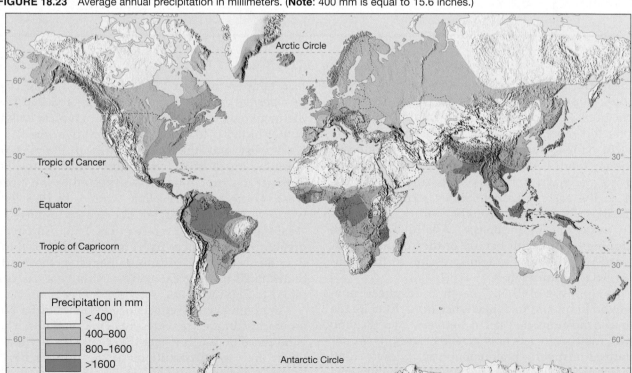

Chapter Summary

- *Air has weight*: At sea level it exerts a pressure of 1 kilogram per square centimeter (14.7 pounds per square inch). *Air pressure* is the force exerted by the weight of air above. With increasing altitude there is less air above to exert a force, and thus air pressure decreases with altitude, rapidly at first, then much more slowly. The unit used by meteorologists to measure atmospheric pressure is the *millibar*. *Standard sea-level pressure* is expressed as 1013.2 millibars. *Isobars* are lines on a weather map that connect places of equal air pressure.

- A *mercury barometer* measures air pressure using a column of mercury in a glass tube that is sealed at one end and inverted in a dish of mercury. As air pressure increases, the mercury in the tube rises; conversely, when air pressure decreases, so does the height of the column of mercury. A mercury barometer measures atmospheric pressure in *inches of mercury*, the height of the column of mercury in the barometer. Standard atmospheric pressure at sea level equals 29.92 inches of mercury. *Aneroid* (without liquid) *barometers* consist of partially evacuated metal chambers that compress as air pressure increases and expand as pressure decreases.

- *Wind* is the horizontal flow of air from areas of higher pressure toward areas of lower pressure. Winds are controlled by the following combination of forces: (1) the *pressure-gradient force* (amount of pressure change over a given distance); (2) *Coriolis effect* (deflective effect of Earth's rotation to the right in the Northern Hemisphere and to the left in the Southern Hemisphere); and (3) *friction* with Earth's surface (slows the movement of air and alters wind direction).

- Upper-air winds, called *geostrophic winds*, blow parallel to the isobars and reflect a balance between the pressure-gradient force and the Coriolis effect. Upper-air winds are faster than surface winds because friction is greatly reduced aloft. Friction slows surface winds, which in turn reduces the Coriolis effect. The result is air movement at an angle across the isobars toward the area of lower pressure.

- The two types of pressure centers are (1) *cyclones*, or *lows* (centers of low pressure), and (2) *anticyclones*, or *highs* (high-pressure centers). In the Northern Hemisphere, winds around a low (cyclone) are counterclockwise and inward. Around a high (anticyclone), winds are clockwise and outward. In the Southern Hemisphere, the Coriolis effect causes winds to move clockwise around a low and counterclockwise around a high. Because air rises and cools adiabatically in low-pressure centers, cloudy conditions and precipitation are often associated with their passage. In high-pressure centers, descending air is compressed and warmed; therefore, cloud formation and precipitation are unlikely, and "fair" weather is usually expected.

- Earth's *global pressure zones* include the *equatorial low, subtropical high, subpolar low,* and *polar high*. The *global surface winds* associated with these pressure zones are the *trade winds, westerlies,* and *polar easterlies*.

- Particularly in the Northern Hemisphere, large seasonal temperature differences over continents disrupt the idealized, or zonal, global patterns of pressure and wind. In winter, large, cold landmasses develop a seasonal high-pressure system from which surface airflow is directed off the land. In summer, landmasses are heated and low pressure develops over them, which permits air to flow onto the land. The seasonal changes in wind direction are known as *monsoons*.

- In the middle latitudes, between 30 and 60 degrees latitude, the general west-to-east flow of the westerlies is interrupted by migrating cyclones and anticyclones. The paths taken by these pressure systems are closely related to upper-level airflow and the polar *jet stream*. The average position of the polar jet stream, and hence the paths followed by cyclones, migrates equatorward with the approach of winter and poleward as summer nears.

- *Local winds* are small-scale winds produced by a locally generated pressure gradient. Local winds include *sea* and *land breezes* (formed along a coast because of daily pressure differences caused by the differential heating of land and water); *valley* and *mountain breezes* (daily wind similar to sea and land breezes except in a mountainous area where the air along slopes heats differently from the air at the same elevation over the valley floor); and *chinook* and *Santa Ana winds* (warm, dry winds created when air descends the leeward side of a mountain and warms by compression).

- The two basic wind measurements are *direction* and *speed*. Winds are always labeled by the direction *from* which they blow. Wind direction is measured with a *wind vane*, and wind speed is measured using a *cup anemometer*.

- *El Niño* is the name given to the periodic warming of the ocean that occurs in the central and eastern Pacific. It is associated with periods when a weakened pressure gradient causes the trade winds to diminish. A major El Niño event triggers extreme weather in many parts of the world. When surface temperatures in the eastern Pacific are colder than average, a *La Niña* event is triggered. A typical La Niña winter blows colder-than-normal air over the Pacific Northwest and the northern Great Plains while warming much of the rest of the United States.

- The global distribution of precipitation is strongly influenced by the global pattern of air pressure and wind, latitude, and distribution of land and water.

Key Terms

air pressure (p. 502)
aneroid barometer (p. 503)
anticyclone (p. 508)
barograph (p. 501)
barometric tendency (p. 510)
chinook (p. 516)
convergence (p. 508)
Coriolis effect (p. 505)
cup anemometer (p. 517)
cyclone (p. 508)
divergence (p. 508)
El Niño (p. 519)
equatorial low (p. 511)

geostrophic wind (p. 506)
high (p. 508)
isobar (p. 504)
jet stream (p. 506)
land breeze (p. 515)
La Niña (p. 521)
low (p. 508)
mercury barometer (p. 503)
monsoon (p. 512)
mountain breeze (p. 516)
polar easterlies (p. 511)
polar front (p. 511)
polar high (p. 511)

pressure gradient (p. 504)
pressure tendency (p. 510)
prevailing wind (p. 510)
Santa Ana (p. 517)
sea breeze (p. 515)
Southern Oscillation (p. 519)
subpolar low (p. 511)
subtropical high (p. 511)
trade winds (p. 511)
valley breeze (p. 516)
westerlies (p. 511)
wind (p. 504)
wind vane (p. 517)

Review Questions

1. What is standard sea-level pressure in millibars? In inches of mercury? In pounds per square inch?

2. Mercury is 13 times heavier than water. If you built a barometer using water rather than mercury, how tall would it have to be to record standard sea-level pressure (in centimeters of water)? In feet?

3. Describe the principle of the *aneroid barometer*.

4. What force is responsible for generating wind?

5. Write a generalization relating the spacing of isobars to the speed of wind.

6. How does the Coriolis effect modify air movement?

7. Contrast surface winds and upper-air winds in terms of speed and direction.

8. Describe the weather that usually accompanies a drop in barometric pressure and a rise in barometric pressure.

9. Sketch a diagram (isobars and wind arrows) showing the winds associated with surface cyclones and anticyclones in both the Northern and Southern hemispheres.

10. If you live in the Northern Hemisphere and are directly west of the center of a cyclone, what most probably will be the wind direction? What will the wind direction be if you are west of an anticyclone?

11. The following questions relate to the global pattern of air pressure and winds.
    (a) The trade winds diverge from which pressure zone?
    (b) Which prevailing wind belts converge in the stormy region known as the polar front?
    (c) Which pressure belt is associated with the equator?

12. What influence does upper-level airflow seem to have on surface-pressure systems?

13. Describe the monsoon circulation of India.

14. What is a local wind? List three examples.

15. A northeast wind is blowing *from* the _____ (direction) *toward* the _____ (direction).

16. Describe the relationship between the Southern Oscillation and a major El Niño event.

17. Which global pressure system is responsible for deserts such as the Sahara and Kalahari in Africa? With which global pressure belt are the tropical rain forests of the Amazon and Congo basins associated?

18. Other than Earth's pressure and wind belts, list two other factors that exert a significant influence on the global distribution of precipitation.

Examining the Earth System

1. Examine the image of Africa in Figure 1.8B (p. 13) and pick out the region dominated by the equatorial low and the areas influenced by the subtropical highs in each hemisphere. What clue(s) did you use? Speculate on the differences in the biosphere between the regions dominated by high pressure and the zone influenced by low pressure.

2. How are global winds related to surface ocean currents? (Try comparing Figure 18.16, p. 513 with Figure 15.2, p. 405). What is the *ultimate* source of energy that drives both of these circulations?

3. Winds and ocean currents change in the tropical Pacific during an El Niño event. How might this impact the biosphere and solid Earth in Peru and Ecuador? How about in Indonesia? (Two excellent Websites that deal with El Niño are the University of Southern California site at **http://www.usc.edu/org/seagrant/elnino/link.html** and NOAA's *El Niño Theme Page* at **http://www.pmel.noaa.gov/toga-tao/el-nino/nino-home.html**).

Online Study Guide

The Earth Science Web site uses the resources and flexibility of the Internet to aid in your study of the topics in this chapter. Written and developed by Earth science instructors, this site will help improve your understanding of Earth science. Visit **http://www.prenhall.com/tarbuck** and click on the cover of Earth Science 11e to find:

- **On-line review quizzes.**
- **Critical thinking exercises.**
- **Links to chapter-specific Web resources.**
- **Internet-wide key term searches.**

http://www.prenhall.com/tarbuck

Tornadoes are intense and destructive local storms of short duration that cause many deaths each year.
(Photo by Alan R. Moller/Getty Images Inc.—Stone Allstock)

CHAPTER 19

Weather Patterns
and Severe Storms

Tornadoes and hurricanes rank among nature's most destructive forces. Each spring, newspapers report the death and destruction left in the wake of a band of tornadoes. During late summer and fall we hear occasional reports about hurricanes in the news. An exceptional example occurred during a six-week span in August and September 2004 when four strong hurricanes hit Florida. Storms named Charley, Frances, Ivan, and Jeanne made front-page headlines (Figure 19.1). Thunderstorms, although less intense and far more common than tornadoes and hurricanes, will also be part of our discussion on severe weather in this chapter. Before looking at violent weather, however, we will study those atmospheric phenomena that most often affect our day-to-day weather: air masses, fronts, and traveling middle-latitude cyclones. Here we will see the interplay of the elements of weather discussed in Chapters 16, 17, and 18.

Air Masses

GEODe Earth's Dynamic Atmosphere
▼ Basic Weather Patterns

For many people who live in the middle latitudes, including much of the United States, summer heat waves and winter cold spells are familiar experiences. In the first instance, several days of high temperatures and oppressive humidities may finally end when a series of thunderstorms pass through the area, followed by a few days of relatively cool relief. By contrast, the clear skies that often accompany a span of frigid subzero days may be replaced by thick gray clouds and a period of snow as temperatures rise to levels that seem mild when compared to those that existed just a day earlier. In both examples, what was experienced was a period of generally constant weather conditions followed by a relatively short period of change, and then the reestablishment of a new set of weather conditions that remained for perhaps several days before changing again.

What Is an Air Mass?

The weather patterns just described result from movements of large bodies of air, called air masses. An **air mass**, as the term implies, is an immense body of air, usually 1600 kilometers (1000 miles) or more across and perhaps several kilometers thick, that is characterized by a similarity of temperature and moisture at any given altitude. When this air moves out of its region of origin, it will carry these temperatures and moisture conditions with it, eventually affecting a large portion of a continent.

▼ **FIGURE 19.1** Hurricane Frances over Florida on September 5, 2004. Frances was the second of four hurricanes that struck Florida over a six week span and caused billions of dollars in damages. (AFP Photo/NOAA/Getty Images)

An excellent example of the influence of an air mass is illustrated in Figure 19.2. Here a cold, dry mass from northern Canada moves southward. With a beginning temperature of −46°C (−51°F) the air mass warms 13 degrees, to −33°C (−27°F) by the time it reaches Winnipeg. It continues to warm as it moves southward through the Great Plains and into Mexico. Throughout its southward journey, the air mass becomes warmer. But it also brings some of the coldest weather of the winter to the places in its path. Thus, the air mass is modified, but it also modifies the weather in the areas over which it moves.

The horizontal uniformity of an air mass is not complete, because it may extend through 20 degrees or more of latitude and cover hundreds of thousands to millions of square kilometers. Consequently, small differences in temperature and humidity are to be expected from one point to another at the same level. Still, the differences observed within an air mass are small in comparison to the rapid changes experienced across air-mass boundaries.

Because it may take several days for an air mass to traverse an area, the region under its influence will probably experience fairly constant weather, a situation called **air-mass weather**. Certainly some day-to-day variations may occur, but the events will be very unlike those in an adjacent air mass. For this reason the boundary between two adjoining air masses that have contrasting characteristics, called a *front*, marks a change in weather.

Source Regions

When a portion of the lower atmosphere moves slowly or stagnates over a relatively uniform surface, the air will assume the distinguishing features of that area, particularly with regard to temperature and moisture conditions.

The area where an air mass acquires its characteristic properties of temperature and moisture is called its **source region**. The source regions that produce air masses influencing North America are shown in Figure 19.3.

Air masses are classified according to their source region. **Polar (P) air masses** originate in high latitudes toward Earth's poles, whereas those that form in low latitudes are called **tropical (T) air masses**. The designation *polar* or *tropical* gives an indication of the temperature characteristics of the air masses. *Polar* indicates cold, and *tropical* indicates warm.

In addition, air masses are classified according to the nature of the surface in the source region. **Continental (c) air masses** form over land, and **maritime (m) air masses** originate over water. The designation *continental* or *maritime* thus suggests the moisture characteristics of the air mass. Continental air is likely to be dry, and maritime air, humid.

The four basic types of air masses according to this scheme of classification are continental polar (cP), continental tropical (cT), maritime polar (mP), and maritime tropical (mT).

Weather Associated with Air Masses

Continental polar and maritime tropical air masses influence the weather of North America most, especially east of the Rocky Mountains. Continental polar air masses originate in northern Canada, interior Alaska, and the Arctic areas that are uniformly cold and dry in winter and cool and dry in summer. In winter, an invasion of continental polar air brings the clear skies and cold temperatures we associate with a cold wave as it moves southward from Canada into the United States. In summer, this air mass may bring a few days of cooling relief.

Although cP air masses are not, as a rule, associated with heavy precipitation, those that cross the Great Lakes during late autumn and winter sometimes bring snow to the leeward shores. These localized storms often form when the surface weather map indicates no apparent cause for a snowstorm to occur. Known as **lake-effect snows**, they make Buffalo and Rochester, New York, among the snowiest cities in the United States (Figure 19.4). What causes lake-effect snow?

▼ **FIGURE 19.2** As this frigid Canadian air mass moved southward, it brought some of the coldest weather of the winter. (After Tom L. McKnight, *Physical Geography*, 5th ed. Upper Saddle River, NJ: Prentice Hall, 1996, p. 174)

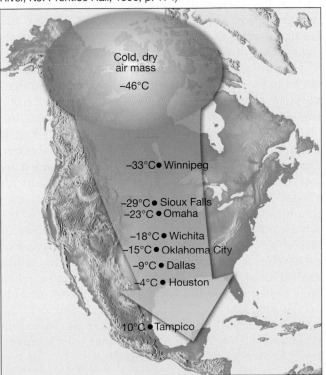

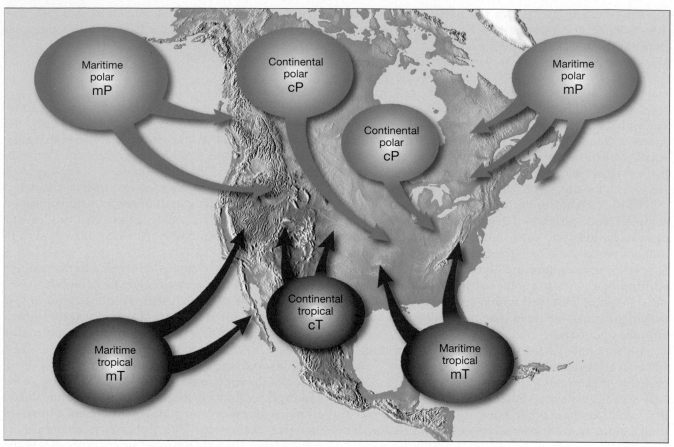

▲ **FIGURE 19.3** Air masses are classified on the basis of their source region. The designation continental (c) or maritime (m) gives an indication of moisture content, whereas polar (P) and tropical (T) indicate temperature conditions.

During late autumn and early winter, the temperature contrast between the lakes and adjacent land areas can be large.* The temperature contrast can be especially great when a very cold cP air mass pushes southward across the lakes. When this occurs, the air acquires large quantities of heat and moisture from the relatively warm lake surface. By the time it reaches the opposite shore, the air mass is humid and unstable, and heavy snow showers are likely.

Maritime tropical air masses affecting North America most often originate over the warm waters of the Gulf of Mexico, the Caribbean Sea, or the adjacent Atlantic Ocean. As you might expect, these air masses are warm, moisture-laden, and usually unstable. Maritime tropical air is the source of much, if not most, of the precipitation received in the eastern two-thirds of the United States. In summer, when an mT air mass invades the central and eastern United States, and occasionally southern Canada, it brings the high temperatures and oppressive humidity typically associated with its source region.

*Recall that land cools more rapidly and to lower temperatures than water. See the discussion of "Land and Water" in the section on "Why Temperatures Vary: The Controls of Temperature" in Chapter 16.

Of the two remaining air masses, maritime polar and continental tropical, the latter has the least influence on the weather of North America. Hot, dry continental tropical air, originating in the Southwest and Mexico during the summer, only occasionally affects the weather outside its source region.

During the winter, maritime polar air masses coming from the North Pacific often originate as continental polar air masses in Siberia. The cold, dry cP air is transformed into relatively mild, humid, unstable mP air during its long journey across the North Pacific (Figure 19.5). As this mP air arrives at the western shore of North America, it is often accompanied by low clouds and shower activity. When this air advances inland against the western mountains, orographic uplift produces heavy rain or snow on the windward slopes of the mountains. Maritime polar air also originates in the North Atlantic off the coast of eastern Canada and occasionally influences the weather of the northeastern United States. In winter, when New England is on the northern or northwestern side of a passing low-pressure center, the counterclockwise cyclonic winds draw in maritime polar air. The result is a storm characterized by snow and cold temperatures, known locally as a *nor'easter*.

▲ **FIGURE 19.4** **(Left)** A remarkable seven-day lake-effect snowstorm in December 2001 dropped more than 200 centimeters (80 inches) of snow on Buffalo, New York. Such snows are associated with the movement of cP air masses across the Great Lakes. (Photo by Don Heupel/AP/Wide World Photos) **(Right)** The snowbelts of the Great Lakes region are the zones that most frequently experience lake-effect snowstorms.

Fronts

Earth's Dynamic Atmosphere
▼ Basic Weather Patterns

Fronts are boundaries that separate air masses of different densities. One air mass is usually warmer and contains more moisture than the other. Fronts can form between any two contrasting air masses. Considering

▼ **FIGURE 19.5** During winter, maritime polar (mP) air masses in the North Pacific usually begin as continental polar (cP) air masses in Siberia. The cP air is modified to mP as it slowly crosses the ocean.

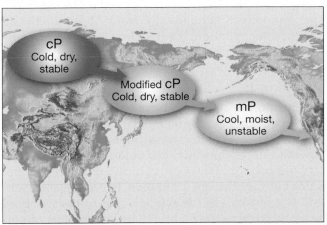

the vast size of the air masses involved, fronts are relatively narrow, being 15- to 200-kilometer- (9- to 120-mile) wide bands of discontinuity. On the scale of a weather map, they are generally narrow enough to be represented by a broad line (as in Figure 19.10, p. 536).

Above Earth's surface, the frontal surface slopes at a low angle so that warmer air overlies cooler air, as shown in Figure 19.6. In the ideal case, the air masses on both sides of the front move in the same direction and at the same speed. Under this condition, the front acts simply as a barrier that travels along with the air masses and that neither mass can penetrate.

Generally, however, the distribution of pressure across a front is such that one air mass moves faster than the other. Thus, one air mass actively advances into another and "clashes" with it. In fact, the boundaries were tagged *fronts* during World War I by Norwegian meteorologists who visualized them as analogous to battle lines between two armies.

As one air mass moves into another, some mixing does occur along the frontal surface, but for the most part the air masses retain their distinct identities as one is displaced upward over the other. No matter which air mass is advancing, *it is always the warmer, less dense air that is forced aloft, whereas the cooler, denser air acts as the wedge upon which lifting takes place.* The term **overrunning** is generally applied to warm air

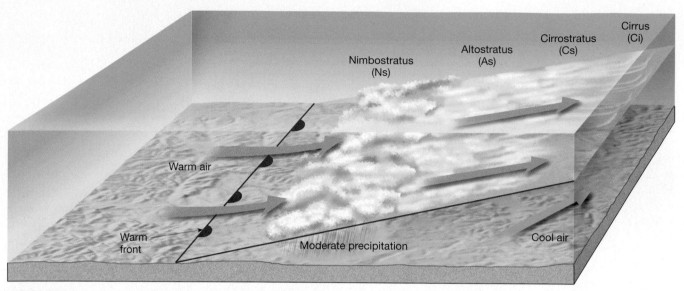

▲ **FIGURE 19.6** Warm front produced as warm air glides up over a cold air mass. Precipitation is moderate and occurs within a few hundred kilometers of the surface front.

gliding up along a cold air mass. We will now take a look at different types of fronts.

Warm Fronts

When the surface (ground) position of a front moves so that warm air occupies territory formerly covered by cooler air, it is called a **warm front** (Figure 19.6). On a weather map, the surface position of a warm front is shown by a red line with semicircles extending into the cooler air.

East of the Rockies, warm tropical air often enters the United States from the Gulf of Mexico and over-runs receding cool air. As the cold air retreats, friction with the ground greatly slows the advance of the sur-face position of the front compared to its position aloft. Stated another way, less dense, warm air has a hard time displacing denser cold air. For this reason, the boundary separating these air masses acquires a very gradual slope. The average slope of a warm front is about 1:200, which means that if you are 200 kilome-ters (120 miles) ahead of the surface location of a warm front, you will find the frontal surface at a height of 1 kilometer (0.6 mile).

As warm air ascends the retreating wedge of cold air, it cools adiabatically to produce clouds, and fre-quently, precipitation. The sequence of clouds shown in Figure 19.6 typically precedes a warm front. The first sign of the approaching warm front is the appear-ance of cirrus clouds. These high clouds form where the overrunning warm air has ascended high up the wedge of cold air, 1000 kilometers (600 miles) or more ahead of the surface front.

As the front nears, cirrus clouds grade into cirro-stratus, which blend into denser sheets of altostratus. About 300 kilometers (180 miles) ahead of the front,

thicker stratus and nimbostratus clouds appear, and rain or snow begins.

Because of their slow rate of advance and very low slope, warm fronts usually produce light-to-moderate precipitation over a large area for an extended period. Warm fronts, however, are occasionally associated with cumulonimbus clouds and thunderstorms. This occurs when the overrunning air is unstable and the tempera-tures on opposite sides of the front contrast sharply. At the other extreme, a warm front associated with a dry air mass could pass unnoticed at the surface.

A gradual increase in temperature occurs with the passage of a warm front. As you would expect, the in-crease is most apparent when a large contrast exists between adjacent air masses. Moreover, a wind shift from the east to the southwest is generally noticeable. (The reason for this shift will become evident later.) The moisture content and stability of the encroaching warm air mass largely determine when clear skies will return. During the summer, cumulus, and occasionally cumulonimbus, clouds are embedded in the warm un-stable air mass that follows the front. These clouds can produce precipitation, but it is usually restricted in ex-tent and of short duration.

Cold Fronts

When cold air actively advances into a region occu-pied by warmer air, the boundary is called a **cold front** (Figure 19.7). As with warm fronts, friction tends to slow the surface position of a cold front more so than its position aloft. However, because of the relative po-sitions of the adjacent air masses, the cold front steep-ens as it moves. On the average, cold fronts are about twice as steep as warm fronts, having a slope of per-haps 1:100. In addition, cold fronts advance more rap-

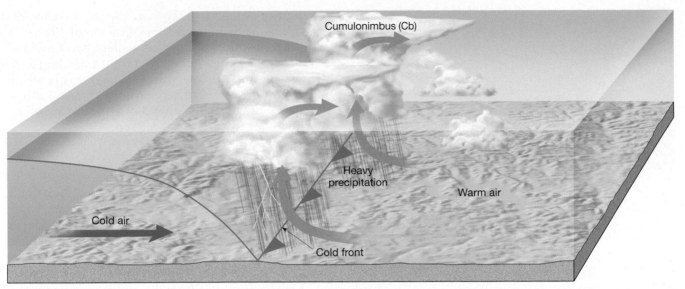

Cumulonimbus (Cb)

Heavy
precipitation

Warm air

Cold air

Cold front

▲ **FIGURE 19.7**   Fast-moving cold front and cumulonimbus clouds. Often thunderstorms occur if the warm air is unstable.

idly than do warm fronts. These two differences—rate of movement and steepness of slope—largely account for the more violent nature of cold-front weather compared to the weather generally accompanying a warm front.

The forceful lifting of air along a cold front is often so rapid that the latent heat released when water vapor condenses will increase the air's buoyancy appreciably. The heavy downpours and vigorous wind gusts associated with mature cumulonimbus clouds frequently result. Because a cold front produces roughly the same amount of lifting as a warm front, but over a shorter distance, the precipitation intensity is greater, but of shorter duration.

As a cold front approaches, generally from the west or northwest, towering clouds often can be seen in the distance. Near the front, a dark band of ominous clouds foretells the ensuing weather. A marked temperature drop and a wind shift usually accompany the passage of the front. (Again, the reason for this shift, from the south or southwest to the west or northwest, will be explained later.)

The sometimes violent weather and sharp temperature contrast along the cold front are symbolized on a weather map by a blue line with triangle-shaped points that extend into the warmer air mass (Figure 19.7).

The weather behind a cold front is dominated by a subsiding and relatively cold air mass. Thus, clearing usually begins soon after the front passes. Although the compression of air due to subsidence causes some adiabatic heating, the effect on surface temperatures is minor. In winter, the long, cloudless nights that often follow the passage of a cold front allow for abundant radiation cooling that reduces surface temperatures. When a cold front moves over a relatively warm area,

surface heating of the air can produce shallow convection. This in turn may generate low cumulus or stratocumulus clouds behind the front.

Stationary Fronts and Occluded Fronts

Occasionally the flow on both sides of a front is neither toward the cold air mass nor toward the warm air mass, but almost parallel to the line of the front. Thus, the surface position of the front does not move. This condition is called a **stationary front**. On a weather map, stationary fronts are shown with blue triangular points on one side of the front and red semicircles on the other. At times, some overrunning occurs along a stationary front. On these occasions, gentle to moderate precipitation is most likely.

The fourth type of front is the **occluded front**. Here an active cold front overtakes a warm front, as shown in Figure 19.8. As the advancing cold air wedges the warm front upward, a new front emerges between the advancing cold air and the air over which the warm front is gliding. The weather of an occluded front is generally complex. Most precipitation is associated with the warm air being forced aloft. When conditions are suitable, however, the newly formed front is capable of initiating precipitation of its own.

A word of caution is in order concerning the weather associated with various fronts. Although the preceding discussion will help you recognize the weather patterns associated with fronts, remember that these descriptions are generalizations. The weather generated along any individual front may or may not conform fully to this idealized picture. Fronts, like all aspects of nature, never lend themselves to classification as nicely as we would like.

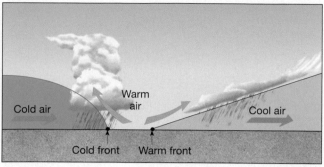

A.

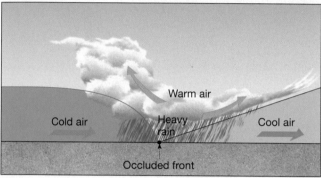

B.

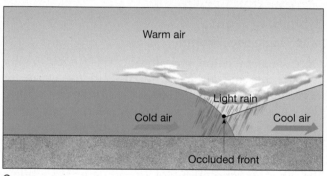

C.

▲ **FIGURE 19.8** Stages in the formation of an occluded front.

The Middle-Latitude Cyclone

 GEODe

Earth's Dynamic Atmosphere
▼ Basic Weather Patterns

So far we have examined the basic elements of weather as well as the dynamics of atmospheric motions. We are now ready to apply our knowledge of these diverse phenomena to an understanding of day-to-day weather patterns in the middle latitudes. For our purposes, *middle latitudes* refers to the region between southern Florida and Alaska. The primary weather producers here are **middle-latitude** or **midlatitude cyclones**. On weather maps, such as those used on the Weather Channel, they are shown by an **L**, meaning *low-pressure center.*

Middle-latitude cyclones are large centers of low pressure that generally travel from west to east. Last-

ing from a few days to more than a week, these weather systems have a counterclockwise circulation with an airflow inward toward their centers. Most middle-latitude cyclones also have a cold front extending from the central area of low pressure, and frequently a warm front as well. Convergence and forceful lifting initiate cloud development and often cause abundant precipitation.

Life Cycle

As early as the 1800s it was known that cyclones were the bearers of precipitation and severe weather. But it was not until the early part of the twentieth century that a model was developed that described how cyclones form and evolve. It was formulated by a group of Norwegian scientists and published in 1918. The model was created primarily from near-surface observations.

Years later, as data from the middle and upper troposphere and from satellite images became available, modifications were necessary. Yet this model is still an accepted working tool for interpreting the weather. If you keep this model in mind when you observe changes in the weather, the changes will no longer come as a surprise. You should begin to see some order in what had appeared to be disorder, and you might even occasionally "predict" the impending weather (see Box 19.1).

Formation: The Clash of Two Air Masses. Cyclones form along fronts and proceed through a generally predictable life cycle. Figure 19.9 shows six stages in the life of a typical midlatitude cyclone. In part A, the stage is set for cyclone formation. Here two air masses of different densities (temperatures) are moving roughly parallel to the front, but in opposite directions. (In the classic model, this would be continental polar air associated with the polar easterlies on the north side of the front and maritime tropical air driven by the westerlies on the south side of the front.)

Under suitable conditions, the frontal surface that separates these two contrasting air masses will take on a wave shape that is usually several hundred kilometers long. These waves are analogous to the waves produced on water by moving air, except that the scale is different. Some waves tend to dampen, or die out, whereas others grow in amplitude. Those storms that intensify develop waves that change in shape over time, much like a gentle ocean swell does as it moves into shallow water and becomes a tall, breaking wave.

Development of Cyclonic Flow. As the wave develops, warm air advances poleward invading the area formerly occupied by colder air, while cold air moves equatorward. This change in the direction of the surface flow is accompanied by a readjustment in the pressure pattern that results in nearly circular isobars, with the low pressure centered at the apex of the wave. The resulting flow is a counterclockwise cyclonic cir-

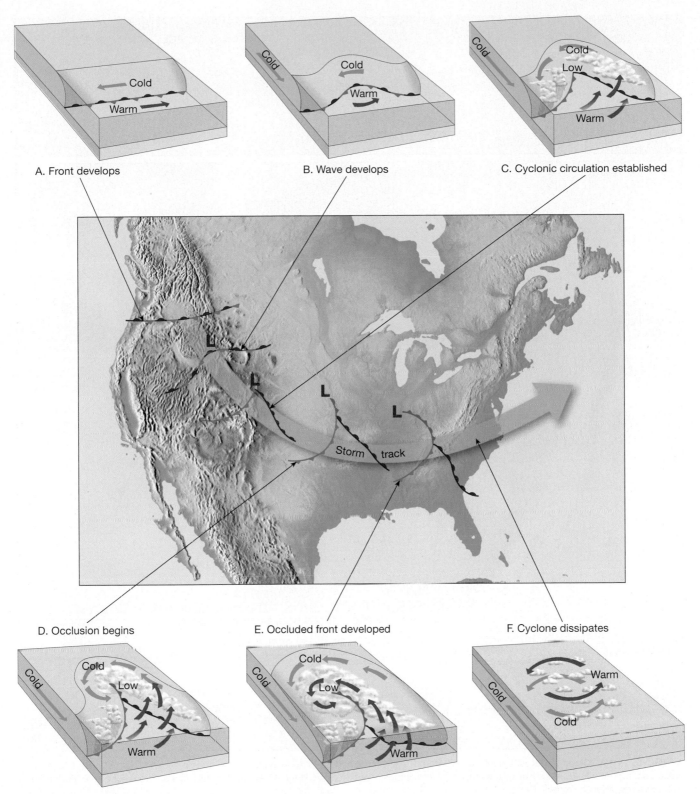

A. Front develops

B. Wave develops

C. Cyclonic circulation established

D. Occlusion begins

E. Occluded front developed

F. Cyclone dissipates

▲ **FIGURE 19.9** Stages in the life cycle of a middle-latitude cyclone.

culation that can be seen clearly on the weather map shown in Figure 19.10. Once the cyclonic circulation develops, we would expect general convergence to result in vertical lifting, especially where warm air is overrunning colder air. You can see in Figure 19.10 that the air in the warm sector (over the southern states) is

flowing northeastward toward colder air that is moving toward the northwest. Because the warm air is moving in a direction perpendicular to this front, we can conclude that the warm air is invading a region formerly occupied by cold air. Therefore, this must be a warm front. Similar reasoning indicates that to the

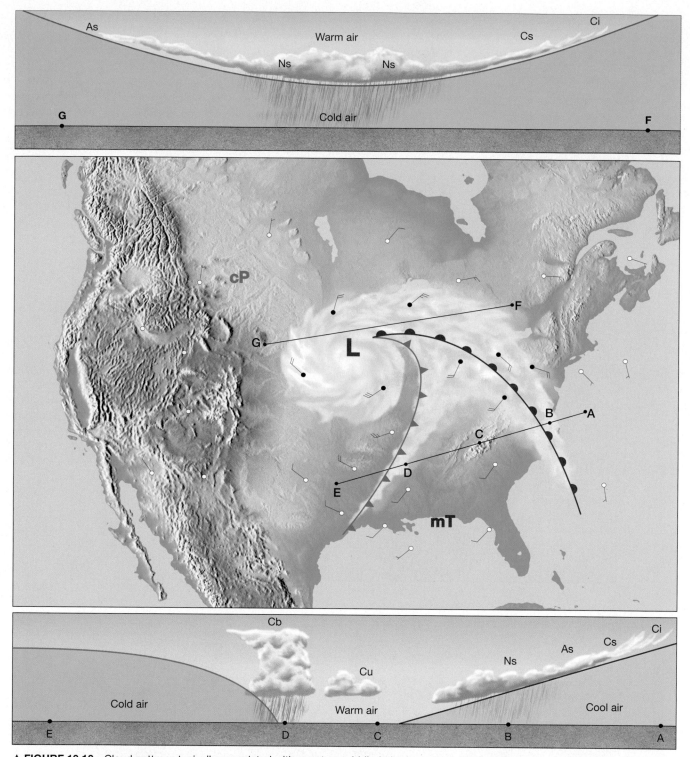

▲ **FIGURE 19.10** Cloud patterns typically associated with a mature middle-latitude cyclone. The middle section is a map view. Note the cross-section lines (F–G, A–E). Above the map is a vertical cross section along line F–G. Below the map is a section along A–E. For cloud abbreviations, refer to Figures 19.6 and 19.7.

left (west) of the cyclonic disturbance, cold air from the northwest is displacing the air of the warm sector and generating a cold front.

Occlusion: The Beginning of the End. Usually, the position of the cold front advances faster than the warm front and begins to close (lift) the warm front, as shown in Figure 19.9D, E. This process, known as **occlusion**, forms an *occluded front*, which grows in length as it displaces the warm sector aloft. As occlusion begins, the storm often intensifies. Pressure at the storm's center falls, and wind speeds increase. In the winter, heavy snowfalls and blizzardlike conditions are possible during this phase of the storm's evolution.

Box 19.1 Understanding Earth:
A Brief Overview of the Weather Business

In the United States, the governmental agency responsible for gathering and disseminating weather-related information is the *National Weather Service (NWS)*. The mission of the NWS is as follows:

The National Weather Service (NWS) provides weather, hydrologic and climate forecasts and warnings for the United States, its territories, adjacent waters and ocean areas, for the protection of life and property and the enhancement of the national economy. NWS data and products form a national information database and infrastructure that can be used by other governmental agencies, the private sector, the public and the global community.

Perhaps the most important services provided by the NWS are forecasts and warnings of hazardous weather including thunderstorms, flooding, hurricanes, tornadoes, winter weather, and extreme heat (Figure 19.A). According to the Federal Emergency Management Agency, 80 percent of all declared emergencies are weather related. In a similar vein, the Department of Transportation reports that more than 6000 fatalities per year can be attributed to the weather.

As the population grows, the economic impact from weather-related phenomena also escalates. Each year weather-related losses are in the billions of dollars. An estimated 90 percent of the U.S. public consults weather forecasts every day. As a result, the pressure on the NWS to provide more accurate and longer-range forecasts continues to grow.

To produce even a short-range forecast is an enormous task. It involves numerous steps, including collecting weather data, transmitting it, and compiling it on a global scale. These data must then be analyzed so that an accurate assessment of the current conditions can be made. From current weather patterns, various methods are used to determine the future state of the atmosphere, a task called *weather forecasting*.

Figure 19.A Meteorologist at the National Hurricane center in Miami, Florida examining satellite imagery of Hurricane Isidore. (Andy Newman/AP Wide World Photos)

The final phase in the weather business is the dissemination of a wide variety of forecasts. Each of the 119 Weather Forecast Offices operated by the NWS issues regional forecasts, aviation forecasts, and warnings covering their forecast area. The *local forecast* seen on The Weather Channel is an unedited version of a forecast issued by one of these offices. Further, all the data and products (maps, charts, and forecasts) produced by the NWS are available at no cost to the general public and to private forecasting services.

The demand for highly visual forecasts containing computer-generated graphics has grown along with the use of personal computers and the Internet. Because it is outside the mission of the National Weather Service, this publicly funded entity is not the source of the animated depictions of the weather that appear on most local newscasts. Instead, the private sector has taken over this task. In addition, private forecast services customize the NWS products to create a variety of specialized weather reports that are tailored for specific audiences. In a farming community, for example, the weather reports might include frost warnings, while winter forecasts in Denver, Colorado, include the snow conditions at ski resorts. Hence, the public receives most of its weather information through local TV stations and newspapers and/or through more widely distributed sources such as The Weather Channel, CNN, and USA Today.

It is important to note that, despite the valuable role that the private sector plays in disseminating weather-related information to the public, the NWS is the *official* voice in the United States for issuing warnings during life-threatening weather situations. Two major weather centers operated by the NWS serve critical functions in this regard. The Storm Prediction Center in Norman, Oklahoma, maintains a constant vigil for severe thunderstorms and tornadoes. Hurricane watches and warnings for the Atlantic, Caribbean, Gulf of Mexico, and eastern Pacific are issued by the National Hurricane Center/Tropical Prediction Center in Miami, Florida.

As more of the sloping discontinuity (front) is forced aloft, the pressure gradient weakens. In a day or two, the entire warm sector is displaced and cold air surrounds the cyclone at low levels (Figure 19.9F). Thus, the horizontal temperature (density) difference that existed between the two contracting air masses has been eliminated. At this point, the cyclone has exhausted its source of energy. Friction slows the surface flow, and the once highly organized counterclockwise flow ceases to exist.

Idealized Weather

The middle-latitude cyclone model provides a useful tool for examining the weather patterns of the middle latitudes. Figure 19.10 illustrates the distribution of clouds and thus the regions of possible precipitation associated with a mature system. Compare this drawing to the satellite image of a cyclone shown in Figure 19.11.

Guided by the westerlies aloft, cyclones generally move eastward across the United States. Therefore, we can expect the first signs of a cyclone's arrival to appear in the western sky. In the region of the Mississippi Valley, however, cyclones often begin a more northeasterly path and occasionally move directly northward. Typically, a midlatitude cyclone requires two to four days to pass over a region. During that brief period, abrupt changes in atmospheric conditions may occur, particularly in the spring of the year when the greatest temperature contrasts occur across the midlatitudes.

Using Figure 19.10 as a guide, let us examine these weather producers and the changes we can expect as they pass in the spring of the year. To facilitate our discussion, Figure 19.10 includes two profiles along lines A–E and F–G.

- Imagine the change in weather as you move from right to left along profile A–E (bottom of figure). At point A, the sighting of high cirrus clouds is the first sign of the approaching cyclone. These high clouds can precede the surface front by 1000 kilometers (600 miles) or more, and they are normally accompanied by falling pressure. As the warm front advances, a lowering and thickening of the cloud deck is noticed.

- Within 12 to 24 hours after the first sighting of cirrus clouds, light precipitation begins (point B). As the front nears, the rate of precipitation increases, a rise in temperature is noticed, and winds begin to change from east or southeast to south or southwest.

- With the passage of the warm front, an area is under the influence of a maritime tropical air mass (point C). Usually the region affected by this sector of the cyclone experiences warm tem-

▲ **FIGURE 19.11** Satellite view of a mature cyclone over the eastern United States. It is easy to see why we often refer to the cloud pattern of a cyclone as having a "comma" shape. (Courtesy of John Jensenius/National Weather Service)

peratures, south or southwest winds, and generally clear skies, although fair-weather cumulus or altocumulus clouds are not uncommon here.

- The relatively warm, humid weather of the warm sector passes quickly in the spring of the year and is replaced by gusty winds and precipitation generated along the cold front. The approach of a rapidly advancing cold front is marked by a wall of dark clouds (point D). Severe weather accompanied by heavy precipitation, hail, and an occasional tornado is a definite possibility at this time of year. The passage of the cold front is easily detected by a wind shift; the southwest winds are replaced by winds from the west to northwest and by a pronounced drop in temperature. Also, rising pressure hints of the subsiding cool, dry air behind the front.

- Once the front passes, skies clear as cooler air invades the region (point E). Often a day or two of almost cloudless deep-blue skies occurs unless another cyclone is edging into the region.

A very different set of weather conditions will prevail in those regions north of the storm's center along profile F–G of Figure 19.10. Often the storm reaches its greatest intensity in this zone, and the area along profile F–G receives the brunt of the storm's fury. Here temperatures remain cold during the passage of the system, and heavy snow, sleet, and/or freezing rain may develop during the winter months.

The Role of Airflow Aloft

When the earliest studies of cyclones were made, little was known about the nature of the airflow in the middle and upper troposphere. Since then, a close relationship has been established between surface disturbances and the flow aloft. Airflow aloft plays an important role in maintaining cyclonic and anticyclonic circulation. In fact, more often than not, these rotating surface wind systems are actually generated by upper-level flow.

Recall that the airflow around a surface low is inward, a fact that leads to mass convergence, or coming together (Figure 19.12). The resulting accumulation of air must be accompanied by a corresponding increase in surface pressure. Consequently, we might expect a low-pressure system to "fill" rapidly and be eliminated, just as the vacuum in a coffee can is quickly dissipated when we open it. However, this does not occur.

On the contrary, cyclones often exist for a week or longer. For this to happen, surface convergence must be offset by a mass outflow at some level aloft (Figure 19.12). As long as divergence (spreading out) aloft is equal to, or greater than, the surface inflow, the low pressure and its accompanying convergence can be sustained.

Because cyclones are bearers of stormy weather, they have received far more attention than anticyclones. Nevertheless, a close relation exists, which makes it difficult to separate any discussion of these two types of pressure systems. The surface air that feeds a cyclone, for example, generally originates as air flowing out of an anticyclone. Consequently, cyclones and anticyclones typically are found adjacent to one another. Like the cyclone, an anticyclone depends

on the flow far above to maintain its circulation. In this instance, divergence at the surface is balanced by convergence aloft and general subsidence of the air column (Figure 19.12).

What's in a Name?

Up to now we have examined the middle-latitude cyclones, which play such an important role in causing day-to-day weather changes. Yet the use of the term *cyclone* is often confusing. To many people, the term implies only an intense storm, such as a tornado or a hurricane. When a hurricane unleashes its fury on India or Bangladesh, for example, it is usually reported in the media as a cyclone (the term denoting a hurricane in that part of the world).

Similarly, tornadoes are referred to as cyclones in some places. This custom is particularly common in portions of the Great Plains of the United States. Recall that in *The Wizard of Oz*, Dorothy's house was carried from her Kansas farm to the land of Oz by a cyclone. Indeed, the nickname for the athletic teams at Iowa State University is the *Cyclones*. Although hurricanes and tornadoes are, in fact, cyclones, the vast majority of cyclones are *not* hurricanes or tornadoes. The term *cyclone* simply refers to the circulation around any low-pressure center, no matter how large or intense it is.

Tornadoes and hurricanes are both smaller and more violent than middle-latitude cyclones. Middle-latitude cyclones can have a diameter of 1600 kilometers

▼ **FIGURE 19.12** Idealized diagram depicting the support that divergence and convergence aloft provide to cyclonic and anticyclonic circulation at the surface. Divergence aloft initiates upward air movement, reduced surface pressure, and cyclonic flow. In contrast, convergence along the jet stream results in general subsidence of the air column, increased surface pressure, and anticyclonic surface winds.

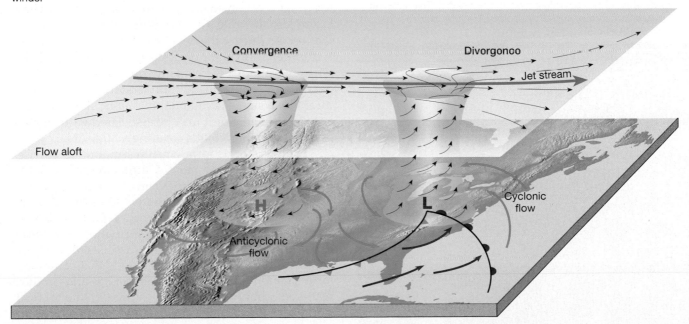

(1000 miles) or more. By contrast, hurricanes average only 600 kilometers (375 miles) across, and tornadoes, with a diameter of just 0.25 kilometer (0.16 mile) are much too small to show up on a weather map.

The thunderstorm, a much more familiar weather event, hardly needs to be distinguished from tornadoes, hurricanes, and midlatitude cyclones. Unlike the flow of air about these latter storms, the circulation associated with thunderstorms is characterized by strong up-and-down movements. Winds in the vicinity of a thunderstorm do not follow the inward spiral of a cyclone, but they are typically variable and gusty.

Although thunderstorms form "on their own" away from cyclonic storms, they also form in conjunction with cyclones. For instance, thunderstorms are frequently spawned along the cold front of a midlatitude cyclone, where on rare occasions a tornado may descend from the thunderstorm's cumulonimbus tower. Hurricanes also generate widespread thunderstorm activity. Thus, thunderstorms are related in some manner to all three types of cyclones mentioned here.

Thunderstorms

This is the first of three severe weather types we will examine in this chapter. Sections on tornadoes and hurricanes follow this look at thunderstorms.

Severe weather has a fascination that everyday weather phenomena cannot provide. The lightning display and booming thunder generated by a severe thunderstorm can be a spectacular event that elicits both awe and fear (Figure 19.13). Of course, hurricanes

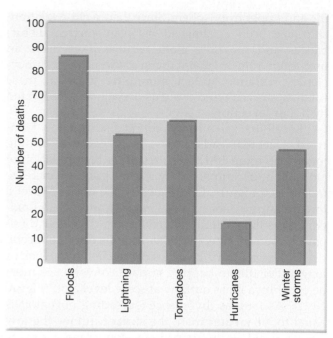

▲ **FIGURE 19.14** Average annual storm-related weather fatalities in the United States over the 10-year span 1993–2002. (Data from NOAA)

and tornadoes also attract a great deal of much-deserved attention. A single tornado outbreak or hurricane can cause billions of dollars in property damage as well as many deaths (Figure 19.14).

Thunderstorm Occurrence

Almost everyone has observed a small-scale phenomenon that is the result of the vertical movements of relatively warm, unstable air. Perhaps you have seen a dust devil over an open field on a hot day whirl its dusty load to great heights, or seen a bird glide effortlessly skyward on an invisible thermal of hot air. These examples illustrate the dynamic thermal instability that occurs during the development of a *thunderstorm*. A **thunderstorm** is simply a storm that generates lightning and thunder. It frequently produces gusty winds, heavy rain, and hail. A thunderstorm may be produced by a single cumulonimbus cloud and influence only a small area or it may be associated with clusters of cumulonimbus clouds covering a large area.

Thunderstorms form when warm, humid air rises in an unstable environment. Various mechanisms can trigger the upward air movement needed to create thunderstorm-producing cumulonimbus clouds. One mechanism, the unequal heating of Earth's surface, significantly contributes to the formation of *air-mass thunderstorms*. These storms are associated with the scattered puffy cumulonimbus clouds that commonly form *within* maritime tropical air masses and produce scattered thunderstorms on summer days. Such storms are usually short-lived and seldom produce strong winds or hail.

▼ **FIGURE 19.13** Cumulonimbus clouds can produce lightning, thunderstorms, and other forms of severe weather. This lightning display occurred over Tucson, Arizona. (Photo by Warren Faidley/DRK Photo)

In contrast, thunderstorms in a second category not only benefit from uneven surface heating but are associated with the lifting of warm air, as occurs along a front or a mountain slope. Moreover, diverging winds aloft frequently contribute to the formation of these storms because they tend to draw air from lower levels upward beneath them. Some of the thunderstorms in this second category may produce high winds, damaging hail, flash floods, and tornadoes. Such storms are described as *severe*.

At any given time there are an estimated 2000 thunderstorms in progress on Earth. As we would expect, the greatest number occur in the tropics where warmth, plentiful moisture, and instability are always present. About 45,000 thunderstorms take place each day, and more than 16 million occur annually around the world. The lightning from these storms strikes Earth 100 times each second (see Box 19.2)!

Annually, the United States experiences about 100,000 thunderstorms and millions of lightning strikes. A glance at Figure 19.15 shows that thunderstorms are most frequent in Florida and the eastern Gulf Coast region, where such activity is recorded between 70 and 100 days each year. The region on the east side of the Rockies in Colorado and New Mexico is next, with thunderstorms occurring on 60 to 70 days each year. Most of the rest of the nation experiences thunderstorms on 30 to 50 days annually. Clearly, the western margin of the United States has little thunderstorm activity. The same is true for the northern tier of states and for Canada, where warm, moist, unstable mT air seldom penetrates.

STUDENTS SOMETIMES ASK...
How far away from a storm can lightning strike?

According to the National Severe Storms Laboratory (NSSL) in Norman, Oklahoma, it is not certain what the maximum possible distance might be. Lightning has been known to strike more than 16 kilometers (10 miles) from a storm in an area of clear skies.

Stages of Thunderstorm Development

All thunderstorms require warm, moist air, which, when lifted, will release sufficient latent heat to provide the buoyancy necessary to maintain its upward

▼ **FIGURE 19.15** Average number of days per year with thunderstorms. Because of its close proximity to the source region for warm, humid, and unstable air masses, the Gulf Coast receives much of its precipitation from thunderstorms. (Source: Environmental Data Service, NOAA)

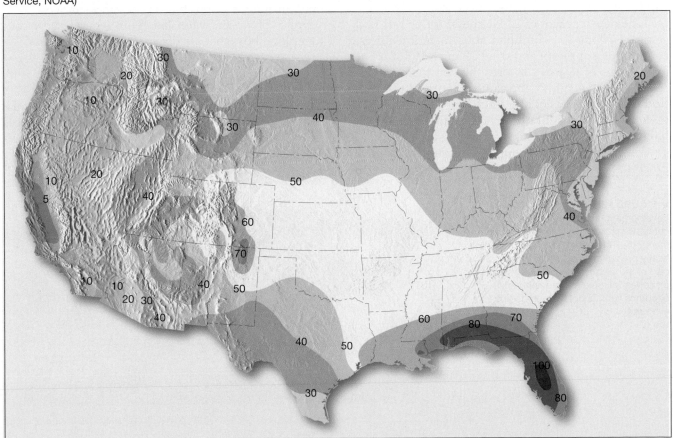

A.

B.

▲ **FIGURE 19.16** **A.** Buoyant thermals often produce fair-weather cumulus clouds that soon evaporate into the surrounding air, making it more humid. As this process of cumulus development and evaporation continues, the air eventually becomes sufficiently humid, so that newly forming clouds do not evaporate but continue to grow. (Photo by Henry Lansford/Photo Researchers, Inc.) **B.** This developing cumulonimbus cloud became a towering August thunderstorm over central Illinois. (Photo by Henry Lansford)

flight. This instability and associated buoyancy are triggered by a number of different processes, yet all thunderstorms have a similar life history.

Because instability and buoyancy are enhanced by high surface temperatures, thunderstorms are most common in the afternoon and early evening (Figure 19.16A). Surface heating is generally not sufficient in itself to cause the growth of towering cumulonimbus clouds. A solitary cell of rising hot air produced by surface heating alone can, at best, produce a small cumulus cloud, which would evaporate within 10 to 15 minutes.

The development of 12-kilometer (or on rare occasions 18-kilometer) cumulonimbus towers requires a continuous supply of moist air (Figure 19.16B). Each new surge of warm air rises higher than the last, adding to the height of the cloud (Figure 19.17). These updrafts must occasionally reach speeds over 100 kilometers (60 miles) per hour to accommodate the size of hailstones they are capable of carrying upward.

Usually within an hour the amount and size of precipitation that has accumulated is too much for the updrafts to support, and in one part of the cloud downdrafts develop, releasing heavy precipitation. This represents the most active stage of the thunderstorm. Gusty winds, lightning, heavy precipitation, and sometimes hail are experienced.

Eventually downdrafts dominate throughout the cloud. The cooling effect of falling precipitation, coupled with the influx of colder air aloft, marks the end of the thunderstorm activity. The life span of a single cumulonimbus cell within a thunderstorm complex is only an hour or two, but as the storm moves, fresh supplies of warm, water-laden air generate new cells to replace those that are dissipating.

Tornadoes

Tornadoes are local storms of short duration that must be ranked high among nature's most destructive forces (see chapter-opening photo). Their sporadic occurrence and violent winds cause many deaths each year. Tornadoes are violent windstorms that take the form

▶ **FIGURE 19.17** Stages in the development of a thunderstorm. During the cumulus stage, strong updrafts act to build the storm. The mature stage is marked by heavy precipitation and cool downdrafts in part of the storm. When the warm updrafts disappear completely, precipitation becomes light, and the cloud begins to evaporate.

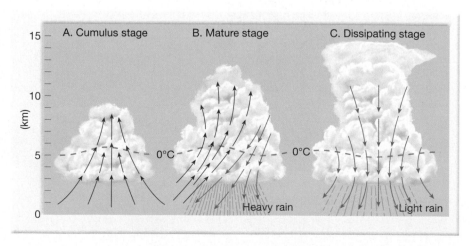

Box 19.2 People and the Environment:
Lightning Safety*

Lightning ranks second only to floods in the number of storm-related deaths each year. Although the number of reported lightning deaths in the United States annually averages about 85, many go unreported. About 100 people are estimated to be killed and more than 500 injured by lightning every year in the United States.

Warnings, statements, and forecasts are routinely issued for floods, tornadoes, and hurricanes but not for lightning. Why is this the case? The answer relates to the wide geographic occurrence and frequency of lightning.

The magnitude of the cloud-to-ground lightning hazard is understood better today than ever before. Lightning occurs in the United States every day in summer and nearly every day during the rest of the year. In an average year the National Lightning Detection Network identifies about 22,000,000 cloud-to-ground flashes. Because lightning is so widespread and strikes the ground with such great frequency, it is not possible to warn each person of every flash. For this reason, lightning can be considered the most dangerous weather hazard that many people encounter each year.

Being aware of and following proven safety guidelines can greatly reduce the risk of injury or death. Individuals are ultimately responsible for their personal safety and should take appropriate action when threatened by lightning.

*Material in this box is based on "Updated Recommendations for Lightning Safety—1998, in *Bulletin of the American Meteorological Society*, Vol. 80, No.10, October 1999, pp. 2035–39.

No place is absolutely safe from the threat of lightning, but some places are safer than others.

- Large enclosed structures (substantially constructed buildings) tend to be much safer than smaller or open structures. The risk for lightning injury depends on whether the structure incorporates lightning protection, the types of construction materials used, and the size of the structure.

- In general, fully enclosed metal vehicles such as cars, trucks, buses, vans, and fully enclosed farm vehicles, etc., with the windows rolled up, provide good shelter from lightning. Avoid contact with metal or conducting surfaces outside or inside the vehicle.

- *Avoid* being in or near high places and open fields, isolated trees, unprotected gazebos, rain or picnic shelters, baseball dugouts, communications towers, flagpoles, light poles, bleachers (metal or wood), metal fences, convertibles, golf carts, and water (ocean, lakes, swimming pools, rivers, etc.).

- When inside a building, *avoid* use of the telephone, taking a shower, washing your hands, doing dishes, or any contact with conductive surfaces with exposure to the outside such as metal door or window frames, electrical wiring, telephone wiring, cable TV wiring, plumbing, and so forth.

Safety guidelines for individuals include the following:

- Generally speaking, if individuals can see lightning and/or hear thunder, they are already at risk. Louder or more frequent thunder indicates that lightning activity is approaching, thus increasing the risk for lightning injury or death. If the time delay between seeing the flash (lightning) and hearing the bang (thunder) is less than 30 seconds, the individual should be in, or seek, a safer location. Be aware that this method of ranging has severe limitations in part due to the difficulty of associating the proper thunder to the corresponding flash.

- High winds, rainfall, and cloud cover often act as precursors to actual cloud-to-ground strikes, and these should motivate individuals to take action. Many lightning casualties occur in the beginning, as the storm approaches, because people ignore these precursors. Also, many lightning casualties occur after the perceived threat has passed. Generally, the lightning threat diminishes with time after the last sound of thunder but may persist for more than 30 minutes. When thunderstorms are in the area but not overhead, the lightning threat can exist even when it is sunny, not raining, or when clear sky is visible.

- When available, pay attention to weather-warning devices such as NOAA (National Oceanic and Atmospheric Administration) weather radio and/or credible lightning detection systems; however, do not let this information override good common sense.

---

of a rotating column of air or *vortex* that extends downward from a cumulonimbus cloud.

Pressures within some tornadoes have been estimated to be as much as 10 percent lower than immediately outside the storm. Drawn by the much lower pressure in the center of the vortex, air near the ground rushes into the tornado from all directions. As the air streams inward, it is spiraled upward around the core until it eventually merges with the airflow of the parent thunderstorm deep in the cumulonimbus tower. Because of the tremendous pressure gradient associated with a strong tornado, maximum winds can sometimes approach 480 kilometers (300 miles) per hour.

Some tornadoes consist of a single vortex, but within many stronger tornadoes are smaller whirls called *suction vortices* that rotate within the main vortex (Figure 19.18). Suction vortices have diameters of only about 10 meters (30 feet) and rotate very rapidly. This structure accounts for occasional observations of virtually total destruction of one building while another one, just 10 or 20 meters away, suffers little damage.

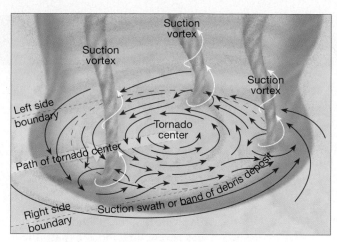

▲ **FIGURE 19.18** Some tornadoes have multiple suction vortices. These small and very intense vortices are roughly 10 meters (30 feet) across and move in a counterclockwise path around the tornado center. Because of this multiple vortex structure, one building might be heavily damaged and another one, just 10 meters away, might suffer little damage. (After Fujita)

Tornado Occurrence and Development

Tornadoes form in association with severe thunderstorms that produce high winds, heavy (sometimes torrential) rainfall, and often damaging hail. Fortunately, less than 1 percent of all thunderstorms produce tornadoes. Nevertheless, a much higher number must be monitored as potential tornado producers. Although meteorologists are still not sure what triggers tornado formation, it is apparent that they are the product of the interaction between strong updrafts in the thunderstorm and the winds in the troposphere.

Tornadoes can form in any situation that produces severe weather, including cold fronts and tropical cyclones (hurricanes). Usually the most intense tornadoes are those that form in association with huge thunderstorms called *supercells*. An important precondition linked to tornado formation in severe thunderstorms is the development of a mesocyclone. A *mesocyclone* is a vertical cylinder of rotating air, typically about 3 to 10 kilometers (2 to 6 miles) across, that develops in the updraft of a severe thunderstorm (Figure 19.19). The formation of this large vortex often precedes tornado formation by 30 minutes or so.

The formation of a mesocyclone does not necessarily mean that tornado formation will follow. Only about half of all mesocyclones produce tornadoes. Because this is the case, forecasters cannot determine in advance which mesocyclones will spawn tornadoes.

General Atmospheric Conditions. Severe thunderstorms and hence tornadoes are most often spawned along the cold front of a middle-latitude cyclone or in

association with a supercell thunderstorm such as the one pictured in Figure 19.19D. Throughout spring, air masses associated with middle-latitude cyclones are most likely to have greatly contrasting conditions. Continental polar air from Canada may still be very cold and dry, whereas maritime tropical air from the Gulf of Mexico is warm, humid, and unstable. The greater the contrast when these air masses meet, the more intense the storm tends to be.

These two contrasting air masses are most likely to meet in the central United States, because there is no significant natural barrier separating the center of the country from the Arctic or the Gulf of Mexico. Consequently, this region generates more tornadoes than any other area of the country or, in fact, the world. Figure 19.20, which depicts tornado incidence in the United States for a 27-year period, readily substantiates this fact.

Tornado Climatology. An average of about 1200 tornadoes were reported annually in the United States between 1990 and 2001. Still, the actual numbers that occur from one year to the next vary greatly. During the 12-year span just mentioned, for example, yearly totals ranged from a low of 1071 in the year 2000 to a high of 1424 in 1998. Tornadoes occur during every month of the year. April through June is the period of greatest tornado frequency in the United States, and December and January are the months of lowest activity (Figure 19.20, graph inset). Of the 40,522 confirmed tornadoes reported over the contiguous 48 states during the 50-year period 1950–1999, an average of almost six per day occurred during May. At the other extreme, a tornado was reported only about every other day in December and January.

Profile of a Tornado. An average tornado has a diameter of between 150 and 600 meters (500 and 2000 feet), travels across the landscape at approximately 45 kilometers (30 miles) per hour, and cuts a path about 10 kilometers (6 miles) long.* Because tornadoes usually occur slightly ahead of a cold front, in the zone of southwest winds, most move toward the northeast. The Illinois example demonstrates this movement (Figure 19.21). Figure 19.21 also shows that many tornadoes do not fit the description of the "average" tornado.

Of the hundreds of tornadoes reported in the United States each year, over half are comparatively weak and short-lived. Most of these small tornadoes have lifetimes of three minutes or less and paths that seldom exceed 1 kilometer (0.6 mile) in length and

---

*The 10-kilometer figure applies to documented tornadoes. Because many small tornadoes go undocumented, the real average path of all tornadoes is unknown but shorter than 10 kilometers.

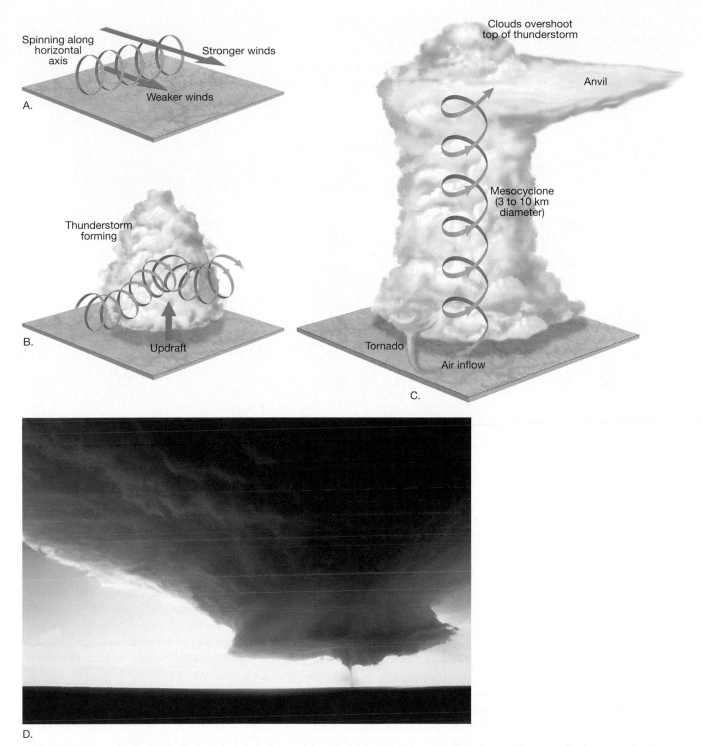

▲ **FIGURE 19.19** The formation of a mesocyclone often precedes tornado formation. **A.** Winds are stronger aloft than at the surface (called *speed wind shear*), producing a rolling motion about a horizontal axis. **B.** Strong thunderstorm updrafts tilt the horizontally rotating air to a nearly vertical alignment. **C.** The mesocyclone, a vertical cylinder of rotating air, is established. **D.** If a tornado develops it will descend from a slowly rotating wall cloud in the lower portion of the mesocyclone. This supercell tornado hit the Texas Panhandle in May 1996. (Photo by Warren Faidley/Weatherstock)

100 meters (330 feet) wide. Typical wind speeds are on the order of 150 kilometers (96 miles) per hour or less. On the other end of the tornado spectrum are the infrequent and often long-lived violent tornadoes. Although large tornadoes constitute only a small percentage of the total reported, their effects are often devastating. Such tornadoes may exist for periods in excess of three hours and produce an essentially continuous damage path more than 150 kilometers (90 miles) long and perhaps a kilometer or more wide. Maximum winds range beyond 500 kilometers (310 miles) per hour.

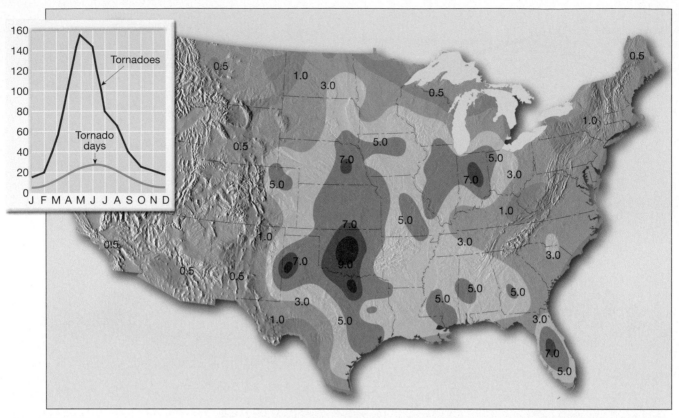

▲ **FIGURE 19.20**   The map shows average annual tornado incidence per 10,000 square miles (26,000 square kilometers) for a 27-year period. The graph shows the average number of tornadoes and tornado days each month in the United States for the same period.

Tornado Destruction

The potential for tornado destruction depends largely on the strength of the winds generated by the storm. Because tornadoes generate the strongest winds in nature, they have accomplished many seemingly impossible tasks, such as driving a piece of straw through a thick wooden plank and uprooting huge trees. Although it may seem impossible for winds to cause some of the fantastic damage attributed to tornadoes, tests in engineering facilities have repeatedly demonstrated that winds in excess of 320 kilometers (200 miles) per hour are capable of incredible feats (Figure 19.22A).

One commonly used guide to tornado intensity is the *Fujita intensity scale*, or simply the *F-scale* (Table 19.1). Because tornado winds cannot be measured directly, a rating on the F-scale is determined by assessing the worst damage produced by a storm.

Although the greatest part of tornado damage is caused by violent winds, most tornado injuries and deaths result from flying debris. For the United States, the average annual death toll from tornadoes is about 60 people. However, the actual number of deaths each year can depart significantly from the average. On April 3–4, 1974, for example, an outbreak of 148 tornadoes brought death and destruction to a 13-state region east of the Mississippi River. More than 300 people died and nearly 5500 people were injured in this worst disaster in half a century. Most tornadoes, however, do not result in a loss of life. In one statistical study that examined a 29-year period, there were 689 tornadoes that resulted in deaths. This figure represents slightly less than 4 percent of the total 19,312 reported storms.

Although the percentage of tornadoes that result in death is small, each tornado is potentially lethal. When tornado fatalities and storm intensities are compared, the results are quite interesting: The majority (63 percent) of tornadoes are weak (F0 and F1), and the number of storms decreases as tornado intensity increases. The distribution of tornado fatalities, however, is just the opposite. Although only 2 percent of tornadoes are classified as violent (F4 and F5), they account for nearly 70 percent of the deaths. If there is some question as to the causes of tornadoes, there certainly is no question about the destructive effects of these violent storms (Figure 19.22B)

TABLE 19.1	Fujita Intensity Scale.		
	Wind Speed		
Scale	**Km/Hr**	**Mi/Hr**	**Damage**
F0	<116	<72	Light damage
F1	116–180	72–112	Moderate damage
F2	181–253	113–157	Considerable damage
F3	254–332	158–206	Severe damage
F4	333–419	207–260	Devastating damage
F5	>419	>260	Incredible damage

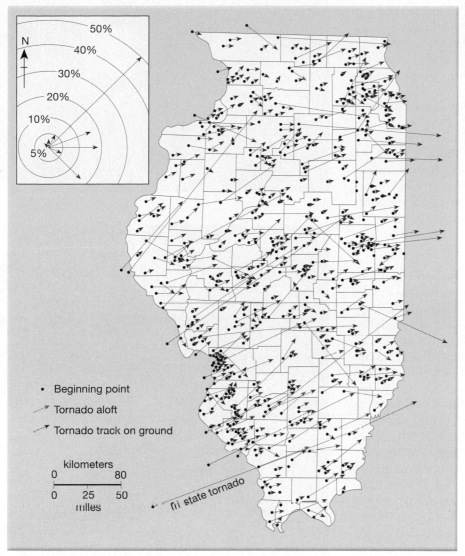

◄ **FIGURE 19.21** Paths of Illinois tornadoes (1916–1969). Because most tornadoes occur slightly ahead of a cold front, in the zone of southwest winds, they tend to move toward the northeast. Tornadoes in Illinois verify this. Over 80 percent exhibited directions of movement toward the northeast through east. (After John W. Wilson and Stanley A. Changnon, Jr., *Illinois Tornadoes*, Illinois State Water Survey Circular 103, 1971, pp. 10, 24)

- Beginning point
- Tornado aloft
- Tornado track on ground

kilometers
0   80

0   25   50
        miles

Tri state tornado

STUDENTS SOMETIMES ASK...

Someone told me that my house could explode if I don't open windows when a tornado is approaching. Is that true?

No. The drop in atmospheric pressure associated with the passage of a tornado plays a minor role in the damage process. Most structures have sufficient venting to allow for the sudden drop in pressure. Opening windows, once thought to be a way of minimizing damage by allowing inside and outside atmospheric pressure to equalize, is no longer recommended. In fact, if a tornado gets close enough to a structure for the pressure drop to be experienced, the strong winds will have already caused significant damage.

Tornado Forecasting

Because severe thunderstorms and tornadoes are small and relatively short-lived phenomena, they are among the most difficult weather features to forecast precisely. Nevertheless, the prediction, the detection, and the monitoring of such storms are among the most important services provided by professional meteorologists. The timely issuance and dissemination of watches and warnings are both critical to the protection of life and property.

STUDENTS SOMETIMES ASK...

What is the most destructive tornado on record?

One tornado easily ranks above all others as the single most dangerous and destructive. Known as the Tri-state tornado, it occurred on March 18, 1925. Its path is labeled on Figure 19.21. Starting in southeastern Missouri, the tornado remained on the ground for 352 kilometers (219 miles), finally ending in Indiana. The losses included 695 dead and 2027 injured. Property losses were also great, with several small towns almost totally destroyed.

The Storm Prediction Center (SPC) located in Norman, Oklahoma, is part of the National Weather Service (NWS) and the National Centers for Environmental Prediction (NCEP). Its mission is to provide timely and accurate forecasts and watches for severe thunderstorms and tornadoes.

Severe thunderstorm outlooks are issued several times daily. *Day 1* outlooks identify those areas likely to be affected by severe thunderstorms during the next 6 to 30 hours, and *day 2* outlooks extend the forecast through the following day. Both outlooks describe the type, coverage, and intensity of the severe weather expected. Many local NWS field offices also issue severe weather outlooks that provide a more local description of the severe weather potential for the next 12 to 24 hours.

Tornado Watches and Warnings **Tornado watches** alert the public to the possibility of tornadoes over a specified area for a particular time interval. Watches serve to fine-tune forecast areas already identified in severe weather outlooks. A typical watch covers an area of about 65,000 square kilometers (25,000 square miles) for a four- to six-hour period. A tornado watch is an important part of the tornado alert system because it sets in motion the procedures necessary to deal adequately with detection, tracking, warning, and response. Watches are generally reserved for organized severe weather events where the tornado threat will affect at least 26,000 square kilometers (10,000 square miles) and/or persist for at least three hours. Watches typically are not issued when the threat is thought to be isolated and/or short-lived.

Whereas a tornado watch is designed to alert people to the possibility of tornadoes, a **tornado warning** is issued by local offices of the National Weather Service when a tornado has actually been sighted in an area or is indicated by weather radar. It warns of a high probability of imminent danger. Warnings are issued for much smaller areas than watches, usually covering portions of a county or counties. In addition, they are in effect for much shorter periods, typically 30 to 60 minutes. Because a tornado warning may be based on an actual sighting, warnings are occasionally issued after a tornado has already developed. However, most warnings are issued prior to tornado formation, sometimes by several tens of minutes, based on Doppler radar data and/or spotter reports of funnel clouds.

If the direction and the approximate speed of the storm are known, an estimate of its most probable path can be made. Because tornadoes often move erratically, the warning area is fan-shaped downwind from the point where the tornado has been spotted. Improved forecasts and advances in technology have contributed to a significant decline in tornado deaths over the last 50 years.

A.

B.

▲ **FIGURE 19.22** **A.** The force of the wind during a tornado near Wichita, Kansas, in April 1991 was enough to drive this piece of metal into a utility pole. (Photo by John Sokich/NOAA) **B.** Damage from a tornado that struck Xenia, Ohio, on September 20, 2000. (Craig Holman/*Columbus Dispatch*/Reuters NewMedia Inc./Corbis/Bettmann).

Doppler Radar Many of the difficulties that once limited the accuracy of tornado warnings have been reduced or eliminated by an advancement in radar technology called **Doppler radar.** Doppler radar not only

◄ **FIGURE 19.23** Satellite image of Hurricane Ivan on September 15, 2004. At the time this image was taken, Ivan was about 275 kilometers (170 miles) south of the Alabama coastline and was an extremely dangerous category 4 storm with maximum sustained winds near 217 kilometers (135 miles) per hour. (NASA image)

performs the same tasks as conventional radar but also has the ability to detect motion directly. Doppler radar can detect the initial formation and subsequent development of a *mesocyclone*. Almost all mesocyclones produce damaging hail, severe winds, or tornadoes. Those that produce tornadoes (about 50 percent) can sometimes be distinguished by their stronger wind speeds and their sharper gradients of wind speeds.

It should also be pointed out that not all tornado-bearing storms have clear-cut radar signatures and that other storms can give false signatures. Detection, therefore, is sometimes a subjective process and a given display could be interpreted in several ways. Consequently, trained observers will continue to form an important part of the warning system in the foreseeable future.

Although some operational problems exist, the benefits of Doppler radar are many. As a research tool, it is not only providing data on the formation of tornadoes but is also helping meteorologists gain new insights into thunderstorm development, the structure and dynamics of hurricanes, and air-turbulence hazards that plague aircraft. As a practical tool for tornado detection, Doppler radar has significantly improved our ability to track thunderstorms and issue warnings.

Hurricanes

Most of us view the weather in the tropics with favor. Places like Hawaii and the islands of the Caribbean are known for their lack of significant day-to-day variations. Warm breezes, steady temperatures, and rains that come as heavy but brief tropical showers are expected. It is ironic that these relatively tranquil regions sometimes produce the most violent storms on Earth.

The whirling tropical cyclones that on occasion have wind speeds attaining 300 kilometers (185 miles) per hour are known in the United States as **hurricanes** and are the greatest storms on Earth (Figure 19.23). Out at sea, they can generate 15-meter (50-foot) waves capable of inflicting destruction hundreds of kilometers from their source. Should a hurricane smash onto land, strong winds coupled with extensive flooding can impose billions of dollars in damage and great loss of life.

Hurricanes are becoming a growing threat because more and more people are living and working along and near coasts. At the close of the twentieth century, more than 50 percent of the U.S. population lived within 75 kilometers (45 miles) of a coast. This number is projected to increase substantially in the early decades of the twenty-first century. The concentration of such large numbers of people near the shoreline means that hurricanes and other large storms place millions at risk. Moreover, the potential costs of property damage are incredible.

Profile of a Hurricane

Most hurricanes form between the latitudes of 5 degrees and 20 degrees over all the tropical oceans except those of the South Atlantic and eastern South Pacific. The North Pacific has the greatest number of storms, averaging 20 per year. Fortunately for those living in the coastal regions of the southern and eastern United States, fewer than five hurricanes, on the average, develop annually in the warm sector of the North Atlantic.

These intense tropical storms are known in various parts of the world by different names. In the western Pacific, they are called *typhoons*, and in the Indian Ocean, including the Bay of Bengal and Arabian Sea, they are simply called *cyclones*. In the following discussion, these storms will be referred to as hurricanes. The term *hurricane* is derived from Huracan, a Carib god of evil.

Although many tropical disturbances develop each year, only a few reach hurricane status. By international agreement, a hurricane has wind speeds in excess of 119 kilometers (74 miles) per hour and a rotary circulation. Mature hurricanes average 600 kilometers (375 miles) in diameter and often extend 12,000 meters (40,000 feet) above the ocean surface. From the outer edge to the center, the barometric pressure has on occasion dropped 60 millibars, from 1010 millibars to 950 millibars. The lowest pressures ever recorded in the Western Hemisphere are associated with these storms.

A steep pressure gradient generates the rapid, inward-spiraling winds of a hurricane (Figure 19.24). As the air rushes toward the center of the storm, its velocity increases. This occurs for the same reason that skaters with their arms extended spin faster as they pull their arms in close to their bodies.

As the inward rush of warm, moist surface air approaches the core of the storm, it turns upward and ascends in a ring of cumulonimbus towers (Figure 19.25). This doughnut-shaped wall of intense convective activity surrounding the center of the storm is called the **eye wall.** It is here that the greatest wind speeds and heaviest rainfall occur. Surrounding the eye wall are curved bands of clouds that trail away in a spiral fashion. Near the top of the hurricane the airflow is outward, carrying the rising air away from the storm center, thereby providing room for more inward flow at the surface.

▼ **FIGURE 19.24**   Weather maps showing Hurricane Fran at 7 a.m. EST on two successive days, September 5 and 6, 1996. On September 5, winds exceeded 190 kilometers (118 miles) per hour. As the storm moved inland, heavy rains caused flash floods, killed 30 people, and caused more than $3 billion in damages. The station information plotted off the Gulf and Atlantic coasts is from data buoys, which are remote floating instrument packages. The small boxes extending southeast from the storm's center show the position of the eye at six-hour intervals.

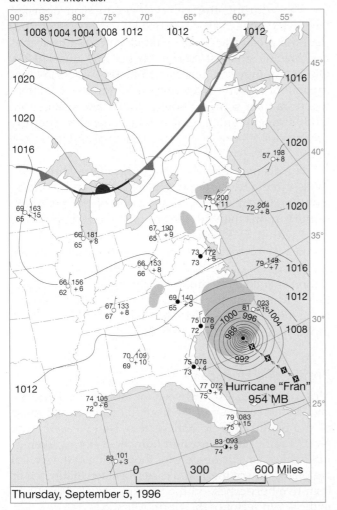

Thursday, September 5, 1996

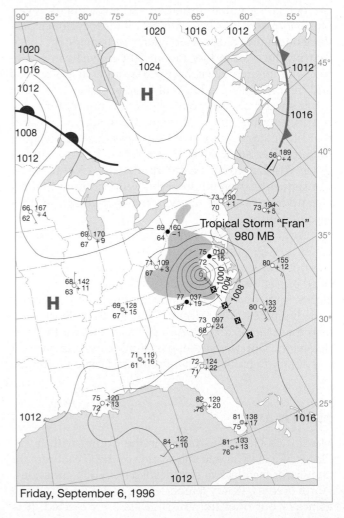

Friday, September 6, 1996

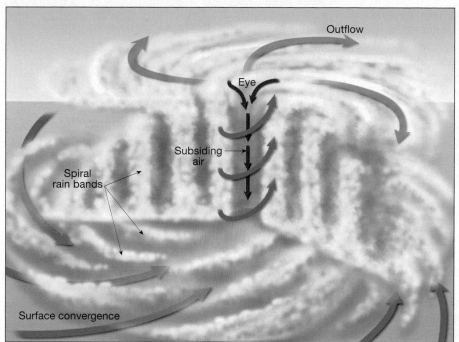

◀ **FIGURE 19.25** Cross section of a hurricane. Note that the vertical dimension is greatly exaggerated. The eye, the zone of relative calm at the center of the storm, is a distinctive hurricane feature. Sinking air in the eye warms by compression. Surrounding the eye is the eye wall, the zone where winds and rain are most intense. Tropical moisture spiraling inward creates rain bands that pinwheel around the storm center. Outflow of air at the top of the hurricane is important because it prevents the convergent flow at lower levels from "filling in" the storm. (After NOAA)

At the very center of the storm is the **eye** of the hurricane (Figure 19.25). This well-known feature is a zone about 20 kilometers (12.5 miles) in diameter where precipitation ceases and winds subside. It offers a brief but deceptive break from the extreme weather in the enormous curving wall clouds that surround it. The air within the eye gradually descends and heats by compression, making it the warmest part of the storm. Although many people believe that the eye is characterized by clear blue skies, such is usually not the case because the subsidence in the eye is seldom strong enough to produce cloudless conditions. Although the sky appears much brighter in this region, scattered clouds at various levels are common.

Hurricane Formation and Decay

A hurricane is a heat engine that is fueled by the latent heat liberated when huge quantities of water vapor condense. The amount of energy produced by a typical hurricane in just a single day is truly immense—roughly equivalent to the entire electrical energy production of the United States in a year. The release of latent heat warms the air and provides buoyancy for its upward flight. The result is to reduce the pressure near the surface, which encourages a more rapid inward flow of air. To get this engine started, a large quantity of warm, moisture-laden air is required, and a continual supply is needed to keep it going.

Hurricanes develop most often in the late summer when water temperatures have reached 27°C (80°F) or higher and thus are able to provide the necessary heat and moisture to the air. This ocean-water temperature requirement accounts for the fact that hurricanes do not form over the relatively cool waters of the South Atlantic and the eastern South Pacific. For the same reason, few hurricanes form poleward of 20 degrees of latitude. Although water temperatures are sufficiently high, hurricanes do not form within 5 degrees of the equator, because the Coriolis effect is too weak to initiate the necessary rotary motion.

Many tropical storms begin as disorganized arrays of clouds and thunderstorms that develop weak pressure gradients but exhibit little or no rotation. Such areas of low-level convergence and lifting are called *tropical disturbances*. Most of the time these zones of convective activity die out. However, occasionally tropical disturbances grow larger and develop a strong cyclonic rotation.

What happens on those occasions when conditions favor hurricane development? As latent heat is released from the clusters of thunderstorms that make up the tropical disturbance, areas within the disturbance get warmer. As a consequence, air density lowers and surface pressure drops, creating a region of weak low pressure and cyclonic circulation. As pressure drops at the storm center, the pressure gradient steepens. If you were watching an animated weather map of the storm, you would see the isobars get closer together. In response, surface wind speeds increase and bring additional supplies of moisture to nurture storm growth. The water vapor condenses, releasing latent heat, and the heated air rises. Adiabatic cooling of rising air triggers more condensation and the release of more latent heat, which causes a further increase in buoyancy. And so it goes.

Meanwhile, at the top of the storm, air is diverging. Without this outward flow up top, the inflow at lower levels would soon raise surface pressures (that is, "fill in" the low) and thwart storm development.

Many tropical disturbances occur each year, but only a few develop into full-fledged hurricanes. By international agreement, lesser tropical cyclones are given different names based on the strength of their winds. When a cyclone's strongest winds do not exceed 61 kilometers (38 miles) per hour, it is called a **tropical depression.** When winds are between 61 and 119 kilometers (38 and 74 miles) per hour, the cyclone is termed a **tropical storm.** It is during this phase that a name is given (Andrew, Mitch, Opal, etc.). Should the tropical storm become a hurricane, the name remains the same. Each year between 80 and 100 tropical storms develop around the world. Of them, usually half or more eventually reach hurricane status.

Hurricanes diminish in intensity whenever they (1) move over ocean waters that cannot supply warm, moist tropical air, (2) move onto land; or (3) reach a location where the large-scale flow aloft is unfavorable. Whenever a hurricane moves onto land, it loses its punch rapidly (see Figure 19.24). The most important reason for this rapid demise is the fact that the storm's source of warm, moist air is cut off. When an adequate supply of water vapor does not exist, condensation and the release of latent heat must diminish. In addition, friction from the increased roughness of the land surface rapidly slows surface wind speeds. This factor causes the winds to move more directly into the center of the low, thus helping to eliminate the large pressure differences.

▲ **FIGURE 19.26** In addition to satellites, reconnaissance aircraft are another important tool used to evaluate hurricanes. When a hurricane is within range, specially instrumented aircraft can fly directly into a threatening storm and accurately measure details of its position and current state of development. Data can be transmitted directly from an aircraft in the midst of a storm to the forecast center. In the Atlantic basin, most operational hurricane reconnaissance is carried out by a special U.S. Air Force squadron based at Keesler AFB, Mississippi. NOAA's P3 Orion aircraft shown here and state-of-the-art high-altitude jet aircraft compliment the reconnaissance function of the U.S. Air Force. (Rafael Perez/ Reuters/Getty Images, Inc.—Liaison)

casts, watches, and warnings. Tropical storms and hurricanes can last a week or longer, and two or more storms can be occurring in the same region at the same time. Thus, names can reduce the confusion about what storm is being described.

The World Meteorological Organization (affiliated with the United Nations) creates the lists of names. The names for Atlantic storms are used again at the end of a six-year cycle unless a hurricane was particularly noteworthy. Such names are retired to prevent confusion when storms are discussed in future years.

STUDENTS SOMETIMES ASK...

Why are hurricanes given names, and who picks the names?

Actually, the names are given once the storms reach tropical-storm status (winds between 61–119 kilometers per hour). Tropical storms are named to provide ease of communication between forecasters and the general public regarding fore-

Hurricane Destruction

A location only a few hundred kilometers from a hurricane—just one day's striking distance away—may experience clear skies and virtually no wind. Prior to the age of weather satellites, such a situation made the task of warning people of impending storms very difficult.

Once a storm develops cyclonic flow and the spiraling bands of clouds characteristic of a hurricane, it receives continuous monitoring (Figure 19.26). When hurricanes form, satellites are able to identify and

TABLE 19.2	Saffir–Simpson hurricane scale.			
Scale Number (category)	Central Pressure (millibars)	Winds (km/hr)	Storm Surge (meters)	Damage
1	≥980	119–153	1.2–1.5	Minimal
2	965–979	154–177	1.6–2.4	Moderate
3	945–964	178–209	2.5–3.6	Extensive
4	920–944	210–250	3.7–5.4	Extreme
5	<920	>250	>5.4	Catastrophic

track the storms long before they make landfall. In the United States, early warning systems have greatly reduced the number of deaths caused by hurricanes. At the same time, however, there has been an astronomical rise in the amount of property damage. The primary reason for this trend has been the rapid population growth in coastal areas.

Although the amount of damage caused by a hurricane depends on several factors, including the size and population density of the area affected and the shape of the ocean bottom near the shore, certainly the most significant factor is the strength of the storm itself. Based on the study of past storms, the *Saffir–Simpson scale* was established to rank the relative intensities of hurricanes (Table 19.2). Predictions of hurricane severity and damage are usually expressed in terms of this scale. When a tropical storm becomes a hurricane, the National Weather Service assigns it a scale (category) number. As conditions change, the category of the storm is reevaluated so that public-safety officials can be kept informed. By using the Saffir–Simpson scale, the disaster potential of a hurricane can be monitored and appropriate precautions can be planned and implemented. A rating of 5 on the scale represents the worst storm possible, and a 1 is least severe. Storms that fall into category 5 are rare. Only three such storms hit the United States in the twentieth century. One was Hurricane Camille in 1969 (Figure 19.27). Another was Hurricane Andrew in August 1992.*

Damage caused by hurricanes can be divided into three categories: (1) storm surge, (2) wind damage, and (3) inland flooding.

Storm Surge The most devastating damage in the coastal zone is caused by the storm surge. It not only accounts for a large share of coastal property losses but is also responsible for 90 percent of all hurricane-caused deaths. A **storm surge** is a dome of water 65 to 80 kilometers (40 to 50 miles) wide that sweeps across the coast near the point where the eye makes landfall. If all wave activity were smoothed out, the storm surge is the height of the water above normal tide level. Thus, a storm surge commonly adds 2 to 3 meters (6 to 10 feet) to normal tide heights—to say nothing of tremendous wave activity superimposed atop the surge.

In the delta region of Bangladesh, for example, the land is mostly less than 2 meters (6.5 feet) above sea level. When a storm surge superimposed upon normal high tide inundated that area on November 13, 1970,

---

*In August 2002, after a reevaluation by the National Hurricane Center, the intensity of Hurricane Andrew was officially changed from category 4 to category 5.

A.

B.

▲ **FIGURE 19.27** In 1969, Hurricane Camille hit the coast of Mississippi. It was a rare category 5 storm. These classic photos document the devastating force of the hurricane's 25-foot storm surge at Pass Christian, Mississippi. **A.** The Richelieu Apartments before the hurricane. This substantial-looking three-story building was directly across the highway from the beach. **B.** The same apartments after the hurricane. (Photos by Chauncey T. Hinman)

the official death toll was 200,000; unofficial estimates ran to 500,000. This was one of the worst disasters of modern times. In May 1991 a similar event again struck Bangladesh. This time the storm took the lives of at least 135,000 people and devastated coastal villages in its path.

Wind Damage Destruction caused by wind is perhaps the most obvious of the classes of hurricane damage. For some structures, the force of the wind is sufficient to cause total ruin. Mobile homes are particularly vulnerable. In addition, the strong winds can create a dangerous barrage of flying debris. In regions with good building codes, wind damage is usually not as catastrophic as storm-surge damage. However, hurricane-force winds affect a much larger area than storm surge and can cause huge economic losses.

For example, winds associated with Hurricane Andrew in 1992 produced more than $20 billion of damage in southern Florida and Louisiana. It was the costliest natural disaster in U.S. history. Tornadoes are spawned by many hurricanes that strike the United States and represent another aspect of wind damage. Although they can be very destructive locally, tornadoes generally account for only a small percentage of the total storm damage.

Inland Flooding. The torrential rains that accompany most hurricanes represent a third significant threat—flooding. Whereas the effects of storm surge and strong winds are concentrated in coastal areas, heavy rains may affect places hundreds of kilometers from the coast for several days after the storm has lost its hurricane-force winds.

Hurricanes weaken rapidly as they move inland, yet the remnants of the storm can still yield 15 to 30 centimeters (6 to 12 inches) or more of rain as they move inland. A good example of such destruction was Hurricane Floyd. In September 1999, this storm dumped more than 48 centimeters (19 inches) of rain on Wilmington, North Carolina, 33.98 centimeters (13.38 inches) of it in a single 24-hour span. In August and September 2004, when the remnants of hurricanes Charley, Frances, Ivan, and Jeanne moved northward from Florida and the Gulf Coast, they brought huge rains and floods from Alabama and Georgia northward through the Carolinas and beyond.

To summarize, extensive damage and loss of life in the coastal zone can result from storm surge, torrential rains, and strong winds. When loss of life occurs, it is commonly caused by the storm surge, which can devastate entire barrier islands and low-lying land along the coast. Although wind damage is usually not as catastrophic as the storm surge, it affects a much larger area. Where building codes are inadequate, economic losses can be especially severe. Because hurricanes weaken as they move inland, most wind damage occurs within 200 kilometers of the coast. Far from the coast a weakening storm can produce extensive flooding long after the winds have diminished below hurricane levels. Sometimes the damage from inland flooding exceeds storm-surge destruction.

STUDENTS SOMETIMES ASK...
What is the deadliest hurricane to strike the United States?

That distinction goes to a hurricane that struck an unprepared Galveston, Texas, on September 8, 1900. In fact, it was the deadliest natural disaster *of any kind* in U.S. history. The strength of the storm, Galveston's vulnerable coastal setting, and the lack of adequate warning were a deadly combination. The storm took the lives of 6000 people in the city and at least 2000 more elsewhere. Fortunately, due to an effective warning system, hurricanes are no longer the unheralded killers they once were.

Chapter Summary

- An *air mass* is a large body of air, usually 1600 kilometers (1000 miles) or more across, which is characterized by a *sameness of temperature and moisture* at any given altitude. When this air moves out of its region of origin, called the *source region*, it will carry these temperatures and moisture conditions elsewhere, perhaps eventually affecting a large portion of a continent.

- Air masses are classified according to (1) the nature of the surface in the source region and (2) the latitude of the source region. *Continental (c)* designates an air mass of land origin, with the air likely to be dry; whereas a *maritime (m)* air mass originates over water, and therefore will be humid. *Polar (P)* air masses originate in high latitudes and are cold. *Tropical (T)* air masses form in low latitudes and are warm. According to this classification scheme, the *four basic types of air masses are continental polar (cP), continental tropical (cT), maritime polar (mP)*, and *maritime tropical (mT)*. Continental polar (cP) and maritime tropical (mT) air masses influence the weather of North America most, especially east of the Rocky Mountains. Maritime tropical air is the source of much, if not most, of the precipitation received in the eastern two-thirds of the United States.

- *Fronts* are boundaries that separate air masses of different densities, one warmer and often higher in moisture content than the other. A *warm front* occurs when the surface position of the front moves so that warm air occupies territory formerly covered by cooler air. Along a warm front, a warm air mass overrides a retreating mass of cooler air. As the warm air ascends, it cools adiabatically to produce clouds and, frequently, light-to-moderate precipitation over a large area. A *cold front* forms where cold air is actively advancing into a region occupied by warmer air. Cold fronts are about twice as steep and move more rapidly than do warm fronts. Because of these two differences, precipitation along a cold front is usually more intense and of shorter duration than is precipitation associated with a warm front.

- The primary weather producers in the middle latitudes are *large centers of low pressure* that generally travel from *west to east*, called *middle-latitude cyclones*. These *bearers of stormy weather*, which last from a few days to a week, have a *counterclockwise circulation* pattern in the Northern Hemisphere, with an *inward flow of air* toward their centers. Most middle-latitude cyclones have a *cold front and frequently a warm front* extending from the central area of low pressure. *Convergence and forceful lifting along the fronts* initiate cloud development and frequently cause precipitation. As a middle-latitude cyclone with its associated fronts passes over a region, it often brings with it abrupt changes in the weather. The particular weather experienced by an area depends on the path of the cyclone.

- *Thunderstorms* are caused by the upward movement of warm, moist, unstable air, triggered by a number of different processes. They are associated with cumulonimbus clouds that generate heavy rainfall, thunder, lightning, and occasionally hail and tornadoes.

- *Tornadoes*, destructive, local storms of short duration, are violent windstorms associated with severe thunderstorms that take the form of a rotating column of air that extends downward from a cumulonimbus cloud. Tornadoes are most often spawned along the cold front of a middle-latitude cyclone, most frequently during the spring months.

- *Hurricanes*, the greatest storms on Earth, are tropical cyclones with wind speeds in excess of 119 kilometers (74 miles) per hour. These complex tropical disturbances develop over tropical ocean waters and are fueled by the latent heat liberated when huge quantities of water vapor condense. Hurricanes form most often in late summer when ocean-surface temperatures reach 27°C (80°F) or higher and thus are able to provide the necessary heat and moisture to the air. Hurricanes diminish in intensity whenever they (1) move over cool ocean water that cannot supply adequate heat and moisture, (2) move onto land, or (3) reach a location where large-scale flow aloft is unfavorable.

Key Terms

air-mass (p. 528)
air-mass weather (p. 529)
cold front (p. 532)
continental (c) air mass (p. 529)
Doppler radar (p. 548)
eye (p. 551)
eye wall (p. 550)
front (p. 531)
hurricane (p. 549)
lake-effect snow (p. 529)

maritime (m) air mass (p. 529)
middle-latitude or midlatitude cyclone (p. 534)
occluded front (p. 533)
occlusion (p. 536)
overrunning (p. 531)
polar (P) air mass (p. 529)
source region (p. 529)
stationary front (p. 533)

storm surge (p. 553)
thunderstorm (p. 540)
tornado (p. 542)
tornado warning (p. 548)
tornado watch (p. 548)
tropical (T) air mass (p. 529)
tropical depression (p. 552)
tropical storm (p. 552)
warm front (p. 532)

Review Questions

1. Describe the weather associated with a continental polar air mass in the winter and in the summer. When would this air mass be most welcome in the United States?

2. What are the characteristics of a maritime tropical air mass? Where are the source regions for the maritime tropical air masses that affect North America? Where are the source regions for the maritime polar air masses?

3. Describe the weather along a cold front where very warm, moist air is being displaced.

4. The formation of an occluded front marks the beginning of the end of a middle-latitude cyclone. Why is this true?

5. For each of the weather elements that follow, describe the changes that an observer experiences when a middle-latitude cyclone passes with its center north of the observer: wind direction, pressure tendency, cloud type, cloud cover, precipitation, temperature.

6. Describe the weather conditions an observer would experience if the center of a middle-latitude cyclone passed to the south.

7. Briefly explain how the flow aloft aids the formation of cyclones at the surface.

8. Compare the wind speeds and sizes of middle-latitude cyclones, tornadoes, and hurricanes. How are thunderstorms related to each?

9. What is the primary requirement for the formation of thunderstorms?

10. Based on your answer to Question 9, where would you expect thunderstorms to be most common on Earth? In the United States?

11. Why do tornadoes have such high wind speeds?

12. What general atmospheric conditions are most conducive to the formation of tornadoes?

13. When is the tornado season? That is, during what months is tornado activity most pronounced?

14. Distinguish between a tornado watch and a tornado warning.

15. Which has stronger winds, a tropical storm or a tropical depression?

16. Refer to Figure 19.24 to answer these questions.
    (a) On which of the two days were Fran's wind speeds probably highest? How were you able to determine this?
    (b) How far (in miles) did the center of the hurricane move in the 24-hour period represented by these maps? At what rate (in miles per hour) did the storm move during this 24-hour span?

17. Hurricane damage can be divided into three broad categories. Name them. Which category is responsible for the highest percentage of hurricane-related deaths?

18. Great damage and significant loss of life can take place a day or more after a hurricane has moved ashore and weakened. When this occurs, what is the likely cause?

19. A hurricane has slower wind speeds than a tornado, yet it inflicts more total damage. How might this be explained?

Examining the Earth System

1. Which parts of the Earth system interact in the Great Lakes region of North America to produce the high snowfalls on the downwind shores of the lakes? What term is applied to these heavy snows? Describe what creates this effect. (A good short review of this process can be found at **http://www.usatoday.com/weather/lakeeffect.htm**

2. Hurricanes are among the most severe storms experienced on Earth. When a hurricane makes landfall in the southeastern United States, what impact might it have on coastal lands, drainage networks (hydrosphere), and natural vegetation (biosphere)?

Online Study Guide

The *Earth Science* Website uses the resources and flexibility of the Internet to aid in your study of the topics in this chapter. Written and developed by Earth science instructors, this site will help improve your understanding of Earth science. Visit **http://www.prenhall.com/tarbuck** and click on the cover of *Earth Science 11e* to find:

- **Online review quizzes.**
- **Critical thinking exercises.**
- **Links to chapter-specific Web resources.**
- **Internet-wide key term searches.**

http://www.prenhall.com/tarbuck

Old-growth forest in Washington's Mount Rainier National Park. The distribution of natural vegetation is strongly influenced by climate. (Photo by Charles Gurche)

CHAPTER 20

Climate

Anyone who has the opportunity to travel around the world will find such an incredible variety of climates that it is hard to believe they could all occur on the same planet. The broad diversity of climates around the globe is an important focus of this chapter. Climate strongly influences the nature of plant and animal life, the soil, and many external geological processes. Climate influences people as well.

Although climate has a significant impact on people, we are learning that people also have a strong influence on climate. The latter portion of this chapter examines the ways in which humans may be changing global climate.

The Climate System

The focus of this chapter is *climate*. In Chapter 16, where this term was introduced, climate was characterized as being an aggregate of weather. It consists not only of average atmospheric values but also involves the variability of elements and information on the occurrence of extreme events.

To understand and appreciate climate, it is important to realize that climate involves more than just the atmosphere.

> The atmosphere is the central component of the complex, connected, and interactive global environmental system upon which all life depends. Climate may be broadly defined as the long-term behavior of this environmental system. To understand fully and to predict changes in the atmospheric component of the climate system, one must understand the Sun, oceans, ice sheets, solid Earth, and all forms of life.*

Indeed, we must recognize that there is a **climate system** that includes the atmosphere, hydrosphere, geosphere, biosphere, and cryosphere. (The *cryosphere* is the ice and snow that exist at Earth's surface.) Powered by energy from the Sun, the climate system involves the exchanges of energy and moisture that occur among the five spheres. These exchanges link the atmosphere to the other parts of the system to produce an integrated and extremely complex interactive unit. The major components of the climate system are shown in Figure 20.1.

The climate system provides a framework for the study of climate. The interactions and exchanges among the parts of the climate system create a complex network that links the five spheres. Changes to the system do not occur in isolation. Rather, when one

part of this interactive unit changes, the other components also react. This well-established relationship is demonstrated often when we study the world's climates and global climate change.

World Climates

Previous chapters have already presented the spatial and seasonal variations of the major elements of weather and climate. Chapter 16 examined the controls of temperature and the world distribution of temperature. In Chapter 18, you studied the general circulation of the atmosphere and the global distribution of precipitation. You are now ready to investigate the *combined* effects of these variations in different parts of the world. The varied nature of Earth's surface and the many interactions that occur among atmospheric processes give every location on our planet a distinctive, even unique, climate. Our intention, however, is not to describe the unique climatic character of countless different locales. Instead, the purpose is to introduce the major climatic regions of the world. The discussion examines large areas and uses particular places only to illustrate the characteristics of these major climatic regions.

Temperature and precipitation are the most important elements in a climatic description because they have the greatest influence on people and their activities and also have an important impact on the distribution of such phenomena as vegetation and soils. Nevertheless, other factors are also important for a complete climatic description. When possible, some of these factors are introduced into our discussion of world climates.

Climate Classification

The distribution of the major atmospheric elements is, to say the least, complex. Because of the many differences from place to place as well as from time to time at a particular locale, it is unlikely that any two sites on Earth experience exactly the same weather conditions. The fact that the number of places on Earth is virtually infinite makes it readily apparent that the number of different climates must be extremely large.

Of course, having a great diversity of information to investigate is not unique to the study of the atmosphere; it is a problem that is basic to all science. To cope with such variety, it is not only desirable but essential to devise some means of classifying the vast array of data to be studied. By establishing groups consisting of items that have certain important characteristics in common, order and simplicity are introduced. Bringing order to large quantities of information not only

---

*The American Meteorological Society and the University Corporation for Atmospheric Research, "Weather and the Nation's Well-Being," *Bulletin of the American Meteorological Society*, 73, no. 12 (December 1991) 2038.

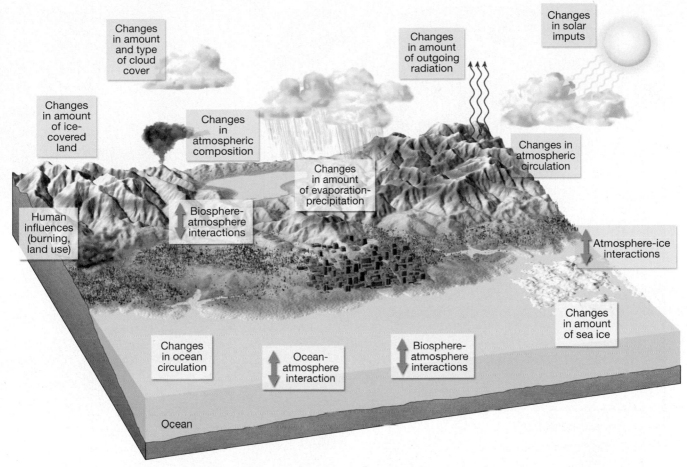

▲ **FIGURE 20.1** Schematic view showing several components of Earth's climate system. Many interactions occur among the various components on a wide range of space and time scales, making the system extremely complex.

aids comprehension and understanding but also facilitates analysis and explanation.

Over the past 100 years, many climate-classification systems have been devised. It should be remembered that the classification of climates (or of anything else) is not a natural phenomenon but the product of human ingenuity. The value of any particular classification is determined largely by its intended use. A system designed for one purpose is not necessarily applicable to another.

In this chapter, we use a classification devised by Russian-born German climatologist Wladimir Köppen (1846–1940). As a tool for presenting the general world pattern of climates, the **Köppen classification** has been the best-known and most used system for decades. It is widely accepted for many reasons. For one, it uses only easily obtained data: mean monthly and annual values of temperature and precipitation. Furthermore, the criteria are unambiguous, relatively simple to apply, and divide the world into climatic regions in a realistic way.

Köppen believed that the distribution of natural vegetation was an excellent expression of the totality of climate. Consequently, the boundaries he chose were largely based on the limits of certain plant associations. Five principal groups were recognized; each group was designated by a capital letter as follows:

A. *Humid tropical.* Winterless climates; all months having a mean temperature above 18°C (64°F).

B. *Dry.* Climates where evaporation exceeds precipitation; there is a constant water deficiency.

C. *Humid middle-latitude.* Mild winters; the average temperature of the coldest month is below 18°C (64°F) but above −3°C (27°F).

D. *Humid middle-latitude.* Severe winters; the average temperature of the coldest month is below −3°C (27°F), and the warmest monthly mean exceeds 10°C (50°F).

E. *Polar.* Summerless climates; the average temperature of the warmest month is below 10°C (50°F).

Notice that four of the major groups (A, C, D, E) are defined on the basis of temperature characteristics, and the fifth, the B group, has precipitation as its primary criterion. Each of the five groups is further subdivided by using the criteria and symbols presented in Table 20.1. In addition, the world distribution of climates according to the Köppen classification is shown in Figure 20.2. You will be referred to this figure several times as Earth's climates are discussed in the following pages.

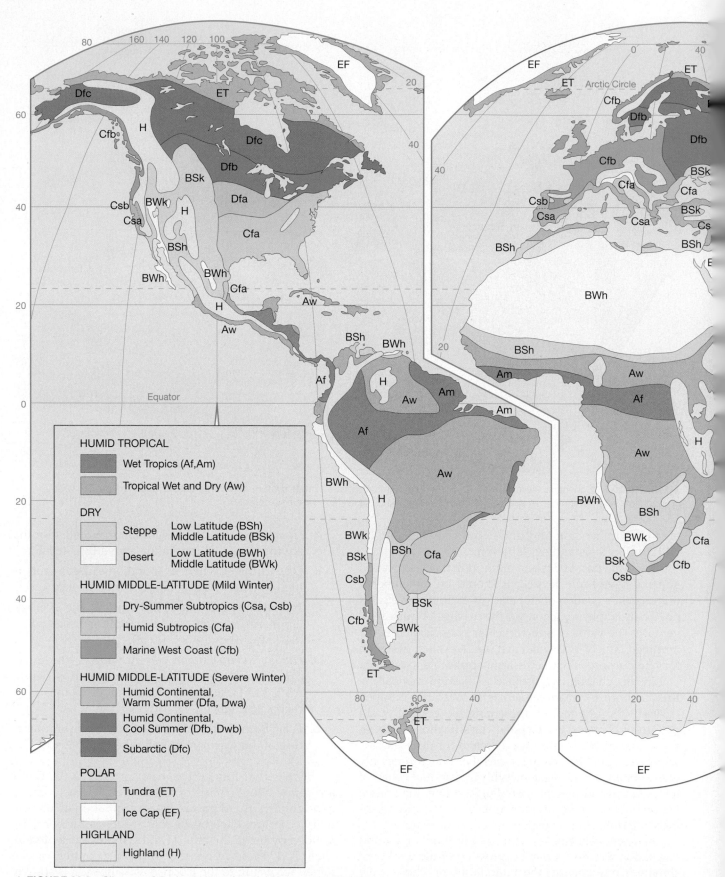

▲ FIGURE 20.2 Climates of the world based on the Köppen classification.

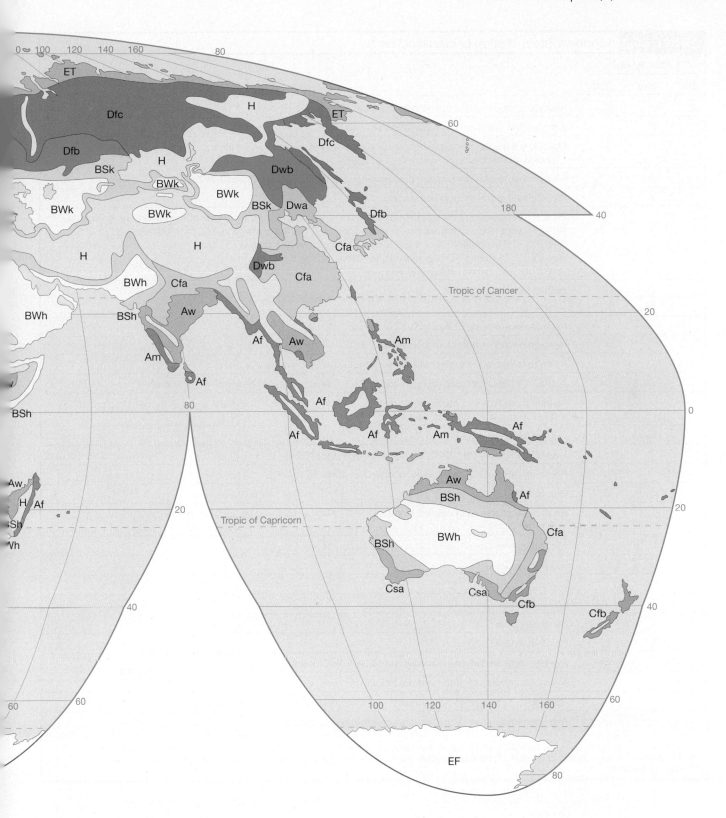

Humid Tropical (A) Climates

Within the A group of climates, two main types are recognized—wet tropical climates (Af and Am) and tropical wet and dry (Aw).

The Wet Tropics

The constantly high temperatures and year-round rainfall in the wet tropics combine to produce the most luxuriant vegetation found in any climatic realm: the **tropical rain forest** (Figure 20.3).

TABLE 20.1 Köppen system of climate classification.*

1st	2nd	3rd	
Letter Symbol			
A			Average temperature of the coldest month is 18°C or higher.
	f		Every month has 6 cm of precipitation or more.
	m		Short, dry season; precipitation in driest month less than 6 cm but equal to or greater than 10–R/25 (R is annual rainfall in cm).
	w		Well-defined winter dry season; precipitation in driest month less than 10–R/25.
	s		Well-defined summer dry season (rare).
B			Potential evaporation exceeds precipitation. The dry–humid boundary is defined by the following formulas: (Note: R is the average annual precipitation in cm and T is the average annual temperature in °C.) R < 2T + 28 when 70% or more of rain falls in warmer 6 months. R < 2T when 70% or more of rain falls in cooler 6 months. R < 2T + 14 when neither half year has 70% or more of rain.
	S		Steppe ___ The BS–BW boundary is 1/2 the dry–humid boundary.
	W		Desert
		h	Average annual temperature is 18°C or greater.
		k	Average annual temperature is less than 18°C.
C			Average temperature of the coldest month is under 18°C and above −3°C.
	w		At least 10 times as much precipitation in a summer month as in the driest winter month.
	s		At least three times as much precipitation in a winter month as in the driest summer month; precipitation in driest summer month less than 4 cm.
	f		Criteria for w and s cannot be met.
		a	Warmest month is over 22°C; at least 4 months over 10°C.
		b	No month above 22°C; at least 4 months over 10°C.
		c	One to 3 months above 10°C.
D			Average temperature of coldest month is −3°C or below; average temperature of warmest month is greater than 10°C.
	s		Same as under C.
	w		Same as under C.
	f		Same as under C.
		a	Same as under C.
		b	Same as under C.
		c	Same as under C.
		d	Average temperature of the coldest month is −38°C or below.
E			Average temperature of the warmest month is below 10°C.
	T		Average temperature of the warmest month is greater than 0°C and less than 10°C.
	F		Average temperature of the warmest month is 0°C or below.

*When classifying climatic data using Table 20.1, you should first determine whether the data meet the criteria for the E climates. If the station is not a polar climate, proceed to the criteria for B climates. If your data do not fit into either the E or B groups, check the data against the criteria for A, C, and D climates, in that order.

The environment of the wet tropics characterizes almost 10 percent of Earth's land area. An examination of Figure 20.2 shows that Af and Am climates form a discontinuous belt astride the equator that typically extends 5° to 10° into each hemisphere. The poleward margins are most often marked by diminishing rainfall, but occasionally decreasing temperatures mark the boundary. Because of the general decrease in temperature with height in the troposphere, this climate region is restricted to elevations below 1000 meters. Consequently, the major interruptions near the equator are principally cooler highland areas.

Data for some representative stations in the wet tropics are shown in Figure 20.4A,B. A brief examination

▲ **FIGURE 20.3** Unexcelled in luxuriance and characterized by hundreds of different species per square kilometer, the tropical rain forest is a broadleaf evergreen forest that dominates the wet tropics. Tutch River region, Sarawak, Malaysia. (Photo by Art Wolfe/Photo Researchers, Inc.)

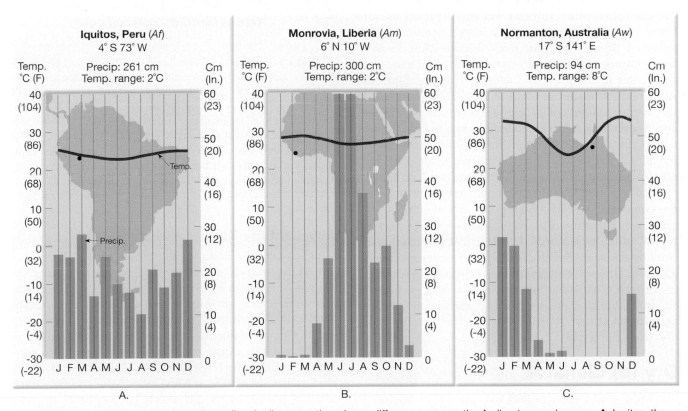

▲ **FIGURE 20.4** By comparing these three climatic diagrams, the primary differences among the A climates can be seen. **A.** Iquitos, the Af station, is wet throughout the year. **B.** Monrovia, the Am station, has a short, dry season. **C.** As is true for all Aw stations, Normanton has an extended dry season and a higher annual temperature range than the others.

reveals the most obvious features that characterize the climate in these areas.

1. Temperatures usually average 25°C (77°F) or more each month. Consequently, not only is the annual mean temperature high, but the annual temperature range is also very small.

2. The total precipitation for the year is high, often exceeding 200 centimeters (80 inches).

3. Although rainfall is not evenly distributed throughout the year, tropical rain forest stations are generally wet in all months. If a dry season exists, it is very short.

Because places with an Af or Am designation lie near the equator, the reason for the uniform temperature rhythm experienced in such locales is clear: The intensity of solar radiation is consistently high. The vertical rays of the Sun are always relatively close, and changes in the length of daylight throughout the year are slight; therefore, seasonal temperature variations are minimal.

The region is strongly influenced by the equatorial low. Its converging trade winds and the accompanying ascent of warm, humid, unstable air produce conditions that are ideal for the formation of precipitation.

Tropical Wet and Dry

In the latitude zone poleward of the wet tropics and equatorward of the subtropical deserts lies a transitional climatic region called **tropical wet and dry**. Here the rain forest gives way to the *savanna*, a tropical grassland with scattered drought-tolerant trees (Figure 20.5). Because temperature characteristics among all A climates are quite similar, the primary factor that distinguishes the Aw climate from Af and Am is precipitation. Although the overall amount of precipitation in the tropical wet and dry realm is often considerably less than in the wet tropics, the most distinctive feature of this climate is not the annual rainfall total but the markedly seasonal character of the rainfall. The climatic diagram for Normanton, Australia (Figure 20.4C), clearly illustrates this trait. As the equatorial low advances poleward in summer, the rainy season commences and features weather patterns typical of the wet tropics. Later, with the retreat of the equatorial low, the subtropical high advances into the region and brings with it pronounced dryness. In some Aw regions such as India, Southeast Asia, and portions of Australia, the alternating periods of rainfall and dryness are associated with a well-established monsoon circulation (see Chapter 18).

Dry (B) Climates

It is important to realize that the concept of dryness is a relative one and refers to any situation in which a water deficiency exists. Climatologists define a dry climate as one in which the yearly precipitation is not as great as the potential loss of water by evaporation. Thus, dryness is not only related to annual rainfall totals but is also a function of evaporation, which in turn is closely dependent upon temperature.

To establish the boundary between dry and humid climates, the Köppen classification uses formulas that involve three variables: average annual precipitation, average annual temperature, and seasonal distribution of precipitation. The use of average annual temperature reflects its importance as an index of evaporation. The amount of rainfall defining the humid–dry boundary increases as the annual mean temperature increases. The use of seasonal precipitation as a variable is also related to this idea. If rain is concentrated in the warmest months, loss to evaporation is greater than if the precipitation were concentrated in the cooler months.

Within the regions defined by a general water deficiency are two climatic types: **arid** or **desert** (BW) and **semiarid** or **steppe** (BS) (Figure 20.6). These two groups have many features in common; their differences are primarily a matter of degree. The semiarid is a marginal and more humid variant of the arid and represents a transition zone that surrounds the desert and separates it from the bordering humid climates.

Low-Latitude Deserts and Steppes

The heart of low-latitude dry climates lies in the vicinities of the Tropics of Cancer and Capricorn. A glance at Figure 20.6 shows a virtually unbroken desert environment stretching for more than 9300 kilometers (nearly 6000 miles) from the Atlantic coast of North Africa to the dry lands of northwestern India. In addition to this single great expanse, the Northern Hemisphere contains another, much smaller area of subtropical desert and steppe in northern Mexico and the southwestern United States. In the Southern Hemisphere, dry climates dominate Australia. Almost 40 percent of the continent is desert, and much of the remainder is steppe. In addition, arid and semiarid areas are found

▲ **FIGURE 20.5** Tropical savanna grassland in Tanzania's Serengeti National Park. This tropical savanna, with its stunted, drought-resistant trees, probably resulted from seasonal burnings carried out by native human populations. (Photo by Stan Osolinski/Dembinsky Photo Associates)

in southern Africa and make a limited appearance in coastal Chile and Peru.

The existence of this dry subtropical realm is primarily the result of the prevailing global distribution of air pressure and winds. Earth's low-latitude deserts and steppes coincide with the subtropical high-

pressure belts (see Figures 6.24, p. 173 and 18.15, p. 512). Here, air is subsiding. When air sinks, it is compressed and warmed. Such conditions are just opposite of what is needed for cloud formation and precipitation. Therefore, clear skies, a maximum of sunshine, and drought are to be expected. The climate diagrams for Cairo,

▼ **FIGURE 20.6** Arid and semiarid climates cover about 30 percent of Earth's land surface. No other climatic group covers so large an area.

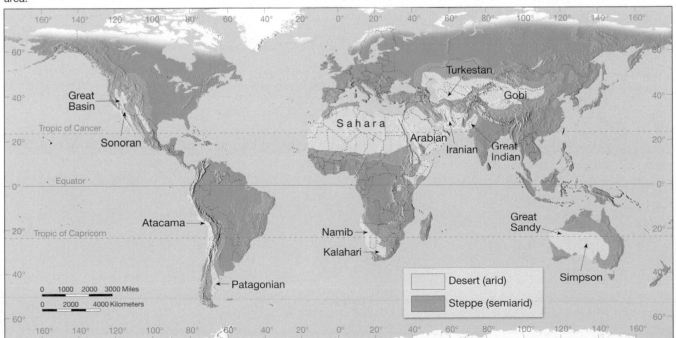

Egypt, and Monterrey, Mexico (Figure 20.7A,B), illustrate the characteristics of low-latitude dry climates.

Actually, this is a misconception. Deserts can certainly be hot places. In fact, the record high temperature for the United States, 57°C (134°F), was set at Death Valley, California. Despite this and other remarkably high figures, deserts also experience cold temperatures. For example, the average daily minimum in January at Phoenix, Arizona, is 1.7°C (35°F), just barely above freezing. At Ulan Bator in Mongolia's Gobi Desert, the average *high* temperature in January is only −19°C (−2°F)! Although subtropical deserts lack a cold season, midlatitude deserts do experience seasonal temperature changes.

Middle-Latitude Deserts and Steppes

Unlike their low-latitude counterparts, middle-latitude deserts and steppes are not controlled by the subsiding air masses associated with high pressure. Instead, these dry lands exist principally because of their positions in the deep interiors of large landmasses far removed from the oceans, which are the ultimate source of moisture for cloud formation and precipitation. In addition, the presence of high mountains across the paths of prevailing winds further acts to separate these areas from waterbearing, maritime air masses.

Windward sides of mountains are often wet. As prevailing winds meet mountain barriers, the air is forced to ascend, producing clouds and precipitation. By contrast, the leeward sides of mountains are usually much dryer and are often arid enough to be referred to as **rain shadow deserts.** Because many middle-latitude deserts occupy sites on the leeward sides of the mountains, they can also be classified as rain shadow deserts (Figure 20.7C). In North America, the Coast Ranges, Sierra Nevada, and Cascades are the foremost mountain barriers. In Asia, the great Himalayan chain prevents the summertime monsoon flow of moist Indian Ocean air from reaching the interior. Because the Southern Hemisphere lacks extensive land areas in the middle latitudes, only a small area of desert and steppe is found in this latitude range, existing primarily in the rain shadow of the towering Andes.

In the case of middle-latitude deserts, we have an example of the impact of tectonic processes on climate. Rain shadow deserts exist by virtue of the mountains produced when plates collide. Without such mountain-building episodes, wetter climates would prevail where many dry regions exist today.

▼ **FIGURE 20.7** Climatic diagrams for representative arid and semiarid stations. Stations **A.** and **B.** are in the subtropics, whereas **C.** is in the middle-latitudes. Cairo and Lovelock are classified as deserts; Monterrey is a steppe. Lovelock, Nevada, may also be called a rain shadow desert.

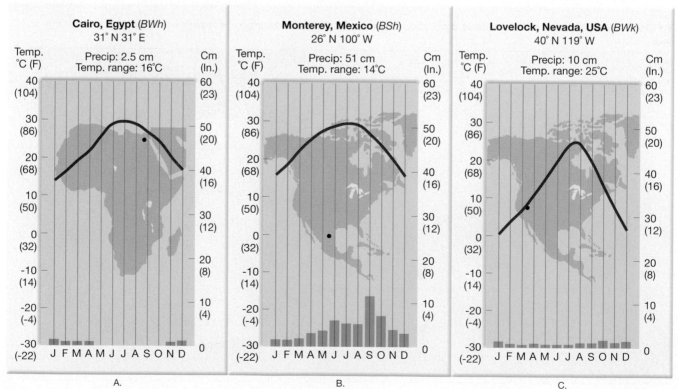

A.

B.

C.

Humid Middle-Latitude Climates with Mild Winters (C Climates)

Although the term *subtropical* is often used for the C climates, it can be misleading. Although many areas with C climates do indeed possess some near-tropical characteristics, other regions do not. For example, we would be stretching the use of the term *subtropical* to describe the climates of coastal Alaska and Norway, which belong to the C group. Within the C group of climates, several subgroups are recognized.

Humid Subtropics

Located on the eastern sides of the continents, in the 25- to 40-degree latitude range, the **humid subtropical climate** dominates the southeastern United States, as well as other similarly situated areas around the world (Figures 20.2 and 20.8A). In the summer, the humid subtropics experience hot, sultry weather of the type one expects to find in the rainy tropics. Daytime temperatures are generally high, and because both mixing ratio and relative humidity are high, the night brings little relief. An afternoon or evening thunderstorm is also possible, for these areas experience such storms on an average of 40 to 100 days each year, the majority during the summer months.

As summer turns to autumn, the humid subtropics lose their similarity to the rainy tropics. Although winters are mild, frosts are common in the higher-latitude Cfa areas and occasionally plague the tropical margins as well. The winter precipitation is also different in character from the summer. Some is in the form of snow, and most is generated along fronts of the frequent middle-latitude cyclones that sweep over these regions.

Marine West Coast

Situated on the western (windward) side of continents, from about 40 to 65 degrees north and south latitude, is a climatic region dominated by the onshore flow of oceanic air. In North America, the **marine west coast climate** extends from near the U.S.–Canadian border northward as a narrow belt into southern Alaska (Figure 20.8B). The largest area of Cfb climate is found in Europe, for here there is no mountain barrier blocking the movement of cool maritime air from the North Atlantic.

The prevalence of maritime air masses means that mild winters and cool summers are the rule, as is an ample amount of rainfall throughout the year. Although there is no pronounced dry period, there is a drop in monthly precipitation totals during the summer. The reason for the reduced summer rainfall is the poleward migration of the oceanic subtropical highs. Although the areas of marine west coast climate are

▼ **FIGURE 20.8**   Each of these climatic diagrams represents one of the three main types of C climates: **A.** humid subtropical, **B.** marine west coast, and **C.** dry-summer subtropical.

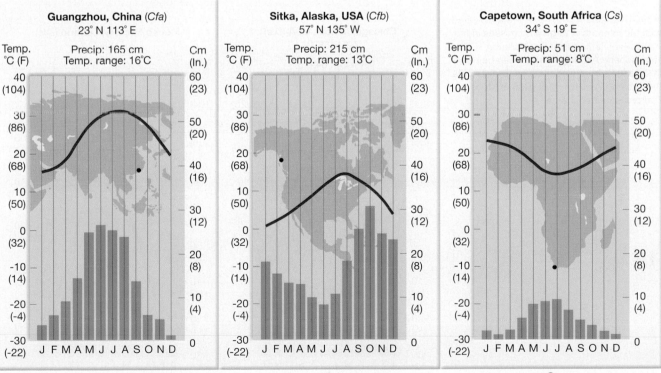

situated too far poleward to be dominated by these dry anticyclones, their influence is sufficient to cause a decrease in warm season rainfall.

Dry-Summer Subtropics

The **dry-summer subtropical climate** is typically located along the west sides of continents between latitudes 30 and 45 degrees. Situated between the marine west coast climate on the poleward side and the subtropical steppes on the equatorward side, this climatic region is best described as transitional in character. It is unique because it is the only humid climate that has a strong winter rainfall maximum, a feature that reflects its intermediate position (Figure 20.8C). In summer, the region is dominated by stable conditions associated with the oceanic subtropical highs. In winter, as the wind and pressure systems follow the Sun equatorward, it is within range of the cyclonic storms of the polar front. Thus, during the course of a year, these areas alternate between becoming a part of the dry tropics and an extension of the humid middle latitudes. Although middle-latitude changeability characterizes the winter, tropical constancy describes the summer.

As was the case for the marine west coast climate, mountain ranges limit the dry-summer subtropics to a relatively narrow coastal zone in both North and South America. Because Australia and southern Africa barely reach to the latitudes where dry-summer climates exist, the development of this climatic type is limited on these continents as well. Consequently, because of the arrangement of the continents, and of their mountain ranges, inland development occurs only in the Mediterranean basin. Here the zone of subsidence extends far to the east in summer; in winter, the sea is a major route of cyclonic disturbances. Because the dry-summer climate is particularly extensive in this region, the name *Mediterranean climate* is often used as a synonym.

Humid Middle-Latitude Climates with Severe Winters (D Climates)

The C climates that were just described characteristically have mild winters. By contrast, D climates experience severe winters. Two types of D climates are recognized, the humid continental and the subarctic. Climatic diagrams of representative locations are shown in Figure 20.9. The D climates are land-controlled climates, the result of broad continents in the middle latitudes. Because continentality is a basic feature, D climates are absent in the Southern Hemisphere where the middle-latitude zone is dominated by the oceans.

Humid Continental

The **humid continental climate** is confined to the central and eastern portions of North America and Eurasia in the latitude range between approximately 40 and 50 degrees north latitude. It may at first seem unusual

▶ **FIGURE 20.9** *D* climates are associated with the interiors of large landmasses in the mid-to-high latitudes of the Northern Hemisphere. Although winters can be harsh in Chicago's humid continental (Dfa) climate, the subarctic environment (Dfc) of Moose Factory is more extreme.

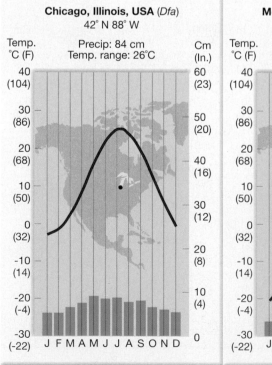

Chicago, Illinois, USA (*Dfa*)
42° N 88° W
Temp. °C (F)   Precip: 84 cm
Temp. range: 26°C   Cm (In.)

A.

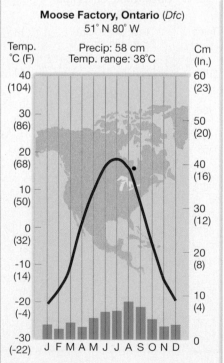

Moose Factory, Ontario (*Dfc*)
51° N 80° W
Temp. °C (F)   Precip: 58 cm
Temp. range: 38°C   Cm (In.)

B.

that a continental climate should extend eastward to the margins of the ocean. However, because the prevailing atmospheric circulation is from the west, deep and persistent incursions of maritime air from the east are not likely to occur.

Both winter and summer temperatures in the humid continental climate can be characterized as relatively severe. Consequently, annual temperature ranges are high throughout the climate.

Precipitation is generally greater in summer than in winter. Precipitation totals generally decrease toward the interior of the continents as well as from south to north, primarily because of increasing distance from the sources of mT air. Furthermore, the more northerly stations are also influenced for a greater part of the year by drier polar air masses.

Wintertime precipitation in humid continental climates is chiefly associated with the passage of fronts connected with traveling middle-latitude cyclones. Part of this precipitation is in the form of snow, the proportion increasing with latitude. Although precipitation is often considerably less during the cold season, it is usually more conspicuous than the greater amounts that fall during summer. An obvious reason is that snow remains on the ground, often for extended periods, and rain, of course, does not.

Subarctic

Situated north of the humid continental climate and south of the polar tundra is an extensive **subarctic climate** region covering broad, uninterrupted expanses from western Alaska to Newfoundland in North America and from Norway to the Pacific coast of Russia in Eurasia. It is often referred to as the *taiga* climate, for its extent closely corresponds to the northern coniferous forest region of the same name (Figure 20.10). Although scrawny, the spruce, fir, larch, and birch trees in the taiga represent the largest stretch of continuous forest on the surface of Earth.

Here in the source regions of continental polar air masses, the outstanding feature is certainly the dominance of winter. Not only is it long but temperatures are also bitterly cold. Winter minimum temperatures are among the lowest ever recorded outside the ice sheets of Greenland and Antarctica. In fact, for many years, the world's coldest temperature was attributed to Verkhoyansk in east central Siberia, where the temperature dropped to $-68°C$ ($-90°F$) on February 5 and 7, 1892. Over a 23-year period, this same station had an average monthly minimum of $-62°C$ ($-80°F$) during January. Although exceptional temperatures, they illustrate the extreme cold that envelops the taiga in winter.

By contrast, summers in the subarctic are remarkably warm, despite their short duration. However, when compared with regions farther south, this short season must be characterized as cool. The extremely cold winters and relatively warm summers combine to produce the highest annual temperature ranges on Earth. Because these far northerly continental interiors are the source regions for cP air masses, there is very limited moisture available throughout the year. Precipitation totals are therefore small, with a maximum occurring during the warmer summer months.

◄ **FIGURE 20.10** The northern coniferous forest is also called the taiga. Denali National Park, Alaska. (Photo © by Carr Clifton)

You might be surprised to learn that Alaska's record high temperature is 38°C (100°F)! The record was set June 27, 1915, at Fort Yukon, a town along the Arctic Circle in the interior of the state. It is also interesting to note that when statistics for the other states are examined, we find that all but one have record high temperatures in *excess* of 38°C. The one state that ties Alaska with the "lowest high" is Hawaii. Panaloa, on the south coast of the big island, recorded 38°C on April 27, 1931.

Polar (E) Climates

Polar climates are those in which the mean temperature of the warmest month is below 10°C (50°F). Thus, just as the tropics are defined by their year-round warmth, the polar realm is known for its enduring cold. As winters are periods of perpetual night, or nearly so, temperatures at most polar locations are understandably bitter. During the summer months temperatures remain cool despite the long days, because the Sun is so low in the sky that its oblique rays are not effective in bringing about a genuine warming. Although polar climates are classified as humid, precipitation is generally meager. Evaporation, of course, is also limited. The scanty precipitation totals are easily understood in view of the temperature characteristics of the region. The amount of water vapor in the air is always small because low mixing ratios must accompany low temperatures. Usually precipitation is most abundant during the warmer summer months when the moisture content of the air is highest.

Two types of polar climates are recognized. The **tundra climate** (ET) is a treeless climate found almost exclusively in the Northern Hemisphere (Figure 20.11). Because of the combination of high latitude and continentality, winters are severe, summers are cool, and annual temperature ranges are high (Figure 20.12A). Further, yearly precipitation is small, with a modest summer maximum.

The **ice cap climate** (EF) does not have a single monthly mean above 0°C (32°F) (Figure 20.12B). Consequently, because the average temperature for all months is below freezing, the growth of vegetation is prohibited and the landscape is one of permanent ice and snow. This climate of perpetual frost covers a surprisingly large area more than 15.5 million square kilometers (6 million square miles), or about 9 percent of Earth's land area. Aside from scattered occurrences in high mountain areas, it is confined to the ice sheets of Greenland and Antarctica (see Figure 6.1, p. 155).

Highland Climates

It is a well-known fact that mountains have climate conditions that are distinctively different from those found in adjacent lowlands. Compared to nearby places at lower elevations, sites with **highland climates** are cooler and usually wetter. Unlike the world climate types already discussed, which consist of large, relatively homogeneous regions, the outstanding characteristic of highland climates is the great diversity of climatic conditions that occur.

The best-known climatic effect of increased altitude is lower temperatures. In addition, an increase in precipitation due to orographic lifting usually occurs at higher elevations. Despite the fact that mountain stations are colder and often wetter than locations at lower elevations, highland climates are often very similar to those in adjacent lowlands in terms of seasonal temperature cycles and precipitation distribution. Figure 20.13 illustrates this relationship.

Phoenix, at an elevation of 338 meters (1109 feet), lies in the desert lowlands of southern Arizona. By contrast, Flagstaff is located at an altitude of 2100 meters (7000 feet) on the Colorado Plateau in northern Arizona (Figure 20.14). When summer averages climb to 34°C (93°F) in Phoenix, Flagstaff is experiencing a pleasant 19°C (66°F), which is a full 15°C (27°F) cooler. Although the temperatures at each city are quite different, the pattern of monthly temperature changes for each place is similar. Both experience their minimum and maximum monthly means in the same months. When precipitation data are examined, both places

▼ **FIGURE 20.11** The tundra in bloom north of Nome, Alaska. It is a region almost completely devoid of trees. Bogs and marshes are common, and plant life frequently consists of mosses, low shrubs, and flowering herbs. (Photo by Fred Bruemmer/DRK Photo)

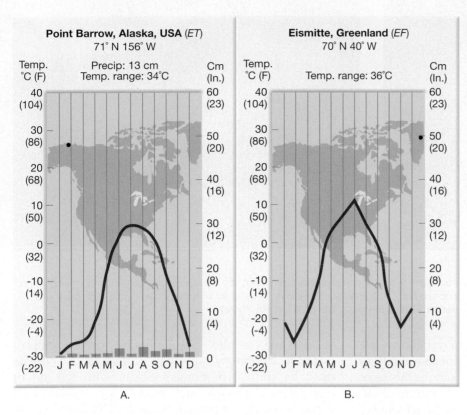

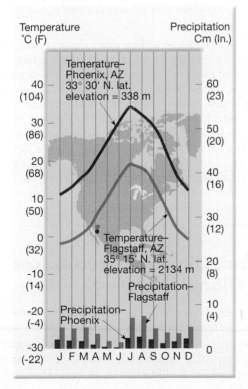

◄ **FIGURE 20.12** These climatic diagrams represent the two basic types of polar climates. **A.** Barrow, Alaska, exhibits a tundra (ET) climate. **B.** Eismitte, Greenland, a station located on a massive ice sheet, is classified as an ice cap (EF) climate.

▼ **FIGURE 20.13** Climate diagrams for two stations in Arizona illustrate the general influence of elevation on climate. Flagstaff is cooler and wetter because of its position on the Colorado Plateau, nearly 1800 meters (6000 feet) higher than Phoenix.

have a similar seasonal pattern, but the amounts at Flagstaff are higher in every month. In addition, owing to its higher altitude, much of Flagstaff's winter precipitation is in the form of snow. By contrast, all of the precipitation at Phoenix is rain.

Perhaps the terms *variety* and *changeability* best describe mountain climates. Because atmospheric conditions fluctuate rapidly with changes in altitude and exposure, a nearly limitless variety of local climates occur in mountainous regions. The climate in a protected valley is very different from that of an exposed peak. Conditions on windward slopes contrast sharply with those on the leeward sides, whereas slopes facing the Sun are unlike those that lie mainly in the shadows.

Human Impact on Global Climate

The proposals to explain global climatic change are many and varied. In Chapter 6, we examined some possible causes for ice-age climates. These hypotheses, which included the movement of lithosphere plates and variations in Earth's orbit, involved natural forcing mechanisms. Another natural forcing mechanism, discussed in Chapter 9, is the possible role of explosive volcanic eruptions in modifying the atmosphere. It is important to remember that these mechanisms, as well as others, not only have contributed to climatic changes in the geologic past but will also be responsible for future shifts in climate. However, when

▲ FIGURE 20.14 **A.** Only scanty drought-tolerant natural vegetation can survive in the hot, dry climate of southern Arizona, near Phoenix. (Photo by Charlie Ott Photography/Photo Researchers, Inc.) **B.** The natural vegetation associated with the cooler, wetter highlands near Flagstaff, Arizona, is much different from the desert lowlands. (Photo by Larry Ulrich/DRK Photo)

relatively recent and future changes in our climate are considered, we must also examine the possible impact of human beings. In this section, we will examine the major way in which humans are likely contributing to global climatic change. This impact results from the addition of carbon dioxide and other gases to the atmosphere.

Human influence of regional and global climate probably did not just begin with the onset of the modern industrial period. There is good evidence that humans have been modifying the environment over extensive areas for thousands of years. The use of fire as well as the overgrazing of marginal lands by domesticated animals have negatively affected the abundance and distribution of vegetation. By altering ground cover, such important climatological factors as surface albedo, evaporation rates, and surface winds have been, and continue to be, modified. Commenting on this aspect of human-induced climatic modification, the authors of one study observed:

> In contrast to the prevailing view that only modern humans are able to alter climate, we believe it is more likely that the human species has made a substantial and continuing impact on climate since the invention of fire.*

It should be pointed out that when any hypothesis of climatic change is examined, whether it depends on natural or human-induced causes, a degree of caution must be exercised. It is safe to say that most, if not all, hypotheses are to some degree controversial and speculative. This is to be expected if we consider the fact that at present all of our models of Earth's climate are far from complete. Because atmospheric processes are

so large and complex, they cannot be physically reproduced in laboratory experiments. Instead, the climate must be simulated mathematically with the aid of computers. Although such models are sophisticated enough to serve as primary tools for climate research, they cannot yet approach the actual complexity of the atmosphere. Computer models are powerful and essential aids, but climate forecasts based on such simulations still contain many uncertainties.

STUDENTS SOMETIMES ASK...
Do cities influence climate?

Yes, in lots of ways, including more fog and rain, as well as less sunshine and slower winds. The most studied and well-documented effect is the *urban heat island,* which refers to the fact that city temperatures are generally higher than surrounding rural areas. Several factors contribute to the heat island. The city's concrete and asphalt surfaces absorb and store more solar energy than the natural rural landscape. At night these stonelike materials help keep the city warmer. Urban temperatures are also higher because cities generate a great deal of waste heat (home heating, factories, cars, etc.). The "blanket" of pollutants also contributes by "trapping" heat that would otherwise escape.

Carbon Dioxide, Trace Gases, and Global Warming

In Chapter 16, we learned that although carbon dioxide (CO_2) represents only about 0.037 percent of the gases that make up clean, dry air, it is nevertheless a meteorologically significant component. The importance of carbon dioxide lies in the fact that it is trans-

*Carl Sagan et al., Anthropogenic Albedo Changes and the Earth's Climate, *Science* 206, no. 4425 (1980), 1367.

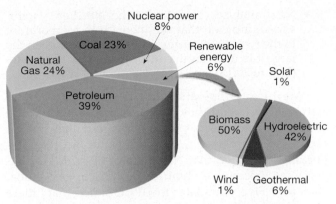

Total = 96.935 quadrillion btu   Total = 5.668 quadrillion btu

▲ **FIGURE 20.15** Paralleling the rapid growth of industrialization, which began in the nineteenth century, has been the combustion of fossil fuels, which has added great quantities of carbon dioxide to the atmosphere. The graph shows energy consumption in the United States, 2001. The total was nearly 97 quadrillion Btu. A quadrillion, by the way, is 10 raised to the 12th power, or a million million—a quadrillion Btu is a convenient unit for referring to U.S. energy use as a whole. Fossil fuels (petroleum, coal, and natural gas) represent about 86 percent of the total. (Source: U.S. Department of Energy, Energy Information Administration)

parent to incoming short-wavelength solar radiation, but it is not transparent to some of the longer-wavelength outgoing terrestrial radiation. A portion of the energy leaving the ground is absorbed by carbon dioxide and subsequently reemitted, part of it toward the surface, thereby keeping the air near the ground warmer than it would be without carbon dioxide.

Thus, along with water vapor, carbon dioxide is largely responsible for the *greenhouse effect* of the atmosphere.* Carbon dioxide is an important heat absorber, and it follows logically that any change in the air's carbon dioxide content could alter temperatures in the lower atmosphere.

CO_2 Levels Are Rising

Earth's tremendous industrialization of the past two centuries has been fueled—and still is fueled—by burning fossil fuels: coal, natural gas, and petroleum (Figure 20.15). Combustion of these fuels has added great quantities of carbon dioxide to the atmosphere.

The use of coal and other fuels is the most prominent means by which humans add CO_2 to the atmosphere, but it is not the only way. The clearing of forests also contributes substantially because CO_2 is released as vegetation, is burned, or decays. Deforestation is particularly pronounced in the tropics, where vast tracts are cleared for ranching and agriculture or subjected to inefficient commercial logging operations. According to U.N. estimates, the destruction of tropical forests exceeded 15 million hectares (38 million acres) per year during the 1990s.

Although some of the excess CO_2 is taken up by plants or is dissolved in the ocean, it is estimated that 45 to 50 percent remains in the atmosphere. Figure 20.16A shows CO_2 concentrations over the past thousand

*To review the greenhouse effect, see Figure 16.21 on p.454.

◄ **FIGURE 20.16 A.** Carbon dioxide (CO_2) concentrations over the past 1000 years. Most of the record is based on data obtained from Antarctic ice cores. Bubbles of air trapped in the glacial ice provide samples of past atmospheres. The record since 1958 comes from direct measurements of atmospheric CO_2 taken at Mauna Loa Observatory, Hawaii. **B.** Fossil fuel CO_2 emissions. The rapid increase in CO_2 concentrations since the onset of industrialization has followed closely the rise in CO_2 emissions from fossil fuels.

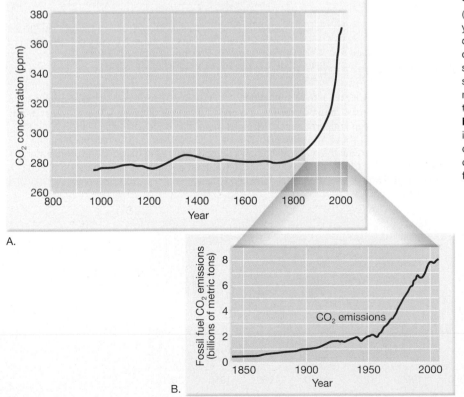

years based on ice-core records and (since 1958) measurements taken at Mauna Loa Observatory, Hawaii. The rapid increase in CO_2 concentration since the onset of industrialization is obvious and has closely followed the increase in CO_2 emissions from burning fossil fuels (Figure 20.16B).

STUDENTS SOMETIMES ASK...

Figure 20.15 shows biomass as a form of renewable energy. What exactly is biomass?

Biomass refers to organic matter that can be burned directly as fuel or converted into a different form and then burned. Biomass is a relatively new name for the oldest human fuels. Examples include firewood, charcoal, crop residues, and animal waste. Biomass burning is especially important in emerging economies.

The Atmosphere's Response

Given the increase in the atmosphere's carbon dioxide content, have global temperatures actually increased? The answer is yes. A report by the Intergovernmental Panel on Climate Change (IPCC)* indicates the following:

- During the twentieth century, the global average surface temperature increased by about 0.6°C (1°F).

- Globally it is very likely that the 1990s was the warmest decade and 1998 the warmest year since 1861 (Figure 20.17).

- New analyses of data for the Northern Hemisphere indicate that the increase in temperature in the twentieth century is likely to have been the largest in any century during the past 1000 years (Figure 20.18).

Are these temperature trends caused by human activities, or would they have occurred anyway? Scientists are cautious but seem convinced that human activities have played a significant role. An IPCC report in 1996 stated that "the balance of evidence suggests a discernible human influence on global climate."* Five years later the IPCC stated that "there is new and stronger evidence that most of the warming observed over the last 50 years is attributable to human activities."[†] What about the future? By the year 2100, models project atmospheric CO_2 concentrations of 540 to 970 ppm. With such an increase, how will global temperatures change? Here is some of what the 2001 IPCC report has to say:[‡]

- The globally averaged surface temperature is projected to increase by 1.4 to 5.8°C by the year 2100.

- The projected rate of warming is much larger than the observed changes during the twentieth century and is very likely to be without precedent during at least the last 10,000 years.

- It is very likely that nearly all land areas will warm more rapidly than the global average, particularly those at northern high latitudes in the cold season.

---

*Intergovernmental Panel on Climate Change, *Climate Change 2001: The Scientific Basis.* Cambridge, UK: Cambridge University Press, 2001, p. 2.

*Intergovernmental Panel on Climate Change, *Climate Change 1995: The Science of Climate Change,* New York: Cambridge University Press, 1996, p. 4.
[†]IPCC. *Climate Change 2001: The Scientific Basis,* p. 10.
[‡]IPCC. *Climate Change 2001: The Scientific Basis,* p. 13.

▼ **FIGURE 20.17** Annual average global temperature variations for the period 1860–2003. The basis for comparison is the average for the 1961–1990 period (the 0.0 line on the graph). Each narrow bar on the graph represents the departure of the global mean temperature from the 1961–1990 average for one year. For example, the global mean temperature for 1862 was more than 0.5°C (1°F) *below* the 1961–1990 average, whereas the global mean for 1998 was more than 0.5°C above. (Specifically, 1998 was 0.56°C warmer.) The bar graph clearly indicates that there can be *significant variations from year to year.* But the graph also shows a trend. Estimated global mean temperatures have been above the 1961–1990 average every year since 1978. Globally the 1990s was the warmest decade, and the years 1998, 2002, and 2003 the warmest years, since 1861. (Modified and updated after G. Bell, et al. "Climate Assessment for 1998," *Bulletin of the American Meteorological Society,* Vol. 80, No. 5, May 1999, p. 54.)

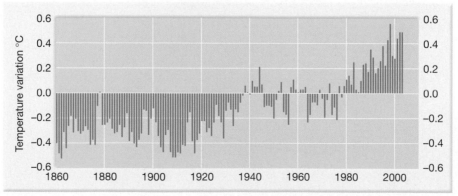

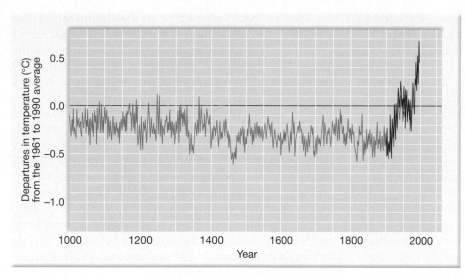

◄ **FIGURE 20.18** Year-by-year variations in average surface temperatures for the Northern Hemisphere for the past 1000 years, reconstructed from tree rings, ice cores, corals, and historical records (blue portion of line), as well as air temperatures directly measured (red portion of line). The warming during the twentieth century was much greater than in any of the previous nine centuries. (After U.S. Global Change Research Program and IPCC)

? STUDENTS SOMETIMES ASK...
What is the Intergovernmental Panel on Climate Change?

Recognizing the problem of potential global climate change, the World Meteorological Organization and the United Nation's Environment Program established the *Intergovernmental Panel on Climate Change (IPCC,* for short). The IPCC assesses the scientific, technical, and socio-economic information that is relevant to an understanding of human-induced climate change. This authoritative group provides advice to the world community through periodic reports that assess the state of knowledge of causes of climate change.

The Role of Trace Gases

Carbon dioxide is not the only gas contributing to a possible global increase in temperature. In recent years, atmospheric scientists have come to realize that the industrial and agricultural activities of people are causing a buildup of several trace gases that may also play a significant role. The substances are called *trace gases* because their concentrations are so much smaller than that of carbon dioxide. The trace gases that appear to be most important are methane (CH_4) (Figure 20.19), nitrous oxide (N_2O), and certain types of chlorofluorocarbons (CFCs). These gases absorb wavelengths of outgoing Earth radiation that would otherwise escape into space. Although individually their impact is modest, taken together the effects of these trace gases may be nearly as great as CO_2 in warming Earth.

Sophisticated computer models of the atmosphere show that the warming of the lower atmosphere caused by CO_2 and trace gases will not be the same everywhere. Rather, the temperature response in polar regions could be two to three times greater than the global average. One reason for such a response is an

expected reduction in sea ice. This topic is explored more fully in the following section.

Climate-Feedback Mechanisms

Climate is a very complex interactive physical system. Thus, when any component of the climate system is altered, scientists must consider many possible outcomes. These possible outcomes are called **climate-feedback mechanisms**. They complicate climate-modeling efforts and add greater uncertainty to climate predictions.

▼ **FIGURE 20.19** Methane is produced by *anaerobic* bacteria in wet places where oxygen is scarce (anaerobic means "without air," specifically oxygen). Such places include swamps, bogs, wetlands, and the guts of termites and grazing animals like cattle and sheep. Methane is also generated in flooded paddy fields ("artificial swamps") used for growing rice. These paddies are in India's Ganges lowlands. Mining of coal and drilling for oil and natural gas are other sources because methane is a product of their formation. (Photo by George Holton/Photo Researchers, Inc.)

What climate-feedback mechanisms are related to carbon dioxide and other greenhouse gases? The most important mechanism is that warmer surface temperatures increase evaporation rates. This in turn increases the water vapor in the atmosphere. Remember that water vapor is an even more powerful absorber of radiation emitted by Earth than is carbon dioxide. Therefore, with more water vapor in the air, the temperature increase caused by carbon dioxide and the trace gases is reinforced.

Recall that the temperature increase at high latitudes may be two to three times greater than the global average. This assumption is based in part on the likelihood that the area covered by sea ice will decrease as surface temperatures rise. Because ice reflects a much larger percentage of incoming solar radiation than does open water, the melting of the sea ice would replace a highly reflecting surface with a relatively dark surface (Figure 20.20). The result would be a substantial increase in the solar energy absorbed at the surface. This in turn would feed back to the atmosphere and magnify the initial temperature increase created by higher levels of greenhouse gases.

So far the climate-feedback mechanisms discussed have magnified the temperature rise caused by the buildup of carbon dioxide. Because these effects reinforce the initial change, they are called *positive-feedback mechanisms*. However, other effects must be classified as *negative-feedback mechanisms* because they produce results that are just the opposite of the initial change and tend to offset it.

▼ FIGURE 20.20 A reduction in sea ice would act as a positive feedback mechanism because surface albedo would decrease and the amount of solar energy absorbed at the surface would increase. This aerial view shows the springtime breakup of sea ice near Antarctica. (Reproduced with permission from Science [Cover, January 25, 2002]. Copyright American Association for the Advancement of Science. Photo: D. N. Thomas)

One probable result of a global temperature rise would be an accompanying increase in cloud cover due to the higher moisture content of the atmosphere. Most clouds are good reflectors of solar radiation. At the same time, however, they are also good absorbers and emitters of radiation emitted by Earth. Consequently, clouds produce two opposite effects. They are a negative-feedback mechanism because they increase albedo and thus diminish the amount of solar energy available to heat the atmosphere. On the other hand, clouds act as a positive-feedback mechanism by absorbing and emitting radiation that would otherwise be lost from the troposphere (see Figure 16.28, p. 459).

Which effect, if either, is stronger? Atmospheric modeling shows that the negative effect of a higher albedo is dominant. Therefore, the net result of an increase in cloudiness should be a decrease in air temperature. The magnitude of this negative feedback, however, is not believed to be as great as the positive feedback caused by added moisture and decreased sea ice. Thus, although increases in cloud cover may partly offset a global temperature increase, climate models show that the ultimate effect of the projected increase in CO_2 and trace gases will still be a temperature increase.

The problem of global warming caused by human-induced changes in atmospheric composition continues to be one of the most studied aspects of climate change. Although no models yet incorporate the full range of potential factors and feedbacks, the scientific consensus is that the increasing levels of atmospheric carbon dioxide and trace gases will lead to a warmer planet with a different distribution of climate regimes.

How Aerosols Influence Climate

Increasing the levels of carbon dioxide and other greenhouse gases in the atmosphere is the most direct human influence on global climate. But it is not the only impact. Global climate is also affected by human activities that contribute to the atmosphere's aerosol content. *Aerosols* are the tiny, often microscopic, liquid and solid particles that are suspended in the air. Atmospheric aerosols are composed of many different materials, including soil, smoke, sea salt, and sulfuric acid. Natural sources are numerous and include such phenomena as dust storms and volcanoes. In Chapter 9 you learned that some explosive volcanoes (such as Mount Pinatubo) emit large quantities of sulfur dioxide gas high into the atmosphere. This gas combines with water vapor to produce clouds of tiny sulfuric acid aerosols that can lower air temperatures near the surface by reflecting solar energy back to space. So it is with sulfuric acid aerosols produced by human activities.

Presently the human contribution of aerosols to the atmosphere *equals* the quantity emitted by natural sources. Most human-generated aerosols come from the sulfur dioxide emitted during the combustion of fossil fuels and as a consequence of burning vegetation to clear agricultural land. Chemical reactions in the atmosphere convert the sulfur dioxide into sulfate aerosols, the same material that produces acid precipitation.

The aerosols produced by human activity act directly by reflecting sunlight back to space and indirectly by making clouds "brighter" reflectors. The second effect relates to the fact that sulfuric acid aerosols attract water and thus are especially effective as cloud condensation nuclei (tiny particles upon which water vapor condenses). The large quantity of aerosols produced by human activities (especially industrial emissions) trigger an increase in the number of cloud droplets that form within a cloud. A greater number of small droplets increases the cloud's brightness—that is, more sunlight is reflected back to space.

By reducing the amount of solar energy available to the climate system, aerosols have a net cooling effect. Studies indicate that the cooling effect of human-generated aerosols could offset a portion of the global warming caused by the growing quantities of greenhouse gases in the atmosphere. Unfortunately, the magnitude and extent of the cooling effect of aerosols is highly uncertain. This uncertainty is a significant hurdle in advancing our understanding of how humans alter Earth's climate.

It is important to point out some significant differences between global warming by greenhouse gases and aerosol cooling. After being emitted, greenhouse gases such as carbon dioxide remain in the atmosphere for many decades. By contrast, aerosols released into the lower atmosphere remain there for only a few days or, at most, a few weeks before they are "washed out" by precipitation. Because of their short lifetime in the atmosphere, human-generated aerosols are distributed unevenly over the globe. As expected, they are concentrated near the areas that produce them, namely industrialized regions that burn fossil fuels and land areas where vegetation is burned.

Because the lifetime of human-generated aerosols in the atmosphere is short, the effect on today's climate is determined by the quantity of material emitted during the preceding couple of weeks. By contrast, the carbon dioxide released into the atmosphere remains for much longer spans and thus influences climate for many decades.

Some Possible Consequences of Global Warming

What consequences can be expected if the carbon dioxide content of the atmosphere reaches a level that is twice what it was early in the twentieth century? Because the climate system is so complex, predicting the distribution of particular regional changes is still very speculative. Nevertheless, plausible scenarios can be given for larger scales of space and time (see Box 20.1).

As noted, the magnitude of the temperature increase will not be the same everywhere. The temperature rise will probably be smallest in the tropics and increase toward the poles. As for precipitation, the models indicate that some regions will experience significantly more precipitation and runoff. However, other regions will experience a decrease in runoff due to reduced precipitation or because of greater evaporation caused by higher temperatures.

Table 20.2 lists some of the more likely effects and their possible consequences. The table also provides

Box 20.1 Understanding Earth
Computer Models of Climate: Important Yet Imperfect Tools

Earth's climate system is amazingly complex. Comprehensive state-of-the-science climate simulation models are among the basic tools used to develop possible climate-change scenarios. Called *General Circulation Models*, or *GCMs*, they are based on fundamental laws of physics and chemistry and incorporate human and biological interactions. GCMs are used to simulate many variables, including temperature, rainfall, snow cover, soil moisture, winds, clouds, sea ice, and ocean circulation over the entire globe through the seasons and over spans of decades.

In many other fields of study, hypotheses can be tested by direct experimentation in the laboratory. However, this is usually not possible in the study of climate. Rather, scientists must construct computer models of how our planet's climate system works. If we understand the climate system correctly and construct the model appropriately, then the behavior of the model climate system should mimic the behavior of Earth's climate system.

What factors influence the accuracy of climate models? Clearly, mathematical models are *simplified* versions of the real Earth and cannot capture its full complexity, especially at smaller geographic scales. Moreover, when computer models are used to simulate future climate change, many assumptions have to be made that significantly influence the outcome. They must consider a wide range of possibilities for future changes in population, economic growth, consumption of fossil fuels, technological development, improvements in energy efficiency, and more.

Despite many obstacles, our ability to use supercomputers to simulate climate continues to improve. Although today's models are far from infallible, they are powerful tools for understanding what Earth's future climate might be like.

TABLE 20.2	Projected changes and effects of global warming in the twenty-first century (estimated probability)*.

Higher maximum temperatures; more hot days and heat waves over nearly all land areas (*very likely*).
Higher minimum temperatures; fewer cold days, frost days, and cold waves over nearly all land areas (*very likely*).
More intense precipitation events (*very likely* over many areas).
Increased summer drying over most midlatitude continental interiors and associated risk of drought (*likely*).
Increase in tropical cyclone peak wind intensities, mean and peak precipitation intensities (*likely* over some areas).
Intensified droughts and floods associated with El Niño events in many different regions (*likely*).
Increased Asian summer monsoon precipitation variability (*likely*).
Increased intensity of midlatitude storms (*uncertain*).

Very likely indicates a probability of 90–99 percent. *Likely* indicates a probability of 67–90 percent.
Source: IPCC; 2001

the IPCC's estimate of the probability of each effect. Levels of confidence for these projections vary from "*likely*" (67 to 90 percent probability) to "*very likely*" (90 to 99 percent probability).

Water Resources and Agriculture

Such changes could profoundly alter the distribution of the world's water resources and hence affect the productivity of agricultural regions that depend on rivers for irrigation water. For example, a 2°C (3.6°F) warming and 10 percent precipitation decrease in the region drained by the Colorado River could diminish the river's flow by 50 percent or more. Because the present flow of the river barely meets current demand for irrigation agriculture, the negative effect would be serious. Many other rivers are the basis for extensive irrigated agriculture, and the projected reduction of their flow could have equally grave consequences. In contrast, large precipitation increases in other areas would increase the flow of some rivers and bring more frequent destructive floods.

Harder to estimate is the effect on nonirrigated crops that depend on direct rainfall for moisture. Some places will no doubt experience productivity loss due to a decrease in rainfall or increase in evaporation. Still, these losses may be offset by gains elsewhere. Warming in the high latitudes could lengthen the growing season, for instance. This in turn could allow expansion of agriculture into areas that are presently not suited to crop production.

Sea-Level Rise

Another impact of a human-induced global warming is a probable rise in sea level. How is a warmer atmosphere related to a global rise in sea level? The most obvious connection, the melting of glaciers, is important, but *not* the only factor. Also significant is that a warmer atmosphere causes an increase in ocean volume due to thermal expansion. Higher air temperatures warm the adjacent upper layers of the ocean, which in turn causes the water to expand and sea level to rise.

Research indicates that sea level has risen between 10 and 25 centimeters (4 and 8 inches) over the past century and that the trend will continue at an accelerated rate. Some models indicate that the rise may approach or even exceed 50 centimeters (20 inches) by the end of the twenty-first century. Such a change may seem modest, but scientists realize that any rise in sea level along a *gently* sloping shoreline, such as the Atlantic and Gulf coasts of the United States, will lead to significant erosion and severe, permanent inland flooding (Figure 20.21). If this happens, many beaches and wetlands will be eliminated and coastal civilization would be severely disrupted.

The Potential for "Surprises"

In summary, you have seen that climate in the twenty-first century, unlike the preceding thousand years, is not expected to be stable. Rather, a constant state of change is very likely. Many of the changes will probably be gradual environmental shifts, imperceptible from year to year. Nevertheless, the effects, accumulated over decades, will have powerful economic, social, and political consequences.

Despite our best efforts to understand future climate shifts, there is also the potential for "surprises." This simply means that due to the complexity of Earth's climate system, we might experience relatively sudden, unexpected changes or see some aspects of climate shift in an unexpected manner. The report on *Climate Change Impacts on the United States* describes the situation like this:

Surprises challenge humans' ability to adapt, because of how quickly and unexpectedly they occur. For example, what if the Pacific Ocean warms in such a way that El Niño events become much more extreme? This could reduce the frequency, but perhaps not the strength, of hurricanes along the East Coast, while on the West Coast, more severe winter storms, extreme precipitation events, and damaging winds could become common. What if large quantities of methane, a potent greenhouse gas currently frozen in icy Arctic tundra and sediments, began to be released to the atmosphere by warming, potentially creating an amplifying "feedback loop" that would cause even more warming? We simply do not know how

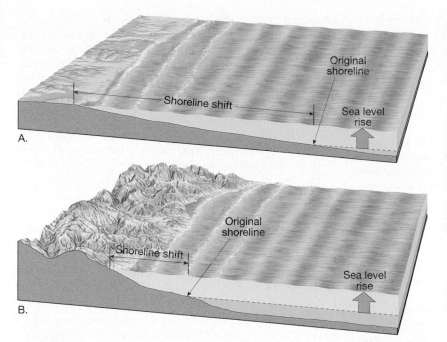

▲ **FIGURE 20.21** The slope of a shoreline is critical to determining the degree to which sea-level changes will affect it. **A.** When the slope is gentle, small changes in sea level cause a substantial shift. **B.** The same sea-level rise along a steep coast results in only a small shoreline shift. **C.** As sea level gradually rises, the shoreline retreats, and structures that were once thought to be safe from wave attack are exposed to the force of the sea. (Photo by Kenneth Hasson)

far the climate system or other systems it affects can be pushed before they respond in unexpected ways.

There are many examples of potential surprises, each of which would have large consequences. Most of these potential outcomes are rarely reported, in this study or elsewhere. Even if the chance of any particular surprise happening is small, the chance that at least one such surprise will occur is much greater. In other words, while we can't know which of these events will occur, it is likely that one or more will eventually occur.*

Clearly, the impact on climate of an increase in atmospheric CO_2 and trace gases is obscured by many unknowns and uncertainties. Policymakers are confronted with responding to the risks posed by emissions of greenhouse gases in the face of significant scientific uncertainties. However, they are also faced with the fact that climate-induced environmental changes cannot be reversed quickly, if at all, owing to the lengthy time scales associated with the climate system. Addressing this issue, the Intergovernmental Panel on Climate Change states:

> Uncertainty does not mean that a nation or the world community cannot position itself better to cope with the broad range of possible climate changes or protect against potentially costly future outcomes. Delaying such measures may leave a nation or the world poorly prepared to deal with adverse changes and may increase the possibility of irreversible or very costly consequences. Options for adapting to change or mitigating change that can be justified for other reasons today (e.g., abatement of air and water pollution) and make society more flexible or resilient to anticipated adverse effects of climate change appear particularly desirable.*

---

*National Assessment Synthesis Team. *Climate Change Impacts on the United States: The Potential Consequences of Climate Variability and Change.* Washington. DC: U.S. Global Research Program, 2000, p. 19.

*Intergovernmental Panel on Climate Change. *Climate Change 1995: Impacts. Adaptations and Mitigation of Climate Change: Scientific–Technical Analysis.* New York: Cambridge University Press. 1996, p. 23.

Chapter Summary

• *Climate* is the aggregate of weather conditions for a place or region over a long period of time. Earth's *climate system* involves the exchanges of energy and moisture that occur among the atmosphere, hydrosphere, solid Earth, biosphere, and *cryosphere* (the ice and snow that exist at Earth's surface).
• Climate classification brings order to large quantities of information, which aids comprehension and understanding, and facilitates analysis and explanation. *Temperature and precipitation are the most important elements in a climatic description.* Many climate classifications have been devised, with the value of each determined by its intended use. The *Köppen classification,* which uses mean monthly and annual values of temperature and precipitation, is a widely used system. The boundaries Köppen chose were largely based on the limits of certain plant

associations. Five principal climate groups, each with subdivisions, were recognized. Each group is designated by a capital letter. Four of the climate groups (A, C, D, and E) are defined on the basis of temperature characteristics, and the fifth, the B group, has precipitation as its primary criterion.

• *Humid tropical (A) climates* are winterless, with all months having a mean temperature above 18°C. *Wet tropical climates* (Af and Am), which lie near the equator, have constantly high temperatures and enough rainfall to support the most luxuriant vegetation (tropical rain forest) found in any climatic realm. *Tropical wet and dry climates* (Aw) are found poleward of the wet tropics and equatorward of the subtropical deserts, where the rain forest gives way to the tropical grasslands and scattered drought-tolerant trees of the savanna. The most distinctive feature of this climate is the seasonal character of the rainfall.

• *Dry (B) climates,* in which the yearly precipitation is less than the potential loss of water by evaporation, are subdivided into two types: *arid* or *desert* (BW) and *semiarid* or *steppe* (BS). Their differences are primarily a matter of degree, with semiarid being a marginal and more humid variant of arid. Low-latitude deserts and steppes coincide with the clear skies caused by subsiding air beneath the subtropical high-pressure belts. Middle-latitude deserts and steppes exist principally because of their position in the deep interiors of large landmasses far removed from the ocean. Because many middle-latitude deserts occupy sites on the leeward sides of mountains, they can also be classified as *rain shadow deserts.*

• *Middle-latitude climates with mild winters* (C climates) occur where the average temperature of the coldest month is below 18°C but above −3°C. Several C climate subgroups exist. *Humid subtropical climates* (Cfa) are located on the eastern sides of the continents, in the 25- to 40-degree latitude range. Summer weather is hot and sultry, and winters are mild. In North America, the *marine west coast climate* (Cfb, Cfc) extends from near the U.S.–Canada border northward as a narrow belt into southern Alaska. The prevalence of maritime air masses means that mild winters and cool summers are the rule. *Dry-summer subtropical climates* (Csa, Csb) are typically located along the west sides of continents between latitudes 30 and 45 degrees. In summer, the regions are dominated by stable, dry conditions associated with the oceanic subtropical highs. In winter they are within range of the cyclonic storms of the polar front.

• *Humid middle-latitude climates with severe winters* (D climates) are land-controlled climates that are absent in the Southern Hemisphere. The D climates have severe winters. The average temperature of the coldest month is −3°C or below, and the warmest monthly mean exceeds 10°C. *Humid continental climates* (Dfa, Dfb, Dwa, Dwb) are confined to the eastern portions of North America and Eurasia in the latitude range between approximately 40 and 50 degrees north latitude. Both winter and summer temperatures can be characterized as relatively severe. Precipitation is generally greater in summer than in winter. *Subarctic climates* (Dfc, Dfd, Dwc, Dwd) are situated north of the humid continental climates and south of the polar tundras. The outstanding feature of subarctic climates is the dominance of winter. By contrast, summers in the subarctic are remarkably warm, despite their short duration. The highest annual temperature ranges on Earth occur here.

• *Polar (E) climates* are summerless, with the average temperature of the warmest month below 10° C. Two types of polar climates are recognized. The *tundra climate* (ET) is a treeless climate found almost exclusively in the Northern Hemisphere. The *ice cap climate* (EF) does not have a single monthly mean above 0°C. As a consequence, the growth of vegetation is prohibited, and the landscape is one of permanent ice and snow.

• Compared to nearby places of lower elevation, *highland climates* are cooler and usually wetter. Because atmospheric conditions fluctuate rapidly with changes in altitude and exposure, these climates are best described by their variety and changeability.

• Humans have been modifying the environment for thousands of years. By altering ground cover with the use of fire and the overgrazing of land, people have modified such important climatological factors as surface albedo, evaporation rates, and surface winds.

• By adding carbon dioxide and other trace gases (methane, nitrous oxide, and chlorofluorocarbons) to the atmosphere, humans may be contributing significantly to global warming.

• When any component of the climate system is altered, scientists must consider the many possible outcomes, called *climate-feedback mechanisms.* Changes that reinforce the initial change are called *positive-feedback mechanisms.* On the other hand, *negative-feedback mechanisms* produce results that are the opposite of the initial change and tend to offset it.

• Because the climate system is very complex, predicting specific regional changes that may occur as the result of increased levels of carbon dioxide in the atmosphere is highly speculative. However, some potential consequences of global warming include: (1) altering the distribution of the world's water resources and therefore the productivity of agricultural regions that depend on rivers for irrigation, (2) a probable rise in sea level, and (3) a change in weather patterns, such as a higher frequency and greater intensity of hurricanes and shifts in the paths of large-scale cyclonic storms.

Key Terms

Review Questions

1. List the five parts of the climate system.

2. Why is classification often a helpful or even necessary task in science?

3. What climatic data are needed in order to classify a climate using the Köppen scheme?

4. **(a)** What primary factor distinguishes Aw climates from Af and Am?
   (b) How is this difference reflected in the vegetation?

5. Describe the influence of the equatorial low and the subtropical high on the precipitation regime in the Aw climate.

6. Why is the amount of precipitation that defines the humid–dry boundary variable?

7. What is the primary reason (control) for the existence of the dry subtropical realm (BWh and BSh)?

8. What is the primary cause for the existence of the middle-latitude deserts and steppes?

9. Describe and explain the differences between summertime and wintertime precipitation in the humid subtropics (Cfa).

10. Why is the marine west coast climate (Cfb and Cfc) represented by only slender strips of land in North and South America, and why is it very extensive in Western Europe?

11. In this chapter the dry-summer subtropics were described as transitional. Explain why this statement is true.

12. Why is the humid continental climate confined to the Northern Hemisphere?

13. Although generally characterized by small precipitation totals, subarctic and polar climates are considered humid. Explain.

14. Describe and explain the annual temperature range one should expect in the realm of the *taiga*.

15. Although polar regions experience extended periods of almost perpetual sunlight in the summer, temperatures remain cool. Explain.

16. The tundra climate is not confined solely to high latitudes. Under what circumstances might the ET climate be found in more equatorward locations?

17. Where are EF climates most extensively developed?

18. The Arizona cities of Flagstaff and Phoenix are relatively close to one another yet have contrasting climates (see Figure 20.13). Briefly explain why the differences occur.

19. Why has the atmosphere's carbon dioxide level been rising for more than 150 years?

20. How are temperatures in the lower atmosphere likely to change as carbon dioxide levels continue to increase? Why?

21. Aside from carbon dioxide, what other trace gases are contributing to a future global temperature change?

22. What are the main sources of human-generated aerosols? What effect do these aerosols have on temperatures in the troposphere? How long do aerosols remain in the lower atmosphere before they are removed?

23. List four possible consequences of a global warming.

Examining the Earth System

1. The Köppen climate classification is based on the fact that there is an excellent association between natural vegetation (biosphere) and climate (atmosphere). Briefly describe the climate conditions (temperature and precipitation) and natural vegetation associated with each of the following Köppen climates: Af, BWh, Dfc, ET.

2. How might the burning of fossil fuels, such as the gasoline to run your car, influence global temperature? If such a temperature change occurs, how might sea level be affected? How might the intensity of hurricanes change? How might these changes impact people who live on a beach or barrier island along the Atlantic or Gulf coasts?

3. What are climate-feedback mechanisms? Give an example of a positive-feedback mechanism and a negative-feedback mechanism.

Online Study Guide

The *Earth Science* Website uses the resources and flexibility of the Internet to aid in your study of the topics in this chapter. Written and developed by Earth science instructors, this site will help improve your understanding of Earth science. Visit **http://www.prenhall.com/tarbuck** and click on the cover of *Earth Science 11e* to find:

- **Online review quizzes.**
- **Critical thinking exercises.**
- **Links to chapter-specific Web resources.**
- **Internet-wide key term searches.**

http://www.prenhall.com/tarbuck

Origin of Modern Astronomy

*Full Moon rising from behind the rugged Chugach Range
near Valdez, Alaska.
(Photo by Paul Souders/Delimont Stock Photography)*

Earth is one of nine planets and numerous smaller bodies that orbit the Sun. The Sun is part of a much larger family of perhaps 100 billion stars that make up the Milky Way, which in turn is only one of billions of galaxies in an incomprehensibly large universe. This view of Earth's position in space is considerably different from that held only a few hundred years ago, when Earth was thought to occupy a privileged position as the center of the universe. This chapter unfolds with some events that led to modern astronomy. In addition, it examines Earth's place in time and space.

Long before recorded history, which began about 5000 years ago, people were aware of the close relationship between events on Earth and the positions of heavenly bodies, the Sun in particular. People noted that changes in the seasons and floods of great rivers like the Nile in Egypt occurred when the celestial bodies, including the Sun, Moon, planets, and stars, reached a particular place in the heavens. Early agrarian cultures, which were dependent on the weather, believed that if the heavenly objects could control the seasons, they must also strongly influence all Earthly events. This belief undoubtedly was the reason that early civilizations began keeping records of the positions of the celestial objects. The Chinese, Egyptians, and Babylonians in particular are noted for this.

These cultures recorded the locations of the Sun, Moon, and the five planets visible to the unaided eye as these objects moved slowly against the background of "fixed" stars. In addition, the Chinese kept quite accurate records of comets and "guest stars" (Figure 21.1). Today we know that a "guest star" is really a normal star, usually too faint to be visible, that increases its brightness as it explosively ejects gases from its surface, a phenomenon we call a *nova* (*novus* = new).

A study of Chinese archives shows that they recorded every appearance of the famous Halley's comet for at least 10 centuries. However, because this comet appears only once every 76 years, they were unable to link these appearances to establish that what they saw was the same object each time. Thus, like most ancients, the Chinese considered comets to be mystical. Comets were seen as bad omens and were blamed for a variety of disasters, from wars to plagues (Figure 21.2).

Ancient Astronomy

The "Golden Age" of early astronomy (600 B.C.–A.D. 150) was centered in Greece. The early Greeks have been criticized, and rightly so, for using philosophical arguments to explain natural phenomena. However, they did rely on observational data as well. The basics of geometry and trigonometry, which they had developed, were used to measure the sizes and distances of the largest-appearing bodies in the heavens—the Sun and the Moon.

Early Greeks

Many astronomical discoveries have been credited to the Greeks. They held the **geocentric** (*geo = Earth, centric = centered*) view, believing that Earth was a sphere that stayed motionless at the center of the universe. Orbiting Earth were the Moon, Sun, and the known planets—Mercury, Venus, Mars, and Jupiter. Beyond the planets was a transparent, hollow sphere (**celestial sphere**) on which the stars traveled daily around Earth (this is how it looks, but of course the effect is actually caused by Earth's rotation about its axis). Some early Greeks realized that the motion of the stars could be explained just as easily by a rotating Earth, but they rejected that idea, because Earth exhibited no sense of motion and seemed too large to be movable. In fact, proof of Earth's rotation was not demonstrated until 1851, a topic we consider in Box 21.1.

▶ **FIGURE 21.1** Comet Hale-Bopp. The two tails are about 10 million to 15 million miles long. (A Peoria Astronomical Society, Inc., photograph by Eric Clifton and Greg Neaveill)

◀ **FIGURE 21.2** The Bayeux Tapestry that hangs in Bayeux, France, shows the apprehension caused by Halley's comet in A.D. 1066. This event preceded the defeat of King Harold by William the Conqueror. ("Sighting of a comet." Detail from Bayeux Tapestry. Musee de la Tapisserie, Bayeux. "With special authorization of the City of Bayeux". Bridgeman-Giraudon/Art Resource, NY)

To the Greeks, all of the heavenly bodies, except seven, appeared to remain in the same relative position to one another. These seven wanderers (*planetai* in Greek) included the Sun, the Moon, Mercury, Venus, Mars, Jupiter, and Saturn. Each was thought to have a circular orbit around Earth. Although this system was incorrect, the Greeks refined it to the point that it explained the apparent movements of all celestial bodies.

As early as the fifth century B.C., the Greeks understood what causes the phases of the Moon. Anaxagoras reasoned that the Moon shines by reflected sunlight, and because it is a sphere, only half is illuminated at one time. As the Moon orbits Earth, that portion of the illuminated half that is visible from Earth is always changing. Anaxagoras also realized that an eclipse of the Moon occurs when it moves into Earth's shadow.

The famous Greek philosopher Aristotle (384–322 B.C.) concluded that Earth is spherical because it always casts a curved shadow when it eclipses the moon. Although most of the teachings of Aristotle were passed along and were considered infallible by many, his belief in a spherical Earth was abandoned during the Middle Ages.

The first Greek to profess a Sun-centered, or **heliocentric**, (*helios* = Sun, *centric* = centered), universe was Aristarchus (312–230 B.C.). Aristarchus also used simple geometric relations to calculate the relative distances from Earth to the Sun and the Moon. He later used these data to calculate their sizes. As a result of observational errors beyond his control, he came up with measurements that were much too small. However, he did learn that the Sun was many times more distant than the Moon and many times larger than Earth. The latter fact may have prompted him to suggest a Sun-centered universe. Nevertheless, because of the strong influence of Aristotle, the Earth-centered view dominated Western thought for nearly 2000 years.

The first successful attempt to establish the size of Earth is credited to Eratosthenes (276–194 B.C.). Eratosthenes observed the angles of the noonday Sun in two Egyptian cities that were roughly north and south of each other—Syene (presently Aswan) and Alexandria (Figure 21.3). Finding that the angles differed by 7 degrees, or 1/50 of a complete circle, he concluded that the circumference of Earth must be 50 times the distance between these two cities. The cities were 5000 *stadia* apart, giving him a measurement of 250,000 *stadia*. Many historians believe the *stadia* was 157.6 meters (517 feet), which would make Eratosthenes' calculation of Earth's circumference—39,400 kilometers (24,428 miles)—a measurement very close to the modern value of 40,075 kilometers (24,902 miles).

▼ **FIGURE 21.3** Orientation of the Sun's rays at Syene (Aswan) and Alexandria in Egypt on June 21 when Eratosthenes calculated Earth's circumference.

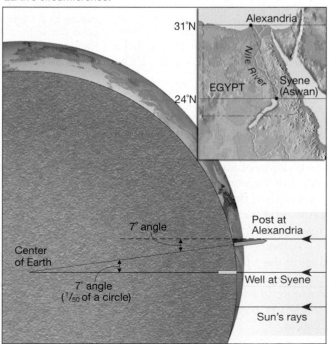

BOX 21.1 UNDERSTANDING EARTH:
Foucault's Experiment

Every schoolchild learns that Earth rotates on its axis once each day to produce periods of daylight and darkness. However, day and night and the apparent motions of the stars can be accounted for equally well by a Sun and celestial sphere that revolve around a stationary Earth. Copernicus realized that a rotating Earth greatly simplified the existing model of the universe and strongly advocated this as the correct view. He was unable, however, to *prove* that Earth rotates. The first substantial proof was presented 300 years after his death by the French physicist Jean Foucault.

In 1851 Foucault used a free-swinging pendulum to demonstrate that Earth does, in fact, turn on its axis. To envision Foucault's experiment, imagine a large pendulum swinging over the North Pole (Figure 21.A). Keep in mind that once a pendulum is put into motion, it continues swinging in the same plane unless acted upon by some outside force. Assume that a sharp stylus is attached to the bottom of this pendulum, marking the snow as it oscillates. When we observe the marks made by the stylus, we note that the pendulum is slowly but continually changing position. At the end of 24 hours it has returned to the starting position (Figure 21.A).

Because no outside force acted on the pendulum to change its position, what we observed must have been

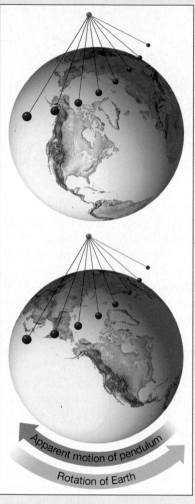

Figure 21.A Apparent movement of a pendulum at the North Pole caused by Earth rotating beneath it.

Earth rotating beneath it. Foucault conducted a similar experiment when he suspended a long pendulum from the dome of the Pantheon in Paris. Today, Foucault pendulums can be found in a number of museums to re-create this famous scientific experiment (Figure 21.B).

Figure 21.B Foucault pendulum housed in Museum of Science and Industry, Chicago. (Courtesy Museum of Science and Industry)

Probably the greatest of the early Greek astronomers was Hipparchus (second century B.C.), best known for his star catalogue. Hipparchus determined the location of almost 850 stars, which he divided into six groups according to their brightness. He measured the length of the year to within minutes of the modern value and developed a method for predicting the times of lunar eclipses to within a few hours.

Although many of the Greek discoveries were lost during the Middle Ages, the Earth-centered view that the Greeks proposed became established in Europe. Presented in its finest form by Claudius Ptolemy, this geocentric outlook became known as the **Ptolemaic system**.

The Ptolemaic System

Much of our knowledge of Greek astronomy comes from a 13-volume treatise, *Almagest* (the great work), which was compiled by Ptolemy in A.D. 141 and survived thanks to the work of Arab scholars. In this work, Ptolemy is credited with developing a model of the universe that accounted for the observable motions of the planets (Figure 21.4). The precision with which his model was able to predict planetary motion is attested to by the fact that it went virtually unchallenged, in principle if not in detail, for nearly 13 centuries.

In the Greek tradition, the Ptolemaic model had the planets moving in circular orbits around a motion-

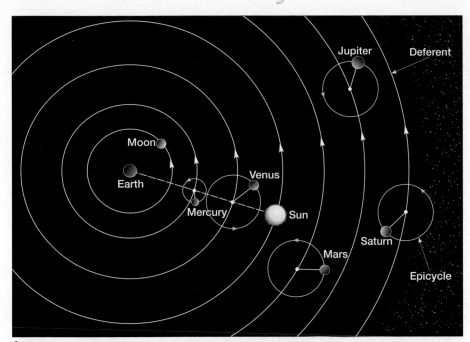

A.

◀ **FIGURE 21.4** The universe according to Ptolemy, second century A.D. **A.** Ptolemy believed that the star-studded celestial sphere made a daily trip around a motionless Earth. In addition he proposed that the Sun, Moon, and planets made trips of various lengths along individual orbits. **B.** Retrograde motion as explained by Ptolemy.

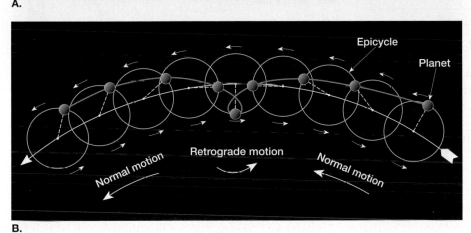

B.

less Earth. (The circle was considered the pure and perfect shape by the Greeks.) However, the motion of the planets, as seen against the background of stars, is not so simple. Each planet, if watched night after night, moves slightly eastward among the stars. Periodically, each planet appears to stop, reverse direction for a period of time, and then resume an eastward motion. The apparent westward drift is called **retrograde** (*retro* = to go back, *gradus* = walking) **motion**. This rather odd *apparent* motion results from the combination of the motion of Earth and the planet's own motion around the Sun.

The retrograde motions of Mars is shown in Figure 21.5. Earth has a faster orbital speed than Mars, so it overtakes its neighbor. While doing so, Mars *appears* to be moving backward, in retrograde motion. This is analogous to what a race-car driver sees out the side window when passing a slower car. The slower planet, like the slower car, appears to be going backward, although its actual motion is in the same direction as the faster-moving body.

It is much more difficult to accurately represent retrograde motion using the incorrect Earth-centered model, but Ptolemy was able to do so (Figure 21.4B). Rather than using a simple circle for each planet's orbit, he showed that the planets orbited on small circles (*epicycles*), revolving along large circles (*deferents*). By trial and error, he found the right combinations of circles to produce the amount of retrograde motion observed for each planet. (An interesting note is that almost any closed curve can be produced by the combination of two circular motions, a fact that can be verified by anyone who has used the Spirograph™ design-drawing toy.)

It is a tribute to Ptolemy's genius that he was able to account for the planets' motions as well as he did, considering that he used an incorrect model. Some suggest that he did not mean his model to represent reality, but only to be used for calculating the positions of heavenly bodies. We probably will never know his intentions. However, the Roman Catholic Church, which dominated European thought for centuries, accepted Ptolemy's theory as the correct representation of the heavens, and this created problems for those who found fault with it.

STUDENTS SOMETIMES ASK...
Why was the discovery of Jupiter's moons important?

According to the Ptolemaic (Earth-centered) model of the universe, all of the heavenly bodies revolved around Earth. When Galileo, using a crude telescope, saw four moons revolving around Jupiter, he knew that Earth was not the center of all motion. Consequently, at least one of the basic tenets of the Ptolemaic model had to be incorrect. Astronomers soon demonstrated that the other tenets of the Earth-centered model were also inconsistent with observations.

The Birth of Modern Astronomy

Modern astronomy was not born overnight. Its development involved a break from deeply entrenched philosophical and religious views that was brought about by the discovery of a new and greater universe governed by discernible laws. Let us now look at the work of five noted scientists involved in this transition: Nicolaus Copernicus, Tycho Brahe, Johannes Kepler, Galileo Galilei, and Sir Isaac Newton.

Nicolaus Copernicus

For almost 13 centuries after the time of Ptolemy, very few astronomical advances were made in Europe. The first great astronomer to emerge after the Middle Ages

▲ **FIGURE 21.6** Polish astronomer Nicolaus Copernicus (1473–1543) believed that Earth was just another planet. (Yerkes Observatory Photograph/University of Chicago)

was Nicolaus Copernicus (1473–1543) from Poland (Figure 21.6). Copernicus became convinced that Earth is a planet, just like the other five then-known planets. The daily motions of the heavens, he reasoned, could be better explained by a rotating Earth. To counter the Ptolemaic objection that Earth would fly apart if it rotated, Copernicus suggested that the much larger celestial sphere would be even more likely to fly apart if it rotated!

▼ **FIGURE 21.5** Retrograde (backward) motion of Mars as seen against the background of distant stars. When viewed from Earth, Mars moves eastward among the stars each day, then periodically appears to stop and reverse direction. This apparent westward drift is a result of the fact that Earth has a faster orbital speed than Mars and overtakes it. As this occurs, Mars appears to be moving backward; that is, it exhibits retrograde motion.

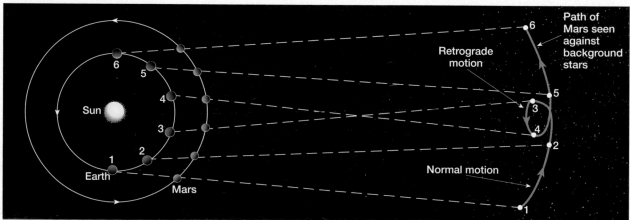

Having concluded that Earth is a planet, Copernicus reconstructed the solar system with the Sun at the center and the planets Mercury, Venus, Earth, Mars, Jupiter, and Saturn orbiting around it. This was a major break from the ancient idea that a motionless Earth lies at the center of all movement. However, Copernicus retained a link to the past and used circles, which were considered to be the perfect geometric shape, to represent the orbits of the planets. Although these circular orbits were close to reality, they didn't quite match what people saw. Unable to get satisfactory agreement between predicted locations of the planets and their observed positions, Copernicus found it necessary to add epicycles like those used by Ptolemy. The discovery that the planets have *elliptical* orbits would wait another century for the insights of Johannes Kepler.

Also, like his predecessors, Copernicus used philosophical justifications, such as the following, to support his point of view.

> . . . In the midst of all stands the Sun. For who could in this most beautiful temple place this lamp in another or better place than that from which it can at the same time illuminate the whole?

Copernicus's monumental work, *De Revolutionibus, Orbium Coelestium (On the Revolution of the Heavenly Spheres)*, which set forth his controversial ideas, was published as he lay on his deathbed. Hence, he never suffered the criticisms that fell on many of his followers.

The Copernican system challenged the primacy of Earth in the universe and was considered heretical by the Roman Catholic Church. Expounding the idea cost at least one person his life. Giordano Bruno was seized by the Inquisition, a Church tribunal, in 1600 and, refusing to denounce the Copernican theory, was burned at the stake.

Tycho Brahe

Tycho Brahe (1546–1601) was born of Danish nobility three years after the death of Copernicus. Reportedly, Tycho became interested in astronomy while viewing a solar eclipse that had been predicted by astronomers. He persuaded King Frederick II to establish an observatory, which he headed, near Copenhagen. There he designed and built pointers (the telescope would not be invented for a few more decades), which he used for 20 years to systematically measure the locations of the heavenly bodies (Figure 21.7). These observations, particularly of Mars, were far more precise than any made previously and are his legacy to astronomy.

Tycho did not believe in the Copernican (Sun-centered) system, because he was unable to observe an apparent shift in the position of stars that would be caused by Earth's motion. His argument went like this: If Earth does revolve along an orbit around the

▲ **FIGURE 21.7** Tycho Brahe (1546–1601) in his observatory, Uraniborg, on the Danish island of Hveen. Tycho (central figure) and the background are painted on the wall of the observatory within the arc of the sighting instrument called a quadrant. In the far right, Tycho can be seen "sighting" a celestial object through the "hole" in the wall. Tycho's accurate measurements of Mars enabled Johannes Kepler to formulate his three laws of planetary motion. (Courtesy of Thomas Clarke, McLaughlin Planetarium/Royal Ontario Museum, copyright ROM)

Sun, the position of a nearby star, when observed from extreme points in Earth's orbit six months apart, should shift with respect to the more distant stars. His *idea* was correct, and this apparent shift of the stars is called *stellar parallax* (see Figure 24.2, p. 661).

The principle of parallax is easy to visualize: Close one eye, and with your index finger vertical, use your eye to line up your finger with some distant object. Now, without moving your finger, view the object with your other eye and notice that the object's position appears to change. The farther away you hold your finger, the less the object's position seems to shift. Herein lay the flaw in Tycho's argument. He was right about parallax, but because the distance to even the nearest stars is enormous compared to the width of Earth's orbit, the shift that occurs is too small to be noticed by using the first primitive telescopes, let alone the unaided eye.

With the death of his patron, the King of Denmark, Tycho was forced to leave his observatory. It was probably his arrogant and extravagant nature that caused a conflict with the next ruler, so Tycho moved to Prague in the Czech Republic. Here, in the last year of his life, he acquired an able assistant, Johannes Kepler. Kepler retained most of the observations made by Tycho and put them to exceptional use. Ironically, the data Tycho collected to refute the Copernican view would later be used by Kepler to support it.

Johannes Kepler

If Copernicus ushered out the old astronomy, Johannes Kepler (1571–1630) ushered in the new (Figure 21.8). Armed with Tycho's data, a good mathematical mind, and, of greater importance, a strong faith in the accuracy of Tycho's work, Kepler derived three basic laws of planetary motion. The first two laws resulted from his inability to fit Tycho's observations of Mars to a circular orbit. Unwilling to concede that the discrepancies were a result of observational error, he searched for another solution. This endeavor led him to discover that the orbit of Mars is not a perfect circle but is elliptical (Figure 21.9). About the same time, he realized that the orbital speed of Mars varies in a predictable way. As it

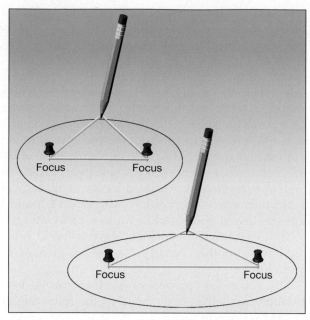

▲ **FIGURE 21.9** Drawing ellipses with various eccentricities. Using two straight pins for foci and a loop of string, trace out a curve while keeping the string taut, and you will have drawn an ellipse. The farther the pins (the foci) are moved apart, the more flattened (more eccentric) is the resulting ellipse.

approaches the Sun, it speeds up, and as it moves away from the Sun, it slows down.

In 1609, after almost a decade of work, Kepler proposed his first two laws of planetary motion:

1. The path of each planet around the Sun is an ellipse, with the Sun at one focus (Figure 21.9). The other focus is symmetrically located at the opposite end of the ellipse.
2. Each planet revolves so that an imaginary line connecting it to the Sun sweeps over equal areas in equal intervals of time (Figure 21.10). This law of equal areas expresses geometrically the variations in orbital speeds of the planets.

Figure 21.10 illustrates the second law. Note that in order for a planet to sweep equal areas in the same amount of time, it must travel more rapidly when it is nearer the Sun and more slowly when it is farther from the Sun.

Kepler was very religious and believed that the Creator made an orderly universe. The uniformity he tried to find eluded him for nearly a decade. Then in 1619, Kepler published his third law in *The Harmony of the Worlds*.

3. The orbital periods of the planets and their distances to the Sun are proportional. In its simplest form, the orbital period of revolution is measured in Earth years, and the planet's distance to the Sun is expressed in terms of Earth's mean

▼ **FIGURE 21.8** German astronomer Johannes Kepler (1571–1630) helped establish the era of modern astronomy by deriving three laws of planetary motion. (National Museum of American History/Smithsonian Institution)

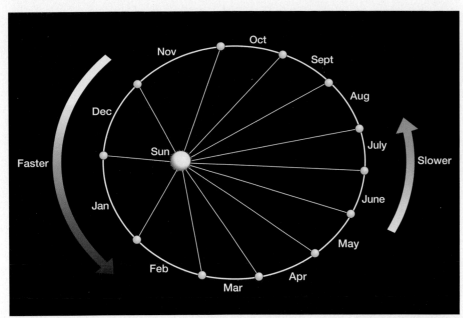

◄ **FIGURE 21.10** Kepler's law of equal areas. A line connecting a planet (Earth) to the Sun sweeps out an area in such a manner that equal areas are swept out in equal times. Thus, Earth revolves slower when it is farther from the Sun (aphelion) and faster when it is closest (perihelion). The eccentricity of Earth's orbit is greatly exaggerated in this diagram.

distance to the Sun. The latter "yardstick" is called the **astronomical unit (AU)** and averages about 150 million kilometers (93 million miles). Using these units, Kepler's third law states that the planet's orbital period squared is equal to its mean solar distance cubed ($p^2 = d^3$). Consequently, the solar distances of the planets can be calculated when their periods of revolution are known. For example, Mars has a period of 1.88 years, which squared equals 3.54. The cube root of 3.54 is 1.52, and that is the distance to Mars in astronomical units (Table 21.1).

Kepler's laws assert that the planets revolve around the Sun and therefore support the Copernican theory. Kepler, however, did fall short of determining the *forces* that act to produce the planetary motion he had so ably described. That task would remain for Galileo Galilei and Sir Isaac Newton.

TABLE 21.1	Period of revolution and solar distances of planets.	
Planet	**Solar Distance(AU)***	**Period (years)**
Mercury	0.39	0.24
Venus	0.72	0.62
Earth	1.00	1.00
Mars	1.52	1.88
Jupiter	5.20	11.86
Saturn	9.54	29.46
Uranus	19.18	84.01
Neptune	30.06	164.80
Pluto	39.44	247.70

*AU = astronomical unit.

Galileo Galilei

Galileo Galilei (1564–1642) was the greatest Italian scientist of the Renaissance (Figure 21.11). He was a contemporary of Kepler and, like Kepler, strongly supported the Copernican theory of a Sun-centered solar system. Galileo's greatest contributions to science were his descriptions of the behavior of moving objects. These he derived from experimentation. The method of using experiments to determine natural laws had essentially been lost since the time of the early Greeks.

All astronomical discoveries before Galileo's time were made without the aid of a telescope. In 1609, Galileo heard that a Dutch lens maker had devised a system of lenses that magnified objects. Apparently without ever seeing a telescope, Galileo constructed his own, which magnified distant objects to three times the size seen by the unaided eye. He immediately made others, the best having a magnification of about 30.

With the telescope, Galileo was able to view the universe in a new way. He made many important discoveries that supported the Copernican view of the universe, such as:

1. The discovery of four satellites, or moons, orbiting Jupiter. Galileo accurately determined their periods of revolution, which range from two to 17 days (Figure 21.12). This find dispelled the old idea that Earth was the only center of motion in the universe; for here, plainly visible, was another center of motion—Jupiter. It also countered the argument, frequently used by those opposed to the Sun-centered system, that the Moon would be left behind if Earth really revolved around the Sun.

▲ **FIGURE 21.11** Italian scientist Galileo Galilei (1564–1642) used a new invention, the telescope, to observe the Sun, Moon, and planets in more detail than ever before. (Yerkes Observatory Photograph/University of Chicago)

2. The discovery, through observation, that the planets are circular disks rather than just points of light, as was previously thought. This indicated that the planets must be Earth-like.

3. The discovery that Venus has phases just like the Moon, demonstrating that Venus orbits its source of light—the Sun. Galileo saw that Venus appears smallest when it is in full phase and thus is farthest from Earth (Figure 21.13B,C). In the Ptolemaic system, as shown in Figure 21.13A, the orbit of Venus lies between Earth and the Sun, which means that only the crescent phase of Venus could be seen from Earth.

4. The discovery that the Moon's surface is not a smooth glass sphere, as the ancients had proclaimed. Rather, Galileo saw mountains, craters, and plains. He thought the plains might be bodies of water, and this idea was strongly promoted by others, as we can tell from the names given to these features (Sea of Tranquility, Sea of Storms, and so forth).

5. The discovery that the Sun (the viewing of which may have caused the eye damage that later blinded him) had sunspots (dark regions caused by slightly lower temperatures). He tracked the movement of these spots and estimated the rota-

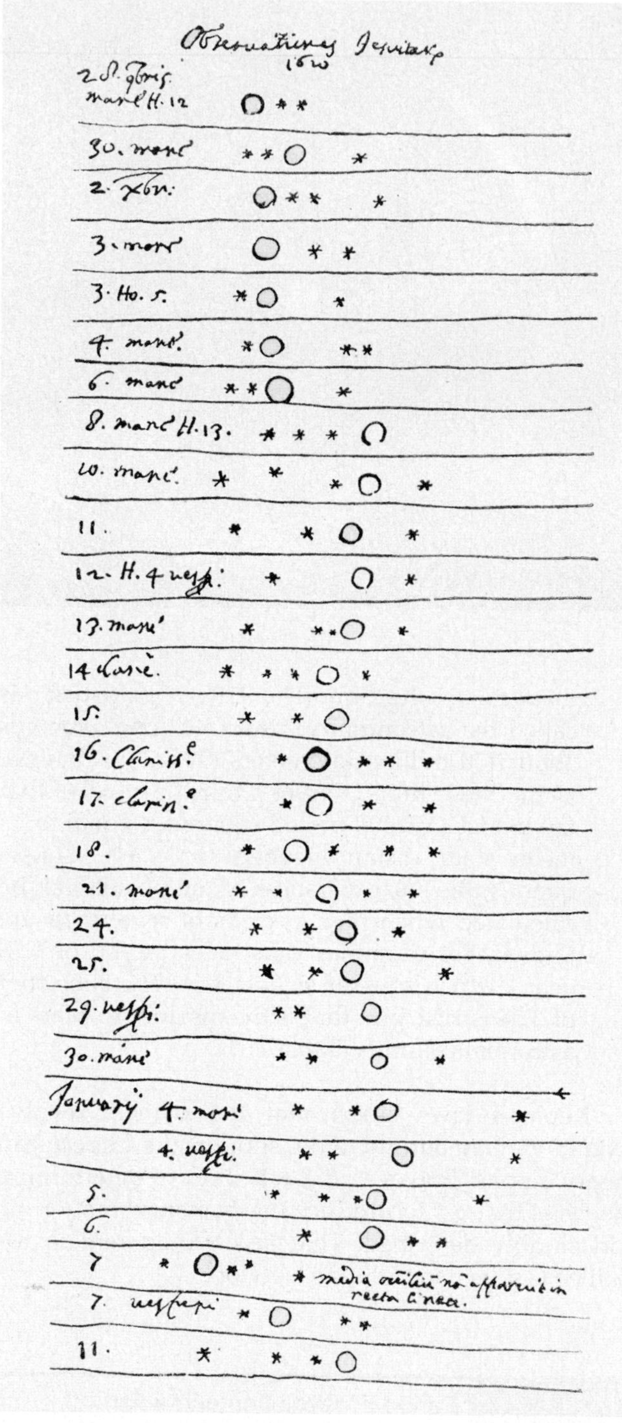

▲ **FIGURE 21.12** Sketch by Galileo of how he saw Jupiter and its four largest satellites through his telescope. The positions of Jupiter's four largest Moons (drawn as stars) change nightly. You can observe these same changes with binoculars. (Yerkes Observatory Photograph/University of Chicago)

tional period of the Sun as just under a month. Hence, another heavenly body was found to have both "blemishes" and rotational motion.

In 1616 the Church condemned the Copernican theory as contrary to Scripture, and Galileo was told to abandon it. Unwilling to accept this verdict, Galileo

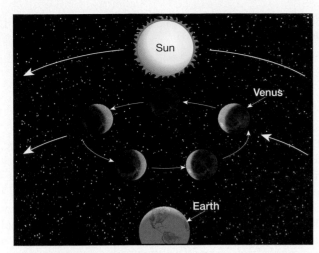

A. Phases of Venus as seen from Earth in the Earth-centered model.

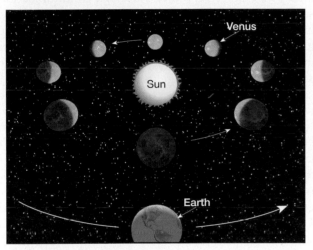

B. Phases of Venus as seen from Earth in the sun-centered model.

◀ **FIGURE 21.13** Using a telescope, Galileo discovered that Venus has phases just like the Moon. **A.** In the Ptolemaic (Earth-centered) system, the orbit of Venus lies between the Sun and Earth as shown in Figure 21.4A. Thus, in an Earth-centered solar system, only the crescent phase of Venus would be visible from Earth. **B.** In the Copernican (Sun-centered) system, Venus orbits the Sun and hence all of the phases of Venus should be visible from Earth. **C.** As Galileo observed, Venus goes through a series of Moonlike phases. Venus appears smallest during the full phase when it is farthest from Earth and largest in the crescent phase when it is closest to Earth. This verified Galileo's belief that the Sun was the center of the solar system. (Photo courtesy of Lowell Observatory)

C.

began writing his most famous work, *Dialogue of the Great World Systems*. Despite poor health, he completed the project and in 1630 went to Rome, seeking permission from Pope Urban VIII to publish. Because the book was a dialogue that expounded both the Ptolemaic and Copernican systems, publication was allowed. However, Galileo's enemies were quick to realize that he was promoting the Copernican view at the expense of the Ptolemaic system. Sales of the book were quickly halted, and Galileo was called before the Inquisition. Tried and convicted of proclaiming doctrines contrary to religious doctrine, he was sentenced to permanent house arrest, under which he remained for the last 10 years of his life.

Despite this restriction, and his age, and his grief after the death of his eldest daughter, Galileo continued to work. In 1637 he became totally blind, yet during the next few years he completed his finest scientific work, a book on the study of motion. When Galileo died in 1642, the Grand Duke of Tuscany wanted to erect a monument in his honor, but fear that it might offend the Church prevailed, and it was never built.

Later, as scientific evidence that supported the Copernican system was discovered, the Catholic Church allowed Galileo's works to be published.

STUDENTS SOMETIMES ASK...
Did Galileo drop balls of iron and wood from the Leaning Tower of Pisa?

Through experimentation, Galileo discovered that the acceleration of falling objects does not depend on their weight. According to some accounts, Galileo made this discovery by dropping balls of iron and wood from the Leaning Tower of Pisa to show that they would fall together and hit the ground at the same time. Despite the popularity of this legend, Galileo probably did not attempt this experiment. In fact, it would have been inconclusive because of the effect of air resistance. However, nearly four centuries later, this experiment was performed most dramatically on the airless Moon when David Scott, an *Apollo 15* astronaut, demonstrated that a feather and a hammer do, indeed, fall at the same rate.

▲ **FIGURE 21.14** English scientist Sir Isaac Newton (1642–1727) explained gravity as the force that holds planets in orbit around the Sun. (Yerkes Observatory Photograph/University of Chicago)

Sir Isaac Newton

Sir Isaac Newton (1642–1727) (Figure 21.14) was born in the year of Galileo's death. His many accomplishments in mathematics and physics led a successor to say that "Newton was the greatest genius that ever existed."

Although Kepler and those who followed attempted to explain the forces involved in planetary motion, their explanations were less than satisfactory. Kepler believed that some force pushed the planets along in their orbits. Galileo, however, correctly reasoned that no force is required to keep an object in motion. Galileo proposed that the natural tendency for a moving object (that is unaffected by an outside force) is to continue moving at a uniform speed and in a straight line. This concept, *inertia*, was later formalized by Newton as his first law of motion.

The problem, then, was not to explain the force that keeps the planets moving but rather to determine the force that *keeps them from going in a straight line out into space*. It was to this end that Newton conceptualized the force of *gravity*. At the early age of 23, he envisioned a force that extends from Earth into space and holds the Moon in orbit around Earth. Although others had theorized the existence of such a force, he was the first to formulate and test the *law of universal gravitation*. It states:

> Every body in the universe attracts every other body with a force that is directly proportional to their masses and inversely proportional to the square of the distance between them.

Thus, the gravitational force decreases with distance, so that two objects 3 kilometers apart have 3^2, or 9, times less gravitational attraction than if the same objects were 1 kilometer apart.

The law of gravitation also states that the greater the mass of the object, the greater its gravitational force. For example, the large mass of the Moon has a gravitational force strong enough to cause ocean tides on Earth, whereas the tiny mass of a communications satellite has no measurable effect on Earth. The mass of an object is a measure of the total amount of matter it contains. But more often mass is measured by determining the resistance an object exhibits in response to any effort made to change its state of motion.

Often we confuse the concept of *mass* with our notion of *weight*. Specifically, weight is the force of gravity acting upon an object. Therefore, weight varies when gravitational forces change. An object weighs less on the Moon than on Earth because the Moon is much less massive than Earth. However, unlike weight, the mass of an object does not change. For example, a person weighing 120 pounds on Earth weighs 1/6 as much, or 20 pounds, on the Moon, but the person's mass remains unchanged.

With his laws of motion, Newton proved that the force of gravity—combined with the tendency of a planet to remain in straight-line motion—results in the elliptical orbits discovered by Kepler. Earth, for example, moves forward in its orbit about 30 kilometers (18.5 miles) each second, and during the same second, the force of gravity pulls it toward the Sun about 0.5 centimeter (1/8 inch). Therefore, as Newton concluded, it is the combination of Earth's forward motion and its "falling" motion that defines its orbit (Figure 21.15). If gravity were somehow eliminated, Earth would move in a straight line out into space. Conversely, if Earth's forward motion suddenly stopped, gravity would pull it directly toward the Sun.

Up to this point, we have discussed Earth as if the only forces involved in its motion were caused by its gravitational relationship with the Sun. However, all bodies in the solar system have gravitational effects on Earth and on each other. For this reason, the orbit of Earth is not the perfect ellipse determined by Kepler. Any variance in the orbit of a body from its predicted path is called **perturbation** (*perturb* = disturb). For example, Jupiter's gravitational pull on Saturn reduces Saturn's orbital period by nearly one week from

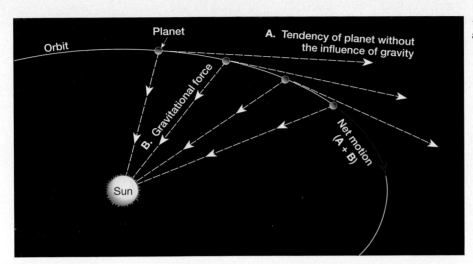

the predicted period. As we shall see, the application of this concept led to the discovery of the planet Neptune, because of Neptune's gravitational effect on the orbit of Uranus.

Newton used the law of universal gravitation to redefine Kepler's third law, which states the relationship between the orbital periods of the planets and their solar distances. When restated, Kepler's third law takes into account the masses of the bodies involved and thereby provides a method for determining the mass of a body when the orbit of one of its satellites is known. For example, the mass of the Sun is known from Earth's orbit, and Earth's mass has been determined from the orbit of the Moon. In fact, the mass of any body with a satellite can be determined. The masses of bodies that do not have satellites can be determined only if the bodies noticeably affect the orbit of a neighboring body or of a nearby artificial satellite.

Constellations

As early as 5000 years ago, people became fascinated with the star-studded skies and began to name the patterns they saw (see Box 21.2). These configurations, called **constellations** (*con* = with, *stella* = star), were named in honor of mythological characters or great heroes, such as Orion. It takes a good bit of imagination to make out the intended subjects, as most constellations were probably not thought of as likenesses in the first place. Although we inherited many of the constellations from Greek mythology, it is believed that Greek astronomers acquired most of theirs from the Babylonians, Egyptians, and Mesopotamians.

Although the stars that make up a constellation all appear to be the same distance from Earth, this is not the case. Some are many times farther away than others. Thus, the stars in a particular constellation are not associated with each other in any physical way. In addition, various cultural groups, including Native Americans, attached their own names, pictures, and stories to the constellations. For example, the constellation Orion, the Hunter, was known as the White Tiger to ancient Chinese astronomers.

Today, 88 constellations are recognized, and they are used to divide the sky into units, just as state boundaries divide the United States. Every star in the sky is in, but is not necessarily part of, one of these constellations. Constellations therefore enable astronomers to roughly identify the area of the heavens they are observing. For the student, the constellations provide a good way to become familiar with the night sky.

Some of the brightest stars were given proper names, such as Sirius, Arcturus, and Betelgeuse. In addition, the brightest stars in a constellation are generally named in order of their brightness by the letters of the Greek alphabet—alpha (α), beta (β), and so on—followed by the name of the parent constellation. For example, Sirius, the brightest star in the constellation Canis Major (Larger Dog), is called Alpha (α) Canis Majoris.

Positions in the Sky

If you gaze away from the city lights on a clear night, you will get the distinct impression that the stars produce a spherical shell surrounding Earth. This impression seems so real that it is easy to understand why many early Greeks regarded the stars as being fixed to a crystalline celestial sphere. Although we realize that no such sphere exists, it is convenient to use this concept for locating stars.

One method for doing this, called the **equatorial system**, divides the celestial sphere into a coordinate system. It is very similar to the latitude-longitude

BOX 21.2 UNDERSTANDING EARTH:
Astrology—the Forerunner of Astronomy

Many people confuse astrology and astronomy to the point of believing these terms to be synonymous. Nothing can be further from the truth. *Astronomy* is a scientific probing of the universe to derive the properties of celestial objects and the laws under which the universe operates. *Astrology*, on the other hand, is based on ancient superstitions that hold that an individual's actions and personality are based on the positions of the planets and stars now, and at the person's birth. Scientists do not accept astrology, regarding it as a pseudoscience ("false science"). It is hoped that most people read horoscopes only as a pastime and do not let them influence daily living.

Apparently astrology had its origin more than 3000 years ago when the positions of the planets were plotted as they regularly migrated against the background of the "fixed" stars. Because the solar system is "flat," like a whirling Frisbee, the planets orbit the Sun along nearly the same plane. Therefore, the planets, Sun, and Moon all appear to move along a band around the sky known as the *zodiac*. Because Earth's Moon cycles through its phases about 12 times each year, the Babylonians divided the zodiac into 12 constellations (Figure 21.C). Thus, each

Figure 21.D Stonehenge, an ancient observatory in England. On June 21–22 (summer solstice), the Sun can be observed rising above the heel stone. (Robin Scagell/Science Photo Library/Photo Researchers, Inc.)

successive full Moon can be seen against the backdrop of the next constellation.

The dozen constellations of the zodiac ("Zone of Animals," so named because some constellations represent animals) are Aries, Taurus, Gemini, Cancer, Leo, Virgo, Libra, Scorpio, Sagittarius, Capricorn, Aquarius, and Pisces. These names may be familiar to you as the astrological signs of the zodiac. When first established, the vernal equinox (first day of spring) occurred when the Sun was viewed against the constellation Aries. However, during each succeeding vernal equinox, the position of the Sun shifts very slightly against the background of stars. Now, over 2000 years later, the vernal equinox occurs when the Sun is in Pisces (Figure 21.C). In several years, it will occur when the Sun appears against Aquarius. (Hence, the "Age of Aquarius" is coming.)

Although astrology is not a science and has no basis in fact, it did contribute to the science of astronomy. The positions of the Moon, Sun, and planets at the time of a person's birth (sign of the zodiac) were considered to have great influence on that person's life.* Even the great astronomer Kepler was required to make horoscopes part of his duties. To make horoscopes for the future, astrologers attempted to predict the future positions of the celestial bodies. Consequently, some of the improvements in astronomical instruments were made because of the desire for more accurate predictions of events such as eclipses, which were considered highly significant in a person's life.

Even prehistoric people built observatories. The structure known as Stonehenge, in England, was undoubtedly an attempt at better solar predictions (Figure 21.D). At the time of midsummer in the Northern Hemisphere (June 21–22—the summer solstice), the rising Sun emerges directly above the heel stone of Stonehenge. Besides keeping this calendar, Stonehenge may also have provided a method of determining eclipses. The remnants of other early observatories exist elsewhere in the Americas, Europe, Asia, and Africa.

*It is interesting to note that 2000 years ago a person born on July 28 was considered a Leo because the Sun was in that constellation. During modern times the Sun appears in the constellation Cancer on this date, but individuals born during this time are still dubbed Leos.

Figure 21.C The 12 constellations of the zodiac. Earth is shown in its autumn (September) position in orbit, from which the Sun is seen against the background of the constellation Virgo.

[Figure 21.C labels: North celestial pole, Celestial sphere, Sun against constellations of the Zodiac, Virgo, Leo, Cancer, Libra, Scorpio, Sagittarius, Sun, Earth, Gemini, Taurus, Aries, Pisces, Aquarius, Capricorn]

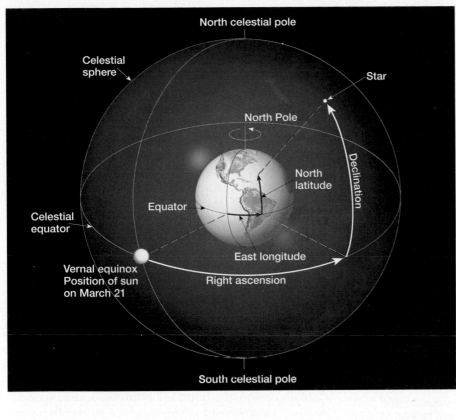

◀ **FIGURE 21.16** Astronomical coordinate system on the celestial sphere.

system used for locations on Earth's surface (Figure 21.16). Because the celestial sphere appears to rotate around an imaginary line extending from Earth's axis, the north and south celestial poles are in line with the terrestrial North Pole and South Pole. The north celestial pole happens to be very near the bright star whose various names reflect its location: "pole star," Polaris, and North Star. To an observer in the Northern Hemisphere, the stars appear to circle Polaris, because it, like the North Pole, is in the center of motion (Figure 21.17). Figure 21.18 shows how to locate the North Star by using two stars in the very visible constellation called the Big Dipper.

Now, imagine a plane through Earth's equator, a plane that extends outward from Earth and intersects the celestial sphere. The intersection of this plane with

the celestial sphere is the *celestial equator* (Figure 21.16). In the equatorial system, the terms *declination*, which is analogous to latitude, and *right ascension*, which is analogous to longitude, are used (Figure 21.16). **Declination** (*delinare* = to turn away), like latitude, is the angular distance north or south of the celestial equator. **Right ascension** (*ascendere* = to climb up) is the angular distance measured eastward along the celestial equator from the position of the vernal equinox. (The *vernal equinox* is at the point in the sky where the Sun crosses the celestial equator, at

▼ **FIGURE 21.18** Locating the North Star (Polaris) from the pointer stars in the Big Dipper, which is part of the constellation Ursa Major. The Big Dipper is shown soon after sunset in December (lower figure), April (upper figure), and August (left).

▼ **FIGURE 21.17** Star trails in the region of Polaris (north celestial pole) on a time exposure. (Courtesy of National Optical Astronomy Observatories)

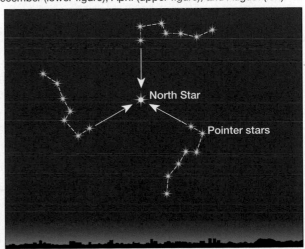

the onset of spring.) While declination is expressed in degrees, right ascension is usually expressed in hours, where each hour is equivalent to 15 degrees. To visualize distances on the celestial sphere, it helps to remember that the Moon and Sun have an apparent width of about 0.5 degree.

Motions of Earth

The two primary motions of Earth are rotation and revolution. **Rotation** is the turning, or spinning, of a body on its axis. **Revolution** is the motion of a body, such as a planet or moon, along a path around some point in space. For example, Earth *revolves* around the Sun, and the Moon *revolves* around Earth. Earth also has another very slow motion known as **precession**, which is the slight movement, over a period of 26,000 years, of Earth's axis.

Rotation

The main consequences of Earth's rotation are day and night. Earth's rotation has become a standard method of measuring time because it is so dependable and easy to use. Each rotation equals about 24 hours. You may be surprised to learn that we can measure Earth's rotation in two ways, making two kinds of days. Most familiar is the **mean solar day**, the time interval from one noon to the next, which averages about 24 hours. Noon is when the Sun has reached its zenith (highest point in the sky).

The **sidereal** (*sider* = star, *at* = pertaining to) **day**, on the other hand, is the time it takes for Earth to make one complete rotation (360 degrees) with respect to a star other than our Sun. The sidereal day is measured by the time required for a star to reappear at the identical position in the sky where it was observed the day before. The sidereal day has a period of 23 hours, 56 minutes, and 4 seconds (measured in solar time), which is almost 4 minutes shorter than the mean solar day. This difference results because the direction to distant stars changes only infinitesimally, owing to Earth's slow revolution along its orbit, whereas the direction to the Sun changes by almost 1 degree each day. This difference is shown in Figure 21.19.

If it is not apparent why we use the mean solar day rather than the sidereal day as a measurement of our day, consider the fact that in sidereal time, "noon" occurs four minutes earlier each day. Therefore, after a span of six months, "noon" occurs at "midnight." Astronomers use sidereal time because the stars appear in the same position in the sky every 24 sidereal hours. Usually, an observatory will begin its sidereal day when the position of the vernal equinox is directly overhead—that is, over the meridian on which the observatory is located. Therefore, when the observatory's sidereal clock is the same as the star's right ascension, the star will be overhead, or at its highest point. For example, the brightest star in the heavens, Sirius, has a right ascension of 6 hours, 42 minutes, and 56 seconds and will be overhead when the clock at the observatory indicates that time.

Revolution

Earth revolves around the Sun in an elliptical orbit at an average speed of 107,000 kilometers (66,000 miles) per hour. Its average distance from the Sun is 150 million kilometers (93 million miles), but because its orbit is an ellipse, Earth's distance from the Sun varies. At **perihelion** (*peri* = near, *helios* = Sun), it is 147 million

▼ **FIGURE 21.19** The difference between a solar day and a sidereal day. Locations X and Y are directly opposite each other. It takes Earth 23 hours and 56 minutes to make one rotation with respect to the stars (sidereal day). However, notice that after Earth has rotated once with respect to the stars, point Y is not yet returned to the "noon position" with respect to the Sun. Earth has to rotate another 4 minutes to complete the solar day.

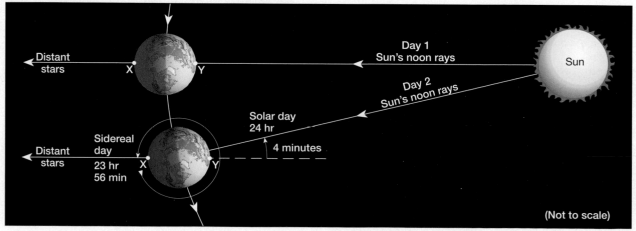

kilometers (91.5 million miles) distant, which occurs about January 3 each year. At **aphelion** (*apo* = away from, *helios* = Sun) Earth is 152 million kilometers (94.5 million miles) distant, which occurs about July 4.

Because of Earth's annual movement around the Sun, each day the Sun appears to be displaced among the constellations at a distance equal to about twice its width, or 1 degree. The apparent annual path of the Sun against the backdrop of the celestial sphere is called the **ecliptic** (Figure 21.20). Generally, the planets and the Moon travel in nearly the same plane as Earth. Hence, their paths on the celestial sphere lie near the ecliptic. (The most notable exception is Pluto, which has an orbit that is tilted 17 degrees to the plane of Earth's orbit.)

The imaginary plane that connects Earth's orbit with the celestial sphere is called the **plane of the ecliptic.** From the reference plane, Earth's axis of rotation is tilted about $23\frac{1}{2}$ degrees. Because of Earth's tilt, the apparent path of the Sun (ecliptic) and the celestial equator intersect each other at an angle of $23\frac{1}{2}$ degrees (Figure 21.20). This angle is very important to Earth's inhabitants. Because of the inclination of Earth's axis to the plane of the ecliptic, Earth exhibits its yearly cycle of seasons, a topic discussed in detail in Chapter 16.

When the *apparent* position of the Sun is plotted on the celestial sphere over a period of a year's time, its path intersects the celestial equator at two points (Figure 21.21). From a Northern Hemisphere point of view, these intersections are called the vernal (spring) equinox (March 20–21) and autumnal equinox (September 22–23). On June 21–22, the date of the summer solstice, the Sun appears $23\frac{1}{2}$ degrees north of the celestial equator, and six months later, on December 21–22, the date of the winter solstice, the Sun appears $23\frac{1}{2}$ degrees south of the celestial equator.

Precession

A third and very slow movement of Earth is called *precession*. Although Earth's axis maintains approximately the same angle of tilt, the direction in which the axis points continually changes. As a result, the axis traces a circle on the sky. This movement is very similar to the movement (wobble) of a spinning top (Figure 21.22A). At the present time, the axis points toward the bright star Polaris. In A.D. 14,000, it will point toward the bright star Vega, which will then become the North Star for a few thousand years (Figure 21.22B). The period of precession is 26,000 years. By the year 28,000, Polaris will once again be the North Star.

Precession has only a minor effect on the seasons, because the angle of tilt changes only slightly. It does, however, cause the positions of the seasons (equinox

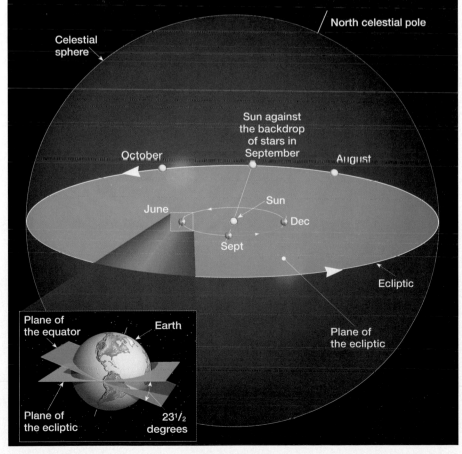

◀ **FIGURE 21.20** Earth's orbital motion causes the apparent position of the Sun to shift about 1 degree each day on the celestial sphere.

▶ **FIGURE 21.21** The apparent position of the Sun plotted on the celestial sphere. The path of the Sun (ecliptic) crosses the celestial equator on two occasions each year, March 20–21 and September 22–23. These are known as the equinox positions because the lengths of daylight and darkness on Earth are equal.

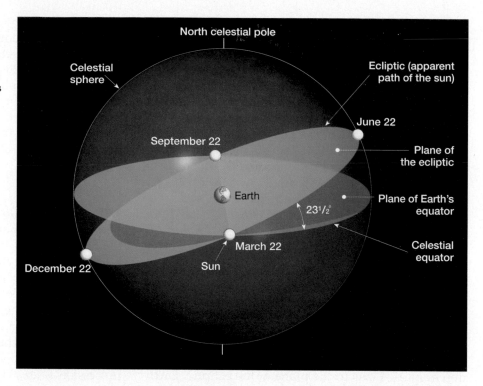

and solstice) to move slightly each year among the constellations.

? STUDENTS SOMETIMES ASK...

Our book states that Earth is farthest from the Sun in July and closest to the Sun in January. What would the seasons be like if this situation were reversed?

The situation you proposed will, in fact, occur in about 13,000 years. (Recall that variations in Earth–Sun distance are not the primary cause of the seasons. Nevertheless, they do affect average seasonal temperatures.) Gradually, Earth's position will change so that in 13,000 years the Northern Hemisphere will experience winter when Earth is farthest from the Sun (aphelion), and summer will occur when our planet is closest to the Sun (perihelion). This will be, of course, just the opposite of the current situation. When perihelion occurs in July, average summer temperatures in the Northern Hemisphere will be warmer than they presently are. A summer in Montreal, Canada, might be more akin to a typical summer in Washington, D.C., today. Likewise, Southern Hemisphere winters will also be warmer than they are currently. When aphelion occurs in January, northern latitudes will experience more extreme winter temperatures whereas average summer temperatures south of the equator will be cooler.

▶ **FIGURE 21.22** **A.** Precession illustrated by a spinning top. **B.** Precession of Earth causes the North Pole to point to different parts of the sky during a 26,000-year cycle. Today, the North Pole points to Polaris (North Star). In 13,000 years, Vega will be the North Star.

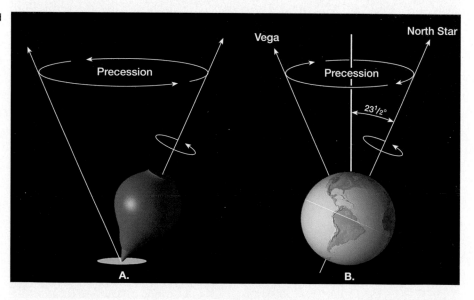

In addition to its own movements, Earth shares numerous motions with the Sun. It accompanies the Sun as the entire solar system speeds in the direction of the bright star Vega at 20 kilometers (12 miles) per second. Also, the Sun, like other nearby stars, revolves around the galaxy, a trip that requires 230 million years to traverse at speeds approaching 250 kilometers (150 miles) per second. In addition, the galaxies themselves are in motion. We are presently approaching one of our nearest galactic neighbors, the Great Galaxy in Andromeda. In summary, the motions of Earth are many and complex, and its speed in space is very great.

Motions of the Earth–Moon System

Earth has one natural satellite, the Moon. In addition to accompanying Earth in its annual trek around the Sun, our Moon orbits Earth within a period of about one month. When viewed from above the North Pole, the direction of this motion is counterclockwise (eastward). Because the Moon's orbit is elliptical, its distance to Earth varies by about 6 percent, averaging 384,401 kilometers (238,329 miles).

The motions of the Earth–Moon system constantly change the relative positions of the Sun, Earth, and Moon. The results are some of the most obvious of astronomical phenomena, namely the *phases of the Moon* and the occasional *eclipses of the Sun and Moon.*

Phases of the Moon

The first astronomical phenomenon to be understood was the regular cycle of the **phases of the Moon.** On a monthly basis, we observe the phases as a systematic change in the amount of the Moon that appears illuminated (Figure 21.23). We will choose the "new-Moon" position in the cycle as a starting point. About two days after the new Moon, a thin sliver (*crescent phase*) appears low in the western sky just after sunset. During the following week, the illuminated portion of the Moon visible from Earth increases (*waxing*) to a half-circle (*first-quarter phase*) and can be seen from about noon to midnight. In another week, the complete disk (*full-Moon phase*) can be seen rising in the east as the Sun is sinking in the west. During the next two weeks, the percentage of the Moon that can be seen steadily declines (*waning*), until the Moon disappears altogether (*new-Moon phase*). The cycle soon begins anew with the reappearance of the crescent Moon.

The lunar phases are a consequence of the motion of the Moon and the sunlight that is reflected from its surface (Figure 21.23B). Half of the Moon is illuminated at all times (note the inner group of Moon sketches in Figure 21.23A). But to an Earthbound observer, the percentage of the bright side that is visible depends on the location of the Moon with respect to the Sun and Earth. When the Moon lies *between* the Sun and Earth, none of its bright side faces Earth, so we see the new-Moon ("no-Moon") phase. Conversely, when the Moon lies on the side of Earth opposite the Sun, all of its lighted side faces Earth, so we see the full Moon. At all positions between these extremes, an intermediate amount of the Moon's illuminated side is visible from Earth.

Lunar Motions

The cycle of the Moon through its phases requires $29\frac{1}{2}$ days, a time span called the **synodic month.** This cycle was the basis for the first Roman calendar. However, this is the *apparent period* of the Moon's revolution around Earth and not the true period, which takes only $27\frac{1}{3}$ days and is known as the **sidereal month.** The reason for the difference of nearly two days each cycle is shown in Figure 21.24. Note that as the Moon orbits Earth, the Earth–Moon system also moves in an orbit around the Sun. Consequently, even after the Moon has made a complete revolution around Earth, it has not yet reached its starting position, which was directly between the Sun and Earth (new-Moon phase). The additional motion to reach the starting point takes another two days.

An interesting fact concerning the motions of the Moon is that its period of rotation about its axis and its revolution around Earth are the same—$27\frac{1}{3}$ days. Because of this, the same lunar hemisphere always faces Earth. All of the manned *Apollo* missions were confined to the Earth-facing side. Only orbiting satellites and astronauts have seen the "back" side of the Moon.

Because the Moon rotates on its axis only once every $27\frac{1}{3}$ days, any location on its surface experiences periods of daylight and darkness lasting about two weeks. This, along with the absence of an atmosphere, accounts for the high surface temperature of 127°C

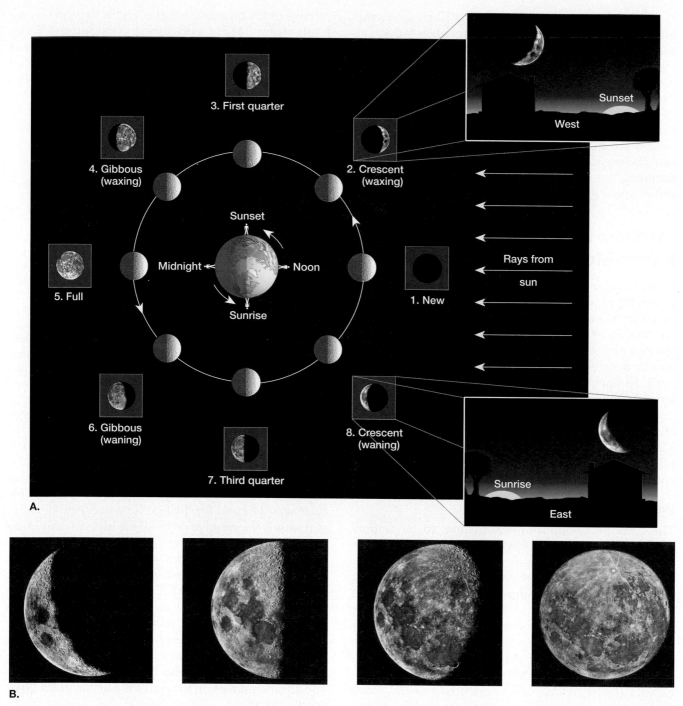

Sunset

West

3. First quarter

4. Gibbous (waxing)

2. Crescent (waxing)

Sunset

Midnight ← → Noon

Rays from

sun

5. Full

1. New

Sunrise

6. Gibbous (waning)

8. Crescent (waning)

Sunrise

East

7. Third quarter

A.

B.

▲ **FIGURE 21.23** Phases of the Moon. **A.** The outer figures show the phases as seen from Earth. **B.** Compare these photographs with the diagram. (Photos © UC Regents/Lick Observatory)

(261°F) on the day side of the Moon and the low surface temperature of −173°C(−280°F) on its night side.

Eclipses

Along with understanding the Moon's phases, the early Greeks also realized that eclipses are simply shadow effects. When the Moon moves in a line directly between Earth and the Sun, which can occur only during the new-Moon phase, it casts a dark shadow on Earth, producing a **solar eclipse** (*eclipsis* = to fail to appear) (Figure 21.25). Conversely, the Moon is eclipsed (**lunar eclipse**) when it moves within Earth's shadow, a situation that is possible only during the full-Moon phases (Figure 21.26).

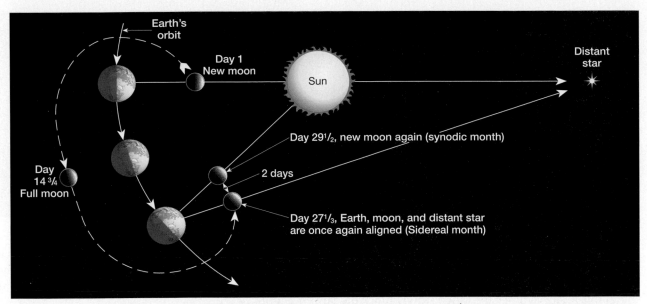

▲ **FIGURE 21.24** The difference between the sidereal month ($27\frac{1}{2}$ days) and the synodic month ($29\frac{1}{2}$ days). Distances and angles are not shown to scale.

Why then does a solar eclipse not occur with every new-Moon phase and a lunar eclipse with every full-Moon phase? They would, if the orbit of the Moon lay exactly along the plane of Earth's orbit (the plane of the ecliptic). However, the Moon's orbit is inclined about 5 degrees to the plane that contains Earth and the Sun (Figure 21.27). Thus, during most new-Moon phases, the shadow of the Moon misses Earth (passes above or below); and during most full-Moon phases, the shadow of Earth misses the Moon. Only when a new- or full-Moon phase occurs where the Moon's orbit crosses the plane of the ecliptic can an eclipse take place.

Because these conditions are normally met only twice a year, the usual number of eclipses is four. These occur as a set of one solar and one lunar eclipse, followed six months later with another set (Figure 21.27). Occasionally the alignment is such that three

eclipses can be squeezed into a one-month period. These occur as a solar eclipse flanked by two lunar eclipses, or vice versa. Furthermore, it occasionally happens that the first set of eclipses occurs at the very beginning of a year, and a third set occurs before the year ends, resulting in six eclipses in that year. More rarely, if one of these sets is a three-eclipse kind, the total number of eclipses in a year can reach seven, which is the maximum.

During a total lunar eclipse, Earth's circular shadow can be seen moving slowly across the disk of the full Moon. When totally eclipsed, the Moon is completely within Earth's shadow but still is visible as a coppery disk, because Earth's atmosphere bends and transmits some long-wavelength light (red) into its shadow. A total eclipse of the Moon can last up to four hours and is visible to anyone on the side of Earth facing the Moon.

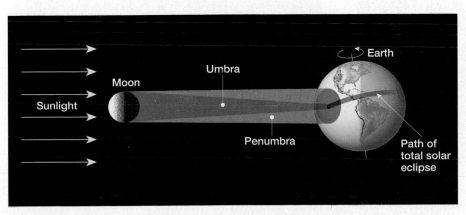

◄ **FIGURE 21.25** Solar eclipse. Observers in the zone of the umbral shadow see a total solar eclipse. Those in the penumbra see a partial eclipse. The path of the solar eclipse moves eastward across the globe.

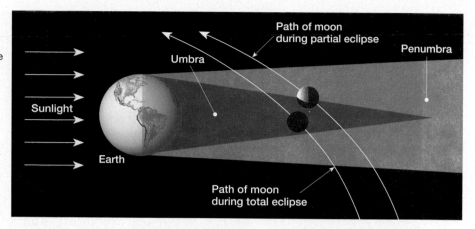

▶ **FIGURE 21.26** Lunar eclipse. During a total lunar eclipse the Moon's orbit carries it into the dark shadow of Earth (umbra). During a partial eclipse only a portion of the Moon enters the umbra.

During a total solar eclipse, the Moon casts a circular shadow that is never wider than 275 kilometers (170 miles), about the size of South Carolina. Anyone observing in this region will see the Moon slowly block the Sun from view and the sky darken (Figure 21.28). Near totality, a sharp drop in temperature of a few degrees is experienced. The solar disk is completely blocked for at most seven minutes, because the Moon's shadow is so small, and then one edge reappears.

At totality, the dark Moon is seen covering the complete solar disk, and only the Sun's brilliant white outer atmosphere is visible (see Figure 23.21). Total solar eclipses are visible only to people in the dark part of the Moon's shadow (*umbra*), while a partial eclipse is seen by those in the light portion (*penumbra*) (see Figure 21.25).

Partial solar eclipses are more common in the polar regions, because it is this zone that the penumbra covers when the dark umbra of the Moon's shadow just misses Earth. A total solar eclipse is a rare event at any given location. The next one that will be visible from the contiguous United States will take place on August 21, 2017.

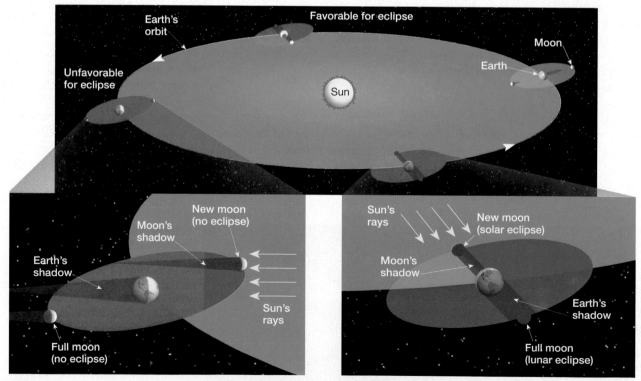

▲ **FIGURE 21.27** The Moon's orbit is inclined about 5 degrees to the plane that contains the Sun and Earth. Thus, during most new-Moon phases, the shadow of the Moon misses Earth (passes above or below), and during most full-Moon phases, the shadow of Earth misses the Moon. Only when a new- or full-Moon phase occurs where the Moon's orbit crosses the Earth-Sun plane can an eclipse take place. These conditions are met at roughly six-month intervals.

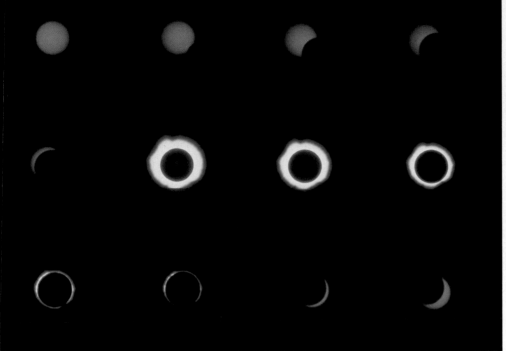

◀ **FIGURE 21.28** This sequence of photos starting from the upper left to the lower right shows the stages of a total solar eclipse. (From *Foundations of Astronomy*, Third Edition, p.54, by Michael Seeds. © 1992. Reprinted with permission of Brooks/Cole Publishing, a division of Thomson Learning)

Chapter Summary

- Early Greeks held the *geocentric* (Earth-centered) view of the universe, believing that Earth was a sphere that stayed motionless at the center of the universe. Orbiting Earth were the seven wanderers (*planetai* in Greek), which included the Moon, Sun, and the known planets Mercury, Venus, Mars, Jupiter, and Saturn. To the early Greeks, the stars traveled daily around Earth on a transparent, hollow sphere called the *celestial sphere.* In A.D. 141, *Claudius Ptolemy* presented the geocentric outlook of the Greeks in its most sophisticated form in a model that became known as the *Ptolemaic system.* The Ptolemaic model had the planets moving in circular orbits around a motionless Earth. To explain the *retrograde motion* of planets (the apparent westward or opposite motion that planets exhibit for a period of time as Earth overtakes and passes them), Ptolemy proposed that the planets orbited in small circles (*epicycles*), revolving along large circles (*deferents*).

- In the fifth century B.C., the Greek *Anaxagoras* reasoned that the Moon shines by reflected sunlight, and because it is a sphere, only half is illuminated at one time. *Aristotle* (384–322 B.C.) concluded that Earth is spherical. The first Greek to profess a Sun-centered, or *heliocentric*, universe was *Aristarchus* (312–230 B.C.). The first successful attempt to establish the size of Earth is credited to *Eratosthenes* (276–194 B.C.). The greatest of the early Greek astronomers was *Hipparchus* (second century B.C.), best known for his star catalogue.

- Modern astronomy evolved through the work of many dedicated individuals during the sixteenth and seventeenth centuries. *Nicolaus Copernicus* (1473–1543) reconstructed the solar system with the Sun at the center and the planets orbiting around it but erroneously continued to use circles to represent the orbits of planets. *Tycho Brahe's* (1546–1601) observations were far more precise than any made previously and are his legacy to astronomy. *Johannes Kepler* (1571–1630) ushered in the new astronomy with his three laws of planetary motion. After constructing his own telescope, *Galileo Galilei* (1564–1642) made many important discoveries that supported the Copernican view of a Sun-centered solar system. *Sir Isaac Newton* (1642–1727) was the first to formulate and test the law of universal gravitation, develop the laws of motion, and prove that the force of *gravity*, combined with the tendency of an object to move in a straight line (*inertia*), results in the elliptical orbits discovered by Kepler.

- As early as 5000 years ago people began naming the configurations of stars, called *constellations*, in honor of mythological characters or great heroes. Today, 88 constellations are recognized that divide the sky into units, just as state boundaries divide the United States.

- One method for locating stars, called the *equatorial system*, divides the celestial sphere into a coordinate system similar to the latitude-longitude system used for locations on Earth's surface. *Declination*, like latitude, is the angular distance north or south of the *celestial equator*. *Right ascension* is the angular distance measured eastward from the position of the *vernal equinox* (the point in the sky where the Sun crosses the celestial equator at the onset of spring).

- The two primary motions of Earth are *rotation* (the turning, or spinning, of a body on its axis) and *revolution* (the motion of a body, such as a planet or moon, along a path around some point in space). Another very slow motion of Earth is *precession* (the slow motion of Earth's axis that traces out a cone over a period of 26,000 years). Earth's rotation can be measured in two ways, making two kinds of days. The *mean solar day* is the time interval from one noon to the next, which averages about 24 hours. In contrast, the *sidereal day* is the time it takes for Earth to make one complete rotation with respect to a star other than the Sun, a period of 23 hours, 56 minutes, and 4 seconds. Earth revolves around the Sun in an elliptical orbit at an average distance from the Sun of 150 million kilometers (93 million miles). At *perihelion* (closest to the Sun), which occurs in January, Earth is 147 million kilometers from the Sun. At *aphelion* (farthest from the Sun), which occurs in July, Earth is 152 million kilometers distant. The imaginary plane that connects Earth's orbit with the celestial sphere is called the *plane of the ecliptic.*

• One of the first astronomical phenomena to be understood was the regular cycle of the phases of the Moon. The cycle of the Moon through its phases requires $29\frac{1}{2}$ days, a time span called the *synodic month*. However, the true period of the Moon's revolution around Earth takes $27\frac{1}{3}$ days and is known as the *sidereal month*. The difference of nearly two days is due to the fact that as the Moon orbits Earth, the Earth–Moon system also moves in an orbit around the Sun.

• In addition to understanding the Moon's phases, the early Greeks also realized that eclipses are simply shadow effects. When the Moon moves in a line directly between Earth and the Sun, which can occur only during the new-Moon phase, it casts a dark shadow on Earth, producing a *solar eclipse*. A *lunar eclipse* takes place when the Moon moves within the shadow of Earth during the full-Moon phase. Because the Moon's orbit is inclined about 5 degrees to the plane that contains the Earth and Sun (the plane of the ecliptic), during most new- and full-Moon phases no eclipse occurs. Only if a new- or full-Moon phase occurs as the Moon crosses the plane of the ecliptic can an eclipse take place. The usual number of eclipses is four per year.

Key Terms

aphelion (p. 601)
astronomical unit (AU) (p. 593)
celestial sphere (p. 586)
constellations (p. 597)
declination (p. 599)
ecliptic (p. 601)
equatorial system (p. 587)
geocentric (p. 596)
heliocentric (p. 597)

lunar eclipse (p. 604)
mean solar day (p. 600)
perihelion (p. 600)
perturbation (p. 596)
phases of the Moon (p. 603)
plane of the ecliptic (p. 601)
precession (p. 600)
Ptolemaic system (p. 588)
retrograde motion (p. 589)

revolution (p. 600)
right ascension (p. 599)
rotation (p. 600)
sidereal day (p. 600)
sidereal month (p. 603)
solar eclipse (p. 604)
synodic month (p. 603)

Review Questions

1. Why did the ancients believe that celestial objects had some control over their lives?

2. Describe what produces the retrograde motion of Mars. What geometric arrangements did Ptolemy use to explain this motion?

3. What major change did Copernicus make in the Ptolemaic system? Why was this change philosophically significant?

4. What was Tycho Brahe's contribution to science?

5. Does Earth move faster in its orbit near perihelion (January) or near aphelion (July)? Keeping your answer to the previous question in mind, is the solar day longest in January or July?

6. Use Kepler's third law ($p^2 = d^3$) to determine the period of a planet whose solar distance is
   (a) 10 AU
   (b) 1 AU
   (c) 0.2 AU

7. Use Kepler's third law to determine the distance from the Sun of a planet whose period is
   (a) 5 years.
   (b) 10 years.
   (c) 10 days.

8. Did Galileo invent the telescope?

9. Explain how Galileo's discovery of a rotating Sun supported the Copernican view of a Sun-centered universe.

10. Using a diagram, explain why the fact that Venus appears full when it is smallest supports the Copernican view and is inconsistent with the Ptolemaic system.

11. Newton learned that the orbits of the planets are the result of two actions. Explain these actions.

12. Of what value are constellations to modern-day astronomers?

13. Explain the difference between the mean solar day and the sidereal day.

14. What is the approximate length of the cycle of the phases of the Moon?

15. What is different about the crescent phase that precedes the new-Moon phase and that which follows the new-Moon phase?

16. What phase of the Moon occurs approximately one week after the new Moon? Two weeks?

17. When you observe the crescent phase early in the evening, is the visible Moon waxing (growing) or waning (declining)? (See Figure 21.23.)

18. What phenomenon results from the fact that the Moon's period of rotation and revolution are the same?

19. The Moon rotates very slowly (once in $27\frac{1}{3}$ days) on its axis. How does this affect the lunar surface temperature?

20. Describe the locations of the Sun, Moon, and Earth during a solar eclipse and during a lunar eclipse.

21. How many eclipses normally occur each year?

22. Solar eclipses are slightly more common than lunar eclipses. Why, then, is it more likely that your region of the country will experience a lunar eclipse?

23. How long can a total eclipse of the Moon last? How about a total eclipse of the Sun?

Examining the Earth System

1. Currently, Earth is closest to the Sun (perihelion) in January (147 million kilometers/91.5 million miles) and farthest from the Sun in July (152 million kilometers/94.5 million miles). As the result of the precession of Earth's axis, 13,000 years from now perihelion (closest) will occur in July and aphelion (farthest) will take place in January. Assuming no other changes, how might this change *average* summer temperatures for your location? What about *average* winter temperatures? What might the impact be on the biosphere and hydrosphere? (To aid your understanding of the effect of Earth's orbital parameters on the seasons, you may want to review the section on "Variations in Earth's Orbit" in Chapter 6.

2. In what ways do the interactions between Earth and its Moon influence the Earth system? If Earth did not have a Moon, would the atmosphere, hydrosphere, geosphere, and biosphere be any different? Explain.

Online Study Guide

The *Earth Science* Website uses the resources and flexibility of the Internet to aid in your study of the topics in this chapter. Written and developed by Earth science instructors, this site will help improve your understanding of Earth science. Visit **http://www.prenhall.com/tarbuck** and click on the cover of *Earth Science* 11e to find:

- **On-line review quizzes.**
- **Critical thinking exercises.**
- **Links to chapter-specific Web resources.**
- **Internet-wide key term searches.**
http://www.prenhall.com/tarbuck

*Panoramic view of the Martian landscape taken by the
Spirit rover in 2004.
(NASA/JPL/Cornell/Peter Arnold, Inc.)*

CHAPTER 22

Touring Our Solar System

The Sun is the hub of a huge rotating system of nine planets, their satellites, and numerous small asteroids, comets, and meteoroids. An estimated 99.85 percent of the mass of our solar system is contained within the Sun. The planets collectively make up most of the remaining 0.15 percent. The planets, traveling outward from the Sun, are Mercury, Venus, Earth, Mars, Jupiter, Saturn, Uranus, Neptune, and Pluto (Figure 22.1).

When people first recognized that the planets resembled Earth more than the stars, excitement grew. Could intelligent life exist on these other planets, or elsewhere in the universe? Space exploration has rekindled this interest. So far, no evidence of extraterrestrial life within our solar system has emerged. Nevertheless, we study the other planets to learn about Earth's formation and early history. Recent space explorations have been organized with this goal in mind. To date, Mercury, Venus, Mars, Jupiter, Saturn, Uranus, and Neptune have been explored by space probes.

The Planets: An Overview

Earth's Place in the Universe
▼ The Planets: An Overview

Under control of the Sun's gravitational force, each planet is tethered in an elliptical orbit, and all of them travel in the same direction. The nearest planet to the Sun—Mercury—has the fastest orbital motion, 48 kilometers per second, and the shortest period of revolution around the Sun, 88 Earth-days. By contrast, the most distant planet, Pluto, has an orbital speed of 5 kilometers per second and requires 248 Earth-years to complete one revolution.

Imagine a planet's orbit drawn on a flat sheet of paper. The paper represents the planet's *orbital plane*. The orbital planes of seven planets lie within 3 degrees of the plane of the Sun's equator. The other two, the innermost and outermost, Mercury and Pluto, are inclined 7 and 17 degrees, respectively.

Careful examination of Table 22.1 shows that the planets fall quite nicely into two groups: the **terrestrial**

▼ **FIGURE 22.1** Orbits of the planets. Positions of the planets are shown to scale along bottom of diagram.

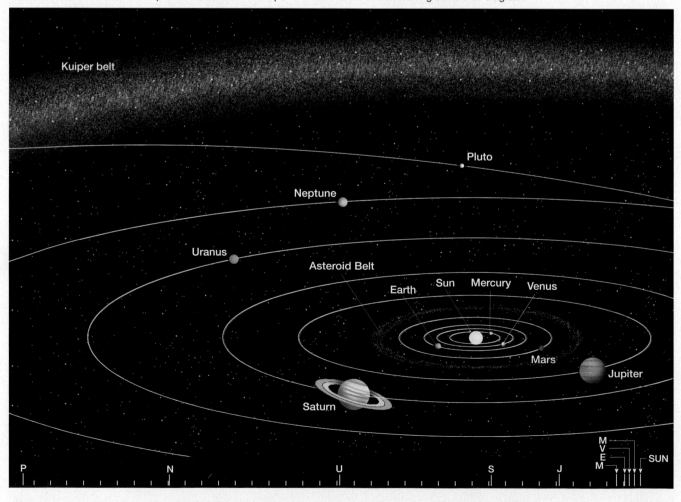

TABLE 22.1 Planetary data.

Planet	Symbol	Mean Distance from Sun: AU*	Mean Distance from Sun: Millions of Miles	Mean Distance from Sun: Millions of Kilometers	Period of Revolution	Inclination of Orbit	Orbital Velocity: mi/s	Orbital Velocity: km/s
Mercury	☿	0.39	36	58	88^d	7°00′	29.5	47.5
Venus	♀	0.72	67	108	225^d	3°24′	21.8	35.0
Earth	⊕	1.00	93	150	365.25^d	0°00′	18.5	29.8
Mars	♂	1.52	142	228	687^d	1°51′	14.9	24.1
Jupiter	♃	5.20	483	778	12^{yr}	1°18′	8.1	13.1
Saturn	♄	9.54	886	1427	29.5^{yr}	2°29′	6.0	9.6
Uranus	⛢	19.18	1783	2870	84^{yr}	0°46′	4.2	6.8
Neptune	♆	30.06	2794	4497	165^{yr}	1°46′	3.3	5.3
Pluto	♇	39.44	3666	5900	248^{yr}	17°12′	2.9	4.7

Planet	Period of Rotation	Diameter: Miles	Diameter: Kilometers	Relative Mass (Earth = 1)	Average Density (g/cm^3)	Polar Flattening (%)	Eccentricity†	Number of Known Satellites**
Mercury	59^d	3015	4878	0.06	5.4	0.0	0.206	0
Venus	244^d	7526	12,104	0.82	5.2	0.0	0.007	0
Earth	$23^h\ 56^m\ 04^s$	7920	12,756	1.00	5.5	0.3	0.017	1
Mars	$24^h\ 37^m\ 23^s$	4216	6794	0.11	3.9	0.5	0.093	2
Jupiter	$9^h\ 50^m$	88,700	143,884	317.87	1.3	6.7	0.048	28
Saturn	$10^h\ 14^m$	75,000	120,536	95.14	0.7	10.4	0.056	31
Uranus	$17^h\ 14^m$	29,000	51,118	14.56	1.2	2.3	0.047	21
Neptune	$16^h\ 03^m$	28,900	50,530	17.21	1.7	1.8	0.009	8
Pluto	6.4^d	≈1425	≈2300	0.002	1.8	0.0	0.250	1

*AU — astronomical unit, Earth's mean distance from the Sun.
**Includes all satellites discovered as of April 2004.
†Eccentricity is a measure of the amount an orbit deviates from a circular shape. The larger the number, the less circular the orbit.

(Earth like) **planets** (Mercury, Venus, Earth, and Mars) and the **Jovian** (Jupiter-like) **planets** (Jupiter, Saturn, Uranus, and Neptune). Pluto is not included in either category. The reason for this will be explored later in the chapter.

The most obvious difference between the terrestrial and the Jovian planets is their size (Figure 22.2). The largest terrestrial planets (Earth and Venus) have diameters only one-quarter as great as the diameter of the smallest Jovian planet (Neptune). Also, their masses are only 1/17 as great as Neptune's. Hence, the Jovian planets are often called *giants.* Because of their relative locations, the four Jovian planets are referred to as the **outer planets,** whereas the terrestrial planets are called the **inner planets.** As we shall see, there appears to be a correlation between the positions of these planets and their sizes.

Other dimensions in which the two groups differ include density, chemical makeup, and rate of rotation. The densities of the terrestrial planets average about five times the density of water, whereas the Jovian planets have densities that average only 1.5 times that of water. One of the outer planets, Saturn, has a density only 0.7 times that of water, which means that Saturn would float if placed in a large enough water tank! Variations in the chemical compositions of the planets are largely responsible for these density differences.

The Interiors of the Planets

The substances that make up the planets are divided into three compositional groups: *gases, rocks,* and *ices,* based on their melting points.

1. The gases, hydrogen and helium, are those with melting points near absolute zero (−273°C or 0 kelvin).
2. The rocks are principally silicate minerals and metallic iron, which have melting points exceeding 700°C.
3. The ices include ammonia (NH_3), methane (CH_4), carbon dioxide (CO_2), and water (H_2O). They have intermediate melting points (for example, H_2O has a melting point of 0°C).

The terrestrial planets are dense, consisting mostly of rocky and metallic substances, with minor amounts of gases and ices. The Jovian planets, on the other hand, contain large amounts of gases (hydrogen and helium) and ices (mostly water, ammonia, and methane). This accounts for their low densities. The outer planets also contain substantial amounts of rocky and metallic materials, which are concentrated in their cores.

Mercury

Venus

Earth

Mars

Sun

Jupiter

Saturn

Uranus

Neptune

Pluto

▲ FIGURE 22.2 The planets drawn to scale.

The Atmospheres of the Planets

The Jovian planets have very thick atmospheres of hydrogen, helium, methane, and ammonia. By contrast, the terrestrial planets, including Earth, have meager

atmospheres at best. The reason is that a planet's ability to retain an atmosphere depends on its mass and temperature.

Simply stated, a gas molecule can evaporate from a planet if it reaches a speed known as the **escape velocity.** For Earth, this velocity is 11 kilometers (7 miles) per second (about 25,000 miles per hour). Any material, including a rocket, must reach this speed before it can escape Earth's gravity and go into space.

The Jovian planets, because of their greater surface gravities, have escape velocities of 21 to 60 kilometers per second (45,000 to 134,000 miles per hour), much higher than the terrestrial planets. Consequently, it is more difficult for gases to evaporate from them. Also, because the molecular motion of a gas is temperature-dependent, at the low temperatures of the Jovian planets even the lightest gases are unlikely to acquire the speed needed to escape.

In contrast, a comparatively warm body with a small surface gravity, like our Moon, is unable to hold even heavy gases, like carbon dioxide and radon, and thus lacks an atmosphere. The slightly larger terrestrial planets of Earth, Venus, and Mars retain some heavy gases such as carbon dioxide, but even their atmospheres make up only an infinitesimally small portion of their total mass.

In the remainder of this chapter, we will consider each planet briefly, plus the minor members of the solar system. First, however, let us visit Earth's companion in space, our own Moon.

? STUDENTS SOMETIMES ASK...
Why are the Jovian planets so much larger than the terrestrial planets?

According to the nebular hypothesis, the planets formed from a rotating disk of dust and gases that surrounded the Sun. The growth of planets began as solid bits of matter began to collide and clump together. In the inner solar system, the temperatures were so high that only metals and silicate minerals could form solid grains. It was too hot for ices of water, carbon dioxide, and methane to form. Thus, the innermost (terrestrial) planets grew mainly from the high melting point substances found in the solar nebula. By contrast, in the frigid outer reaches of the solar system, it was cold enough for ices of water and other substances to form. Consequently, the outer planets grew not only from accumulations of solid bits of metals and silicate minerals but also from large quantities of ices. Eventually, the outer planets became large enough to gravitationally capture even the lightest gases (hydrogen and helium), and thus grow to become "giant" planets.

Earth's Moon

Earth's Place in the Universe
▼ Earth's Moon

Earth now has hundreds of satellites, but only one natural satellite, the Moon, accompanies us on our annual journey around the Sun. Although other planets have moons, our planet-satellite system is unique in the solar system, because Earth's Moon is unusually large compared to its parent planet. The diameter of the Moon is 3475 kilometers (2150 miles), about one-fourth of Earth's 12,756 kilometers (7920 miles).

From calculation of the Moon's mass, its density is 3.3 times that of water. This density is comparable to that of *mantle* rocks on Earth but is considerably less than Earth's average density, which is 5.5 times that of water. Geologists have suggested that this difference can be accounted for if the Moon's iron core is small. The gravitational attraction at the lunar surface is one-sixth of that experienced on Earth's surface (a 150-pound person on Earth weighs only 25 pounds on the Moon). This difference allows an astronaut to carry a heavy life-support system with relative ease. If not burdened with such a load, an astronaut could jump six times higher than on Earth.

The Lunar Surface

The Moon has no atmosphere or flowing water. Therefore, the processes of weathering and erosion that continually modify Earth's surface are virtually lacking on the Moon. In addition, tectonic forces are not active on the Moon, so earthquakes and volcanic eruptions no longer occur. However, because the Moon is unprotected by an atmosphere, a different kind of erosion occurs: tiny particles from space (micrometeorites) continually bombard its surface and gradually smooth the landscape.

Most of the Moon's surface was shaped by meteoroid impacts and ancient volcanic processes that largely ceased nearly 3 billion years ago. Since then, only scattered meteorite impacts have modified the lunar surface.

Craters. The most obvious features of the lunar surface are craters. They are so profuse that craters-within-craters are the rule! The larger ones in the lower portion of Figure 22.3 are about 250 kilometers (150 miles) in diameter, roughly the width of Indiana. Most craters were produced by the impact of rapidly moving debris (meteoroids), a phenomenon that was considerably more common in the early history of the solar system than it is today.

By contrast, Earth has only about a dozen easily recognized impact craters. This difference can be attributed

◄ **FIGURE 22.3** Telescopic view from Earth of the lunar surface. The major features are the dark "seas" (maria) and the light highly cratered lunar highlands (terrae). (UCO/Lick Observatory Image)

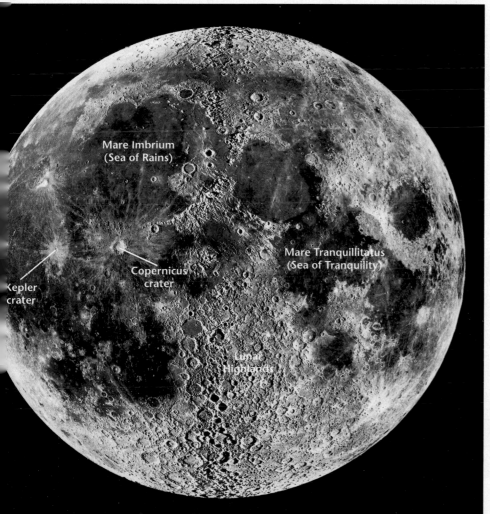

Mare Imbrium
(Sea of Rains)

Copernicus
crater

Kepler
crater

Mare Tranquillitatus
(Sea of Tranquility)

Lunar
Highlands

to Earth's atmosphere. Friction with the air burns up small debris before it reaches the ground. In addition, evidence for most of the craters that formed in Earth's history has been obliterated by erosion or tectonic processes.

The formation of an impact crater is illustrated in Figure 22.4. Upon impact, the high-speed meteoroid

▼ **FIGURE 22.4** Formation of an impact crater. The energy of the rapidly moving meteoroid is transformed into heat energy and compressional waves. The rebound of the compressed rock causes debris to be ejected from the crater, and the heat melts some material, producing shock-melted glass. Small secondary craters are often formed by the material "splashed" from the impact crater. (After E. M. Shoemaker)

compresses the material it strikes. Then almost instantaneously the compressed rock rebounds, ejecting material from the crater. This process is analogous to the splash that occurs when a rock is dropped into water. Large craters that are created by objects several kilometers across often exhibit a central peak, as seen in Figure 22.5. Most of the ejected material (*ejecta*) lands near the crater, building a rim around it. The heat generated by the impact is sufficient to melt some of the impacted rock. Astronauts have brought back samples of glass beads produced in this manner, as well as rock formed when broken fragments and dust were welded together by the impact.

A meteoroid only 3 meters (10 feet) in diameter can blast out a 150-meter- (500-foot-) wide crater. A few of the large craters, such as Kepler and Copernicus, shown in Figure 22.3, formed from the impact of bodies 1 kilometer or more in diameter. These two large craters are thought to be relatively young because of the bright *rays* (splash marks) that radiate outward for hundreds of kilometers.

Highlands and "Seas." When Galileo first pointed his telescope toward the Moon, he saw two different types of terrain (see Figure 22.3). The bright areas as seen from Earth are called **terrae** (plural for *terra*, the Latin word for land), or **lunar highlands**. By contrast, the dark areas, called **maria** (plural for *mare*, the Latin word for sea), are flat, lowland regions. (The use of the name maria came from the mistaken belief that these features were seas similar to those on Earth.) Together the arrangement of terrae and maria result in the well-known "face of the Moon."

The densely pockmarked highland areas make up most of the lunar surface (Figure 22.6). In fact, most of the far side of the Moon is characterized by such to-

▼ **FIGURE 22.5** The 20-kilometer-wide lunar crater Euler in the southwestern part of Mare Imbrium. Clearly visible are the bright rays, central peak, secondary craters, and the large accumulation of ejecta near the crater rim. (Courtesy of NASA)

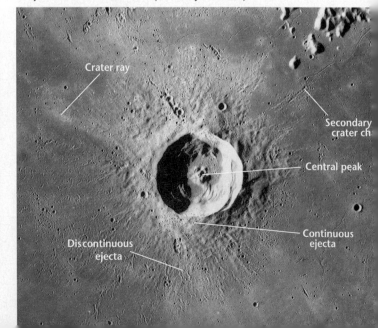

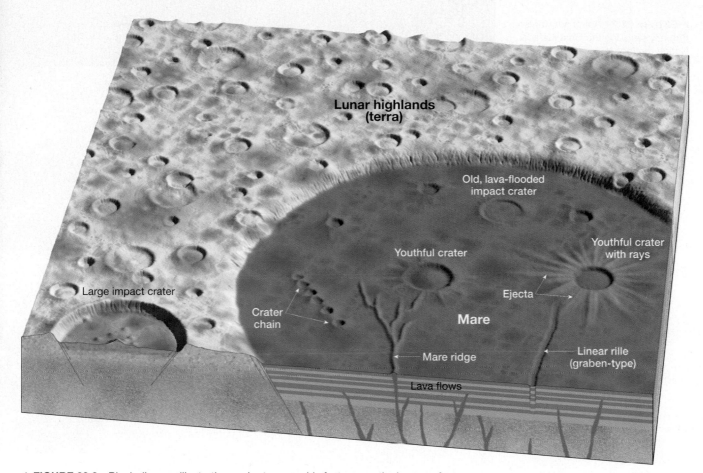

▲ **FIGURE 22.6** Block diagram illustrating major topographic features on the lunar surface.

pography. (Only astronauts have directly observed the far side, because the Moon rotates on its axis once with each revolution around Earth, always keeping the same side facing Earth.) As seen in Figure 22.3, the highlands consist of an apparently endless sequence of overlapping craters. Indeed, the great number of impact craters is evidence of the Moon's early violent history. This activity crushed and repeatedly mixed at least the upper few kilometers of the lunar crust. As a result, the terrae consist of very rugged topography.

The dark, flat maria make up only about 16 percent of the Moon's landscape and are concentrated on the side of the Moon facing Earth (Figure 22.6). Each basin formed when a meteorite, perhaps having a diameter as large as Rhode Island, excavated a huge crater. Because the crust was sufficiently fractured, magma began to bleed out. Apparently, the craters were flooded with layer upon layer of very fluid basaltic lava, forming lava plains that resemble the Columbia Plateau in the northwestern United States. The lava flows are often over 30 meters (100 feet) thick, and the total thickness of the material that fills the maria may approach a few thousand meters.

Regolith. Both the maria and terrae are mantled with a layer of gray, unconsolidated debris derived from a few billion years of meteoric bombardment (Figure 22.7). This soil-like layer, called **lunar regolith** (*rhegos* = blanket, *lithos* = stone), is composed of igneous rocks, breccia, glass beads, and fine *lunar dust*. In the maria that have been explored by *Apollo* astronauts, the lunar regolith appears to be just over 3 meters (10 feet) thick.

Lunar History

Although the Moon is our nearest planetary neighbor and astronauts have sampled its surface directly, much is still unknown about its early history. The most widely accepted model for the origin of the Moon suggests that during the formative period of the solar system, a Mars-size body collided with Earth. According to this hypothesis, the off-center blow ejected huge quantities of crustal and mantle rock from both bodies. A portion of this ejected debris entered an orbit around Earth, where it rapidly coalesced to form the Moon.

The giant-impact hypothesis is consistent with a number of facts we know about the Moon. The ejected material was mostly iron-poor mantle and crustal rocks, which would account for the lack of a sizable iron core on the Moon. Further, the impact would have vaporized much of the ejected debris, thereby

► **FIGURE 22.7** Astronaut Harrison Schmitt sampling the lunar surface. Notice the footprints (inset) in the lunar "soil." (Courtesy of NASA)

driving off the volatile materials, such as water. This explains the relative absence of water on the Moon.

Planetary geologists have also worked out the basic details of the Moon's more recent history. One of the methods of dating topographic features on the lunar surface is to observe variations in crater density (quantity per unit area). The greater the crater density, the longer the feature must have existed. From such evidence, scientists concluded that the Moon's surface evolved in four phases that produced: (1) the original crust, (2) the lunar highlands, (3) the maria basins, and (4) the rayed craters.

During its formation about 4.5 billion years ago, the Moon was continually impacted as it accumulated debris. This continuous bombardment, and perhaps radioactive decay, generated enough heat to melt the Moon's outer shell and quite possibly some of the interior as well. Based on samples recovered from the lunar highlands, the outer shell of the Moon solidified early in its history (about 4.3 billion years ago) to form a crust composed largely of plagioclase feldspar. Once formed, the lunar crust went through a period of heavy bombardment that lasted at least 500 million years. This period of intense cratering destroyed the upper crust and created the rugged highland topography.

The next major event was the formation of the large maria basins. Based on radiometric dates of samples obtained by the *Apollo* and *Luna* missions, three of the maria basins were created at about the same time—3.9 billion years ago. It has been proposed that a very large body in the asteroid belt broke apart 3.9 billion years ago and the debris was swept toward the Earth–Moon system. Although an interesting idea, this catastrophic period of cratering has yet to be verified by sampling of the other maria basins. What is known is that once formed, the basins were flooded with basaltic lavas over a time span of perhaps a billion years or more.

The last prominent features to form were the rayed craters, as exemplified by the Copernicus crater in Figure 22.3. Material ejected from these more recently formed depressions is clearly seen blanketing the surface of the maria and many older rayless craters. Even a relatively young crater like Copernicus must be millions of years old. Had it formed on Earth, erosional forces would have long since obliterated it.

If photos of the Moon taken several hundreds of millions of years ago were available, they would reveal that the Moon changed little in the intervening years. By all measures, the Moon is a tectonically dead body wandering through space and time.

The Planets: A Brief Tour

GEODe

Earth's Place in the Universe
▼ A Brief Tour of the Planets

Mercury: The Innermost Planet

Mercury, the innermost and second smallest planet, is hardly larger than Earth's Moon and is smaller than three other moons in the solar system. Like our own Moon, it absorbs most of the sunlight that strikes it, reflecting only 6 percent into space (Figure 22.8). This is characteristic of terrestrial bodies that have no atmosphere. (Earth reflects about 30 percent of the light that strikes it, most of it from clouds.)

Mercury has cratered highlands, much like the Moon, and vast smooth terrains that resemble maria. However, unlike the Moon, Mercury is a very dense planet, which implies that it contains a very large iron core for its size. Also, Mercury has very long scarps that cut across the plains and craters alike. These scarps may have resulted from crustal shortening as the planet cooled and shrank.

Mercury revolves quickly but rotates slowly. One full day–night cycle on Earth takes 24 hours, but on Mercury it requires 179 Earth-days. Thus, a night on Mercury lasts for about three months and is followed by three months of daylight. Nighttime temperatures drop as low as −173°C (−280°F), and noontime temperatures exceed 427°C (800°F), hot enough to melt tin and lead. Mercury has the greatest extremes of any planet. The odds of life as we know it existing on Mercury are nil.

STUDENTS SOMETIMES ASK...
Do any nearby stars have planets?

Yes. Although this was long suspected, it was not until recently that the presence of extrasolar planets has been verified. Astronomers have found these bodies by measuring the telltale wobbles of nearby stars. The first apparent planet outside the solar system was discovered in 1995, orbiting the star 51 Pegasi, 42 light-years from Earth. Since that time, over two dozen Jupiter-size bodies have been identified, most of them surprisingly close to the stars they orbit.

Venus: The Veiled Planet

Venus, second only to the Moon in brilliance in the night sky, is named for the goddess of love and beauty. It orbits the Sun in a nearly perfect circle once every 255 Earth-days. Venus is similar to Earth in size, density, mass, and location in the solar system. Thus, it has been referred to as "Earth's twin." Because of these similarities, it is hoped that a detailed study of Venus will provide geologists with a better understanding of Earth's evolutionary history.

Venus is shrouded in thick clouds impenetrable to visible light. Nevertheless, radar mapping by the unmanned *Magellan* spacecraft and by instruments on Earth has revealed a varied topography with features somewhat between those of Earth and Mars (Figure 22.9). Simply, radar pulses in the microwave range are sent toward the Venusian surface, and the heights of plateaus and mountains are measured by timing the return of the radar echo. These data have confirmed that basaltic volcanism and tectonic deformation are the dominant processes operating on Venus. Further, based on the low density of impact craters, volcanism and tectonic deformation must have been very active during the recent geologic past.

About 80 percent of the Venusian surface consists of subdued plains that are mantled by volcanic flows. Some lava channels extend hundreds of kilometers; one meanders 6800 kilometers across the planet. Thousands of volcanic structures have been identified, mostly small shield volcanoes, although over 1500 volcanoes greater than 20 kilometers (12 miles) across have been mapped (Figure 22.10). One is Sapas Mons,

▼ **FIGURE 22.8** Photomosaic of Mercury. This view of Mercury is remarkably similar in appearance to the "far side" of Earth's Moon. (Courtesy of NASA)

▲ FIGURE 22.9 This global view of the surface of Venus is computer-generated. It resulted from two years of Magellan Project radar mapping. The twisting bright features that cross the globe are highly fractured mountains and canyons of the eastern Aphrodite highland. (Courtesy of NASA/JPL)

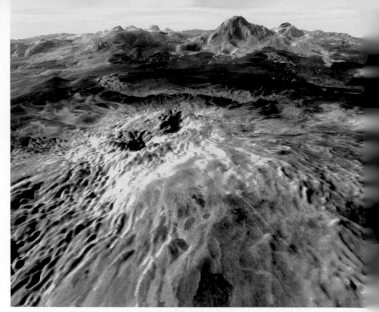

▲ FIGURE 22.10 Computer-generated image of Venus. Near the horizon is Maat Mons, a large volcano. The bright feature below is the summit of Sapas Mons. (Courtesy David P. Anderson/SMU/NASA/Science Photo Library/Photo Researchers, Inc.)

400 kilometers (250 miles) across and 1.5 kilometers (0.9 mile) high. Many flows from this volcano erupted from its flanks rather than the summit, in the manner of Hawaiian shield volcanoes.

Only 8 percent of the Venusian surface consists of highlands that may be likened to continental areas on Earth. Tectonic activity on Venus seems to be driven by upwelling and downwelling of material in the planet's interior. Although mantle convection still operates on Venus, the processes of plate tectonics, which recycle rigid lithosphere, do not appear to have contributed to the present Venusian topography.

Before the advent of space vehicles, Venus was considered to be a potentially hospitable site for living organisms. However, evidence from space probes indicates otherwise. The surface of Venus reaches temperatures of 475°C (900°F), and the Venusian atmosphere is 97 percent carbon dioxide. Only scant water vapor and nitrogen have been detected. The Venusian atmosphere contains an opaque cloud layer about 25 kilometers (15 miles) thick, and has an atmospheric pressure that is 90 times that at Earth's surface. This hostile environment makes it unlikely that life as we know it exists on Venus.

Mars: The Red Planet

Mars has evoked greater interest than any other planet, for both scientists and nonscientists (see Box 22.1). When one imagines intelligent life on other worlds, little green Martians may come to mind. Interest in Mars

stems mainly from this planet's accessibility to observation. All other planets within telescopic range have their surfaces hidden by clouds, except for Mercury, whose nearness to the Sun makes viewing difficult. Through the telescope, Mars appears as a reddish ball interrupted by some permanent dark regions that change intensity during the Martian year. The most prominent telescopic features of Mars are its brilliant white polar caps, resembling Earth's.

The Martian Atmosphere. The Martian atmosphere has only 1 percent the density of Earth's, and it is primarily carbon dioxide with tiny amounts of water vapor. Data from Mars probes confirm that the polar caps of Mars are made of water ice, covered by a thin layer of frozen carbon dioxide. As winter nears in either hemisphere, we see the equatorward growth of that hemisphere's ice cap as temperatures drop to −125°C (−193°F), and additional carbon dioxide is deposited.

Although the atmosphere of Mars is very thin, extensive dust storms occur and may cause the color changes observed from Earth-based telescopes. Hurricane-force winds up to 270 kilometers (170 miles) per hour can persist for weeks. Images from *Viking 1* and *Viking 2* revealed a Martian landscape remarkably similar to a rocky desert on Earth, with abundant sand dunes and impact craters partially filled with dust.

Mars' Dramatic Surface. *Mariner 9*, the first spacecraft to orbit another planet, reached Mars in 1971 amid a raging dust storm. When the dust cleared, images of Mars's northern hemisphere revealed numerous large volcanoes. The biggest, Mons Olympus, is the size of Ohio and is 23 kilometers (75,000 feet) high, over two and a half times higher than Mount Everest.

BOX 22.1 UNDERSTANDING EARTH:
Pathfinder—The First Geologist on Mars

On July 4, 1997, the *Mars Pathfinder* bounced onto the rock-littered surface of Mars and deployed its wheeled companion, *Sojourner*. For the next three months the lander sent back to Earth three gigabits of data, including 16,000 images and 20 chemical analyses. The landing site was a vast rolling landscape carved by ancient floods. The flood-deposit locale was selected in hope that a variety of rock types would be available for the rover *Sojourner* to examine.

Sojourner carried an alpha photon X-ray spectrometer (APXS) used to determine the chemical composition of rocks and Martian "soil" (regolith) at the landing site (Figure 22.A). In addition, the rover was able to take close-up images of the rocks. From these images, researchers concluded that the rocks were igneous. However, one hard, white, flat object named Scooby Doo was originally thought to be sedimentary rock, but the APXS data suggest its chemistry is like that of the soil found at the site. Thus, Scooby Doo is probably a well-cemented soil.

During its first week on Mars, *Sojourner*'s APXS obtained data for a patch of windblown soil and a medium-sized rock, known affectionately as Barnacle Bill. Preliminary evaluation of the APXS data on Barnacle Bill shows that it contains over 60 percent silica. If these data

Figure 22.A *Pathfinder*'s rover *Sojourner* (left) obtaining data on the chemical composition of a Martian rock known as Yogi. (Photo courtesy of NASA)

are confirmed, it could indicate that Mars contains the volcanic rock andesite. Researchers had expected that most volcanic rocks on Mars would be basalt, which is lower in silica (less than 50 percent). On Earth, andesites are associated with tectonically active regions where oceanic crust is subducted into the mantle. Examples include the volcanoes of South America's Andes Mountains and the Cascades of North America.

Sojourner analyzed eight rocks and seven soils. The results thus far are only prelimary. Because these rocks are covered with a reddish dust that is high in sulfur, some controversy has arisen as to the exact composition of these Martian rocks. Some researchers believe they are all of the same composition. The differences in measurements, they claim, are the result of varying thickness of dust.

This gigantic volcano and others resemble Hawaiian shield volcanoes on Earth (Figure 22.11).

Most Martian surface features are old by Earth standards. The highly cratered southern hemisphere is probably similar in age to the lunar highlands (3.5 billion to 4.5 billion years old). Even the relatively fresh-appearing volcanic features of the northern hemisphere may be older than 1 billion years. This fact and the absence of Mars quake recordings by *Viking* seismographs point to a tectonically dead planet.

Another surprising find made by *Mariner 9* was the existence of several canyons that dwarf even Earth's Grand Canyon of the Colorado River. One of the largest, Valles Marineris, is thought to have formed by slippage of material along huge faults in the crustal layer. In this respect, it would be comparable to the rift valleys of Africa (Figure 22.12).

Water on Mars? Today liquid water does not exist anywhere on the Martian surface. However, poleward of about 30° latitude, ice can be found within a meter of the surface, and in the polar regions, it forms small permanent ice caps. In addition, considerable evidence indicates that in the first billion years of the planet's history, liquid water flowed on the surface, creating valleys and related features. In particular, some areas of Mars exhibit branching drainage patterns similar to those created by streams on Earth. Furthermore, *Viking* orbiter images revealed unmistakable ancient islands in what are now dry streambeds. When these stream-like channels were first discovered, some observers speculated that a thick water-rich atmosphere capable of generating torrential downpours once existed on Mars. Other researchers, however, were skeptical about this scenario.

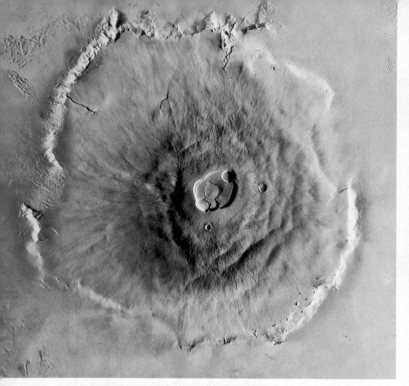

▲ **FIGURE 22.11** Image of Mons Olympus, an inactive shield volcano on Mars that covers an area about the size of the state of Ohio. (Courtesy of the U.S. Geological Survey)

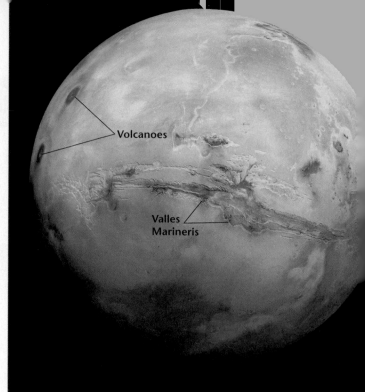

▲ **FIGURE 22.12** This image shows the entire Valles Marineris canyon system, over 5000 kilometers long and up to 8 kilometers deep. The dark spots on the left edge of the image are huge volcanoes, each about 25 kilometers high. (Courtesy of U.S. Geological Survey)

Today, most planetary geologists agree that running water was involved in carving at least some of the valleys on Mars. Some examples can be seen on an image from the *Mars Global Surveyor* in Figure 22.13. Researchers have proposed that melting of subsurface ice caused springlike seeps to emerge along this valley wall, slowly creating the gullies shown.

Other channels have streamlike banks and contain numerous teardrop-shaped islands. These valleys appear to have been cut by catastrophic floods that had discharge rates that were more than 1000 times greater than the Mississippi River. Most of these large flood channels emerge from areas of chaotic topography that appear to have formed when the surface collapsed. The most likely source of water for these flood valleys is the melting of subsurface ice. If the meltwater was trapped beneath a thick layer of permafrost, pressure could mount until a massive release of groundwater occurred. As a result, the overlying surface layer would collapse, creating the chaotic terrain.

Not all Martian valleys appear to be the result of water released by the melting of subsurface ice. Some valley systems consist of branching treelike channels that closely resemble the river valley networks found on Earth. Such features are clear evidence that Mars once had an active hydrologic cycle similar to Earth's. Further evidence to support this view came in 2004 when the *Mars Opportunity* rover discovered salt-laden sediments on Meridiai Planum. These deposits are precisely the kind of materials associated with shallow evaporating lakes or seas.

Because water is an essential ingredient for life, astrobiologists are also interested in the role flowing water may have played in the development of the Martian landscape.

The Martian Satellites. Tiny Phobos and Deimos, the two satellites of Mars, were not discovered until 1977 because they are only 24 and 15 kilometers in diameter. Phobos is nearer to its parent than any other natural satellite in the solar system—only 5500 kilometers (3400 miles)—and requires just 7 hours and

▼ **FIGURE 22.13** Image from *Mars Global Surveyor* showing valley wall with large gullies that may have been cut by liquid water, mixed with soil, rocks, and ice. (NASA/JPL/Malin Space Science Systems/NGS Image Collection)

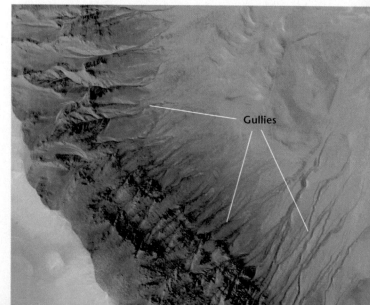

39 minutes for one revolution. *Mariner 9* revealed that both satellites are irregularly shaped and have numerous impact craters.

It is likely that these moons are asteroids captured by Mars. A most interesting coincidence in astronomy and literature is the close resemblance between Phobos and Deimos and two fictional satellites of Mars described by Jonathan Swift in *Gulliver's Travels,* written about 150 years before these satellites were actually discovered.

? STUDENTS SOMETIMES ASK...
*Why are the volcanoes on Earth
so much smaller than those on Mars?*

The largest shield volcanoes form where plumes of hot rock rise from deep within a planet's interior. Earth is tectonically active, with moving plates that keep the crust in constant motion. For example, the Hawaiian Islands consist of a chain of shield volcanoes that formed as the Pacific plate moved over a relatively stationary mantle plume. On Mars, volcanoes such as Mons Olympus have grown to great size because the crust there remains stationary. Successive eruptions occur at the same location and add to the bulk of a single volcano rather than producing several smaller structures, as occurs on Earth.

Jupiter: Lord of the Heavens

Jupiter, truly a giant among planets, has a mass two and a half times greater than the combined mass of all the remaining planets, satellites, and asteroids. In fact, had Jupiter been about 10 times larger, it would have evolved into a small star. Despite its great size, however, it is only 1/800 as massive as the Sun. Jupiter also rotates more rapidly than any other planet, completing one rotation in slightly less than 10 Earth-hours. The effect of this fast spin is to make the equatorial region bulge and to make the polar dimension flatten (see the Polar Flattening column in Table 22.1).

When viewed through a telescope or binoculars, Jupiter appears to be covered with alternating bands of multicolored clouds aligned parallel to its equator (Figure 22.14). The most striking feature is the *Great Red Spot* in the southern hemisphere (Figure 22.14). The Great Red Spot has been a prominent feature since it was first discovered more than three centuries ago. When *Voyager 2* swept by Jupiter in 1979, the Great Red Spot was the size of two Earth-size circles placed side by side. On occasion, it has grown even larger.

Images from *Pioneer 11* as it moved within 42,000 kilometers of Jupiter's cloud tops in 1974 indicated that the Great Red Spot is a counterclockwise-rotating (cyclonic) storm. It is caught between two jetstream-like bands of atmosphere flowing in opposite directions. This huge hurricane-like storm rotates once every 12 Earth-days. Although several smaller storms have been

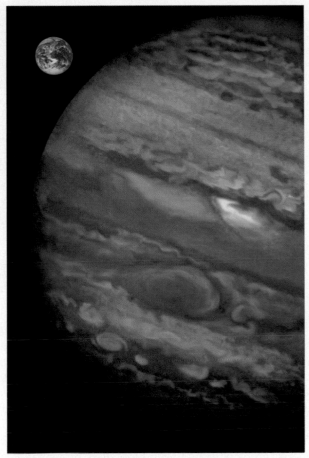

▲ **FIGURE 22.14** Artist's view of Jupiter with Great Red Spot visible in its southern hemisphere. Earth for scale.

observed in other regions of Jupiter's atmosphere, none have survived for more than a few days.

Structure of Jupiter. Jupiter's hydrogen-helium atmosphere also has methane, ammonia, water, and sulfur compounds as minor constituents. The wind systems generate the light- and dark-colored bands that encircle this giant (Figure 22.15). Unlike the winds on Earth, which are driven by solar energy, Jupiter itself gives off nearly twice as much heat as it receives from the Sun. Thus, it is the *interior* heat from Jupiter that produces huge convection currents in the atmosphere.

Atmospheric pressure at the top of the clouds is equal to sea-level pressure on Earth. Because of Jupiter's immense gravity, the pressure increases rapidly toward its surface. At 1000 kilometers below the clouds, the pressure is great enough to compress hydrogen gas into a liquid. Consequently, Jupiter's surface is thought to be a gigantic ocean of liquid hydrogen. Less than halfway into Jupiter's interior, extreme pressures cause the liquid hydrogen to turn into *liquid metallic* hydrogen. Jupiter is also believed to contain as much rocky and metallic material as is found in the terrestrial planets, probably in a central core.

▶ **FIGURE 22.15** The structure of Jupiter's atmosphere. The areas of light clouds (zones) are regions where gases are ascending and cooling. Sinking dominates the flow in the darker cloud layers (belts). This convective circulation, along with the rapid rotation of the planet, generates the high-speed winds observed between the belts and zones.

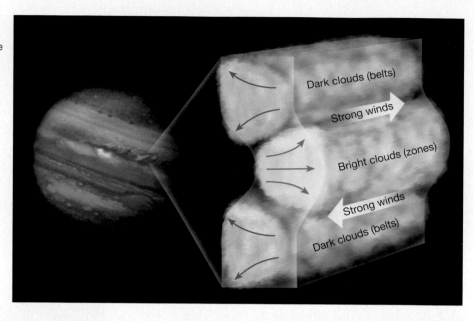

Jupiter's Moons. Jupiter's satellite system, consisting of 28 moons discovered so far, resembles a miniature solar system. The four largest satellites, discovered by Galileo, travel in nearly circular orbits around the parent, with periods of from 2 to 17 Earth-days (Figure 22.16). The two largest Galilean satellites, Callisto and Ganymede, surpass Mercury in size, while the two smaller ones, Europa and Io, are about the size of Earth's Moon. The Galilean moons can be observed with binoculars or a small telescope and are interesting in their own right.

By contrast, Jupiter's four outermost satellites are very small (20 kilometers in diameter), revolve in a direction that is opposite (*retrograde motion*) the largest moons, and have orbits that are steeply inclined to the Jovian equator. These satellites appear to be asteroids that passed near enough to be captured gravitationally by Jupiter.

Images from *Voyagers 1* and *2* in 1979 revealed, to the surprise of almost everyone, that each of the four Galilean satellites is a unique geological world, as shown in Figure 22.16. The innermost of the Galilean moons, Io (Figure 22.16A), is one of three volcanically active bodies discovered in our solar system, the others being Earth and Neptune's moon Triton. To date, numerous active sulfurous volcanic centers have been discovered. Umbrella-shaped plumes have been seen rising from the surface of Io to heights approaching 200 kilometers (120 miles) (Figure 22.17). The heat source for volcanic activity on Io is thought to be tidal energy generated by a relentless "tug of war" between Jupiter and the other Galilean satellites. Because Io is gravitationally locked to Jupiter, the same side always faces the giant planet, like Earth's Moon. The gravitational power of Jupiter and the other nearby satellites pulls and pushes on Io's tidal bulge as its slightly eccentric orbit takes it alternately closer to and farther from Jupiter. This gravitational flexing of Io is trans-

formed into heat energy (similar to the back-and-forth bending of a paper clip) and results in Io's spectacular sulfurous volcanic eruptions.

One of the most unexpected discoveries made by *Voyager 1* was Jupiter's ring system. By analyzing how these rings scatter light, researchers concluded that the rings are composed of fine, dark particles, similar in size to smoke particles. Further, the faint nature of the rings indicates that these minute fragments are widely dispersed. The particles are thought to be fragments blasted by meteorite impacts from the surfaces of Metis and Adrastea, two small moons of Jupiter.

STUDENTS SOMETIMES ASK...

Besides Earth, do any other bodies in the solar system have liquid water?

The planets closer to the Sun than Earth are considered too warm to contain liquid water, and those farther from the Sun are generally too cold to have water in the liquid form (although some features on Mars suggest that it may have had abundant liquid water at some point in its history). However, the best prospects of finding liquid water within our solar system lie beneath the icy surfaces of some of Jupiter's moons. For instance, Europa is suspected to have an ocean of liquid water hidden under its outer covering of ice. Detailed images sent back to Earth from the *Galileo* spacecraft have revealed that Europa's icy surface is quite young and exhibits cracks apparently filled with dark fluid from below. This suggests that under its icy shell, Europa must have a warm, mobile interior—and perhaps an ocean. Because the presence of water in the liquid form is a necessity for life as we know it, there has been much interest in sending an orbiter to Europa—and eventually a lander capable of launching a robotic submarine—to determine if it too may harbor life.

| A. Io | B. Europa | C. Ganymede | D. Callisto |

▲ FIGURE 22.16   Jupiter's four largest moons (from left to right) called the Galilean moons because they were discovered by Galileo. **A.** The innermost moon, Io, is one of only three volcanically active bodies in the solar system. **B.** Europa, smallest of the Galilean moons, has an icy surface that is criss-crossed by many linear features. **C.** Ganymede, the largest Jovian satellite, contains cratered areas, smooth regions, and areas covered by numerous parallel grooves. **D.** Callisto, the outermost of the Galilean satellites, is densely cratered, much like Earth's moon. (Courtesy of NASA/NGS Image Collection)

Saturn: The Elegant Planet

Requiring 29.46 Earth-years to make one revolution, Saturn is almost twice as far from the Sun as Jupiter, yet its atmosphere, composition, and internal structure are thought to be remarkably similar to Jupiter's. The most prominent feature of Saturn is its system of rings (Figure 22.18), discovered by Galileo in 1610. With his primitive telescope, the rings appeared as two small bodies adjacent to the planet. Their ring nature was discovered 50 years later by the Dutch astronomer Christian Huygens.

Saturn Close up.   In 1980 and 1981, fly-by missions of the nuclear-powered *Voyagers 1* and *2* space vehicles came within 100,000 kilometers of Saturn. More information was gained in a few days than had been acquired since Galileo first viewed this elegant planet telescopically.

1. Saturn's atmosphere is very dynamic, with winds roaring at up to 1500 kilometers (930 miles) per hour.
2. Large cyclonic "storms" similar to Jupiter's Great Red Spot, although smaller, occur in Saturn's atmosphere.
3. Eleven additional moons were discovered.
4. The rings of Saturn were discovered to be more complex than expected.

More recently, observations from ground-based telescopes and the Hubble Space Telescope have added to our knowledge of Saturn's ring system. In July of 2004, the Cassini spacecraft passed through Saturn's rings before swinging into orbit around the planet. Cassini's path was midway between the F and G rings because the mission planners thought that zone was empty. However, the spacecraft ran into about 100,000 tiny ring particles, the size of smoke particles, which have been labeled "dirt" for lack of a better name.

Planetary Ring Systems.   Until the recent discovery that Jupiter, Uranus, and Neptune also have ring systems, this phenomenon was thought to be unique to

▼ FIGURE 22.17   A volcanic eruption on Io. This plume of volcanic gases and debris is rising more than 100 kilometers (60 miles) above Io's surface. (Courtesy of NASA)

▶ **FIGURE 22.18** A view of the dramatic ring system of Saturn.

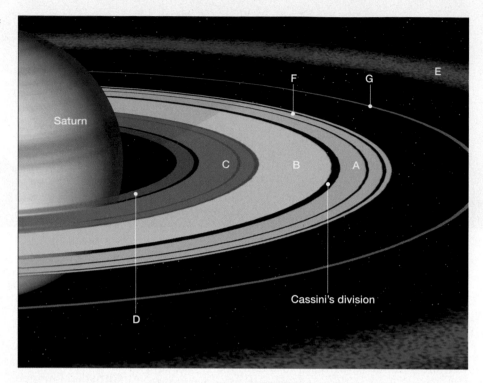

Saturn. Although the four known ring systems differ in detail, they share many attributes. They all consist of multiple concentric rings separated by gaps of various widths. In addition, each ring is composed of individual particles—"moonlets" of ice and rock—that circle the planet while regularly impacting one another.

Most rings fall into one of two categories based on particle density. Saturn's main rings (designated A and B in Figure 22.18) and the bright rings of Uranus are tightly packed and contain "moonlets" that range in size from a few centimeters (pebble-size) to several meters (house-size). These particles are thought to collide frequently as they orbit the parent planet. Despite the fact that Saturn's dense rings stretch across several hundred kilometers, they are very thin, perhaps less than 100 meters (300 feet) from top to bottom.

At the other extreme, the faintest rings, such as Jupiter's ring system and Saturn's outermost rings (designated E in Figure 22.18), are composed of very fine (smoke-size) particles that are widely dispersed. In addition to having very low particle densities, these rings tend to be thicker than Saturn's bright rings.

Recent studies have shown that the moons that coexist with the rings play a major role in determining their structure. In particular, the gravitational influence of these moons tends to shepherd the ring particles by altering their orbits. The narrow rings appear to be the work of satellites located on either side that confine the ring by pushing back particles that try to escape.

More important, the ring particles are thought to be debris ejected from these moons. According to this

view, material is continually being recycled between the rings and the ring moons. The moons gradually sweep up particles, which are subsequently ejected by collisions with large chunks of ring material, or perhaps by energetic collisions with other moons. It seems, then, that planetary rings are not the timeless features that we once thought; rather, they continually reinvent themselves.

The origin of planetary ring systems is still being debated. Did the rings form out of a flattened cloud of dust and gases that encircled the parent planet? In this scenario, the rings formed at the same time and of the same material as the planets and moons. Or, did the rings form later, when a moon or large asteroid was gravitationally pulled apart after straying too close to a planet? Still another hypothesis suggests that a foreign body blasted apart one of the planet's moons. The fragments from this impact would tend to jostle one another and form a flat, thin ring. Researchers expect more light to be shed on the origin of planetary rings as the Cassini spacecraft continues its four-year tour of Saturn.

Saturn's Moons. The Saturnian satellite system consists of 30 named moons (Figure 22.19). (If you count the "moonlets" that comprise Saturn's rings, this planet has millions of satellites.) The largest, Titan, is bigger than Mercury and is the second-largest satellite in the solar system (after Jupiter's Ganymede). Titan and Neptune's Triton are the only satellites in the solar system known to have a substantial atmosphere. Because of its dense gaseous cover, the atmospheric pressure at Titan's surface is about 1.5 times

▲ **FIGURE 22.19** Montage of the Saturnian satellite system. The moon Dione is in foreground; Tethys and Mimas are at lower right; Enceladus and Rhea are off ring's left; and Titan is upper right. (Image courtesy of NASA)

that at Earth's surface. Another satellite, Phoebe, exhibits retrograde motion. It, like other moons with retrograde orbits, is most likely a captured asteroid or large planetesimal left over from the major episode of planetary formation.

Uranus and Neptune: The Twins

Earth and Venus have similar traits, but Uranus and Neptune are nearly twins. Only 1 percent different in diameter, they are both bluish in appearance, which is attributable to the methane in their atmospheres (Figures 22.20 and 22.21). Their structure and composition are believed to be similar. Neptune, however, is colder, because it is half again as distant from the Sun's warmth as is Uranus.

Uranus: The Sideways Planet. A unique feature of Uranus is that it rotates "on its side." Its axis of rotation, instead of being generally perpendicular to the plane of its orbit, like the other planets, lies nearly parallel with the plane of its orbit. Its rotational motion, therefore, has the appearance of rolling, rather than the toplike spinning of the other planets. Because the axis of Uranus is inclined almost 90 degrees, the Sun is nearly overhead at one of its poles once each revolu-

tion, and then half a revolution later it is overhead at the other pole.

A surprise discovery in 1977 revealed that Uranus has a ring system. This find occurred as Uranus passed in front of a distant star and blocked its view, a process called *occultation* (*occult* = hidden). Observers saw the star "wink" briefly five times (meaning five rings) before the primary occultation and again five times afterward. Later studies indicate that Uranus has at least nine distinct belts of debris orbiting its equatorial region.

Spectacular views from *Voyager 2* of the five largest moons of Uranus show quite varied terrains. Some have long, deep canyons and linear scars, whereas others possess large, smooth areas on otherwise crater-riddled surfaces. The Jet Propulsion Laboratory described Miranda, the innermost of the five largest moons, as having a greater variety of landforms than any body yet examined in the solar system.

Neptune: The Windy Planet Even when the most powerful telescope is focused on Neptune, it appears as a bluish fuzzy disk. Until the 1989 *Voyager 2* encounter, astronomers knew very little about this planet. However, the 12-year, nearly 3-billion-mile journey of *Voyager 2* provided investigators with so much new information

627

▲ **FIGURE 22.20** This image of Uranus was sent back to Earth by *Voyager 2* as it passed by this planet on January 24, 1986. Taken from a distance of nearly 1 million kilometers, little detail of its atmosphere is visible, except a few streaks (clouds) in the Northern Hemisphere (Courtesy of NASA)

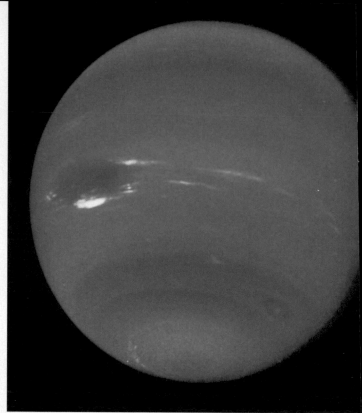

▲ **FIGURE 22.21** This image of Neptune shows the Great Dark Spot (left center). Also visible are bright cirruslike clouds that travel at high speeds around the planet. A second oval spot is at 54° south latitude on the east side of the planet. (Courtesy of the Jet Propulsion Laboratory)

about Neptune and its satellites that years will be needed to analyze it all.

Neptune has a dynamic atmosphere, much like those of Jupiter and Saturn (Figure 22.21). Winds exceeding 1000 kilometers per hour (600 miles per hour) encircle the planet, making it one of the windiest places in the solar system. In Figure 22.1, which is an image of Neptune sent back by the *Voyager* spacecraft in 1986, Neptune has an Earth-size blemish called the *Great Dark Spot* that is reminiscent of Jupiter's Great Red Spot and is assumed to be a large rotating storm. About five years later, when the Hubble Space Telescope viewed Neptune, the spot had vanished, only to be replaced by another dark spot in the planet's northern hemisphere.

Perhaps most surprising are white, cirruslike clouds that occupy a layer about 50 kilometers above the main cloud deck, probably frozen methane. Six new satellites were discovered in the *Voyager* images, bringing Neptune's family to eight. All of the newly discovered moons orbit the planet in a direction opposite that of the two larger satellites. *Voyager* images also revealed a ring system around Neptune.

Triton, Neptune's largest moon, is a most interesting object. Its diameter is nearly that of Earth's Moon. Triton is the only large moon in the solar system that exhibits retrograde motion. This indicates that Triton formed independently of Neptune and was gravitationally captured.

Triton also has the lowest surface temperature yet measured on any body in the solar system, −200°C (−391°F). Its atmosphere is mostly nitrogen with a little methane. Despite low surface temperatures, Triton displays volcanic-like activity. In 1989, *Voyager 2* de-

tected active plumes that extended to an altitude of 8 kilometers and were blown downwind for more than 100 kilometers. Presumably, the surface layers of darker methane ice absorb solar energy more readily. Such surface warming vaporizes some of the underlying nitrogen ice. As subsurface pressures increase, explosive eruptions result.

? STUDENTS SOMETIMES ASK...
Why does Uranus spin on its side?

The most likely explanation for the unusual sideways spin of Uranus is that it started out spinning the same way as the other planets, but then its spin was altered by a giant impact—which was probably very common when the planets were first formed. However, a giant impact would be very difficult to verify because it would not have left any crater on Uranus, which has no solid surface. Like many events that occurred early on in the formation of our solar system, the reason for Uranus's sideways spin may never be known for sure.

Pluto: The Outermost Planet?

Pluto lies on the fringe of the solar system, almost 40 times farther from the Sun than Earth. It is 10,000 times too dim to be visible to the unaided eye. Because of its great distance and slow orbital speed, it takes Pluto 248 Earth-years to orbit the Sun. Ever since its discovery in 1930, it has completed about one-fourth of a revolution. Pluto's orbit is noticeably elongated (highly eccentric), causing it to occasionally travel inside the orbit of Neptune, where it resided from 1979 through February 1999. There is no chance that Pluto and Neptune will ever collide, because their orbits are inclined to each other and do not actually cross (see Figure 22.1).

In 1978 the moon Charon was discovered orbiting Pluto. Because of its close proximity to the planet, the best ground-based images of Charon show it only as an elongated bulge. In 1990 the Hubble Space Telescope produced an image that clearly resolves the separation between these two icy worlds. Charon orbits Pluto once every 6.4 Earth-days at a distance 20 times closer to Pluto than our Moon is to Earth (Figure 22.22).

The average temperature of Pluto is estimated at −210°C (−340°F) cold enough to solidify most gases that might be present. Thus, Pluto might best be described as a dirty iceball of frozen gases with lesser amounts of rocky substances.

Is Pluto Really a Planet?

Ever since Pluto's discovery in 1930, it has been a mystery on the edge of the solar system. At first, Pluto was thought to be about as large as Earth, but as better images were obtained, Pluto's diameter was estimated to be a little less than one half that of Earth. Recent images obtained by the Hubble Space Telescope established the diameter of Pluto at only 2300 kilometers (1425 miles). This is about one-fifth that of Earth and less than half that of Mercury, long considered the "runt" of the solar system. In fact, seven moons, including Earth's moon, are larger than Pluto (Figure 22.22).

Even more attention was given to Pluto's status as a planet, when in 1992 astronomers discovered another icy body in orbit beyond Neptune. Soon hundreds of these objects were discovered forming a band similar to the asteroid belt between Mars and Jupiter. However, these orbiting bodies are made up of dust and ices, like comets, rather than metallic and rocky substances, like asteroids. Some astronomers believe that planetary objects even larger than Pluto may exist in this belt of icy worlds found in the outermost reaches of the solar system. In fact, one body larger than Pluto's moon, Charon, has already been discovered.

A growing number of astronomers assert that Pluto's small size and its location within a swarm of similar icy objects means that it should be reclassified as a minor planet, like asteroids and comets. Others insist that regardless of how Pluto's identity changes, demoting Pluto to a minor planet would dishonor astronomical history and confuse the public.

For now it seems that the International Astronomical Union, a group that has the power to vote on whether or not Pluto is a planet, is content with the status quo. Nevertheless, Pluto's planetary status will never be the same. It is now clear that Pluto is unique among the planets, being very different from the four rocky innermost planets, and unlike the four gaseous giants. Perhaps Pluto is best described as one of the largest members of a belt of millions of small, icy worlds (comets) that orbit in the outer reaches of our solar system called the *Kuiper belt* (see the section on comets).

▼ **FIGURE 22.22** Pluto and its moon Charon. Earth is shown for scale.

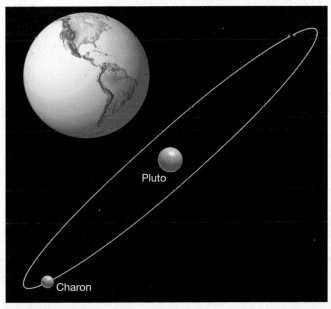

Minor Members of the Solar System

Asteroids: Microplanets

Asteroids are smaller bodies that have been likened to "flying mountains." The largest, Ceres, is about 1000 kilometers (600 miles) in diameter, but most of the 50,000 that have been observed are only about 1 kilometer across. The smallest asteroids are assumed to be no larger than grains of sand. Most lie between the orbits of Mars and Jupiter and have periods of three to six years (Figure 22.23). Some have very eccentric orbits and travel very near the Sun, and a few larger ones regularly pass close to Earth and the Moon. Many of the most recent impact craters on the Moon and Earth were probably caused by collisions with asteroids. Inevitably, future Earth–asteroid collisions will occur (see Box 22.2).

▶ **FIGURE 22.23** The orbits of most asteroids lie between Mars and Jupiter. Also shown are the orbits of a few known near-Earth asteroids. Perhaps a thousand or more asteroids have near-Earth orbits. Luckily, only a few dozen are thought to be larger than 1 kilometer in diameter.

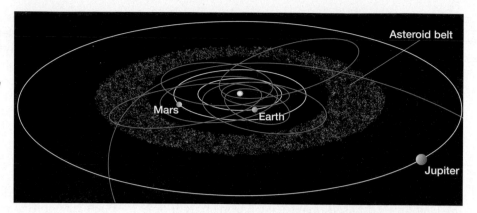

Because many asteroids have irregular shapes, planetary geologists first speculated that they might be fragments of a broken planet that once orbited between Mars and Jupiter (Figure 22.24). However, the total mass of the asteroids is estimated to be only 1/1000 that of Earth, which itself is not a large planet. What happened to the remainder of the original planet? Others have hypothesized that several larger bodies once coexisted in close proximity and that their collisions produced numerous smaller ones. The existence of several families of asteroids has been used to support this explanation. However, no conclusive evidence has been found for either hypothesis.

In February 2001 an American spacecraft became the first visitor to an asteroid. Although it was not designed for landing, *NEAR Shoemaker* landed successfully and generated information that has planetary geologists intrigued and perplexed. Images obtained as the spacecraft drifted at the rate of 6 kilometers (4 miles) per hour toward the surface of Eros revealed a barren, rocky surface composed of particles ranging in size from fine dust to boulders up to 8 meters (26 feet) across. Researchers unexpectedly discovered that fine debris is concentrated in the low areas that form flat deposits that resemble ponds. Surrounding the low areas, the landscape is marked by an abundance of large boulders.

One of several hypotheses being considered as an explanation for the boulder-laden topography is seismic shaking, which would move the boulders upward. Analogous to what happens when a can of mixed nuts is shaken, the larger materials rise to the top while the smaller materials settle to the bottom.

Comets: Dirty Snowballs

Comets are among the most interesting and unpredictable bodies in the solar system. They have been compared to dirty snowballs, because they are made of frozen gases (water, ammonia, methane, carbon dioxide, and carbon monoxide) that hold together small pieces of rocky and metallic materials. Many comets travel in very elongated orbits that carry them far beyond Pluto. These comets take hundreds of thousands of years to complete a single orbit around the Sun. However, a few *short-period comets* (those having orbital periods of less than 200 years), such as Halley's comet, make regular encounters with the inner solar system.

When first observed, a comet appears very small, but as it approaches the Sun, solar energy begins to vaporize the frozen gases, producing a glowing head called the **coma** (Figure 22.25). The size of the coma varies greatly from one comet to another. Extremely rare ones exceed the size of the Sun, but most approximate the size of Jupiter. Within the coma, a small glowing nucleus with a diameter of only a few kilometers can sometimes be detected. As comets approach the Sun, some, but not all, develop a tail that extends for millions of kilometers. Despite the enormous size of their tails and comas, comets are relatively small members of the solar system.

▼ **FIGURE 22.24** Image of asteroid 951 (Gaspra) obtained by the Jupiter-bound *Galileo* spacecraft. Like other asteroids, Gaspra is probably a collision-produced fragment of a larger body. (Courtesy of NASA)

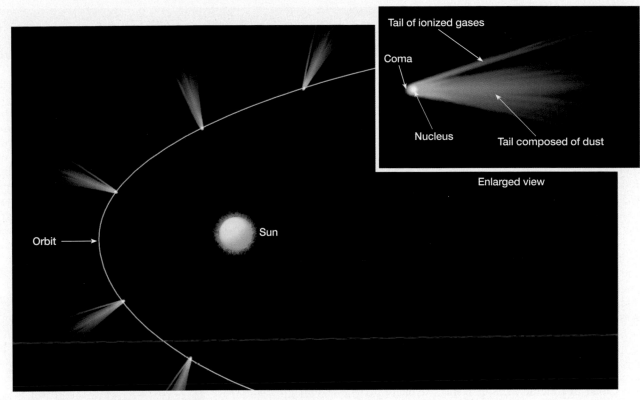

▲ **FIGURE 22.25** Orientation of a comet's tail as it orbits the Sun.

The fact that the tail of a comet points away from the Sun in a slightly curved manner (Figure 22.25) led early astronomers to propose that the Sun has a repulsive force that pushes the particles of the coma away, thus forming the tail. Today, two solar forces are known to contribute to this formation. One, *radiation pressure*, pushes dust particles away from the coma. The second, known as *solar wind*, is responsible for moving the ionized gases, particularly carbon monoxide. Sometimes a single tail composed of both dust and ionized gases is produced, but often two tails are observed (Figure 22.26).

As a comet moves away from the Sun, the gases forming the coma recondense, the tail disappears, and the comet returns to cold storage. Material that was blown from the coma to form the tail is lost from the comet forever. Consequently, it is believed that most comets cannot survive more than a few hundred close orbits of the Sun. Once all the gases are expelled, the remaining material—a swarm of tiny metallic and

◀ **FIGURE 22.26** Comet Hale-Bopp. The two tails seen in the photograph are between 10 million and 15 million miles long. (Peoria Astronomical Society Photograph by Eric Clifton and Craig Neaveill)

BOX 22.2 EARTH AS A SYSTEM:
Is Earth on a Collision Course?

The solar system is cluttered with meteoroids, asteroids, active comets, and extinct comets. These fragments travel at great speeds and can strike Earth with the explosive force of a powerful nuclear weapon.

In the last few decades, it has become increasingly clear that comets and asteroids have collided with Earth far more frequently than was previously known. The evidence is giant impact structures. More than 100 have been identified (Figure 22.B). Many were once misunderstood to result from some volcanic process. Although most impact structures are so old that they no longer resemble impact craters, evidence of their intense impact remains (Figure 22.C). One notable exception is a very fresh-looking crater near Winslow, Arizona, known as Meteor Crater (see Figure 22.28, p. 634).

Evidence is mounting that about 65 million years ago a large asteroid about 10 kilometers (6 miles) in diameter collided with Earth. This impact may have caused the extinction of the dinosaurs, as well as nearly 50 percent of all plant and animal species (see Box 12.2, p. 352).

More recently, a spectacular explosion has been attributed to the collision of our planet with a comet or asteroid. In 1908, in a remote region of Siberia, a "fireball" that appeared more brilliant than the Sun exploded with a violent force. The shock waves rattled windows and triggered reverberations heard up to 1000 kilometers away. The "Tunguska event," as it is called, scorched, delimbed, and flattened trees to 30 kilometers from the epicenter. But expeditions to the area found no evidence of an impact crater nor any metallic fragments. Evidently, the explosion, which equaled at least a 10-megaton nuclear bomb, occurred a few kilometers above the surface. Most likely it was the demise of a comet or perhaps a stony asteroid. Why it exploded prior to impact is uncertain.

The dangers of living with these small but deadly objects from space again came to public attention in 1989 when an asteroid nearly 1 kilometer across shot past Earth. It was a near miss, about twice the distance to the Moon. Traveling at 70,000 kilometers (44,000 miles) per hour, it could have produced a crater 10 kilometers (6 miles) in diameter and perhaps 2 kilometers (1.2 miles) deep. As an observer noted, "Sooner or later it will be back." As it was, it crossed our orbit just six hours ahead of Earth. Statistics show that collisions of this tremendous magnitude should take place every few hundred million years and could have drastic consequences for life on Earth.

Figure 22.C Manicouagan, Quebec, is a 200-million-year-old eroded impact structure. The lake outlines the crater remnant, which is 70 kilometers (42 miles) across. Fractures related to this event extend outward for an additional 30 kilometers. (Courtesy of U.S. Geological Survey)

Figure 22.B World map of major impact structures. Others are being identified every year. (Data from Griffith Observatory)

stony particles—continues the orbit without a coma or a tail.

Comets apparently originate in two regions of the outer solar system. Most short period comets are thought to orbit beyond Neptune in a region called the **Kuiper belt**, in honor of astronomer Gerald Kuiper, who had predicted their existence. (During the past decade over a hundred of these icy bodies have been discovered.) Like the asteroids in the inner solar system, most Kuiper belt comets move in nearly circular orbits that lie roughly in the same plane as the planets. A chance collision between two Kuiper belt comets, or the gravitational influence of one of the Jovian planets, may occasionally alter the orbit of a comet enough to send it to the inner solar system, and into our view.

Unlike Kuiper belt comets, long-period comets have orbits that are *not* confined to the plane of the solar system. These comets appear to be distributed in all directions from the Sun, forming a spherical shell around the solar system, called the **Oort cloud**, after the Dutch astronomer Jan Oort. Millions of comets are believed to orbit the Sun at distances greater than 10,000 times the Earth-Sun distance. The gravitational effect of a distant passing star is thought to send an occasional Oort cloud comet into a highly eccentric orbit that carries it toward the Sun. However, only a tiny portion of the Oort cloud comets have orbits that bring them into the inner solar system.

The most famous short-period comet is Halley's comet. Its orbital period averages 76 years, and every one of its 29 appearances since 240 B.C. has been recorded by Chinese astronomers. This record is a testimonial to their dedication as astronomical observers and to the endurance of their culture. When seen in 1910, Halley's comet had developed a tail nearly 1.6 million kilometers (1 million miles) long and was visible during the daylight hours.

In 1986 the unspectacular showing of Halley's comet was a disappointment to many people in the Northern Hemisphere. Yet it was during this most recent visit to the inner solar system that a great deal of new information was learned about this most famous of comets. The new data were gathered by space probes sent to rendezvous with the comet. Most notably, the European probe *Giotto* approached to within 600 kilometers of the comet's nucleus and obtained the first images of this elusive structure.

We now know that the nucleus is potato-shaped, 16 kilometers by 8 kilometers. The surface is irregular and full of craterlike pits. Gases and dust that vaporize from the nucleus to form the coma and tail appear to gush from its surface as bright jets or streams. Only about 10 percent of the comet's total surface was emitting these jets at the time of the rendezvous. The remaining surface area of the comet appeared to be covered with a dark layer that may consist of organic material.

In 1997 the comet Hale-Bopp made for spectacular viewing around the globe. As comets go, the nucleus of Hale-Bopp was unusually large, about 40 kilometers (25 miles) in diameter. As shown in Figure 22.26, two tails nearly 15 million miles long extended from this comet. The bluish gas-tail is composed of positively charged ions, and it points almost directly away from the Sun. The yellowish tail is composed of dust and other rocky debris. Because the rocky material is more massive than the ionized gases, it is less affected by the solar wind and follows a different trajectory away from the comet.

Meteoroids: Visitors to Earth

Nearly everyone has seen a **meteor**, popularly (but inaccurately) called a "shooting star." This streak of light lasts from an eye blink to a few seconds and occurs when a small solid particle, a **meteoroid**, enters Earth's atmosphere from interplanetary space. The friction between the meteoroid and the air heats both and produces the light we see. Most meteoroids originate from any one of the following three sources: (1) interplanetary debris that was not gravitationally swept up by the planets during the formation of the solar system, (2) material that is continually being displaced from the asteroid belt, or (3) the solid remains of comets that once traveled near Earth's orbit. A few meteoroids are believed to be fragments of the Moon, or possibly Mars, that were ejected when an asteroid impacted these bodies.

Although a rare meteoroid is as large as an asteroid, most are the size of sand grains and weigh less than 1/100 gram. Consequently, they vaporize before reaching Earth's surface. Some, called *micrometeorites*, are so tiny that their rate of fall becomes too slow to cause them to burn up, so they drift down as space dust. Each day, the number of meteoroids that enter Earth's atmosphere must reach into the thousands. After sunset on a clear night, a half dozen or more are bright enough to be seen with the naked eye each hour from anywhere on Earth.

Occasionally, meteor sightings increase dramatically to 60 or more per hour. These displays, called **meteor showers**, result when Earth encounters a swarm of meteoroids traveling in the same direction and at nearly the same speed as Earth. The close association of these swarms to the orbits of some short-term comets strongly suggests that they represent material lost by these comets (Table 22.2). Some

TABLE 22.2	Major meteor showers.	
Shower	**Approximate Dates**	**Associated Comet**
Quadrantids	January 4–6	
Lyrids	April 20–23	Comet 1861 I
Eta Aquarids	May 3–5	Halley's comet
Delta Aquarids	July 30	
Perseids	August 12	Comet 1862 III
Draconids	October 7–10	Comet Giacobini-Zinner
Orionids	October 20	Halley's comet
Taurids	November 3–13	Comet Encke
Andromedids	November 14	Comet Biela
Leonids	November 18	Comet 1866 I
Geminids	December 4–16	

▲ FIGURE 22.27 Iron meteorite found near Meteor Crater, Arizona. (Courtesy of Meteor Crater Enterprises, Inc., northern Arizona, USA).

ground. Most meteoroids large enough to survive the heated fall are thought to originate among the asteroids, where chance collisions modify their orbits and send them toward Earth. Earth's gravitational force does the rest.

The remains of meteoroids, when found on Earth, are referred to as **meteorites** (Figure 22.27). A few very large meteoroids have blasted out craters on Earth's surface that strongly resemble those on the lunar surface. The most famous is Meteor Crater in Arizona (Figure 22.28). This huge cavity is about 1.2 kilometers (0.75 miles) across, 170 meters (560 feet) deep, and has an upturned rim that rises 50 meters (165 feet) above the surrounding countryside. Over 30 tons of iron fragments have been found in the immediate area, but attempts to locate the main body have been unsuccessful. Based on erosion, the impact likely occurred within the last 50,000 years.

Prior to Moon rocks brought back by lunar explorers, meteorites were the only extraterrestrial materials that could be directly examined. Meteorites are classified by their composition: (1) **irons**—mostly iron, with 5 percent to 20 percent nickel; (2) **stony**—silicate minerals with inclusions of other minerals; and (3) **stony-irons**—mixtures. Although stony meteorites are probably more common, people find mostly irons. This is understandable, for irons withstand the impact better, weather more slowly, and are much easier for a layperson to distinguish from terrestrial rocks. Iron

swarms not associated with orbits of known comets are probably the remains of the nucleus of a long-defunct comet. The notable Perseid meteor shower that occurs each year around August 12 is believed to be the remains of the Comet 1862 III, which has a period of 110 years.

Meteoroids that are thought to be the remains of comets are small and only occasionally reach the

▼ FIGURE 22.28 Meteor Crater, near Winslow, Arizona. This cavity is about 1.2 kilometers (0.75 mile) across and 170 meters (560 feet) deep. The solar system is cluttered with meteoroids and other objects that can strike Earth with explosive force. (Photo by Michael Collier)

meteorites are probably fragments of once-molten cores of large asteroids or small planets.

One rare kind of meteorite, called a *carbonaceous chondrite,* was found to contain simple amino acids and other organic compounds, which are the basic building blocks of life. This discovery confirms similar findings in observational astronomy, which indicate that numerous organic compounds exist in the frigid realm of outer space.

If meteorites represent the makeup of Earth-like planets, as some planetary geologists suggest, then Earth must contain a much larger percentage of iron than is indicated by surface rocks. This is one reason why geologists suggest that Earth's core may be mostly iron and nickel. In addition, meteorite dating indicates that our solar system's age certainly exceeds 4.5 billion years. This "old age" has been confirmed by data from lunar samples.

Chapter Summary

- The planets can be arranged into two groups: the *terrestrial* (Earth-like) *planets* (Mercury, Venus, Earth, and Mars) and the *Jovian* (Jupiter-like) *planets* (Jupiter, Saturn, Uranus, and Neptune). Pluto is not included in either group. *When compared to the Jovian planets, the terrestrial planets are smaller, more dense, contain proportionally more rocky material, have slower rates of rotation, and lower escape velocities.*

- The lunar surface exhibits several types of features. Most *craters* were produced by the impact of rapidly moving interplanetary debris (*meteoroids*). Bright, densely cratered *highlands* (*terrae*) make up most of the lunar surface. The dark, fairly smooth lowlands are called *maria* (singular, *mare*). Maria basins are enormous impact craters that have been flooded with layer upon layer of very fluid basaltic lava. All lunar terrains are mantled with a soil-like layer of gray unconsolidated debris, called *lunar regolith*, which has been derived from a few billion years of meteoric bombardment. Much is still unknown about the Moon's origin. One hypothesis suggests that a giant asteroid collided with Earth to produce the Moon. Scientists conclude that the *lunar surface evolved in four phases*: (1) *the original crust*; (2) *lunar highlands*; (3) *maria basins*; and (4) *youthful rayed craters.*

- *Mercury* is a small, dense planet that has no atmosphere and exhibits the greatest temperature extremes of any planet. *Venus,* the brightest planet in the sky, has a thick, heavy atmosphere composed of 97 percent carbon dioxide, a surface of relatively subdued plains and inactive volcanic features, a surface atmospheric pressure 90 times that of Earth's, and surface temperatures of 475°C (900°F). *Mars,* the Red Planet, has a carbon dioxide atmosphere only 1 percent as dense as Earth's, extensive dust storms, numerous inactive volcanoes, many large canyons, and several valleys of debatable origin exhibiting drainage patterns similar to stream valleys on Earth. *Jupiter,* the largest planet, rotates rapidly, has a banded appearance caused by huge convection currents driven by the planet's interior heat, a *Great Red Spot* that varies in size, a thin ring system, and at least 16 moons (one of the moons, *Io*, is a volcanically active body). *Saturn* is best known for its system of rings. It also has a dynamic atmosphere with winds up to 930 miles per hour and storms similar to Jupiter's Great Red Spot. *Uranus* and *Neptune* are often called the twins because of similar structure and composition. A unique feature of Uranus is the fact that it rotates on its side. Neptune has white, cirruslike clouds above its main cloud deck and an Earth-size *Great Dark Spot,* assumed to be a large rotating storm similar to Jupiter's Great Red Spot. *Pluto* is a small frozen world with one moon (Charon). Pluto's noticeably elongated orbit causes it to occasionally travel inside the orbit of Neptune, but with no chance of collision.

- The minor members of the solar system include the *asteroids, comets,* and *meteoroids*. Most asteroids lie between the orbits of Mars and Jupiter. No conclusive evidence has been found to explain their origin. Comets are made of frozen gases (water, ammonia, methane, carbon dioxide, and carbon monoxide) with small pieces of rocky and metallic material. Many travel in very elongated orbits that carry them beyond Pluto, and little is known about their origin. Meteoroids, small solid particles that travel through interplanetary space, become *meteors* when they enter Earth's atmosphere and vaporize with a flash of light. *Meteor showers* occur when Earth encounters a swarm of meteoroids, probably material lost by a comet. *Meteorites* are the remains of meteoroids found on Earth. The *three types of meteorites* (classified by their composition) are (1) *irons*, (2) *stony*, and (3) *stony-irons*. One rare kind of meteorite, called a *carbonaceous chondrite*, was found to contain amino acids and other organic compounds.

Key Terms

asteroid (p. 629)
coma (p. 630)
comet (p. 630)
escape velocity (p. 614)
inner planets (p. 613)
iron meteorite (p. 634)
Jovian planet (p. 613)

Kuiper belt (p. 632)
lunar highlands (p. 616)
lunar regolith (p. 617)
maria (p. 616)
meteor (p. 633)
meteorite (p. 634)
meteoroid (p. 633)

meteor shower (p. 633)
Oort cloud (p. 633)
outer planets (p. 613)
stony-iron meteorite (p. 634)
stony meteorite (p. 634)
terrae (p. 613)
terrestrial planet (p. 612)

Review Questions

1. By what criteria are the planets placed into either the Jovian or terrestrial group?

2. What are the three types of materials thought to make up the planets? How are they different? How does their distribution account for the density differences between the terrestrial and Jovian planetary groups?

3. Explain why different planets have different atmospheres.

4. How is crater density used in the relative dating of features on the Moon?

5. Briefly outline the history of the Moon.

6. How are the maria of the Moon thought to be similar to the Columbia Plateau?

7. Why has Mars been the planet most studied telescopically?

8. What surface features does Mars have that are also common on Earth?

9. What evidence supports a water cycle on Mars?

10. Why are astrobiologists intrigued about evidence that groundwater has seeped onto the surface of Mars?

11. The two "moons" of Mars were once suggested to be artificial. What characteristics do they have that would cause such speculation?

12. What is the nature of Jupiter's Great Red Spot?

13. Why are the Galilean satellites of Jupiter so named?

14. What is distinctive about Jupiter's satellite Io?

15. Why are the four *outer* satellites of Jupiter thought to have been captured rather than having been formed with the rest of the satellite system?

16. How are Jupiter and Saturn similar?

17. What two roles do ring moons play in the nature of planetary ring systems?

18. How are Saturn's satellite Titan and Neptune's Triton similar?

19. What three bodies in the solar system exhibit volcanic-like activity?

20. Where are most asteroids found?

21. What do you think would happen if Earth passed through the tail of a comet?

22. Where are most comets thought to reside? What eventually becomes of comets that orbit close to the Sun?

23. Compare meteoroid, meteor, and meteorite.

24. What are the three main sources of meteoroids?

25. Why are meteorite craters more common on the Moon than on Earth, even though the Moon is a much smaller target?

26. It has been estimated that Halley's comet has a mass of 100 billion tons. Further, this comet is estimated to lose 100 million tons of material during the few months that its orbit brings it close to the Sun. With an orbital period of 76 years, what is the maximum remaining life span of Halley's comet?

Examining the Earth System

1. On Earth the four major spheres (atmosphere, hydrosphere, geosphere, and biosphere) interact as a system with occasional influence from our near-space neighbors. Which of these spheres are absent, or nearly absent, on the Moon? Because the Moon lacks these spheres, list at least five processes that operate on Earth but are absent on the Moon.

2. Among the planets in our solar system, Earth is unique because water exists in all three states (solid, liquid, and gas) on and near its surface. In what state(s) of matter is water found on Mercury, Venus, and Mars? How would Earth's hydrologic cycle be different if (a) its orbit was inside the orbit of Venus? (b) its orbit was outside the orbit of Mars?

3. If a large meteorite were to strike Earth in the near future, what effect might this event have on the atmosphere (in particular, average temperatures and climate)? Assuming the conditions persisted, speculate about how the changes might influence the biosphere?

Online Study Guide

The *Earth Science* Website uses the resources and flexibility of the Internet to aid in your study of the topics in this chapter. Written and developed by Earth science instructors, this site will help improve your understanding of Earth science. Visit **http://www.prenhall.com/tarbuck** and click on the cover of *Earth Science 11e* to find:

- **On-line review quizzes.**
- **Critical thinking exercises.**
- **Links to chapter-specific Web resources.**
- **Internet-wide key term searches.**

http://www.prenhall.com/tarbuck

Time lapse photograph of the night sky over the Keck Telescope on Hawaii's Mauna Kea volcano. (Photo by Roger Ressmeyer/CORBIS)

CHAPTER 23

Light, Astronomical Observations, and the Sun

stronomers are in the business of gathering and studying light. Almost everything that is known about the universe beyond Earth comes by analyzing light from distant sources. Consequently, an understanding of the nature of light is basic to modern astronomy. This chapter deals with the study of light and the tools used by astronomers to gather light in order to probe the universe. In addition, we will examine the nearest source of light, our Sun. By understanding the processes that operate on the Sun, astronomers can better grasp the nature of more distant celestial objects.

The Study of Light

The vast majority of our information about the universe is obtained from the study of the light emitted from celestial bodies (Figure 23.1). Although visible light is most familiar to us, it constitutes only a small part of an array of energy generally referred to as **electromagnetic radiation.** Included in this array are gamma rays, X-rays, ultraviolet light, visible light, infrared radiation, and radio waves (see Figure 16.17, p. 450). All radiant energy travels through the vacuum of space in a straight line at the rate of 300,000 kilo-

meters (186,000 miles) per second.* Over a 24-hour day, this equals a staggering 26 billion kilometers.

Nature of Light

Experiments have demonstrated that light can be described in two ways. In some instances light behaves like waves, and in others like discrete particles. In the wave sense, light is analogous to swells in the ocean. This motion is characterized by the wavelength, which is the distance from one wave crest to the next. Wavelengths vary from several kilometers for radio waves to less than a billionth of a centimeter for gamma rays. Most of these waves are either too long or too short for our eyes to detect.

The narrow band of electromagnetic radiation we can see is sometimes referred to as *white light*. However, white light consists of an array of waves having various wavelengths, a fact easily demonstrated with a prism (Figure 23.2). As white light passes through a prism, the color with the shortest wavelength, violet, is bent more than blue, which is bent more than green, and so forth (Table 23.1). Thus, white light can be separated into its component colors in the order of their

---

*Light rays are "bent" slightly when they pass nearby a very massive object such as the Sun.

▼ FIGURE 23.1 The Trifid Nebula, in the constellation Sagittarius. This colorful nebula is a cloud of dust, plus hydrogen and helium gases. These gases are excited by the radiation of the hot, young stars within and produce a pink glow. (Courtesy of National Optical Astronomy Observatories)

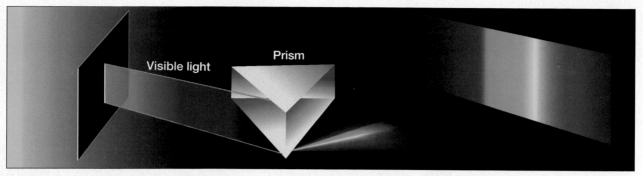

▲ **FIGURE 23.2** A spectrum is produced when sunlight (white light) is passed through a prism, which bends each wavelength at different angles.

wavelengths, producing the familiar rainbow of colors (Figure 23.2).

Wave theory, however, cannot explain some effects of light. In some cases, light acts like a stream of particles, analogous to infinitesimally small bullets fired from a machine gun. These particles, called **photons,** can exert a pressure (push) on matter, which is called **radiation pressure.** Photons from the Sun are responsible for pushing material away from a comet to produce its tail. Each photon has a specific amount of energy, which is related to its wavelength in a simple way: *Shorter wavelengths* correspond to *more energetic* photons. Thus, blue light has more energetic photons than does red light.

Which theory of light—the wave theory or the particle theory—is correct? Both, because each will predict the behavior of light for certain phenomena. As George Abell, a prominent astronomer, stated about all scientific laws, "The mistake is only to apply them to situations that are outside their range of validity."

Spectroscopy

When Sir Isaac Newton used a prism to disperse white light into its component colors, he unknowingly initiated the field of **spectroscopy,** which is the study of properties of light that depend on wavelength. The rainbow of colors Newton produced is called a "continuous spectrum," because all wavelengths of light are included. It was later learned that two other types

of spectra exist and that all three are generated under somewhat different conditions (Figure 23.3).

1. A **continuous spectrum** is produced by an incandescent solid, liquid, or gas under high pressure. It consists of an uninterrupted band of color (Figure 23.3A). One example would be light generated by a common lightbulb. (*Incandescent* means "to emit light when hot.")

2. A **dark-line spectrum (absorption spectrum)** is produced when white light is passed through a comparatively cool gas under low pressure. The gas absorbs selected wavelengths of light, so the spectrum that is produced appears as a continuous spectrum, but with a series of dark lines running through it (Figure 23.3B).

3. A **bright-line spectrum (emission spectrum)** is produced by a hot (incandescent) gas under low pressure. It is a series of bright lines of particular wavelengths, depending on the gas that produces them (Figure 23.3C). These bright lines appear in the exact location as the dark lines that are produced by this gas in a dark-line (absorption) spectrum.

The spectra of most stars are of the dark-line type. The importance of these spectra is that each element or compound that is in gaseous form (in a star, material is usually in the gaseous form) produces a unique set of spectral lines. When the spectrum of a star is studied, the spectral lines act as "fingerprints," which identify the elements present.

In an admittedly oversimplified manner, we can imagine how the Sun and other stars create a dark-line spectrum. A continuous spectrum is produced in the interior of the Sun, where the gases are under very high pressure. When this light passes through the less dense gases of the solar atmosphere, they absorb selected wavelengths, which appear as dark lines in the spectrum.

When Newton studied solar light, he obtained a continuous spectrum. However, when a prism is used

TABLE 23.1	Colors and corresponding wavelengths.
Color	**Wavelength (nanometers*)**
Violet	380–440
Blue	440–500
Green	500–560
Yellow	560–590
Orange	590–640
Red	640–750

*One nanometer is 10^{-9} meter.

▶ **FIGURE 23.3** Formation of the three types of spectra. **A.** Continuous spectrum. **B.** Dark-line spectrum. **C.** Bright-line spectrum.

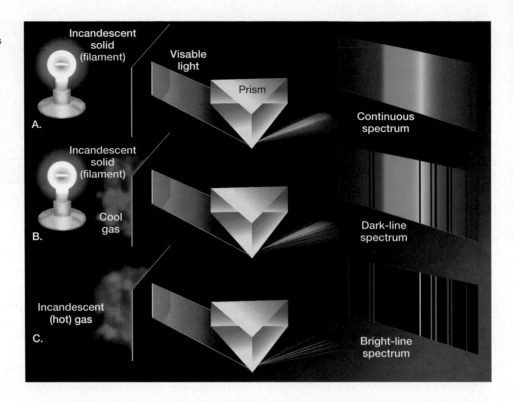

in conjunction with lenses, the solar spectrum can be dispersed even further. An instrument that does this is called a **spectroscope** (*spectro* = to look, *scope* = instrument). The spectrum of the Sun contains thousands of dark lines. Over 60 elements have been identified by matching these lines with those of elements known on Earth.

Two other facts concerning a radiating body are important. First, if the temperature of a radiating surface is increased, the total amount of energy emitted increases. The rate of increase is stated by the Stefan-Boltzmann law: The energy radiated by a body is directly proportional to the fourth power of its absolute temperature. For example, if the temperature of a star is doubled, the total radiation emitted increases by 2^4 $(2 \times 2 \times 2 \times 2)$, or 16 times. Second, as the temperature of an object increases, a larger proportion of its energy is radiated at shorter wavelengths. To illustrate this, imagine a metal rod that is heated slowly.

The rod first appears dull red (longer wavelengths) and later bluish white (shorter wavelengths). From this, it follows that blue stars are hotter than yellow stars, which are hotter than red stars.

The Doppler Effect

You may have heard the change in pitch of a car horn or ambulance siren as it passes by. When it is approaching, the sound seems to have a higher-than-normal pitch, and when it is moving away, the pitch sounds lower than normal. This effect, which occurs for both sound and light waves, was first explained by Christian Doppler in 1842 and is called the **Doppler effect.** The reason for the difference in pitch is that it takes time for the wave to be emitted. If the source is moving away, the beginning of the wave is emitted nearer to you than the end, which stretches the wave—that is, gives it a longer wavelength (Figure 23.4). The opposite is true for an approaching source.

▶ **FIGURE 23.4** The Doppler effect, illustrating the apparent lengthening and shortening of wavelengths caused by the relative motion between a source and an observer.

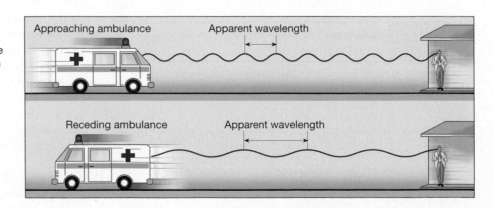

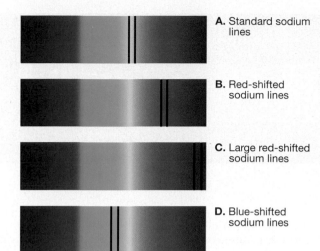

A. Standard sodium lines

B. Red-shifted sodium lines

C. Large red-shifted sodium lines

D. Blue-shifted sodium lines

▲ **FIGURE 23.5** **A.** An illustration of standard sodium lines produced in a laboratory compared to sodium lines as they would appear when the source is receding (red shift, **B.** and **C.**) or approaching (blue shift, **D.**).

In the case of light, when a source is moving away, its light appears redder than it actually is because its waves appear lengthened. Objects approaching have their light waves shifted toward the blue (shorter wavelength). Thus, if a source of red light approached you at a very high speed (near the speed of light), it would actually appear blue. The same effect would be produced if you moved and the light were stationary.

Therefore, the Doppler effect reveals whether Earth is approaching or receding from a star or another celestial body. In addition, the amount of shift allows us to recalculate the rate at which the relative movement is occurring. Larger Doppler shifts indicate higher velocities; smaller Doppler shifts indicate slower velocities. Doppler shifts are generally measured from the dark lines in the spectra of stars by comparing them with a standard spectrum produced in the laboratory (Figure 23.5).

Astronomical Tools

Having examined the nature of light, we will now turn our attention to the tools astronomers use to intercept and study the energy emitted by distant objects in the universe (Figure 23.6). Because the basic principles of detecting radiation were originally developed through visual observations, we will consider optical telescopes first.

Refracting Telescopes

Galileo is considered to be the first person to use telescopes for astronomical observations. Having learned about the newly invented instrument, Galileo built one of his own that was capable of magnifying objects 30 times. Because this early instrument, as well as its modern counterparts, used a lens to bend or refract light, it is known as a **refracting telescope**.

The most important lens in a refracting telescope, the **objective lens,** produces an image by bending light from a distant object in such a way that the light converges at an area called the **focus** (*focus* = central point) (Figure 23.7). For an object such as a star, the image appears as a point of light, but for nearby objects it appears as an inverted replica of the original.

You can easily demonstrate the latter case by holding a lens in one hand and, with the other hand, placing a white card behind the lens. Now vary the distance between them until an image appears on the card. The distance between the focus (where the image appears) and the lens is called the **focal length** of the lens.

Astronomers usually study an image from a telescope by first photographing the image. However, if a telescope is used to examine an image directly, a second lens, called an **eyepiece,** is required (Figure 23.7). The eyepiece magnifies the image produced by the objective lens. In this respect, it is similar to a magnifying glass. Thus, the objective lens produces a very small,

◀ **FIGURE 23.6** The 10-meter Keck Telescope in located at the summit of Hawaii's Mauna Kea volcano. (Photo by Roger Ressmeyer/CORBIS)

▶ **FIGURE 23.7** Simple refracting telescope.

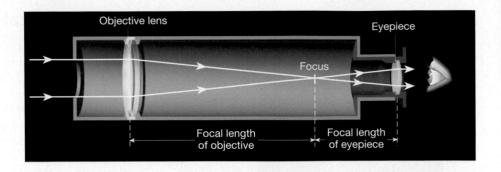

bright image of an object, and the eyepiece enlarges the image so that details can be seen.

Although used extensively in the nineteenth century, refracting telescopes suffer a major optical defect. As light passes through any lens, the shorter wavelengths of light are bent more than the longer wavelengths. (Recall the effect of a prism in separating the colors of the spectrum.) Consequently, when a refracting telescope is in focus for red light, blue and violet light are out of focus. The troublesome effect, known as **chromatic** (*chroma* = color), **aberration** (*aberrare* = to go astray), weakens the image and produces a halo of color around it. When blue light is in focus, a reddish halo appears, and vice versa. Although this effect cannot be eliminated completely, it is reduced by using a second lens made of a different type of glass.

Reflecting Telescopes

Newton was so bothered by chromatic aberration that he built and used telescopes that reflected light from a shiny surface (mirror). Because reflected light is not dispersed into its component colors, the problem is avoided. **Reflecting telescopes** use a concave mirror that focuses the light in front of the objective (the mirror), rather than behind it, like a lens (Figure 23.8). The mirror is generally made of glass that is finely ground. In the case of the 5-meter (200-inch) Hale Telescope, the grinding is accurate to about 1 millionth of a centimeter. The surface is then coated with a highly reflective material, usually an aluminum compound.

To focus parallel incoming rays onto one spot, the mirror is ground to a special curved surface called a *paraboloid*. This is the same shape as that used for the reflector in the headlights of automobiles. In the case of the auto bulb, however, the light source is at the focus, and the light goes out in parallel rays rather than coming in.

Because the focus of a reflecting telescope is in front of the mirror, provisions have to be made to view the image without blocking too much of the

incoming light. Figure 23.9 illustrates the most common arrangements. Most large telescopes employ more than one type. When using a very large reflecting telescope, the observer enters a viewing cage positioned at the focus to make the observations. The viewing cage blocks only about 10 percent of the total incoming light, and this is more than compensated for by the large objective (mirror) that is used.

In the good old days, an astronomer would spend numerous long nights outdoors in the cold mountain air, perched in a viewing cage. But advances in photographic materials and computer-enhancing technologies now allow indoor work and reduce the time required to obtain an image.

Nearly all large optical telescopes built today are reflectors. Among the reasons is the monumental task of producing a large piece of high-quality, bubble-free glass for refracting telescopes. Because light does not pass through a mirror, the glass for a reflecting telescope does not have to be of optical quality, nor does the instrument suffer from chromatic aberration. In addition, a lens can be supported only around the edge, so it sags. Mirrors, on the other hand, can be supported fully from behind.

Large reflecting telescopes of about 4 meters (150 inches) are located at Kitt Peak, Arizona (Figure 23.10); Mauna Kea, Hawaii; Cerro Tololo, Chile; and Siding Spring, Australia (see Box 23.1). By comparison, the largest refractor in the world is the 1-meter (40-inch) telescope at Yerkes Observatory in Williams Bay, Wisconsin. This instrument was built before the turn of the twentieth century.

Properties of Optical Telescopes

Telescopes have three properties that aid astronomers in their work. They provide the observer with light-gathering power, resolving power, and magnifying power.

Light-Gathering Power. Because most celestial objects are very faint sources of light, astronomers are *most* interested in improving the *light-gathering* power

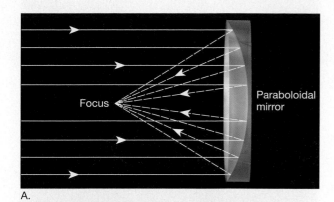

A.

B.

▲ **FIGURE 23.8** **A.** Diagram illustrating how concave mirrors, like those used in reflector telescopes, gather light. **B.** Preparation of the 2.4-meter mirror for the Hubble Space Telescope. (Courtesy of Space Telescope Science Institute)

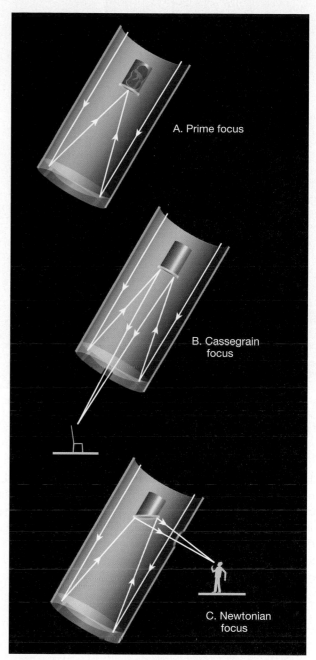

▲ **FIGURE 23.9** Viewing methods used with reflecting telescopes. **A.** Prime focus method (only used in very large telescopes). **B.** Cassegrain method (most common). **C.** Newtonian method.

of their instruments. As seen in Figure 23.11, a telescope with a large lens (or mirror) intercepts more light from distant objects and thereby produces brighter images. As very distant stars appear very dim as well, a great deal of light must be collected before the image is bright enough to be seen. Consequently, telescopes with large objectives "see" farther into space than do those with small objectives.

Resolving Power. Another advantage of telescopes with large-diameter objectives is their greater *resolving power* (Figure 23.12), which allows for sharper images and finer detail. For example, with the unaided eye, the Milky Way appears as a vague band of light in the night sky, but even a small telescope is capable of resolving (separating it into) individual stars. Even so, the condition of Earth's atmosphere (which is known as *seeing*) greatly limits the resolving power of telescopes on Earth. On a night when the

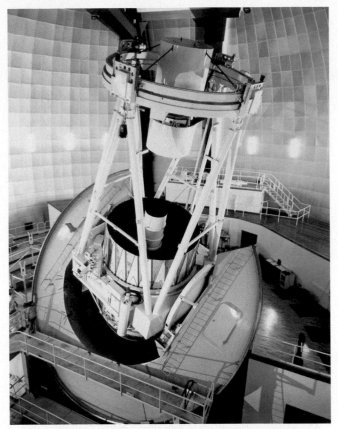

▲ **FIGURE 23.10** The 3.9-meter Anglo-Australian telescope. This modern reflecting telescope is one of the world's most advanced. (Royal Observatory Edinburgh/Photo Researchers, Inc.)

A.

B.

▲ **FIGURE 23.12** Appearance of the galaxy in the constellation Andromeda using telescopes of different resolution. (Courtesy of AURA)

stars twinkle, the seeing is poor because the air is moving rapidly. This causes the image to move about, blurring it. Conversely, when the stars shine steadily, the seeing is described as good. Even under ideal conditions, however, some blurring occurs, eliminating the fine details. Thus, even the largest telescopes cannot photograph lunar features less than 0.5 kilometer (0.3 mile) in size.

To eliminate the problems of Earthbound viewing, the United States built the Hubble Space Telescope, which was put into Earth orbit in April 1990 (Figure 23.13). This 2.4-meter (94-inch) space telescope has 10 billion times more light-gathering power than the human eye. As the technical problems that hampered the early operation of this instrument have been corrected, many spectacular images have been received. For example, the Hubble Space Telescope has provided images that clearly resolve the separation between Pluto and its moon, Charon.

▶ **FIGURE 23.11** Comparison of the light-gathering ability of two lenses.

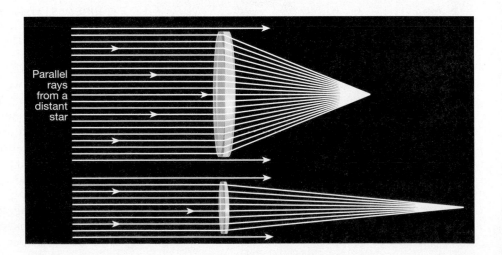

Parallel rays from a distant star

▲ **FIGURE 23.13** The deployment of the Hubble Space Telescope in Earth orbit, April 24, 1990, from Space Shuttle *Discovery.* (Courtesy of NASA)

? STUDENTS SOMETIMES ASK...
Why do astronomers build observatories on mountaintops?

Observatories are most often located on mountaintops because sites above the densest part of the atmosphere provide better conditions for "seeing." At high elevations, there is less air to scatter and dim the incoming light, and less water vapor to absorb infrared radiation. Further, the thin air on mountaintops causes less distortion of the images being observed. (Think of when you have seen a blurring effect caused by movement in the atmosphere on a hot summer day.) This also helps explain why the Hubble Space Telescope is such a valuable instrument.

Magnifying Power. When you think of the power of a telescope, you probably think of its *magnifying power,* the ability to make an object larger. Magnification is calculated by dividing the focal length of the objective by the focal length of the eyepiece. Thus, the magnification of a telescope can be changed by simply changing the eyepiece. However, increased magnification does not necessarily improve the clarity of the image. What can be viewed telescopically is limited by atmospheric conditions and the resolving power of the telescope. Any part of an image that is not clear at low magnification will appear only as a larger blur at higher magnification. Furthermore, increasing magnification spreads out the light and decreases the brightness of the object. Thus, astronomers describe telescopes not in terms of their magnification but by the diameter of the objective mirror or lens, because it is this factor that determines both the light-gathering power and the resolving power of a telescope.

Most modern telescopes have supplemental devices that enhance the image. A simple but important example is a photographic plate that can be exposed for long periods of time, thereby collecting enough light from a star to make an image that otherwise would be undetectable. One of the latest advances makes use of high-speed computers that adjust the optics to partially remove the distortion caused by the atmosphere. This process greatly enhances the image sharpness.

Detecting Invisible Radiation

As we said, sunlight is made up of more than just the radiation that is visible to our eyes. Gamma rays, X-rays, ultraviolet radiation, infrared radiation, and radio waves are also produced by stars. Photographic film that is sensitive to ultraviolet and infrared radiation has been developed, thereby extending the limits of our vision. However, most of this radiation cannot penetrate our atmosphere, so balloons, rockets, and satellites must transport cameras "above" the atmosphere to record it.

Of great importance is a narrow band of radio radiation that does penetrate the atmosphere. One particular wavelength is the 21-centimeter (8-inch) line produced by neutral hydrogen (hydrogen atoms that lack an electrical charge). Measurement of this radiation has permitted us to map the galactic distribution of hydrogen, the material from which stars are made.

The detection of radio waves is accomplished by big dishes called **radio telescopes** (Figure 23.14A). In principle, the dish of one of these telescopes operates in the same manner as the mirror of an optical telescope. It is parabolic in shape and focuses the incoming radio waves on an antenna, which absorbs and transmits these waves to an amplifier, just like any radio antenna.

Because radio waves are about 100,000 times longer than visible radiation, the surface of the dish need not be as smooth as a mirror. In fact, except for the shortest radio waves, a wire mesh is a good reflector. On the other hand, because radio signals from celestial sources are very weak, large dishes are necessary to intercept an adequate signal. The largest radio telescope is a bowl-shaped antenna hung in a natural depression in Puerto Rico (Figure 23.15). It is 300 meters (1000 feet) in diameter and has some directional flexibility in its movable antenna. The largest steerable types have about 100-meter (330-foot) dishes like that currently being constructed at the National Radio Astronomy Observatory in Green Bank, West Virginia.

Radio telescopes also have rather poor resolution, making it difficult to pinpoint the radio source. Pairs or groups of telescopes are used to reduce this problem. When several radio telescopes are wired together, the resulting network is called a **radio interferometer** (Figure 23.14B).

A.

B.

▲ **FIGURE 23.14** **A.** The 43-meter (140-foot) steerable radio telescope at Green Bank, West Virginia. The dish acts like the mirror of a reflector-type optical telescope to focus radio waves onto the antenna. The antenna is the small round object supported on four "legs" above the dish. **B.** Twenty-seven identical antennas operate together to form the Very Large Array near Socorro, New Mexico. (Courtesy of National Radio Astronomy Observatory)

Box 23.1 Understanding Earth:
The Largest Optical Telescopes

The main focus of a telescope is to collect as much light as possible. The larger the telescope's lens or mirror, the more light it collects, allowing for viewing of fainter objects. Because an important focus of astronomy involves observing very distant and hence very faint cosmic sources, very large telescopes are essential.

Until recently, the largest telescopes were limited to mirrors about 5 meters (200 inches) in diameter because the task of casting, cooling, and polishing large glass mirrors to very fine tolerances (less than the thickness of a human hair) was enormously time consuming and expensive. For example, the construction of the 5-meter mirror for the Hale Telescope on Mount Palomar, California, began 1934 and was not completed until 1948. However, during the last decade, with the aid of high-tech manufacturing techniques, several large-diameter telescopes have been built, and several more are being planned.

Recently, two twin 8.1-meter telescopes went into operation. Gemini North is located in the Northern Hemisphere, on Hawaii's Mauna Kea. Located at an altitude of nearly 4,200 meters (13,800 feet), the Mauna Kea Observatory is the world's highest. Its twin, Gemini South, is located in Chile, on the western slopes of the Andes Mountains. Mauna Kea also houses the Japanese 8.3-meter Subaru telescope. (Subaru is the Japanese name for the Pleiades.)

A different innovative design is employed in the recently completed twin 10-meter Keck Telescope, also located on Mauna Kea (Figure 23.A). These instruments, operated by the California Institute of Technology and the University of California, use a mosaic of 36 six-sided 1.8-meter mirrors carefully positioned by the computer to give the optical effect of a 10-meter (400-inch) mirror.

To date, the largest optical telescope in terms of total light-gathering is the European Southern Observatory's Very Large Telescope (VLT), located at Cerro Paranal, Chile. It consists of four separate 8.2-meter instruments that work independently, or in conjunction with each other. When working in tandem, these telescopes have 10 times the light-gathering capacity of the 5-meter Hale Telescope and therefore can "see" cosmic objects that are 10 times dimmer.

Figure 23.A Mirror of the 10-meter Keck Telescope. The mirror was constructed from 36 hexagonal segments. (Photo by Roger Ressmeyer/CORBIS)

▲ FIGURE 23.15 The 300-meter (1000-foot) radio telescope at Arecibo, Puerto Rico. (Courtesy of National Astronomy and Ionosphere Center's Arecibo Observatory, operated by Cornell University under contract with the National Science Foundation)

▲ FIGURE 23.16 The Sun is the source of more than 99 percent of all energy on Earth. (Photo by Thomas Dimock/Corbis/Stock Market)

Radio telescopes have some advantages over optical telescopes. They are much less affected by turbulence in the atmosphere, clouds, and the weather in general. No protective dome is required, which reduces the cost of construction, and "viewing" is possible 24 hours a day. More important, radio telescopes can "see" through interstellar dust clouds that obscure our view at visible wavelengths. Radio signals from distant points in the universe pass unhindered through the dust, giving us an unobstructed view. Furthermore, radio telescopes can detect clouds of gases too cool to emit visible light. These cold gas clouds are important because they are the sites of star formation.

Radio telescopes are, however, hindered by human-made radio interference. Thus, while optical telescopes are placed on remote mountaintops to reduce interference from city lights, radio telescopes are often hidden in valleys to block human-made radio interference.

Radio telescopes have revealed such spectacular events as the collision of two galaxies. Of even greater interest was the discovery of *quasars* (quasi-stellar radio sources). These perplexing objects are the most distant things known in the universe and will be examined further in Chapter 24.

The Sun

The Sun is one of the 200 billion stars that make up the Milky Way Galaxy. Although the Sun is of no significance to the universe as a whole, to us on Earth it is the primary source of energy (see Box 23.2). Every-

thing from the fossil fuels we burn in our automobiles and power plants to the food we eat is ultimately derived from solar energy (Figure 23.16). The Sun is also important to astronomers, since it is the only star close enough to permit study of the surface. Even with the largest telescopes, other stars appear only as points of light.

Because of the Sun's brightness and its damaging radiation, it is not safe to observe it directly. However, a small telescope will project its image on a piece of cardboard held behind the telescope's eyepiece, and the Sun can be studied safely in this manner. This basic method is used in several telescopes around the world, which keep a constant vigil of the Sun. One of the finest is at the Kitt Peak National Observatory in southern Arizona (Figure 23.17). It consists of a 150-meter sloped enclosure that directs sunlight to a mirror situated below ground. From the mirror, an 85-centimeter (33-inch) image of the Sun is projected to an observing room, where it is studied.

Compared to other stars of the universe, many of which are larger, smaller, hotter, cooler, more red, or more blue, the Sun is an "average star." However, on the scale of our solar system, it is truly gigantic, having a diameter equal to 109 Earth diameters (1.35 million kilometers) and a volume of 1.25 million times as great as that of Earth. Yet, because of its gaseous nature, the density is only one quarter that of solid Earth, nearly the density of water.

Structure of the Sun

For convenience of discussion, we divide the Sun into four parts: the solar interior; the visible surface, or photosphere; and the two layers of its atmosphere, the

▲ FIGURE 23.17 The unique Robert J. McMath Solar Telescope at Kitt Peak, near Tucson, Arizona. Movable mirrors at the top follow the Sun, reflecting its light down the sloping tunnel. (Photo by Kent Wood/Photo Researchers, Inc.).

chromosphere and the corona (Figure 23.18). Because the Sun is gaseous throughout, no sharp boundaries exist between these layers. The Sun's interior makes up all but a tiny fraction of the solar mass, and unlike the outer three layers, it is not accessible to direct observation. We shall discuss the visible layers first.

Photosphere. The **photosphere** (*photos* = light, *sphere* = a ball) is aptly named, for it radiates most of the sunlight we see and therefore appears as the bright disk of the Sun. Although it is considered to be the Sun's "surface," it is unlike most surfaces to which we are accustomed. The photosphere consists of a layer of incandescent gas less than 500 kilometers (300 miles) thick, having a pressure less than 1/100 of our atmosphere. Furthermore, it is neither smooth nor uniformly bright, as the ancients had imagined. It has numerous blemishes.

When viewed through a telescope under ideal conditions, the photosphere's grainy texture is apparent. This is the result of numerous relatively small, bright markings called **granules** (*granum* = small grain), which are surrounded by narrow, dark regions (Figure 23.19). Granules are typically the size of Texas, and they owe their brightness to hotter gases that are rising from below. As this gas spreads laterally, cooling causes it to darken and sink back into the interior. Each granule survives only 10 to 20 minutes, while the combined motion of new

BOX 23.2 EARTH AS A SYSTEM:
Variable Sun and Climatic Change

Among the most persistent hypotheses of climatic change have been those based on the idea that the Sun is a variable star and that its output of energy varies through time. The effect of such changes would seem direct and easily understood: Increases in solar output would cause the atmosphere to warm, and reductions would result in cooling. This notion is appealing because it can be used to explain climatic changes of any length or intensity. Still, there is at least one major drawback: No major long-term variations in the total intensity of solar radiation have yet been measured outside the atmosphere. Such measurements were not even possible until modern satellite technology became available. Now that it is possible, we will need many years of records before we begin to sense how variable (or invariable) energy from the Sun really is.

Several proposals for climatic change based on a variable Sun relate to *sunspot cycles*. The most conspicuous and best-known features on the surface of the Sun are the dark blemishes called *sunspots*. Although their origin is uncertain, it has been established that sunspots are huge magnetic storms that extend from the

Sun's surface deep into the interior. Moreover, these spots are associated with the Sun's ejection of huge masses of particles that, on reaching Earth's atmosphere, interact with gases there to produce auroral displays.

Along with other solar activity, the number of sunspots seem to increase and decrease in a regular way, creating a cycle of about 11 years. The graph in Figure 23.B shows the annual number of sunspots, beginning in the early 1700s. However, this pattern is not always as regular as it appears.

There have been prolonged periods when sunspots have been absent or nearly so. Moreover, these events correspond closely with cold periods in Europe and

North America. Conversely, periods characterized by plentiful sunspots have correlated well with warmer times in these regions.

Referring to these matches, some scientists have suggested that such correlations make it appear that changes on the Sun are an important cause of climate change. But other scientists seriously question this notion. Their hesitation stems in part from subsequent investigations using different climate records from around the world that failed to find a significant correlation between solar activity and climate. Even more troubling is that no testable physical mechanism exists to explain the purported effect.

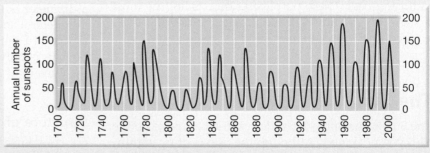

Figure 23.B Mean annual sunspot numbers.

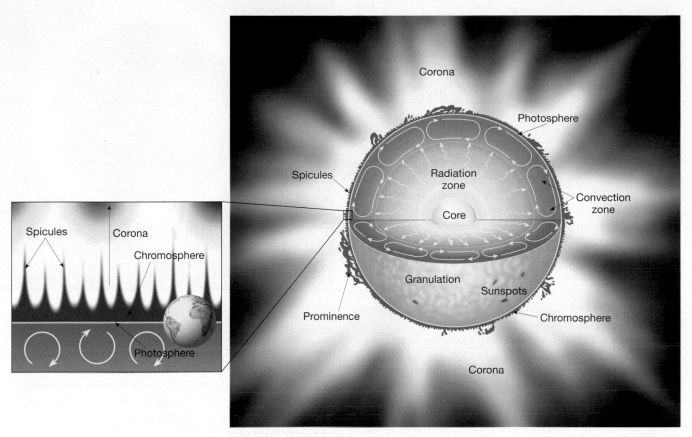

▲ **FIGURE 23.18** Diagram of solar structure in cutaway view. Earth is shown for scale.

granules replacing old ones gives the photosphere the appearance of boiling. This up-and-down movement of gas is called *convection.* Besides producing the grainy appearance of the photosphere, convection is believed to be responsible for the transfer of energy in the uppermost part of the Sun's interior (Figure 23.18).

The composition of the photosphere is revealed by the dark lines of its absorption spectrum (see Fig-

▼ **FIGURE 23.19** Granules of the solar photosphere. The granules appear as yellowish-orange patches, while the dark areas shown are small sunspots. Each granule is about the size of Texas and lasts for only 10 to 20 minutes before being replaced by a new granule. (Courtesy of National Optical Astronomy Observatories)

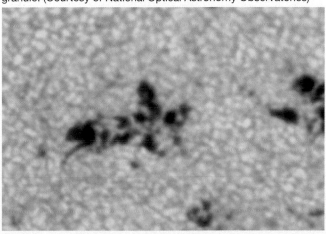

ure 23.3). When these fingerprints are compared to the spectra of known elements, they indicate that most of the elements found on Earth also occur on the Sun. When the strengths of the absorption lines are analyzed, the relative abundance of the elements can be determined. These studies reveal that 90 percent of the Sun's surface atoms are hydrogen, almost 10 percent are helium, and only minor amounts of the other detectable elements are present. Other stars also indicate similar disproportionate percentages of these two lightest elements, a fact we shall consider later.

STUDENTS SOMETIMES ASK...

If the Sun is an enormous ball of hot gas, how can it have a surface?

That's a good observation. Actually, the visible surface of the Sun, the photosphere, is a layer of gas about 500 kilometers (300 miles) thick. It is from this layer that we receive most sunlight. Although more light is emitted from the layer below the photosphere, that light is absorbed in the overlying layers of gas. Further, above the photosphere, the gas is less dense and thus unable to radiate much light. The photosphere is the layer that is dense enough to emit ample light yet has a density low enough to allow light to escape. Since the photosphere emits most of the light we see, it appears as the outermost surface of the Sun.

Chromosphere. Just above the photosphere lies the **chromosphere** (color sphere), a relatively thin layer of hot, incandescent gases a few thousand kilometers thick. The chromosphere is observable for a few moments during a total solar eclipse or by using a special instrument that blocks out the light from the photosphere. Under such conditions, it appears as a thin red rim around the Sun. Because the chromosphere consists of hot, incandescent gases under low pressure, it produces a bright-line spectrum that is nearly the reverse of the dark-line spectrum of the photosphere. One of the bright lines of hydrogen contributes a good portion of its total light and accounts for this sphere's red color.

A study of the chromospheric spectrum conducted in 1868 revealed the existence of an unknown element. It was named helium, from *helios*, the Greek word for Sun. Originally, helium was thought to be an element unique to the stars, but 27 years later it was discovered in natural-gas wells on Earth.

The top of the chromosphere contains numerous **spicules** (*spica* = point), flamelike structures that extend upward as much as 10,000 kilometers into the lower corona, almost like trees that reach into our atmosphere (Figure 23.20). Spicules are produced by the turbulent motion of the granules below.

Corona. The outermost portion of the solar atmosphere, the **corona** (*corona* = crown), is very tenuous and, as with the chromosphere, is visible only when the brilliant photosphere is covered (Figure 23.21). This envelope of ionized gases normally extends a million kilometers from the Sun and produces a glow about half as bright as the full Moon.

At the outer fringe of the corona, the ionized gases have speeds great enough to escape the gravitational pull of the Sun. The streams of protons and electrons that boil from the corona constitute the **solar wind**. They travel outward through the solar system at very high speeds (250 to 800 kilometers a second) and even-

▲ **FIGURE 23.21** Solar corona photographed during a total eclipse. (Photo by Lawrence Burr/Getty Images, Inc—Liaison)

tually are lost to interstellar space. During its journey, the solar wind interacts with the bodies of the solar system, continually bombarding lunar rocks and altering their appearance. Although Earth's magnetic field prevents the solar winds from reaching our surface, these winds do affect our atmosphere, as we shall discuss later.

Studies of the energy emitted from the photosphere indicate that its temperature averages about 6000 K (10,000°F). Upward from the photosphere, the temperature unexpectedly increases, exceeding 1 million K at the top of the corona. It should be noted that although the coronal temperature exceeds that of the photosphere by many times, it radiates much less energy overall because of its very low density.

Surprisingly, the high temperature of the corona is probably caused by sound waves generated by the convective motion of the photosphere. Just as boiling water makes noise, the energetic sound waves generated in the photosphere are believed to be absorbed by the gases of the corona and thereby raise their temperatures.

The Active Sun

Sunspots. The most conspicuous features on the surface of the Sun are the dark blemishes called **sunspots** (Figure 23.22A). Although large sunspots were occasionally observed before the advent of the telescope, they were generally regarded as opaque objects located somewhere between the Sun and Earth. In 1610 Galileo concluded that they were residents of the solar surface, and from their motion he deduced that the Sun rotates on its axis about once a month.

▼ **FIGURE 23.20** Spicules of the chromosphere on the edge of the solar disk. (National Solar Observatory/Sacramento Peak)

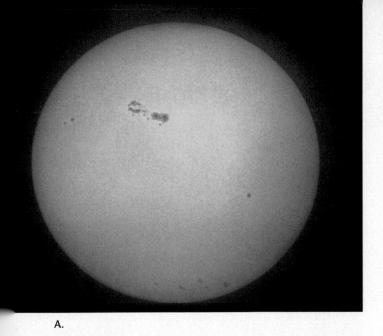

A.

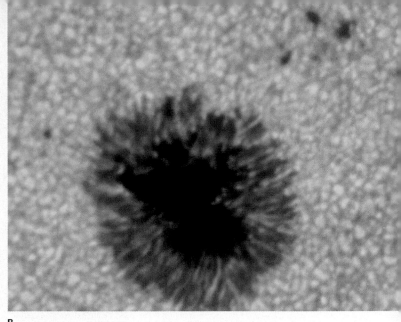

B.

▲ **FIGURE 23.22** **A.** Large sunspot group on the solar disk. (Celestron 8 photo courtesy of Celestron International) **B.** Sunspots having visible umbra (dark central area) and penumbra (lighter area surrounding umbra). (Courtesy of National Optical Astronomy Observatories)

Observations made later indicated that not all parts of the Sun rotate at the same speed. The Sun's equator rotates once in 25 days, while a place located 70 degrees from the solar equator, whether north or south, requires 33 days for one rotation. If Earth rotated in a similar disjointed manner, imagine the consequences! The Sun's nonuniform rotation is a testimonial to its gaseous nature.

Sunspots begin as small dark pores about 1600 kilometers (1000 miles) in diameter. Although most pores last for only a few hours, some grow into blemishes many times larger than Earth and last for a month or more. The largest spots often occur in pairs surrounded by several smaller spots. An individual spot contains a black center, the *umbra*, (*umbra* = shadow), which is rimmed by a lighter region, the *penumbra* (*paene* − almost, *umbra* = shadow) (Figure 23.22B). Sunspots appear dark only by contrast with the brilliant photosphere, a fact accounted for by their temperature, which is about 1500 K less than that of the solar surface. If these dark spots could be observed away from the Sun, they would appear many times brighter than the full moon.

During the early nineteenth century, it was believed that a tiny planet named Vulcan orbited between Mercury and the Sun. In the search for Vulcan an accurate record of sunspot occurrences was kept. Although the planet was never found, the sunspot data collected did reveal that the number of sunspots observable on the solar disk varies in an 11-year cycle.

First, the number of sunspots increases to a maximum, with perhaps a hundred or more visible at a given time. Then, over a period of five to seven years, their numbers decline to a minimum, when only a few or even none are visible. At the beginning of each cycle, the first sunspots form about 30 degrees from

the solar equator, but as the cycle progresses and their numbers increase, they form nearer the equator. During the period when sunspots are most abundant, the majority form about 15 degrees from the equator. They rarely occur more than 40 degrees away from the Sun's equator, or within 5 degrees of it.

Another interesting characteristic of sunspots was discovered by astronomer George Hale, for whom the Hale Telescope is named. Hale deduced that the large spots are strongly magnetized, and when they occur in pairs, they have opposite magnetic poles. For instance, if one member of the pair is a north magnetic pole, then the other member is a south magnetic pole, as with the north and south poles of Earth's magnetic field. Also, every pair located in the same hemisphere is magnetized in the same manner. However, all pairs in the *other* hemisphere are magnetized in the opposite manner. At the beginning of each sunspot cycle, the situation reverses, and the polarity of these sunspot pairs is opposite those of the previous cycle. The cause of this change in polarity—in fact, the cause of sunspots themselves—is not fully explained. However, other solar activity varies in the same cyclic manner as sunspots do, indicating a common origin.

Prominences. Among the more spectacular features of the active Sun are **prominences** (*prominere* = to jut out). These huge cloudlike structures, consisting of concentrations of chromospheric gases, are best observed when they are on the edge, or limb, of the Sun, where they often appear as great arches that extend well into the corona (Figure 23.23). Many prominences have the appearance of a fine tapestry and seem to hang motionless for days at a time, but motion pictures reveal that the material within them is continually falling like luminescent rain. On the other hand,

653

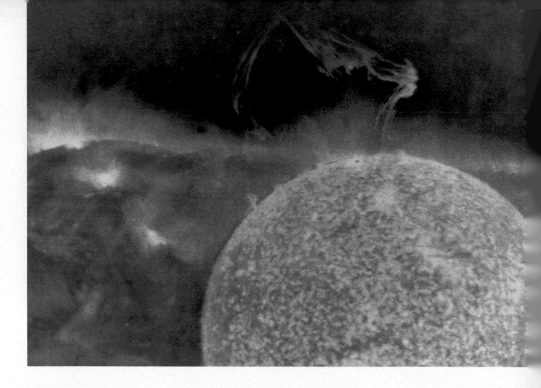

► **FIGURE 23.23** A huge solar prominence. (PHOTRI/Corbis/Stock Market)

eruptive prominences rise almost explosively away from the Sun. These active prominences reach velocities up to 1000 kilometers (620 miles) per second and may leave the Sun entirely. Whether eruptive or quiescent, prominences are ionized chromospheric gases trapped by magnetic fields that extend from regions of intense solar activity.

Solar Flares. These are the most explosive events associated with sunspots. **Solar flares** are brief outbursts that normally last an hour or so and appear as a sudden brightening of the region above a sunspot cluster. During their existence, enormous quantities of energy are released, much of it in the form of ultraviolet, radio, and X-ray radiation. Simultaneously, fast-moving atomic particles are ejected, causing the solar wind to intensify noticeably. Although a major flare could conceivably endanger a manned space flight, they are relatively rare. About a day after a large outburst, the ejected particles reach Earth and disturb the ionosphere,* affecting long-distance radio communications.

The most spectacular effects of solar flares, however, are the **auroras**, also called the Northern and Southern Lights (Figure 23.24). Following a strong solar flare, Earth's upper atmosphere near its magnetic poles is set aglow for several nights. The auroras appear in a wide variety of forms. Sometimes the display consists of vertical streamers with considerable movement. At other times, the auroras appear as a series of luminous expanding arcs or as a quiet, almost foglike, glow. Auroral displays, like other solar activities, vary in intensity with the 11-year sunspot cycle.

---

*The ionosphere is a complex atmospheric zone of ionized gases extending between about 80 and 400 kilometers (50 and 250 miles) above Earth's surface.

The Solar Interior

The interior of the Sun cannot be observed directly. For that reason, all we know about it is based on information acquired from the energy it radiates and from theoretical studies. The source of the Sun's energy, **nuclear fusion** (*fusus* = to melt), was not discovered until the late 1930s.

Deep in its interior, a nuclear reaction called the **proton-proton chain** converts four hydrogen nuclei (protons) into the nucleus of a helium atom. The energy released from the proton-proton reaction results because some of the matter involved is actually converted to energy. This can be illustrated by noting that four hydrogen atoms have a combined atomic mass of 4.032 (4 × 1.008), whereas the atomic mass of helium is 4.003, or 0.029 less than the combined mass of the hydrogen. The tiny missing mass is emitted as energy according to Einstein's formula $E = mc^2$, where E equals energy, m equals mass, and c equals the speed of light. Because the speed of light is very great, the amount of energy released from even a small amount of mass is enormous.

The conversion of just one pinhead's worth of hydrogen to helium generates more energy than burning thousands of tons of coal. Most of this energy is in the form of high-energy photons that work their way toward the solar surface, being absorbed and reemitted many times until they reach an opaque layer just below the photosphere. Here, convection currents serve to transport this energy to the solar surface, where it radiates through the transparent chromosphere and corona (see Figure 23.18).

Only a small percentage (0.7 percent) of the hydrogen in the proton-proton reaction is actually converted to energy. Nevertheless, the Sun is consuming an estimated 600 million tons of hydrogen each second, with

▲ **FIGURE 23.24** Aurora borealis (Northern Lights) as seen from Alaska. The same phenomenon occurs toward the South Pole, where it is called the Aurora australis (Southern Lights). (Photo by Michio Hoshino/Minden Pictures)

about 4 million tons of it being converted to energy. As hydrogen is consumed, the product of this reaction, helium, forms the solar core, which continually grows in size.

Just how long can the Sun produce energy at its present rate before all of its fuel (hydrogen) is consumed? Even at the enormous rate of consumption, the Sun has enough fuel to last easily another 100 billion years. However, evidence from other stars indicates that the Sun will grow dramatically and engulf Earth long before all of its hydrogen is gone. It is thought that a star the size of the Sun can exist in its present stable state for 10 billion years. As the Sun is already 5 billion years old, it is "middle-aged."

To initiate the proton-proton reaction, the Sun's internal temperature must have reached several million degrees. What was the source of this heat? As previously noted, the solar system is believed to have formed from an enormous cloud of dust and gases (mostly hydrogen) that condensed gravitationally. The consequence of squeezing (compressing) a gas is to increase its temperature. Although all of the bodies in the solar system were compressed, the Sun was the only one, because of its size, that became hot enough to trigger the proton-proton reaction. Astronomers currently estimate its internal temperature at 15 million K.

The planet Jupiter is basically a hydrogen-rich gas ball; if it were about 10 times more massive, it too might have become a star. The idea of one star orbiting another may seem odd, but recent evidence indicates that about 50 percent of the stars in the universe probably occur in pairs or multiples!

Chapter Summary

- Visible light constitutes only a small part of an array of energy, generally referred to as *electromagnetic radiation*. Light, a type of electromagnetic radiation, can be described in two ways: (1) as waves and (2) as a stream of particles, called *photons*. The wavelengths of electromagnetic radiation vary from several kilometers for *radio waves* to less than a billionth of a centimeter for *gamma rays*. The shorter wavelengths correspond to more energetic photons.

- *Spectroscopy* is the study of the properties of light that depend on wavelength. When a prism is used to disperse visible light into its component parts (wavelengths), one of three possible types of *spectra* is produced (a *spectrum*, the singular form of *spectra*, is the light pattern produced by passing light through a prism). The three types of spectra are (1) *continuous spectrum*, (2) *dark-line (absorption) spectrum*, and (3) *bright-line (emission) spectrum*. The spectra of most stars are of the dark-line type. Spectroscopy can be used to determine (1) the state of matter of an object (solid, liquid, high- or low-pressure gas; (2) the composition of gaseous objects; (3) the temperature of a radiating body; and (4) the motion of an object. Motion (direction toward or away and velocity) is determined using the *Doppler effect*—the apparent change in the wavelength of radiation emitted by an object caused by the relative motions of the source and the observer.

- There are two types of optical telescopes: (1) the *refracting telescope*, which uses a *lens* as its *objective* to bend or refract light, so that it converges at an area called the *focus*; and (2) the *reflecting telescope*, which uses a *concave mirror* to focus (gather) the light. When examining an image directly, both types of telescopes require a second lens, called an *eyepiece*, which magnifies the image produced by the objective.

- Telescopes have three properties that aid astronomers: (1) *light-gathering power*, which is a function of the size of the objective—large objectives gather more light and therefore "see" farther into space; (2) *resolving power*, which allows for sharper images and finer details, is the ability of a telescope to separate objects that are close together, e.g., Pluto and its moon, Charon; and (3) *magnifying power*, the ability to make an object larger. Most modern telescopes have supplemental devices that enhance the image.

- Invisible radio-wave radiation is detected by "big dishes" called *radio telescopes*. A parabolic-shaped dish, often consisting of a wire mesh, operates in the same manner as the mirror of a reflecting telescope. Radio telescopes have poor resolution, making it difficult to pinpoint a radio source. To reduce this problem, several can be wired together into a network called a *radio interferometer*. The advantages of radio telescopes over optical telescopes are that radio telescopes are less affected by the weather, they are less expensive to construct, "viewing" is possible 24 hours a day, they can detect material in the universe too cool to emit visible radiation, and they can "see" through interstellar dust clouds.

- The *Sun* is one of the 200 billion stars that make up the Milky Way Galaxy. The Sun can be divided into four parts: (1) the *solar interior*, (2) the *photosphere* (visible surface), and the two layers of its atmosphere, (3) the *chromosphere* and (4) *corona*. The photosphere radiates most of the light we see. Unlike most surfaces, it consists of a layer of incandescent gas less than 500 kilometers (300 miles) thick, with a grainy texture consisting of numerous relatively small, bright markings called *granules*. Just above the photosphere lies the chromosphere, a relatively thin layer of hot, incandescent gases a few thousand kilometers thick. At the edge of the uppermost portion of the solar atmosphere, called the *corona*, ionized gases escape the gravitational pull of the Sun and stream toward Earth at high speeds, producing the *solar wind*.

- Numerous features have been identified on the active Sun. *Sunspots* are dark blemishes with a black center, the *umbra*, which is rimmed by a lighter region, the *penumbra*. The number of sunspots observable on the solar disk varies in an 11-year cycle. *Prominences*, huge cloudlike structures best observed when they are on the edge, or limb, of the Sun, are produced by ionized chromospheric gases trapped by magnetic fields that extend from regions of intense solar activity. The most explosive events associated with sunspots are *solar flares*. Flares are brief outbursts that release enormous quantities of energy that appear as a sudden brightening of the region above sunspot clusters. During the event, radiation and fast-moving atomic particles are ejected, causing the solar wind to intensify. When the ejected particles reach Earth and disturb the ionosphere, radio communication is disrupted and the *auroras*, also called the Northern and Southern Lights, occur.

- The source of the Sun's energy is *nuclear fusion*. Deep in the solar interior, at a temperature of 15 million K, a nuclear reaction called the *proton-proton chain* converts four hydrogen nuclei (protons) into the nucleus of a helium atom. During the reaction some of the matter is converted to the energy of the Sun. A star the size of the Sun can exist in its present stable state for 10 billion years. As the Sun is already 5 billion years old, it is a "middle-aged" star.

Key Terms

aurora (p. 654)
bright-line (emission) spectrum (p. 641)
chromatic aberration (p. 644)
chromosphere (p. 652)
continuous spectrum (p. 641)

corona (p. 652)
dark-line (absorption) spectrum (p. 641)
Doppler effect (p. 642)
electromagnetic radiation (p. 640)
eyepiece (p. 643)

focal length (p. 643)
focus (p. 643)
granules (p. 650)
nuclear fusion (p.654)
objective lens (p. 643)
photon (p. 641)

photosphere (p. 650)
prominence (p. 653)
proton-proton chain (p. 654)
radiation pressure (p. 641)
radio interferometer (p. 647)

radio telescope (p. 647)
reflecting telescope (p. 644)
refracting telescope (p. 643)
solar flare (p. 654)
solar wind (p. 652)

spectroscope (p. 642)
spectroscopy (p. 641)
spicule (p. 652)
sunspot (p. 652)

Review Questions

1. What term is used to describe the collection that includes gamma rays, X-rays, ultraviolet light, visible light, infrared radiation, and radio waves?

2. Which color has the longest wavelength? The shortest?

3. What is spectroscopy?

4. Describe a continuous spectrum. Give an example of a natural phenomenon that exhibits a continuous spectrum.

5. What produces emission lines (bright lines) in a spectrum?

6. What can be learned about a star (or other celestial objects) from a dark-line (absorption) spectrum?

7. How can astronomers determine whether a star is moving toward or away from Earth?

8. True or false: The primary objective of a refracting telescope is a mirror.

9. What three properties do telescopes have that aid astronomers?

10. Of the three properties listed in Question 9, which is the least important to astronomers?

11. What is the advantage of a space telescope over a similar instrument on Earth?

12. Why do astronomers seek to design telescopes with larger and larger objectives?

13. Why do all large optical telescopes use mirrors to collect light rather than lenses?

14. With reflecting telescopes, why are special viewing systems needed?

15. Explain the following statement: "Photography has extended the limits of our vision."

16. Why would the Moon make a good site for an observatory?

17. Why are radio telescopes much larger than optical telescopes?

18. What are some of the advantages of radio telescopes over optical telescopes?

19. Compare the diameter of the Sun to that of Earth.

20. Describe the photosphere, chromosphere, and corona.

21. List the features associated with the active Sun and describe each.

22. Explain how a sunspot can be very hot and yet appear dark.

23. What is the solar wind?

24. What "fuel" does the Sun consume?

25. What happens to the matter that is consumed in the proton-proton chain reaction?

Examining the Earth System

1. Of the two sources of energy that power the Earth system, the Sun is the main driver of Earth's external processes. If the Sun increased its energy output by 10 percent, what would happen to global temperatures? What effect would this temperature change have on the percentage of water that exists as ice? What would be the impact on the position of the ocean shoreline?

Speculate as to whether the change in temperature might produce an increase or decrease in the amount of surface vegetation? In turn, what impact might this change in vegetation have on the level of atmospheric carbon dioxide? How would such a change in the amount of carbon dioxide in the atmosphere affect global temperatures?

Online Study Guide

The *Earth Science* Website uses the resources and flexibility of the Internet to aid in your study of the topics in this chapter. Written and developed by Earth science instructors, this site will help improve your understanding of Earth science. Visit **http://www.prenhall.com/tarbuck** and click on the cover of *Earth Science 11e* to find:

- **On-line review quizzes.**
- **Critical thinking exercises.**
- **Links to chapter-specific Web resources.**
- **Internet-wide key term searches.**
http://www.prenhall.com/tarbuck

Stars embedded in clouds of dust and gases produce colorful
emission nebulae.
(Royal Observatory, Edinburgh, Scotland/Anglo-Australian
Telescope Board/Science Photo Library/Photo Researchers, Inc.)

CHAPTER 24

Beyond Our Solar System

The star Proxima Centauri is about 4.3 light-years away, roughly 100 million times farther than the Moon. Yet, other than our own Sun, it is the closest star to Earth. This fact suggests that the universe is incomprehensibly large. What is the nature of this vast cosmos beyond our solar system? Are the stars distributed randomly, or are they organized into distinct clusters? Do stars move, or are they permanently fixed features, like lights strung out against the black cloak of outer space? Does the universe extend infinitely in all directions, or is it bounded? To consider these questions, this chapter will examine the universe by taking a census of the stars—the most numerous objects in the night sky.

Properties of Stars

The Sun is the only star whose surface we can observe. Yet a great deal is known about the universe beyond our solar system. In fact, more is known about the stars than about our outermost planet, Pluto. This knowledge hinges on the fact that stars, and even gases in the "empty" space between stars, radiate energy in all directions into space (Figure 24.1). The key to understanding the universe is to collect this radiation and unravel the secrets it holds. Astronomers have devised many ingenious methods to do just that. We will begin by examining stellar distances and some intrinsic properties of stars, including *color, brightness, mass, temperature,* and *size.*

Measuring Distances to the Stars

Measuring the distance to a star is very difficult. Obviously, we cannot journey to the star. Nevertheless, astronomers have developed some *indirect* methods of measuring stellar distances. The most basic of these measurements is called stellar parallax.

Recall from Chapter 21 that **stellar parallax** is the extremely slight back-and-forth shifting in the apparent position of a nearby star due to the orbital motion of Earth. To review, the principle of parallax is easy to visualize. Close one eye, and with your index finger in a vertical position, use your eye to line up your finger with some distant object. Now, without moving your finger, view the object with your other eye and notice that its position appears to have changed. The farther away you hold your finger, the less its position seems to shift.

In practice, parallax is determined by photographing a nearby star against the background of distant stars. Then, six months later, when Earth has moved halfway around its orbit, a second photograph is taken. When these photographs are compared, the position of the nearby star appears to have shifted with respect to the background stars. Figure 24.2 illustrates this shift and the parallax angle determined from it. The nearest stars have the largest parallax angles, while those of distant stars are too slight to measure. Recall that the sixteenth-century astronomer Tycho Brahe was unable to detect stellar parallax, leading him to reject the idea that Earth orbits the Sun.

It should be emphasized that parallax angles are very small. The parallax angle to the nearest star, Proxima Centauri, is less than 1 second of arc, which equals 1/3600 of a degree. To put this in perspective, fully extend your arm and raise your little finger. Your finger is roughly 1 degree wide. Try doing this on a moonlit night, covering the Moon with your finger. The Moon is only about 1/2 degree wide. Now imagine detecting a movement that is only

▶ **FIGURE 24.1** Lagoon Nebula. It is in glowing clouds like these that gases and dust particles become concentrated into stars. (Courtesy of National Optical Astronomy Observatories)

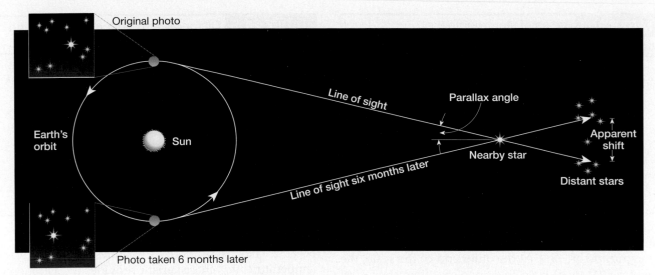

▲ **FIGURE 24.2** Geometry of stellar parallax. The parallax angle shown here is enormously exaggerated to illustrate the principle. Because the distance to even the nearest stars is thousands of times greater than the Earth–Sun distance, the triangles that astronomers work with are quite long and narrow, making the angles that are measured very small.

1/3600 as wide as your finger. It should be apparent why Tycho Brahe, without the aid of a telescope, was unable to observe stellar parallax.

The distances to stars are so large that the conventional units, such as kilometers or astronomical units, are often too cumbersome to use. A better unit to express stellar distance is the **light-year**, which is the distance light travels in one Earth-year—about 9.5 trillion kilometers (5.8 trillion miles).

In principle, the method used to measure stellar distances is elementary and was known to the ancient Greeks. But in practice, measurements are greatly complicated because of the tiny angles involved and because the Sun, as well as the star being measured, also have actual motion through space. The first accurate stellar parallax was not determined until 1838. Even today, parallax angles for only a few thousand of the nearest stars are known with certainty.

Most stars have such small parallax shifts that accurate measurement is not possible. Fortunately, a few other methods have been derived for estimating distances to these stars. Also, the Hubble Space Telescope, which is not hindered by Earth's turbulent, light-distorting atmosphere, is expected to obtain accurate parallax distances for many additional remote stars.

Stellar Brightness

Three factors control the apparent brightness of a star as seen from Earth: *how big* it is, *how hot* it is, and *how far away* it is. The stars in the night sky are a grand assortment of sizes, temperatures, and distances, so their brightnesses vary widely.

Apparent Magnitude. Stars have been classified according to their brightness since the second century B.C., when Hipparchus placed about 1000 of them into six categories. The measure of a star's brightness is called its **magnitude**. Some stars may appear dimmer than others only because they are farther away. Therefore, a star's brightness, *as it appears when viewed from Earth*, has been termed its **apparent magnitude**.

When numbers are employed to designate relative brightness, the larger the magnitude number, the dimmer the star. *Stars that appear the brightest are of the first magnitude, whereas the faintest stars visible to the unaided eye are of the sixth magnitude.* With the invention of the telescope, many stars fainter than the sixth magnitude were discovered.

In the mid-1800s, a method was developed for comparing the brilliance of stars using magnitude. Just as we can compare the brightness of a 50 watt light-bulb to that of a 100-watt bulb, we can compare the brightness of stars having different magnitudes. It was determined that a first-magnitude star was about 100 times brighter than a sixth-magnitude star. Therefore, on the scale that was devised, two stars that differ by 5 magnitudes have a ratio in brightness of 100 to 1. Hence, a seventh-magnitude star is 100 times brighter than a twelfth-magnitude star.

It follows, then, that the brightness ratio of two stars differing by only one magnitude is about 2.5.* Thus, a star of the first magnitude is about 2.5 times brighter than a star of the second magnitude. Table 24.1 shows some magnitude differences and the corresponding brightness ratios.

Because some stars are brighter than first-magnitude stars, zero and negative magnitudes were introduced. The brightest star in the night sky, Sirius, has an apparent

---

*Calculations: 2.512 × 2.512 × 2.512 × 2.512 × 2.512, or 2.512 raised to the fifth power, equals 100.

TABLE 24.1	Ratios of star brightness.
Difference in Magnitude	**Brightness Ratio**
0.5	1.6 : 1
1	2.5 : 1
2	6.3 : 1
3	16 : 1
4	40 : 1
5	100 : 1
10	10,000 : 1
20	100,000,000 : 1

TABLE 24.2	Distance, apparent magnitude, and absolute magnitude of some stars.		
Name	**Distance (Light-years)**	**Apparent Magnitude***	**Absolute Magnitude***
Sun	NA	−26.7	5.0
Alpha Centauri	4.27	0.0	4.4
Sirius	8.70	−1.4	1.5
Arcturus	36	−0.1	−0.3
Betelgeuse	520	0.8	−5.5
Deneb	1600	1.3	−6.9

*The more negative, the brighter; the more positive, the dimmer.

magnitude of −1.4, about 10 times brighter than a first-magnitude star. On this scale, the Sun has an apparent magnitude of −26.7. At its brightest, Venus has a magnitude of −4.3. At the other end of the scale, the 5-meter (200-inch) Hale Telescope can view stars with an apparent magnitude of 23, approximately 100 million times dimmer than stars that are visible to the unaided eye.

Absolute Magnitude. Astronomers are also interested in the "true" brightness of stars, called their **absolute magnitude**. Stars of the same luminosity, or brightness, usually do not have the same *apparent* magnitude, because their distances from us are not equal. To compare their true, or intrinsic, brightness, astronomers determine what magnitude the stars would have if they were at a standard distance of about 32.6 light-years. For example, the Sun, which has an apparent magnitude of −26.7, would, if located at a distance of 32.6 light-years, have an absolute magnitude of about 5. Thus, stars with absolute magnitudes greater than 5 (smaller numerical value) are intrinsically brighter than the Sun, but because of their distance, they appear much dimmer.

Table 24.2 lists the absolute and apparent magnitudes of some stars as well as their distances from Earth. Most stars have an absolute magnitude between −5 and 15, which puts the Sun near the middle of this range.

Stellar Color and Temperature

The next time you are outdoors on a clear night, take a good look at the stars and note their color (Figure 24.3). Some that are quite colorful can be found in the constellation Orion. Of the two brightest stars in Orion, Betelgeuse (α Orionis) is definitely red, whereas Rigel (β Orionis) appears blue.

Very hot stars with surface temperatures above 30,000 K emit most of their energy in the form of short-wavelength light and therefore appear blue. Red stars, on the other hand, are much cooler, generally less than 3000 K, and most of their energy is emitted as longer-wavelength red light. Stars with temperatures between 5000 and 6000 K appear yellow, like the Sun.

Because color is primarily a manifestation of a star's temperature, this characteristic provides the astronomer with useful information about a star.

Binary Stars and Stellar Mass

One of the night sky's best-known constellations, the Big Dipper, appears at first glance to consist of seven stars. But anyone with good eyesight can resolve the second star in the handle of the Big Dipper (Mizar) as two stars. During the eighteenth century, astronomers used their new tool, the telescope, to discover numerous such star pairs. One of the stars in the pair is usually fainter than the other, and for this reason it was considered to be farther away. In other words, the stars were not considered true pairs but were thought only to lie in the same line of sight.

In the early nineteenth century, careful examination of numerous star pairs by William Herschel revealed that many stars actually orbit one another. The two stars are in fact united by their mutual gravitation. These pairs of stars, in which the members are far enough apart to be resolved telescopically, are called

▼ **FIGURE 24.3** Time-lapse photograph of stars in the constellation Orion. These star trails show some of the various star colors. It is important to note that the eye sees color somewhat differently than does photographic film. (Courtesy of National Optical Astronomy Observatories)

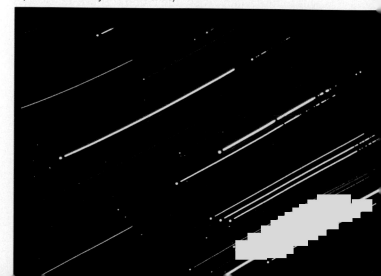

visual binaries (*binaries* = double). The idea of one star orbiting another may seem unusual, but evidence indicates that more than 50 percent of the stars in the universe occur in pairs or multiples.

Binary stars are used to determine the star property most difficult to calculate—its mass. The mass of a body can be established if it is gravitationally attached to a partner, which is the case for any binary star system. Binary stars orbit each other around a common point called the *center of mass* (Figure 24.4). For stars of equal mass, the center of mass lies exactly halfway between them. If one star is more massive than its partner, their common center will be located closer to the more massive one. Thus, if the sizes of their orbits can be observed, a determination of their individual masses can be made. You can experience this relationship on a seesaw by trying to balance a person who has a much greater mass.

For illustration, when one star has an orbit half the size (radius) of its companion, it is twice as massive. If their combined masses are equal to three times the mass of the Sun, then the larger will be twice as mas-

▼ **FIGURE 24.4**   Binary stars orbit each other around their common center of mass. **A.** For stars of equal mass, the center of mass lies exactly halfway between them. **B.** If one star is twice as massive as its companion, it is twice as close to their common center of mass. Therefore, more massive stars have proportionately smaller orbits than do their less massive companions.

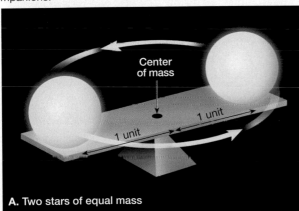

A. Two stars of equal mass

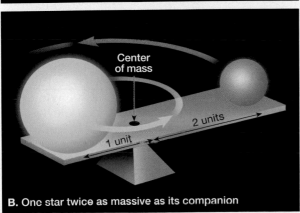

B. One star twice as massive as its companion

sive as the Sun, and the smaller will have a mass equal to that of the Sun. Most stars have masses that range between 1/10 and 50 times the mass of the Sun.

Hertzsprung-Russell Diagram

Early in the twentieth century, Einar Hertzsprung and Henry Russell independently studied the relation between the true brightness (absolute magnitude) and temperature of stars. From this research each developed a graph, now called a **Hertzsprung-Russell diagram, (H-R diagram)**, that exhibits these intrinsic stellar properties. By studying H-R diagrams, we learn a great deal about the sizes, colors, and temperatures of stars.

To produce an H-R diagram, astronomers survey a portion of the sky and plot each star according to its luminosity (brightness) and temperature (Figure 24.5). Notice that the stars in Figure 24.5 are not uniformly distributed. Rather, about 90 percent of all stars fall along a band that runs from the upper-left corner to the lower-right corner of an H-R diagram. These "ordinary" stars are called **main-sequence stars**. As shown in Figure 24.5, *the hottest main-sequence stars are intrinsically the brightest, and vice versa.*

The luminosity of the main-sequence stars is also related to their mass. The hottest (blue) stars are about 50 times more massive than the Sun, whereas the coolest (red) stars are only 1/10 as massive. Therefore, on the H-R diagram, the main-sequence stars appear in a decreasing order, from *hotter, more massive* blue stars to *cooler, less massive* red stars.

Note the location of the Sun in Figure 24.5. The Sun is a yellow main-sequence star with an absolute magnitude of about 5. Because the magnitude of a vast majority of main-sequence stars lie between −5 and 15, and because the Sun falls midway in this range, the Sun is often considered an average star. However, more main-sequence stars are cooler and less massive than our Sun.

Just as all humans do not fall into the normal size range, some stars do not fit in with the main-sequence stars. Above and to the right of the main sequence in the H-R diagram in Figure 24.5 lies a group of very luminous stars called *giants,* or, on the basis of their color, **red giants**. The size of these giants can be estimated by comparing them with stars of known size that have the same surface temperature. We know that objects having equal surface temperatures radiate the same amount of energy per unit area. Therefore, any difference in the brightness of two stars having the same surface temperature is attributable to their relative sizes.

As an example, let us compare the Sun, which has been assigned a luminosity of 1, with another yellow star that has a luminosity of 100. Because both stars

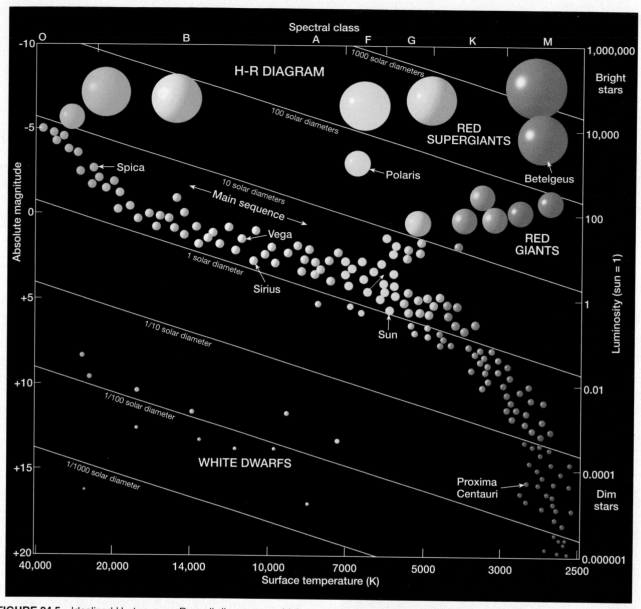

▲ **FIGURE 24.5** Idealized Hertzsprung-Russell diagram on which stars are plotted according to temperature and absolute magnitude.

have the same surface temperature, they both radiate the same amount of energy per unit area. Therefore, for the more luminous star to be 100 times brighter than the Sun, it must have 100 times more surface area. It should be clear why stars whose plots fall in the upper-right position of an H-R diagram are called *giants.*

Some stars are so large that they are called **supergiants.** Betelgeuse, a bright red supergiant in the constellation Orion, has a radius about 800 times that of the Sun. If this star were at the center of our solar system, it would extend beyond the orbit of Mars, and Earth would find itself inside the star! Other red giants that are easy to locate in our sky are Arcturus in the constellation Bootes and Antares in Scorpius.

In the lower-central portion of the H-R diagram, the opposite situation occurs. These stars are much

fainter than main-sequence stars of the same temperature, and by using the same reasoning as before, they must be much smaller. Some probably approximate Earth in size. This group has come to be called **white dwarfs**, although not all are white.

Soon after the first H-R diagrams were developed, astronomers realized their importance in interpreting stellar evolution. Just as with living things, a star is born, ages, and dies. Owing to the fact that almost 90 percent of the stars lie on the main sequence, we can be relatively certain that stars spend most of their active years as main-sequence stars. Only a few percent are giants, and perhaps 10 percent are white dwarfs.

After considering some variable stars and the nature of interstellar matter, we will return to the topic of stellar evolution.

March 10, 1935

May 6, 1935

▲ **FIGURE 24.6** Photographs of Nova Herculis (a nova in the constellation Hercules), taken about 2 months apart, showing the decrease in brightness. (Courtesy of Lick Observatory)

Variable Stars

Stars that fluctuate in brightness are known as variables. Some, called **pulsating variables**, fluctuate regularly in brightness by expanding and contracting in size. The importance of one member of this group (cepheid variables) in determining stellar distances is discussed in Box 24.1. Other pulsating variables have irregular periods and are of no value for this purpose.

The most spectacular variables belong to a group known as **eruptive variables**. When one of these explosive events associated with these stars occurs, it appears as a sudden brightening of a star, called a **nova** (Figure 24.6). The term *nova*, meaning "new," was used by the ancients because these stars were unknown to them before their abrupt increase in luminosity.

During the outburst, the outer layer of the star is ejected at high speed. The "cloud" of ejected material occasionally can be captured photographically

---

![Earth]

BOX 24.1 UNDERSTANDING EARTH:
Determining Distance from Magnitude

For a star too distant for parallax measurements, knowing its absolute and apparent bright-ness provides astronomers with a tool for approximating its distance. The apparent magnitude is measured with a photometer (light meter) attached to a telescope. If we also know a star's true brightness, we can determine just how far away that star would have to be for it to have the brightness we observe.

You use this same principle when you drive at night and estimate the distance to an oncoming car from the brightness of its headlights. You can do this because you know the true brightness of an automobile headlight. But how do astronomers determine the intrinsic brightness of a star? Fortunately, some stars have characteristics that provide the necessary data.

One important star group is called *cepheid variables*. These are pulsating stars that get brighter and fainter in a rhythmic

fashion. The interval between two successive occurrences of maximum brightness of a pulsating variable is termed its *light period*. Most cepheid variables pulsate with periods of between two and 50 days. For example, the North Pole Star (Polaris) varies about 10 percent in brightness over a period of four days. In

general, the longer the light period of a cepheid, the greater is its absolute magnitude (Figure 24.A). Thus, by determining the light period of a cepheid, its absolute magnitude can be calculated. When this absolute magnitude is compared to the apparent magnitude, a good approximation of its distance can be made.

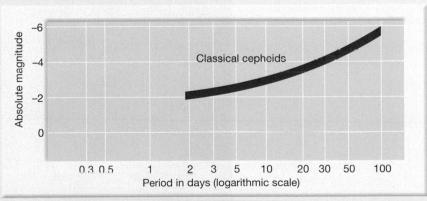

Figure 24.A Relationship between the light period (two successive occurrences of maximum brightness) and absolute magnitude of pulsating stars (cepheid variables).

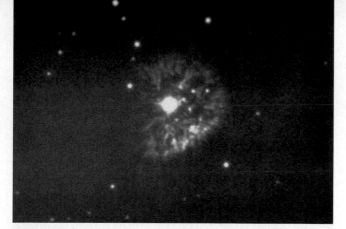

▲ **FIGURE 24.7** Expanding "cloud" around Nova Persei. (Courtesy of Hale Observatories. Copyright by the California Institute of Technology and Carnegie Institution of Washington)

(Figure 24.7). A nova generally reaches maximum brightness in a couple of days, remains bright for only a few weeks, then slowly returns in a year or so to its original brightness. Because the star returns to its prenova brightness, we can assume that only a small amount of its mass is lost during the flareup. Some stars have experienced more than one such event. In fact, the process probably occurs repeatedly.

The modern explanation of novae proposes that they occur in binary systems consisting of an expanding red giant and a hot white dwarf that are in close proximity. Hydrogen-rich gas from the oversized giant encroaches near enough to the white dwarf to be gravitationally transferred. Eventually, enough of the hydrogen-rich gas is added to the dwarf to cause it to ignite explosively. Such a thermonuclear reaction rapidly heats and expands the outer gaseous envelope of the hot dwarf to produce a nova event. In a relatively short time, the white dwarf returns to its prenova state, where it remains inactive until the next buildup occurs.

Interstellar Matter

Between existing stars is "the vacuum of space." However, it is far from being a pure vacuum, for it is populated with accumulations of dust and gases. The name applied to these concentrations of interstellar matter is **nebula** (*nebula* = cloud). If this interstellar matter is close to a very hot (blue) star, it will glow and is called a **bright nebula**. The two main types of bright nebulae are known as *emission nebulae* and *reflection nebulae*.

Emission nebulae are gaseous masses that consist largely of hydrogen. They absorb *ultraviolet radiation* emitted by an embedded or nearby hot star. Because these gases are under very low pressure, they reradiate, or emit, this energy as *visible light*. This conversion of ultraviolet light to visible light is known as *fluorescence,* an effect you observe daily in fluorescent lights. A well-known emission nebula easily seen with binoculars is the sword of the hunter in the constellation Orion (Figure 24.8).

Reflection nebulae, as the name implies, merely reflect the light of nearby stars (Figure 24.9). Reflection nebulae are thought to be composed of rather dense clouds of large particles called **interstellar dust**. This view is supported by the fact that atomic gases with low densities could not reflect light sufficiently to produce the glow observed.

When a dense cloud of interstellar material is not close enough to a bright star to be illuminated, it is referred to as a **dark nebula**. Exemplified by the Horsehead Nebula in Orion, dark nebulae appear as opaque objects silhouetted against a bright background (Figure 24.10). Dark nebulae can also easily be seen as starless regions—holes in the heavens when viewing the Milky Way.

▶ **FIGURE 24.8** The Orion Nebula is a well-known emission nebula. Bright enough to be seen by the naked eye, the Orion Nebula is located in the sword of the hunter in the constellation of the same name. (Courtesy of National Optical Astronomy Observatories)

Although nebulae appear very dense, they actually consist of very thinly scattered matter. Because of their enormous size, however, the total mass of rarefied particles and molecules may be many times that of the Sun. Interstellar matter is of great interest to astronomers because it is from this material that stars and planets are formed.

Stellar Evolution

The idea of describing how a star is born, ages, and then dies may seem a bit presumptuous, for many of these objects have life spans that surely exceed billions of years. However, by studying stars of different ages, astronomers have been able to piece together a plausible model for stellar evolution. The method that was used to create this model is analogous to what an alien being, upon reaching Earth, might do to determine the developmental stages of human life. By examining a large number of humans, this stranger would be able to observe the birth of human life, the activities of children and adults, and the death of the elderly. From this information, the alien would attempt to put the stages of human development into their proper sequence. Based on the relative abundance of humans in each stage of development, it would even be possible to conclude that humans spend more of their lives as adults than as toddlers. In a similar

▲ FIGURE 24.9 A faint blue reflection nebula, in the Pleiades star cluster, is caused by the reflection of starlight from dust in the nebula. The Pleiades star cluster, just visible to the naked eye in the constellation Taurus, is spectacular when viewed through binoculars or a small telescope. (Palomar Observatories/California Institute of Technology [Caltech])

▼ FIGURE 24.10 The Horsehead Nebula, a dark nebula in a region of glowing nebulosity in Orion. (© Anglo-Australian Observatory/Royal Observatory Edinburgh, photography by David Malin)

fashion, astronomers have pieced together the story of the stars.

Simply, stars exist because of gravity. The mutual gravitational attraction of particles in a thin, gaseous cloud causes the cloud to collapse. As the cloud is squeezed to unimaginable pressures, its temperature rises, igniting its nuclear furnace, and a star is born. A star is a ball of very hot gases, caught between the opposing forces of gravity trying to contract it and thermal nuclear energy trying to expand it. Eventually, all of a star's nuclear fuel will be exhausted and gravity takes over, collapsing the stellar remnant into a small, dense body.

Star Birth

The birthplaces of stars are dark, cool interstellar clouds, which are comparatively rich in dust and gases (Figure 24.11). In the neighborhood of the Milky Way, these gaseous clouds consist of 92 percent hydrogen, 7 percent helium, and less than 1 percent of the remaining heavier elements. By some mechanism not yet fully understood, these thin gaseous clouds become concentrated enough to begin to contract gravitationally. One proposal to explain the triggering of stellar formation is a shock wave traveling from a catastrophic explosion (supernova) of a nearby star. But, regardless of the force that initiates the concentration of

interstellar matter, once it is accomplished, mutual gravitational attraction of the particles squeeze the cloud, pulling every particle toward the center. As the cloud shrinks, gravitational energy (potential energy) is converted into energy of motion, or heat energy.

The initial contraction spans a million years or so. With the passage of time, the temperature of this gaseous body slowly rises, eventually reaching a temperature sufficiently high to cause it to radiate energy from its surface in the form of long-wavelength red light. Because this large red object is not hot enough to engage in nuclear fusion, it is not yet a star. The name **protostar** is applied to these bodies.

Protostar Stage

During the protostar stage, gravitational contraction continues, slowly at first, then much more rapidly (Figure 24.11). This collapse causes the core of the developing star to heat much more intensely than the outer envelope. When the core has reached at least 10 million K, the pressure within is so great that groups of four hydrogen nuclei are fused together into single helium nuclei. Astronomers refer to this nuclear reaction as **hydrogen burning** because an enormous amount of energy is released. However, keep in mind that thermonuclear "burning" is not burning in the usual chemical sense.

▼ **FIGURE 24.11** H-R diagram showing stellar evolution for a star about as massive as the Sun.

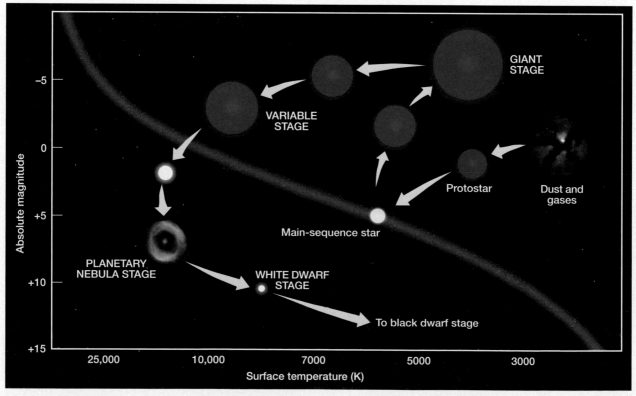

Heat from hydrogen fusion causes the stellar gases to increase their motion. This in turn results in an increase in the outward gas pressure. At some point, this outward pressure exactly balances the inward force of gravity. When this balance is reached, the star becomes a stable main-sequence star (Figure 24.11). Stated another way, a stable main-sequence star is balanced between two forces: *gravity*, which is trying to squeeze it into a smaller sphere, and *gas pressure*, which is trying to expand it.

Main-Sequence Stage

From this point in the evolution of a main-sequence star until its death, the internal gas pressure struggles to offset the unrelenting force of gravity. Typically, hydrogen burning continues for a few billion years and provides the outward pressure required to support the star from gravitational collapse.

Different stars age at different rates. Hot, massive blue stars radiate energy at such an enormous rate that they substantially deplete their hydrogen fuel in only a few million years. By contrast, the very smallest (red) main-sequence stars may remain stable for hundreds of billions of years. A yellow star, such as the Sun, remains a main-sequence star for about 10 billion years. As the solar system is about 5 billion years old, it is comforting to know that the Sun will remain stable for another 5 billion years.

An average star spends 90 percent of its life as a hydrogen-burning main-sequence star. Once the hydrogen fuel in the star's core is depleted, it evolves rapidly and dies. However, with the exception of the least-massive (red) stars, a star can delay its death by burning other fuels and becoming a giant.

Red Giant Stage

The evolution to the red giant stage results because the zone of hydrogen burning continually migrates outward, leaving behind an inert helium core. Eventually, all of the hydrogen in the star's core is consumed. While hydrogen fusion is still progressing in the star's outer shell, no fusion is taking place in the core. Without a source of energy, the core no longer has enough pressure to support itself against the inward force of gravity. As a result, the core begins to contract.

Although the core cannot generate nuclear energy, it does grow hotter by converting gravitational energy into heat energy. Some of this energy is radiated outward, initiating a more vigorous level of hydrogen fusion in the star's outer shell. This energy in turn heats and enormously expands the star's outer envelope, producing a giant body hundreds to thousands of times its main-sequence size (Figure 24.11).

As the star expands, its surface cools, which explains the star's reddish appearance. Eventually the star's gravitational force will stop this outward expansion. Once again, the two opposing forces, gravity and gas pressure, will be in balance, and this gaseous mass will be a stable but much larger star. Some red giants overshoot the equilibrium point and rebound like an overextended spring. Such stars continue to oscillate in size, becoming variable stars.

While the envelope of a red giant expands, the core continues to collapse and heat until it reaches 100 million K. At this incredible temperature, it is hot enough to initiate a nuclear reaction in which helium is converted to carbon. Thus, a red giant consumes both hydrogen and helium to produce energy. In stars more massive than the Sun, still other thermonuclear reactions occur that generate all the elements on the periodic table up to number 26, iron. Nuclear "burning" of elements heavier than iron requires an additional source of energy to keep the reaction progressing. Hence, these elements are not produced in ordinary stars.

Eventually, all the usable nuclear fuel in these giants will be consumed. The Sun, for example, will spend less than a billion years as a giant, and the more massive stars will pass through this stage even more rapidly. The force of gravity will again control the star's destiny as it squeezes the star into the smallest, most dense piece of matter possible.

Burnout and Death

Most of the events of stellar evolution discussed thus far are well documented. What happens to a star after the red giant phase is more speculative. We do know that a star, regardless of its size, must eventually exhaust all of its usable nuclear fuel and collapse in response to its immense gravitational force. With this in mind, we will now consider the final stage of stars in three different mass categories.

Death of Low-Mass Stars. Stars less than one half the mass of the Sun (0.5 solar mass) consume their fuel at a comparatively slow rate (Figure 24.12A). Consequently, these small, *cool red stars* may remain on the main sequence for up to 100 billion years. Because the interior of a low-mass star never attains sufficiently high temperatures and pressures to fuse helium, its only energy source is hydrogen fusion. Thus, low-mass stars never evolve to become bloated red giants. Rather, they remain as stable main-sequence stars until they consume their hydrogen fuel and collapse into a hot, dense *white dwarf*. As we shall see, white dwarfs are small, compact objects unable to support nuclear burning.

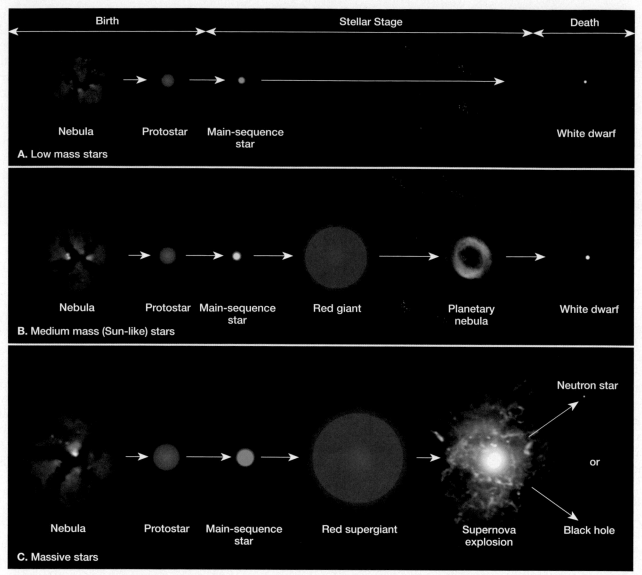

▲ FIGURE 24.12 The evolutionary stages of stars having various masses.

Death of Medium-Mass (Sun-like) Stars. Main-sequence stars with masses ranging between half that of the Sun and three times that of the Sun evolve in essentially the same way (Figure 24.12B). During their giant phase, Sun-like stars fuse hydrogen and helium fuel at an accelerated rate. Once this fuel is exhausted, these stars (like low-mass stars) collapse into an Earth-size body of great density—a white dwarf. The density of a white dwarf is as great as physics will allow, short of destroying protons and electrons.

The gravitational energy supplied to a collapsing white dwarf is reflected in its high surface temperature. However, without a source of nuclear energy, a white dwarf becomes cooler and dimmer as it continually radiates its remaining thermal energy into space.

During their collapse from red giants to white dwarfs, medium-mass stars are thought to cast off their bloated outer atmosphere, creating an expanding spherical cloud of gas. The remaining hot, central white dwarf heats the gas cloud, causing it to glow. These often beautiful, gleaming spherical clouds are called **planetary nebulae**. A good example of a planetary nebula is the Helix Nebula in the constellation Aquarius (Figure 24.13). This nebula appears as a ring because our line of sight through the center traverses less gaseous material than at the nebula's edge. It is, nevertheless, spherical in shape.

Death of Massive Stars. In contrast to Sun-like stars, which expire gracefully, stars exceeding three solar masses have relatively short life spans and terminate in a brilliant explosion called a **supernova** (Figure 24.12C). During a supernova event, a star becomes millions of times brighter than its prenova stage (see Box 24.2). If one of the nearest stars to Earth produced such an outburst, its brilliance would surpass that of the Sun. Supernovae are rare; none have been observed in our galaxy since the advent of the telescope,

◀ **FIGURE 24.13** The Helix Nebula, the nearest planetary nebula to our solar system. A planetary nebula is the ejected outer envelope of a Sun-like star that formed during the star's collapse from a red giant to a white dwarf. (© Anglo-Australian Observatory, photography by David Malin)

although Tycho Brahe and Galileo each recorded one about 30 years apart. An even larger supernova was recorded in A.D. 1054 by the Chinese. Today, the remnant of this great outburst is the Crab nebula, shown in Figure 24.14.

A supernova event is thought to be triggered when a massive star consumes most of its nuclear fuel. Without a heat engine to generate the gas pressure required to balance its immense gravitational field, it collapses. This implosion is of cataclysmic proportion, resulting in a shock wave that moves out from the star's interior. This energetic shock wave destroys the star and blasts the outer shell into space, generating the supernova event.

Theoretical work predicts that during a supernova, the star's interior condenses into a very hot object, possibly no larger than 20 kilometers in diameter (see Box 24.3). These incomprehensibly dense bodies have been named *neutron stars*. Some supernovae events are thought to produce even smaller and most intriguing objects called *black holes*. We will consider the nature of neutron stars and black holes in the Stellar Remnants section.

H-R Diagrams and Stellar Evolution

The Hertzsprung-Russell diagrams have been very helpful in formulating and testing models of stellar evolution. They are also useful for illustrating the changes that take place in an individual star during its life span. Figure 24.11 shows on an H-R diagram the evolution of a star about the size of the Sun. Keep in mind that the star does not physically move along this path, but rather that its position on the H-R diagram represents the color (temperature) and absolute magnitude (brightness) of the star at various stages in its evolution.

STUDENTS SOMETIMES ASK...
How will the Sun die?

In a few billion years, the Sun will exhaust the remaining hydrogen fuel in its core, an event that will trigger hydrogen fusion in the surrounding shell. As a result, the Sun's outer envelope will expand, producing a red giant that is hundreds of times larger and more luminous. The intense solar radiation will cause Earth's oceans to boil, and the solar winds will drive away Earth's atmosphere. In another billion years, the Sun will expel its outermost layer, producing a spectacular planetary nebula, while its interior will collapse to form a dense, small (planet size), white dwarf. Because of its small size, the Sun's energy output will be less than 1 percent of its current level. Gradually, the Sun will emit its remaining thermal energy, eventually becoming a cold, nonluminous body.

For example, on the H-R diagram, a protostar would be located to the right and above the main sequence (Figure 24.11). It is formed to the right because of its relatively cool surface temperature (red color), and above because it would be more luminous than a main-sequence star of the same color, a fact attributable to its large size. Careful examination of Figure 24.11 should help you visualize the evolutionary changes experienced by a star the size of our Sun. In addition, Table 24.3 provides a summary of the evolutionary history of stars having various masses.

Box 24.2 Understanding Earth:
Supernova 1987A

The first naked-eye supernova in 383 years was discovered in the southern sky in February 1987 (Figure 24.B). This stellar explosion was officially named SN 1987A (SN stands for supernova, and 1987A indicates that it was the first supernova observed in 1987). Naked-eye supernovae are rare. Only a few have been recorded in historic times. Arab observers saw one in 1006, and the Chinese recorded one in 1054 at the present location of the Crab Nebula. In addition, the astronomer Tycho Brahe observed a supernova in 1572, and Kepler saw one shortly thereafter in 1604.

Prior to this event, researchers could only test their hypotheses on dim supernovae seen in distant galaxies. Thus, when SN 1987A occurred, astronomers quickly focused every available telescope in the Southern Hemisphere on this spectacular event. As one astronomer remarked, "This supernova is better studied by far than any supernova in history." More important, this event has allowed astronomers to use observational data to test their theoretical models of stellar evolution.

Supernova 1987A occurred about 170,000 light-years away in the Large Magellanic Cloud, a satellite galaxy to our own Milky Way. As expected, the supernova rapidly increased in brightness to a peak magnitude of 2.4, outshining all the other stars in the Large Magellanic Cloud. Also as predicted, within a few weeks it began to fade. However, SN 1987A did provide some surprises.

From old photographs taken of the area, researchers identified the exploded star as Sanduleak. Astronomers were surprised to find that the parent star was a hot blue star about 15 times the mass of our Sun. Recall that only cool red giants are thought to die as a supernova event. Further, the Hubble Space Telescope made another unexpected discovery. Its camera revealed a very large shell of gas that predates the supernova explosion by about 40,000 years.

Astronomers now think that Sanduleak was once a red supergiant that had blown away its outer shell, exposing a hot blue core. It is this ejected outer shell that appears in the image produced by the Hubble Space Telescope. Then, some 40,000 years later, the remaining hot core of the red supergiant collapsed, producing the supernova of 1987.

Despite these twists, the theory of stellar evolution has held up very well. Theory predicts that the expanding remnants to Supernova 1987A will be large enough to be observed by the turn of the twenty-first century. Thus, astronomers continue to monitor SN 1987A to unravel its secrets and to confirm or refute their ideas about the final stages of stellar evolution.

Figure 24.B The great Supernova 1987A. The photo on the left was made prior to the supernova and the one on the right was made following the event. (© Anglo-Australian Observatory, photography by David Malin)

Stellar Remnants

Eventually, all stars consume their nuclear fuel and collapse into one of three final states—white dwarf, neutron star, or black hole. Although different in some ways, these small, compact objects are all composed of incomprehensibly dense material and all have extreme surface gravity.

White Dwarfs

White dwarfs are extremely small stars with densities greater than any known terrestrial material. It is believed that white dwarfs once were low-mass or medium-mass stars whose internal heat was able to keep these gaseous bodies from collapsing under their own gravitational force.

Although some white dwarfs are no larger than Earth, the mass of such a dwarf can equal 1.4 times that of the Sun. Thus, their densities may be a million times greater than water. A spoonful of such matter would weigh several tons. Densities this great are possible only when electrons are displaced inward from their regular orbits, around an atom's nucleus, allowing the atoms to take up less than the "normal" amount of space. Material in this state is called **degenerate matter**.

In degenerate matter, the atoms have been squeezed together so tightly that the electrons are dis-

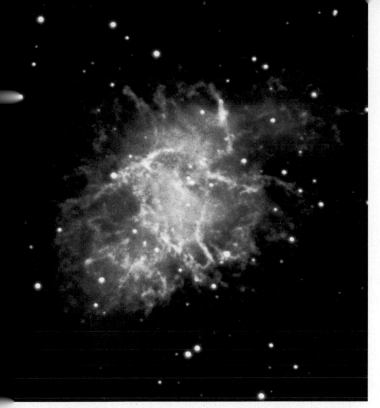

▲ **FIGURE 24.14** Crab Nebula in the constellation Taurus: the remains of the supernova of A.D. 1054. (UCO Lick Observatory Image)

placed much nearer to the nucleus. Degenerate matter uses electrical repulsion rather than molecular motion to support itself from total collapse. Although atomic particles in degenerate matter are much closer together than in normal Earth matter, they still are not packed as tightly as possible. Stars made of matter that has an even greater density are thought to exist.

As a star contracts into a white dwarf, its surface becomes very hot, sometimes exceeding 25,000 K. Even so, without a source of energy, it can only become cooler and dimmer. Although none have been observed, the terminal stage of a white dwarf must be a small, cold, nonluminous body called a *black dwarf*.

Neutron Stars

A study of white dwarfs produced what might at first appear to be a surprising conclusion. The smallest white dwarfs are the most massive, and the largest are the least massive. The explanation for this is that a more massive star, because of its greater gravitational force, is able to squeeze itself into a smaller, more densely packed object than can a less massive star. Thus, the smallest white dwarfs were produced from the collapse of larger, more massive stars than were the larger white dwarfs.

This conclusion led to the prediction that stars smaller and more massive than white dwarfs must exist. Named **neutron stars**, these objects are thought to be the remnants of supernova events. In a white dwarf, the electrons are pushed close to the nucleus, whereas in a neutron star the electrons are forced to combine with protons to produce neutrons (hence the name). If Earth were to collapse to the density of a neutron star, it would have a diameter equivalent to the length of a football field. A pea-size sample of this matter would weigh 100 million tons. This is approximately the density of an atomic nucleus; thus, neutron stars can be thought of as large atomic nuclei.

During a supernova implosion, the envelope of the star is ejected (Figure 24.15), while the core collapses into a very hot neutron star about 20 kilometers (12.4 miles) in diameter. Although neutron stars have high surface temperatures, their small size would greatly limit their luminosity. Consequently, locating one visually would be extremely difficult.

However, theory predicts that a neutron star would have a very strong magnetic field. Further, as a star collapses, it will rotate faster, for the same reason ice skaters rotate faster as they pull in their arms. If the Sun were to collapse to the size of a neutron star, it would increase its rate of rotation from once every 25 days to nearly 1000 times per second. Radio waves generated by these rotating stars would be concentrated into two narrow zones that would align with the star's magnetic poles. Consequently, these stars would resemble a rapidly rotating beacon emitting strong radio waves. If Earth happened to be in the path of these beacons, the star would appear to blink on and off, or pulsate, as the waves swept past.

In the early 1970s, a source that radiates short pulses of radio energy, called a **pulsar** (pulsating radio source), was discovered in the Crab Nebula. Visual inspection of this radio source revealed it to be a small

TABLE 24.3	Summary of evolution for stars of various masses.			
Initial Mass of Interstellar Cloud (Sun = 1)	**Main-Sequence Stage**	**Giant Phase**	**Evolution After Giant Phase**	**Terminal State (Final Mass)***
0.001	None (Planet)	No	None	Planet (0.001)
0.1	Red	No	None	White dwarf (0.1)
1–3	Yellow	Yes	Planetary Nebula	White dwarf (<1.4)
6	White	Yes	Supernova	Neutron star (1.4–3)
20	Blue	Yes (Supergiant)	Supernova	Black hole (>3.0)

*These mass numbers are estimates.

673

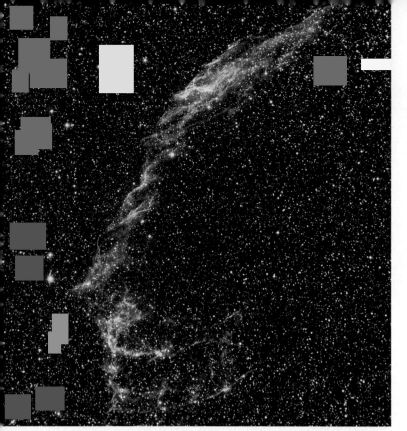

▲ FIGURE 24.15 Veil Nebula in the constellation Cygnus is the remnant of an ancient supernova implosion. (Palomar Observatories/California Institute of Technology [Caltech])

star centered in the nebula. The pulsar found in the Crab Nebula is undoubtedly the remains of the supernova of A.D. 1054 (see Figure 24.14). Thus, the first neutron star had been discovered.

Black Holes

Are neutron stars made of the most dense materials possible? No. During a supernova event, remnants of stars greater than three solar masses apparently collapse into objects even smaller and denser than neutron stars. Even though these objects would be very hot, their surface gravity would be so immense that even light could not escape the surface. Consequently, they would literally disappear from sight. These incredible bodies have appropriately been named **black holes**. Anything that moved too near a black hole would be swept in by its irresistible gravity and be devoured forever.

How can astronomers find an object whose gravitational field prevents the escape of all matter and energy? One strategy is to seek evidence of matter being rapidly swept into a region of apparent nothingness. Theory predicts that as matter is pulled into a black hole, it should become very hot and emit a

BOX 24.3 EARTH AS A SYSTEM:
From Stardust to You

During a supernova implosion, the internal temperature of a star may reach 1 billion K, a condition thought to produce very heavy elements such as gold and uranium. These heavy elements, plus the debris of novae and the planetary nebulae, are continually returned to interstellar space where they are available for the formation of other stars (Figure 24.C).

Astronomers believe that the earliest stars were made of nearly pure hydrogen. Fusion during the life and death of stars in turn produced heavier elements, some of which were returned to space. Because the Sun contains some heavy elements but has not yet reached the stage in its evolution where it could have produced them, it must be at least a second-generation star. Thus, our Sun, as well as the rest of the solar system, is believed to have formed from debris scattered from preexisting stars. If this is the case, the atoms in your body were produced billions of years ago inside a star, and the gold in your jewelry was formed during a supernova event that occurred trillions of kilometers away. Without such events, the development of life on Earth would not have been possible.

Figure 24.C Eagle Nebula in the constellation Serpens. This gaseous nebula is the site of a recent star formation. (Courtesy of National Optical Astronomy Observatories)

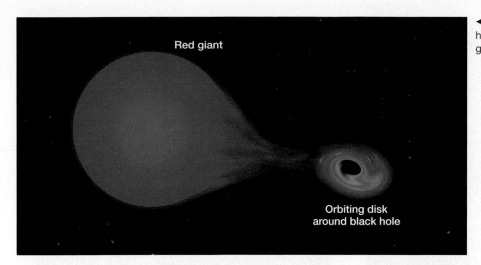

Red giant

Orbiting disk
around black hole

◄ **FIGURE 24.16** This illustration shows how astronomers believe a binary pair (red giant/black hole) might function.

flood of X-rays before being engulfed. Because isolated black holes would not have a source of matter to engulf, astronomers first looked at binary-star systems.

A likely candidate for a black hole is Cygnus X-1, a strong x-ray source in the constellation Cygnus. In this case, the x-ray source can be observed orbiting a supergiant companion with a period of 5.6 days. It appears that gases are pulled from this companion and spiral into the disc-shaped structure around the black hole (Figure 24.16). The result is a stream of X-rays. Because X-rays cannot penetrate our atmosphere efficiently, the existence of black holes was not confirmed until recently. The first x-ray sources were discovered in 1971 by detectors on satellites. Cygnus X-1 is such a source.

The Milky Way Galaxy

On a clear and moonless night away from city lights, you can see a truly marvelous sight—our own Milky Way Galaxy (Figure 24.17). With his telescope, Galileo discovered that this band of light was produced by countless individual stars that the unaided eye is unable to resolve. Today we realize that the Sun is actually a part of this vast system of stars, which number about 100 billion (Figure 24.18A). The milky appear-

ance of our galaxy results because the solar system is located within the flat *galactic disk*. Thus, when it is viewed from the inside, a higher concentration of stars appears in the direction of the galactic plane than in any other direction. You can see this in the edge-on view in Figure 24.18B.

When astronomers began telescopically to survey the stars located along the plane of the Milky Way, it appeared that equal numbers lay in every direction. Could Earth actually be at the center of the galaxy? A better explanation was put forth. Imagine that the trees in an enormous forest represent the stars in the galaxy. After hiking into this forest a short distance, you look around. What you see is an equal number of trees in every direction. Are you really in the center of the forest? Not necessarily; anywhere in the forest, except at the very edge, you will seem to be in the middle.

Structure of the Milky Way Galaxy

Attempts to visually inspect the Milky Way Galaxy are hindered by the large quantities of interstellar matter that lie in our line of sight. Nevertheless, with the aid of radio telescopes, the gross structure of our galaxy has been determined. The Milky Way is a rather large spiral galaxy whose disk is about 100,000

▼ **FIGURE 24.17** Panorama of our galaxy, the Milky Way. Notice the dark band caused by the presence of interstellar dark nebulae. (Courtesy of Axel Mellinger)

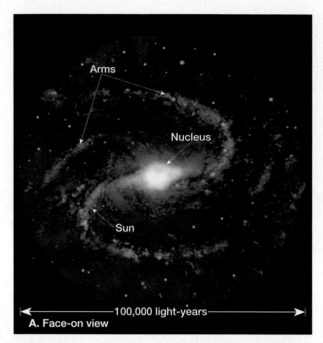

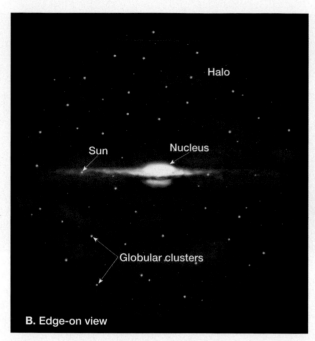

A. Face-on view

B. Edge-on view

▲ FIGURE 24.18   Structure of the visible portion of the Milky Way Galaxy.

light-years wide and about 10,000 light-years thick at the nucleus (Figure 24.18). As viewed from Earth, the center of the galaxy lies beyond the constellation Sagittarius.

Radio telescopes reveal the existence of at least three distinct *spiral arms,* with some showing splintering (Figure 24.19). The Sun is positioned in one of these arms about two-thirds of the way from the center, at a distance from the hub of about 30,000 light-years. The stars in the arms of the Milky Way rotate around the *galactic nucleus,* with the most outward ones moving the slowest, such that the ends of the arms appear to trail. The Sun and the arm it is in require about 200 million years for each orbit around the nucleus.

Surrounding the galactic disk is a nearly spherical *halo* made of very tenuous gas and numerous globular clusters. These star clusters do not participate in the rotating motion of the arms but rather have their own orbits that carry them through the disk. Although some clusters are very dense, they pass among the stars of the arms with plenty of room to spare.

Galaxies

In the mid-1700s, German philosopher Immanuel Kant proposed that the telescopically visible fuzzy patches of light scattered among the stars were actually distant

▶ FIGURE 24.19   If the Milky Way were photographed from a distance, it might appear like the spiral galaxy NGC 2997. (© Anglo-Australian Observatory, photography by David Malin)

▲ FIGURE 24.20 Great Galaxy, a spiral galaxy, in the constellation Andromeda. The two bright spots to the left and right are dwarf elliptical galaxies. (Palomar Observatories/California Institute of Technology [Caltech])

galaxies like the Milky Way. Kant described them as island universes. Each galaxy, he believed, contained billions of stars and, as such, was a universe in itself. The weight of opinion, however, favored the hypothesis that they were dust and gas clouds (nebulae) within our galaxy.

This matter was not resolved until the 1920s, when American astronomer Edwin Hubble was able to locate, within one of these fuzzy patches, some unique stars that are known to be intrinsically very bright. Because these very bright stars appeared only very faintly in the telescope, Hubble believed they must lie outside the Milky Way.

This fuzzy patch, which lies over a million light-years away, was named the Great Galaxy in Andromeda (Figure 24.20). Hubble had extended the universe far beyond the limits of our imagination, to include hundreds of billions of galaxies, each containing hundreds of billions of stars. It has been said that a million galaxies are found in that portion of the sky bounded by the cup of the Big Dipper. There are more stars in the heavens than grains of sand in all the beaches on Earth.

Types of Galaxies

From the hundreds of billions of galaxies, several basic types have been identified: spiral, elliptical, and irregular.

Spiral Galaxies The Milky Way and the Great Galaxy in Andromeda are examples of fairly large **spiral galaxies** (Figure 24.21). Andromeda can be

seen with the unaided eye as a fuzzy fifth-magnitude object. Typically, spiral galaxies are disk-shaped, with a somewhat greater concentration of stars near their centers, but there are numerous variations. Viewed broadside, arms are often seen extending from the central nucleus and sweeping gracefully away. The outermost stars of these arms rotate most slowly, giving the galaxy the appearance of a fireworks pinwheel.

One type of spiral galaxy, however, has the stars arranged in the shape of a bar, which rotates as a rigid system. This requires that the outer stars move faster than the inner ones, a fact not easy for astronomers to reconcile with the laws of motion. Attached to each end of these bars are curved spiral arms. These have become known as **barred spiral galaxies** (Figure 24.22). Spiral galaxies are generally quite large, ranging from 20,000 to about 125,000 light-years in diameter. About 10 percent of all galaxies are thought to be barred spirals, and another 20 percent are regular spiral galaxies like the Milky Way.

Elliptical Galaxies The most abundant group, making up 60 percent of the total, is the **elliptical galaxies**. These are generally smaller than spiral galaxies. Some are so much smaller, in fact, that the term *dwarf* has been applied. Because these dwarf galaxies are not visible at great distances, a survey of the sky reveals more of the conspicuous large spiral galaxies. Although most elliptical galaxies are small, the very largest known galaxies (200,000 light-years in diameter) are also elliptical. As their name implies, elliptical galaxies have an ellipsoidal shape that ranges to nearly spherical, and they lack spiral arms. The two dwarf companions of Andromeda shown in Figure 24.20 are elliptical galaxies.

Irregular Galaxies Only 10 percent of the known galaxies lack symmetry and are classified as **irregular galaxies**. The best-known irregular galaxies, the Large and Small Magellanic Clouds in the Southern Hemisphere, are easily visible with the unaided eye. Named after the explorer Ferdinand Magellan, who observed them when he circumnavigated Earth in 1520, they are our nearest galactic neighbors—only 150,000 light-years away.

One of the major differences among the galactic types is the age of the stars that make them up. The irregular galaxies are composed mostly of young stars, whereas the elliptical galaxies contain old stars. The Milky Way and other spiral galaxies consist of both young and old stars, with the youngest located in the arms.

Galactic Clusters

Once astronomers discovered that stars were associated in groups, they set out to determine whether galaxies also were grouped or just randomly distributed

A.

B.

▲ **FIGURE 24.21** Two views illustrating the idealized structure of spiral galaxies. (Courtesy of Hansen Planetarium/U.S. Naval Observatory)

throughout the universe. They found that, like stars, galaxies are grouped in **galactic clusters** (Figure 24.23). Some abundant clusters contain thousands of galaxies. Our own, called the **Local Group**, contains at least 28 galaxies. Of these, 3 are spirals, 11 are irregulars, and 14 are ellipticals. Galactic clusters also reside in huge swarms called superclusters. From visual observations, it appears that superclusters may be the largest entities in the universe.

Red Shifts

You probably have noticed the change in pitch of a car horn or ambulance siren as it passes by. When it is approaching, the sound seems to have a higher-than-normal pitch, and when it is moving away, the pitch sounds lower than normal. This effect, which occurs for all wave motion, including sound and light waves, was first explained by Christian Doppler in 1842 and is called the *Doppler effect*. The reason for the difference in pitch is that it takes time for the wave to be emitted. If the source of the wave is moving away, the beginning of the wave is emitted nearer to you than the end of the wave, effectively "stretching" the wave. This gives it a longer wavelength (see Figure 23.14). The opposite is true for an approaching source.

In the case of light, when a source is moving away, its light appears redder than it actually is, because its waves appear lengthened. Objects approaching have their light waves shifted toward the blue (shorter wavelength). Therefore, the Doppler effect reveals whether Earth and another celestial body are approaching or leaving one another. In addition, the amount of shift allows us to calculate the *rate* at which the relative movement is occurring. Large Doppler

shifts indicate higher velocities; smaller Doppler shifts indicate lower velocities.

Expanding Universe

One of the most important discoveries of modern astronomy was made in 1929 by Edwin Hubble. Observations completed several years earlier revealed that most galaxies have Doppler shifts toward the red end of the spectrum. Recall that red shift occurs because the light waves are "stretched," indicating that Earth and the source are moving away from each other. Hubble set out to explain the predominance of red shift.

Hubble realized that dimmer galaxies were probably farther away than were brighter galaxies. Thus, he tried to determine whether a relation existed between the distances to galaxies and their red shifts. Using estimated distances based on relative brightness and the observed Doppler red shifts, Hubble discovered that galaxies that exhibit the greatest red shifts are the most distant.

A consequence of the universal red shift is that it predicts that most galaxies (except for a few nearby)

▼ **FIGURE 24.22** Barred spiral galaxy. (Palomar Observatories/California Institute of Technology [Caltech])

▲ FIGURE 24.23 A cluster of galaxies located about 1 million light-years from Earth. (Courtesy of NASA)

are receding from us. Recall that the amount of Doppler red shift is dependent on the velocity at which the object is moving away. Greater red shifts indicate faster recessional velocities. Because more distant galaxies have greater red shifts, Hubble concluded that they must be retreating from us at greater velocities. This idea is currently termed **Hubble's law** and states that galaxies are receding from us at a speed that is proportional to their distance.

Hubble was surprised at this discovery because it implied that the most distant galaxies are moving away from us many times faster than those nearby. What type of cosmological theory can explain this fact? It was soon realized that an *expanding universe* can adequately account for the observed red shifts.

To help visualize the nature of this expanding universe, we will employ a popularly used analogy. Imagine a loaf of raisin bread dough that has been set out to rise for a few hours (Figure 24.24). As the dough doubles in size, so does the distance between all of the raisins. However, the raisins that were originally farther apart traveled a greater distance in the same time span than those located closer together. We therefore conclude that in an expanding universe, as in our analogy, those objects located farther apart move away from each other more rapidly.

Another feature of the expanding universe can be demonstrated using the raisin bread analogy. No matter which raisin you select, it will move away from all the other raisins. Likewise, no matter where one is located in the universe, every other galaxy (except those in the same cluster) will be receding. Edwin Hubble had indeed advanced our understanding of the universe. The Hubble Space Telescope is named in his honor.

The Big Bang

Did the universe have a beginning? Will it have an end? Cosmologists are trying to answer these questions, and that makes them a rare breed.

First and foremost, any viable theory regarding the origin of the universe must account for the fact that all galaxies (except for the very nearest) are moving away from us. Because all galaxies appear to be moving away from Earth, is our planet in the center of the universe? Probably not, because if we are not even in the center of our own solar system, and our solar system is not even in the center of the galaxy, it seems unlikely that we could be in the center of the universe.

A more probable explanation exists: Imagine a balloon with paper-punch "dots" glued to its surface. When the balloon is inflated, each dot spreads apart from every other dot. Similarly, if the universe is expanding, every galaxy would be moving away from every other galaxy.

This concept of an expanding universe led to the widely accepted **big bang** theory. According to this theory, the entire universe was at one time confined to a dense, hot, supermassive ball. Then, almost 14 billion years ago, a cataclysmic explosion occurred, hurling this material in all directions. The big bang marks the inception of the universe; all matter and space were created at that instant. The ejected masses of gas cooled and condensed, forming the stars that compose the galactic systems we now observe fleeing from their birthplace.

If the universe began with a big bang, how will it end? One view is that the universe will last forever. In this scenario, the stars will slowly burn out, being replaced by invisible degenerate matter and black holes

A. Raisin bread dough before it rises.

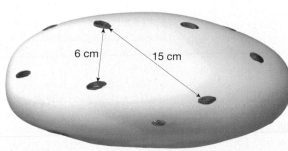

B. Raisin bread dough a few hours later.

◄ FIGURE 24.24 Illustration of the raisin bread analogy of an expanding universe. As the dough rises, raisins originally farther apart travel a greater distance in the same time span as those located closer together. Thus, raisins (like galaxies in a uniform expanding universe) that are located farther apart move away from each other more rapidly than those located nearer to each other.

that will travel outward through an endless, dark cold universe. The other possibility is that the outward flight of the galaxies will slow and eventually stop. Gravitational contraction would follow, causing the galaxies to collide and coalesce into the high-energy, high-density mass from which the universe began. This fiery death of the universe, the big bang operating in reverse, has been called the "big crunch."

Whether or not the universe will expand forever, or eventually collapse upon itself, depends on its average density. If the average density of the universe is more than an amount known as its *critical density* (about one atom for every cubic meter), the gravitational field is sufficient to stop the outward expansion and cause the universe to contract. On the other hand, if the density of the universe is less than the critical value, it will expand forever. Current estimates of the density of the universe place it below the critical density, which predicts an ever expanding, or *open universe*. Additional support for an open universe comes from studies that indicate the universe is expanding faster now than in the past. Hence, the view currently favored by most cosmologists is an expanding universe with no ending point.

It should be noted, however, that the methods used to determine the ultimate fate of the universe have substantial uncertainties. It is possible that previ-ously undetected matter (*dark matter*) exists in great quantities in the universe. If this is so, the galaxies could, in fact, collapse in the "big crunch."

Absence of evidence is not evidence of absence.

—Anonymous

? STUDENTS SOMETIMES ASK...
I have a hard time buying into the universe starting as a "big bang." Did it really happen?

You're not the first to have this doubt. In fact, the name Big Bang was originally coined by cosmologist Fred Hoyle as a sarcastic comment on the believability of the theory. The Big Bang Theory proposes that our universe began as a violent explosion, from which the universe continues to expand, evolve, and cool. Through decades of experimentation and observation, scientists have gathered substantial evidence that supports this theory. Despite this fact, the Big Bang Theory, like all other scientific theories, can never be proved. It is always possible that a future observation will disprove a previously accepted theory. Nevertheless, the Big Bang has replaced all alternative theories and remains the only widely accepted scientific model for the origin of the universe.

Chapter Summary

• One method for determining the distance to a star is to use a measurement called *stellar parallax*, the extremely slight back-and-forth shifting in a nearby star's position due to the orbital motion of Earth. *The farther away a star is, the less its parallax.* A unit used to express stellar distance is the *light-year*, which is the distance light travels in a year about 9.5 trillion kilometers (5.8 trillion miles).

• The intrinsic properties of stars include *brightness, color, temperature, mass,* and *size*. Three factors control the brightness of a star as seen from Earth: how big it is, how hot it is, and how far away it is. *Magnitude* is the measure of a star's brightness. *Apparent magnitude* is how bright a star appears when viewed from Earth. *Absolute magnitude* is the "true" brightness of a star if it were at a standard distance of about 32.6 light-years. The difference between the two magnitudes is directly related to a star's distance. Color is a manifestation of a star's temperature. Very hot stars (surface temperatures above 30,000 K) appear blue; red stars are much cooler (surface temperatures generally less than 3000 K). Stars with surface temperatures between 5000 and 6000 K appear yellow, like our Sun. The center of mass of orbiting *binary stars* (two stars revolving around a common center of mass under their mutual gravitational attraction) is used to determine the mass of the individual stars in a binary system.

• A *Hertzsprung-Russell diagram* is constructed by plotting the absolute magnitudes and temperatures of stars on a graph. A great deal about the sizes of stars can be learned from H-R diagrams. Stars located in the upper-right position of an H-R diagram are called *giants*, luminous stars of large radius. *Supergiants* are very large. Very small *white dwarf* stars are located in the lower-central portion of an H-R diagram. Ninety percent of all stars, called *main-sequence stars*, are in a band that runs from the upper-left corner to the lower-right corner of an H-R diagram.

• *Variable stars* fluctuate in brightness. Some, called *pulsating variables*, fluctuate regularly in brightness by expanding and contracting in size. When a star explosively brightens, it is called a *nova*. During the outburst, the outer layer of the star is ejected at high speed. After reaching maximum brightness in a few days, the nova slowly returns in a year or so to its original brightness.

• New stars are born out of enormous accumulations of dust and gases, called *nebula*, that are scattered between existing stars. A *bright nebula* glows because the matter is close to a very hot (blue) star. The two main types of bright nebulae are *emission nebulae* (which derive their visible light from the fluorescence of the ultraviolet light from a star in or near the nebula) and *reflection nebulae* (relatively dense dust clouds in interstellar space that are illuminated by reflecting the light of nearby stars). When a nebula is not close enough to a bright star to be illuminated, it is referred to as a *dark nebula*.

• Stars are born when their nuclear furnaces are ignited by the unimaginable pressures and temperatures in collapsing nebulae. New stars not yet hot enough for nuclear fusion are called *protostars*. When collapse causes the core of a protostar to reach a temperature of at least 10 million K, the fusion of hydrogen nuclei into helium nuclei begins in a process called *hydrogen burning*. The opposing forces acting on a star are *gravity* trying to contract it and *gas pressure (thermal nuclear energy)* trying to expand it. When the two forces are balanced, the star becomes a stable *main-sequence star*. When the hydrogen in a star's core is consumed, its outer envelope expands enormously and a *red giant* star, hundreds to thousands of times larger than its main-sequence size, forms. When all the usable nuclear fuel in these giants is exhausted and gravity takes over, the stellar remnant collapses into a small dense body.

• The *final fate of a star is determined by its mass*. Stars with less than one half the mass of the Sun collapse into hot, dense *white dwarf* stars. Medium-mass stars (between 0.5 and 3.0 times the mass of the Sun) become red giants, collapse, and end up as white dwarf stars, often surrounded by expanding spherical clouds of glowing gas called *planetary nebulae*. Stars more than three times the mass of the Sun terminate in a brilliant explosion called a *supernova*. Supernovae events can produce small, extremely dense *neutron stars*, composed entirely of subatomic particles called neutrons; or even smaller and more dense *black holes*, objects that have such immense gravity that light cannot escape their surface.

• The *Milky Way Galaxy* is a large, disk-shaped *spiral galaxy* about 100,000 light-years wide and about 10,000 light-years thick at the center. There are three distinct *spiral arms* of stars, with some showing splintering. The Sun is positioned in one of these arms about two-thirds of the way from the galactic center, at a distance of about 30,000 light-years. Surrounding the galactic disk is a nearly spherical halo made of very tenuous gas and numerous *globular clusters* (nearly spherically shaped groups of densely packed stars).

• The various types of galaxies include (1) *irregular galaxies*, which lack symmetry and account for only 10 percent of the known galaxies; (2) *spiral galaxies*, which are typically disk-shaped with a somewhat greater concentration of stars near their centers, often containing arms of stars extending from their central nucleus; and (3) *elliptical galaxies*, the most abundant type, which have an ellipsoidal shape that ranges to nearly spherical and that lack spiral arms.

• Galaxies are not randomly distributed throughout the universe. They are grouped in *galactic clusters*, some containing thousands of galaxies. Our own, called the *Local Group*, contains at least 28 galaxies.

• By applying the *Doppler effect* (the apparent change in wavelength of radiation caused by the motions of the source and the observer) to the light of galaxies, galactic motion can be determined. Most galaxies have Doppler shifts toward the red end of the spectrum, indicating increasing distance. The amount of Doppler shift is dependent on the velocity at which the object is moving. Because the most distant galaxies have the greatest red shifts, Edwin Hubble concluded in the early 1900s that they were retreating from us with greater recessional velocities than were more nearby galaxies. It was soon realized that an *expanding universe* can adequately account for the observed red shifts.

• The belief in the expanding universe led to the widely accepted *Big Bang Theory*. According to this theory, the entire universe was at one time confined in a dense, hot, supermassive concentration. Almost 14 billion years ago, a cataclysmic explosion hurled this material in all directions, creating all matter and space. Eventually the ejected masses of gas cooled and condensed, forming the stellar systems we now observe fleeing from their place of origin.

Key Terms

absolute magnitude (p. 662)
apparent magnitude (p. 661)
barred spiral galaxy (p. 677)
big bang (p. 679)
black hole (p. 674)
bright nebula (p. 666)
dark nebula (p. 666)
degenerate matter (p. 672)
elliptical galaxy (p. 677)
emission nebula (p. 666)
eruptive variables (p. 665)
galactic cluster (p. 678)

Hertzsprung-Russell (H-R) diagram (p. 663)
Hubble's law (p. 679)
hydrogen burning (p. 668)
interstellar dust (p. 666)
irregular galaxy (p. 677)
light-year (p. 661)
Local Group (p. 678)
magnitude (p. 661)
main-sequence stars (p. 663)
nebula (p. 666)
neutron star (p. 673)

nova (p. 665)
planetary nebula (p. 670)
protostar (p. 668)
pulsar (p. 673)
pulsating variables (p. 665)
red giant (p. 663)
reflection nebula (p. 666)
spiral galaxy (p. 677)
stellar parallax (p. 660)
supergiant (p. 664)
supernova (p. 660)
white dwarf (p. 664)

Review Questions

1. How far away in light-years is our nearest stellar neighbor, Proxima Centauri? Convert your answer to kilometers.

2. What is the most basic method of determining stellar distances?

3. Explain the difference between a star's apparent and absolute magnitudes. Which one is an intrinsic property of a star?

4. What is the ratio of brightness between a twelfth-magnitude and fifteenth-magnitude star?

5. What information about a star can be determined from its color?

6. What color are the hottest stars? Medium-temperature stars? Coolest stars?

7. Which property of a star can be determined from binary-star systems?

8. Make a generalization relating the mass and luminosity of main-sequence stars.

9. The disk of a star cannot be resolved telescopically. Explain the method that astronomers have used to estimate the size of stars.

10. Where on an H-R diagram does a star spend most of its lifetime?

11. How does the Sun compare in size and brightness to other main-sequence stars?

12. Why is interstellar matter important to stellar evolution?

13. Compare a bright nebula and a dark nebula.

14. What element is the fuel for main-sequence stars? Red giants?

15. What causes a star to become a giant?

16. Why are less massive stars thought to age more slowly than more massive stars, even though they have much less "fuel"?

17. Enumerate the steps thought to be involved in the evolution of Sun-like stars.

18. What is the final state of a low-mass (red) main-sequence star?

19. What is the final state of a medium-mass (Sun-like) star?

20. How do the "lives" of the most massive stars end? What are the two possible products of this event?

21. Describe the general structure of the Milky Way Galaxy.

22. Compare the three general types of galaxies.

23. Explain why astronomers consider elliptical galaxies more abundant than spiral galaxies, even though more spiral galaxies have been sighted.

24. How did Edwin Hubble determine that the Great Galaxy in Andromeda is located beyond our galaxy?

25. What evidence supports the Big Bang Theory?

Examining the Earth System

1. Briefly describe how the atmosphere, hydrosphere, solid Earth, and biosphere are each related to the death of stars that occurred billions of years ago.

2. If a supernova explosion were to occur within the immediate vicinity of our solar system, what might be some possible consequences of the intense x-ray and gamma radiation that would reach Earth?

3. Scientists are continuously searching the Milky Way Galaxy for other stars that may have planets. What types of stars would most likely have a planet or planets suitable for life as we know it? If you would like to investigate extra-solar planets on-line, you might find these two Web-sites helpful: *NOVA Online* at **http://www.pbs.org/wgbh/nova/worlds/** and the *Electronic Universe Project* at **http://zebu.uoregon.edu/galaxy.html**

4. Based upon your knowledge of the Earth system, the planets in our solar system, and the cosmos in general, speculate about the likelihood that extra-solar planets exist with atmospheres, hydrospheres, lithospheres, and biospheres, similar to Earth's. Explain your speculation.

Online Study Guide

The *Earth Science* Website uses the resources and flexibility of the Internet to aid in your study of the topics in this chapter. Written and developed by Earth science instructors, this site will help improve your understanding of Earth science. Visit **http://www.prenhall.com/tarbuck** and click on the cover of *Earth Science 11e* to find:

- **On-line review quizzes.**
- **Critical thinking exercises.**
- **Links to chapter-specific Web resources.**
- **Internet-wide key term searches.**

http://www.prenhall.com/tarbuck

Appendix A

Metric and English Units Compared

Units

1 kilometer (km)	= 1000 meters (m)
1 meter (m)	= 100 centimeters
1 centimeter (cm)	= 0.39 inch (in.)
1 mile (mi)	= 5280 feet (ft)
1 foot (ft)	= 12 inches (in.)
1 inch (in.)	= 2.54 centimeters (cm)
1 square mile (mi^2)	= 640 acres (a)
1 kilogram (kg)	= 1000 grams (g)
1 pound (lb)	= 16 ounces (oz)
1 fathom	= 6 feet (ft)

Conversions

Length

When you want to convert:	multiply by:	to find:
inches	2.54	centimeters
centimeters	0.39	inches
feet	0.30	meters
meters	3.28	feet
yards	0.91	meters
meters	1.09	yards
miles	1.61	kilometers
kilometers	0.62	miles

Area

When you want to convert:	multiply by:	to find:
square inches	6.45	square centimeters
square centimeters	0.15	square inches
square feet	0.09	square meters
square meters	10.76	square feet
square miles	2.59	square kilometers
square kilometers	0.39	square miles

Volume

When you want to convert:	multiply by:	to find:
cubic inches	16.38	cubic centimeters
cubic centimeters	0.06	cubic inches
cubic feet	0.028	cubic meters
cubic meters	35.3	cubic feet
cubic miles	4.17	cubic kilometers
cubic kilometers	0.24	cubic miles
liters	1.06	quarts
liters	0.26	gallons
gallons	3.78	liters

Masses and Weights

When you want to convert:	multiply by:	to find:
ounces	20.33	grams
grams	0.035	ounces
pounds	0.45	kilograms
kilograms	2.205	pounds

Temperature

When you want to convert degrees Fahrenheit (°F) to degrees Celsius (°C), subtract 32 degrees and divide by 1.8.

When you want to convert degrees Celsius (°C) to degrees Fahrenheit (°F), multiply by 1.8 and add 32 degrees.

When you want to convert degrees Celsius (°C) to kelvins (K), delete the degree symbol and add 273. When you want to convert kelvins (K) to degrees Celsius (°C), add the degree symbol and subtract 273.

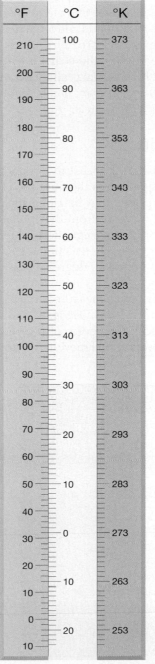

◀ FIGURE A.1 Temperature scales.

Appendix B
Earth's Grid System

A glance at any globe reveals a series of north-south and east-west lines that together make up Earth's grid system, a universally used scheme for locating points on Earth's surface. The north-south lines of the grid are called **meridians** and extend from pole to pole (Figure B.1). All are halves of great circles. A **great circle** is the largest possible circle that may be drawn on a globe; if a globe were sliced along one of these circles, it would be divided into two equal parts called **hemispheres**. By viewing a globe or Figure B.1, you can see that meridians are spaced farthest apart at the equator and converge toward the poles. The east-west lines (circles) of the grid are called **parallels**. As their name implies, these circles are parallel to one another (Figure B.1). Whereas all meridians are parts of great circles, all parallels are not. In fact, only one parallel, the equator, is a great circle.

Latitude and Longitude

Latitude may be defined as distance, measured in degrees, *north* and *south* of the equator. Parallels are used to show latitude. Because all points that lie along the same parallel are an identical distance from the equator, they all have the same latitude designation. The latitude of the equator is 0 degrees, whereas the north and south poles lie 90 degrees N and 90 degrees S, respectively.

Longitude is defined as distance, measured in degrees, *east* and *west* of the zero or *prime meridian*. Because all meridians are identical, the choice of a zero line is obviously arbitrary. However, the meridian that passes through the Royal Observatory at Greenwich, England, is universally accepted as the reference meridian. Thus, the longitude for anyplace on the globe is measured east or west from this line. Longitude can vary from 0 degrees along the prime meridian to 180 degrees, halfway around the globe.

It is important to remember that when a location is specified, directions must be given—that is, north or south latitude and east or west longitude (Figure B.2). If this is not done, more than one point on the globe is being designated. The only exceptions, of course, are places that lie along the equator, the prime meridian, or the 180-degree meridian. It should also be noted that while it is not incorrect to use fractions, a degree of latitude or longitude is usually divided into minutes and seconds. A minute (') is 1/60th of a degree, and a second (") is 1/60th of a minute. When locating a place on a map, the degree of exactness will depend on the scale of the map. When using a small-scale world map or globe, it may be difficult to estimate latitude and longitude to the nearest whole degree or two. On the other hand, when a large-scale map of an area is used, it is often possible to estimate latitude and longitude to the nearest minute or second.

Distance Measurement

The length of a degree of longitude depends on where the measurement is taken. Along the equator, which is a great circle, a degree of east-west distance is equal to approximately 111 kilometers (69 miles). This figure is found by dividing Earth's circumference—40,075 kilometers (24,900 miles)— by 360. However, with an increase in latitude, the parallels become smaller, and the length of a degree of longitude diminishes (see Table B.1). Thus, at about latitude 60 degrees N and S, a degree of longitude has a value equal to about half of what it was at the equator.

As all meridians are halves of great circles, a degree of latitude is equal to about 111 kilometers (69 miles), just as a degree of longitude along the equator is. However, Earth is not a perfect sphere but is slightly flattened at the poles and bulges slightly at the equator. Because of this, there are small differences in the length of a degree of latitude.

Determining the shortest distance between two points on a globe can be done easily and fairly accurately using the globe-and-string method. It should be noted here that the arc of a great circle is the shortest distance between two points on a sphere. To determine

▼ **FIGURE B.1** Earth's grid system.

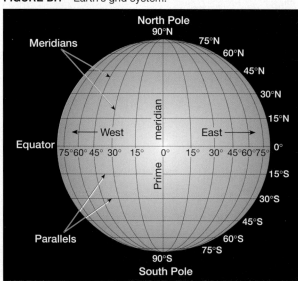

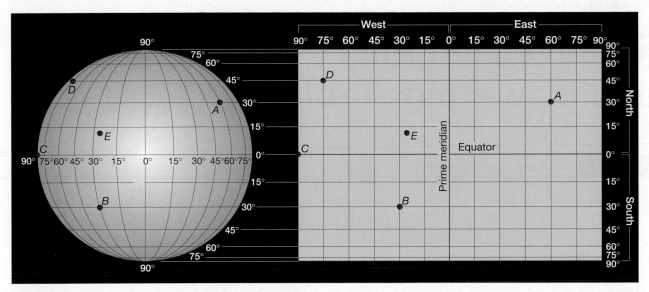

▲ **FIGURE B.2**   Locating places using the grid system. For both diagrams: Point *A* is latitude 30 degrees N, longitude 60 degrees E; Point *B* is latitude 30 degrees S, longitude 30 degrees W; Point *C* is latitude 0 degrees, longitude 90 degrees W; Point *D* is latitude 45 degrees N, longitude 75 degrees W; Point *E* is approximately latitude 10 degrees N, longitude 25 degrees W.

the great circle distance (as well as observe the great circle route) between two places, stretch a piece of string between the locations in question. Then, measure the length of the string along the equator (since it is a great circle with degrees marked on it) to determine the number of degrees between the two points. To calculate the distance in kilometers or miles, simply multiply the number of degrees by 111 or 69, respectively.

TABLE B.1	Longitude as distance.							
	Length of 1° Longitude			Length of 1° Longitude			Length of 1° Longitude	
° Latitude	km	miles	° Latitude	km	miles	° Latitude	km	miles
0	111.367	69.172	30	96.528	59.955	60	55.825	34.674
1	111.349	69.161	31	95.545	59.345	61	54.131	33.622
2	111.298	69.129	32	94.533	58.716	62	52.422	62.560
3	111.214	69.077	33	93.493	58.070	63	50.696	31.488
4	111.096	69.004	34	92.425	57.407	64	48.954	30.406
5	110.945	68.910	35	91.327	56.725	65	47.196	29.314
6	110.760	68.795	36	90.203	56.027	66	45.426	28.215
7	110.543	68.660	37	89.051	55.311	67	43.639	27.105
8	110.290	68.503	38	87.871	54.578	68	41.841	25.988
9	110.003	68.325	39	86.665	53.829	69	40.028	24.862
10	109.686	68.128	40	85.431	53.063	70	38.204	23.729
11	109.333	67.909	41	84.171	52.280	71	36.368	22.589
12	108.949	67.670	42	82.886	51.482	72	34.520	21.441
13	108.530	67.410	43	81.575	50.668	73	32.662	20.287
14	108.079	67.130	44	80.241	49.839	74	30.793	19.126
15	107.596	66.830	45	78.880	48.994	75	28.914	17.959
16	107.079	66.509	46	77.497	48.135	76	27.029	16.788
17	106.530	66.168	47	76.089	47.260	77	25.134	15.611
18	105.949	65.807	48	74.659	46.372	78	23.229	14.428
19	105.337	65.427	49	73.203	45.468	79	21.320	13.242
20	104.692	65.026	50	71.727	44.551	80	19.402	12.051
21	104.014	64.605	51	70.228	43.620	81	17.480	10.857
22	103.306	64.165	52	68.708	42.676	82	15.551	9.659
23	102.565	63.705	53	67.168	41.719	83	13.617	8.458
24	101.795	63.227	54	65.604	40.748	84	11.681	7.255
25	100.994	62.729	55	64.022	39.765	85	9.739	6.049
26	100.160	62.211	56	62.420	38.770	86	7.796	4.842
27	99.297	61.675	57	60.798	37.763	87	5.849	3.633
28	98.405	61.121	58	59.159	36.745	88	3.899	2.422
29	97.481	60.547	59	57.501	35.715	89	1.950	1.211
30	96.528	59.955	60	55.825	34.674	90	0.000	0.000

Relative Humidity and Dew-Point Tables*

TABLE C.1	Relative humidity (percent).																					

Dry bulb (°C) — Dry-Bulb (Air) Temperature

Depression of Wet-Bulb Temperature
(Dry-Bulb Temperature Minus Wet Bulb Temperature = Depression of the Wet Bulb)

Dry bulb (°C)	1	2	3	4	5	6	7	8	9	10	11	12	13	14	15	16	17	18	19	20	21	22
−20	28																					
−18	40																					
−16	48	0																				
−14	55	11																				
−12	61	23																				
−10	66	33	0																			
−8	71	41	13																			
−6	73	48	20	0																		
−4	77	54	32	11																		
−2	79	58	37	20	1																	
0	81	63	45	28	11																	
2	83	67	51	36	20	6																
4	85	70	56	42	27	14																
6	86	72	59	46	35	22	10	0														
8	87	74	62	51	39	28	17	6														
10	88	76	65	54	43	38	24	13	4													
12	88	78	67	57	48	38	28	19	10	2												
14	89	79	69	60	50	41	33	25	16	8	1											
16	90	80	77	62	54	45	37	29	21	14	7	1										
18	91	81	72	64	56	48	40	33	26	19	12	6	0									
20	91	82	74	66	58	51	44	36	30	23	17	11	5									
22	92	83	75	68	60	53	46	40	33	27	21	15	10	4	0							
24	92	84	76	69	62	55	49	42	36	30	25	20	14	9	4	0						
26	92	85	77	70	64	57	51	45	39	34	28	23	18	13	9	5						
28	93	86	78	71	65	59	53	45	42	36	31	26	21	17	12	8	4					
30	93	86	79	72	66	61	55	49	44	39	34	29	25	20	16	12	8	4				
32	93	86	80	73	68	62	56	51	46	41	36	32	27	22	19	14	11	8	4			
34	93	86	81	74	69	63	58	52	48	43	38	34	30	26	22	18	14	11	8	5		
36	94	87	81	75	69	64	59	54	50	44	40	36	32	28	24	21	17	13	10	7	4	
38	94	87	82	76	70	66	60	55	51	46	42	38	34	30	26	23	20	16	13	10	7	5
40	94	89	82	76	71	67	61	57	52	48	44	40	36	33	29	25	22	19	16	13	10	7

Relative Humidity Values

* To determine the relative humidity, find the air (dry-bulb) temperature on the vertical axis (far left) and the depression of the wet bulb on the horizontal axis (top). Where the two meet, the relative humidity is found. For example, when the dry-bulb temperature is 20°C and a wet-bulb temperature is 14°C, then the depression of the wet bulb is 6°C (20°C − 14°C). From Table C-1, the relative humidity is 51 percent and from Table C-2, the dew point is 10°C.

TABLE C.2 | Dew-point temperature (°C)

Dry bulb (°C)	1	2	3	4	5	6	7	8	9	10	11	12	13	14	15	16	17	18	19	20	21	22
−20	−33																					
−18	−28																					
−16	−24																					
−14	−21	−36																				
−12	−18	−28																				
−10	−14	−22																				
−8	−12	−18	−29																			
−6	−10	−14	−22																			
−4	−7	−12	−17	−29																		
−2	−5	−8	−13	−20																		
0	−3	−6	−9	−15	−24																	
2	−1	−3	−6	−11	−17																	
4	1	−1	−4	−7	−11	−19																
6	4	1	−1	−4	−7	−13	−21															
8	6	3	1	2	−5	−9	−14															
10	8	6	4	1	−2	−5	−9	−14	−18													
12	10	8	6	4	1	−2	−5	−9	−16													
14	12	11	9	6	4	1	−2	−5	−10	−17												
16	14	13	11	9	7	4	1	−1	−6	−10	−17											
18	16	15	13	11	9	7	4	2	−2	−5	−10	−19										
20	19	17	15	14	12	10	7	4	2	−2	−5	−10	−19									
22	21	19	17	16	14	12	10	8	5	3	−1	−5	−10	−19								
24	23	21	20	18	16	14	12	10	8	6	2	−1	−5	−10	−18							
26	25	23	22	20	18	17	15	13	11	9	6	3	0	−4	−9	−18						
28	27	25	24	22	27	19	17	16	14	11	9	7	4	1	−3	−9	16					
30	29	27	26	24	23	21	19	18	16	14	12	10	8	5	1	−2	−8	−15				
32	31	29	28	27	25	24	22	21	19	17	15	13	11	8	5	2	−2	−7	−14			
34	33	31	30	29	27	26	24	23	21	20	18	16	14	12	9	6	3	−1	−5	−12	−29	
36	35	33	32	31	29	28	27	25	24	22	20	19	17	15	13	10	7	4	0	−4	−10	
38	37	35	34	33	32	30	29	28	26	25	23	21	19	17	15	13	11	8	5	1	−3	9
40	39	37	36	35	34	32	31	30	28	27	25	24	22	20	18	16	14	12	9	6	2	−2

Column header group: (Dry-Bulb Temperature Minus Wet-Bulb Temperature = Depression of the Wet Bulb)

Left axis label: Dry-Bulb (Air) Temperature

Diagonal label: Dew-Point Values

Appendix D

Landforms of the Conterminous United States

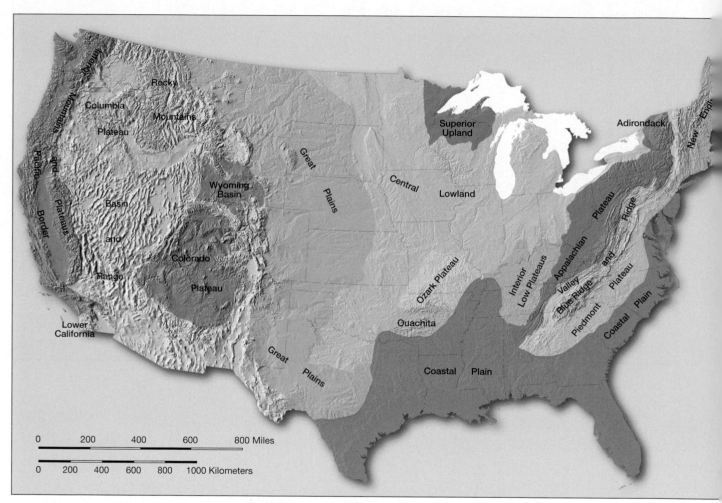

▲ **FIGURE D.1** Outline map showing major physiographic provinces of the United States.

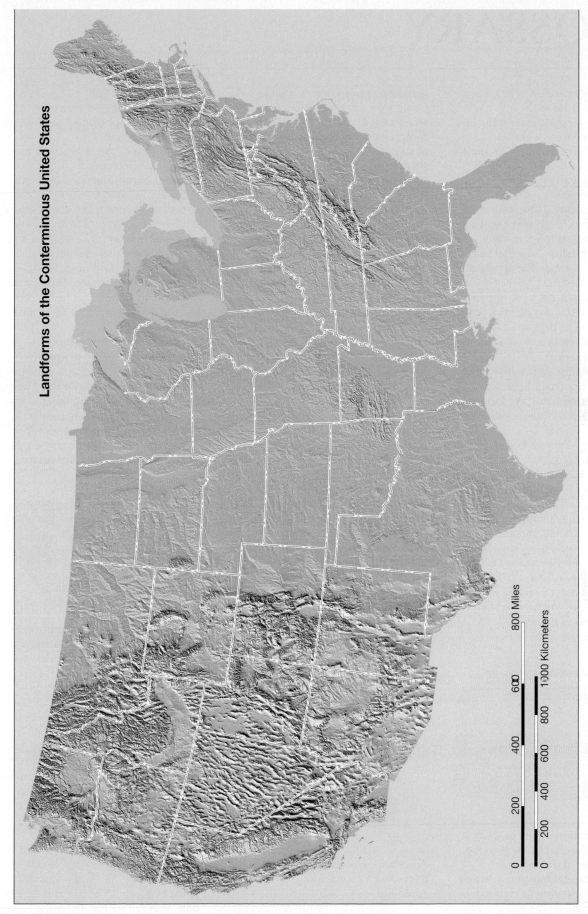

Landforms of the Conterminous United States

▲ **FIGURE D.2** Digital shaded relief landform map of the United States. (Data provided by the U.S. Geological Survey)

GLOSSARY

Aa A type of lava flow that has a jagged blocky surface.

Ablation A general term for the loss of ice and snow from a glacier.

Abrasion The grinding and scraping of a rock surface by the friction and impact of rock particles carried by water, wind, or ice.

Absolute humidity The weight of water vapor in a given volume of air (usually expressed in grams/m^3).

Absolute instability Air that has a lapse rate greater than the dry adiabatic rate.

Absolute magnitude The apparent brightness of a star if it were viewed from a distance of 10 parsecs (32.6 light-years). Used to compare the true brightness of stars.

Absolute stability Air with a lapse rate less than the wet adiabatic rate.

Absorption spectrum A continuous spectrum with dark lines superimposed.

Abyssal plain Very level area of the deep-ocean floor, usually lying at the foot of the continental rise.

Abyssal zone A subdivision of the benthic zone characterized by extremely high pressures, low temperatures, low oxygen, few nutrients, and no sunlight.

Accretionary wedge A large wedge-shaped mass of sediment that accumulates in subduction zones. Here sediment is scraped from the subducting oceanic plate and accreted to the overriding crustal block.

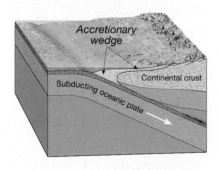

Acid precipitation Rain or snow with a pH value that is less than the pH of unpolluted precipitation.

Active continental margin Usually narrow and consisting of highly deformed sediments. They occur where oceanic lithosphere is being subducted beneath the margin of a continent.

Adiabatic temperature change Cooling or warming of air caused when air is allowed to expand or is compressed, not because heat is added or subtracted.

Advection Horizontal convective motion, such as wind.

Advection fog A fog formed when warm, moist air is blown over a cool surface.

Aerosols Tiny solid and liquid particles suspended in the atmosphere.

Aftershocks Smaller earthquakes that follow the main earthquake.

Air A mixture of many discrete gases, of which nitrogen and oxygen are most abundant, in which varying quantities of tiny solid and liquid particles are suspended.

Air mass A large body of air that is characterized by a sameness of temperature and humidity.

Air-mass weather The conditions experienced in an area as an air mass passes over it. Because air masses are large and fairly homogenous, air-mass weather will be fairly constant and may last for several days.

Air pollutants Airborne particles and gases that occur in concentrations that endanger the health and well-being of organisms or disrupt the orderly functioning of the environment.

Air pressure The force exerted by the weight of a column of air above a given point.

Albedo The reflectivity of a substance, usually expressed as a percentage of the incident radiation reflected.

Alluvial fan A fan-shaped deposit of sediment formed when a stream's slope is abruptly reduced.

Alluvium Unconsolidated sediment deposited by a stream.

Alpine glacier A glacier confined to a mountain valley, which in most instances had previously been a stream valley.

Altitude (of the Sun) The angle of the Sun above the horizon.

Andesitic composition *See* Intermediate composition.

Anemometer An instrument used to determine wind speed.

Aneroid barometer An instrument for measuring air pressure that consists of evacuated metal chambers very sensitive to variations in air pressure.

Angle of repose The steepest angle at which loose material remains stationary without sliding downslope.

Angular unconformity An unconformity in which the strata below dip at an angle different from that of the beds above.

Annual mean temperature An average of the 12 monthly temperature means.

Annual temperature range The difference between the highest and lowest monthly temperature means.

Anthracite A hard, metamorphic form of coal that burns clean and hot.

Anticline A fold in sedimentary strata resembling an arch.

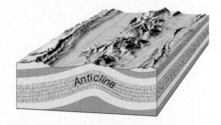

Anticyclone A high-pressure center characterized by a clockwise flow of air in the Northern Hemisphere.

Aphelion The place in the orbit of a planet where the planet is farthest from the Sun.

Aphotic zone That portion of the ocean where there is no sunlight.

Apparent magnitude The brightness of a star when viewed from Earth.

Aquifer Rock or soil through which groundwater moves easily.

Aquitard Impermeable beds that hinder or prevent groundwater movement.

Archean eon The second eon of Precambrian time, following the Hadean and preceding the Proterozoic. It extends between 3.8 billion and 2.5 billion years before the present.

Arête A narrow knifelike ridge separating two adjacent glaciated valleys.

Arid   *See* Desert.

Arid climate   *See* Dry climate.

Arkose   A feldspar-rich sandstone.

Artesian well   A well in which the water rises above the level where it was initially encountered.

Asteroids   Thousands of small planetlike bodies, ranging in size from a few hundred kilometers to less than a kilometer, whose orbits lie mainly between those of Mars and Jupiter.

Asthenosphere   A subdivision of the mantle situated below the lithosphere. This zone of weak material exists below a depth of about 100 kilometers and in some regions extends as deep as 700 kilometers. The rock within this zone is easily deformed.

Astronomical theory   A theory of climatic change first developed by the Yugoslavian astronomer Milankovitch. It is based upon changes in the shape of Earth's orbit, variations in the obliquity of Earth's axis, and the wobbling of Earth's axis.

Astronomical Unit (AU)   Average distance from Earth to the Sun; 1.5×10^8 km, or 93×10^6 miles.

Astronomy   The scientific study of the universe; it includes the observation and interpretation of celestial bodies and phenomena.

Atmosphere   The gaseous portion of a planet; the planet's envelope of air. One of the traditional subdivisions of Earth's physical environment.

Atoll   A continuous or broken ring of coral reef surrounding a central lagoon.

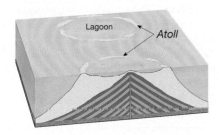

Atom   The smallest particle that exists as an element.

Atomic number   The number of protons in the nucleus of an atom.

Atomic weight   The average of the atomic masses of isotopes for a given element.

Aurora   A bright display of ever-changing light caused by solar radiation interacting with the upper atmosphere in the region of the poles.

Autumnal equinox   The equinox that occurs on September 21–23 in the Northern Hemisphere and on March 21–22 in the Southern Hemisphere.

Backshore   The inner portion of the shore, lying landward of the high-tide shoreline. It is usually dry, being affected by waves only during storms.

Backswamp   A poorly drained area on a floodplain that results when natural levees are present.

Bar   Common term for sand and gravel deposits in a stream channel.

Barchan dune   A solitary sand dune shaped like a crescent with its tips pointing downward.

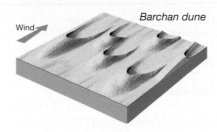

Barchan dune

Wind→

Barchanoid dune   Dunes forming scalloped rows of sand oriented at right angles to the wind. This form is intermediate between isolated barchans and extensive waves of transverse dunes.

Barograph   A recording barometer.

Barometer   An instrument that measures atmospheric pressure.

Barometric tendency   *See* Pressure tendency.

Barred spiral   A galaxy having straight arms extending from its nucleus.

Barrier island   A low, elongate ridge of sand that parallels the coast.

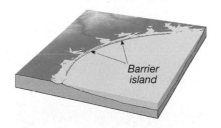

Barrier island

Basalt   A fine-grained igneous rock of mafic composition.

Basaltic composition   A compositional group of igneous rocks indicating that the rock contains substantial dark silicate minerals and calcium-rich plagioclase feldspar.

Base level   The level below which a stream cannot erode.

Basin   A circular downfolded structure.

Batholith   A large mass of igneous rock that formed when magma was emplaced at depth, crystallized, and subsequently exposed by erosion.

Bathymetry   The measurement of ocean depths and the charting of the shape or topography of the ocean floor.

Baymouth bar   A sandbar that completely crosses a bay, sealing it off from the open ocean.

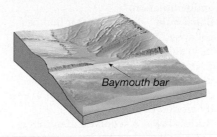

Baymouth bar

Beach   An accumulation of sediment found along the landward margin of the ocean or a lake.

Beach drift   The transport of sediment in a zigzag pattern along a beach caused by the uprush of water from obliquely breaking waves.

Beach face   The wet, sloping surface that extends from the berm to the shoreline.

Beach nourishment   The process by which large quantities of sand are added to the beach system to offset losses caused by wave erosion.

Bed load   Sediment that is carried by a stream along the bottom of its channel.

Benioff zone   Zone of inclined seismic activity that extends from a trench downward into the asthenosphere.

Benthic zone   The marine life zone that includes *any* seabottom surface regardless of its distance from shore.

Benthos   The forms of marine life that live on or in the ocean bottom.

Bergeron process   A theory that relates the formation of precipitation to supercooled clouds, freezing nuclei, and the different saturation levels of ice and liquid water.

Berm   The dry, gently sloping zone on the backshore of a beach at the foot of the coastal cliffs or dunes.

Big Bang Theory   The theory that proposes that the universe originated as a single mass, which subsequently exploded.

Binary stars   Two stars revolving around a common center of mass under their mutual gravitational attraction.

Biogenous sediment   Seafloor sediments consisting of material of marine-organic origin.

Biomass   The total mass of a defined organism or group of organisms in a particular area or ecosystem.

Biosphere   The totality of life on Earth; the parts of the solid Earth, hydrosphere, and atmosphere in which living organisms can be found.

Bituminous   The most common form of coal, often called soft, black coal.

691

Black dwarf   A final state of evolution for a star, in which all of its energy sources are exhausted and it no longer emits radiation.

Black hole   A massive star that has collapsed to such a small volume that its gravity prevents the escape of all radiation.

Blowout (deflation hollow)   A depression excavated by the wind in easily eroded deposits.

Bode's law   A sequence of numbers that approximates the mean distances of the planets from the Sun.

Body waves   Seismic waves that travel through Earth's interior.

Bowen's reaction series   A concept proposed by N. L. Bowen that illustrates the relationships between magma and the minerals crystallizing from it during the formation of igneous rocks.

Braided stream   A stream consisting of numerous intertwining channels.

Breakwater   A structure protecting a nearshore area from breaking waves.

Breccia   A sedimentary rock composed of angular fragments that were lithified.

Bright-line spectrum   The bright lines produced by an incandescent gas under low pressure.

Bright nebula   A cloud of glowing gas excited by ultraviolet radiation from hot stars.

Brittle failure (deformation)   Deformation that involves the fracturing of rock. Associated with rocks near the surface.

Cactolith   A quasi-horizontal chonolith composed of anastomosing ductoliths, whose distal ends curl like a harpolith, thin like a sphenolith, or bulge discordantly like an akmolith or ethmolith.

Caldera   A large depression typically caused by collapse or ejection of the summit area of a volcano.

Calorie   The amount of heat required to raise the temperature of one gram of water 1°C.

Calving   Wastage of a glacier that occurs when large pieces of ice break off into water.

Capacity   The total amount of sediment a stream is able to transport.

Carbonate group   Mineral group whose members contain the carbonate ion (CO_2^{-2}) and one or more kinds of positive ions. Calcite is a common example.

Cassini division   A wide gap in the ring system of Saturn between the A ring and the B ring.

Catastrophism   The concept that Earth was shaped by catastrophic events of a short-term nature.

Cavern   A naturally formed underground chamber or series of chambers most commonly produced by solution activity in limestone.

Celestial sphere   An imaginary hollow sphere upon which the ancients believed the stars were hung and carried around Earth.

Cenozoic era   A span on the geologic time scale beginning about 65 million years ago following the Mesozoic era.

Cepheid variable   A star whose brightness varies periodically because it expands and contracts. A type of pulsating star.

Chemical sedimentary rock   Sedimentary rock consisting of material that was precipitated from water by either inorganic or organic means.

Chemical weathering   The processes by which the internal structure of a mineral is altered by the removal and/or addition of elements.

Chinook   A wind blowing down the leeward side of a mountain and warming by compression.

Chromatic aberration   The property of a lens whereby light of different colors is focused at different places.

Chromosphere   The first layer of the solar atmosphere found directly above the photosphere.

Cinder cone   A rather small volcano built primarily of pyroclastics ejected from a single vent.

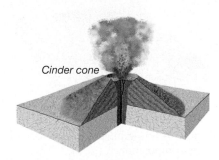

Cinder cone

Circle of illumination   The great circle that separates daylight from darkness.

Cirque   An amphitheater-shaped basin at the head of a glaciated valley produced by frost wedging and plucking.

Cirrus   One of three basic cloud forms; also one of the three high cloud types. They are thin, delicate ice-crystal clouds often appearing as veil-like patches or thin, wispy fibers.

Clastic rock   A sedimentary rock made of broken fragments of preexisting rock.

Cleavage   The tendency of a mineral to break along planes of weak bonding.

Climate   A description of aggregate weather conditions; the sum of all statistical weather information that helps describe a place or region.

Climate system   The exchanges of energy and moisture that occur among the atmosphere, hydrosphere, solid Earth, biosphere, and cryosphere.

Climate-feedback mechanism   Because the atmosphere is a complex interactive physical system, several different possible outcomes may result when one of the system's elements is altered. These various possibilities are called *climate-feedback mechanisms*.

Climatology   The scientific study of climate.

Closed system   A system that is self contained with regards to matter—that is, no matter enters or leaves.

Cloud   A form of condensation best described as a dense concentration of suspended water droplets or tiny ice crystals.

Clouds of vertical development   A cloud that has its base in the low-height range but extends upward into the middle or high altitudes.

Cluster (star)   A large group of stars.

Coarse-grained texture   An igneous rock texture in which the crystals are roughly equal in size and large enough so that individual minerals can be identified with the unaided eye.

Coast   A strip of land that extends inland from the coastline as far as ocean-related features can be found.

Coastline   The coast's seaward edge. The landward limit of the effect of the highest storm waves on the shore.

Col   A pass between mountain valleys where the headwalls of two cirques intersect.

Cold front   A front along which a cold air mass thrusts beneath a warmer air mass.

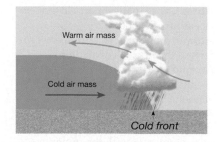

Collision-coalescence process   A theory of raindrop formation in warm clouds (above 0°C) in which large cloud droplets (giants) collide and join together with smaller droplets to form a raindrop. Opposite electrical charges may bind the cloud droplets together.

Column A feature found in caves that is formed when a stalactite and stalagmite join.

Columnar joints A pattern of cracks that form during cooling of molten rock to generate columns that are generally six-sided.

Coma The fuzzy, gaseous component of a comet's head.

Comet A small body that generally revolves about the Sun in an elongated orbit.

Competence A measure of the largest particle a stream can transport; a factor dependent on velocity.

Composite cone A volcano composed of both lava flows and pyroclastic material.

Compound A substance formed by the chemical combination of two or more elements in definite proportions and usually having properties different from those of its constituent elements.

Condensation The change of state from a gas to a liquid.

Condensation nuclei Tiny bits of particulate matter that serve as surfaces on which water vapor condenses.

Conditional instability Moist air with a lapse rate between the dry and wet adiabatic rates.

Conduction The transfer of heat through matter by molecular activity. Energy is transferred through collisions from one molecule to another.

Conduit A pipelike opening through which magma moves toward Earth's surface. It terminates at a surface opening called a vent.

Cone of depression A cone-shaped depression in the water table immediately surrounding a well.

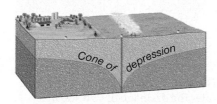

Conformable Layers of rock that were deposited without interruption.

Conglomerate A sedimentary rock composed of rounded gravel-size particles.

Constellation An apparent group of stars originally named for mythical characters. The sky is presently divided into 88 constellations.

Contact metamorphism Changes in rock caused by the heat from a nearby magma body.

Continental (c) air mass An air mass that forms over land; it is normally relatively dry.

Continental drift theory A theory that originally proposed that the continents are rafted about. It has essentially been replaced by the plate tectonics theory.

Continental margin That portion of the seafloor adjacent to the continents. It may include the continental shelf, continental slope, and continental rise.

Continental rise The gently sloping surface at the base of the continental slope.

Continental shelf The gently sloping submerged portion of the continental margin, extending from the shoreline to the continental slope.

Continental slope The steep gradient that leads to the deep-ocean floor and marks the seaward edge of the continental shelf.

Continental volcanic arc Mountains formed in part by igneous activity associated with the subduction of oceanic lithosphere beneath a continent.

Continuous spectrum An uninterrupted band of light emitted by an incandescent solid, liquid, or gas under pressure.

Convection The transfer of heat by the movement of a mass or substance. It can take place only in fluids.

Convergence The condition that exists when the distribution of winds within a given area results in a net horizontal inflow of air into the area. Because convergence at lower levels is associated with an upward movement of air, areas of convergent winds are regions favorable to cloud formation and precipitation.

Convergent plate boundary A boundary in which two plates move together, causing one of the slabs of lithosphere to be consumed into the mantle as it descends beneath on an overriding plate.

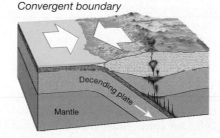

Convergent boundary

Coral reef Structure formed in a warm, shallow, sunlit ocean environment that consists primarily of the calcite-rich remains of corals as well as the limy secretions of algae and the hard parts of many other small organisms.

Core Located beneath the mantle, it is the innermost layer of Earth. The core is divided into an outer core and an inner core.

Coriolis force (effect) The deflective force of Earth's rotation on all free-moving objects, including the atmosphere and oceans. Deflection is to the right in the Northern Hemisphere and to the left in the Southern Hemisphere.

Corona The outer, tenuous layer of the solar atmosphere.

Correlation Establishing the equivalence of rocks of similar age in different areas.

Covalent bond A chemical bond produced by the sharing of electrons.

Crater The depression at the summit of a volcano, or that which is produced by a meteorite impact.

Creep The slow downhill movement of soil and regolith.

Crevasse A deep crack in the brittle surface of a glacier.

Cross-bedding Structure in which relatively thin layers are inclined at an angle to the main bedding. Formed by currents of wind or water.

Cross-cutting A principle of relative dating. A rock or fault is younger than any rock (or fault) through which it cuts.

Crust The very thin outermost layer of Earth.

Crystal An orderly arrangement of atoms.

Crystal form The external appearance of a mineral as determined by its internal arrangement of atoms.

Crystallization The formation and growth of a crystalline solid from a liquid or gas.

Crystal settling During the crystallization of magma, the earlier-formed minerals are denser than the liquid portion and settle to the bottom of the magma chamber.

Cumulus One of three basic cloud forms; also the name given one of the clouds of vertical development. Cumulus are billowy individual cloud masses that often have flat bases.

Cup anemometer *See* Anemometer.

Curie point The temperature above which a material loses its magnetization.

Cutoff A short channel segment created when a river erodes through the narrow neck of land between meanders.

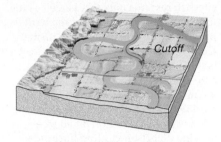
Cutoff

Cyclone A low-pressure center characterized by a counterclockwise flow of air in the Northern Hemisphere.

Daily mean The mean temperature for a day that is determined by averaging the 24 hourly readings or, more commonly, by averaging the maximum and minimum temperatures for a day.

Daily temperature range The difference between the maximum and minimum temperatures for a day.

Dark-line spectrum *See* Absorption spectrum.

Dark nebula A cloud of interstellar dust that obscures the light of more distant stars and appears as an opaque curtain.

Daughter product An isotope resulting from radioactive decay.

Debris flow A relatively rapid type of mass wasting that involves a flow of soil and regolith containing a large amount of water. Also called *mudflows*.

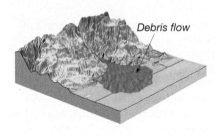
Debris flow

Declination (stellar) The angular distance north or south of the celestial equator denoting the position of a celestial body.

Decompression melting Melting that occurs as rock ascends due to a drop in confining pressure.

Deep-ocean basin The portion of seafloor that lies between the continental margin and the oceanic ridge system. This region comprises almost 30 percent of Earth's surface.

Deep-ocean trench *See* Trench.

Deep-sea fan A cone-shaped deposit at the base of the continental slope. The sediment is transported to the fan

by turbidity currents that follow submarine canyons.

Deflation The lifting and removal of loose material by wind.

Deformation General term for the processes of folding, faulting, shearing, compression, or extension of rocks as the result of various natural forces.

Delta An accumulation of sediment formed where a stream enters a lake or ocean.

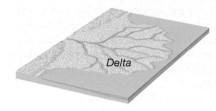
Delta

Dendritic pattern A stream system that resembles the pattern of a branching tree.

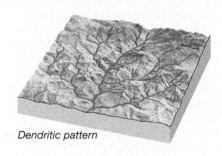
Dendritic pattern

Density Mass per unit volume of a substance, usually expressed as grams per cubic centimeter (g/cm^3).

Deposition The process by which water vapor is changed directly to a solid without passing through the liquid state.

Desalination The removal of salts and other chemicals from seawater.

Desert One of the two types of dry climate; the driest of the dry climates.

Desert pavement A layer of coarse pebbles and gravel created when wind removed the finer material.

Detrital sedimentary rock Rock formed from the accumulation of material that originated and was transported in the form of solid particles derived from both mechanical and chemical weathering.

Dew-point temperature The temperature to which air has to be cooled in order to reach saturation.

Differential weathering The variation in the rate and degree of weathering caused by such factors as mineral make-up, degree of jointing, and climate.

Diffused light Solar energy scattered and reflected in the atmosphere that

reaches Earth's surface in the form of diffuse blue light from the sky.

Dike A tabular-shaped intrusive igneous feature that cuts through the surrounding rock.

Dip-slip fault A fault in which the movement is parallel to the dip of the fault.

Discharge The quantity of water in a stream that passes a given point in a period of time.

Disconformity A type of unconformity in which the beds above and below are parallel.

Disseminated deposit Any economic mineral deposit in which the desired mineral occurs as scattered particles in the rock but in sufficient quantity to make the deposit an ore.

Dissolved load That portion of a stream's load carried in solution.

Distributary A section of a stream that leaves the main flow.

Diurnal tidal pattern A tidal pattern exhibiting one high tide and one low tide during a tidal day; a daily tide.

Divergence The condition that exists when the distribution of winds within a given area results in a net horizontal outflow of air from the region. In divergence at lower levels the resulting deficit is compensated for by a downward movement of air from aloft; hence, areas of divergent winds are unfavorable to cloud formation and precipitation.

Divergent plate boundary A region where the rigid plates are moving apart, typified by the mid-oceanic ridges.

Divergent boundary

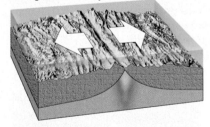

Divide An imaginary line that separates the drainage of two streams; often found along a ridge.

Dome A roughly circular upfolded structure similar to an anticline.

Doppler effect The apparent change in wavelength of radiation caused by the relative motions of the source and the observer.

Doppler radar In addition to the tasks performed by conventional radar, this new generation of weather radar can detect motion directly and hence greatly improve tornado and severe storm warnings.

Drainage basin The land area that contributes water to a stream.

Drawdown The difference in height between the bottom of a cone of depression and the original height of the water table.

Drift *See* Glacial drift.

Drumlin A streamlined asymmetrical hill composed of glacial till. The steep side of the hill faces the direction from which the ice advanced.

Dry adiabatic rate The rate of adiabatic cooling or warming in unsaturated air. The rate of temperature change is 1°C per 100 meters.

Dry climate A climate in which yearly precipitation is not as great as the potential loss of water by evaporation.

Dry-summer subtropical climate A climate located on the west sides of continents between latitudes 30° and 45°. It is the only humid climate with a strong winter precipitation maximum.

Ductile deformation A type of solid state flow that produces a change in the size and shape of a rock body without fracturing. Occurs at depths where temperatures and confining pressures are high.

Dune A hill or ridge of wind-deposited sand.

Earthflow The downslope movement of water-saturated, clay-rich sediment. Most characteristic of humid regions.

Earthquake The vibration of Earth produced by the rapid release of energy.

Earth science The name for all the sciences that collectively seek to understand Earth. It includes geology, oceanography, meteorology, and astronomy.

Earth system science An interdisciplinary study that seeks to examine Earth as a system composed of numerous interacting parts or subsystems.

Ebb current The movement of a tidal current away from the shore.

Eccentricity The variation of an ellipse from a circle.

Echo sounder An instrument used to determine the depth of water by measuring the time interval between emission of a sound signal and the return of its echo from the bottom.

Eclipse The cutting off of the light of one celestial body by another passing in front of it.

Ecliptic The yearly path of the Sun plotted against the background of stars.

Elastic rebound The sudden release of stored strain in rocks that results in movement along a fault.

Electromagnetic radiation *See* Radiation.

Electromagnetic spectrum The distribution of electromagnetic radiation by wavelength.

Electron A negatively charged subatomic particle that has a negligible mass and is found outside an atom's nucleus.

Element A substance that cannot be decomposed into simpler substances by ordinary chemical or physical means.

Elements of weather and climate Those quantities or properties of the atmosphere that are measured regularly and that are used to express the nature of weather and climate.

Elliptical galaxy A galaxy that is round or elliptical in outline. It contains little gas and dust, no disk or spiral arms, and few hot, bright stars.

El Niño The name given to the periodic warming of the ocean that occurs in the central and eastern Pacific. A major El Niño episode can cause extreme weather in many parts of the world.

Eluviation The washing out of fine soil components from the A horizon by downward-percolating water.

Emergent coast A coast where land that was formerly below sea level has been exposed either because of crustal uplift or a drop in sea level or both.

Emission nebula A gaseous nebula that derives its visible light from the fluorescence of ultraviolet light from a star in or near the nebula.

End moraine A ridge of till marking a former position of the front of a glacier.

Energy The capacity to do work.

Energy levels Spherically shaped negatively charged zones that surround the nucleus of an atom.

Environment Everything that surrounds and influences an organism.

Environmental lapse rate The rate of temperature decrease with increasing height in the troposphere.

Eon The largest time unit on the geologic time scale, next in order of magnitude above era.

Ephemeral stream A stream that is usually dry because it carries water only in response to specific episodes of rainfall. Most desert streams are of this type.

Epicenter The location on Earth's surface that lies directly above the focus of an earthquake.

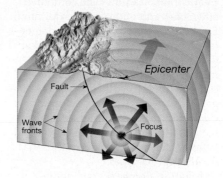

Epoch A unit of the geologic calendar that is a subdivision of a period.

Equatorial low A belt of low pressure lying near the equator and between the subtropical highs.

Equatorial system A method of locating stellar objects much like the coordinate system used on Earth's surface.

Equinox The time when the vertical rays of the Sun are striking the equator. The length of daylight and darkness is equal at all latitudes at equinox.

Era A major division on the geologic calendar; eras are divided into shorter units called periods.

Erosion The incorporation and transportation of material by a mobile agent, such as water, wind, or ice.

Eruptive variable A star that varies in brightness.

Escape velocity The initial velocity an object needs to escape from the surface of a celestial body.

Esker Sinuous ridge composed largely of sand and gravel deposited by a stream flowing in a tunnel beneath a glacier near its terminus.

Estuary A partially enclosed coastal water body that is connected to the ocean. Salinity here is measurably reduced by the freshwater flow of rivers.

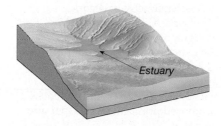

Euphotic zone The portion of the photic zone near the surface, where light is bright enough for photosynthesis to occur.

Evaporation The process of converting a liquid to a gas.

Evaporite A sedimentary rock formed of material deposited from solution by evaporation of the water.

Evapotranspiration The combined effect of evaporation and transpiration.

Evolution, (Theory of) A fundamental theory in biology and paleontology that sets forth the process by which members of a population of organisms come to differ from their ancestors. Organisms evolve by means of mutations, natural selection, and genetic factors. Modern species are descended from related but different species that lived in earlier times.

Exfoliation dome Large, dome-shaped structure, usually composed of granite, formed by sheeting.

Exotic stream A permanent stream that traverses a desert and has its source in well-watered areas outside the desert.

External process Process such as weathering, mass wasting or erosion that is powered by the Sun and transforms solid rock into sediment.

Extrusive Igneous activity that occurs outside the crust.

Eye A zone of scattered clouds and calm averaging about 20 kilometers in diameter at the center of a hurricane.

Eyepiece A short-focal-length lens used to enlarge the image in a telescope. The lens nearest the eye.

Eye wall The doughnut-shaped area of intense cumulonimbus development and very strong winds that surrounds the eye of a hurricane.

Fall A type of movement common to mass-wasting processes that refers to the free falling of detached individual pieces of any size.

Fault A break in a rock mass along which movement has occurred.

Fault-block mountain A mountain formed by the displacement of rock along a fault.

Fault creep Displacement along a fault that is so slow and gradual that little seismic activity occurs.

Fault scarp A cliff created by movement along a fault. It represents the exposed surface of the fault prior to modification by weathering and erosion.

Felsic The group of igneous rocks composed primarily of feldspar and quartz.

Fetch The distance that the wind has traveled across the open water.

Filaments Dark, thin streaks that appear across the bright solar disk.

Fine-grained texture A texture of igneous rocks in which the crystals are too small for individual minerals to be distinguished with the unaided eye.

Fiord A steep-sided inlet of the sea formed when a glacial trough was partially submerged.

Fissure eruption An eruption in which lava is extruded from narrow fractures or cracks in the crust.

Flare A sudden brightening of an area on the Sun.

Flood basalts Flows of basaltic lava that issue from numerous cracks or fissures and commonly cover extensive areas to thicknesses of hundreds of meters.

Flood current The tidal current associated with the increase in the height of the tide.

Floodplain The flat, low-lying portion of a stream valley subject to periodic inundation.

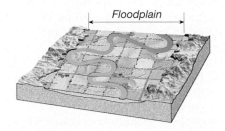

Floodplain

Flow A type of movement common to mass-wasting processes in which water-saturated material moves downslope as a viscous fluid.

Fluorescence The absorption of ultraviolet light, which is reemitted as visible light.

Focal length The distance from the lens to the point where it focuses parallel rays of light.

Focus (earthquake) The zone within Earth where rock displacement produces an earthquake.

Focus (light) The point where a lens or mirror causes light rays to converge.

Fog A cloud with its base at or very near Earth's surface.

Fold A bent rock layer or series of layers that were originally horizontal and subsequently deformed.

Foliated A texture of metamorphic rocks that gives the rock a layered appearance.

Food chain A succession of organisms in an ecological community through which food energy is transferred from producers through herbivores and on to one or more carnivores.

Food web A group of interrelated food chains.

Foreshocks Small earthquakes that often precede a major earthquake.

Foreshore That portion of the shore lying between the normal high and low water marks; the intertidal zone.

Fossil fuel General term for any hydrocarbon that may be used as a fuel, including coal, oil, and natural gas.

Fossils The remains or traces of organisms preserved from the geologic past.

Fossil succession Fossil organisms that succeed one another in a definite and determinable order, and any time period can be recognized by its fossil content.

Fracture Any break or rupture in rock along which no appreciable movement has taken place.

Freezing The change of state from a liquid to a solid.

Freezing nuclei Solid particles that serve as cores for the formation of ice crystals.

Front The boundary between two adjoining air masses having contrasting characteristics.

Frontal fog Fog formed when rain evaporates as it falls through a layer of cool air.

Frontal wedging Lifting of air resulting when cool air acts as a barrier over which warmer, lighter air will rise.

Frost wedging The mechanical breakup of rock caused by the expansion of freezing water in cracks and crevices.

Fumarole A vent in a volcanic area from which fumes or gases escape.

Galactic cluster A system of galaxies containing from several to thousands of member galaxies.

Geocentric The concept of an Earth-centered universe.

Geologic time scale The division of Earth history into blocks of time—eons, eras, periods, and epochs. The time scale was created using relative dating principles.

Geology The science that examines Earth, its form and composition, and the changes it has undergone and is undergoing.

Geosphere The solid Earth, the largest of Earth's four major spheres.

Geostrophic wind A wind, usually above a height of 600 meters (2000 feet), that blows parallel to the isobars.

Geothermal energy Natural steam used for power generation.

Geothermal gradient The gradual increase in temperature with depth in the crust. The average is 30°C per kilometer in the upper crust.

Geyser A fountain of hot water ejected periodically.

Giant (star) A luminous star of large radius.

Glacial drift An all-embracing term for sediments of glacial origin, no matter how, where, or in what shape they were deposited.

Glacial erratic An ice-transported boulder that was not derived from bedrock near its present site.

Glacial striations Scratches and grooves on bedrock caused by glacial abrasion.

Glacial trough A mountain valley that has been widened, deepened, and straightened by a glacier.

Glacier A thick mass of ice originating on land from the compaction and recrystallization of snow that shows evidence of past or present flow.

Glassy A term used to describe the texture of certain igneous rocks, such as obsidian, that contain no crystals.

Glaze A coating of ice on objects formed when supercooled rain freezes on contact.

Globular cluster A nearly spherically shaped group of densely packed stars.

Globule A dense, dark nebula thought to be the birthplace of stars.

Gondwanaland The southern portion of Pangaea consisting of South America, Africa, Australia, India, and Antarctica.

Graben A valley formed by the downward displacement of a fault-bounded block.

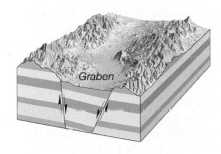

Graben

Graded bed A sediment layer that is characterized by a decrease in sediment size from bottom to top.

Gradient The slope of a stream; generally measured in feet per mile.

Granitic composition A compositional group of igneous rocks that indicates a rock is composed almost entirely of light-colored silicates.

Granules The fine structure visible on the solar surface caused by convective cells below.

Gravitational collapse The gradual subsidence of mountains caused by

lateral spreading of weak material located deep within these structures.

Greenhouse effect The transmission of short-wave solar radiation by the atmosphere, coupled with the selective absorption of longer-wavelength terrestrial radiation, especially by water vapor and carbon dioxide.

Groin A short wall built at a right angle to the shore to trap moving sand.

Ground moraine An undulating layer of till deposited as the ice front retreats.

Groundwater Water in the zone of saturation.

Guyot A submerged flat-topped seamount.

Gyre The large circular surface current pattern found in each ocean.

Hadean eon The first eon on the geologic time scale; this eon ended 3.8 billion years ago and preceded the Archean eon.

Hail Nearly spherical ice pellets having concentric layers and formed by the successive freezing of layers of water.

Half-life The time required for one half of the atoms of a radioactive substance to decay.

Halocline A layer of water in which there is a high rate of change in salinity in the vertical dimension.

Hanging valley A tributary valley that enters a glacial trough at a considerable height above its floor.

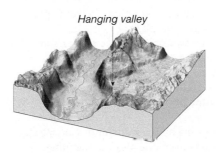

Hanging valley

Hardness The resistance a mineral offers to scratching.

Hard stabilization Any form of artificial structure built to protect a coast or to prevent the movement of sand along a beach. Examples include groins, jetties, breakwaters, and seawalls.

Heliocentric The view that the Sun is at the center of the solar system.

Heat The kinetic energy of random molecular motion.

Hertzsprung-Russell diagram *See* H-R diagram.

High A center of high pressure characterized by anticyclonic winds.

High cloud A cloud that normally has its base above 6000 meters; the base may be lower in winter and at high-latitude locations.

Highland climate Complex pattern of climate conditions associated with mountains. Highland climates are characterized by large differences that occur over short distances.

Hogback A narrow, sharp-crested ridge formed by the upturned edge of a steeply dipping bed of resistant rock.

Horizon A layer in a soil profile.

Horn A pyramid-like peak formed by glacial action in three or more cirques surrounding a mountain summit.

Horst An elongate, uplifted block of crust bounded by faults.

Horst

Hot spot A concentration of heat in the mantle capable of producing magma, which in turn extrudes onto Earth's surface. The intraplate volcanism that produced the Hawaiian Islands is one example.

Hot spring A spring in which the water is 6–9°C (10–15°F) warmer than the mean annual air temperature of its locality.

H-R diagram A plot of stars according to their absolute magnitudes and spectral types.

Hubble's law Relates the distance to a galaxy and its velocity.

Humid continental climate A relatively severe climate characteristic of broad continents in the middle latitudes between approximately 40 and 50 degrees north latitude. This climate is not found in the Southern Hemisphere, where the middle latitudes are dominated by the oceans.

Humidity A general term referring to water vapor in the air but not to liquid droplets of fog, cloud, or rain.

Humid subtropical climate A climate generally located on the eastern side of a continent and characterized by hot, sultry summers and cool winters.

Humus Organic matter in soil produced by the decomposition of plants and animals.

Hurricane A tropical cyclonic storm having winds in excess of 119 kilometers (74 miles) per hour.

Hydrogen burning The conversion of hydrogen through fusion to form helium.

Hydrogenous sediment Seafloor sediments consisting of minerals that crystallize from seawater. An important example is manganese nodules.

Hydrosphere The water portion of our planet; one of the traditional subdivisions of Earth's physical environment.

Hydrothermal solution The hot, watery solution that escapes from a mass of magma during the later stages of crystallization. Such solutions may alter the surrounding country rock and are frequently the source of significant ore deposits.

Hygrometer An instrument designed to measure relative humidity.

Hygroscopic nuclei Condensation nuclei having a high affinity for water, such as salt particles.

Hypothesis A tentative explanation that is tested to determine if it is valid.

Ice cap A mass of glacial ice covering a high upland or plateau and spreading out radially.

Ice cap climate A climate that has no monthly means above freezing and supports no vegetative cover except in a few scattered high mountain areas. This climate, with its perpetual ice and snow, is confined largely to the ice sheets of Greenland and Antarctica.

Ice sheet A very large, thick mass of glacial ice flowing outward in all directions from one or more accumulation centers.

Igneous rock A rock formed by the crystallization of molten magma.

Immature soil A soil lacking horizons.

Incised meander Meandering channel that flows in a steep, narrow valley. They form either when an area is uplifted or when base level drops.

Inclination of the axis The tilt of Earth's axis from the perpendicular to the plane of Earth's orbit.

Inclusion A piece of one rock unit contained within another. Inclusions are used in relative dating. The rock mass adjacent to the one containing the inclusion must have been there first in order to provide the fragment.

Index fossil A fossil that is associated with a particular span of geologic time.

Inertia A property of matter that resists a change in its motion.

Infiltration The movement of surface water into rock or soil through cracks and pore spaces.

Infrared Radiation with a wavelength from 0.7 to 200 micrometers.

Inner core The solid innermost layer of Earth, about 1300 kilometers (800 miles) in radius.

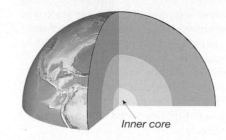

Inner core

Inner planets *See* Terrestrial planets.

Inselberg An isolated mountain remnant characteristic of the late stage of erosion in an arid region.

Intensity (earthquake) A measure of the degree of earthquake shaking at a given locale based on the amount of damage.

Interface A common boundary where different parts of a system interact.

Interior drainage A discontinuous pattern of intermittent streams that do not flow to the ocean.

Intermediate composition The composition of igneous rocks lying between felsic and mafic.

Interstellar matter Dust and gases found between stars.

Intertidal zone The area where land and sea meet and overlap; the zone between high and low tides.

Intraplate volcanism Igneous activity that occurs within a tectonic plate away from plate boundaries.

Intrusive rock Igneous rock that formed below Earth's surface.

Ion An atom or molecule that possesses an electrical charge.

Ionic bond A chemical bond between two oppositely charged ions formed by the transfer of valence electrons from one atom to the other.

Ionosphere A complex zone of ionized gases that coincides with the lower portion of the thermosphere.

Iron meteorite One of the three main categories of meteorites. This group is composed largely of iron with varying amounts of nickel (5–20 percent). Most meteorite finds are irons.

Irregular galaxy A galaxy that lacks symmetry.

Island arc *See* Volcanic island arc.

Isobar A line drawn on a map connecting points of equal atmospheric pressure, usually corrected to sea level.

Isostacy The concept that Earth's crust is floating in gravitational balance upon the material of the mantle.

Isotherms Lines connecting points of equal temperature.

Isotopes Varieties of the same element that have different mass numbers; their nuclei contain the same number of protons but different numbers of neutrons.

Jet stream Swift (120–240 kilometers per hour), high-altitude winds.

Jetties A pair of structures extending into the ocean at the entrance to a harbor or river that are built for the purpose of protecting against storm waves and sediment deposition.

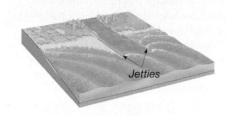

Jetties

Joint A fracture in rock along which there has been no movement.

Jovian planet The Jupiter-like planets: Jupiter, Saturn, Uranus, and Neptune. These planets have relatively low densities.

Kame A steep-sided hill composed of sand and gravel originating when sediment is collected in openings in stagnant glacial ice.

Karst A topography consisting of numerous depressions called *sinkholes.*

Kettle holes Depressions created when blocks of ice became lodged in glacial deposits and subsequently melted.

Köppen classification A system for classifying climates devised by Wladimir Köppen that is based on mean monthly and annual values of temperature and precipitation.

Kuiper belt A region outside the orbit of Neptune where most short-period comets are thought to originate.

Laccolith A massive igneous body intruded between preexisting strata.

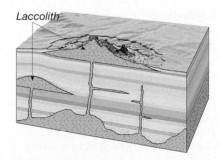

Laccolith

Lahar Mudflows on the slopes of volcanoes that result when unstable layers of ash and debris become saturated and flow downslope, usually following stream channels.

Lake-effect snow   Snow showers associated with a cP air mass to which moisture and heat are added from below as the air mass traverses a large and relatively warm lake (such as one of the Great Lakes), rendering the air mass humid and unstable.

Laminar flow   The movement of water particles in straight-line paths that are parallel to the channel. The water particles move downstream without mixing.

Land breeze   A local wind blowing from land toward the water during the night in coastal areas.

La Niña   An episode of strong trade winds and unusually low sea-surface temperatures in the central and eastern Pacific. The opposite of *El Niño.*

Lapse rate (normal)   The average drop in temperature (6.5°C per kilometer; 3.5°F per 1000 feet) with increased altitude in the troposphere.

Latent heat   The energy absorbed or released during a change in state.

Lateral moraine   A ridge of till along the sides of an alpine glacier composed primarily of debris that fell to the glacier from the valley walls.

Laurasia   The northern portion of Pangaea consisting of North America and Eurasia.

Lava   Magma that reaches Earth's surface.

Law of conservation of angular momentum   The product of the velocity of an object around a center of rotation (axis), and the distance squared of the object from the axis is constant.

Leaching   The depletion of soluble materials from the upper soil by downward-percolating water.

Lightning   A sudden flash of light generated by the flow of electrons between oppositely charged parts of a cumulonimbus cloud or between the cloud and the ground.

Light-year   The distance light travels in a year; about 6 trillion miles.

Liquefaction   A phenomenon, sometimes associated with earthquakes, in which soils and other unconsolidated materials containing abundant water are turned into a fluidlike mass that is not capable of supporting buildings.

Lithification   The process, generally cementation and/or compaction, of converting sediments to solid rock.

Lithosphere   The rigid outer layer of Earth, including the crust and upper mantle.

Local group   The cluster of 20 or so galaxies to which our galaxy belongs.

Localized convective lifting Unequal surface heating that causes localized pockets of air (thermals) to rise because of their buoyancy.

Loess   Deposits of windblown silt, lacking visible layers, generally buff-colored, and capable of maintaining a nearly vertical cliff.

Longitudinal (seif dunes)   Long ridges of sand oriented parallel to the prevailing wind; these dunes form where sand supplies are limited.

Longitudinal dunes

Longshore current   A nearshore current that flows parallel to the shore.

Low   A center of low pressure characterized by cyclonic winds.

Low cloud   A cloud that forms below a height of 2000 meters.

Lower mantle   The part of the mantle that extends from the core-mantle boundary to a depth of 660 kilometers.

Low-velocity zone   *See* Asthenosphere.

Luminosity   The brightness of a star. The amount of energy radiated by a star.

Lunar breccia   A lunar rock formed when angular fragments and dust are welded together by the heat generated by the impact of a meteoroid.

Lunar eclipse   An eclipse of the Moon.

Lunar highlands   *See* Terrae.

Lunar regolith   A thin, gray layer on the surface of the Moon, consisting of loosely compacted, fragmented material believed to have been formed by repeated meteoritic impacts.

Luster   The appearance or quality of light reflected from the surface of a mineral.

Mafic   Igneous rocks with a low silica content and a high iron-magnesium content.

Magma   A body of molten rock found at depth, including any dissolved gases and crystals.

Magmatic differentiation   The process of generating more than one rock type from a single magma.

Magnitude (earthquake)   The total amount of energy released during an earthquake.

Magnitude (stellar)   A number given to a celestial object to express its relative brightness.

Main sequence   A sequence of stars on the Hertzsprung-Russell diagram, containing the majority of stars, that runs diagonally from the upper left to the lower right.

Manganese nodules   Rounded lumps of hydrogenous sediment scattered on the ocean floor, consisting mainly of manganese and iron and usually containing small amounts of copper, nickel, and cobalt.

Mantle   The 2900-kilometer (1800-mile)-thick layer of Earth located below the crust.

Mantle plume   A mass of hotter-than-normal mantle material that ascends toward the surface, where it may lead to igneous activity. These plumes of solid yet mobile material may originate as deep as the core-mantle boundary.

Maria   The Latin name for the smooth areas of the Moon formerly thought to be seas.

Marine terrace   A wave-cut platform that has been exposed above sea level.

Marine west coast climate   A climate found on windward coasts from latitudes 40 to 65 degrees and dominated by maritime air masses. Winters are mild, and summers are cool.

Maritime (m) air mass   An air mass that originates over the ocean. These air masses are relatively humid.

Mass number   The number of neutrons and protons in the nucleus of an atom.

Mass wasting   The downslope movement of rock, regolith, and soil under the direct influence of gravity.

Meander   A looplike bend in the course of a stream.

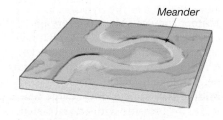

Meander

Mean solar day   The average time between two passages of the Sun across the local celestial meridian.

Mechanical weathering   The physical disintegration of rock, resulting in smaller fragments.

Medial moraine A ridge of till formed when lateral moraines from two coalescing alpine glaciers join.

Melt The liquid portion of magma, excluding the solid crystals.

Melting The change of state from a solid to a liquid.

Mercalli intensity scale *See* Modified Mercalli intensity scale.

Mercury barometer A mercury-filled glass tube in which the height of the mercury column is a measure of air pressure.

Mesocyclone An intense, rotating wind system in the lower part of a thunderstorm that precedes tornado development.

Mesopause The boundary between the mesosphere and the thermosphere.

Mesosphere The layer of the atmosphere immediately above the stratosphere and characterized by decreasing temperatures with height.

Mesozoic era A span on the geologic time scale between the Paleozoic and Cenozoic eras from about 248 million to 65 million years ago.

Metamorphic rock Rocks formed by the alteration of preexisting rock deep within Earth (but still in the solid state) by heat, pressure, and/or chemically active fluids.

Metamorphism The changes in mineral composition and texture of a rock subjected to high temperature and pressure within Earth.

Meteor The luminous phenomenon observed when a meteoroid enters Earth's atmosphere and burns up; popularly called a "shooting star."

Meteorite Any portion of a meteoroid that survives its traverse through Earth's atmosphere and strikes Earth's surface.

Meteoroid Small solid particles that have orbits in the solar system.

Meteorology The scientific study of the atmosphere and atmospheric phenomena; the study of weather and climate.

Meteor shower Many meteors appearing in the sky caused when Earth intercepts a swarm of meteoritic particles.

Middle cloud A cloud occupying the height range from 2000 to 6000 meters.

Middle-latitude cyclone Large center of low pressure with an associated cold front and often a warm front. Frequently accompanied by abundant precipitation.

Mid-ocean ridge *See* Oceanic ridge.

Mineral A naturally occurring, inorganic crystalline material with a unique chemical composition.

Mineralogy The study of minerals.

Mineral resource All discovered and undiscovered deposits of a useful mineral that can be extracted now or at some time in the future.

Mixed tidal pattern A tidal pattern exhibiting two high tides and two low tides per tidal day with a large inequality in high water heights, low water heights, or both. Coastal locations that experience such a tidal pattern may also show alternating periods of diurnal and semidiurnal tidal patterns. Also called mixed semidiurnal.

Mixing depth The height to which convectional movements extend above Earth's surface. The greater the mixing depth, the better the air quality.

Mixing ratio The mass of water vapor in a unit mass of dry air; commonly expressed as grams of water vapor per kilogram of dry air.

Model A term often used synonymously with hypothesis but is less precise because it is sometimes used to describe a theory as well.

Modified Mercalli intensity scale A 12-point scale developed to evaluate earthquake intensity based on the amount of damage to various structures.

Mohorovičić discontinuity (Moho) The boundary separating the crust from the mantle, discernible by an increase in seismic velocity.

Mohs scale A series of 10 minerals used as a standard in determining hardness.

Moment magnitude A more precise measure of earthquake magnitude than the Richter scale that is derived from the amount of displacement that occurs along a fault zone.

Monocline A one-limbed flexure in strata. The strata are unusually flat-lying or very gently dipping on both sides of the monocline.

Monsoon Seasonal reversal of wind direction associated with large continents, especially Asia. In winter, the wind blows from land to sea; in summer, from sea to land.

Monthly mean temperature The mean temperature for a month that is calculated by averaging the daily means.

Mountain breeze The nightly downslope winds commonly encountered in mountain valleys.

Natural leeves The elevated landforms that parallel some streams and act to confine their waters, except during floodstage.

Neap tide Lowest tidal range, occurring near the times of the first- and third-quarter phases of the Moon.

Nearshore zone The zone of beach that extends from the low-tide shoreline seaward to where waves break at low tide.

Nebula A cloud of interstellar gas and/or dust.

Nebular hypothesis The basic idea that the Sun and planets formed from the same cloud of gas and dust in interstellar space.

Nekton Pelagic organisms that can move independently of ocean currents by swimming or other means of propulsion.

Negative feedback mechanism A feedback mechanism that tends to maintain a system as it is—that is, maintain the status quo.

Neritic zone The marine-life zone that extends from the low tideline out to the shelf break.

Neutron A subatomic particle found in the nucleus of an atom. The neutron is electrically neutral and has a mass approximately that of a proton.

Neutron star A star of extremely high density composed entirely of neutrons.

Nonconformity An unconformity in which older metamorphic or intrusive igneous rocks are overlain by younger sedimentary strata.

Nonfoliated Metamorphic rocks that do not exhibit foliation.

Nonmetallic mineral resource Mineral resource that is not a fuel or processed for the metals it contains.

Nonrenewable resource Resource that forms or accumulates over such long time spans that it must be considered as fixed in total quantity.

Normal fault A fault in which the rock above the fault plane has moved down relative to the rock below.

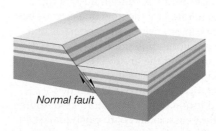

Normal fault

Normal polarity A magnetic field that is the same as that which exists at present.

Nova A star that explosively increases in brightness.

Nucleus The small heavy core of an atom that contains all of its positive charge and most of its mass.

Nuée ardente Incandescent volcanic debris buoyed up by hot gases that moves downslope in an avalanche fashion.

Numerical date Date that specifies the actual number of years that have passed since an event occurred.

Objective lens In a refracting telescope, the long-focal-length lens that forms an image of the object viewed. The lens closest to the object.

Obliquity The angle between the planes of Earth's equator and orbit.

Obsidian A volcanic glass of felsic composition.

Occluded front A front formed when a cold front overtakes a warm front. It marks the beginning of the end of a middle-latitude cyclone.

Occlusion The overtaking of one front by another.

Occultation An eclipse of a star or planet by the Moon or a planet.

Ocean basin floor Area of the deep-ocean floor between the continental margin and the mid-ocean ridge.

Oceanic plateau An extensive region on the ocean floor composed of thick accumulations of pillow basalts and other mafic rocks that in some cases exceed 30 kilometers in thickness.

Oceanic ridge A continuous elevated zone on the floor of all the major ocean basins and varying in width from 500 to 5000 kilometers (300 to 3000 miles). The rifts at the crests of ridges represent divergent plate boundaries.

Oceanic zone The marine-life zone beyond the continental shelf.

Oceanography The scientific study of the oceans and oceanic phenomena.

Offshore zone The relatively flat submerged zone that extends from the breaker line to the edge of the continental shelf.

Oort cloud A spherical shell composed of comets that orbit the Sun at distances generally greater than 10,000 times the Earth-Sun distance.

Open cluster A loosely formed group of stars of similar origin.

Open system One in which both matter and energy flow into and out of the system. Most natural systems are of this type.

Orbit The path of a body in revolution around a center of mass.

Ore Usually a useful metallic mineral that can be mined at a profit. The term is also applied to certain non-metallic minerals such as fluorite and sulfur.

Original horizontality Layers of sediments are generally deposited in a horizontal or nearly horizontal position.

Orogenesis The processes that collectively result in the formation of mountains.

Orographic lifting Mountains acting as barriers to the flow of air, forcing the air to ascend. The air cools adiabatically, and clouds and precipitation may result.

Outer core A layer beneath the mantle about 2200 kilometers (1364 miles) thick that has the properties of a liquid.

Outer planets *See* Jovian planets.

Outgassing The escape of gases that had been dissolved in magma.

Outwash plain A relatively flat, gently sloping plain consisting of materials deposited by meltwater streams in front of the margin of an ice sheet.

Overrunning Warm air gliding up a retreating cold air mass.

Oxbow lake A curved lake produced when a stream cuts off a meander.

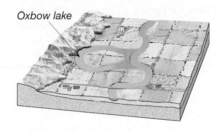

Oxbow lake

Ozone A molecule of oxygen containing three oxygen atoms.

Pahoehoe A lava flow with a smooth-to-ropey surface.

Paleomagnetism The natural remnant magnetism in rock bodies. The permanent magnetization acquired by rock that can be used to determine the location of the magnetic poles and the latitude of the rock at the time it became magnetized.

Paleontology The systematic study of fossils and the history of life on Earth.

Paleozoic era A span on the geologic time scale between the eons of the Precambrian and Mesozoic era from about 540 million to 248 million years ago.

Pangaea The proposed supercontinent that 200 million years ago began to break apart and form the present landmasses.

Parabolic dunes The shape of these dunes resembles barchans, except their tips point into the wind; they often form along coasts that have strong on-shore winds, abundant sand, and vegetation that partly covers the sand.

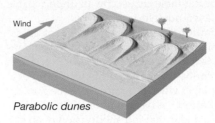

Parabolic dunes

Paradigm A theory that is held with a very high degree of confidence and is comprehensive in scope.

Parallax The apparent shift of an object when viewed from two different locations.

Parasitic cone A volcanic cone that forms on the flank of a larger volcano.

Parcel An imaginary volume of air enclosed in a thin elastic cover. Typically it is considered to be a few hundred cubic meters in volume and is assumed to act independently of the surrounding air.

Parent material The material upon which a soil develops.

Parsec The distance at which an object would have a parallax angle of 1 second of arc (3.26 light-years).

Partial melting The process by which most igneous rocks melt. Since individual minerals have different melting points, most igneous rocks melt over a temperature range of a few hundred degrees. If the liquid is squeezed out after some melting has occurred, a melt with a higher silica content results.

Passive continental margin Margins that consist of a continental shelf, continental slope, and continental rise. They are *not* associated with plate boundaries and therefore experience little volcanism and few earthquakes.

Pegmatite A very coarse-grained igneous rock (typically granite) commonly found as a dike associated with a large mass of plutonic rock that has smaller crystals. Crystallization in a water-rich environment is believed to be responsible for the very large crystals.

Pelagic zone Open ocean of *any* depth. Animals in this zone swim or float freely.

Penumbra The portion of a shadow from which only part of the light source is blocked by an opaque body.

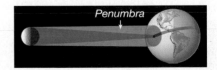

Penumbra

Perched water table   A localized zone of saturation above the main water table created by an impermeable layer (aquiclude).

Peridotite   An igneous rock of ultramafic composition thought to be abundant in the upper mantle.

Perihelion   The point in the orbit of a planet where it is closest to the Sun.

Period   A basic unit of the geologic calendar that is a subdivision of an era. Periods may be divided into smaller units called epochs.

Periodic table   The tabular arrangement of the elements according to atomic number.

Permeability   A measure of a material's ability to transmit water.

Perturbation   The gravitational disturbance of the orbit of one celestial body by another.

Phanerozoic eon   That part of geologic time represented by rocks containing abundant fossil evidence. The eon extending from the end of the Proterozoic eon (about 540 million years ago) to the present.

Phases of the Moon   The progression of changes in the Moon's appearance during the month.

Pheoncryst   Conspicuously large crystals embedded in a matrix of finer-grained crystals.

Photic zone   The upper part of the ocean into which any sunlight penetrates.

Photochemical reaction   A chemical reaction in the atmosphere that is triggered by sunlight, often yielding a secondary pollutant.

Photon   A discrete amount (quantum) of electromagnetic energy.

Photosphere   The region of the Sun that radiates energy to space. The visible surface of the Sun.

Photosynthesis   The process by which plants and algae produce carbohydrates from carbon dioxide and water in the presence of chlorophyll, using light energy and releasing oxygen.

pH scale   A common measure of the degree of acidity or alkalinity of a solution, it is a logarithmic scale ranging from 0 to 14. A value of 7 denotes a neutral solution, values below 7 indicate greater acidity, and numbers above 7 indicate greater alkalinity.

Physical environment   The part of the environment that encompasses water, air, soil, and rock, as well as conditions such as temperature, humidity, and sunlight.

Phytoplankton   Algal plankton, which are the most important community of primary producers in the ocean.

Piedmont glacier   A glacier that forms when one or more valley glaciers emerge from the confining walls of mountain valleys and spread out to create a broad sheet in the lowlands at the base of the mountains.

Pipe   A vertical conduit through which magmatic materials have passed.

Placer   Deposit formed when heavy minerals are mechanically concentrated by currents, most commonly streams and waves. Placers are sources of gold, tin, platinum, diamonds, and other valuable minerals.

Plane of the ecliptic   The imaginary plane that connects Earth's orbit with the celestial sphere.

Planetary nebula   A shell of incandescent gas expanding from a star.

Plankton   Passively drifting or weakly swimming organisms that cannot move independently of ocean currents. Includes microscopic algae, protozoa, jellyfish, and larval forms of many animals.

Plate   One of numerous rigid sections of the lithosphere that moves as a unit over the material of the asthenosphere.

Plate tectonics   The theory that proposes that Earth's outer shell consists of individual plates that interact in various ways and thereby produce earthquakes, volcanoes, mountains, and the crust itself.

Playa   A flat area on the floor of an undrained desert basin. Following heavy rain, the playa becomes a lake.

Playa lake   A temporary lake in a playa.

Pleistocene epoch   An epoch of the Quaternary period beginning about 1.8 million years ago and ending about 10,000 years ago. Best known as a time of extensive continental glaciation.

Plucking (quarrying)   The process by which pieces of bedrock are lifted out of place by a glacier.

Pluton   A structure that results from the emplacement and crystallization of magma beneath the surface of Earth.

Pluvial lake   A lake formed during a period of increased rainfall. During the Pleistocene epoch this occurred in some nonglaciated regions during periods of ice advance elsewhere.

Polar (P) air mass   A cold air mass that forms in a high-latitude source region.

Polar easterlies   In the global pattern of prevailing winds, winds that blow from the polar high toward the subpolar low. These winds, however, should not be thought of as persistent winds, such as the trade winds.

Polar front   The stormy frontal zone separating air masses of polar origin from air masses of tropical origin.

Polar high   Anticyclones that are assumed to occupy the inner polar regions and are believed to be thermally induced, at least in part.

Polar wandering   As the result of paleomagnetic studies in the 1950s, researchers proposed that either the magnetic poles migrated greatly through time or the continents had gradually shifted their positions.

Population I   Stars rich in atoms heavier than helium. Nearly always relatively young stars found in the disk of the galaxy.

Population II   Stars poor in atoms heavier than helium. Nearly always relatively old stars found in the halo, globular clusters, or nuclear bulge.

Porosity   The volume of open spaces in rock or soil.

Porphyritic   An igneous texture consisting of large crystals embedded in a matrix of much smaller crystals.

Positive feedback mechanism   A feedback mechanism that enhances or drives change.

Precambrian   All geologic time prior to the Paleozoic era.

Precession   A slow motion of Earth's axis that traces out a cone over a period of 26,000 years.

Precipitation fog   Fog formed when rain evaporates as it falls through a layer of cool air.

Pressure gradient   The amount of pressure change occurring over a given distance.

Pressure tendency   The nature of the change in atmospheric pressure over the past several hours. It can be a useful aid in short-range weather prediction.

Prevailing wind   A wind that consistently blows from one direction more than from another.

Primary pollutants   Those pollutants emitted directly from identifiable sources.

Primary productivity   The amount of organic matter synthesized by organisms from inorganic substances through photosynthesis or chemosynthesis within a given volume of water or habitat in a unit of time.

Primary (P) wave A type of seismic wave that involves alternating compression and expansion of the material through which it passes.

Principal shells *See* Energy levels.

Prominence A concentration of material above the solar surface that appears as a bright archlike structure.

Proterozoic eon The eon following the Archean and preceding the Phanerozoic. It extends between about 2500 million (2.5 billion) and 540 million years ago.

Proton A positively charged subatomic particle found in the nucleus of an atom.

Proton-proton chain A chain of thermonuclear reactions by which nuclei of hydrogen are built up into nuclei of helium.

Protostar A collapsing cloud of gas and dust destined to become a star.

Psychrometer A device consisting of two thermometers (wet bulb and dry bulb) that is rapidly whirled and, with the use of tables, yields the relative humidity and dew point.

Ptolemaic system An Earth-centered system of the universe.

Pulsar A variable radio source of small size that emits radio pulses in very regular periods.

Pulsating variable A variable star that pulsates in size and luminosity.

Pycnocline A layer of water in which there is a rapid change of density with depth.

Pyroclastic An igneous rock texture resulting from the consolidation of individual rock fragments that are ejected during a violent eruption.

Pyroclastic flow A highly heated mixture, largely of ash and pumice fragments, traveling down the flanks of a volcano or along the surface of the ground.

Pyroclastic material The volcanic rock ejected during an eruption, including ash, bombs, and blocks.

Radial pattern A system of streams running in all directions away from a central elevated structure, such as a volcano.

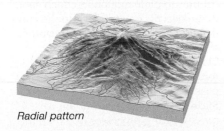

Radial pattern

Radiation The transfer of energy (heat) through space by electromagnetic waves.

Radiation fog Fog resulting from radiation heat loss by Earth.

Radiation pressure The force exerted by electromagnetic radiation from an object such as the Sun.

Radioactivity The spontaneous decay of certain unstable atomic nuclei.

Radiocarbon (carbon-14) The radioactive isotope of carbon, which is produced continuously in the atmosphere and is used in dating events from the very recent geologic past (the last few tens of thousands of years).

Radio interferometer Two or more radio telescopes that combine their signals to achieve the resolving power of a larger telescope.

Radiometric dating The procedure of calculating the absolute ages of rocks and minerals that contain radioactive isotopes.

Radio telescope A telescope designed to make observations in radio wavelengths.

Rainshadow desert A dry area on the lee side of a mountain range. Many middle-latitude deserts are of this type.

Rainshadow desert

Rapids A part of a stream channel in which the water suddenly begins flowing more swiftly and turbulently because of an abrupt steepening of the gradient.

Ray (lunar) Any of a system of bright elongated streaks, sometimes associated with a crater on the Moon.

Recessional moraine An end moraine formed as the ice front stagnated during glacial retreat.

Rectangular pattern A drainage pattern characterized by numerous right-angle bends that develops on jointed or fractured bedrock.

Red giant A large, cool star of high luminosity; a star occupying the upper-right portion of the Hertzsprung-Russell diagram.

Reflecting telescope A telescope that concentrates light from distant objects by using a concave mirror.

Reflection The process whereby light bounces back from an object at the same angle at which it encounters a surface and with the same intensity.

Reflection nebula A relatively dense dust cloud in interstellar space that is illuminated by starlight.

Refracting telescope A telescope that employs a lens to bend and concentrate the light from distant objects.

Refraction The process by which the portion of a wave in shallow water slows, causing the wave to bend and tend to align itself with the underwater contours.

Regional metamorphism Metamorphism associated with large-scale mountain-building processes.

Regolith The layer of rock and mineral fragments that nearly everywhere covers Earth's surface.

Relative dating Rocks are placed in their proper sequence or order. Only the chronological order of events is determined.

Relative humidity The ratio of the air's water-vapor content to its water-vapor capacity.

Renewable resource A resource that is virtually inexhaustible or that can be replenished over relatively short time spans.

Reserve Already identified deposits from which minerals can be extracted profitably.

Residual soil Soil developed directly from the weathering of the bedrock below.

Resolving power The ability of a telescope to separate objects that would otherwise appear as one.

Retrograde motion The apparent westward motion of the planets with respect to the stars.

Reverse fault A fault in which the material above the fault plane moves up in relation to the material below.

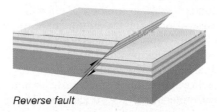

Reverse fault

Reverse polarity A magnetic field opposite to that which exists at present.

Revolution The motion of one body about another, as Earth about the Sun.

Richter scale A scale of earthquake magnitude based on the motion of a seismograph.

Ridge push A mechanism that may contribute to plate motion. It involves the oceanic lithosphere sliding down the oceanic ridge under the pull of gravity.

Rift zone A region of Earth's crust along which divergence is taking place.

Right ascension An angular distance measured eastward along the celestial equator from the vernal equinox. Used with declination in a coordinate system to describe the position of celestial bodies.

Rime A thin coating of ice on objects produced when supercooled fog droplets freeze on contact.

Rock A consolidated mixture of minerals.

Rock cycle A model that illustrates the origin of the three basic rock types and the interrelatedness of Earth materials and processes.

Rock flour Ground-up rock produced by the grinding effect of a glacier.

Rockslide The rapid slide of a mass of rock downslope along planes of weakness.

Rockslide

Rotation The spinning of a body, such as Earth, about its axis.

Runoff Water that flows over the land rather than infiltrating into the ground.

Salinity The proportion of dissolved salts to pure water, usually expressed in parts per thousand (‰).

Saltation Transportation of sediment through a series of leaps or bounces.

Santa Ana The local name given a chinook wind in southern California.

Saturation The maximum quantity of water vapor that the air can hold at any given temperature and pressure.

Scattering The redirecting (in all directions) of light by small particles and gas molecules in the atmosphere. The result is diffused light.

Scoria Hardened lava that has retained the vesicles produced by escaping gases.

Scoria cone *See* Cinder cone.

Sea arch An arch formed by wave erosion when caves on opposite sides of a headland unite.

Sea breeze A local wind blowing from the sea during the afternoon in coastal areas.

Seafloor spreading The process of producing new seafloor between two diverging plates.

Seamount An isolated volcanic peak that rises at least 1000 meters (3000 feet) above the deep-ocean floor.

Sea stack An isolated mass of rock standing just offshore, produced by wave erosion of a headland.

Seawall A barrier constructed to prevent waves from reaching the area behind the wall. Its purpose is to defend property from the force of breaking waves.

Secondary enrichment The concentration of minor amounts of metals that are scattered through unweathered rock into economically valuable concentrations by weathering processes.

Secondary pollutants Pollutants that are produced in the atmosphere by chemical reactions that occur among primary pollutants.

Secondary (S) wave A seismic wave that involves oscillation perpendicular to the direction of propagation.

Sediment Unconsolidated particles created by the weathering and erosion of rock, by chemical precipitation from solution in water, or from the secretions of organisms and transported by water, wind, or glaciers.

Sedimentary rock Rock formed from the weathered products of preexisting rocks that have been transported, deposited, and lithified.

Seismic sea wave A rapidly moving ocean wave generated by earthquake activity capable of inflicting heavy damage in coastal regions.

Seismogram The record made by a seismograph.

Seismograph An instrument that records earthquake waves.

Seismology The study of earthquakes and seismic waves.

Semiarid *See* Steppe.

Semidiurnal tidal pattern A tidal pattern exhibiting two high tides and two low tides per tidal day with small inequalities between successive highs and successive lows; a semidaily tide.

Shadow zone The zone between 104 and 143 degrees distance from an earthquake epicenter in which direct waves do not arrive because of refraction by Earth's core.

Sheeting A mechanical weathering process characterized by the splitting off of slablike sheets of rock.

Shelf break The point where a rapid steepening of the gradient occurs, marking the outer edge of the continental shelf and the beginning of the continental slope.

Shield A large, relatively flat expanse of ancient metamorphic rock within the stable continental interior.

Shield volcano A broad, gently sloping volcano built from fluid basaltic lavas.

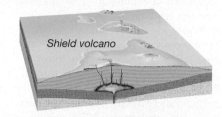

Shield volcano

Shore Seaward of the coast, this zone extends from the highest level of wave action during storms to the lowest tide level.

Shoreline The line that marks the contact between land and sea. It migrates up and down as the tide rises and falls.

Sidereal day The period of Earth's rotation with respect to the stars.

Sidereal month A time period based on the revolution of the Moon around Earth with respect to the stars.

Silicate Any one of numerous minerals that have the oxygen and silicon tetrahedron as their basic structure.

Silicon-oxygen tetrahedron A structure composed of four oxygen atoms surrounding a silicon atom that constitutes the basic building block of silicate minerals.

Sill A tabular igneous body that was intruded parallel to the layering of preexisting rock.

Sinkhole A depression produced in a region where soluble rock has been removed by groundwater.

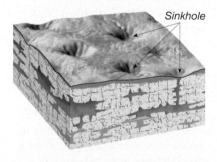

Sinkhole

Slab pull A mechanism that contributes to plate motion in which cool, dense oceanic crust sinks into the mantle and "pulls" the trailing lithosphere along.

Sleet Frozen or semifrozen rain formed when raindrops freeze as they pass through a layer of cold air.

Slide A movement common to mass-wasting processes in which the material moving downslope remains fairly coherent and moves along a well-defined surface.

Slip face   The steep, leeward slope of a sand dune; it maintains an angle of about 34 degrees.

Slump   The downward slipping of a mass of rock or unconsolidated material moving as a unit along a curved surface.

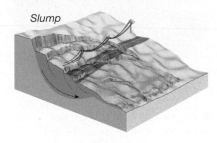

Slump

Snow   A solid form of precipitation produced by sublimination of water vapor.

Snowfield   An area where snow persists year-round.

Snowline   Lower limit of perennial snow.

Soil   A combination of mineral and organic matter, water, and air; that portion of the regolith that supports plant growth.

Soil horizon   A layer of soil that has identifiable characteristics produced by chemical weathering and other soil-forming processes.

Soil profile   A vertical section through a soil showing its succession of horizons and the underlying parent material.

Soil Taxonomy   A soil classification system consisting of six hierarchical categories based on observable soil characteristics. The system recognizes 12 soil orders.

Soil texture   The relative proportions of clay, silt, and sand in a soil. Texture strongly influences the soil's ability to retain and transmit water and air.

Solar constant   The rate at which solar radiation is received outside Earth's atmosphere on a surface perpendicular to the Sun's rays when Earth is at an average distance from the Sun.

Solar eclipse   An eclipse of the Sun.

Solar flare   A sudden and tremendous eruption in the solar chromosphere.

Solar winds   Subatomic particles ejected at high speed from the solar corona.

Solifluction   Slow, downslope flow of water-saturated materials common to permafrost areas.

Solstice   The time when the vertical rays of the Sun are striking either the Tropic of Cancer or the Tropic of Capricorn. Solstice represents the longest or shortest day (length of daylight) of the year.

Solum   The O, A, and B horizons in a soil profile. Living roots and other plant and animal life are largely confined to this zone.

Sorting   The process by which solid particles of various sizes are separated by moving water or wind. Also, the degree of similarity in particle size in sediment or sedimentary rock.

Source region   The area where an air mass acquires its characteristic properties of temperature and moisture.

Specific gravity   The ratio of a substance's weight to the weight of an equal volume of water.

Spectral class   A classification of a star according to the characteristics of its spectrum.

Spectroscope   An instrument for directly viewing the spectrum of a light source.

Spectroscopy   The study of spectra.

Spheroidal weathering   Any weathering process that tends to produce a spherical shape from an initially blocky shape.

Spicule   A narrow jet of rising material in the solar chromosphere.

Spiral galaxy   A flattened, rotating galaxy with pinwheel-like arms of interstellar material and young stars winding out from its nucleus.

Spiral galaxy

Spit   An elongate ridge of sand that projects from the land into the mouth of an adjacent bay.

Spring   A flow of groundwater that emerges naturally at the ground surface.

Spring equinox   The equinox that occurs on March 21–22 in the Northern Hemisphere and on September 21–23 in the Southern Hemisphere.

Spring tide   Highest tidal range that occurs near the times of the new and full moons.

Stable air   Air that resists vertical displacement. If it is lifted, adiabatic cooling will cause its temperature to be lower than the surrounding environment; if it is allowed, it will sink to its original position.

Stable platform   That part of the craton that is mantled by relatively undeformed sedimentary rocks and underlain by a basement complex of igneous and metamorphic rocks.

Stalactite   The icicle-like structure that hangs from the ceiling of a cavern.

Stalagmite   The columnlike form that grows upward from the floor of a cavern.

Star dune   Isolated hill of sand that exhibits a complex form and develops where wind directions are variable.

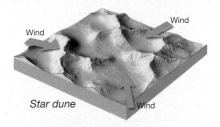

Star dune

Stationary front   A situation in which the surface position of a front does not move; the flow on either side of such a boundary is nearly parallel to the position of the front.

Steam fog   Fog having the appearance of steam, produced by evaporation from a warm water surface into the cool air above.

Stellar parallax   A measure of stellar distance.

Steppe   One of the two types of dry climate. A marginal and more humid variant of the desert that separates it from bordering humid climates.

Stony-iron meteorite   One of the three main categories of meteorites. This group, as the name implies, is a mixture of iron and silicate minerals.

Stony meteorite   One of the three main categories of meteorites. Such meteorites are composed largely of silicate minerals with inclusions of other minerals.

Storm surge   The abnormal rise of the sea along a shore as a result of strong winds.

Strata   Parallel layers of sedimentary rock.

Stratified drift   Sediments deposited by glacial meltwater.

Stratopause   The boundary between the stratosphere and the mesosphere.

Stratosphere   The layer of the atmosphere immediately above the troposphere, characterized by increasing temperatures with height, owing to the concentration of ozone.

Stratovolcano   *See* Composite cone.

Stratus One of three basic cloud forms; also, the name given one of the flow clouds. They are sheets or layers that cover much or all of the sky.

Streak The color of a mineral in powdered form.

Stream valley The channel, valley floor, and sloping valley walls of a stream.

Striations (glacial) Scratches or grooves in a bedrock surface caused by the grinding action of a glacier and its load of sediment.

Strike-slip fault A fault along which the movement is horizontal.

Stromatolite Structures that are deposited by algae and consist of layered mounds of calcium carbonate.

Subarctic climate A climate found north of the humid continental climate and south of the polar climate and characterized by bitterly cold winters and short cool summers. Places within this climatic realm experience the highest annual temperature ranges on Earth.

Subduction The process of thrusting oceanic lithosphere into the mantle along a convergent boundary.

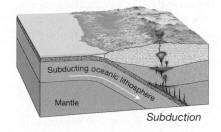

Subduction

Subduction zone A long, narrow zone where one lithospheric plate descends beneath another.

Sublimation The conversion of a solid directly to a gas without passing through the liquid state.

Submarine canyon A seaward extension of a valley that was cut on the continental shelf during a time when sea level was lower, or a canyon carved into the outer continental shelf, slope, and rise by turbidity currents.

Submergent coast A coast with a form that is largely the result of the partial drowning of a former land surface either because of a rise of sea level or subsidence of the crust or both.

Subpolar low Low pressure located at about the latitudes of the Arctic and Antarctic circles. In the Northern Hemisphere the low takes the form of individual oceanic cells; in the Southern Hemisphere there is a deep and continuous trough of low pressure.

Subsoil A term applied to the B horizon of a soil profile.

Subtropical high Not a continuous belt of high pressure but rather several semipermanent, anticyclonic centers characterized by subsidence and divergence located roughly between latitudes 25 and 35 degrees.

Summer solstice The solstice that occurs on June 21–22 in the Northern Hemisphere and on December 21–22 in the Southern Hemisphere.

Sunspot A dark spot on the Sun, which is cool by contrast to the surrounding photosphere.

Supercooled The condition of water droplets that remain in the liquid state at temperatures well below 0°C.

Supergiant A very large star of high luminosity.

Supernova An exploding star that increases in brightness many thousands of times.

Superposition In any undeformed sequence of sedimentary rocks, each bed is older than the layers above and younger than the layers below.

Supersaturation The condition of being more highly concentrated than is normally possible under given temperature and pressure conditions. When describing humidity, it refers to a relative humidity that is greater than 100 percent.

Surf A collective term for breakers; also, the wave activity in the area between the shoreline and the outer limit of breakers.

Surface soil The uppermost layer in a soil profile: the A horizon.

Surface waves Seismic waves that travel along the outer layer of Earth.

Suspended load The fine sediment carried within the body of flowing water.

Swells Wind-generated waves that have moved into an area of weaker winds or calm.

Syncline A linear downfold in sedimentary strata; the opposite of anticline.

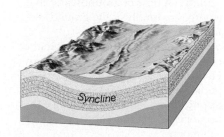

Syncline

Synodic month The period of revolution of the Moon with respect to the Sun, or its cycle of phases.

System Any size group of interacting parts that form a complex whole.

Talus An accumulation of rock debris at the base of a cliff.

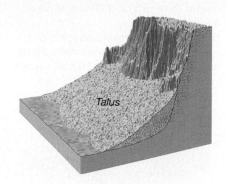

Talus

Tarn A small lake in a cirque.

Tectonics The study of the large-scale processes that collectively deform Earth's crust.

Temperature A measure of the degree of hotness or coldness of a substance; a measure of the *average* kinetic energy of individual atoms or molecules in a substance.

Temperature inversion A layer in the atmosphere of limited depth where the temperature increases rather than decreases with height.

Temporary (local) base level The level of a lake, resistant rock layer, or any other base level that stands above sea level.

Terminal moraine The end moraine marking the farthest advance of a glacier.

Terrace A flat, benchlike structure produced by a stream, which was left elevated as the stream cut downward.

Terrae The extensively cratered highland areas of the Moon.

Terrane A crustal block bounded by faults, whose geologic history is distinct from the histories of adjoining crustal blocks.

Terrestrial planets Any of the Earth-like planets, including Mercury, Venus, Mars, and Earth.

Terrigenous sediment Seafloor sediments derived from terrestrial weathering and erosion.

Texture The size, shape, and distribution of the particles that collectively constitute a rock.

Theory A well-tested and widely accepted view that explains certain observable facts.

Thermal gradient The increase in temperature with depth. It averages 1°C per 30 meters (1–2°F per 100 feet) in the crust.

Thermal metamorphism *See* Contact metamorphism.

Thermocline   A layer of water in which there is a rapid change in temperature in the vertical dimension.

Thermohaline circulation   Movements of ocean water caused by density differences brought about by variations in temperature and salinity.

Thermosphere   The region of the atmosphere immediately above the mesosphere and characterized by increasing temperatures due to absorption of very shortwave solar energy by oxygen.

Thrust fault   A low-angle reverse fault.

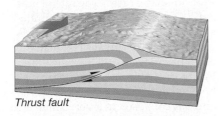
Thrust fault

Thunder   The sound emitted by rapidly expanding gases along the channel of lightning discharge.

Thunderstorm   A storm produced by a cumulonimbus cloud and always accompanied by lightning and thunder. It is of relatively short duration and usually accompanied by strong wind gusts, heavy rain, and sometimes hail.

Tidal current   The alternating horizontal movement of water associated with the rise and fall of the tide.

Tidal delta   A deltalike feature created when a rapidly moving tidal current emerges from a narrow inlet and slows, depositing its load of sediment.

Tidal flat   A marshy or muddy area that is covered and uncovered by the rise and fall of the tide.

Tide   Periodic change in the elevation of the ocean surface.

Till   Unsorted sediment deposited directly by a glacier.

Tombolo   A ridge of sand that connects an island to the mainland or to another island.

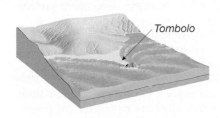
Tombolo

Tornado   A small, very intense cyclonic storm with exceedingly high winds, most often produced along cold fronts in conjunction with severe thunderstorms.

Tornado warning   A warning issued when a tornado has actually been sighted in an area or is indicated by radar.

Tornado watch   A warning issued for areas of about 65,000 square kilometers (25,000 square miles), indicating that conditions are such that tornadoes may develop; it is intended to alert people to the possibility of tornadoes.

Trade winds   Two belts of winds that blow almost constantly from easterly directions and are located on the equatorward sides of the subtropical highs.

Transform fault   A major strike-slip fault that cuts through the lithosphere and accommodates motion between two plates.

Transform fault boundary   A boundary in which two plates slide past one another without creating or destroying lithosphere.

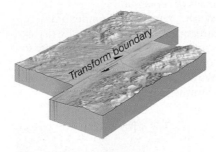
Transform boundary

Transpiration   The release of water vapor to the atmosphere by plants.

Transported soil   Soils that form on unconsolidated deposits.

Transverse dunes   A series of long ridges oriented at right angles to the prevailing wind; these dunes form where vegetation is sparse and sand is very plentiful.

Travertine   A form of limestone ($CaCO_3$) that is deposited by hot springs or as a cave deposit.

Trellis pattern   A system of streams in which nearly parallel tributaries occupy valleys cut in folded strata.

Trench   An elongated depression in the seafloor produced by bending of oceanic crust during subduction.

Trophic level   A nourishment level in a food chain. Plant and algae producers constitute the lowest level, followed by herbivores and a series of carnivores at progressively higher levels.

Tropical depression   By international agreement, a tropical cyclone with maximum winds that do not exceed 61 kilometers (38 miles) per hour.

Tropical rain forest   A luxuriant broadleaf evergreen forest; also, the name given the climate associated with this vegetation.

Tropical storm   By international agreement, a tropical cyclone with maximum winds between 61 and 119 kilometers (38 and 74 miles) per hour.

Tropical wet and dry   A climate that is transitional between the wet tropics and the subtropical steppes.

Tropic of Cancer   The parallel of latitude, 23 1/2 degrees north latitude, marking the northern limit of the Sun's vertical rays.

Tropic of Capricorn   The parallel of latitude, 23 1/2 degrees south latitude, marking the southern limit of the Sun's vertical rays.

Tropopause   The boundary between the troposphere and the stratosphere.

Troposphere   The lowermost layer of the atmosphere. It is generally characterized by a decrease in temperature with height.

Tsunami   The Japanese word for a seismic sea wave.

Tundra climate   Found almost exclusively in the Northern Hemisphere or at high altitudes in many mountainous regions. A treeless climatic realm of sedges, grasses, mosses, and lichens that is dominated by a long, bitterly cold winter.

Turbidite   Turbidity current deposit characterized by graded bedding.

Turbidity current   A downslope movement of dense, sediment-laden water created when sand and mud on the continental shelf and slope are dislodged and thrown into suspension.

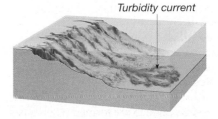
Turbidity current

Turbulent flow   The movement of water in an erratic fashion often characterized by swirling, whirlpool-like eddies. Most streamflow is of this type.

Ultimate base level   Sea level; the lowest level to which stream erosion could lower the land.

Ultramafic   Igneous rocks composed mainly of iron and magnesium-rich minerals.

Ultraviolet   Radiation with a wavelength from 0.2 to 0.4 micrometer.

Umbra   The central, completely dark part of a shadow produced during an eclipse.

Unconformity A surface that represents a break in the rock record, caused by erosion or nondeposition.

Uniformitarianism The concept that the processes that have shaped Earth in the geologic past are essentially the same as those operating today.

Unstable air Air that does not resist vertical displacement. If it is lifted, its temperature will not cool as rapidly as the surrounding environment, so it will continue to rise on its own.

Upslope fog Fog created when air moves up a slope and cools adiabatically.

Upwelling The rising of cold water from deeper layers to replace warmer surface water that has been moved away.

Urban heat island The fact that temperatures within a city are generally higher than in surrounding rural areas.

Valence electron The electrons involved in the bonding process; the electrons occupying the highest-principal energy level of an atom.

Valley breeze The daily upslope winds commonly encountered in a mountain valley.

Valley glacier *See* Alpine glacier.

Valley train A relatively narrow body of stratified drift deposited on a valley floor by meltwater streams that issue from a valley glacier.

Vapor pressure That part of the total atmospheric pressure attributable to water-vapor content.

Vein deposit A mineral filling a fracture or fault in a host rock. Such deposits have a sheetlike, or tabular, form.

Ventifact A cobble or pebble polished and shaped by the sandblasting effect of wind.

Vesicular A term applied to igneous rocks that contain small cavities called vesicles, which are formed when gases escape from lava.

Viscosity A measure of a fluid's resistance to flow.

Visible light Radiation with a wavelength from 0.4 to 0.7 micrometer.

Volatiles Gaseous components of magma dissolved in the melt. Volatiles will readily vaporize (form a gas) at surface pressures.

Volcanic bomb A streamlined pyroclastic fragment ejected from a volcano while molten.

Volcanic island arc A chain of volcanic islands generally located a few hundred kilometers from a trench where active subduction of one oceanic slab beneath another is occurring.

Volcanic neck An isolated, steep-sided, erosional remnant consisting of lava that once occupied the vent of a volcano.

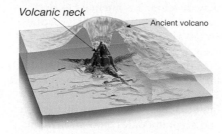

Volcanic neck
Ancient volcano

Volcano A mountain formed of lava and/or pyroclastics.

Warm front A front along which a warm air mass overrides a retreating mass of cooler air.

Wash A common term for a desert stream course that is typically dry except for brief periods immediately following a rain.

Water table The upper level of the saturated zone of groundwater.

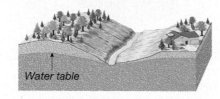

Water table

Wave-cut cliff A seaward-facing cliff along a steep shoreline formed by wave erosion at its base and mass wasting.

Wave-cut platform A bench or shelf in the bedrock at sea level, cut by wave erosion.

Wave height The vertical distance between the trough and crest of a wave.

Wavelength The horizontal distance separating successive crests or troughs.

Wave of oscillation A water wave in which the wave form advances as the water particles move in circular orbits.

Wave of translation The turbulent advance of water created by breaking waves.

Wave period The time interval between the passage of successive crests at a stationary point.

Wave refraction *See* Refraction.

Weather The state of the atmosphere at any given time.

Weathering The disintegration and decomposition of rock at or near Earth's surface.

Welded tuff A pyroclastic rock composed of particles that have been fused together by the combination of heat still contained in the deposit after it has come to rest and by the weight of overlying material.

Well An opening bored into the zone of saturation.

Westerlies The dominant west-to-east motion of the atmosphere that characterizes the regions on the poleward side of the subtropical highs.

Wet adiabatic rate The rate of adiabatic temperature change in saturated air. The rate of temperature change is variable, but it is always less than the dry adiabatic rate.

White dwarf A star that has exhausted most or all of its nuclear fuel and has collapsed to a very small size; believed to be near its final stage of evolution.

White frost Ice crystals instead of dew that form on surfaces when the dew point is below freezing.

Wind Air flowing horizontally with respect to Earth's surface.

Wind vane An instrument used to determine wind direction.

Winter solstice The solstice that occurs on December 21–22 in the Northern Hemisphere and on June 21–22 in the Southern Hemisphere.

Yazoo tributary A tributary that flows parallel to the main stream because a natural levee is present.

Zodiac A band along the ecliptic containing the 12 constellations of the zodiac.

Zone of accumulation The part of a glacier characterized by snow accumulation and ice formation. Its outer limit is the snowline.

Zone of aeration Area above the water table where openings in soil, sediment, and rock are not saturated but filled mainly with air.

Zone of fracture The upper portion of a glacier consisting of brittle ice.

Zone of saturation Zone where all open spaces in sediment and rock are completely filled with water.

Zone of wastage The part of a glacier beyond the zone of accumulation where all of the snow from the previous winter melts, as does some of the glacial ice.

Zooplankton Animal plankton.

Index

Edward J. Tarbuck/Frederick K. Lutgens/Dennis Tasa
GEODe: Earth Science version 2.0
0-13-149814-2
© 2006 Pearson Education, Inc.
Pearson Prentice Hall
Pearson Education, Inc.
Upper Saddle River, NJ 07458
All rights Reserved.
Pearson Prentice Hall™ is a trademark
of Pearson Education, Inc.

YOU SHOULD CAREFULLY READ THE TERMS AND CONDITIONS BEFORE USING THE CD-ROM PACKAGE. USING THIS CD-ROM PACKAGE INDICATES YOUR ACCEPTANCE OF THESE TERMS AND CONDITIONS.

Pearson Education, Inc. provides this program and licenses its use. You assume responsibility for the selection of the program to achieve your intended results, and for the installation, use, and results obtained from the program. This license extends only to use of the program in the United States or countries in which the program is marketed by authorized distributors.

LICENSE GRANT

You hereby accept a nonexclusive, nontransferable, permanent license to install and use the program ON A SINGLE COMPUTER at any given time. You may copy the program solely for backup or archival purposes in support of your use of the program on the single computer. You may not modify, translate, disassemble, decompile, or reverse engineer the program, in whole or in part.

TERM

The License is effective until terminated. Pearson Education, Inc. reserves the right to terminate this License automatically if any provision of the License is violated. You may terminate the License at any time. To terminate this License, you must return the program, including documentation, along with a written warranty stating that all copies in your possession have been returned or destroyed.

LIMITED WARRANTY

THE PROGRAM IS PROVIDED "AS IS" WITHOUT WARRANTY OF ANY KIND, EITHER EXPRESSED OR IMPLIED, INCLUDING, BUT NOT LIMITED TO, THE IMPLIED WARRANTIES OR MERCHANTABILITY AND FITNESS FOR A PARTICULAR PURPOSE. THE ENTIRE RISK AS TO THE QUALITY AND PERFORMANCE OF THE PROGRAM IS WITH YOU. SHOULD THE PROGRAM PROVE DEFECTIVE, YOU (AND NOT PEARSON EDUCATION, INC. OR ANY AUTHORIZED DEALER) ASSUME THE ENTIRE COST OF ALL NECESSARY SERVICING, REPAIR, OR CORRECTION. NO ORAL OR WRITTEN INFORMATION OR ADVICE GIVEN BY PEARSON EDUCATION, INC., ITS DEALERS, DISTRIBUTORS, OR AGENTS SHALL CREATE A WARRANTY OR INCREASE THE SCOPE OF THIS WARRANTY. SOME STATES DO NOT ALLOW THE EXCLUSION OF IMPLIED WARRANTIES, SO THE ABOVE EXCLUSION MAY NOT APPLY TO YOU. THIS WARRANTY GIVES YOU SPECIFIC LEGAL RIGHTS AND YOU MAY ALSO HAVE OTHER LEGAL RIGHTS THAT VARY FROM STATE TO STATE.

Pearson Education, Inc. does not warrant that the functions contained in the program will meet your requirements or that the operation of the program will be uninterrupted or error-free. However, Pearson Education, Inc. warrants the CD-ROM(s) on which the program is furnished to be free from defects in material and workmanship under normal use for a period of ninety (90) days from the date of delivery to you as evidenced by a copy of your receipt. The program should not be relied on as the sole basis to solve a problem whose incorrect solution could result in injury to person or property. If the program is employed in such a manner, it is at the user's own risk and Pearson Education, Inc. explicitly disclaims all liability for such misuse.

LIMITATION OF REMEDIES

Pearson Education, Inc.'s entire liability and your exclusive remedy shall be:

1. the replacement of any CD-ROM not meeting Pearson Education, Inc.'s "LIMITED WARRANTY" and that is returned to Pearson Education, or 2. if Pearson Education is unable to deliver a replacement CD-ROM that is free of defects in materials or workmanship, you may terminate this agreement by returning the program.

IN NO EVENT WILL PEARSON EDUCATION, INC. BE LIABLE TO YOU FOR ANY DAMAGES, INCLUDING ANY LOST PROFITS, LOST SAVINGS, OR OTHER INCIDENTAL OR CONSEQUENTIAL DAMAGES ARISING OUT OF THE USE OR INABILITY TO USE SUCH PROGRAM EVEN IF PEARSON EDUCATION, INC. OR AN AUTHORIZED DISTRIBUTOR HAS BEEN ADVISED OF THE POSSIBILITY OF SUCH DAMAGES, OR FOR ANY CLAIM BY ANY OTHER PARTY.

SOME STATES DO NOT ALLOW FOR THE LIMITATION OR EXCLUSION OF LIABILITY FOR INCIDENTAL OR CONSEQUENTIAL DAMAGES, SO THE ABOVE LIMITATION OR EXCLUSION MAY NOT APPLY TO YOU.

GENERAL

You may not sublicense, assign, or transfer the license of the program. Any attempt to sublicense, assign or transfer any of the rights, duties, or obligations hereunder is void.

This Agreement will be governed by the laws of the State of New York.

Should you have any questions concerning this Agreement, you may contact Pearson Education, Inc. by writing to:
ESM Media Development
Higher Education Division
Pearson Education, Inc.
1 Lake Street
Upper Saddle River, NJ 07458

Should you have any questions concerning technical support, you may write to:
New Media Production
Higher Education Division
Pearson Education, Inc.
1 Lake Street
Upper Saddle River, NJ 07458

YOU ACKNOWLEDGE THAT YOU HAVE READ THIS AGREEMENT, UNDERSTAND IT, AND AGREE TO BE BOUND BY ITS TERMS AND CONDITIONS. YOU FURTHER AGREE THAT IT IS THE COMPLETE AND EXCLUSIVE STATEMENT OF THE AGREEMENT BETWEEN US THAT SUPERSEDES ANY PROPOSAL OR PRIOR AGREEMENT, ORAL OR WRITTEN, AND ANY OTHER COMMUNICATIONS BETWEEN US RELATING TO THE SUBJECT MATTER OF THIS AGREEMENT.

Windows
Minimum System Requirements:
Processor: 200 MHz Intel Pentium II processor or greater (Windows XP requires a 233 MHz processor)

RAM:
Windows 98—32 MB of RAM
Windows NT—32 MB of RAM
Windows 2000—64 MB of RAM
Windows Me—32 MB of RAM
Windows XP—128 MB of RAM

Operating System: Windows 98, NT4, 2000, Me, XP
Monitor resolution: 640 x 480 pixels; thousands of colors (millions of colors recommended)

Required third-party software: QuickTime™ (version 5.0.2)
Required hardware: CD drive, sound card, speakers

Installation Instructions:
GEODe Earth Science has no installation or setup program.

GEODe Earth Science requires QuickTime™ version 5.0.2.
To check or install QuickTime™, visit:

http://browsertuneup.pearsoncmg.com/browser_tuneup.html

Program Instructions:
To start the GEODe Earth Science program simply insert GEODe Earth Science CD-ROM into your CD-ROM drive. The GEODe Earth Science program will automatically start.

It is recommended that the display be set to "millions of colors."

If your display is larger than 640 x 480 pixels, there will be a black border around the 640 x 480 display window.

This CD-ROM is intended for use on stand alone computers. Network installation is not supported.

CD Content:

This CD-ROM also contains a version of GEODe Earth Science appropriate for use on Macintosh computers.

Mac

Minimum System Requirements:
Processor: 180 MHz PowerPC (G3 required for OS X)
RAM: In addition to the RAM required by the operating system, this application requires:
        10 MBytes of RAM (16 MBytes recommended) for OS 8.6-9.2.
        28 MBytes of available RAM for OS X.
Operating System: Mac OS 8.6 to 9.2 and Mac OS X (10.1.5 or later)
Monitor resolution: 640 x 480 pixels; thousands of colors (millions of colors recommended)
Required third-party software: QuickTime™ (version 5.0.2)
Required hardware: CD drive, speakers

Installation Instructions
GEODe Earth Science has no installation or setup program.

GEODe Earth Science requires QuickTime™ version 5.0.2. To check or update your version of QuickTime, visit:

http://browsertuneup.pearsoncmg.com/browser_tuneup.html

Program Instructions
To start the GEODe Earth Science program simply:
1) insert GEODe Earth Science CD-ROM into your CD-ROM drive.
2) double-click the GEODe Earth Science CD-ROM icon on your desktop,
3) double-click the GEODe Earth Science application for your operating system version

It is recommended that the display be set to "millions of colors."

If your display is larger than 640 x 480 pixels, there will be a black border around the 640 x 480 display window.

This CD-ROM is intended for use on stand alone computers. Network installation is not supported.

CD Content:
This CD-ROM also contains a version of GEODe Earth Science appropriate for use on computers running Microsoft Windows as well.

Technical Support Information:
If you are having problems with this software, call (800) 677-6337 between 8:00 a.m. and 8:00 p.m. EST, Monday through Friday, and 5:00 p.m. through Midnight EST onSundays. You can also get support by filling out the web form located at:

http://247.prenhall.com/mediaform.html

Our technical staff will need to know certain things about your system in order to help us solve your problems more quickly and efficiently. If possible, please be at your computer when you call for support. You should have the following information ready:

• product and title and product ISBN
• computer make and model
• operating system version
• RAM available
• hard disk space available
• graphics card type
• sound card type
• printer make and model
• network connection
• detailed description of the problem, including the exact wording of any error messages.

Third Party Software:
Pearson does not handle technical support for third-party software. The web site for QuickTime™ technical support is:

http://www.info.apple.com/usen/quicktime/